# Handbook of Energy Transitions

The global energy scenario is undergoing an unprecedented transition. In the wake of enormous challenges—such as increased population, higher energy demands, increasing greenhouse gas emissions, depleting fossil fuel reserves, volatile energy prices, geopolitical concerns, and energy insecurity issues—the energy sector is experiencing a transition in terms of energy resources and their utilization. This modern transition is historically more dynamic and multidimensional compared to the past considering the vast technological advancements, socioeconomic implications and political responses, and ever-evolving global policies and regulations. Energy insecurity in terms of its critical dimensions—access, affordability, and reliability—remains a major problem hindering the socioeconomic progress in developing countries.

The *Handbook of Energy Transitions* presents a holistic account of the 21st-century energy transition away from fossil fuels. It provides an overview of the unfolding transition in terms of overall dimensions, drivers, trends, barriers, policies, and geopolitics, and then discusses transition in terms of particular resources or technologies, such as renewable energy systems, solar energy, hydropower, hydrogen and fuel cells, electric vehicles, energy storage systems, batteries, digitalization, smart grids, blockchain, and machine learning. It also discusses the present energy transition in terms of broader policy and developmental perspectives. Further, it examines sustainable development, the economics of energy and green growth, and the role of various technologies and initiatives like renewables, nuclear power, and electrification in promoting energy security and energy transition worldwide.

**Key Features**

- Includes technical, economic, social, and policy perspectives of energy transitions
- Features practical case studies and comparative assessments
- Examines the latest renewable energy and low-carbon technologies
- Explains the connection between energy transition and global climate change

# Handbook of Energy Transitions

Edited by
## Muhammad Asif

**CRC Press**
Taylor & Francis Group
Boca Raton London New York

CRC Press is an imprint of the
Taylor & Francis Group, an **informa** business

First edition published 2023
by CRC Press
6000 Broken Sound Parkway NW, Suite 300, Boca Raton, FL 33487-2742

and by CRC Press
4 Park Square, Milton Park, Abingdon, Oxon, OX14 4RN

*CRC Press is an imprint of Taylor & Francis Group, LLC*

*Library of Congress Cataloging-in-Publication Data*
Names: Asif, Muhammad, editor.
Title: Handbook of energy transitions / edited by Muhammad Asif.
Description: First edition. | Boca Raton : CRC Press, 2023. | Includes
bibliographical references and index.
Identifiers: LCCN 2022016063 (print) | LCCN 2022016064 (ebook) |
ISBN 9780367688592 (hbk) | ISBN 9781032324982 (pbk) |
ISBN 9781003315353 (ebk)
Subjects: LCSH: Renewable energy sources.
Classification: LCC TJ808 .H356 2023  (print) | LCC TJ808  (ebook) |
DDC 333.79/4—dc23/eng/20220713
LC record available at https://lccn.loc.gov/2022016063
LC ebook record available at https://lccn.loc.gov/2022016064

ISBN: 9780367688592 (hbk)
ISBN: 9781032324982 (pbk)
ISBN: 9781003315353 (ebk)

DOI: 10.1201/9781003315353

Typeset in Times
by codeMantra

The Open Access version of chapter 3 was funded by FWO – Research Foundation-Flanders.

*To my daughter*
*Hareem Zainab*

# Contents

## SECTION I   Introduction to the 21st-Century Energy Transition

## SECTION II   Energy Transition: Resources and Technologies

## SECTION III    Energy Transition: Policies and Prospects

Contents                                                                                                ix

# Preface

The global energy scenario, led by fossil fuels, is at a crossroads as climate change is recognized as one of the biggest challenges facing the planet. Fossil fuels, which propelled the industrial revolution, and the technological and socioeconomic advancements ever since, are being increasingly criticized for their greenhouse gas emissions. Challenges like growing energy demand, depleting fossil fuel reserves, and mounting energy prices are adding to the complexities of the energy landscape. Energy insecurity in terms of its critical dimensions—access, affordability, and reliability—remains to be a major problem hindering the socioeconomic progress in the developing countries, as globally around 1 billion people lack access to electricity and nearly 3 billion people have to rely on raw biomass to meet cooking and heating requirements. Developments like fuel shortages, supply disruptions, and soaring energy prices are issues not just for the developing countries but also for the emerging and developed economies. Energy affordability is becoming a global problem as the post-COVID economic recovery is linked to the escalating energy prices around the world. Some of the EU member states, for example, have experienced electricity and gas prices increase by up to 500% in 2021. The Russia-Ukraine crisis has also sent shockwaves throughout the world with far-reaching implications for the global energy outlook.

Climate change is already upon us with implications like rising sea level, seasonal disorder, a trend of more frequent and intense weather-driven disasters such as flooding, droughts, heatwaves, wildfires, storms, and the consequent loss of lives and economy. The energy sector, for its severe environmental footprint, owes to lead the fight against climate change as also echoed by the Paris Agreement and more recently by COP 26. In response to climate change and to ensure a supply of energy compatible with the demands of a sustainable future for the planet, the energy sector is embracing a transition.

The unfolding energy transition is evolving in dynamics and scope to tackle the emerging energy and environmental challenges. This is not the first time the world is witnessing a major transformation in the energy sector. The present energy transition is, however, far more vibrant and meaningful compared to the earlier ones. The 18th- and the 20th-century switchovers of energy systems, respectively, from biomass to coal, and from coal to oil and gas—also regarded as energy transitions—primarily sought more efficient fuels in terms of logistics and utilization. In terms of energy resources and utilization technologies, the 21st-century energy transition, though predominantly pursues decarbonization of the energy sector, has several other important dimensions, such as distributed energy generation and digitalization of energy systems. Energy conservation and management is also an imperative part of this transition.

The *Handbook of Energy Transitions* presents a holistic account of the 21st-century energy transition. In terms of structure, the Handbook is divided into three sections. The first section—"Introduction to the 21st-Century Energy Transition"—consists of three chapters. These chapters provide a broader overview of the unfolding energy transition in terms of topics like dimensions, drivers, trends, barriers, policies, and geopolitics. The second section—"Environmental Transition: Resources and Technologies"—contains 12 chapters. Each of the chapters in this section discusses the ongoing energy transition in terms of a particular resource or technology. Chapters in this section have focused on technologies like renewable energy systems, solar energy, hydropower, hydrogen and fuel cells, electric vehicles, energy storage systems, batteries, digitalization, smart grids, blockchain, and machine learning. The third section—"Energy Transition: Policies and Prospects"—consisting of 10 chapters, discusses the present energy transition in terms of broader policy and developmental perspectives. Chapters in this section discuss topics like energy and sustainable development, the economics of energy and green growth, effects of the energy transition on economic growth, and role of various technologies and initiatives like renewables, nuclear power, and electrification in promoting energy security and energy transition.

# Acknowledgments

This book is a teamwork, and I am grateful to the chapter contributors for helping me accomplish it. I would like to thank reviewers for their time and efforts in reviewing chapter abstracts and manuscripts. I would also acknowledge the King Fahd University of Petroleum and Minerals (KFUPM) for the appreciative support.

# Editor

**Dr Muhammad Asif** is a Professor at the King Fahd University of Petroleum and Minerals. He is a Chartered Engineer, a Certified Energy Manager, and a Member of the Energy Institute. He has 20 years of teaching and research experience at various European and Middle Eastern Universities. His areas of research interests include energy transition, renewable energy, energy policy, energy security, sustainable buildings, and life cycle assessment. He has authored/edited six books and has published more than 100 journal and conference papers.

# Contributors

**Abdulazeez Abdulraheem**
Department of Petroleum Engineering
King Fahd University of
    Petroleum & Minerals
Dhahran, KSA

**Muhammad Abid**
Department of Mechanical Engineering
Interdisciplinary Research Center
COMSATS University Islamabad
Wah Campus, Pakistan

**Zaineb Abid**
School of Water, Energy and Environment
    (SWEE)
Cranfield University
United Kingdom

**Manzoor Khan Afridi**
Department of Politics and International
    Relations
International Islamic University
Islamabad, Pakistan

**Khaled H. M. Al-Hamed**
Department of Mechanical and Manufacturing
    Engineering
Faculty of Engineering and Applied Science
Ontario Tech. University
Oshawa, Ontario, Canada

**Ghulam Ali**
USPCASE, National University of Sciences
    and Technology (NUST)
Islamabad, Pakistan

**Waqas Ali**
Department of Electrical Engineering (RCET)
University of Engineering & Technology
Lahore, Pakistan

**Tarek Alskaif**
Information Technology group (INF)
Wageningen University & Research (WUR)
Wageningen, The Netherlands

**Ewere Evelyn Anyokwu**
Center for Development Research (ZEF)
University of Bonn
Bonn, Germany

**Hüseyin Turan Arat**
Department of Mechanical Engineering
Faculty of Engineering and architecture
Sinop University
Sinop, Turkey

**Muhammad Asif**
Department of Architectural Engineering
King Fahd University of Petroleum & Minerals
Dhahran, KSA

**Muhammad Asim**
School of Professional Education and Executive
    Development
The Hong Kong Polytechnic University
Kowloon, Hong Kong

**Niccolò Aste**
Department of Architecture, Built Environment
    and Construction Engineering
Politecnico di Milano
Milan, Italy

**Tabbi Wilberforce**
Department of Mechanical, Biomedical and
    Design Engineering
College of Engineering & Physical Science
Aston University
Birmingham, UK

**Young-Jin Baik**
Thermal Energy Systems Laboratory
Korea Institute of Energy Research
Daejeon, South Korea

**Tri Ratna Bajracharya**
Centre for Energy Studies (CES)
Institute of Engineering
Tribhuvan University
Lalitpur, Nepal

**Mustafa Kaan Baltacioğlu**
Department of Mechatronics Engineering
Faculty of Engineering and natural sciences
İskenderun Technical University
İskenderun, Turkey

**J. C. Barbosa**
Center of Physics
University of Minho
Braga, Portugal
and
Department of Chemistry and CQ-VR
University of Trás-os-Montes e Alto Douro
Vila Real, Portugal

**Maksud Bekchanov**
Department of Mathematics, Informatics and
    Natural Science (MIN)
University of Hamburg
and
Center for Earth System Research and
    Sustainability (CEN)
Research Unit Sustainability and Global
    Change (FNU)
Cluster of Excellence Climate, Climate
    Change, and Society (CLICCS)
Hamburg, Germany

**Cağlar Conker**
Department of Mechatronics Engineering
Faculty of Engineering and natural sciences
İskenderun Technical University
İskenderun, Turkey

**C. M. Costa**
Center of Physics
Institute of Science and Innovation for
    Bio-Sustainability (IB-S)
University of Minho
Braga, Portugal

**Sandip Deshmukh**
Department of Mechanical Engineering,
    Hyderabad Campus
Birla Institute of Technology & Science, Pilani
Hyderabad, India

**Ibrahim Dincer**
Department of Mechanical and Manufacturing
    Engineering
Faculty of Engineering and Applied Science
Ontario Tech. University
Oshawa, Ontario, Canada

**Salaheldin Elkatatny**
Department of Petroleum Engineering
King Fahd University of
    Petroleum & Minerals
Dhahran, KSA

**Haroon Farooq**
Department of Electrical Engineering (RCET)
University of Engineering & Technology
Lahore, Pakistan

**S. Ferdov**
Center of Physics
University of Minho
Braga, Portugal

**Anton Ming-Zhi Gao**
The Institute of Law for Science and
    Technology (ILST)
National Tsing-Hua University
Hsinchu, Taiwan

**Marco Giuli**
Department of Energy, Climate and Resources
    Programme
Brussels School of Governance
Vrije Universiteit Brussel (VUB)
Rome, Italy

**R. Gonçalves**
Center of Chemistry
University of Minho
Braga, Portugal

**Mabroor Hassan**
Department of Environmental Science
International Islamic University
Green Environ Sol (Private) Limited
Islamabad, Pakistan

**Muhammad Haseeb Hassan**
Hydrogen Energy Research Division
Korea Institute of Energy Research
and
Department of Advanced Energy and System
    Engineering, University of Science and
    Technology (UST)
Daejeon, South Korea

**Muhammad Imran**
Department of Mechanical, Biomedical
    and Design Engineering, College of
    Engineering & Physical Science
Aston University
Birmingham, UK

**Hafiz Ahmad Ishfaq**
Hydrogen Energy Research Division
Korea Institute of Energy Research
and
Department of Advanced Energy and System
    Engineering, University of Science and
    Technology (UST)
Daejeon, South Korea

**Vikrant Katekar**
Department of Mechanical Engineering
S. B. Jain Institute of Technology, Management
    and Research
Nagpur, India

**Muhammad Irfan Khan**
Department of Environmental Science
International Islamic University
Islamabad, Pakistan

**Muhammad Zubair Khan**
Department of Material Science and
    Engineering
Pak-Austria Fachhochschule: Institute of
    Applied Sciences and Technology
Haripur, Pakistan

**Zafar Ali Khan**
Department of Mechanical, Biomedical
    and Design Engineering, College of
    Engineering & Physical Science
Aston University
Birmingham, UK
and
Department of Electrical Engineering
Mirpur University of Science and Technology
Mirpur, Pakistan

**S. Lanceros-Mendez**
BCMaterials, Basque Center for Materials,
    Applications, and Nanostructures
UPV/EHU Science Park, Spain
Ikerbasque, Basque Foundation for Science
Bilbao, Spain

**Gijs Van Leeuwen**
Department of Human-Centered Design
Delft University of Technology
Delft, The Netherlands

**Fabrizio Leonforte**
Department of Architecture, Built Environment
    and Construction Engineering
Politecnico di Milano
Milan, Italy

**Rehan Liaqat**
Department of Electrical Engineering
Government College University (GCU)
    Faisalabad
Faisalabad, Pakistan

**Ahmed Abdulhamid Mahmoud**
Department of Petroleum Engineering
King Fahd University of Petroleum & Minerals
Dhahran, KSA

**Zubair Masaud**
Hydrogen Energy Research Division
Korea Institute of Energy Research
and
Department of Advanced Energy and System
    Engineering, University of Science and
    Technology (UST)
Daejeon, South Korea

**Hafiz Ali Muhammad**
Thermal Energy Systems Laboratory
Korea Institute of Energy Research
and
Department of Renewable Energy Engineering,
University of Science and Technology (UST)
Daejeon, South Korea

**Ghulam M. Mustafa**
Department of Physics
University of Education Lahore
Faisalabad Campus
Faisalabad, Pakistan

**Tahira Nazir**
Department of Management Sciences
COMSATS University Islamabad
Wah Campus, Pakistan

**Chinedu Miracle Nevo**
The Open University Business School
Milton Keynes, United Kingdom

**Chigozie Nweke-Eze**
Integrated Africa Power (IAP),
and
Institute of Geography, University of Bonn
Bonn, Germany

**Claudio Del Pero**
Department of Architecture, Built Environment
and Construction Engineering
Politecnico di Milano
Milan, Italy

**Liliana N. Proskuryakova**
Research Laboratory for Science and
Technology Studies
National Research University Higher School of
Economics
Moscow, Russia

**Saeed-ur-Rehman**
Hydrogen Energy Research Division
Korea Institute of Energy Research
Daejeon, South Korea

**Guller Sahin**
Department of Administrative and Financial
Affairs
Kutahya Health Sciences University
Kütahya, Turkey

**Intisar Ali Sajjad**
Department of Electrical Engineering
University of Engineering & Technology
Taxila, Pakistan

**Nobuhiro Sawamura**
Energy Statistics & Training Office
Asia Pacific Energy Research Centre
Tokyo, Japan

**Shree Raj Shakya**
Centre for Energy Studies (CES), Institute of
Engineering
Tribhuvan University
Lalitpur, Nepal
and
Institute for Advanced Sustainability Studies
(IASS)
Potsdam, Germany

**Anzoo Sharma**
Centre for Energy Studies (CES), Institute of
Engineering
Tribhuvan University
Lalitpur, Nepal
and
Center for Rural Technology (CRT/N)
Kathmandu, Nepal

**Bedir Tekinerdogan**
Information Technology Group (INF)
Wageningen University & Research (WUR)
Wageningen, The Netherlands

**Meltem Ucal**
Department of Economics
Kadir Has University
Istanbul, Turkey

**Muhammad Umer**
Department of Management Sciences
COMSATS University Islamabad
Wah Campus, Pakistan

# Section I

## Introduction to the 21st-Century Energy Transition

# 1 Dynamics of a Sustainable Energy Transition

*Muhammad Asif*
King Fahd University of Petroleum & Minerals

## CONTENTS

## 1.1 INTRODUCTION

Energy is the backbone of modern societies. The 18th-century industrial revolution transformed the human-energy relationship. Ever since, the extensive and efficient utilization of energy has played an instrumental role in human development. Energy is becoming an increasingly critical commodity on multiple fronts including technological, socio-economic, and geopolitical. Energy has attained the status of a prerequisite for all crucial aspects of societies, that is, mobility, agriculture, industry, health, education, and trade and commerce [1]. Energy resources exist in a wide range of physical states, which can be harnessed and capitalized through various technologies. Energy resources can be broadly classified into two categories: renewables and nonrenewables. Renewable energy resources are the ones that are naturally replenished or renewed. Examples of renewable resources include solar energy, wind power, hydropower, and wave and tidal power. Energy resources that are finite and exhaustible are termed nonrenewable such as coal, oil, and natural gas.

An important dimension of the human use of energy is its contribution to climate change. Unchecked emissions of GHS are leading to global warming. Climate change, as a result of global warming, is regarded as the biggest challenge facing the world. Different types of energy resources, especially fossil fuels, contribute to greenhouse gas (GHG) emissions. Fossil fuels are considered to be the primary reason for the anthropogenic emission of carbon dioxide ($CO_2$)—the 18th-century industrial revolution is considered to have triggered the rapid growth in the release of GHG [2]. The carbon dioxide ($CO_2$) concentration in the atmosphere, for example, has increased from the pre-industrial age level of 280 parts-per-million (ppm) to 415 ppm. The acceleration in the growth of $CO_2$ concentration can be gauged from the fact that almost 100 ppm of the total 135 ppm increment has occurred since 1960 [3].

DOI: 10.1201/9781003315353-2

Climate change is leading to a wide range of consequences such as seasonal disorder, a pattern of intense and more frequent weather-related events such as floods, droughts, storms, heatwaves and wildfires, financial loss, and health problems [4–6]. Climate change is also adversely affecting the water and food supplies around the world. Warmer temperatures are increasing sea level as a result of melting of glaciers. During the 20th century, the global sea level rose by around 20 cm. The pace of rise in sea level is accelerating every year—over the last two decades, it has been almost double that of the last century [7]. As a result of warmer temperatures, glaciers are shrinking across the world including in the Himalayas, Alps, Alaska, Rockies, and Africa. An extremely alarming dimension of climate change is that it is growing in momentum. Most of the temperature rise since the industrial revolution has occurred since the 1960s. Extreme weather conditions and climate abnormalities are becoming so frequent that the situation is already being widely dubbed as climate crisis. With the recorded acceleration in the accumulation of GHG and consequent increase in atmospheric temperature, the climate change-driven, weather-related disasters are becoming more intense and recurrent. The past 7 years have been observed to be the warmest since records began with 2020 being the hottest year ever [7]. The year 2021 set new records of natural disasters including heatwaves, wildfires, storms, and floods. Extreme weather events are now considered to be a new normal as experts predict more intense natural calamities including wildfires, storms, floods, and droughts.

The global energy scenario also faces a number of other challenges such as rapid growth in energy demand, depletion of fossil fuel reserves, volatile energy prices, and a lack of universal access to energy. Fast growth in the global energy demand—owing to factors such as surging population, economic and infrastructural development, and urbanization—is adding pressures on the energy supply chain. According to the Energy Information Administration (EIA), between 2018 and 2050, the world energy requirements are projected to increase by 50% [8]. Access to refined energy resources remains to be a major challenge for significant proportions of population in the developing countries.

The energy sector is experiencing a major transformation to address the energy and environmental challenges. The primary aim of this transition is to shift the global energy system away from fossil fuels. Renewable and low-carbon technologies are at the heart of this energy transition. This chapter presents an overview of the key dynamics of the ongoing energy transition. It defines the energy transition to have four main dimensions: decarbonization, energy efficiency, decentralization, and digitalization. In addition, this chapter discusses the prospects of these four dimensions (4Ds) of the energy transition, especially in terms of relevant technological and policy advancements.

## 1.2  KEY DIMENSIONS OF THE ENERGY TRANSITION

The use of energy is closely linked to the environment [9–11]. It is estimated that despite the pledges and efforts by the global community to tackle climate change, $CO_2$ emissions from energy and industry have increased by 60% since the United Nations Framework Convention on Climate Change (UNFCCC) signed in 1992 [12]. Climate change is also observed with its implications such as seasonal disorder, rising sea levels, a trend of more frequent and intense weather-driven disasters such as flooding, droughts, heat waves, wildfires, and storms, and associated financial losses [13,14]. The situation calls for an urgent paradigm shift across the entire energy sector. Responding to the challenges on hand, and to ensure a supply of energy compatible with the demands of a sustainable future for the planet, the global energy sector is going through a transformation. This energy transition can be defined as "The energy transition is a pathway toward transformation of the global energy sector from fossil-based to zero-carbon by the second half of this century." At the heart of the ongoing energy transition is the need to reduce energy-related $CO_2$ emissions to limit climate change [15].

In the recent history, mankind has witnessed two major energy transitions. The first energy transition propelled the industrial revolution mainly attributed to coal replacing biomass and wood as a more efficient and effective fuel to drive machines. The second energy transition was a shift from coal to more refined forms of fossil fuels—oil and gas—in the later part of the 20th century. The world is now experiencing the third energy transition. This energy transition is much more vibrant, intriguing, and impactful compared to the earlier ones. It is fundamentally a sustainability-driven energy pathway with the focus on decarbonization of the energy sector by shifting away from fossil fuels. This energy transition, therefore, can also be termed as "sustainable energy transition" or "low-carbon energy transition." Holistically, however, the ongoing energy transition is not just about going low carbon or shifting away from fossil fuels. It is rather much more diverse and comprehensive in terms of scope and impacts. The 21st-century energy transition is being propelled by unprecedented developments on the fronts of energy resources and their consumption, technological advancements, socio-economic and political response, and evolving policy landscape. This energy transition has four key dimensions: decarbonization, energy efficiency, decentralization, and digitalization [16].

## 1.3 DECARBONIZATION

Decarbonization of the energy sector is at the heart of the energy transition as reduction in $CO_2$ and other GHG emissions is fundamental in the fight against climate change. The energy sector can be decarbonized through a range of technologies and solutions such as renewable energy, electric vehicles (EVs), hydrogen and fuel cells, carbon capture and storage (CCS), and phasing out of fossil fuels. Renewable energy is the single most critical component of the decarbonization drive.

Through the Paris Agreement, the world has adopted the first-ever universally legally binding global climate deal to avoid dangerous climate changes by limiting global warming to well below 2°C. Alarm bells are, however, being frequently rung by the concerned circles including the United Nations Intergovernmental Panel on Climate Change (IPCC), that the world is seriously overshooting this target. In order to be anywhere closer to achieving this target, the world needs to make major changes in four big global systems: energy, land use, cities, and industry. The energy sector is where the greatest challenges and opportunities exist [17]. Energy systems have a huge variation in terms of their associated environmental emissions. Table 1.1, for example, shows the carbon dioxide emissions from different types of power generation technologies [18].

### TABLE 1.1
### Comparison of $CO_2$ Emissions from Different Energy Systems

| Power Plant | Type of Fuel/Energy | $CO_2$/(kg/kWh) |
| --- | --- | --- |
| Steam power | Lignite | 1.04–1.16 |
| Steam power | Hard coal | 0.83 |
| Gas turbine | Pit coal | 0.79 |
| Thermal power | Fuel oil (heavy) | 0.76 |
| Gas turbine | Natural gas | 0.58 |
| Nuclear power | Uranium | 0.025 |
| Solar thermal | Solar energy | 0.1–0.15 |
| Photovoltaic | Solar energy | 0.1–0.2 |
| Wind power | Solar/wind energy | 0.02 |
| Hydro-electric | Hydropower | 0.004 |

## 1.3.1 RENEWABLE ENERGY

Renewable energy is the primary pathway for the energy sector's low- or zero-carbon transition. Over the last couple of decades, renewable technologies, especially solar PV and wind turbines, have made a great progress in terms of technological developments and economic maturity. The global installed capacity of renewables increased from 2,581 GW in 2019 to 2,838 GW in 2020, exceeding expansion in the previous year by almost 50%. For several years now in a row, renewable energy is adding more power generation capacity compared to the combined addition of fossil fuels and nuclear power. In the year 2020, for example, renewables contributed to more than 80% of all the new power generation capacity added worldwide. The growth of the renewable sector is primarily being propelled by solar and wind power, with the two technologies accounting for 91% of the new renewables added during the year [19]. The annual growth and the cumulative installed capacity of solar PV and wind power over the last 10 years are shown in Figures 1.1 and 1.2,

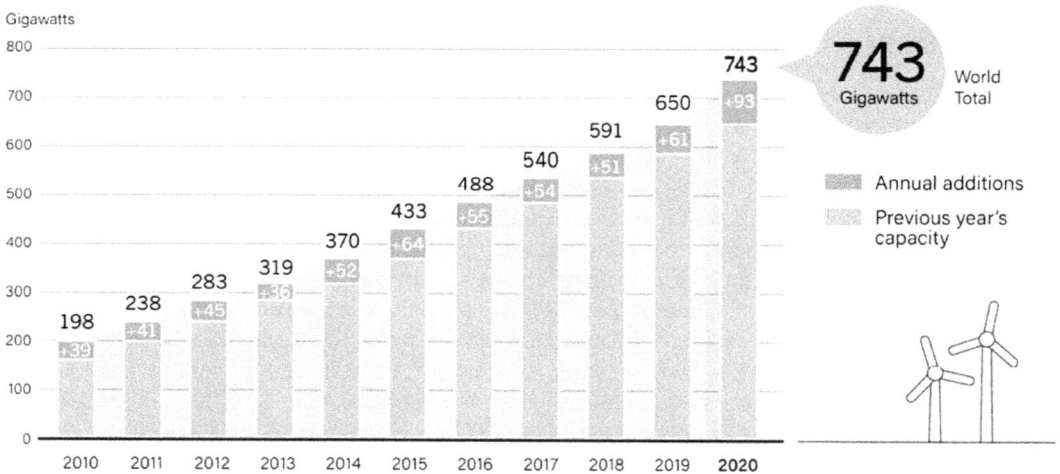

**FIGURE 1.1**  Growth in the solar PV sector between 2010 and 2020.

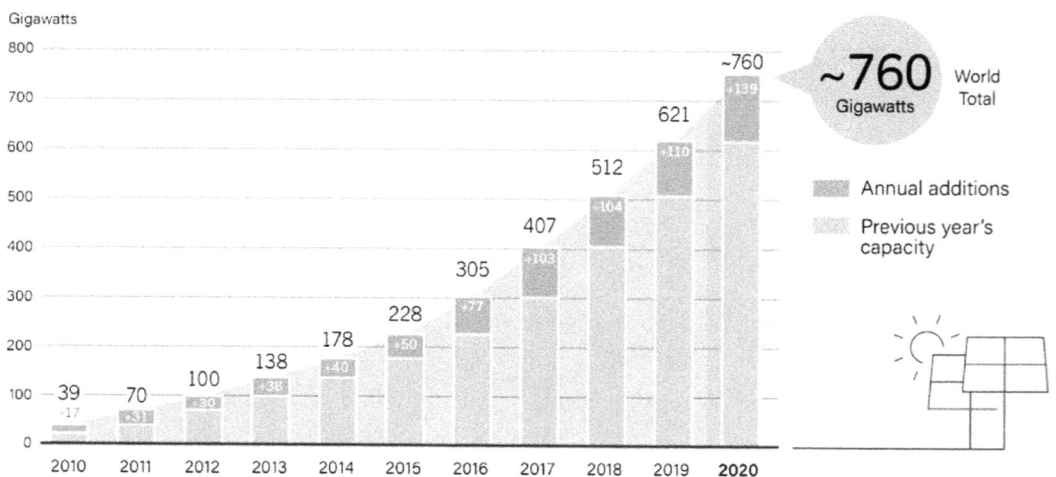

**FIGURE 1.2**  Growth in wind power between 2010 and 2020.

**FIGURE 1.3**  Concentrated solar PV.

respectively [19]. Renewable energy is already supplying 26% of the global electricity needs. According to IEA, to achieve the net-zero emissions by 2050, almost 90% of global electricity generation is to be supplied from renewables. There was over US$ 303 billion invested in renewable energy projects during the year [20]. The upward scale of the renewables development can be gauged from the fact that China has started developing the first 100 GW phase of massive solar and wind power initiative. The initiative is likely to be expanded to several hundreds of GW in capacity as China aims to develop 1,200 GW of renewables by 2030 [21]. The renewables growth trends are projected to continue, as the annual capacity addition of solar and wind power is set to grow four-fold between 2020 and 2030 [17]. It is also expected that by 2026, solar and wind power will account for around 95% of the total new capacity addition in the power sector.

The success of renewables has been propelled by technological advancements, economy of scale, and supportive policies. Solar and wind power industries are massively benefiting from the scientific and engineering advancements. Solar PV cells, for example, are becoming more efficient and reliable. Concentrated solar PV cells, as shown in Figure 1.3 [22], have achieved efficiency figures of over 40%. Figure 1.4 shows the advancements in the PV cell technology [23]. The progress of renewable technologies, especially solar energy systems, is significantly being helped by their broadening application domains. The building sector has been a vital area of application for solar PV and solar water heating systems [24–28]. PV systems are also being installed over agricultural farms, termed as Agrivoltaics. For PV and wind turbines, the issue of low power density is being addressed by off-shore applications. Application of PV on water bodies, that is, lakes, canals, rivers, and oceans, termed as floating PV, is becoming popular with the added advantage of higher system efficiency. Wind turbines are witnessing improvements both at the manufacturing and installation ends. Besides improvements in aerodynamic designs, advanced and sophisticated materials are helping develop larger, lighter, and stronger wind blades. These developments have enabled wind turbines grow rapidly in size, as shown in Figure 1.5. The off-shore application of wind turbines has significantly boosted the capacity factor. Within a couple of decades, larger and more sophisticated wind turbines and better site selection have resulted in an increase in

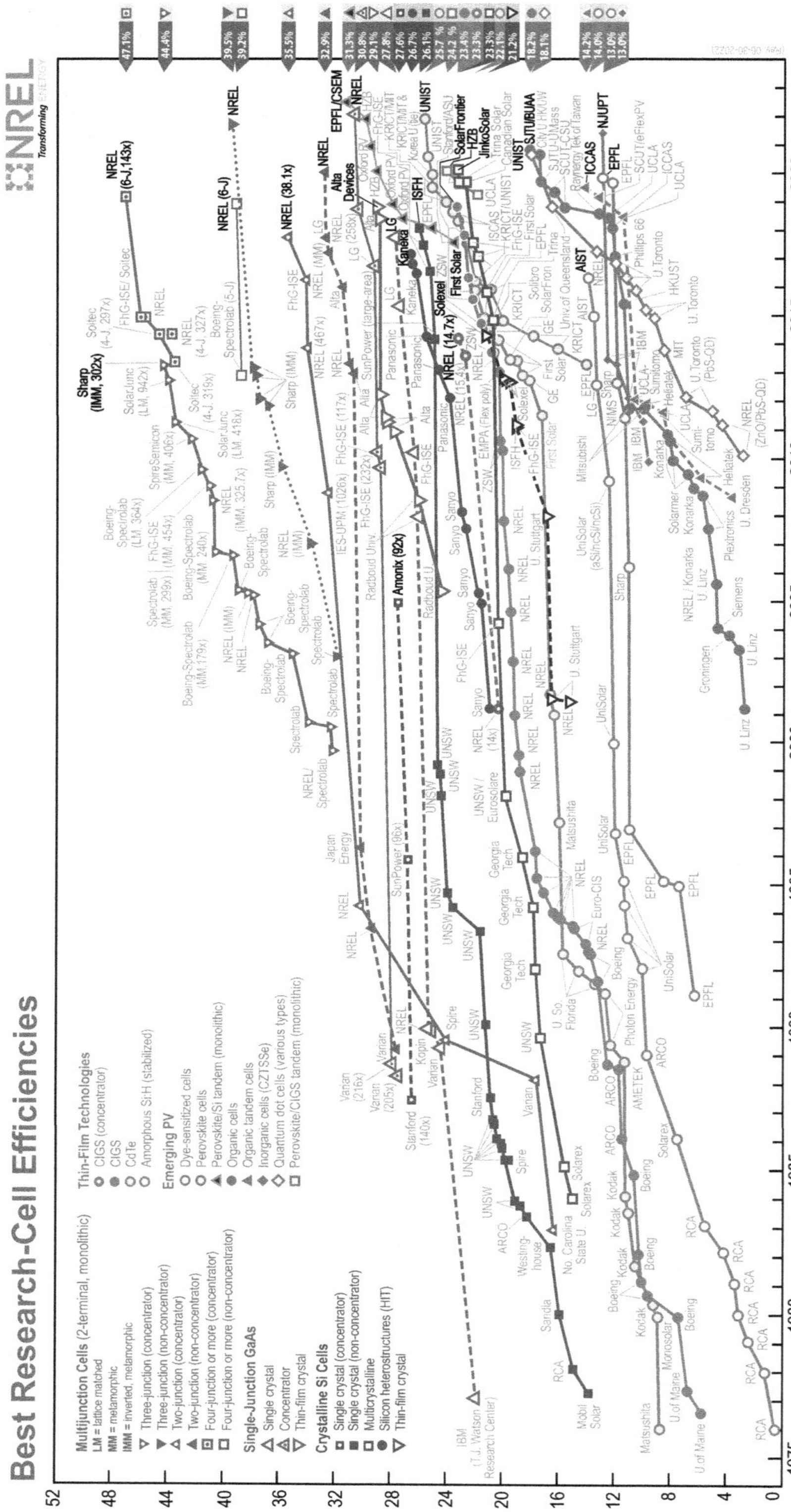

**FIGURE 1.4** Growth of solar cells in terms of types and efficiency.

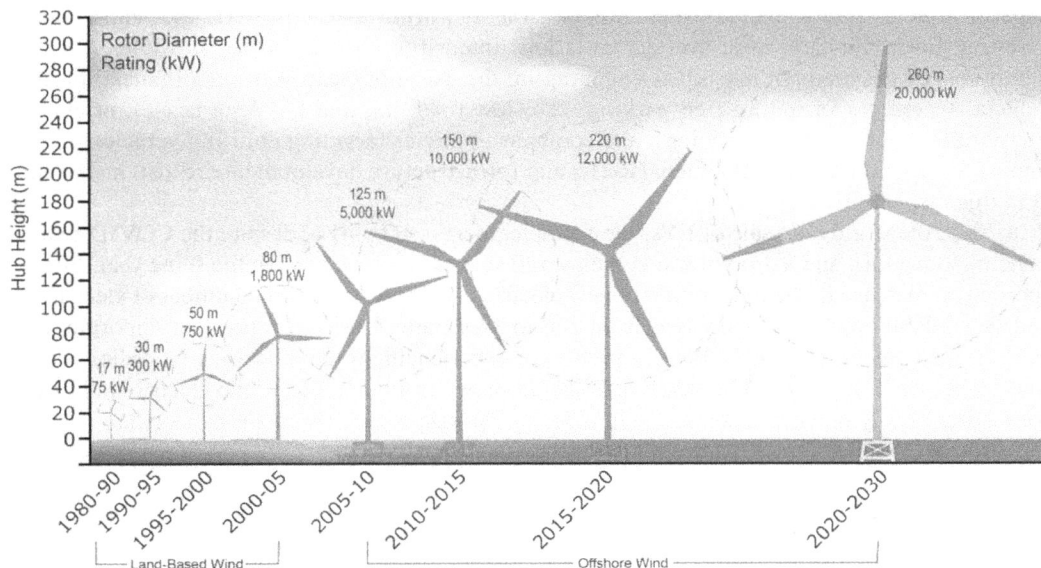

**FIGURE 1.5**    Growth in size of wind turbines.

the average annual capacity factor from 20%–30% to 40%–50%. Some off-shore wind turbines are now claiming to have a capacity factor over 60%.

### 1.3.2 Decarbonization in the Fossil Fuel Sector

Fossil fuels have traditionally led the energy sector, presently contributing to almost 80% of the total supplies. Oil and gas are the main fuels being used in the transport sector. Industrial sector also heavily relies on fossil fuels. Despite the grounds being gained by renewables, in a foreseeable future, fossil fuels are projected to be a major part of the global energy outlook. The emissions associated with fossil fuels, however, need to be curbed in order to avoid irreparable damage to the planet's ecosystem. Among the three main types of fossil fuels, coal is the most polluting one, while natural gas has the least environmental burdens. Within the fossil fuel sector, there is a gradual shift away from coal, especially in power generation, which can also be considered as a decarbonization trend. There is, however, need to have more profound decarbonization efforts. There can be two major pathways in this respect, which are carbon capture and storage, and transformation of fossil fuels into hydrogen or hydrogen-rich fuels. Removal of carbon from fossil fuels—which are fundamentally hydrocarbons—allows for the generation of hydrogen as a secondary energy source that does not produce $CO_2$. Carbon capture and storage (CCS), also termed as carbon capture and sequestration, involves removal of $CO_2$ from the direct combustion of fossil fuels for power generation or industrial processes. The idea is to prevent the release of $CO_2$ into the atmosphere, and instead store it in underground geological formations for the long term.

### 1.3.3 Electric Vehicles

Electric vehicles are leading the decarbonization efforts in the transport sector. Electric vehicles are environment-friendly, require low maintenance due to fewer components, are noise-free, and are convenient to use in urban areas. The growth of electric vehicles is being supported by a wide range of policies. These include standards (such as requiring a certain share of clean vehicles or setting limits for fleet-wide average emissions intensity); purchase price subsidies (i.e., tax exemptions or tax credits); incentives encouraging the usage of clean vehicles (financial or non-financial incentives including free parking, zero-low road tax, and bus lane usage); pricing of externalities (such as carbon pricing); and scrappage policies targeting emitting vehicles. Active support for research and development (R&D) and infrastructure development are also instrumental in this respect [3].

In 2020, the worldwide sale of EVs, for example, increased by 41%, despite the COVID-related economic downturn and a drop of 6% in the overall sale of vehicles. During the same year, Europe recorded an increase in the registration of new electric cars by 100% and the number of electric car models available worldwide increased from 260 to 370. Battery price has been an important cost factor in the economics of EVs. Battery prices are experiencing a rapid decline, recording a drop from $1,191/kWh in 2010 to $137/kWh in 2020, as shown in Figure 1.6. While electric mobility is also paving its way in the aviation and ship industry, the sale of electric cars is expected to increase from around 3.5 million in 2020 to over 23 million by 2030 [29].

### 1.3.4 ENERGY STORAGE

Energy storage is an important aspect of the energy value chain. The modern electricity infrastructure, especially in the wake of increasing supplies from renewable technologies, is finding energy storage to be critically important for its smooth operations. Energy storage is becoming of particular interest to power sector stakeholders, including utilities, end-users, grid system operators, and regulators.

Modern renewables like solar energy and wind power, being dependent on weather conditions, have the intermittency drawback. Solar radiation, for example, is available only during the daytime. While the daytime availability of solar radiation can be hindered by multiple weather conditions such as rain, snow, fog, and overcast conditions. Issues like dust storms, smog, haze, and smoke from wildfires also affect the intensity of solar radiation. Similarly, availability of wind is not a constant phenomenon either. Furthermore, even during their spells of availability, solar radiation and wind speed can fluctuate quickly and heavily, accordingly affecting the output from the respective systems. Renewable energy thus needs a backup storage to serve as a reliable source of energy.

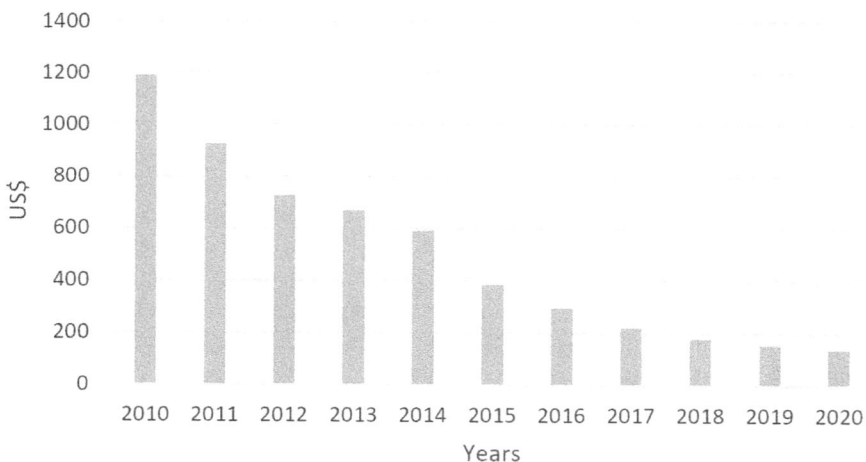

**FIGURE 1.6**   Declining price trend of lithium batteries.

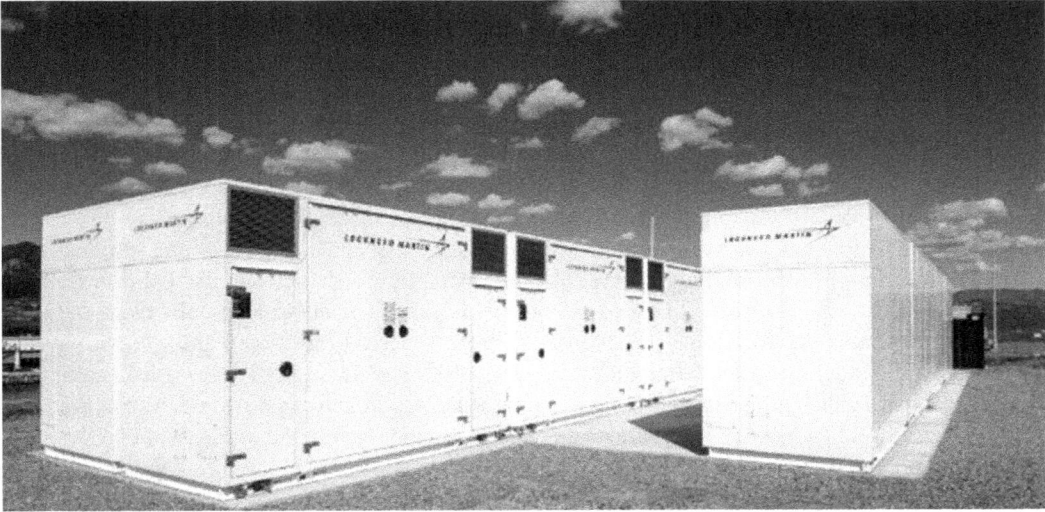

**FIGURE 1.7**   Utility-scale lithium-ion battery. (Photo: 56318 by Dennis Schroeder, NREL. With permission.)

Pumped storage hydropower projects have traditionally been used as the optimum large-scale energy storage solution. In recent years, battery technology has seen major techno-economic breakthroughs to become another option for large-scale energy storage. The 100 MW lithium battery storage developed by Tesla in Australia in 2017 has been a turning point in the field of large-scale battery storage. Large- and utility-scale battery storage systems, as shown in Figure 1.7 [30], have become a viable option. The United States has planned a 1,500 MW/6,000 MWh lithium-ion battery project, the first phase of which with a capacity of 300 MW/1,200 MWh started to operate in December 2020. Australia and the United Kingdom are also developing major battery storage projects. Some of these projects include a 1,200 MW project in New South Wales, a 700 MW system by Origin Energy Ltd., a 500 MW system in New South Wales, and a 300 MW facility in Victoria. The United Kingdom has over 1.1 GW of battery storage capacity in operation, while projects of 600 MW of cumulative capacity are under construction. An overview of the leading battery storage projects currently in operation around the world is provided in Table 1.2.

**TABLE 1.2**
**World's Largest Battery Storage Projects**

| Project | Capacity (MW) | Battery Technology | Country |
|---|---|---|---|
| Vistra | 300 | Li-ion | USA |
| Hornsdale power | 150 | Li-ion | UK |
| Stocking pelham | 50 | Li-ion | Australia |
| Jardelund | 48 | Li-ion | Germany |
| Minamisoma substation | 40 | Li-ion | Japan |
| Nishi-Sendai substation | 40 | Li-ion | Japan |
| Laurel AES | 32 | Li-ion | USA |
| Escondido substation | 30 | Li-ion | USA |

### 1.3.5  HYDROGEN AND FUEL CELLS

Hydrogen as a fuel has unique characteristics. Its use does not release any toxic emissions, and it has the highest calorific value compared to other commonly used fuels. Hydrogen is the simplest and one of the most plentiful elements in the universe. Despite its abundance, however, hydrogen does not occur independently—it exists bonded with other elements and is available in compound forms such as water, hydrocarbons, and carbohydrates [31]. Hydrogen can be produced from fossil fuels as well as renewables. In the former case, hydrogen can be produced through various routes such as reformation of natural gas, partial oxidation of heavy fossil fuels, and coal gasification. Production of hydrogen from fossil fuels—making up over 90% of the current hydrogen supplies—however, leads to GHG emissions, and is typically termed as "blue hydrogen." The environmentally clean option to produce hydrogen is electrolysis which involves splitting water into hydrogen and oxygen with the help of electricity. Ideally, the electricity used for electrolysis should be from renewables, making it a "clean hydrogen," which is also termed as "green hydrogen." Hydrogen can be stored, transported, and used for energy applications through various technologies. Hydrogen in the capacity of energy vector also has the potential to become the optimum solution for the renewables' intermittency by providing energy storage solutions. The vision of building an energy infrastructure that uses hydrogen as a fuel and energy carrier, a concept called hydrogen economy, is the path toward the full commercial application of hydrogen energy technologies [32]. Fuel cell is an important complimentary technology. It is a device that converts hydrogen directly into electricity through electrochemical oxidation while generating pure water as its byproduct. Fuel cells come in many types and are classified mainly in terms of the kind of electrolyte they use which in turn influences other features such as the involved electrochemical reaction, required catalysts, cell's operating temperature range, and the needed fuel [33].

## 1.4  ENERGY EFFICIENCY

The fast growth in the global demand for energy is pressurizing the entire energy value chain. A one-dimensional approach, of matching the growing energy demands with corresponding capacity addition, is not a sustainable solution, especially when the planet is already overshooting its bio-capacity by almost 70%. A sustainable energy pathway, aiming to satisfy the global energy requirements while protecting the environment, must begin by decreasing the consumption of energy by taking energy efficiency measures. Energy efficiency is regarded as a better solution to address energy shortages than adding new capacity. To industrial and commercial entities, energy efficiency delivers economic and environmental gains, besides offering competitive edge.

Energy efficiency is a broad domain in terms of the nature of facilities, types of energy losses and wastes, and the range of solutions available to improve the efficiency. Given the technological and policy advancements, the field of energy efficiency is continuously evolving. In terms of fundamental approach, energy efficiency solutions can be broadly classified into three types, as shown in Figure 1.8. An energy efficiency program, typically involving eradication of the unnecessary use of energy and improvement in the efficiency of the required energy, starts with an energy audit exercise. The type of energy audit to be carried out primarily depends upon the scope and objectives of the intended energy efficiency program. The energy audit process is also influenced by factors like available resources (funding, manpower, and time), type of facility, and provision of data and support. Detailed energy management programs can also include execution of the recommended solutions and post-implementation measurement and verification work to ensure the desired energy-saving goals are achieved.

The use of energy can be reduced across all major sectors including buildings, industry, and transport. Buildings account for over one-third of the global energy consumption [34–37]. Energy use in buildings can be reduced through a range of energy efficiency measures. Energy-efficient solutions for buildings can be broadly classified as active and passive energy-saving measures [38]. The choice of energy efficiency solutions depends on factors like the nature of the facility, site condition and local climate, desired levels of comfort and improvement, and financial situation. Through energy efficiency

Change of
Attitude

Energy
Efficiency

New
Technologies

Performance
Enhancement

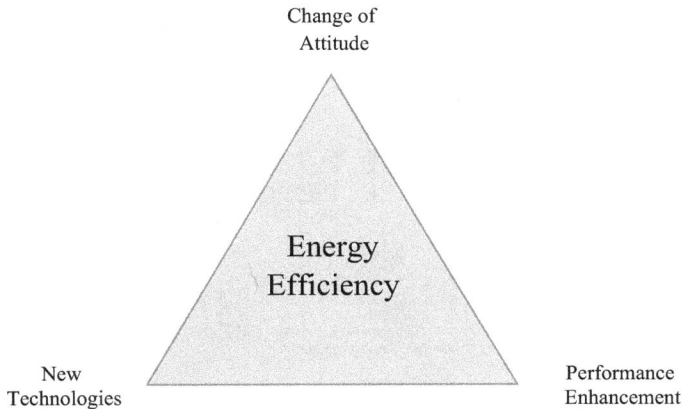

**FIGURE 1.8**  Classification of energy efficiency approaches.

measures, energy demand in existing and new buildings can be reduced by 30%–80% [39–42]. Energy efficiency in the transport sector can be improved through measures like incorporating fuel economy standards and eco-driving [43]. Digital technologies can also help save on fuel across the road, air, and sea transportation through optimization of routes. The industrial sector also offers a significant potential for energy efficiency, especially in the energy-intensive industry. Improvement in energy efficiency enables industrial entities to enhance their productivity and competitiveness, besides contributing to addressing energy and environmental problems locally, nationally, and globally. The energy efficiency drive in the industrial sector is also being helped by digital energy management technologies. It is estimated that with the help of proven and commercially viable technologies, energy use in the manufacturing industry can be reduced by 18%–26% [44].

Energy efficiency is beneficial both at the micro- and macro-levels. While it offers saving on bills to individual consumers, it also helps commercial and industrial entities in terms of economic competitiveness. While energy efficiency helps the utility and energy sector in terms of demand-side management and peak-load shaving, it also helps foster national security. The global economy could increase by $18 trillion by 2035 if energy efficiency is adopted as the "first choice" for new energy supplies, which would also achieve the emission reductions required to limit global warming to 2°C [45]. Besides enabling economic growth and improving energy security, energy efficiency can also play a vital role in the fight against climate change as it can deliver more than 40% of the reduction in energy-related GHG emissions over the next 20 years [46].

## 1.5 DECENTRALIZATION

Decentralized or distributed generation is the energy generated close to the point of use, as shown in Figure 1.9. Decentralized generation (DG) avoids/minimizes transmission and distribution setup, thus saving on cost and losses. It offers better efficiency, flexibility, and economy as compared to large and centralized generation systems. There are several energy technologies that can be used in DG systems depending on the application and type of project. Based on the type of energy resource, DG technologies can be classified into two categories renewables-based systems and nonrenewables-based systems. Renewables-based DG systems employ technologies like solar energy, wind power, hydropower, biomass, and geothermal energy. Some of these technologies can be further classified into different types. Solar technologies, for example, can be categorized into solar PV, solar thermal power, and solar water heating. Similarly, biomass can be used to deliver solid fuels, liquid fuels such as biodiesel and bioethanol, and gaseous fuels. Renewables-based DG systems offer several benefits such as reduced GHG emissions, and lower operation and maintenance costs.

## Centralized Generation                    Distributed Generation

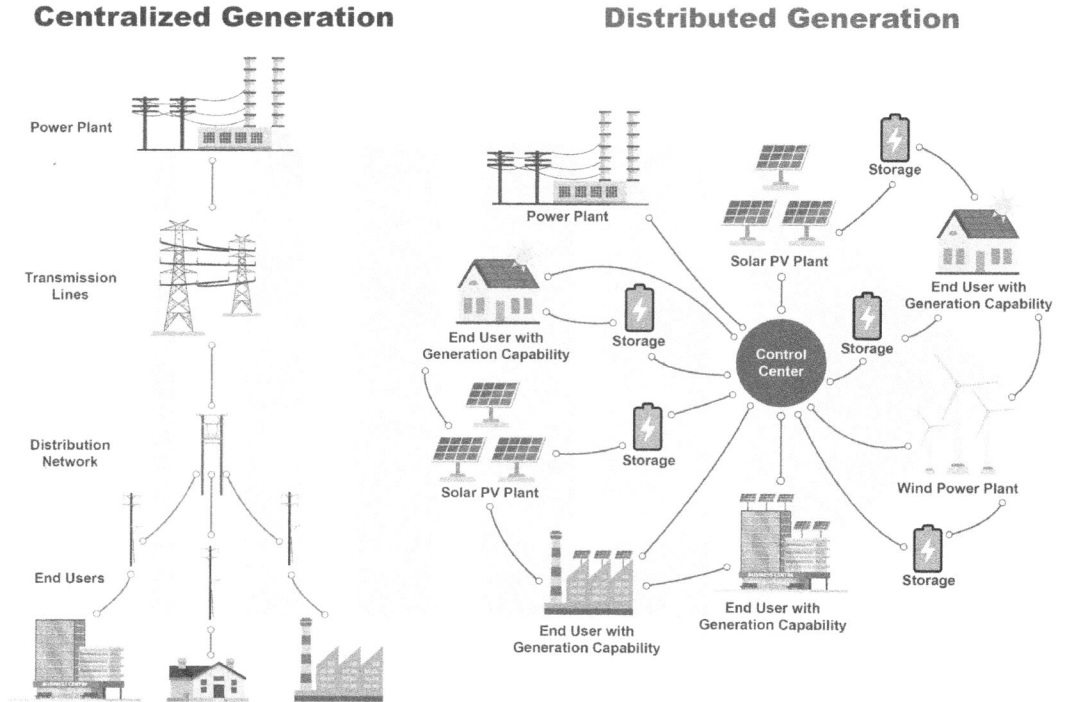

**FIGURE 1.9**   Overview of central and distributed generation systems.

**FIGURE 1.10**   Rooftop PV systems 45231 and 45180.

These systems, however, are typically intermittent and need energy storage to offer reliable solutions. Nonrenewables-based DG technologies are also available in a wide range and may include internal combustion engine, combined heat and power (CHP), gas turbines, micro-turbines, Stirling engines, and fuel cells. These technologies can use different types of fossil fuels.

Renewables like solar and wind power systems are leading the DG landscape. DG is playing an important role in the global electrification efforts and is presenting viable solutions for meeting modern energy needs and enabling the livelihoods of hundreds of millions who still lack access to electricity or clean cooking solutions [47]. Solar PV is one of the most successful DG technologies, especially at small-scale and off-grid levels. The building sector offers a tremendous potential for DG PV systems [27–28]. It is estimated that since 2010, over 180 million off-grid solar systems have been installed, including 30 million solar home systems. In 2019, the market for off-grid solar systems grew by 13%, with sales totaling 35 million units. Rooftop PV systems, as shown in Figure 1.10 [48], make up 40% of the total PV installations worldwide. Renewable energy had also supplied around half of the 19,000 mini-grids installed worldwide by the end of 2019. Efficient biomass systems such as improved cooking stoves and biogas systems are also helping the global efforts toward clean energy access. In 2020, the installed capacity of off-grid DG systems grew by 365 MW to reach 10.6 GW. Solar systems alone added 250 MW to have a total installed capacity of 4.3 GW.

## 1.6   DIGITALIZATION

Digitalization, also referred to as the Fourth Industrial Revolution, is driving the needed fundamental shift in the energy industry, which is also disrupting traditional market players [49]. Digitalization is a broad term in the context of energy sector. An important dimension of digitalization is collection and analysis of energy data to optimize energy demand and supply, and to achieve system efficiency and cost-effectiveness. While decarbonization, decentralization, and decreasing use of energy are transforming the energy sector, digitalization—through the proliferation of sensors, computing, communication, and predictive and control techniques—is also set to change the way energy services are realized and delivered. This is accomplished through a range of established and emerging technologies, above all artificial intelligence (AI). Digitalization in the perspective of business opportunities created in the energy sector can be regarded as the use of digital technologies to change a business process and enhance efficiency and revenue; it is the process of moving to a digital business. However it is defined, digitalization is having a profound impact on the global energy scenario. While digitalization is leading to new business models, it is also disrupting the existing models of generation, consumption, markets, and businesses and employment, potentially pushing some of the established ones on their way out [50]. Digitalization of the energy sector employs technologies like artificial intelligence, machine learning, big data and data analytics, Internet of Things, cloud computing, blockchain, and robotics and automation. These technologies are at various degrees of techno-economic maturity for their application in the energy sector. Digitalization is revolutionizing the energy sector by improving the productivity, safety, accessibility, and overall sustainability of energy systems. New, smarter ways of modeling, monitoring, analyzing, and forecasting energy production and consumption are helping the sustainable energy transition. With the range of advantages it offers, digitalization is also posing several challenges. Most importantly, the digital transformation heavily relies on large datasets, handling of which is increasingly exposing utilities and the energy industry to cyber security risks.

## 1.7   POLICY AND INVESTMENT TRENDS

Decarbonization is becoming the central part of the national and international energy policy frameworks across the world. As part of GHG emission targets, which mandate a reduction in overall emissions and can include net-zero and carbon-neutral targets, there have been new emission reduction commitments covering around 47% of total global emissions. Some examples in this respect are of China aiming to become carbon neutral by 2060, Japan by 2050, and the Republic of Korea by 2050. A major development is from the European Union (EU) committing to reduce carbon emissions by 55% from the 1990 levels by 2030 and to become net zero by 2050. The United Kingdom has planned to reduce carbon emissions by 68% by 2030 compared to the 1990 level to go net zero by 2050. A major decarbonization boost with far-reaching impacts at the global level has been the United States' rejoining of the Paris Agreement in 2021. The United States, after having walked out of the Paris Agreement in 2020, under the Nationally Determined Contribution (NDC) has committed to reduce emissions by 50%–52% below the 2005 levels by 2030, which is equivalent to a reduction of 40%–43% below the 1990 levels [51]. Overall, over 30 nations have incorporated climate neutrality by 2050 (or 2060) in laws, proposed legislation, or a national policy document [52]. There have also been significant decarbonization efforts on the part of other stakeholders including the energy, banking, and corporate sectors. World's leading corporations are becoming

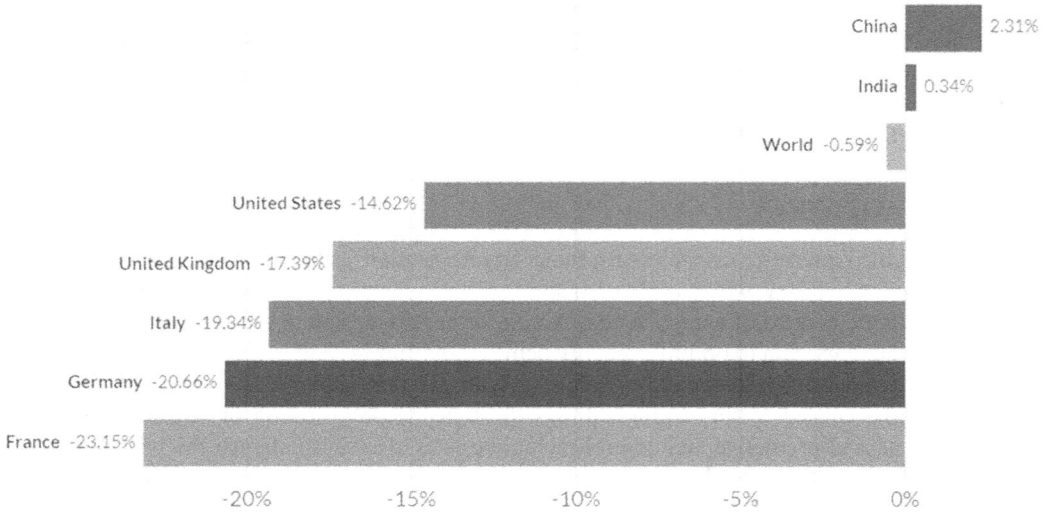

**FIGURE 1.11**  Annual changes in the coal power capacity.

increasingly aware of the threats associated with climate change and the business opportunities in taking action. It will help not only reduce their own emissions but also from their business associates. For example, the British insurer Prudential is working on an "energy transition mechanism" (ETM), with the Asian Development Bank on a scheme to buy out coal-fired power plants in Asia to shut them down within 15 years. The initiative involves insurance groups, Asian governments, and multilateral banks [53]. As a result of these developments, globally, with the exception of mainly China and India, all other major coal-consuming economies are reducing dependence on coal, with France, Germany, Italy, the United Kingdom, and the United States recording annual decline in coal consumption by 23.2%, 20.7%, 19.3%, 17.4%, and 14.6%, respectively, as shown in Figure 1.11 [54].

Despite the fact that coal has been a critical part of their energy mix, several countries have plans in place to phase out coal, while many others have decided to reduce its use. The United Kingdom, for example, has ambitious targets toward shifting away from coal. In 2020, it completely avoided the use of coal for more than 2 months in a row, for the first time in history [55]. The country has decided to close down all coal power plants by 2024. It means that within a decade, the country brings down its reliance on coal for power generation from nearly 40% to 0%. It is a major step toward the transition away from fossil fuels and decarbonization of power sector to eliminate contributions to climate change by 2050. Germany, one of the leading economies of the world which has traditionally heavily relied on coal, has also plans to phase it out by 2038. Similarly, France, Canada, and Denmark have plans to go coal-free before 2030. While the phasing out of coal is underway in many parts of the world, the use of carbon capture and storage (CSS) is deemed as a key incremental technology on the path to net-zero emissions. Decarbonization of the power sector, in particular, is also critical from the fact that majority of the energy used by any country is wasted. In the United States, for example, 61% of the total energy goes into wastes [56]. While energy waste exists across all sectors, over 90% of the losses are associated with power generation and transport. Currently, both these sectors overwhelmingly rely on fossil fuels; thus, hefty GHG emission is an integral part of the process. Switching power generation to renewable energy and transport to electric vehicles can help control the emissions associated with energy waste.

The decarbonization efforts are being crucially helped by the vital advancements in the renewable sector on technical, economic, and policy fronts. There have been significant technological improvements, for example, in terms of efficiency gains in the solar industry and improving capacity factors in the wind power sector. These technological breakthroughs, coupled with economy of scale, are helping renewables become economically competitive with the conventional energy options. The renewable energy policy landscape has also steadily improved over the years. From 2010 to 2020, the number of countries with regulatory incentives/mandates policies in the areas of power generation, transport, and heating and cooling has, respectively, increased from 81 to 145, 35 to 64, and 11 to 24 [20]. Getting to net-zero emissions by 2050 is, however, a mammoth task. According to the IPCC, a paradigm shift is needed across four major global systems: energy, industry, land use, and urban development. Besides firm policy commitments and technological advancements, the zero-emission target is estimated to require an investment of around $50 trillion by 2050 [57].

Renewable and low-carbon technologies are making significant progress in attracting investment. Global investment in the energy sector is set to rebound in 2021, reversing most of the drop in 2020 caused by the Covid-19 pandemic. Investment in the power sector is also set to rise. Over the last 10 years, investment in the power sector has been relatively stable compared with significant fluctuations in the oil and gas industry, mainly due to renewables. Worldwide, since 2010, the annual investment in renewable energy technologies has been over the US$ 200 billion mark. Global new investment on renewable projects (excluding hydropower projects larger than 50 MW) totaled US$ 301.7 billion in 2019, up by 5% from 2018. Within the renewable sector, the main focus of investment has been on wind and solar power. In 2019, investment in small-scale solar PV installations (less than 1 MW) increased by 43.5% to US$ 52.1 billion worldwide [20]. Renewables are dominating investments in the power sector, accounting for 70% of the total new investments in 2020. Importantly, investment in renewables is becoming more impactful. Money now goes further than ever in financing clean electricity, with a dollar spent on solar PV deployment today resulting in four times more electricity than 10 years ago, thanks to greatly improved technology and falling costs. The International Energy Agency (IEA) warns that not enough is going into clean energy, especially in emerging markets and developing economies, a new IEA report finds. The anticipated investment of US$ 750 billion in renewable and energy efficiency technologies in 2021 remains far below what's required for shifting the energy sector to a sustainable path [58].

## 1.8 CONCLUSIONS

In the backdrop of the fight against climate change, the global energy sector is experiencing an unprecedented transition. This energy transition is regarded as a pathway of shifting away from fossil fuel-based energy systems by the middle of the century. It is fundamentally a sustainable or low-carbon energy transition, which is having profound impacts across the entire energy value chain. The transition is not just about becoming carbon neutral or zero carbon; it is rather much more vibrant and impactful, thanks to the changes and advancements occurring on the fronts of energy resources and their consumption, technological solution, socio-economic adjustments, and political and policy response. The 21st-century energy transition has four main dimensions: decarbonization, energy efficiency, decentralization, and digitalization. Decarbonization of the energy sector is the most important dimension of the energy transition. Reduction in $CO_2$ and other GHG emissions is fundamental to the fight against climate change. The energy sector can be decarbonized through a range of technologies and solutions including renewable energy, electric vehicles, hydrogen and fuel cells, carbon capture and storage, and phasing out of fossil fuels. Replacement of fossil fuels with renewable energy is the most critical part of the decarbonization drive. Renewable energy is already supplying over 25% of the global electricity needs. To achieve the net-zero emissions by 2050, almost 90% of global electricity generation is to be supplied from renewables. Renewable energy has already become an important stakeholder in the energy sector, accounting for over 80% of the global newly added power generation capacity in 2020. Energy efficiency is an imperative part of the energy transition with massive scope across various sectors, especially in buildings, industry, and transport. Global economy could increase by $18 trillion by 2035 if energy efficiency is prioritized as a solution toward addressing energy supply issues. Decentralized or distributed generation offers efficiency, flexibility, and economy, and is thus regarded as an integral part of the unfolding energy transition. Digitalization is revolutionizing the energy sector by improving the productivity, safety, accessibility, and overall sustainability of energy systems. New, smarter ways of modeling, monitoring, analyzing, and forecasting energy production and consumption are helping the sustainable energy transition. While the advanced countries like EU and OECD member states are relatively well positioned to decarbonize their economies, developing countries critically lack the needed resources including finance, technical infrastructure, knowledge and awareness, and policy frameworks. There is thus a need for strong international partnerships and support toward the developing nations to achieve the global decarbonization targets.

## REFERENCES

1. M. Asif, *Energy and Environmental Security in Developing Countries*, Springer, ISBN: 978-3-030-63653-1, 2021.
2. M. Asif & T. Muneer, Energy supply, its demand and security issues for developed and emerging economies, *Renewable & Sustainable Energy Reviews*, Vol. 11, No. 7, September 2007, pp. 1388–1413.
3. M. Asif, *The 4Ds of Energy Transition: Decarbonization, Decentralization, Decreasing Use, and Digitalization*, Wiley, ISBN: 978-3-527-34882-4, 2020.
4. M. Asif, *Energy and Environmental Outlook for South Asia*, CRC Press, ISBN: 978-0-367-67343-7, 2021.
5. H. Qudratullah & M. Asif, *Dynamics of Energy, Environment and Economy: A Sustainability Perspective*, Springer, ISBN: 978-3-030-43578-3, 2020.
6. M. Asif, *Energy and Environmental Security, Handbook of Environmental Management*, Taylor & Francis, Florida, 2019.
7. NASA, Climate change: How do we know, facts, National Aeronautics and Space Administration, Evidence | Facts – Climate Change: Vital Signs of the Planet (nasa.gov).
8. EIA, *EIA Projects Nearly 50% Increase in World Energy Usage by 2050, Led by Growth in Asia*, Today in Energy, U.S. Energy Information Administration (EIA), 24 September 2019.
9. M Asif, Energy and Environmental Security, *Encyclopaedia of Environmental Management*, Taylor & Francis: New York, 2013; Vol. II, pp. 833–842.
10. Kh. Nahiduzaman, A. Al-Dosary, A. Abdallah, M. Asif, H. Kua & A. Alqadhib, Change-agents driven interventions for energy conservation at the Saudi households: Lessons learnt, *Journal of Cleaner Production*, Vol. 185, 1 June 2018, pp. 998–1014.
11. M. Asif, An empirical study on life cycle assessment of double-glazed aluminium-clad timber windows, *International Journal of Building Pathology and Adaptation*, Vol. 37, No. 5, pp. 547–564, Doi: 10.1108/IJBPA-01-2019-0001.
12. IEA, *Net Zero by 2050: A Roadmap for the Global Energy Sector*, Flagship report, International Energy Agency, May 2021.
13. M. Asif, A. Dehwa, F. Ashraf, M. Shaukat, H. Khan & M. Hassan, Life cycle assessment of a three-bedroom house in Saudi Arabia, *Environments*, Vol. 4, No. 3, 2017, p. 52, Doi: 10.3390/environments4030052.
14. M. Khan, M. Asif & E. Stach, Rooftop PV potential in the residential sector of the kingdom of Saudi Arabia, *Buildings*, Vol. 7, No. 2, 2017, p. 46, Doi: 10.3390/buildings7020046.
15. IRENA, *Energy Transition*, International Renewable Energy Agency, Energy Transition (irena.org).
16. M. Asif, Role of energy conservation and management in the 4D sustainable energy transition, *Sustainability*, Vol. 12, 2020, p. 10006, Doi: 10.3390/su122310006.
17. E. Gillam and R. Asplund, *Will Solar Take the Throne*, Invesco, August 2021.
18. M. Asif, *Energy Crisis in Pakistan: Origins, Challenges and Sustainable Solutions*, Oxford University Press, ISBN: 978-0-19-547876-1, 2011.
19. IRENA, *World Adds Record New Renewable Capacity in 2020*, Press Release, 5 April 2021, International Renewable Energy Agency, World Adds Record New Renewable Energy Capacity in 2020 (irena.org).
20. REN21, *Renewables 2020 Global Status Report*, Renewable Energy Network, 2020.
21. J. Scully, China signals construction start of 100GW, first phase of desert renewables rollout, PV-Tech, 12 October, 2021, (pv-tech.org).
22. NREL, *Concentrated Solar PV*, Photo by Dennis Schroeder, National Renewable Energy Lab, 2021.
23. NREL, *Best Research-Cell Efficiency*, National Renewable Energy Laboratory, 2021.
24. A. Dehwah & M. Asif, Assessment of net energy contribution to buildings by rooftop PV systems in hot-humid climates, *Renewable Energy*, Vol. 131, February 2019, pp. 1288–1299, Doi: 10.1016/j.renene.2018.08.031.

25. A. Dehwah, M. Asif & M. Tauhidurrahman, Prospects of PV application in unregulated building rooftops in developing countries: A perspective from Saudi Arabia, *Energy and Buildings*, Vol. 171, 2018, pp. 76–87.
26. A. Mahmood, M. Asif, M. Hassanain & M. Babsail, Energy and economic evaluation of green roof for residential buildings in hot humid climates, buildings,
27. M. Asif, Urban scale application of solar PV to improve sustainability in the building and the energy sectors of KSA, *Sustainability*, Vol. 8, 2016, p. 1127, Doi: 10.3390/su8111127.
28. M. Asif, Growth and sustainability trends in the GCC countries with particular reference to KSA and UAE, *Renewable & Sustainable Energy Reviews Journal*, Vol. 55, 2016, pp. 1267–1273.
29. K. Adler, Global electric vehicle sales grew 41% in 2020, more growth coming through decade: IEA, HIS Markit, 3 May 2021, https://ihsmarkit.com/research-analysis/global-electric-vehicle-sales-grew-41-in-2020-more-growth-comi.html.
30. NREL, *Utility Scale Lithium Ion Battery*, Photo: 56318 by Dennis Schroeder, National Renewable Energy Lab, 2021.
31. M. Asif, T. Muneer & J. Kubie, Security assessment of importing solar electricity for the EU, *Journal of Energy Institute*, No 1, March 2009, 102–105.
32. T. Muneer, M. Asif & S. Munawwar, Sustainable production of solar electricity with particular reference to the Indian economy, *Renewable & Sustainable Energy Reviews*, Vol. 9, No. 5, October 2005, 444–473.
33. DOE, *Types of Fuel Cells, Energy Efficiency and Renewable Energy, Types of Fuel Cells*, Department of Energy.
34. M. Hamida, W. Ahmed, M. Asif & F. Almaziad, Techno-economic assessment of energy retrofitting educational, *Sustainability*, Vol. 13, No. 1, 2021, p. 179, Doi: 10.3390/su13010179.
35. W. Ahmed & M. Asif, BIM-based techno-economic assessment of energy retrofitting residential buildings in hot humid climate, *Energy and Buildings*, Vol. 227, Doi: 10.1016/j.enbuild.2020.110406.
36. R. Alawneh, F.E.M. Ghazali, H. Ali & M. Asif, A new index for assessing the contribution of energy efficiency in LEED certified green buildings to achieving UN sustainable development goals in Jordan, *International Journal of Green Energy*, Vol. 6, 2019, pp. 490–499, Doi: 10.1080/15435075.2019.1584104.
37. A. Alazameh & M. Asif, Commercial building retrofitting: Assessment of improvements in energy performance and indoor air quality, *Case Studies in Thermal Engineering*, Doi: 10.1016/j.csite.2021.100946.
38. W. Ahmed & M. Asif, A critical review of energy retrofitting trends in residential buildings with particular focus on the GCC countries, *Renewable and Sustainable Energy Reviews*, Vol. 144, July 2021, p. 111000, Doi: 10.1016/j.rser.2021.111000.
39. H. Khan & M. Asif, Impact of green roof and orientation on the energy performance of buildings: A case study from Saudi Arabia, *Sustainability*, Vol. 9, No. 4, 2017, p. 640, Doi:10.3390/su9040640.
40. H. Khan, M. Asif & M. Mohammed, Case study of a nearly zero energy building in Italian climatic conditions, *Infrastructures*, Vol. 2, No. 4, 2017, p. 9, Doi: 10.3390/infrastructures2040019.
41. A. Mahmood, M. Asif, M. Hassanain & M. Babsail, Energy and economic evaluation of green roof for residential buildings in hot humid climates,
42. W. Ahmed, M. Asif & F. Alrashed, Application of building performance simulation to design energy-efficient homes: Case study from Saudi Arabia, *Sustainability*, Vol. 11, No. 21, 2019, p. 6048, Doi: 10.3390/su11216048.
43. K. Kojima & L. Ryan, *Transport Energy Efficiency, Energy Efficiency Series*, International Energy Agency, September 2010.
44. M.T. Hassan, S. Burek & M. Asif, Barriers to industrial energy efficiency improvement- Manufacturing SMEs of Pakistan, *Energy Procedia*, Vol. 113, 2017, pp. 135–142.
45. UNEP, *Energy Efficiency: The Game Changer*, United Nations Environment, https://www.unep.org/explore-topics/energy/what-we-do/energy-efficiency.
46. IEA, *Energy Efficiency 2020*, Final Report, International Energy Agency, December 2020.
47. M. Asif, *Energy and Environmental Outlook for South Asia*, CRC Press, ISBN: 978-0-367-67343-7, 2021.
48. NREL, *Rooftop PV Systems*, Photos: 45231 and 45180 by Dennis Schroeder, National Renewable Energy Lab, 2021.
49. DNV-GL, *Digitalization and the Future of Energy*, DNV-GL, 2019, Digitalization_report_pages.pdf (smartenergycc.org).
50. IEA, *Digitalization and Energy*, technical report, International Energy Agency, November 2017.
51. CAT, *Ambitious US Target Upgrade Reduces the 2030 Global Emissions Gap by 5–10%*, Climate Action Tracker, 23 April 2021.

52. J.D. Sachs, C. Kroll, G. Lafortune, G. Fuller & F. Woelm, *Sustainable Development Report 2021, The Decade of Action for the SDGs*, Cambridge University Press, 2021, Doi: 10.1017/9781009106559, https://s3.amazonaws.com/sustainabledevelopment.report/2021/2021-sustainable-development-report.pdf.
53. J. Ambrose, Prudential in talks to buy out and shut coal-fired plants in Asia, *Guardian*, 3 August 2021.
54. OWID, *Annual Percentage Change in Coal Energy Consumption*, Our World in Data, 2019 (ourworld-indata.org).
55. M. Farmer, What does Britain's two months without coal power mean? *Power Technology*, 11 June 2020.
56. CT, *US Wastes 61–86% of Its Energy*, CleanTechnica, 26 August 2013.
57. M. Stanley, *Decarbonization: The Race to Zero Emissions*, Morgan Stanley, 25 November 2019.
58. IEA, *Global Energy Investments Set to Recover in 2021 But Remain Far from a Net Zero Pathway*, International Energy Agency, Press Release, 2 June 2021.

# 2 Energy Transitions
## *Trend, Drivers, Barriers, and Policies*

*Tri Ratna Bajracharya*
Tribhuvan University

*Shree Raj Shakya*
Tribhuvan University
Institute for Advanced Sustainability Studies (IASS)

*Anzoo Sharma*
Tribhuvan University
Center for Rural Technology (CRT/N)

## CONTENTS

DOI: 10.1201/9781003315353-3

## 2.1  TREND AND TRANSITIONS OF GLOBAL ENERGY

The leaps of humanity are well defined by the transition of energy uses in different eras. Tens of millions of years ago, the life relied on solar energy. The harnessing of fire changed the genetic and cultural trajectory of mankind. Maritime trade and population dispersion over vast land was made possible with the harnessing of wind power for sailing ships. Horsepower eased distance, meanwhile enabling the exchange of ideas, enriching science, art, and religion, and also determined the rise and fall of empires. James Watt's coal-fired steam engine in 1776 commenced the "Fossil Fuel Age." With the ability to harness coal, oil, and gas, humanity tapped into a seemingly limitless reserve of power, both defined by Physics and Politics. Had the climate crisis such as intense storms, heatwaves, droughts, massive forest fires, rising sea levels, floods, and emerging infectious diseases, triggered by $CO_2$ emissions not been extensively detectable, the fossil fuel era would continue for several centuries. The Earth's average temperature has risen more than 1.2°C since the late 19th century and if the trend continues, the temperature can be expected to rise 3.2°C above preindustrial levels by the end of this century (UNEP, 2019). For global warming to stay within 1.5°C, the world must achieve net-zero greenhouse gas (GHG) emissions by 2050, that is, the known fossil reserves have to stay in the ground. The scientists and engineers have signaled that decarbonization by 2050 is feasible by shifting power generation from fossil fuels to zero-carbon energy sources (renewable sources, nuclear or fossil fuel with carbon capture and storage) supply energy to downstream uses notably transport, building, and industry.

For approximately over half a century (1971–2018), the world has experienced a significant shift within fossil fuels and other sources (Figure 2.1). The lighting market that used to be dominated by oil, is captured by electricity while oil captured the transportation market. Transportation and home heating markets transitioned from coal to oil, meanwhile, coal dominates the fastest-growing electricity market. According to IEA (2020c), in 2018, almost two-thirds of the transportation sector is fueled by oil (65.63%) while the industrial sector takes 80% share of coal consumption and 37.1% share of natural gas (NG) consumption and 30% of NG is consumed in the residential sector

Electricity has the fastest growth rate among energy sectors (Figure 2.2): its demand is increasing at an annual rate of 4.6%, while the global total final energy consumption was increasing at 1.94% per annum from 2000 to 2018. The electricity demand is likely to increase further due to the rapid increase in energy access in developing countries, and electrification is noted as a strong tool for decarbonization. Though the electric end-uses (services) are carbon-free but not necessarily the generation process. High-income countries have largely stopped using coal, peat, and biomass for heat and mechanical work but extensively use them to generate electricity (Figure 2.3). Since most of the electricity is generated by burning fossil fuel (64%) (IEA, 2020b), the decarbonization of the power sector through higher penetration of low-carbon sources, renewable energy sources, and demand-side management are crucial. Electricity and heat producers contributed to 39.34% of the total 33.5Gt of global $CO_2$ emission by combustion, followed by transport (23.24%), and industry

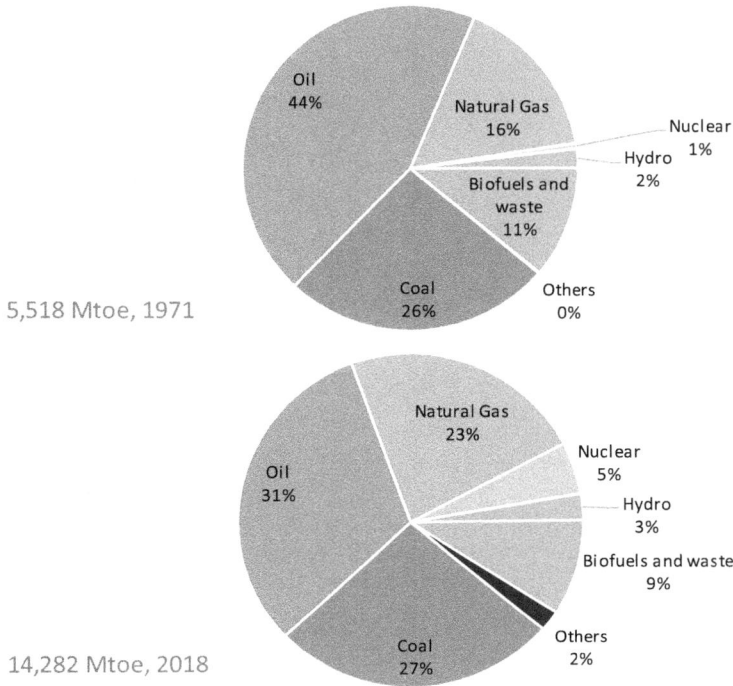

**FIGURE 2.1** World total energy supply 1971–2018. (From IEA, 2020c, *Key World Energy Statistics 2020*, Retrieved from https://www.iea.org/reports/key-world-energy-statistics-2020. With permission.)

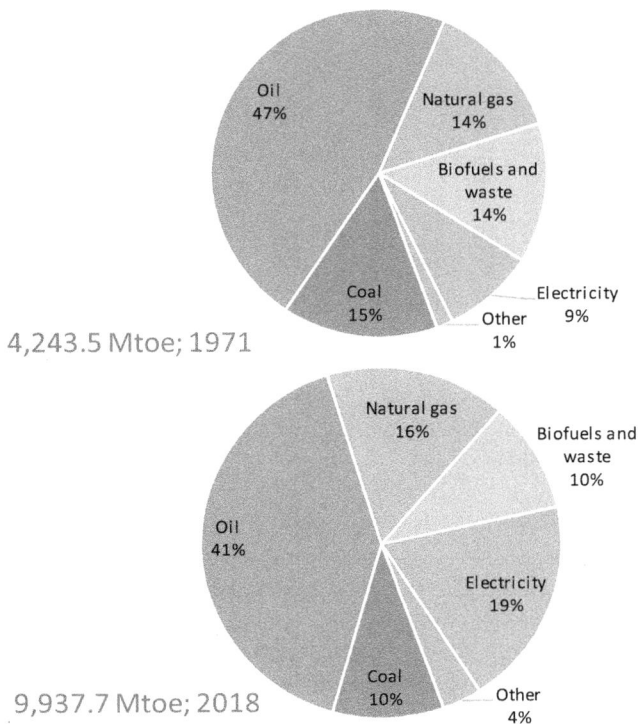

**FIGURE 2.2** World total final consumption by source 1971–2018.

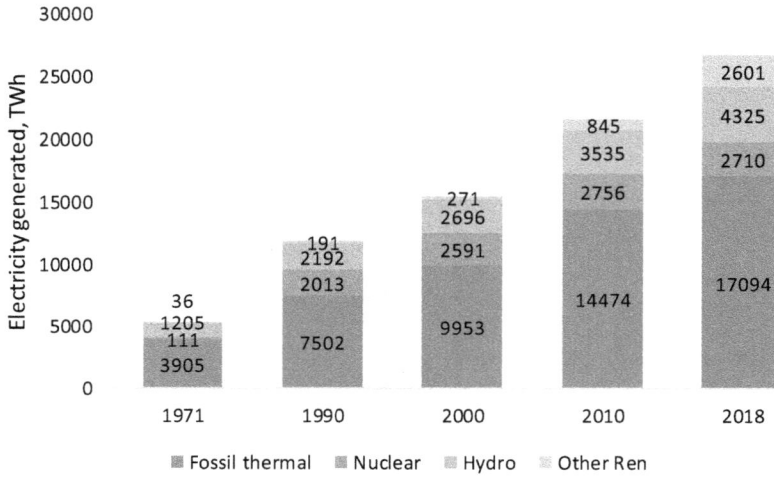

**FIGURE 2.3** World electricity generation by fuel, 1971–2018 (TWh). (From IEA, 2020c, *Key World Energy Statistics 2020*, Retrieved from https://www.iea.org/reports/key-world-energy-statistics-2020. With permission.)

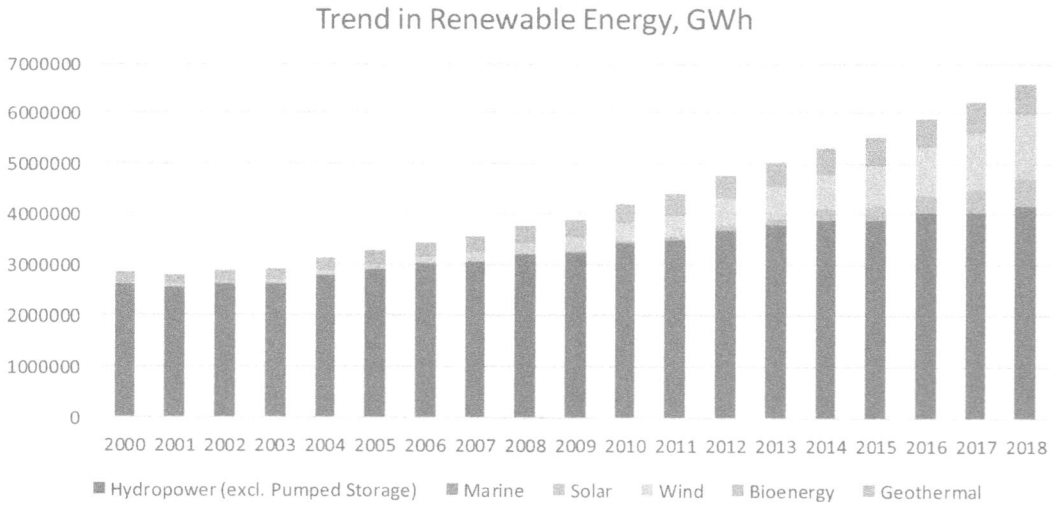

**FIGURE 2.4** Global trend in renewable Energy, GWh. (From IRENA, 2020b. *Statistics Time Series*, Retrieved February 18, 2021, from https://irena.org/Statistics/View-Data-by-Topic/Capacity-and-Generation/Statistics-Time-Series. With permission.)

(17.33%) (IEA, 2020a). From 2010 to 2018, the supply of coal, oil, and NG have a growth rate of 0.63%, 1.08%, and 2.22%, respectively, while that of renewable sources is 16.43%

The technologies for generating electricity using renewable sources and zero carbon emission are at various stages of maturity (Figure 2.4). Hydropower is the most mature and most versatile source, its application ranging from baseload to peak power matched with reservoir or pumped storage plants. The next promising renewable sources are wind and solar. According to IRENA (2020b), the total global installed capacity of renewables is 2,537 GW: hydropower is 1,190 GW, wind is 623 GW, solar PV is 586 GW, and other 638 GW.

## 2.2 DRIVERS OF ENERGY TRANSITIONS

Demand and supply situations and enabling environment are the major factors affecting an energy transition in the country. The availability and cost of energy resources, the invention of energy conversion technologies, and their interaction with the environment can change the pattern of energy use to fulfill a demand for energy services. For example, over the past century, there is a significant transition in the demand for illumination starting from the use of coal oil in a lamp which is followed by the use of coal gas in a gaslight to presently coal-produced electricity in a lightbulb, using the same resource but changing the conversion technologies for resources or end-use devices (O'connor, 2010). But after post-1900, issues related to the climate change, environmental pollution, and sustainable development have led to the use of cleaner source and renewable form of energy. On top of that, energy efficiency has played an important role in optimizing the use of scarce energy resources. Today, increasing concerns about the impacts of global climate change have brought together the international community to limit the increase in the global average temperature within 1.5°C relative to preindustrial levels by end of this century as per the COP 21 Paris Agreement. To achieve the target, studies have shown that net-zero emission should be reached by 2050 which requires a substantial change in energy generation, transformation, and consumption technologies (IRENA, 2021d).

There are certain factors that govern the direction of the energy transition as pointed out by O'connor (2010).

### 2.2.1 SUPPLY CONSTRAINTS

It occurs when energy demand exceeds the resource available. Sometimes, extraction of energy resources at a finite level can be maintained at a low cost, but it may be economically infeasible for large-scale expansion resulting in a relative decline in share over time. In other cases, maintaining a fixed level of extraction may lead to resource depletion, resulting in a rise in prices and a decline in demand. During fast economic growth, supply constraints are often seen as demand growth surpasses regular supply. In some cases, other external factors like environmental, social, and economic issues also create a constraint on the extraction and supply of energy resources.

Examples of energy transitions under supply constraints include the following:

- coal replacing firewood in the United Kingdom;
- decrease in the share of hydropower in electricity generation mix since 1949 in the United States;
- decrease in the use of whale oil despite its rising prices after 1846;
- limited use of geothermal energy which may be cost-effective at a low scale application but not possible to expand economically on large-scale use;
- reduction in the use of kerosene in the urban areas of Nepal despite government subsidy.

### 2.2.2 COST ADVANTAGES

Choice of fuel is not only governed by the actual cost of the fuel, but it is also determined by the associated labor cost, energy converting devise cost, and other externality costs from economic impacts. Even though the energy source is not facing supply constraints, it may be replaced by alternative fuels due to economic cost disadvantages. The effect of cost advantages can be seen even in a slowly rising market when emerging alternative fuels commercialize and gain market share.

Examples of cost advantage factors in energy transition include the following:

- replacement of electricity from rural distributed generation plants (e.g., micro-hydro plant, wind turbine) by the national grid as it extends to those areas;
- use of coal replaced by NG for building heating applications;

- wood replaced by coal in locomotive applications in the United States;
- limited use of nuclear power plants in the U.S. market (not supply-constrained, but cost-disadvantaged);
- oil sand petroleum mining operation in Canada during 1973 Arab oil embargo and 1990, 2001, 2003 middle east crisis.

### 2.2.3  Performance Advantages

It includes an accounting of all performance benefits except resource price such as better speed and acceleration to cleanliness and safety. These can be considered as a subcategory of cost advantages and consist of an implicit value established by consumer behavior (though these qualities are not explicitly valued in some cases). They can be observed when introducing a new fuel which is often linked to the developments in energy conversion technologies.

Examples of performance advantage factors are as follows:

- evolution in naval ships from wind-powered to coal-driven steam engines and then to diesel combustion engines;
- early acceptance of electric lighting, even though it was more expensive than oil or gas lamps;
- iceboxes replaced by refrigerators;
- streetcars and trolleys being replaced by the modern automobile;
- Choice of a higher-power engine despite its cost disadvantage while buying a car;
- Fuel switching from kerosene and LPG to electric cooking.

### 2.2.4  Policy Decisions

They play an instrumental role in driving the energy market of the particular fuel. The energy supply and demand can be affected by the policy-based actions taken by governments like subsidies, tariffs, infrastructure development, taxes, regulations, and other measures.

Examples of policy decision factors include the following:

- OPEC supply restrictions resulting in price shocks during 1973 and 1979;
- Public sector investment in the hydropower projects, transmission systems, distribution systems, and so on;
- Land grants provided to support the development of railroad network;
- Subsidies provided for various appropriate energy technologies (for rural electrification, solar rooftop programs, etc.);
- Energy and carbon taxes on the use of various fuels;
- Ban on the selling of incandescent bulbs in the European market;
- Development of charging station and parking area for promoting electric vehicles.

### 2.2.5  Energy and Environment Standard Ratings and Codes

Standardization of performance and control of environmental impacts for various energy generating and end-use technologies plays a crucial role in the transition of the energy system toward an efficient and low-carbon path.

Examples of standard ratings and codes affecting energy transitions include the following:

- Energy efficiency rating of energy-using technologies;
- Building codes limiting energy intensity;
- Use of Carbon footprint for grading of energy-using products.

## 2.3 BARRIERS OF ENERGY TRANSITIONS

Poor access to clean technology and value chains, absence of financing mechanisms, and trade restrictions are the major barriers to a successful transition toward a low-carbon or sustainable energy system. These barriers are highlighted by Goldthau, Eicke, and Weko (2020) in the case of the Global South as follows.

### 2.3.1 TECHNOLOGY TRANSFER AND VALUE CHAINS

Access to technology for clean and affordable energy and inclusion of the participating country in the value chains of the energy supply plays a crucial role in the ability of the country for the low-carbon energy transition. Though renewable energy resources like solar and wind are free, the technology required to convert them into final energy commodities is expensive and accessible to limited developed countries. Generally, they are developed and manufactured by a few developed and emerging economy countries which can invest in research and development for innovation and are well established in the supply value chains directly supporting their national economy. Developing countries are becoming only the final consumers of such imported technologies with limited capability to diversify technology as per the local need. Besides, the absence of the role in the supply value chain of clean energy technology will create minimal effects on the local economy, thus demotivating policymakers to adopt low-carbon energy transition. Many developing countries are deprived of participation in the international clean technology flows (Glachant & Dechezleprêtre, 2016) due to the lack of supportive government policy, subsidies to fossil fuels, limited financing ability, and low capacity to absorb and manage technology (Kirchherr & Urban, 2018). This has created the carbon lock-in at technological, infrastructural, institutional, and behavioral levels. The possibility for such countries to get out from the carbon lock-in depends on their capacity to bear the costs of switching away from carbon-intensive systems and availability of alternative options, institutional reform capability, and behavior and cultural change (Seto et al., 2016).

So far, many nationally determined contributions (NDCs) submitted to the UNFCCC secretariat mostly by developing countries are strongly conditional on the availability of climate finance and the accessibility of low-carbon technologies to meet their committed emission mitigation targets. In addition, most of the small developing countries have limited technological capacity and small domestic markets, making them less attractive to local and foreign investments (Pueyo, Mendiluce, Naranjo, & Lumbreras, 2012). Another aspect of the poor technology transfer may create market distortion in the low-carbon technology due to the monopolies (Goldthau et al., 2020).

### 2.3.2 FINANCIAL RISK AND PATH DEPENDENCY

Energy infrastructure in many countries is still fossil fuel-based with huge capital lock-in in such infrastructure. Today, many developing countries are found at the transition phase between leapfrog from a carbon-intensive economy and copying the carbon lock-in economy that was once followed by the developed countries (Goldthau et al., 2020). As many countries and emerging economies are in the process of determining their future national energy systems, the strategic decisions made today on the energy infrastructure development will determine the energy mix of the countries for decades to follow (Seto et al., 2016). An emerging economy like China has invested in overseas fossil fuels (reported to be USD 128 bn) significantly higher than that made in renewables (reported to be USD 32 bn) between 2000 and 2018 (Li, Gallagher, & Mauzerall, 2018, 2020; Wright, 2018). Such investments are responsible for creating path dependency on the fossil infrastructure due to the economy of scale and capital lock-in, which may result in a delay in the adoption of low-carbon technologies despite their economic viability for decades in the investment recipient country as well (Unruh & Carrillo-Hermosilla, 2006). As decarbonization in the energy sector is one of the major options to mitigate global climate change impacts, delay in the departure from such a lock-in will result in high transaction costs due to the requirement of changes on already built infrastructure and established rules, and (political and economic) institutional mechanisms (Seto et al., 2016).

One of the implications of achieving the low-carbon transition to limit global warming to well below 2°C as per the 2015 Paris Agreement is that almost two-thirds of already identified fossil fuel reserves will have to remain untapped (IEA, 2018a). Much of the fossil fuels (oil, coal, or gas) which are currently listed in the records will not be monetized, creating economic vulnerability. It has been estimated that loss in the fossil fuel assets value will reach up to $100 trillion by 2050 (Curmi et al., 2015). It has been suggested that the resulting global "carbon bubble" could lead to losses similar to the financial crisis in 2008 (Mercure et al., 2018). As per estimates of IEA (2018b), there would be a risk of losing approximately USD 7 trillion by the oil- and gas-producing states under the Paris Agreement scenario. This risk can be further enhanced, resulting in economic vulnerability for many developing countries which are forced to invest in infrastructure to meet the rising fossil fuel energy demand.

### 2.3.3 TRADE BARRIER

Globally, the energy transition is also related to the energy and electricity trade practices. One school of thought suggests that countries will become more self-sufficient through the indigenous renewable energy installations and global energy trade will slow down (Scholten, 2018). An opposing view suggests that the energy trade volumes can be expected to stay at the same level or rise further (Schmidt et al., 2019). The technological innovations in the renewable energy power sector and fuel technologies along with resource availability will create price variations among countries ultimately resulting in international trade (Goldthau et al., 2020). Some of the energy trading infrastructures like gas pipelines might be upgraded and used for hydrogen trade, helping to lessen the risk associated with stranded investment and hedge the probable financial losses from the decarbonization process (Schmidt et al., 2019). These possible future pathways will decide the affairs between the countries that are present energy net-importers, transit, and net exporters (Goldthau et al., 2020). During the development of renewable infrastructures, it is also anticipated that a regional electricity trade will increase with the construction of more sophisticated trans-border power grids to balance variations in the generation of renewables energy supply (Bahar & Sauvage, 2013). It is also suggested that the regional integration in such networked grids might happen around power centers (Goldthau, Westphal, Bazilian, & Bradshaw, 2019). As an example, China has been observed investing heavily in the strategic grid infrastructures as part of its increasing Belt and Road Initiative. Past evidence from the European Union (EU) and other regions of the world suggests that in addition to administrative capacity, political stability and mutual trust are vital prerequisites for effective governance structures of cross-border energy trading infrastructures (Goldthau et al., 2019). As such energy and power trading infrastructure need huge investment and lead time which will ultimately affect the pace of low-carbon energy transition. Besides, the future geopolitical situation will also play a major role in the effective implementation of low-carbon energy trade.

## 2.4 POLICIES IN ENERGY TRANSITIONS

Energy transition implies a significant structural change in an energy system. The world has gone through various energy resource transitions, energy carrier transitions, energy service transitions, and energy converter transitions. In the contemporary world, the energy transition refers to the pathway toward the transformation of the global energy sector from fossil-based to zero carbon by the second half of this century. Even though energy transitions historically have been driven by supply constraints, cost, and performance advantages, policy decisions are observed to be the most pivotal considering the urgency to limit global warming. Policy decisions such as tariffs, subsidies, vehicle standards for cars, energy efficiency labeling for appliances, regulations, infrastructural development, and other measures revolving around climate and sustainability have been implemented by many governments shifting the way energy is produced and consumed. Energy policies are and will continue to be a fundamental driver in the clean energy transition, accelerating renewables deployments, incentivizing adaptation of energy efficiency, strengthening changes in the energy systems, and concreting the role of Carbon Capture, Utilization, and Storage (CCUS). Solar PV and wind have succeeded in being the frontier on green power technologies with the continuous decline in the cost and the policies have been evolving over the past two decades in order to tackle various market-entry barriers.

Energy transition policies around the globe can be classified as below (Daszkiewicz, 2020).

### 2.4.1 Targets Setting and Strategic Planning Policies

Target setting is the foremost step in any strategy and policymaking. IRENA (2015) defines renewable energy targets as "numerical goals established by governments to achieve a specific amount of renewable energy production or consumption in the electricity, heating/cooling or transport sectors." Targets are complemented by a roadmap or action plan that is developed through analyses and consultations on the availability of resources, potential technologies, cost-competitiveness, stability of power grid, and energy demand evolution. As of the end of 2021, 156 countries have regulatory renewable power policies and some 135 countries have renewable power targets (REN21, 2022). In order to reduce dependency on Russian fossil fuels, European Commission is assessing the possibility of increasing renewable targets from 40% to 45% by 2030. As of 2020, the share of renewals in EU final energy consumption is 22.1%, around 2% above the 2020 target (Renewable energy statistics,2022).

### 2.4.2 Policies Targeting Upfront Investment Costs

The high capital cost of most of the low-carbon technologies acts as a barrier to their deployment. Fiscal and financial policies are introduced to reduce the upfront costs and foster a cheaper financial environment for investors. These policies can be implemented under various schemes, for example, tax benefits or tax waivers, soft loans, capital grants, and rebates, and can be revised, altered, or withdrawn with the government's regular budgetary plan (Daszkiewicz, 2020).

### 2.4.3 Policies Targeting Energy Generation

In 2019, a total of 244 GW of power plants were added globally, of which 72% were renewables. Much of the renewable policies are aimed at increasing installation in the power sector. *Feed-in tariffs* (FITs) refer to a policy mechanism developed toward the late 1970s for the support of renewable energy technologies. It is an administratively set price per kWh injected into the electricity grid, and usually has a contract duration of 10–15 years. For the high learning rates and decreasing technology cost of renewables (Table 2.1), other price-finding mechanisms are being adopted in developed countries while it is still an important mechanism to promote renewable energy generation where the financing opportunities are limited, and investment is generally riskier. *Feed-in premiums* (FIPs) are similar to FITs but are also exposed to electricity markets. The overall price per unit electricity generated by a project is the sum of the market electricity and the FIP which can be set to constant over time. FITs, auctions, and premiums have all been key in enabling the massive increase of renewable energy capacity over the last decade. According to IEA (2019) analysis, two-third of renewable capacity to be commissioned from 2019 to 2024 will have fixed remuneration, set competitively, mostly through *auctions*. For utility-scale renewable electricity technologies (>5 MW), mostly solar PV and onshore wind, many European, Latin American, North America, African and Asian countries are shifting from administratively set tariffs to competitive auctions for long-term power purchase agreements. Auction is an emerging price-discovery tool.

**TABLE 2.1**

**Global Weighted-Average Total Installed Cost and LCOE of Utility-Scale Renewable Power Plants**

| Types of Utility-Scale Power Plant | Global Weighted-Average Installed Cost (USD/kW) | | Global Weighted-Average LCOE (USD/kWh) | |
|---|---|---|---|---|
| | **2010** | **2019** | **2010** | **2019** |
| Fossil fuel-powered | | | | 0.05–0.177 |
| Solar PV | 4,702 | 995 | 0.378 | 0.068 |
| Onshore wind | 1,949 | 1,473 | 0.086 | 0.053 |
| Offshore wind | 4,650 | 3,800 | 0.161 | 0.115 |
| Concentrated solar power | 8,987 | 5,774 | 0.35 | 0.182 |
| Hydro power | 1,254 | 1,704 | 0.037 | 0.047 |
| Geothermal | 2,588 | 3,916 | 0.049 | 0.073 |
| Biofuel | 2,588 | 2,141 | 0.076 | 0.066 |

*Source:* From IEA, 2020d, *Projected Costs of Generating Electricity 2020*, Retrieved July 22, 2021, from https://www.iea.org/reports/projected-costs-of-generating-electricity-2020. With permission; From IRENA, 2020a, *Renewable Power Generation Costs in 2019*, International Renewable Energy Agency, Abu Dhabi, Retrieved June 30, 2021, from https://www.iea.org/data-and-statistics/data-browser?country=WORLD&fuel=CO2 emissions&indicator=CO2BySector. With permission.

### 2.4.4 REGULATORY POLICIES

Reduction in technology cost of renewable energy has increased its penetration in the grid. However, the intermittent nature of ocean power, solar PV, and wind adds challenges to overall grid operation and management. Therefore, the grid integration of variable renewable energy (VRE) must follow certain rules on connection and dispatch to ensure adequacy, reliability, quality, and affordability of electricity. To combat the effect of variability on the system, the policymakers should adopt a long-term energy transition vision and set regulatory frameworks that foster: (i) integration of VRE into diverse electricity resources, (ii) smart grid infrastructure, and (iii) demand-side management mechanisms like standards, labels, mandates, and obligations.

### 2.4.5 OTHER POLICIES

Renewable energy delivers multifaceted benefits and must be quantified in the energy system. Apart from training, research, development, and deployment of programs, policymakers must also prioritize education and information dissemination to reach the grassroots level so that no one is left behind in the journey of the energy transition.

Though scientists have pointed that renewables energy and energy efficiency measures can potentially achieve 90% of the required carbon reduction (IRENA, 2021b), decarbonization of the energy sector is a complex process as its management and governance are characterized by large uncertainties and ambiguities. The energy transition is not only about the technology transition from fossil fuels to renewables but must undertake the social, economic, and environmental aspects of the development.

## 2.5  GLOBAL INITIATIVE

Climate change, the consequence of global warming activated by GHG emissions from energy production and utilization, is a global threat and must be dealt with globally. The climate change discussion started with the establishment of the Intergovernmental Panel on Climate Change (IPCC)4 in 1988. The United Nations General Assembly at its 45th session decided to start the process of formulation of the framework convention on climate change. As a result, the text of the United Nations Framework Convention on Climate Change was prepared which was adopted in May 1992 and entered into force in 1994. The adoption of the UNFCCC embarks on a new history in an effort of the global community to address challenges of climate change. After the UNFCCC enters into force until now 25th Conference of Parties (COP) to the UNFCCC have been organized and several decisions were made. Various clean energy transition policies are formulated at a global level, binding the global members for sustainable development. One of the major ones is the Paris Agreement.

## 2.5.1 THE PARIS AGREEMENT

At COP 21, on December 12, 2015, a groundbreaking agreement was attained, and it was called the Paris Agreement. The Paris Agreement (2015) is a legally binding international treaty on climate change. The parties agree for accelerating and intensifying the actions and investment required for a sustainable future, thereby limiting global warming preferably below 1.5°C compared to preindustrial levels. The agreement is ratified by 196 parties. The agreement works on a 5-year cycle. By 2020, the member countries are supposed to submit nationally determined contributions (NDCs) that contain the targets and a list of actions to cut down GHG emissions and to build resilience to adapt to the impacts of rising temperatures including the finance flow required. The Paris Agreement also provides a framework for financial, technical, and capacity-building support to those countries who need it. Starting from 2024, countries are obligated to transparently report on actions taken and progress made in climate change mitigation, adaptation measures, and support provided/received. The achievement so far is that more and more countries, regions, cities, and companies have established carbon neutrality targets and have created low-carbon solutions and markets, especially in the power and transport sectors which is responsible for over 70% of global emissions. The energy transition is the fundamental transformational effort to achieve the SDGs and Paris Agreement targets. According to IRENA's World Energy Transitions Outlook (2021c), by 2050, an investment of USD 33 trillion is required into efficiency, renewables, end-use electrification, power grids, flexibility, hydrogen, and innovation. The benefits through the revitalization of economies, saving on human health, sustainable environment, and adaptation and mitigation of climate changes supersede the cost of investment.

## 2.5.2 INTERNATIONAL BEST PRACTICES ON LONG-TERM STRATEGY ON NET-ZERO EMISSION

Launched in 2019, UNFCCC's Race to Zero campaign has brought "net zero targets" to mainstream. 128 countries and self governing territories have set targets for net zero, 10 of them have set target years after 2050 (Mooldijk et al., 2022). Bhutan is the first carbon-neutral country and plans to become net-zero GHG emission and zero waste country. Bhutan's intended nationally determined contribution (INDC) suggests the adoption of climate-smart agriculture and livestock farming as potential measures to reduce GHG emissions from the agriculture sector. Besides, sustainable forest management and conservation of biodiversity are deemed to be options to increase sequestration potential from the land-use sector. Singapore and Japan have planned to achieve net-zero GHG emissions in the second half of the century. Singapore aims to have 80% green building by 2030 and zero private vehicle growth to reach the target. The circular economy approach in water and waste management is important to reduce waste and increase recycling. The adoption of Carbon Capture Utilization and Storage (CCUS) technologies that are still not matured would have a significant role in the future (NCCS, 2020). The EU and remaining other countries aim to achieve net-zero by 2050. In the United Kingdom, resource and energy efficiency, CCS technology, reducing methane venting or leakage, and fuel switching from fossil fuels to hydrogen, electricity, and bioenergy are considered to be the long-term strategy to become carbon neutral. The use of heat pumps, biomass-based boilers, and hydrogen boilers are considered as the potential mitigation measures in the industry as well as buildings. In transport, phasing out diesel trains and replacing IC engine vehicles with BEV and plug-in hybrid electric vehicles (PHEV), improving logistic efficiency along with investment in cycling and walking infrastructure are considered as potential options (CCC, 2019). Similarly, in other countries, the low-carbon strategy includes energy efficiency improvement, fuel switching, use of advanced low-emission technologies (existing and under development phase), afforestation/reforestation, and CCUS as potential options.

Long-term net-zero target is a challenge to every country as they need to transform their economies and is also an opportunity toward sustainable economic growth that will avoid possible climate impacts. In addition to setting net-zero targets and identifying viable options, a broad range of policies and measures are also needed such as framework legislation and strategies (e.g., climate laws and long-term strategies), economic instruments (e.g., carbon taxes, subsidy reform, trade policy, and tax incentives), regulatory instruments (e.g., emissions, technology, and product standards), and other approaches, such as information policies, procurement policies, voluntary agreements, and valuation and accountability mechanisms (Levin et al., 2020).

## 2.6  EXAMPLES OF ENERGY TRANSITIONS IN DEVELOPED COUNTRIES

The EU has been playing an instrumental role in the international efforts while devising Paris Agreement to fight climate change. It had submitted its enhanced Nationally Determined Contribution (NDC) target to decrease GHG emissions by at least 55% by 2030 from 1990 levels and establish information to facilitate clarity, transparency, and understanding (ICTU) of the NDC in December 2020. The EU and its member states are dedicated to a binding target of a net domestic GHG emissions reduction by at least 55% by 2030 compared to 1990 (European Commission, 2021). EU's 2030 climate and energy targets pursue an energy transition by gradually shifting from fossil fuels to low-emission and carbon-neutral energy sources and adopting for greater energy efficiency. The national energy transition for each Member State will be based on their country's energy mix and the renewable energy potential available. Regardless of differences among the EU member states, all member states are expected to adopt solutions over the next decade for pursuing the same set of objectives. Energiewende (2019) has listed several issues related to solidarity, competitiveness and innovation, and security of energy supply and energy systems that need to be addressed for the just and clean European energy transition, as shown in Figure 2.5.

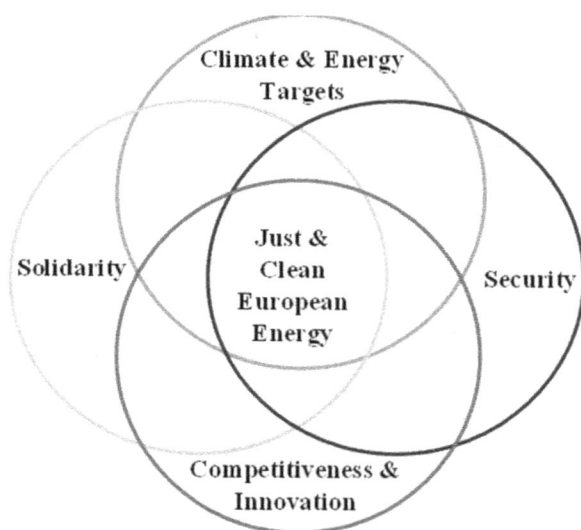

**FIGURE 2.5**  Overlapping priorities and effective policies for a just and clean European energy transition. (Adapted from Energiewende, *European Energy Transition 2030: The Big Picture*, 2019.)

### 2.6.1 SOLIDARITY

Energy transition policies can affect the overall socio-economic transformation thus the actual benefits that would result from the energy transition (e.g., cleaner air, new jobs) must be spelled out so that it will increase legitimacy and public acceptance. Some examples of addressing issues related to solidarity are as follows.

#### 2.6.1.1 Building Energy Efficiency Retrofitting

Many homeowners with lower income cannot afford the high upfront investment costs of building energy efficiency retrofitting. Therefore, they will require public financial incentives and supports to undertake the retrofitting. The EU funds support the member states with lower economies by contributing a substantial share of public investment, sometimes the aid is as high as two-thirds of the cost.

#### 2.6.1.2 Electric Vehicle Penetration

It is estimated that under the decarbonization path there will be the need for at least 40 mn electric vehicles supported by 4 mn publicly accessible recharging points in Europe by 2030. To realize this, 2.7–3.8 bn Euros are estimated to be invested annually from 2021. Based on the current plans of member states, the penetration of electric vehicles will likely happen in three distinct phases: the initiation is expected to happen in western Europe and Scandinavian countries in the early 2020s, and then it will be followed by southern Europe (Spain, Italy, Portugal, and Slovenia) in the mid-2020s and later phase in central and eastern Europe by the end of the decade (with the possibility of the appearance of a second-hand market for electric vehicles).

#### 2.6.1.3 Fuel Switching in Power Plants

Due to the dominance of coal-based power plants in the electricity sector of the EU, there will be a need to reduce its consumption by two-thirds. Presently, about 41 regions in 12 member states rely heavily on economic revenues generated from coal mining and coal uses with direct employment of around 185,000 people. To address the socio-economic impacts of a coal phase-out, the European Commission launched the Coal Regions in Transition Platform in December 2017 to foster dialogue between governments and stakeholders in affected regions and to facilitate the modernization of the required structural and economic changes.

### 2.6.2 ENERGY SECURITY

Ensuring energy security at a high level is a fundamental objective of all national or regional energy policies. Energy security can be seen from two aspects: the security of energy supply (i.e. the reliable access to energy) and the security of the power system (i.e. the reliable operation of the power system). In the case of the EU, energy security has to be addressed from the regional level due to the integrated nature of the European energy market. Some examples are as follows.

#### 2.6.2.1 Reducing Fossil Fuel Import Dependency

Dependence on imported fossil fuel is increasing in the EU due to declining domestic production resulting in susceptibility to volatile energy market prices and supply shortages. It is particularly true of NG supply. For example, Russia is the sole supplier providing 42% of the gas imports to nine EU member states. After the gas crisis between Russia and Ukraine in 2014, the EU started investing in the infrastructure projects, developing more than one entry point for pipeline gas supply in each member state, upgrading the internal gas network, creating provisions for the reverse flow of gas (from west to east), and adding more terminals for importing liquefied NG. It is envisaged that the EU energy transition will shift toward the use of more renewable energy sources, efficiency improvement, and lessen dependency on the imported fossil fuels in the medium to long term.

### 2.6.2.2   Increasing Regional Grid Integration

Maintaining electricity supply security at the least possible cost to consumers and taxpayers is expected to be achieved in Europe through the use of expanding interconnected power markets in the region. The "software" for the cross-border trade of electricity and integration of intermittent renewable electricity (e.g., wind, solar PV) has been established by the EU laws. The EU laws and financial support mechanisms have been influential for upgrading the power system "hardware." The actual targets for cross-border interconnection capacity development have been established and the building of transmission lines for operating an integrated trans-European power network has been expedited. "Clean Energy for All Europeans" initiative states additional steps which include the use of a European resource adequacy assessment for knowing whether each member states have enough electricity generating capacity to ensure the reliability of power supply during peak load demand and prevent power outages.

### 2.6.2.3   Smart Energy Systems

The present trend of digitalization has enabled the evolution of more integrated and smarter energy systems. But at the same time, it also increases the risk of cyberattacks and the theft of private personal data. In May 2018, the EU's General Data Protection Regulation entered into force and is generally regarded as one of the most comprehensive and advanced privacy protection systems in the world. It empowered EU citizens to determine who can access their data and how it can be used. This must be respected by the service providers while implementing an emerging "smart" energy system throughout Europe. In line with this, EU laws established a common framework to ensure that the member states and other actors cooperate across borders for preventing cyberattacks and working toward certifying products and services as "cyber secure."

### 2.6.3   COMPETITIVENESS AND INNOVATION

Standardization and innovation in low-carbon technologies and services play a major role in the economic and political success of the EU energy transition. In some selected areas of the energy transition (e.g. power system integration of solar and wind), Europe is considered as the leading international actor, creating opportunities for exporting new technologies and services. Some examples are as follows.

### 2.6.3.1   Standardization of Energy Technology and Supply

With more than 500 mn consumers, the EU is considered the largest market in the world. In addition, European institutions have started to develop common European standards to ensure that all companies in the regional market comply with the same basic requirements. The process of early standardization in Europe (e.g., for charging stations, smart electricity meters, wind blade design, cybersecurity design) is expected to make the EU the main reference point for those overseas companies doing business in Europe. It becomes important for the EU, particularly when dealing with major competitors like the United States of America or China.

### 2.6.3.2   Common Energy Regulation

Under the European treaties, the European Commission is given an influential mandate to regulate the EU's single market. As an example, energy-intensive companies are often exempted from levies if they finance renewable energy infrastructure to compensate for significant additional costs through energy transition policies and become competitive in the international market. Such state aid needs to be approved by the European Commission under EU State Aid rules. EU competition laws and EU State Aid frequently come into play when scaling up nascent markets for energy transition products and services like green hydrogen.

### 2.6.3.3   Advancing Innovation

The EU's multiannual budget has played a vital role in advancing innovation that will ultimately help in enabling and accelerating the European energy transition. As an example, the EU research budget for the 2014–2020 period (Horizon 2020) was 77 bn euros. For the 2021–2027 period (Horizon Europe), the proposed EU research budget is expected to increase around 100 bn euros. In addition, the EU will (i) allocate a significant amount of research grants for energy transition and climate protection, (ii) emphasize dedicated "research missions" to focus on specific global challenges (including climate change), and (iii) widen the scope of funding beyond academic research by introducing a new financing instrument that supports those companies that move highly innovative products from an advanced development stage to the market (i.e., bridging the so-called valley of death).

## 2.7 EXAMPLES OF ENERGY TRANSITIONS IN DEVELOPING COUNTRIES

Energy transitions in developing countries are mostly focused on modern energy access, electrification, and energy efficiency. In line with this, UN SDG goal 7 has, especially, highlighted promoting renewable energy and improving energy efficiency for ensuring universal access to affordable, reliable, and modern energy services by 2030 (UNDP, 2021).

Energy consumption per capita tends to rise in developing countries and the emerging economy at a faster rate compared to the developed countries mostly due to the ongoing economic development activities and energy access programs. Table 2.2 shows the energy consumption per capita and electricity consumption per capita for different countries for different periods. It can be seen that the rate of growth in electricity consumption per capita is more than the energy consumption per capita partly due to the increase in the share of service-based economy and electrification programs. In the case of developed countries, energy consumption per capita decreases but electricity consumption per capita increases partly due to energy efficiency programs, economic structure change, and fuel switching. It can be seen that there exists a huge gap between developed and developing countries in terms of the scale of energy consumption.

In terms of energy intensity of economic activities, it can be observed that energy efficiency is improving in almost all the countries due to the technological improvement and change in economic

**TABLE 2.2**
**Energy Consumption Per Capita during 1990–2013**

| Country | Energy per Capita (MJ/capita) | | | Electricity per Capita (MJ/capita) | | |
|---|---|---|---|---|---|---|
| | 1990 | 2000 | 2013 | 1990 | 2000 | 2013 |
| Germany | 4,421 | 4,094 | 3,940 | 6,640 | 6,635 | 7,218 |
| Japan | 3,553 | 4,084 | 3,568 | 6,809 | 8,299 | 7,989 |
| USA | 7,672 | 8,057 | 6,906 | 11,713 | 13,671 | 13,004 |
| Brazil | 941 | 1,072 | 1,461 | 1,461 | 1,897 | 2,569 |
| China | 767 | 899 | 2,214 | 511 | 993 | 3,773 |
| India | 350 | 417 | 606 | 272 | 394 | 764 |
| Russia | 5,942 | 4,224 | 5,079 | 6,688 | 5,198 | 6,539 |
| South Africa | 2,472 | 2,425 | 2,603 | 4,239 | 4,580 | 4,286 |
| Nigeria | 698 | 704 | 780 | 87 | 74 | 143 |
| Saudi Arabia | 3,573 | 4,736 | 6,395 | 4,018 | 5,665 | 8,785 |
| Venezuela | 2,016 | 2,119 | 2,309 | 2,478 | 2,668 | 3,299 |
| Paraguay | 728 | 723 | 759 | 504 | 884 | 1,463 |
| Nepal | 306 | 339 | 417 | 35 | 59 | 137 |
| Mozambique | 456 | 405 | 428 | 42 | 126 | 451 |
| World | 1,664 | 1,637 | 1,896 | 2,128 | 2,387 | 3,110 |

*Source:* From World Bank, 2021, *World Bank Open Data*, Retrieved July 22, 2021, from https://data.worldbank.org/indicator/. With permission.

**TABLE 2.3**

**Energy Intensity and CO$_2$ Intensity during 1990–2013**

| Country | Energy Intensity Level of Primary Energy (MJ/$2011 PPP GDP) | | | CO$_2$ Intensity (kg/kg of Oil Equivalent Energy Use) | | |
|---|---|---|---|---|---|---|
| | 1990 | 2000 | 2013 | 1990 | 2000 | 2013 |
| Germany | 5.88 | 4.64 | 3.83 | 2.72 | 2.47 | 2.45 |
| Japan | 5.03 | 5.31 | 4.21 | 2.49 | 2.28 | 2.77 |
| USA | 8.67 | 7.34 | 5.68 | 2.53 | 2.54 | 2.33 |
| Brazil | 3.81 | 3.95 | 3.94 | 1.41 | 1.67 | 1.64 |
| China | 21.18 | 10.23 | 7.85 | 2.50 | 2.95 | 3.31 |
| India | 8.29 | 6.95 | 4.99 | 1.84 | 2.13 | 2.53 |
| Russia | 12.03 | 12.59 | 8.46 | 2.46 | 2.41 | 2.23 |
| South Africa | 10.44 | 10.45 | 8.84 | 2.72 | 2.61 | 3.13 |
| Nigeria | 9.60 | 10.34 | 5.96 | 1.02 | 0.88 | 0.83 |
| Saudi Arabia | 4.20 | 5.42 | 5.44 | 2.87 | 2.50 | 2.60 |
| Venezuela | 5.77 | 6.08 | 5.38 | 2.60 | 2.62 | 2.73 |
| Paraguay | 5.06 | 5.00 | 3.89 | 0.67 | 0.91 | 1.07 |
| Nepal | 10.79 | 9.29 | 7.77 | 0.16 | 0.40 | 0.54 |
| Mozambique | 49.44 | 29.60 | 16.73 | 0.19 | 0.20 | 0.32 |
| World | 7.96 | 6.85 | | 2.40 | 2.39 | 2.51 |

*Source:* From World Bank, 2021, *World Bank Open Data*, Retrieved July 22, 2021, from https://data.worldbank.org/indicator/. With permission.

structure as shown in Table 2.3. However, the emission intensity of fuel used is increasing in developing and emerging countries mostly due to the growing infrastructure development, transportation, and economic activities using fossil fuels. In contrast, developed countries show a decreasing trend in both energy intensity of GDP and emission intensity of energy, indicating a move toward a low-carbon economy. It indicates that the role of developing and emerging economies will be very crucial in the coming days for achieving global climate change mitigation targets.

The major trend in the energy transition especially in the case of developing countries are as follows.

## 2.7.1 ELECTRIFICATION

Most of the developing countries that are not 100% electrified have set targets and established programs for addressing access to electricity for all. Still, 789 mn people do not have access to electricity in the world. This drive is also impetus by the UN SDG goal 7 highlighting the need for access to electricity for all by 2030 (UNDP, 2021). Electricity provides a higher-quality service than other energy options thus improving the quality of life. Some of the examples are listed below:

- Switching of fuel for lighting application from kerosene to electricity produced by off-grid electrification like solar PV systems, micro-hydro, micro-wind systems, etc. in off-grid areas.
- Electrification by extension of grid or development of off-grid electrification projects for increasing the access to electricity.
- Use of electricity for agriculture irrigation and agro-processing.
- Electrification of the industries and service sectors.

### 2.7.2 EFFICIENCY IMPROVEMENT AND FUEL SWITCHING

Around 2.4 bn people in developing countries rely on traditional biofuels for cooking and heating (Modi, McDade, Lallement, & Saghir, 2005). The use of improved efficiency cookstoves replacing less-efficient traditional stoves can have several benefits, including health benefits due to reduction in emissions of carbon monoxide (CO) and fine particulate matter (PM10) and possible ecosystem improvement (Modi et al., 2005). The use of efficient devices and fuel switching from lower-grade fuels like traditional biomass to commercial fuels like LPG and electricity is the major transition taking place in the residential and commercial sectors. Some of the examples are as follows:

- Promotion of the use of efficient cooking and heating devices by providing subsidy and technical support (O'connor, 2010).
- Use of LPG and charcoal for cooking and heating in the urban area of the country.
- Use of electricity replacing LPG and charcoal for residential and commercial activities.
- Cogeneration and heat recovery systems are used in the industrial process.

## 2.8 ENERGY TRANSITION IN ASIAN EMERGING COUNTRIES

The energy transition policies of two economically emerging countries are presented below.

### 2.8.1 INDIA'S ENERGY TRANSITION

India is the world's third-largest producer and consumer of electricity with an installed capacity of 382.15 GW. India imports 82% of its oil needs and aims to bring that down to 67% by 2022 by replacing it with local exploration, renewable energy, and indigenous ethanol fuel. Its energy transition is characterized by ambitious targets. By 2022, India seeks to provide all households in the country 24×7 power (Waray, 2018). By 2022, India also seeks to install 175 GW of new renewable energy (RE) in the country. India plans to reduce its emissions intensity by 33%–35% between 2005 and 2030. However, India's actions toward climate change mitigation have a strong development impact. To this effect, it is focusing on accelerating the use of clean and renewable energy by 40% by 2030, and on promoting the efficient use of energy. By 2030, India also intends to increase its carbon sinks by creating an additional capacity equivalent to 2.5–3 bn tons of $CO_2$ through significant efforts (Government of India, 2016). Over the past few years, India has set itself ambitious targets for expanding its renewable energy capacity, which has led to dynamic changes in the energy ecosystem in India comprising utilities, RE producers, and electricity consumers.

### 2.8.2 CHINA'S ENERGY TRANSITION

China's energy mix was dominated by fossil fuels. In 2014, coal accounted for 66% of total energy consumption, then followed by oil which is 18%, hydroelectricity 8%, NG 6%, and nuclear power 1%. Other renewable excluding hydroelectricity only accounted for 2% of the energy mix (Isoaho, Goritz, & Schulz, 2017). Yet, this rather bleak picture is contrasted with recent developments to push renewable energy. The goals in the roadmap, such as increasing non-fossil fuel in the nation's energy mix by 4% points and nuclear power capacity by 40% over the next 5 years, are "underwhelming" and unlikely sufficient to reach its goal of reducing carbon emissions by 18% by 2025. According to its 14th 5-year plan, which was announced recently (March 11, 2021), China pledged to achieve net-zero emissions by 2060. Under the 5-year plan, China will reduce its emissions intensity—the amount of GHG produced per unit of gross domestic product—by 18% over the period of 2021–2025, the same as in 2016. Non-fossil fuel sources are targeted to make up 20% of China's energy mix by 2025, up from about 16% in 2020 (Yamaguchi, 2021).

### 2.8.3  INDONESIA'S ENERGY TRANSITION

Indonesia is geographically located in the thermohaline circulation, also referred to as the ocean's "conveyor belt" zone, and is the largest archipelagic country, which makes it highly vulnerable to natural disasters that are aggravated by climate change. In addition, its location in the ring of fires challenges the resources management which includes rehabilitation and reconstruction. In 2005, 63% of the $CO_2e$ emissions resulted from land-use change and peat and forest fires while the combustion of fossil fuels was responsible for approximately 19% of total emissions. In 2016, the emission from the energy sector increased to 36.91% of total emission (1.45 Gt $CO_2e$), and the share of LUCF decreased to 43.58% (Republic of Indonesia, 2021). In 2005, Indonesia's energy portfolio (2093.4 PJ) consisted of 62% fossil, 1% hydro, 5% renewable (wind and solar), and 32% biofuels and waste. In 2019, the portfolio (1557.6 PJ) has changed to 73% fossil, 1% hydro, 8% renewables, and 18% biofuels and waste (IEA, 2021). Though the country has an ambitious energy mix target of at least 23% renewable energy by 2025, the country failed to envision the green recovery stimulus after the economic recession due to COVID-19 pandemic. The total installed capacity in 2020 is 59.833 GW, of which 10.554 is renewable and the coal-fired power plants capacity is 35.2 GW (Deon et al., 2021; IRENA, 2021a). The coal power plant is expected to reach 57 GW by 2028. However, the Indonesian oil and coal companies have shown interest in investments in renewables. An annual investment of approximately US$ 5 bn is estimated to meet the renewable target (Deon et al., 2021).

## 2.9  CONCLUSIONS

The energy transition is governed by the demand and supply situations and enabling environment prevailing in the country. Studies have shown that factors like supply constraints, cost advantages, performance advantages, policy decisions, and energy and environment standard ratings and codes were instrumental in determining the direction of energy transitions. It was observed that poor access to clean technology and value chains, absence of financing mechanisms, and trade restrictions were the major barriers for a successful transition to a low-carbon or sustainable energy system mostly in developing countries. Developed countries are at the forefront in the decarbonization of the energy systems and following a low-carbon development path. As a representative case, the EU submitted its updated and enhanced Nationally Determined Contribution (NDC) target in December 2020 which set the target to reduce GHG emissions by at least 55% by 2030 from 1990 levels and establish information to facilitate clarity, transparency, and understanding (ICTU) of the NDC. EU's 2030 climate and energy targets seek a European energy transition based on greater energy efficiency and the gradual phase-out of fossil fuels in favor of low-emission and carbon-neutral energy sources. In order to achieve these targets several issues related to solidarity, security of energy supply and energy systems, competitiveness, and innovation need to be addressed for the just and clean energy transition. In the case of developing countries, the energy transitions were mostly focused on modern energy access, electrification, and energy efficiency. In line with this, UN SDG goal 7 has especially highlighted promoting renewable energy and improving energy efficiency for ensuring universal access to affordable, reliable, and modern energy services by 2030.

## REFERENCES

Bahar, H., & Sauvage, J. (2013). *Cross-Border Trade in Electricity and the Development of Renewables-Based Electric Power*. OECD Trade and Environment Working Papers, (2013/02). https://doi.org/10.1787/5k4869cdwnzr-en.

CCC. (2019). *Net Zero Technical Report*. London. Retrieved from https://www.theccc.org.uk/publication/net-zero-technical-report/.

Curmi, E., Rahbari, E., Prior, E., Kleinman, S. M., Channell, J., Jansen, H. R., … Kruger, T. (2015). *Energy Darwinism II Why a Low Carbon Future Doesn't Have to Cost the Earth*. Retrieved from https://ir.citi.com/E8%2B83ZXr1vd%2Fqyim0DizLrUxw2FvuAQ2jOlmkGzr4ffw4YJCK8s0q2W58AkV%2FypGoKD74zHfji8%3D

Daszkiewicz, K. (2020). Policy and regulation of energy transition. In M. Hafner & S. Tagliapietra (Eds.), *The Geopolitics of the Global Energy Transition. Lecture Notes in Energy* (Vol. 73, pp. 203–226). Springer. Doi: 10.1007/978-3-030-39066-2_9.

Deon, A., Prasojo, H., Tampubolon, A. P., Simamora, P., Kurniawan, D., Marciano, I., & Adiatma, J. C. (2021). *Indonesia Energy Transition Outlook 2021.* Jakarta. Retrieved from https://iesr.or.id/download/ieto-2021.

Energiewende, A. (2019). *European Energy Transition 2030: The Big Picture,* Ten Priorities for the next European Commission to meet the EU's 2030 targets and accelerate towards 2050." Retrieved from https://www.agora-energiewende.de/fileadmin/Projekte/2019/EU_Big_Picture/153_EU-Big-Pic_WEB.pdf

European Commission. (2021). *Paris Agreement | Climate Action.*

Glachant, M., & Dechezleprêtre, A. (2016). What role for climate negotiations on technology transfer? *Climate Policy, 17*(8), 962–981. Doi: 10.1080/14693062.2016.1222257.

Goldthau, A., Eicke, L., & Weko, S. (2020). The global energy transition and the global south. *Lecture Notes in Energy, 73,* 319–339. Springer. Doi: 10.1007/978-3-030-39066-2_14.

Goldthau, A., Westphal, K., Bazilian, M., & Bradshaw, M. (2019, May). Model and manage the changing geopolitics of energy. *Nature.* Nature Publishing Group. Doi: 10.1038/d41586-019-01312-5.

IEA. (2018a). *World Energy Outlook 2018, International Energy Agency. Retrieved from* https://www.iea.org/reports/world-energy-outlook-2018

IEA. (2018b). *World Energy Outlook Special Report Outlook for Producer Economies What Do Changing Energy Dynamics Mean for Major Oil and Gas Exporters?, IEA,* Paris. Retrieved from https://www.iea.org/reports/outlook-for-producer-economies

IEA. (2019, November). *What Are the Trends for Remunerating Renewable Electricity in the Next Five Years? –* Analysis - IEA. Retrieved March 25, 2021, from https://www.iea.org/articles/what-are-the-trends-for-remunerating-renewable-electricity-in-the-next-five-years.

IEA. (2020a). $CO_2$ *Emission from Fuel Combustion.* Retrieved June 30, 2021, from https://www.iea.org/data-and-statistics/data-browser?country=WORLD&fuel=CO2 emissions&indicator=CO2BySector.

IEA. (2020b). *Electricity Information 2020.* Retrieved June 30, 2021, from https://www.iea.org/data-and-statistics/data-browser/?country=WORLD&fuel=Electricity and heat&indicator=ElecGenByFuel.

IEA. (2020c). *Key World Energy Statistics 2020.* Retrieved from https://www.iea.org/reports/key-world-energy-statistics-2020.

IEA. (2020d, December). *Projected Costs of Generating Electricity 2020.* Retrieved July 22, 2021, from https://www.iea.org/reports/projected-costs-of-generating-electricity-2020.

IEA. (2021). *Data & Statistics - Indonesia.* Retrieved October 17, 2021, from https://www.iea.org/data-and-statistics/data-browser/?country=INDONESIA&fuel=Energy supply&indicator=TESbySource.

Government of India (2016). India's Intended Nationally Determined Contribution, Ministry of Environment, Forest and Climate Change. Retrieved from https://moef.gov.in/wp-content/uploads/2018/04/revised-PPT-Press-Conference-INDC-v5.pdf.

IRENA. (2015). *Renewable Energy Target Setting.* Retrieved from https://www.irena.org/-/media/Files/IRENA/Agency/Publication/2015/IRENA_RE_Target_Setting_2015.pdf.

IRENA. (2020a). *Renewable Power Generation Costs in 2019.* International Renewable Energy Agency. Abu Dhabi. Retrieved from https://www.irena.org/-/media/Files/IRENA/Agency/Publication/2018/Jan/IRENA_2017_Power_Costs_2018.pdf.

IRENA. (2020b). *Statistics Time Series.* Retrieved February 18, 2021, from https://irena.org/Statistics/View-Data-by-Topic/Capacity-and-Generation/Statistics-Time-Series.

IRENA. (2021a). *Energy Profile-Indonesia.* Retrieved October 17, 2021, from https://www.irena.org/IRENADocuments/Statistical_Profiles/Asia/Indonesia_Asia_RE_SP.pdf.

IRENA. (2021b). *Energy Transition.* Retrieved March 10, 2021, from https://www.irena.org/energytransition.

IRENA. (2021c). *World Energy Transitions Outlook: 1.5°C Pathway.* Abu Dhabi. Retrieved from www.irena.org/publications.

IRENA. (2021d). *World Energy Transitions Outlook: 1.5°C Pathway.* Abu Dhabi: International Renewable Energy Agency. Retrieved from https://www.irena.org/-/media/Files/IRENA/Agency/Publication/2021/March/IRENA_World_Energy_Transitions_Outlook_2021.pdf.

Isoaho, K., Goritz, A., & Schulz, L. (2017). Governing clean energy transitions in China and India. *The Political Economy of Clean Energy Transitions,* (February 2021), 231–249. Doi:10.1093/oso/9780198802242.003.0012.

Kirchherr, J., & Urban, F. (2018). Technology transfer and cooperation for low carbon energy technology: Analysing 30 years of scholarship and proposing a research agenda. *Energy Policy, 119,* 600–609. Doi: 10.1016/j.enpol.2018.05.001.

Levin, K., Rich, D., Ross, K., Fransen, T., & Elliott, C. (2020). *Designing and Communicating Net-Zero Targets*. Washington, DC. Retrieved from www.wri.org/design-net-zero.

Li, Z., Gallagher, K. P., & Mauzerall, D. (2018). Estimating Chinese Foreign investment in the electric power sector. In *GCI Working Paper 003(10/2018)*.Retrieved from https://www.bu.edu/gdp/files/2018/12/Li-Gallagher-Mauzerall-2018.pdf

Li, Z., Gallagher, K. P., & Mauzerall, D. L. (2020). China's global power: Estimating Chinese foreign direct investment in the electric power sector. *Energy Policy, 136*, 111056. Doi: 10.1016/j.enpol.2019.111056.

Mercure, J. F., Pollitt, H., Viñuales, J. E., Edwards, N. R., Holden, P. B., Chewpreecha, U., ... Knobloch, F. (2018). Macroeconomic impact of stranded fossil fuel assets. *Nature Climate Change, 8*(7), 588–593. Doi: 10.1038/s41558-018-0182-1.

Modi, V., McDade, S., Lallement, D., & Saghir, J. (2005). *Energy Services for the Millennium Development Goals*. New York: Energy Sector Management Assistance Programme, United Nations Development Programme, UN Millenium Project, and World Bank. Retrieved from https://www.undp.org/sites/g/files/zskgke326/files/publications/MP_Energy2006.pdf

Mooldijk, S., Beuerle, J., Höhne, N., Chalkley, P., Smith, S., Hyslop, C., ... Axelsson, K. (2022). Net Zero Stocktake 2022: Assessing the status and trends of net zero target setting across countries, sub-national governments and companies. Retrieved from www.zerotracker.net/analysis/

NCCS. (2020). *Charting Singapore's Low-Carbon and Clim Ate RE Silient Future*. Singapore. Retrieved from https://www.nccs.gov.sg/docs/default-source/publications/nccsleds.pdf.

O'connor, P. A. (2010). Energy transitions. *The Frederick S. Pardee Center for the Study of the Longer-Range Future, 12*. Retrieved from www.bu.edu/pardee.

Pueyo, A., Mendiluce, M., Naranjo, M. S., & Lumbreras, J. (2012). How to increase technology transfers to developing countries: A synthesis of the evidence. *Climate Policy*. Taylor and Francis Ltd. Doi: 10.1080/14693062.2011.605588.

REN21. (2020). *Renewables 2020 Global Status Report*. Paris: REN21.

REN21. (2022). Renewables 2022 *Global Status Report*. Paris:REN21- Renewables Now. Retrieved from https://www.ren21.net/gsr-2022/

Renewable energy statistics. (2022). *Eurostat Statistics Explained*. Retrieved from https://ec.europa.eu/eurostat/statistics-explained/index.php?title=Renewable_energy_statistic

Republic of Indonesia. (2021). *Updated Nationally Determined Contribution*, Republic of Indonesia. Retrieved October 17, 2021, from https://www4.unfccc.int/sites/ndcstaging/PublishedDocuments/Indonesia First/Updated NDC Indonesia 2021 - corrected version.pdf.

Schmidt, J., Gruber, K., Klingler, M., Klöckl, C., Ramirez Camargo, L., Regner, P., ... Wetterlund, E. (2019, July). A new perspective on global renewable energy systems: Why trade in energy carriers matters. *Energy and Environmental Science*. Royal Society of Chemistry. Doi: 10.1039/c9ee00223e.

Scholten, D. (2018). The geopolitics of renewables - An introduction and expectations. In D. Scholten (Ed.), *The Geopolitics of Renewables* (pp. 1–36). London: Springer.

Seto, K. C., Davis, S. J., Mitchell, R. B., Stokes, E. C., Unruh, G., & Ürge-Vorsatz, D. (2016). Carbon lock-in: Types, causes, and policy implications. *Annual Review of Environment and Resources, 41*, 425–452. Doi: 10.1146/annurev-environ-110615-085934.

UNDP. (2021). *Goal 7 Affordable and Clean Energy*. United Nations Development Programme (UNDP). Retrieved from https://www.unep.org/explore-topics/sustainable-development-goals/why-do-sustainable-development-goals-matter/goal-7#:~:text=Target%207.,infrastructure%20and%20clean%20energy%20technology

UNEP. (2019). *Emissions Gap Report 2019*. Nairobi. Retrieved from http://www.un.org/Depts/Cartographic/english/htmain.htm

United Nations. (2015). *Paris Agreement*. Paris: United Nations. Retrieved from https://unfccc.int/sites/default/files/english_paris_agreement.pdf.

Unruh, G. C., & Carrillo-Hermosilla, J. (2006). Globalizing carbon lock-in. *Energy Policy, 34*(10), 1185–1197. Doi: 10.1016/j.enpol.2004.10.013.

Waray, S. (2018). Bringing 24×7 power to all by 2022 - The Hindu BusinessLine. Retrieved from https://www.thehindubusinessline.com/opinion/bringing-24x7-power-to-all-by-2022/article22136414.ece1

World Bank. (2021). *World Bank Open Data*. Retrieved July 22, 2021, from https://data.worldbank.org/indicator/.

Wright, C. (2018). *Chinese Overseas Investments in Fossil Fuel 100x bigger than Renewables Since Paris*. Climatetracker Retrieved from https://climatetracker.org/chinese-overseas-investments-in-fossil-fuel-100x-bigger-than-renewables/

Yamaguchi, Y. (2021). *China's Energy Transition Plan "Underwhelming," May Cool Green Financing Hype*. S&P Global Market Intelligence. Retrieved from https://www.spglobal.com/marketintelligence/en/news-insights/latest-news-headlines/china-s-energy-transition-plan-underwhelming-may-cool-green-financing-hype-63163141

# 3 Geopolitics of the Energy Transition

*Marco Giuli*
Vrije Universiteit Brussel
Istituto Affari Internazionali

## CONTENTS

## 3.1 INTRODUCTION

For more than a century, oil and gas have been critical components of international relations. Since Winston Churchill, in his capacity as the First Lord of the Admiralty, pushed the British Navy to move from coal to oil—a fuel that allows greater speed, greater radius of action, and at-sea refueling—at the cost of external dependence, the control over hydrocarbons has become a matter of national security (Yergin 1992). In parallel, after the Second World War, oil seduced global capital with the promise of a higher energy density and a lower need for unionized manpower (Mitchell 2011) and paved the way for an age of mass consumption (Pirani 2018), becoming essential to industrialized nations' economic development and social reproduction (Di Muzio 2015). This gave rise to important geopolitical developments—out of which the most evident has been the strategic relevance of oil provinces such as the Persian Gulf to large powers' economies and military might. The energy transition—implying a gradual, but ultimately massive reduction in the global consumption of fossil fuels and a qualitative change toward decarbonized forms of energy—is expected to bring about a structural change in many socio-technical foundations of our civilization (Bridge et al. 2018), including the way in which power and influence are allocated and exerted on the international stage.

In this respect, different aspects raise the attention of scholars and policy communities. First, a reduction in the role of fossil fuels in the global economy, although slow and geographically uneven, is expected to alter the strategic balance between importers, exporters, and transit countries. Second, a rise in renewable energy will make most countries able to source energy domestically, blurring the boundaries between energy exporters and importers, and giving rise to new energy actors and new patterns of international interdependencies (Scholten et al. 2020). At the same time, while change in geographical and socio-technical features of energy systems affects international relations, it also takes place in an international system in flux. Concepts such as the rise of China, the neorevisionism

DOI: 10.1201/9781003315353-4

**41**

of emerging and regional powers, or a sense of "westlessness" (Kennedy 2020) and decline of the liberal international order—summarized by some as a "return of geopolitics" (Mead 2014)—interrogate expectations on the pace and patterns of the energy transition and call the geopolitical analysis of the transition to take into consideration shifts in the surrounding political context.

To dissect this complex puzzle, this chapter—based on a dominant review element—will first offer an understanding of what does it mean to look at energy systems through geopolitical lenses. Then, it will explore the geography of geopolitical "winners" and "losers" emerging from a declining importance of fossil fuels. Section 3.4 discusses which emerging interdependencies are expected to rise with the energy transition, figuring out their geopolitical dimension. Section 3.5 will then reflect on the evolving quality of energy geopolitics, focusing on shifting balances in technological mastery, before proceeding to the conclusions.

## 3.2 FRAMING ENERGY IN GEOPOLITICAL TERMS

Some definitional clarification is necessary with respect to what does it mean to think at the energy transition through geopolitical lenses. Geopolitics is referred to as a descriptive approach to the international relations, looking at inter-state patterns of amity and enmity through the prism of a dynamic interaction between power and geographical features that are supposed to be fixed—such as geographical positions, natural borders, demography, resource endowments (Dodds 2007). The geopolitics of energy looks, in particular, at how the geographical and technical features of energy systems contribute to shaping patterns of inter-state relations (Scholten 2018; Högselius 2019). Despite the abundance of historical, descriptive accounts on the interplay between energy and geopolitics, international politics, and security (Yergin 1992; Brown 1999; Yergin 2020), systematic attempts to conceptualize such an interplay are scarce. According to Van de Graaf and Sovacool (2020), looking at energy in geopolitical terms means to consider energy resources as "a currency of power, a strategic tool in foreign policy, and a source of conflict." To this extent, the focus is placed on three main notions associated to energy that are geopolitically relevant: inter-state relations of dependence, geoeconomic statecraft, and resource conflicts.

Few states are entirely independent as for energy and energy systems. Oil and gas, in particular, are the subject of a dense international trade. In geopolitical terms, dependence on foreign resources is typically associated to vulnerability (Mearsheimer 2001; Krasner 1978). As hydrocarbons are essential to the economy and to mechanized warfare, energy dependence is ultimately a matter of national security. Nevertheless, states establish relations of energy dependence because of the uneven distribution of energy resources, economic opportunities—in case imported fuels are cheaper than domestically sourced energy—or environmental concerns—in case importing fuels helps preserve the domestic environment. In some cases, they can even establish dependence for geopolitical reasons, i.e. to expand their influence over exporters. To counter the risks associated to dependence, importing states adopt energy security policies (Deese 1979; APERC 2007; Chester 2010; Ang et al. 2015; Cherp and Jewell 2014) such as fuel switch to domestic supply, diversification of importing routes, demand reduction, or energy diplomacy (Verrastro and Ladislaw 2017). Depending on different strategies to contain the risks associated to dependence, relations between importers and suppliers, or among importers, can evolve in cooperative or antagonistic manners.

Notably, energy dependence affects importers as much as exporters, giving rise to an interdependence. While symmetric interdependence can be expected—especially under liberal institutionalist assumptions—to give rise to peaceful relations and the buildup of institutions to manage it and prevent its politicization, when interdependence grows asymmetrical it can be used by actors as a source of power (Keohane and Nye 2001; Blackwill and Harris 2016; Szabo 2015), bringing about opportunities of geoeconomic instrumentalization of energy by states or non-state actors such as regions, international organizations, or terrorist organizations and rebel groups (Van de Graaf and Sovacool 2020). Different declinations of energy statecraft include the manipulation of flows, infrastructures or system-building, prices, discourses (Högselius 2019), the rent associated to energy exports, and the instrumentalization of large import markets. The *control of energy flows* affords geostrategic centrality to actors holding it. Actors such as producers or terrorist organizations can leverage on the threat of interrupting flows, as much as a large military power can achieve influence and establish relations of dependence by protecting flows in critical chokepoints (Yergin 1992; Glaser and Kelanic 2016; Krane and Medlock 2018). Similarly, the control of *infrastructure or system-building* can be instrumentalized to entrench involved countries into long-term interdependencies or marginalize adversaries, excluding them from certain energy routes (Jentleson 1986; Little 1990; Kemp 2000; Rashid 2002; Sziklai et al. 2020; Gustafson 2020). *Price manipulation* became a less viable tool of energy statecraft after the restructuring of global physical and financial markets that followed the '70s oil crisis. However, the practice of politically motivated discounts on hydrocarbons is still present (Colgan and Van de Graaf 2017). More recent tendencies based on critical geopolitics (Dodds 2007; Power and Campbell 2010) address how energy is politicized through *discursive practices* (Kratochvíl and Tichý 2013; Kuzemko 2014a, b). Very often, disputes that originate on legal or technical grounds are presented in political terms, and inter-state pipeline projects crossing contested regions become symbols in the context of political rivalries (Baev and Øverland 2010; Meierding 2020). Another—perhaps more indirect and possibly disputed—form of fossil energy statecraft relates to a foreign policy-motivated use of the *oversea investment of oil rent*. Oil money can be re-invested by petrostates in foreign assets in more or less transparent ways and with different degrees of political implications, i.e. supporting the influence of corrupt networks aimed at promoting a petrostate's interests in other countries (Shaxson 2008). A final declination of fossil energy statecraft refers to importing countries. Large importers can instrumentalize their *energy demand* to establish influence over third countries, and keep producers locked into bilateral relations and outside the influence of rival powers.

A third notion of interest for the geopolitics of energy is the "resource conflicts." These can stem from asymmetric distribution and global scarcity of resources so that actors perceive them as something worth fighting for (Klare 2007; Månsson 2015; Klare 2016); in other circumstances, resource conflicts could be expected to take place where the environmental impact of extractive activities causes the eruption of violence involving local communities (Watts 2016). The fact that conflicts appear to be frequent in resource-rich regions has nourished popular geopolitical imaginary, and for this reason, the idea of resource conflicts is frequently played in political narratives. Indeed, these events are rare (Meierding 2020), no less because the prospect of conquering access to oil rarely offsets the costs of initiating a conflict. For this reason, great powers are more used to recur to anticipatory strategies (Kelanic 2020) such as operating to keep energy in friendly hands and denying access to enemies (Van de Graaf and Sovacool 2020) rather than utilizing military means to achieve direct control of resources. Conflict frequency in oil-rich regions was explained by Colgan (2013) through the notion of "petro-aggression"—notably, the tendency of revolutionary petrostates to rely upon their resources to support the regional revisionism of revolutionary leaders. However, regardless of whether energy can originate conflicts or not, access to oil is fundamental to warring parties for its centrality to mechanized warfare, so that oil can play a critical role in orienting military tactics during non-resource conflicts and even contribute to deciding the fate of some of them (Yergin 1992).

All these declinations of energy geopolitics have been analyzed and associated for decades to global and regional energy systems dominated by fossil fuels, and their application to clean energy resources and technologies is not exempt from criticism (Øverland 2019). Indeed, it is a difficult task to balance the risk of geopolitical reductionism and conceptual overstretches. Nevertheless, they constitute a useful reference framework to understand the magnitude of the geopolitical change the energy transition is bringing. This is the objective of the remainder of this chapter.

## 3.3   THE (SLOW) DEMISE OF FOSSIL FUELS

Attaining the objectives of the Paris Agreement implies a significant reduction in the consumption of fossil fuels in the course of the XIX century, as well as the "unburnability" of vast amounts of known fossil fuel reserves (Van de Graaf 2018). Reversing traditional assumptions about peak oil supply—according to which the world would have moved toward a structural trend of declining production and a "race to what's left" driven by growing scarcity (Klare 2012)—the climate agenda helped introduce the notion of "peak demand," according to which carbon constraints and structurally declining demand would move the world toward a "geopolitics of too much oil" (Verbruggen and Van de Graaf 2015).

In this context, indicators focusing on geopolitical gains and losses from the energy transition uniformly reveal that the main losses would affect fuel exporters, which would find themselves holding "stranded geopolitical assets" (Øverland et al. 2019; Smith Stegen 2018). As the distribution of the oil rent is central to the social contract of exporters of hydrocarbons (Eisenstadt 1973; Beblawi 1987; Gelb 1988; Noreng 2005; Krane 2019; Bradshaw et al. 2019), the loss in oil revenues associated to the global energy transition is expected to compromise authoritarian petrostates' output legitimacy (Goldthau and Westphal 2019) and trigger processes of reform and/or domestic instability (O'Sullivan et al. 2017; Tagliapietra 2017, 2019; IRENA Global Commission 2019; Krane 2019; Bradshaw et al. 2019), which can, in some cases, affect regional security. Besides compromising neo-patrimonial petrostates' model of buying consensus through patronage and strategic allocation of the oil rent, a decline in oil demand would hit oil suppliers also on ideational grounds, as the possession of natural resources is often discursively constructed by petrostates' elites as elements of great power status, i.e. in Russia (Sakwa 2017), post-colonial identity or sovereignty, i.e. in Algeria, Mexico, and Venezuela (Entelis 1999; Colgan 2013), or generous social models, i.e. in Norway (Austvik 2019).

However, not all oil suppliers are made equal. Their exposure to energy transition can vary in function of many variables. Vulnerability can be, for instance, being understood as directly related to the weight of the public sector on total employment, as a measure of oil's importance for domestic political legitimacy; the level of export dependence on fossil fuels, as a measure of oil's importance to keep the economy afloat; and costs of extraction—as low-cost producers are likely to continue supplying a shrinking demand during the transition. Considering these aspects, geography of vulnerability would reveal a deep exposure in Angola, Nigeria, Libya, Venezuela, and Turkmenistan, while the UAE, Australia, and Mexico seem to be the best equipped for the transition. Different studies on petrostates' vulnerability may use further indicators, generally reaching similar results in terms of nations' grouping (IRENA Global Commission 2019; Carbon Tracker Initiative 2020).

Confronting a critical situation and the prospect of a weakened external and internal position, fossil fuel exporters are expected to undertake strategies to reduce their vulnerability to the energy transition, which may constitute, therefore, a further factor of differentiation among petrostates in terms of stability and geopolitical alignments in the age of the transition.

A frequently discussed short-term coping strategy is a push to extract oil at the quickest possible pace. If oil left in the ground is no longer perceived as a financially safe asset, producers may see an incentive to extract as much oil as possible (Van de Graaf and Verbuggen 2015), calculating that resources that can go stranded are less valuable than oil sold underprice (Van de Graaf 2018). Such a "cash in" approach would be especially rational among high-cost producers and producers presenting high fiscal break-evens—notably the oil price required to balance state budgets—as they are those that would be eventually squeezed out of the market in a low prices environment (Goldthau and Westphal 2019). Yet, this is not a strategy without risks. It may imply the acceleration of the transition toward a structurally low oil prices environment. Risks may provide incentives for an opposite strategy, such as engaging in some form of multilateral—formal or informal—energy governance to manage a process of structural fossil fuel decline. The Organization of Petroleum Exporting Countries plus (OPEC+) scheme, emerged in 2016 to coordinate OPEC supply policy with the policy of Russia and other non-OPEC producers, and the US' lobbying toward OPEC to cut production in the wake of the 2020 price collapse, constitute signals in this direction. On the other hand, oil suppliers may undertake aggressive military and geoeconomic action to deny rivals' production capacity (Van de Graaf 2018). The civil war in Libya took the country's production off the global markets in a context where warring parties are influenced by several petrostates' interests. The US sanctions limited, to different extent, supply or supply prospects in Venezuela, Iran, and even Russia. Iran has made a permanent threat to Saudi oil facilities and Gulf shipping lanes a key feature of its foreign policy toolbox. Under a carbon constrain, these tendencies may consolidate and aggravate, coexisting with or replacing cooperative attempts to manage the oil decline.

A second coping strategy, currently taking place among several petrostates and arguably valid in the short-to-medium term, is the energy re-specialization and export re-orientation of fossil fuels producers. Some petrostates seem willing to expand downstream in the energy-intensive sectors. Especially Gulf countries are well-placed to hedge the decline of hydrocarbons with a rise in petrochemical production.[1] For petrostates wishing to maximize their resources' value in the transition phase, control over refining means also to enhance demand security in times of potentially tougher competition. In recent years, petrostates' national oil companies (NOCs) such as Aramco, ADNOC, or Rosneft expanded their control over refining and distribution assets in emerging economies (Øverland 2015; Braekhus and Øverland 2007). In parallel, crude oil and gas export capacities expanded eastwards based on the assumption that emerging Asian economies may ensure at least a certain level of demand security during the transition[2] (Griffiths 2019). A major geopolitical implication would be an expanded Chinese influence over exporters such as Russia and the Gulf countries, which would consolidate these countries' "detachment" from the historical west (Sakwa 2017) in a phase where US-Saudi, US-Russia, and EU-Russia political relations are under strain. At the same time, it remains an open geopolitical question whether this may ultimately imply that emerging Asian economies could grow more involved in the Gulf security issues.

Finally, petrostates may attempt to diversify their economies. The capacity to undergo such a change, however, is unevenly distributed. The UAE is usually portrayed as a success story. In two decades, oil went from accounting for almost the whole UAE GDP to 40%, while non-oil exports rose in the last decade from 13% of total exports to almost 60%. This evolution—largely leveraged, however, on oil money—took place without major social and political unrest, while the country's foreign policy grew over time increasingly ambitious and assertive (Krane 2019; Yergin 2020). Saudi Arabia engaged in an even more ambitious path of diversification, as Crown Prince Mohamed bin Salman's Vision 2030 plan conflates economic diversification with plans of major socio-economic transformation aimed at changing the neo-patrimonial nature of the state (Bradshaw et al. 2019). These petrostates show unparalleled advantages in their quest for diversification—notably, a combination of large financial availabilities and relatively stable governments. As for many other large oil producers, the journey has not even started. Notably, fragile petrostates (i.e. in Iraq, Nigeria, Libya, or South Sudan), or petrostates promoting revolutionary or externally revisionist policies, (i.e. Iran, Venezuela, or Russia) lag behind in their political commitment to diversification.

As the demise of fossil fuels is expected to weaken exporters, it is also well expected to benefit large importers, whose dependence on oil volumes or prices constrains policy options and affects

---

[2] In 2019, Saudi Aramco merged with Saudi petrochemical giant SABIC, expanding its refining assets to 50 countries and doubling its refining capacity (Todd 2019).

[3] In 2014, Russia and China signed a $400 bn, 30-year gas deal to be supplied to China by way of the new Power of Siberia pipeline, financed by China for $400 bn. The deal makes China the second largest recipient of Russian gas after Germany. In addition, Chinese CNPC participate to Novatek's Yamal LNG project, providing critical financing to complete a complex project in an age of sanctions that cut off Novatek's access to western finance (Yergin 2020). In 2017, Russian oil company Rosneft expanded its downstream presence in India through the acquisition of Indian refiner Essar Oil (Rosneft 2017). Similarly, Saudi Aramco took over 17% of Korean refiner Hyundai Oilbank (Todd 2019).

economic performance. However, the picture is ambiguous with "winners" as well. The US is generally seen as a geopolitical "winner" in the energy transition, due to the role that oil dependence has played in constraining Washington's foreign policy options and enabling adversaries such as the USSR/Russia, post-revolutionary Iran, Saddam Hussein's Iraq, Gaddafi's Libya, or the Bolivarian Venezuela. However, for the US—a large fossil fuel producer itself—the oil should not just be considered as a geopolitical constraint, but also as a geopolitical asset. As much as coal was central in building up the British hegemony, the history of US hegemony is deeply intertwined with the emergence of the global oil order (Yergin 1992). The role of the US Navy in patrolling critical chokepoints, prominently emerged with the "Carter doctrine," can be considered an international public good whose provision—in line with the basic tenets of the hegemonic stability theory (Gilpin 1987; Keohane 1984)—has been playing as an integral part of the US hegemony on the international system (Glaser and Kelanic 2016) by establishing a strategic dependence of both allies and rivals—such as Europe, Japan, the Gulf Monarchies, or more recently China—on Washington (Högselius 2019). Oil has also played a central role in the US' dominance of the global financial architecture. The dollar-denomination of oil transactions contributes to the primacy of the dollar as an international reserve currency, facilitating the enforcement of an ever more defining instrument of US geoeconomic statecraft such as the application of extraterritorial sanctions. Yet, things change. The energy transition, spurring an era of newfound fossil fuel abundance and redefinition of energy security, conveniently occurs in a time when the US leaders are undergoing a process of reflection on the costs and benefits of certain aspects of its hegemonic engagement (Mead 2014; Ikenberry 2018). To this extent, it remains an open question whether the demise of the oil order will coincide with the emergence of a more multipolar international system, where a more mixed, decentralized, and deglobalized energy system will require less engagement from a global hegemon (Sivaram and Saha 2018).

In contrast, Europeans unambiguously find themselves in the camp of the geopolitical "winners" of the transition. Dependence on fossil fuel imports turned out to be a chapter of tension between the buildup of EU actorness in external energy relations and member states' commercial priorities. Nowhere has this proved more evident than in the case of energy relations with Russia, where understandings of energy security and Russia's reliability as a supplier diverge among the EU member states (Schmidt-Felzmann 2011; Sharples 2013; Haukkala 2015; Krickovic 2015; Thaler 2020; Gustafson 2020), complicating the emergence of a coherent EU external actorness instead of facilitating it (Herranz Surralles 2015). To this extent, the process of decarbonization could be expected to remove a divisive element in the crafting of unified external actorness toward fuel suppliers, and potentially an obstacle to the improvement of their relations with the EU. Similarly, risks associated to vanishing European influence on exporters of fossil fuels as a result of the energy transition should not be overestimated. In a distinctive exercise of normative, transformative geopolitics, the EU has sought to leverage on its energy relations to expand its regulatory frameworks beyond its borders (Lavenex and Schimmelfennig 2009). However, except for those fuel suppliers or transit countries embedded in a highly institutionalized relation with Europe—i.e. Norway or Ukraine—the majority of hydrocarbon suppliers to Europe and transit countries have not moved in the regulatory or political direction desired by the EU (Aalto and Korkmaz Temel 2014; Thaler 2020; Darbouche 2010). To this extent, bilateral energy relations appear to be a poor predictor of the EU's ability to shape outcomes in supplier countries, especially if insulated from other factors.

Large Asian importers are also likely to geopolitically benefit from a reduced centrality of fossil fuels in their economy. China is the biggest consumer of oil flowing out of the Persian Gulf through the Strait of Hormuz, making the country dependent on flow security that can only be provided by its systemic rival, the US. Half of the oil shipped by tankers passes through the South China Sea—one of the most contentious geopolitical hotspots in the world (Holslag 2015)—fueling not only China but also South Korea and Japan. To this extent, oil dependence is for China and other Asian powers a factor of major strategic fragility. As the three of them are now committed to carbon neutrality plans for 2050–2060, their geopolitical position is expected to be stronger.

Yet, such a map of winning and losing sides—limited, in its exemplifying purposes, to few geo-political actors—in the geopolitics of fossil fuel decline requires some nuance. A transition should not necessarily be seen as a quick adjustment from a steady-state to another. According to the IEA Sustainable Development Scenario, oil supply is expected to still amount to 3000 Mt in 2040, while natural gas supply is expected to amount to 2800 Mtoe (IEA 2020). As coal is expected to be phased out more urgently than oil and gas, especially emerging economies are expected to raise their consumption of natural gas in the near future. This means that in all likelihood the phase-out of fossil fuels will take the form of a protracted process, lasting decades, with elements of non-linearity. A generalized, gradual decline of fossil fuels may imply a temporary re-concentration of fossil fuel supply in low-cost production provinces, such as Russia and the Gulf, as global demand decline would drive a reduction in prices (Van de Graaf 2018; O'Sullivan et al. 2017; Goldthau and Westphal 2019).

## 3.4 GEOPOLITICS OF RENEWABLE ENERGY

The geopolitical dimension of renewable energy is currently the most explored aspect of the geo-politics of the energy transition (Vakulchuk et al. 2020; Scholten et al. 2020; Øverland 2019; O'Sullivan et al. 2017; Criekemans 2018). Many forms of renewable energy—such as wind, solar, geothermal, and tidal waves—show characteristics that are fundamentally different from those of hydrocarbons in geographical and socio-technical terms (Paltsev 2016). First, renewable energy is, by definition, not scarce. Second, renewable resources are not geographically concentrated, although some areas can produce renewable energy at lower costs than others. This would ideally imply a lower density of inter-state energy dependence, interdependence, and therefore less geo-political contention around issues such as access to resources (O'Sullivan et al. 2017; Paltsev 2016; Criekemans 2018). Yet, while a renewable energy sources (RES)-based energy system is allegedly less internationally connected—and eventually less contested—than the fossil fuel-based one, cross-border interdependencies will not necessarily be over. This section will identify inter-state interdependencies associated to decarbonizing and decarbonized energy systems, exploring their geopolitical dimension.

### 3.4.1 GEOPOLITICS OF ELECTRIFICATION

Electricity is going to become a dominant carrier of renewable energy, so that the attention of geopolitics is likely to shift toward the availability and control of electricity infrastructures, stor-age capacity, and cross-border interconnections (Casier 2015; O'Sullivan et al. 2017). In addition, a decarbonized energy mix requires the replacement of fossil fuels with electricity in many compo-nents of final energy consumption, from mobility to residential and industrial heating. Deep decar-bonization scenarios are ultimately electrification scenarios.

In contrast to oil and gas, electricity is an efficient energy carrier only if traded across relatively short distances. As such, a switch toward electricity as a carrier of choice is likely to produce a *deglobalization* of energy trading. Unless technological breakthroughs happen with utility-scale storage or fuel cells, electricity's centers of production will need proximity to centers of consump-tion. Scholten and Bosman (2016) identify a "continental" or regional scenario, where inter-state integration continues between producers, consumers, and transit countries through high-voltage power lines within a regional dimension. Such a scenario is more likely in regions that already show a high level of economic and regulatory integration, such as Europe, where the integration of power grids was initiated well before the massive penetration of renewable energy technol-ogy, with early attempts dating back to the immediate post-war period. In the meantime, several regional high-voltage direct current (HVDC) projects are emerging. The Sun Cable project aims at providing Australian renewable electricity to the ASEAN power grid. In other contexts, however, the direct current—implying one-way trade—raised geopolitical concerns. It is, for instance, the case of China's State Grid plans to provide renewable electricity from Xinjiang to Central Asia or even Europe (Högselius 2019), or the Desertec Industrial Initiative (DII) promising to provide electricity from the sunny Sahara Desert to Europe (Lilliestam and Ellenbeck 2016). In theory, inter-state HVDC raises the same geoeconomic implications of inter-state trading of fossil fuels—especially gas—in terms of manipulation of flows, infrastructure (Smith Stegen 2018) and discourses in politically charged environments. As such, a "regional scenario" is less likely to emerge in regions characterized by inter-state tensions. Wherever geopolitical framings of energy security take precedence over commercial framings or efficiency considerations, states may profit from the distributed nature of renewable energy to pursue energy self-sufficiency instead of cross-border integration, adopting a "westphalian" approach to energy systems (Aalto 2008). While case studies are generally rare, Escribano (2018) looked at Spain-Morocco RES cooperation, arguing that fossil fuel-related concepts such as the quest for "energy independence" or "RES mercantilism" continue to affect inter-state energy relations, acting as a barrier to strategic opportunities. A third electrification scenario would be a "neomedieval" one (Aalto 2008), where security is ensured neither by cross-border interconnection or national self-sufficiency, but instead by a decentralized model of self-sufficient communities. Such a scenario is geopolitically relevant as for the possibility it raises for sub-national entities to assert more autonomy from central governments. Powerful sub-national entities with ambitious decarbonization agendas such as California operate their own energy diplomacy, i.e. by autonomously establishing carbon markets and linking them with those of foreign entities. On the other hand, decentralization can also coexist with certain levels of nationalization/regionalization of the grid, ideally providing system resilience to possible attacks (Øverland 2019) – especially in light of the growing cybersecurity threats implied by the growing digitalization of electricity systems (O'Sullivan et al. 2017; Hielscher and Sovacool 2018).

### 3.4.2 HYDROGEN GEOPOLITICS

Electricity is not the only possible energy carrier under deep decarbonization scenarios. Especially in recent years and in some contexts such as Europe and Japan, the potential role of hydrogen in the energy transition has been gaining traction, so that optimistic forecasts foresee hydrogen to serve up to one-quarter of the world's energy needs by mid-century (Bloomberg NEF 2020). Hydrogen offers the possibility to store and transport renewable electricity over long distances in liquid or gaseous form via ships or pipelines, potentially giving rise to transnational connectivity. With its large availability of wind and hydro power, large network of transmission infrastructure, gas reserves, and natural sites for storage, Norway shows the ideal profile for a supplier of hydrogen to the industrial clusters of north-western European countries. Similarly, North African countries, Chile, the Gulf countries or Australia could generate renewable hydrogen at the lowest cost and make the most of abundant gas export infrastructure (De Blasio and Pflugmann 2020), potentially becoming large suppliers for the Asian markets (Van de Graaf et al. 2020). Similar considerations could be made about Russia, although an expansion of the energy partnership beyond natural gas may remain exposed to the political vagaries of Russia's relations with the transatlantic community, not to mention the poor Russian record with renewable energy deployment (Khrushcheva and Maltby 2016).

In contrast to fossil fuel importers, hydrogen importers could retain the choice of procuring hydrogen domestically, although potentially at higher costs. Such a possibility reduces the room for geoeconomic statecraft by hydrogen producers. In all cases, questions remain unanswered with respect to the likely uptake and spread of hydrogen technologies. Hydrogen would remain in many sectors in direct competition with other green technologies that could be currently deployed at lower cost with less complex logistics (i.e. in the residential and transport sectors). This would make it a much less pervasive carrier for energy systems—and ultimately economies and societies—than hydrocarbons. Such a lower pervasivity, or even niche utilization, points in the direction of a relatively low geopolitical relevance.

### 3.4.3   A Race for Critical Materials?

Geopolitical re-articulations driven by the expansion of renewable energy also revolve around the role of raw materials that are critical to the decarbonization of energy systems (Bazilian 2018; O'Sullivan et al. 2017; Månberger and Johansson 2019; Kalantzakos 2020; Smith Stegen 2015; 2018). Materials like lithium, copper, nickel, cobalt, graphite, indium, platinum group metals, and rare earth elements (REE) such as neodymium, dysprosium, and yttrium are essential components for renewable energy hardware such as solar panels and wind turbines, batteries, fuel cells, and electrolyzers. The level of concentration in critical materials extraction and processing is remarkable. In the case of cobalt, the Democratic Republic of Congo (DRC) accounts for more than 50% of production and slightly less than half the world's reserves. Lithium production is concentrated in Australia and Chile, with a prominent role of Argentina and China. The production of natural graphite and REE is currently concentrated in China, mostly due to unparalleled processing capabilities—with a prominent role of Mozambique and Brazil for the former, and the US and Australia for the latter. All in all, out of many metals and metalloids critical for renewable energy, only copper, tellurium, and silicon show a geographical concentration that is less accentuated than oil (Månberger and Johansson 2019). Such a concentration implies some shift with geostrategic centralities. Latin America, China, and Sub-Saharan Africa would achieve strategic importance, while the Middle East and North Africa are largely out of the picture—potentially as both suppliers and customers. On the other hand, the EU and Japan—likely large buyers—continue to be confronted with external dependency, while Russia would confirm its role as an exporter of raw materials. Around such a concentration, three main factors of geopolitical risk emerge. The potential for a confrontational race among major manufacturing powers for the access to critical raw materials, the manipulation of flows and systems, and the fragility of some critical raw materials (CRM) exporters (Pitron 2018; Gulley et al. 2018; Kalantzakos 2020; Church and Crawford 2020).

Rivalry for access among the three largest CRMs consumers—the US, China, and the EU—could emerge especially for access to lithium, for which external dependence accounts for 50% for the US, 75% for China, and 100% for Europe (Gulley et al. 2018). Especially China moved assertively, given the large processing capacity and its leadership in battery production, in trying to acquire production assets abroad. With respect to bilateral competition, China and Europe are heavily dependent on cobalt imports, while Europe and the US depend on REE imports. However, in this latter case, common over-dependence caused cooperation between the two importers rather than competition. Notably, Washington and Brussels aligned their action to contrast Chinese REE export restrictions, bringing the dispute to the WTO. In 2016, the EU undertook legal action against China before the WTO over Chinese restrictions on the export of raw materials such as graphite, cobalt, copper, lead, chromium, magnesia, talcum, tantalum, tin, antimony, and indium (European Commission 2016) following similar legal actions in 2012 and 2014.

As for the risks associated to the instrumentalization of interdependence, concerns have mounted since 2010 among the US foreign policy and security bureaucracies with respect to potential coercive use by China of its quasi-monopolistic position in the REE extraction (US Department of State 2010), on the occasion of an export embargo that China imposed on Japan following an incident concerning the disputed Senkaku/Diayou islands. Australia had already blocked a Chinese attempt to take over Australian REE assets (Kalantzakos 2020). To be sure, Chinese activism on this front cannot be reduced to merely strategic considerations. Instead, one needs to consider China's quest for primacy in finite products—hence the need to secure access to raw materials for an increasingly sophisticated industrial-technological complex (Freeman 2018). This should not be surprising, considering that REE markets accounted in 2010 for 1.3 bn USD value, against 4,800 bn USD of end-use industries (Smith Stegen 2015).

The third factor of geopolitical risk concerns the potential for a "resource curse"—notably, the paradox where countries rich in raw materials are particularly exposed to intra-state violence, authoritarianism, and economic backwardness (Ross 1999). Månsson (2015) notices that despite a lower propensity of RES to provide incentives to conflicts to secure control, increased competition for land possession may trigger conflictual dynamics involving non-state actors—depending on specific technologies and the implementation of sustainability side policies. Månberger and Johansson (2019) suggest that risks remain country-specific. Some exporters show a very diversified economy, where raw materials supercycles or downturns are unlikely to heavily affect political or macroeconomic stability. It is, for instance, the case with China or the US. The second group of countries—such as Russia or Australia—export CRM as well as hydrocarbons. Due to the foreseeable contraction in the trade of hydrocarbons and the relatively smaller CRMs trade volume, it is predictable that in aggregate terms this second group will become less dependent on raw materials—and therefore less exposed to the resource curse. Major distortions may however be expected in the present and future CRMs exporters with small economies and an already dominant mining sector such as Chile, Cuba, or the DRC (World Bank 2020). These contexts risk a growing exposure to the fluctuation of CRMs' prices, which may amplify political or economic instability that were already latent or present (Church and Crawford 2020).

However, risks connected to the concentration could be easily overestimated in a context where China's domination of supply chains is conducive to a securitization (Buzan et al. 1997) of CRMs access in western discursive practices. Indeed, there are important factors that mitigate CRMs' geopolitical sensibility, especially with respect to fossil fuels. *First*, clean technologies need different CRMs and are frequently in competition with each other. Greater electrical interconnectivity would imply greater demand for copper while mitigating the need for storage - and with it the demand for lithium and cobalt for batteries or PMG for fuel cells. Greater use of hydrogen in electric mobility would increase the demand for PMG, but mitigate the demand for lithium. The shift toward wind turbines of ever-larger capacity involves a shift from neodymium to yttrium.[3] Other materials such as aluminum, zinc, or selenium would still be employed in a large number of non-energy sectors, implying that the demand for wind turbines or solar panels could be offset by fewer uses of the raw materials needed in other sectors. *Secondly*, it is appropriate to relativize the notions of scarcity and concentration for many of the elements necessary for the transition (Øverland 2019). The production of REE underwent a de-concentration over the last 10 years, so that the Chinese market share dropped from 97% to 64%. Lithium also seems to show no major risks in terms of depletion. The projections concern only reserves currently being exploited, while the available resources seem much more abundant. In the long term, any shortage in conventional sites would favor the development of alternative mining technologies.[4] *Thirdly*, in contrast to fossil fuels, CRMs do not need to flow daily to importers. They have a flexible logistics and low storage costs, which act as important factors mitigating flow manipulation risks. *Fourthly*, in the medium and long terms—intervals in which the results of technological innovation take over—opportunities arise from recycling, efficiency, and substitutability of CRMs. These interventions reduce both the CRMs' environmental and geopolitical criticalities. While advanced economies have for a long time looked at CRMs through the lenses of trade policy principles and instruments—i.e. pushing for open global supply chains and attacking export restrictions—the new geopolitical environment is shifting toward import-substitution policies (European Commission 2020). In the light of the cases examined, it is concluded that environmental and geostrategic risks can mainly affect in short term, and the role of politics and technological development remains essential in their mitigation.

---

[4] Neodymium is employed for magnets in turbines below 10 MW of capacity, while yttrium is necessary for superconductors in turbines above 10 MW.

[5] An example is lithium's extraction from sea water. Already now the prices of lithium carbonate, at \$ 12–15,000/t, are approaching the \$ 16,000/t needed to make marine extraction profitable, capable of expanding current reserves from 15,000 to 2 mn tons (Greim et al. 2020).

### 3.4.4  HYDRO, NUCLEAR, AND BIOMASS

By concentrating on the geopolitical implications of renewable energy, one should not neglect the role of more traditional low-carbon sources—which currently provide most of the carbon-free energy and are often found in competition with each other (Paltsev 2016). Based on these sources, some states such as Brazil, France, Sweden, or Norway already rely upon a relatively decarbonized energy mix. One of the major difficulties in assessing the geopolitics of the transition is factoring in the relevance of these sources on a long-term horizon, as their permanence or removal would have significant consequences on the volume of other sources—fossil or not—that the world will need in the future.

*Biomass* accounts for a large proportion of renewable energy supply. It is central to the decarbonization strategies of several countries such as Brazil, Sweden, or Finland, where bioenergy accounts for 20–30% of the total primary energy supply (IEA Bioenergy 2018). Yet, their future role is constrained by growing reconsideration of their environmental sustainability. Despite its critical role in decarbonization patterns, biomass attracts little attention in terms of geopolitical analysis. Biomass trade can be hardly weaponized. Despite serious social issues associated to processes of land-use change and commodity prices (Dalby 2020), there is limited record of direct connection to meaningful cooperative or confrontational transnational patterns, with the notable exception of trade complaints against restrictive practices (Ghosh 2016). Only a small proportion of biofuels are traded internationally, as large producers of biofuels are also the largest consumers, while tariff and non-tariff barriers play an important role in limiting the prospects for an enhanced trade expansion sought by producers such as Indonesia and Malaysia (IRENA Global Commission 2019).

*Hydropower* is currently one of the major sources of carbon-free electricity. In 2019, hydroelectricity accounted for 15.6% of power generation, and 60% of electricity produced through renewable energy.[5] For some countries—i.e. Brazil, Canada, Norway, and many other countries in South America and Sub-Saharan Africa—it even accounts as the main source of electricity. Yet, the production of hydroelectricity has been stagnating since the mid-90s everywhere except for China. As for the future, a sustained decline of capacity additions is putting hydropower off track with respect to the Sustainable Development Scenario of the IEA. Hydropower prospects are constrained by the challenge of climate change itself, which—acting on water's availability and variability—works as an amplifier of the elements of political contentiousness that already affect large hydropower infrastructure with a cross-border dimension (De Stefano et al. 2017). As such, regardless of uncertain future prospects, tensions around its use are likely to increase in certain geographical contexts (Hancock and Sovacool 2018). A relevant example is the current tension between Egypt and Ethiopia with respect to the Grand Ethiopian Renaissance Dam (GERD), which puts in conflict vital needs of the two countries: water and food security for Egypt and fighting energy poverty for Ethiopia, in a context complicated by expanding populations, domestic and regional tensions, and the impact of climate change on water resources. At the same time, cooperative patterns may emerge given hydropower's value for regional balancing markets. Notably, Norwegian and Swiss water reservoirs are likely to become increasingly central to lower the costs of the EU's grid stabilization as the continent increasingly relies on intermittent sources, or serve the production and export of green hydrogen.

*Nuclear energy* has been also frequently portrayed as potentially experiencing a rebirth thanks to decarbonization (Ramana 2016). Yet, the consumption of nuclear energy remained stable over the last decades—accounting for about 10% of power generation—with a slight contraction in Europe compensated by a rise in Asia. Nuclear energy has always been geopolitically relevant due to the dual use of nuclear technology and the concerns associated to nuclear proliferation (Van de Graaf and Sovacool 2020). The geography of its current development shows a dual dynamic. Stringent regulation, rising costs for decommissioning and waste disposal, and low popularity are making

---

[6] Author calculation on BP (2020).

nuclear energy a lesser attractive option for decarbonization with respect to easier-to-deploy renewable energy in western markets. On the other hand, the growing power demand of developing and emerging economy in a carbon-constrained world translates into a renewed interest for nuclear (Markard et al. 2020). Largely due to the US' opposition to nuclear proliferation, many emerging and developing countries are looking at the investment schemes proposed by state-owned nuclear actors from Russia and China—which benefit from state subsidies and are not bound by OECD export credit rules (Nakano 2021). Russia's Rosatom, already a key actor in the Iranian nuclear program, is building power plants in Turkey, Egypt, Belarus, Finland, and Hungary (Giuli 2017; Aalto et al. 2017; Schepers 2019), while China is active in the international marketing of its Hualong-1 type reactor by way of its realization in Pakistan, a country chronically affected by energy poverty and deeply entrenched in the network of China's infrastructure diplomacy and debt schemes (Small 2015). From the US perspective, strategic risks connected to the emergence of new dependencies of several countries on Russian and Chinese technology, as much as the commercial decline of western nuclear industrial actors on the world stage are reasons for concern. In response, the US is in a process of bipartisan reconsideration of its non-proliferation policy, raising the prospects of commercial and geopolitical competition in the definition of civil nuclear standards (Nakano 2020).

## 3.5 REDEFINING ENERGY MASTERY

The previous sections interpreted clean energy systems through the lenses of the geopolitical categories usually applied to fossil energy. Still, it needs to be acknowledged that the geopolitics of the energy transition will be played less and less on the chessboard of access to resources and increasingly on the mastery of innovation in processes and products, supply and value chains, integrated infrastructure, and influence on processes of standardization and regulation (Casier 2015; O'Sullivan et al. 2017; Ghosh 2016), so that those actors that are better placed in terms of innovative ecosystems, market size, financial availabilities, and administrative capacity will be more likely to achieve energy mastery in the XXI century (Criekemans 2018) and therefore geopolitical dividends in terms of acquiring revenues, status, and establishing relations of asymmetric dependence. Research, intellectual property, industrial (Rodrik 2014; Johnstone and Kivimaa 2018), trade (Jebe et al. 2012; Ghosh 2016; Marhold 2017) and tax and carbon pricing policies (Condon and Ignaciuk 2013) and their surrounding bureaucratic-industrial ecosystems are becoming the central instruments and actors in a geopolitics of energy where dominating technology is a more important determinant of power than accessing resources (Casier 2015; O'Sullivan et al. 2017), in contrast to the traditional role played by foreign policy and security bureaucracies and oil and gas companies in the framings and pursuit of energy security. Innovation agencies and public procurement authorities are going, as well, to provide resources, design policies, and practices, and create lead markets critical to the achievement of global primacy in clean energy systems.

Existing literature identified the number of RES patents, the weight of capital investment, and the density of commercial clean energy companies as indicators of power in the technological aspects of the geopolitics of renewable energy (Criekemans 2018; Månsson 2015; Paltsev 2016; Smith Stegen 2018; Escribano 2018). Results show significant geographical concentration—notably among economies with large domestic markets and innovation capacity such as the US, the EU, and China, with the addition of mid-size countries such as Japan, South Korea, and the UK. In the early 2000s, Germany, the US, and Japan were the only powers holding more than 10% share in global patents for wind energy, solar energy, and fuel cells. In Europe, investment in renewable energy – driven by the adoption of renewable energy targets—occupied more than 75% of clean technology investments, while clean energy companies were concentrated in the US, Europe—driven by Germany—and Japan (Criekemans 2018). In few years, the distribution rebalanced eastwards. In 2014, the value added of Chinese clean energy manufacturing[6] amounted to almost $40 bn,

[7] The data refers to wind turbine components, crystalline silicon PV modules, LED packages and lithium-ion battery cells (IRENA Global Commission 2019).

in contrast to Japan, Germany, and the US, which totaled each slightly more than $5 bn. In 2016, China cumulated almost 30% share of RES patents, followed by the US (18%), the EU, and Japan (14% each). In 2019, China was the top investor in clean energy globally ($83 bn), followed by the US ($55.5 bn) and Japan ($16.5 bn). As for firms, Chinese giants, such as Suntech and BYD, promise to become the clean energy challengers in a similar way Huawei and ZTE have established themselves in the digital sector. As such, besides almost monopolizing the extraction, processing, and exports of CRMs, China established itself as a leading exporter of clean energy products. China controls 77% of battery cell capacity and 60% of components, and 60% of global manufacturing in the solar supply chain. Based on a vast array of measures such as long-term planning, subsidization, government procurement, and local content requirements, combined with a large domestic market and social and environmental dumping, China achieved dominance over first-generation clean energy technologies, such as lithium-ion batteries, onshore wind, and crystalline silicon solar (Finamore 2021).

While the rise of China's clean energy technological and industrial capacity has contributed to rendering the early phases of the energy transition affordable for the climate frontrunners in the west, China's industrial primacy is no longer considered by other powers as just a neoliberalism and multilateral trade institutions' success story, but also—and increasingly—as a strategic challenge (Blackwill and Harris 2016; Allison 2017). The intensification of great power rivalry is, therefore, expected to inform the race for the next generation of clean energy technologies—i.e. floating offshore wind, perovskite solar cells, solid-state batteries. In the US, the race to clean energy is increasingly framed in antagonizing terms. At the beginning of 2021, President Joe Biden ordered a review of four critical foreign supply chains to address dependence on geopolitical rivals in essential sectors, including batteries for electric vehicles (The White House 2021). Without adopting a confrontational rhetoric, the EU is also showing interest to shorten strategic value chains, including those related to decarbonization. The European Green Deal strategy (European Commission 2019) and side-strategies underpinning it foresee a range of industrial policy measures aimed at reshoring battery manufacturing and enhancing the domestic processing and recycling capacity for CRMs, ranging from the relaxation of state aid discipline to the support to research and development, to a carbon border adjustment mechanism—notably, a tariff on the embedded carbon content of imports, whose prospects are generating acrimonious reactions among Europe's trade partners (Hook 2021). In the context of these competitive patterns, the big prize in terms of power and influence is the definition of clean energy standards for processes, products, and infrastructures' interoperability. In China's understanding, standards guarantee institutional influence on standardization agencies, originally created and dominated by the historical west; economic revenues, as mastering standards-essential patents provides royalties and reduces adaptation costs (Fägersten and Rühlig 2019).

It would be indeed too extreme and early to claim that the new clean energy powers are on the verge of technological decoupling (Meidan 2021). The complexity of emerging energy value chains lie also in the fact that the reshoring of certain productions (i.e. batteries) implies the rise of the dependence on other products (i.e. CRMs). Yet, signals are pointing toward a clean energy technology race entangled within a confrontational environment, marked by mercantilist tendencies and little appetite for multilateral frameworks and governance forms.

## 3.6   CONCLUSIONS

In this short summary, the geopolitical dimension of the transition was interpreted as the impact of shifting technical and geographical features of energy systems on inter-state patterns of cooperation ad conflict. The overview has shown that the energy transition is expected to have important geopolitical implications, raising strategic opportunities and risks.

Oil and gas exporters' foreign policy options are likely to shrink as the geostrategic importance of an asset they possess is diminished, while the foreign policy options of large importers are going to expand. Questions remain about the extent the transition will result in political and economic instability among petrostates, and whether oil suppliers and the large powers of the XXI century will undergo geopolitical realignments as a result of the transition. To be sure, energy is not the only chapter of interaction between large powers and oil suppliers. A vast array of issues including security, migration, climate change, and economic interlinkages will continue to bound actors together. Then, a question to explore is whether a fuel "disconnection" is likely to result in a factor that facilitates addressing other geopolitical issues, or instead as an amplifier of geopolitical risks. At least, in part, the answer to such a question will depend on the pace of the energy transition—as in the short-to-medium term low-cost producers and their NOCs may see their geostrategic centrality to temporarily increase before declining—on individual petrostates' characteristics, and their choices on how to deal with the transition.

On the other hand, a different geography of interdependences—smaller in economic volumes but bigger in complexity, actors, and articulations—is likely to emerge. Such a geography will be defined by the control on critical raw materials, processing capacities, and decarbonized systems' infrastructure, mastery over advanced technologies, and influence on the definition of regulation and standards. Whether these emerging patterns of interdependencies will lead *per se* toward cooperative or competitive patterns of inter-state interactions will largely depend on their interplay with broader geopolitical dynamics, mainly related to the rise of China and geoeconomic declinations of great power rivalry.

## REFERENCES

Aalto, P. 2008. ed. *The EU-Russian Energy Dialogue: Europe's Future Energy Security.* Burlington, VT: Routledge, 2008.

Aalto, P., and D. Korkmaz Temel. 2014. European energy security: Natural gas and the integration process. *JCMS: Journal of Common Market Studies* 52, no. 4: 758–74. Doi: 10.1111/jcms.12108.

Aalto, P., H. Nyyssönen, M. Kojo, and P. Pal. 2017, November. Russian nuclear energy diplomacy in Finland and Hungary. *Eurasian Geography and Economics*: 1–32. Doi: 10.1080/15387216.2017.1396905.

Allison, G. 2017. *Destined for War: Can America and China Escape Thucydides's Trap?* London, Scribe: Mariner Books.

Ang, B. W., W. L. Choong, and T. S. Ng. 2015. Energy security: Definitions, dimensions and indexes. *Renewable and Sustainable Energy Reviews* 42: 1077–93. Doi: 10.1016/j.rser.2014.10.064.

Asia Pacific Energy Research Centre (APERC). 2007. ed. *A Quest for Energy Security in the 21st Century: Resources and Constraints.* Tokyo: Inst. of Energy Economics.

Austvik, O. G. 2019. Norway: Small state in the great European energy game. In *New Political Economy of Energy in Europe: Power to Project, Power to Adapt*, ed. J. M. Godzimirski, 139–64. Cham: Springer International Publishing. Doi: 10.1007/978-3-319-93360-3_6.

Baev, P. K., and I. Øverland. 2010. The south stream versus Nabucco pipeline race: Geopolitical and economic (Ir)rationales and political stakes in mega-projects. *International Affairs* 86, no. 5: 1075–90. Doi: 10.1111/j.1468-2346.2010.00929.x.

Bazilian, M. D. 2018. The mineral foundation of the energy transition. *The Extractive Industries and Society* 5, no. 1: 93–97. Doi: 10.1016/j.exis.2017.12.002.

Beblawi, H. 1987. The rentier state in the Arab world. *Arab Studies Quarterly* 9, no. 4: 383–98.

Braekhus, K., and I. Øverland. 2007. A match made in heaven? Strategic convergence between China and Russia. *China and Eurasia Forum Quarterly* 5 no. 2: 41–61. https://www.researchgate.net/publication/265455681_A_Match_Made_in_Heaven_Strategic_Convergence_between_China_and_Russia.

Blackwill, R.D., and J.M. Harris. *War by Other Means: Geoeconomics and Statecraft.* Cambridge, MA: Harvard University Press, 2016.

Bridge, G., S. Barr, S. Bouzarovski, M. Bradshaw, E. Brown, H. Bulkeley, and G. Walker. 2018 *Energy and Society: A Critical Perspective.* London: Routledge.

Bloomberg NEF. 2020. *Hydrogen Economy Outlook 2020.* Accessed December 12, 2020. https://data.bloomberglp.com/professional/sites/24/BNEF-Hydrogen-Economy-Outlook-Key-Messages-30-Mar-2020.pdf.

BP. 2020. *Statistical Review of World Energy 2020.* https://www.bp.com/content/dam/bp/business-sites/en/global/corporate/pdfs/energy-economics/statistical-review/bp-stats-review-2020-full-report.pdf.

Bradshaw, M., T. Van de Graaf, and R. Connolly. 2019. Preparing for the new oil order? Saudi Arabia and Russia. *Energy Strategy Reviews* 26: 100374. Doi: 10.1016/j.esr.2019.100374.

Brown, A. C. 1999. *Oil, God and Gold: The Story of Aramco and the Saudi Kings.* Boston, MA: Houghton Mifflin.

Buzan, B., O. Wver, and J. De Wilde. 1997. *Security: A New Framework for Analysis.* Boulder, CO: Lynne Rienner Publishers.

Carbon Tracker Initiative. 2020. *Beyond Petrostates.* Accessed March 6, 2021. https://carbontracker.org/reports/petrostates-energy-transition-report/.

Casier, T. 2015. The geopolitics of the EU's decarbonization strategy: A bird's eye perspective. In *Decarbonization in the European Union: Internal Policies and External Strategies*, eds. C. Dupont, and S. Oberthür, 159–79. London: Palgrave Macmillan UK. Doi: 10.1057/9781137406835_8.

Cherp, A., and J. Jewell. 2014. The concept of energy security: Beyond the four as. *Energy Policy* 75: 415–21. Doi: 10.1016/j.enpol.2014.09.005.

Chester, L. 2010. Conceptualising energy security and making explicit its polysemic nature. *Energy Policy* 38, no. 2: 887–95. Doi: 10.1016/j.enpol.2009.10.039.

Church, C., and A. Crawford. 2020. Minerals and the metals for the energy transition: Exploring the conflict implications for mineral-rich, fragile states. In *The Geopolitics of the Global Energy Transition*, ed. M. Hafner and S. Tagliapietra, 279–304. Cham: Springer International Publishing. Doi: 10.1007/978-3-030-39066-2_12.

Colgan, J. D. 2013. *Petro-Aggression: When Oil Causes War.* Cambridge: Cambridge University Press. Doi: 10.1017/CBO9781139342476.

Colgan, J. D., and T. Van de Graaf. 2017. A crude reversal: The political economy of the United States crude oil export policy. *Energy Research & Social Science* 24: 30–35. Doi: 10.1016/j.erss.2016.12.012.

Condon, M., and A. Ignaciuk. 2013. Border carbon adjustment and international trade: A literature review. *OECD Working Paper n. 6.* Doi: 10.2139/ssrn.2693236.

Criekemans, D. 2018. Geopolitics of the renewable energy game and its potential impact upon global power relations. In *The Geopolitics of Renewables*, ed. D. Scholten, 37–73. Cham: Springer International Publishing. Doi: 10.1007/978-3-319-67855-9_2.

Dalby, S. 2020. *Anthropocene Geopolitics: Globalization, Security, Sustainability.* Ottawa: University of Ottawa Press.

Darbouche, H. 2010. Energising' EU–Algerian relations. *The International Spectator* 45, no. 3: 71–83. Doi: 10.1080/03932729.2010.504624.

De Blasio, N. and F. Plugmann. 2020. *Geopolitical and Market Implications of Renewable Hydrogen: New Dependencies in a Low-Carbon Energy World.* Belfer Center for Science and International Affairs Report Accessed March 6, 2021. https://www.belfercenter.org/publication/geopolitical-and-market-implications-renewable-hydrogen-new-dependencies-low-carbon.

Deese, D. A. 1979. Energy: Economics, politics, and security. *International Security* 4, no. 3: 140–53. Doi: 10.2307/2626698.

De Stefano, L., J. D. Petersen-Perlman, E. A. Sproles, J. Eynard, and A. T. Wolf. 2017. Assessment of transboundary river basins for potential hydro-political tensions. *Global Environmental Change* 45: 35–46. Doi: 10.1016/j.gloenvcha.2017.04.008.

Di Muzio, T. 2015. *Carbon Capitalism: Power, Social Reproduction and World Order.* London: Rowman & Littlefield International.

Dodds, K. 2007. *Geopolitics: A Very Short Introduction.* Oxford: Oxford University Press.

Eisenstadt, S. N. 1973. *Traditional Patrimonialism and Modern Neopatrimonialism.* Beverly Hills, CA: Sage Publications.

Entelis, J. P. 1999. Sonatrach: The political economy of an Algerian state institution. *Middle East Journal* 53, no. 1: 9–27.

Escribano, G. 2019. The geopolitics of renewable and electricity cooperation between Morocco and Spain. *Mediterranean Politics* 24, no. 5: 674–81. Doi: 10.1080/13629395.2018.1443772.

European Commission. 2016. EU takes legal action against export restrictions on Chinese raw materials. Press release, 19 July 2016. Text. Accessed March 6, 2021. https://ec.europa.eu/commission/presscorner/detail/en/IP_16_2581.

European Commission. 2019, December 11. The European green deal. Communication from the Commission to the European Parliament, the European Council, the Council, the European Economic and Social Committee and the Committee of the Regions. COM/2019/640. https://eur-lex.europa.eu/legal-content/EN/TXT/?qid=1588580774040&uri=CELEX%3A52019DC0640.

European Commission. 2020, March 10. A new industrial strategy for Europe. Communication from the Commission to the European Parliament, the European Council, the Council, the European Economic and Social Committee and the Committee of the Regions. COM/2020/102. https://eur-lex.europa.eu/legal-content/EN/TXT/?qid=1593086905382&uri=CELEX%3A52020DC0102.

Fägersten, B., and T. Rühlig. 2019. *China's Standard Power and Its Geopolitical Implications for Europe.* Swedish Institute of International Affairs, UI Brief 2/2019. https://www.beltroadresearch.com/wp-content/uploads/2019/07/Chinas-standard-power-and-its-geopolitical-implications-for-europe.pdf.

Finamore, B. A. 2021. Clean tech innovation in China and its impact on the geopolitics of the energy transition. *OIES Forum* 126: 18-22. https://www.oxfordenergy.org/wpcms/wp-content/uploads/2021/02/OEF-126.pdf.

Freeman, D. 2018. China and renewables: The priority of economics over geopolitics. In *The Geopolitics of Renewables*, ed. D. Scholten, 187–201. Cham: Springer International Publishing, 2018. Doi: 10.1007/978-3-319-67855-9_7.

Gelb, A. H. 1988. *Oil Windfalls: Blessing or Curse?* Washington: World Bank, 1988.

Ghosh, A. 2016. Clean energy trade conflicts: The political economy of a future energy system. In *The Palgrave Handbook of the International Political Economy of Energy*, eds. T. Van de Graaf, B. K. Sovacool, A. Ghosh, F. Kern, and M. T. Klare, 175–204. London: Palgrave Macmillan UK. Doi: 10.1057/978-1-137-55631-8_7.

Gilpin, R. 1987. *The Political Economy of International Relations.* Princeton, NJ: Princeton University Press.

Giuli, M. 2017. Russia's nuclear energy diplomacy. *EPC Policy Brief.* https://euagenda.eu/upload/publications/untitled-92744-ea.pdf.

Glaser, C. L., and R. A. Kelanic, eds. 2016. *Crude Strategy: Rethinking the US Military Commitment to Defend Persian Gulf Oil.* Washington, DC: Georgetown University Press.

Goldthau, A., and K. Westphal. 2019. Why the global energy transition does not mean the end of the petro-state. *Global Policy* 10, no. 2: 279–83. Doi: 10.1111/1758-5899.12649.

Greim, P., A. A. Solomon, and C. Breyer. 2020. Assessment of lithium criticality in the global energy transition and addressing policy gaps in transportation. *Nature Communications* 11, no. 1: 4570. Doi: 10.1038/s41467-020-18402-y.

Griffiths, S. 2019. Energy diplomacy in a time of energy transition. *Energy Strategy Reviews* 26: 100386. Doi: 10.1016/j.esr.2019.100386.

Gulley, A. L., N. T. Nassar, and S. Xun. 2018. China, the United States, and competition for resources that enable emerging technologies. *Proceedings of the National Academy of Sciences* 115, no. 16: 4111–15. Doi: 10.1073/pnas.1717152115.

Gustafson, T. 2020. *The Bridge: Natural Gas in a Redivided Europe.* Cambridge, MA: Harvard University Press.

Hancock, K. J., and B. K. Sovacool. 2018. International political economy and renewable energy: Hydroelectric power and the resource curse. *International Studies Review* 20, no. 4: 615–32. Doi: 10.1093/isr/vix058.

Haukkala, H. 2015. From cooperative to contested Europe? The conflict in ukraine as a culmination of a long-term crisis in EU–Russia relations. *Journal of Contemporary European Studies* 23, no. 1: 25–40. Doi: 10.1080/14782804.2014.1001822.

Herranz-Surralles, A. 2015. European external energy policy: Governance, diplomacy and sustainability, In *The SAGE Handbook of European Foreign Policy*, Eds K.E. Jrgensen, A. K. Aarstad, E. Drieskens, K. Laatikainen, and B. Tonra, 913–27. Doi: 10.4135/9781473915190.n63.

Hielscher, S., and B. K. Sovacool. 2018. Contested smart and low-carbon energy futures: Media discourses of smart meters in the United Kingdom. *Journal of Cleaner Production* 195: 978–90. Doi: 10.1016/j.jclepro.2018.05.227.

Högselius, P. 2019. *Energy and Geopolitics.* New York: Routledge, 2018. Doi: 10.4324/9781315177403.

Holslag, J. 2015. *China's Coming War with Asia.* Malden, MA: Polity, 2015.

Hook, L. 2021, March 12. John Kerry warns EU against carbon border tax. *Financial Times.* https://www.ft.com/content/3d00d3c8-202d-4765-b0ae-e2b212bbca98.

IEA. 2020. World energy outlook 2020. Accessed March 6, 2021. https://www.iea.org/reports/world-energy-outlook-2020.

IEA Bioenergy. 2018. Bioenergy countries' report – Update 2018. https://www.ieabioenergy.com/wp-content/uploads/2018/10/IEA-Bioenergy-Countries-Report-Update-2018-Bioenergy-policies-and-status-of-implementation.pdf.

IRENA Global Commission. 2019. A new world: The geopolitics of the energy transformation. Accessed November 10, 2020. /publications/2019/Jan/A-New-World-The-Geopolitics-of-the-Energy-Transformation.

Ikenberry, G. J. 2018. The end of liberal international order? *International Affairs* 94, no. 1: 7–23. Doi: 10.1093/ia/iix241.

Jebe, R., D. Mayer, and Yong-Shik L. 2012. China's export restrictions of raw materials and rare earths: A new balance between free trade and environmental protection. *George Washington International Law Review* 44, no. 4: 579–642.

Jentleson, B. W. 1986. *Pipeline Politics: The Complex Political Economy of East-West Energy Trade*. Ithaca, NY: Cornell University Press.

Johnstone, P., and P. Kivimaa. 2018. Multiple dimensions of disruption, energy transitions and industrial policy. *Energy Research & Social Science* 37: 260–65. Doi: 10.1016/j.erss.2017.10.027.

Kalantzakos, S. 2020. The race for critical minerals in an era of geopolitical realignments. *The International Spectator* 55, no. 3: 1–16. Doi: 10.1080/03932729.2020.1786926.

Kelanic, R. A. 2020. *Black Gold and Blackmail: Oil and Great Power Politics*. Ithaca, NY: Cornell University Press.

Kemp, G. 2000. US-Iranian relations: Competition and cooperation in the Caspian Basin. In *Energy and Conflict in Central Asia and the Caucasus*, eds R. Ebel and A. Menon: 145–162. New York: Rowman & Littlefield.

Keohane, R. O. 1984. *After Hegemony: Cooperation and Discord in the World Political Economy*. Princeton: Princeton University Press.

Keohane, R. O., and J.S. Nye. 2001. *Power and Interdependence*. London: Longman Publishing Group.

Kennedy, L. 2020. 'This is not America': States of emergency in Europe. *European Journal of American Studies* 15, no. 15-4. Doi: 10.4000/ejas.16368.

Klare, M. T. 2007. *Blood and Oil: The Dangers and Consequences of America's Growing Dependency on Imported Petroleum*. New York: Henry Holt and Company.

Klare, M. T. 2012. *The Race for What's Left: The Global Scramble for the World's Last Resources*. New York: Picador.

Klare, M. T. 2016. No blood for oil? Hydrocarbon abundance and international security. In *The Palgrave Handbook of the International Political Economy of Energy*, Eds. T. Van de Graaf, B. K. Sovacool, A. Ghosh, F. Kern, and M. T. Klare, 419–439. London: Palgrave Macmillan UK.

Krane, J. 2019. *Energy Kingdoms: Oil and Political Survival in the Persian Gulf*. New York: Columbia University Press.

Krane, J., and K. B. Medlock. 2018. Geopolitical dimensions of US oil security. *Energy Policy* 114: 558–65. Doi: 10.1016/j.enpol.2017.12.050.

Krasner, S. D. 1978. *Defending the National Interest: Raw Materials Investments and U.S. Foreign Policy*. Princeton, NJ: Princeton University Press.

Kratochvíl, P., and L. Tichý. EU and Russian discourse on energy relations. *Energy Policy* 56: 391–406. Doi: 10.1016/j.enpol.2012.12.077.

Krickovic, A. 2015. When interdependence produces conflict: EU–Russia energy relations as a security dilemma. *Contemporary Security Policy* 36, no. 1: 3–26. Doi: 10.1080/13523260.2015.1012350.

Khrushcheva, O., and T. Maltby. 2016. The future of EU-Russia energy relations in the context of decarbonisation. *Geopolitics* 21, no. 4: 799–830. Doi: 10.1080/14650045.2016.1188081.

Kuzemko, C. 2014a. Politicising UK energy: What 'speaking energy security' can do. *Policy & Politics* 42, no. 2: 259–74. Doi: 10.1332/030557312X655990.

Kuzemko, C. 2014b. Ideas, power and change: Explaining EU–Russia energy relations. *Journal of European Public Policy* 21, no. 1: 58–75. Doi: 10.1080/13501763.2013.835062.

Lavenex, S., and F. Schimmelfennig. 2009. EU rules beyond EU borders: Theorizing external governance in European politics. *Journal of European Public Policy* 16, no. 6: 791–812. Doi: 10.1080/13501760903087696.

Lilliestam, J., and S. Ellenbeck. 2011. Energy security and renewable electricity trade—Will desertec make Europe vulnerable to the 'energy weapon'? *Energy Policy* 39, no. 6: 3380–91. Doi: 10.1016/j.enpol.2011.03.035.

Little, D. 1990. Pipeline politics: America, TAPLINE, and the Arabs. *The Business History Review* 64, no. 2: 255–85. Doi: 10.2307/3115583.

Månberger, A., and B. Johansson. 2019. The geopolitics of metals and metalloids used for the renewable energy transition. *Energy Strategy Reviews* 26: 100394. Doi: 10.1016/j.esr.2019.100394.

Månsson, A. 2015. A resource curse for renewables? Conflict and cooperation in the renewable energy sector. *Energy Research & Social Science* 10: 1–9. Doi: 10.1016/j.erss.2015.06.008.

Marhold, A. 2017. EU state aid law, WTO subsidy disciplines and renewable energy support schemes: Disconnected paradigms in decarbonizing the grid. *TILEC Discussion Paper No 2017-029*. https://papers.ssrn.com/abstract=3009124.

Markard, J., N. Bento, N. Kittner, and A. Nuñez-Jimenez. 2020. Destined for decline? examining nuclear energy from a technological innovation systems perspective. *Energy Research & Social Science* 67: 101512. Doi: 10.1016/j.erss.2020.101512.

Mead, W. R. 2014. The return of geopolitics: The revenge of the revisionist powers. *Foreign Affairs* 93, no. 3: [i]–79.

Mearsheimer, J. J. 2001. *The Tragedy of Great Power Politics*. New York: W. W. Norton & Company.

Meidan, M. 2021. China's emergence as a powerful player in the old and new geopolitics of energy. *OIES Forum* 126: 12–15. https://www.oxfordenergy.org/wpcms/wp-content/uploads/2021/02/OEF-126.pdf.

Meierding, E. 2020. *The Oil Wars Myth: Petroleum and the Causes of International Conflict*. Ithaca, NY: Cornell University Press.

Mitchell, T. 2011. *Carbon Democracy: Political Power in the Age of Oil*. London: Verso Books.

Nakano, J. 2020. The changing geopolitics of nuclear energy: A look at the United States, Russia and China. *CSIS Report*. https://www.csis.org/analysis/changing-geopolitics-nuclear-energy-look-united-states-russia-and-china.

Nakano, J. 2021. The geopolitics of nuclear energy in the era of energy transition. *OIES Forum* 126: 34-36. https://www.oxfordenergy.org/wpcms/wp-content/uploads/2021/02/OEF-126.pdf.

Noreng, Ø. 2005. *Crude Power: Politics and the Oil Market*. London: I.B. Tauris.

O'Sullivan, M., I. Overland, and D. Sandalow. 2017. *The Geopolitics of Renewable Energy*. *SSRN Scholarly Paper*. Rochester, NY: Social Science Research Network. Doi: 10.2139/ssrn.2998305.

Øverland, I. 2015. Future petroleum geopolitics: consequences of climate policy and unconventional oil and gas. In *Handbook of Clean Energy Systems*, 1–29. American Cancer Society, 2015. Doi: 10.1002/9781118991978.hces203.

Øverland, I. 2019. The geopolitics of renewable energy: Debunking four emerging myths. *Energy Research & Social Science* 49: 36–40. Doi: 10.1016/j.erss.2018.10.018.

Øverland, I., M. Bazilian, T. I. Uulu, R. Vakulchuk, and K. Westphal. 2019. The GeGaLo index: Geopolitical gains and losses after energy transition. *Energy Strategy Reviews* 26: 100406. Doi: 10.1016/j.esr.2019.100406.

Paltsev, S. 2016. The complicated geopolitics of renewable energy. *Bulletin of the Atomic Scientists* 72, no. 6: 390–95. Doi: 10.1080/00963402.2016.1240476.

Pirani, S. 2018. *Burning Up: A Global History of Fossil Fuel Consumption*. London: Pluto Press.

Pitron, G. 2018. *La guerre des métaux rares: La face cachée de la transition énergétique et numérique*. Paris: Liens qui Liberent.

Power, M. and D. Campbell. 2010. The state of critical geopolitics. *Political Geography*, 29, no. 5: 243–46. Doi: 10.1016/j.polgeo.2010.06.003.

Ramana, M.V. 2016. Second life or half-life? The contested future of nuclear power and its potential role in sustainable energy transition. In *The Palgrave Handbook of the International Political Economy of Energy*, Eds. T. Van de Graaf, B. K. Sovacool, A. Ghosh, F. Kern, and M. T. Klare, 363–96. London: Palgrave Macmillan UK

Rashid, A. 2002. *Taliban: Islam, Oil and the New Great Game in Central Asia*. London: I.B. Tauris.

Rodrik, D. 2014. Green industrial policy. *Oxford Review of Economic Policy* 30, no. 3: 469–91. Doi: 10.1093/oxrep/gru025.

Rosneft. 2017. *Rosneft Successfully Closes Strategic Deal for the Acquisition of 49% of Essar Oil Limited*. Press Release, 21 August 2017. Accessed March 6, 2021. https://www.rosneft.com/press/releases/item/187527/.

Ross, M. 2001. Does oil hinder democracy? *World Politics* 53: 325–61. Doi: 10.1353/wp.2001.0011.

Sakwa, R. 2017. *Russia Against the Rest: The Post-Cold War Crisis of World Order*. Cambridge: Cambridge University Press.

Schmidt-Felzmann, A. 2011. EU member states' energy relations with Russia: Conflicting approaches to securing natural gas supplies. *Geopolitics* 16, no. 3: 574–99. Doi: 10.1080/14650045.2011.520864.

Schepers, N. 2019. Russia's nuclear energy exports: Status, prospects and implications. *SIPRI Non-Proliferation and Disarmament Papers No. 61*. https://www.sipri.org/sites/default/files/2019-02/eunpdc_no_61_final.pdf.

Scholten, D., M. Bazilian, I. Overland, and K. Westphal. 2020. The geopolitics of renewables: New board, new game. *Energy Policy* 138: 111059. Doi: 10.1016/j.enpol.2019.111059.

Scholten, D., and R. Bosman. 2016. The geopolitics of renewables; exploring the political implications of renewable energy systems. *Technological Forecasting and Social Change* 103, no. C: 273–83.

Scholten, D., D. Criekemans, and T. Van de Graaf. 2019. An energy transition amidst great power rivalry. *Journal of International Affairs* 73, no. 1: 195–204. Doi: 10.2307/26872789.

Sharples, J. D. 2013 "Russian approaches to energy security and climate change: Russian gas exports to the EU." *Environmental Politics* 22, no. 4: 683–700. Doi: 10.1080/09644016.2013.806628.

Shaxson, N. 2008. *Poisoned Wells*. New York: Griffin.

Sivaram, V., and S. Saha. 2018. The geopolitical implications of a clean energy future from the perspective of the United States. In *The Geopolitics of Renewables*, ed. D. Scholten, 125–62. Cham: Springer International Publishing. Doi: 10.1007/978-3-319-67855-9_5.

Small, A. 2015. *The China-Pakistan Axis: Asia's New Geopolitics.* Oxford: Oxford University Press.

Smith Stegen, K. 2015. Heavy rare earths, permanent magnets, and renewable energies: An imminent crisis. *Energy Policy* 79: 1–8. Doi: 10.1016/j.enpol.2014.12.015.

Smith Stegen, K. 2018. Redrawing the geopolitical map: International relations and renewable energies. In *The Geopolitics of Renewables*, ed. D. Scholten, 75–95. Cham: Springer International Publishing. Doi: 10.1007/978-3-319-67855-9_3.

Szabo, S. F. 2015. *Germany, Russia, and the Rise of Geo-Economics.* London: Bloomsbury Academic.

Sziklai, B. R., L. Á. Kóczy, and D. Csercsik. 2020. The impact of Nord stream 2 on the European gas market bargaining positions. *Energy Policy* 144: 111692. Doi: 10.1016/j.enpol.2020.111692.

Tagliapietra, S. 2017. *Energy Relations in the Euro-Mediterranean: A Political Economy Perspective.* Palgrave Macmillan. Doi: 10.1007/978-3-319-35116-2.

Tagliapietra, S. 2019. The impact of the global energy transition on MENA oil and gas producers. *Energy Strategy Reviews* 26: 100397. Doi: 10.1016/j.esr.2019.100397.

Thaler, P. 2020. *Shaping EU Foreign Policy Towards Russia: Improving Coherence in External Relations.* Cheltenham: Edward Elgar Pub.

The White House. 2021, February 24. Fact sheet: Securing America's critical supply chains. *Statements and Releases.* https://www.whitehouse.gov/briefing-room/statements-releases/2021/02/24/fact-sheet-securing-americas-critical-supply-chains/

Todd, F. 2019, June 6. Analysing all Saudi Aramco's acquisitions since oil giant issued debut bond. *NS Energy.* Accessed March 6, 2021. https://www.nsenergybusiness.com/features/saudi-aramco-acquisition-oil-debut-bond/.

US Department of State. 2010, October 27. Joint press availability with Japanese Foreign Minister Seiji Maehara. https://2009-2017.state.gov/secretary/20092013clinton/rm/2010/10/150110.htm.

Vakulchuk, R., I. Overland, and D. Scholten. 2020. Renewable energy and geopolitics: A review. *Renewable and Sustainable Energy Reviews* 122: 109547. Doi: 10.1016/j.rser.2019.109547.

Van de Graaf, T. 2018. Battling for a shrinking market : Oil producers, the renewables revolution, and the risk of stranded assets. In *The Geopolitics of Renewables*, ed. D. Scholten, 97–121. Springer. http://hdl.handle.net/1854/LU-8544872.

Van de Graaf, T. 2020. Is OPEC dead? Oil exporters, the Paris agreement and the transition to a post-carbon world. In *Beyond Market Assumptions: Oil Price as a Global Institution*, ed. A.V. Belyi, 63–77. Cham: Springer International Publishing, 2020. Doi: 10.1007/978-3-030-29089-4_4.

Van de Graaf, T., I. Overland, D. Scholten, and K. Westphal. 2020. The new oil? The geopolitics and international governance of hydrogen. *Energy Research & Social Science* 70: 101667. Doi: 10.1016/j.erss.2020.101667.

Van de Graaf, T., and B. K. Sovacool. 2020. *Global Energy Politics.* Cambridge: Polity.

Van de Graaf, T., and A. Verbruggen. 2015. The oil endgame: Strategies of oil exporters in a carbon-constrained world. *Environmental Science & Policy* 54: 456–62. Doi: 10.1016/j.envsci.2015.08.004.

Verbruggen, A., and T. Van de Graaf. 2015. The geopolitics of oil in a carbon-constrained world. *IAEE Energy Forum* 2: 21–24.

Verrastro, F., and S. Ladislaw. 2007. Providing energy security in an interdependent world. *The Washington Quarterly* 30, no. 4: 95–104. Doi: 10.1162/wash.2007.30.4.95.

Watts, M. 2016. The political ecology of oil and gas in West Africa's Gulf of Guinea: State, Petroleum, and the conflict in Nigeria. In *The Palgrave Handbook of the International Political Economy of Energy*, eds. T. Van de Graaf, B. K. Sovacool, A. Ghosh, F. Kern, and M. T. Klare, 585–619. London: Palgrave Macmillan UK.

World Bank. 2020. *Mineral Rents (% of GDP) | Data.* Accessed March 6, 2021. https://data.worldbank.org/indicator/NY.GDP.MINR.RT.ZS.

Yergin, D. 1992. *The Prize: The Epic Quest for Oil, Money, and Power.* London: Simon & Schuster.

Yergin, D. 2020. *The New Map: Energy, Climate, and the Clash of Nations.* New York: Penguin Press.

# Section II

*Energy Transition: Resources and Technologies*

# 4 Renewable Energy Systems

*Hafiz Ali Muhammad*
University of Science and Technology
Korea Institute of Energy Research

*Muhammad Asim*
The Hong Kong Polytechnic University

*Muhammad Imran*
Aston University

*Zafar Ali Khan*
Aston University
Mirpur University of Science and Technology

*Tabbi Wilberforce*
Aston University

*Young-Jin Baik*
Korea Institute of Energy Research

## CONTENTS

DOI: 10.1201/9781003315353-6

## 4.1  INTRODUCTION

With a tremendous increase in the global population, the world reaches new heights in terms of energy needs. The acceleration in the population growth with accompanying higher energy needs poses new issues. The world population was 2.6 billion in 1950, which now tops 7.6 billion and is expected to become 9–10 billion by 2050, which means that the population-doubling period of the world is half a century. The primary energy demand has increased more than 45% in the last 20 years, and this is just the beginning. By 2035, it is expected that the global energy consumption will increase by 35% and 50% by 2050. The end-use energy consumption by industry, in buildings, and in transportation is set to grow by 45%, 30%, and 40%, respectively [1]. Therefore, an exponential growth in energy need driven by the increasing population and increased use of technology dependent on electrical energy is imminent.

The renewable energy resources have made a significant progress in the past 25 years. The environmentalists are drawn to the importance of renewable energy by the emerging threat of climate change. On the other hand, the economists fret about the cost of replacing the fossil fuels. Despite all the controversies, the nonrenewable fuels (fossils) are already being replaced by the renewables as world's primary energy resources, which is an inevitable and impactful development in this modern world. The change from petroleum products to renewables in the 21$^{st}$-century mirrors past energy advances in human history, that is, from wood to coal for heat, and from animal feed to oil for transportation [2]. Recent trends show a significant increase in the use of renewable energy with a global increase of up to 2%–3% in 2020 alone. Moreover, the renewables' share in global electricity generation increased to 29% in 2020 as compared to 27% in 2019. Therefore, the renewables are on a track to set the new benchmarks in 2021 [3].

In 2021, the electricity generation from renewables is set to expand by more than 8% (to reach 8,300 TWh), which is the fastest year-on-year growth since the 1970s. Two-thirds of this renewable growth is shared by solar photovoltaics (PVs) and wind. Moreover, China accounts for about half of the global increase in electricity generation in 2021 followed by the USA, the European Union, and India.

### 4.1.1  ECONOMIC PERSPECTIVE

Nowadays, the renewable energy sources are the lowest-cost sources of electricity in many markets. The decade has been a remarkable decade of change for solar PVs and wind power technologies in terms of renewable electricity generation. The global average levelized cost of energy (LCOE) of solar PV fell by 85%, from 0.381 to 0.057 US$/kWh among the newly commissioned projects between 2010 and 2020. Similarly, the global LCOE from onshore wind projects fell by 56% from 0.089 to 0.039 US$/kWh over the same period. For biomass, the LCOE remains almost the same, that is, 0.076 US$/kWh, whereas there is an increase of 45% in the LCOE for geothermal projects from 0.049 US$/kWh in 2010 to 0.071 US$/kWh in 2020 [4]. The global LCOE of newly commissioned utility-scale renewable power generation technologies from 2010 to 2020 is shown in Figure 4.1.

With reducing costs of renewables, the investments steadily grew over the span of last 15 years, from US$ 70 billion in 2005 to over US$ 300 billion in 2019. Despite the dramatic COVID-19 pandemic impacts in 2020, the investments in the renewables crossed US$ 320 billion. The largest share of renewable energy investment is contributed by the Asia-Oceanic region led by China (55% during 2005–2019). In the same duration, Europe and the USA shared 20% and 16%, respectively [4].

#### 4.1.1.1  Cost Comparison of 2019 and 2020

The development of renewable energy systems has been aggressively pursued across the globe. The production of electricity through renewable resources has increased by 2.9%/year from renewable resources since the last decade [6]. The two critical merits of any power generation system are (i) LCOE and (ii) the capacity factor which represents the period in which a plant is running at its maximum capacity [7]. The renewable resources were initially characterized as expensive resources, but the trend has been shifting since the last decade (as shown in Figure 4.2) where a significant reduction in LCOE of renewable resources is evident from 2010 to 2019.

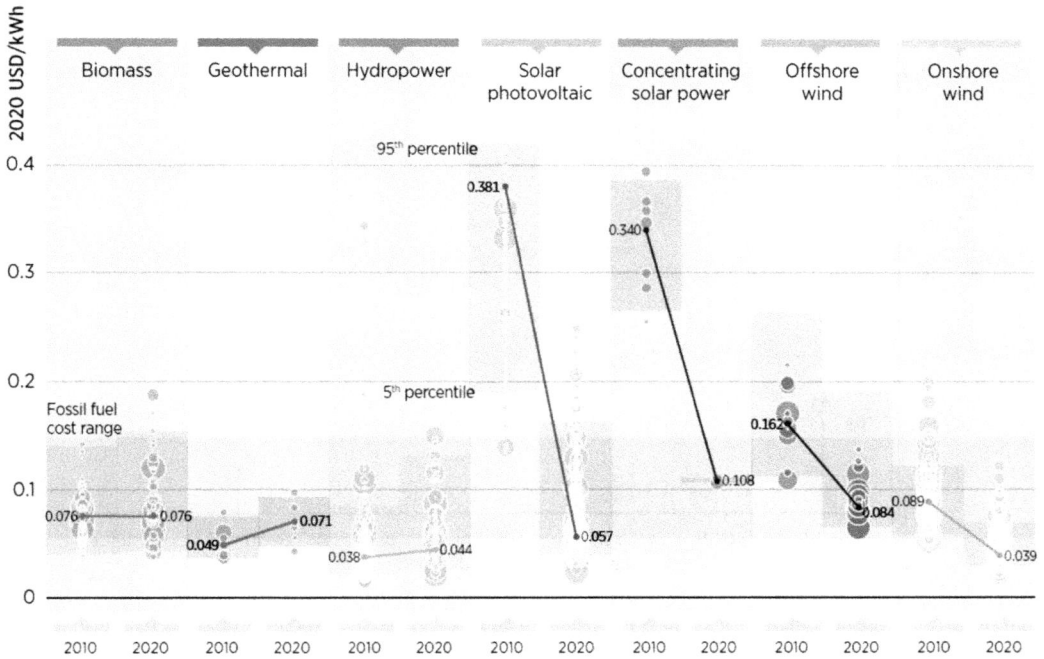

**FIGURE 4.1**  Global LCOE of newly commissioned utility-scale renewable power generation technologies. (From Ref. [5]. With permission.)

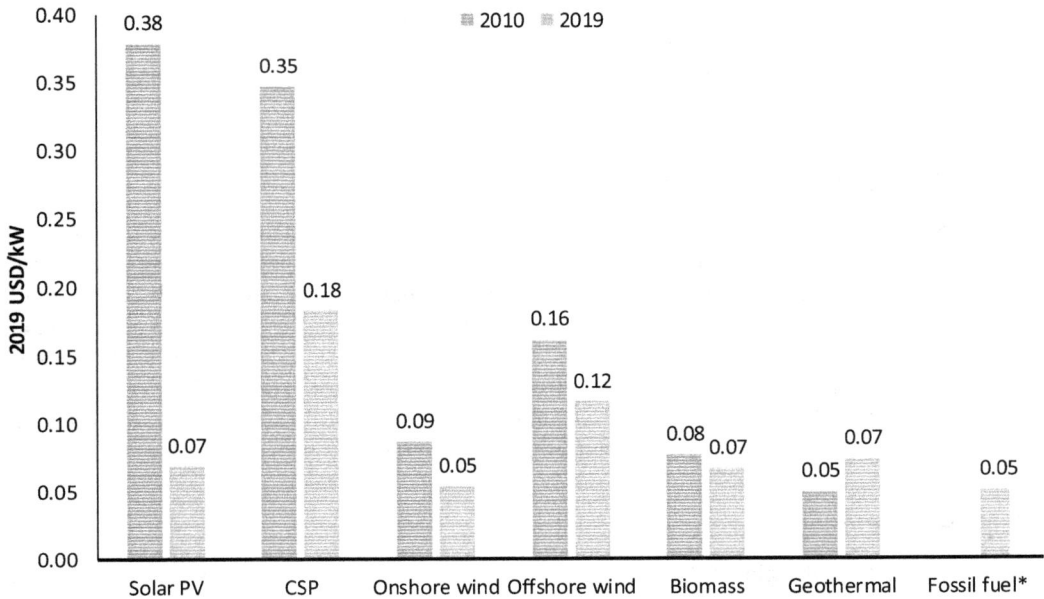

**FIGURE 4.2**  Global weighted LCOE for various resources. *Based on the new coal pants commissioned in coal producing region. https://www.energy.gov/ne/articles/what-generation-capacity. (From Ref. [8]. With permission.)

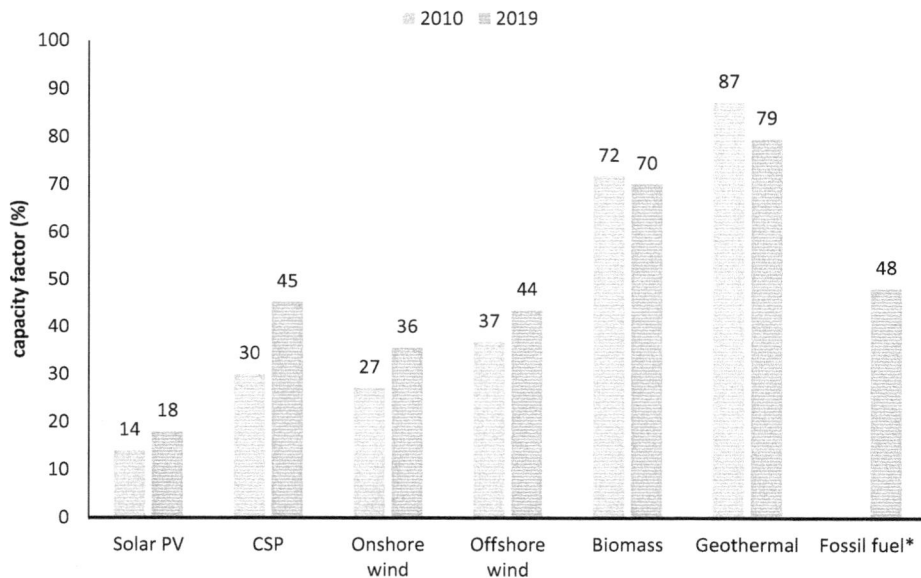

**FIGURE 4.3**  Capacity factor for various renewable resources. *Based on the new coal pants commissioned in coal producing region. https://www.energy.gov/ne/articles/what-generation-capacity. (From Ref. [8]. With permission.)

The second critical factor is the capacity factor of renewable resources, which normally fell short of fossil-fueled plants due to the fickle nature of these resources. The capacity factor of various resources is presented in Figure 4.3.

The solar and wind renewable resources are weather dependent, which is random in nature, so they have the capacity factor in the lower range. However, the biomass and geothermal resources are toward the upper range since the feedstock can be sustained in a continuous manner. Thus, the aspiration in advancements of renewable energy systems is to develop the cost-competitive and reliable renewable energy-based systems.

## 4.2  SOLAR ENERGY TRANSITION

Solar energy is the infinite source of renewable energy as it is available naturally. Due to its versatility and sustainability, it has gained popularity in recent years. With rapidly depleting conventional energy reserves, the solar energy ensures massive financial benefits and reliability in the future [9].

The recent advancements in the field of power electronics triggered scientists and researchers to explore solar PVs and concentrated solar power (CSP) systems. For any system (renewable or nonrenewable) to be adopted for power generation, the most important motivation and the reason behind this is the cost. If the levelized cost of the system is low with successive years, then we can say that the renewable system is viable to be adopted for power generation.

### 4.2.1  SOLAR PVS

In case of solar PVs, the global decrease in the LCOE motivates the market as well as customers to adopt this technology as a prominent replacement for conventional fossil fuels and to adopt this technology to achieve the zero-emission targets. Second, because of its zero marginal cost characteristics, solar PV output is always prioritized for power generation [10]. According to the recent market trends of solar PV, more than 707 GW of solar PV systems have been installed globally as of the end of 2020. This technological growth is about 16 times more since 2010. Asia is a leading contributor to new solar PV installations since 2013 and contributed about 60% of the new PV installation in that year. Meanwhile, Vietnam is an emerging solar PV market where the installation doubled between 2019 and 2020. Vietnam has become the second leading market shareholder of new solar PV installations (after China) in Asia and commissioned more than 11.6 GW of solar PV projects. Japan, India, and Korea also contributed around 13.7 GW of new PV capacity projects following China and Vietnam. In the United States, the new installed capacity of the solar PV projects doubled compared to 2019. The United States, Australia, and Germany installed 24 GW of solar PV during 2020, whereas Brazil and the Netherlands added up 3 GW in their new installations making up the combined growth of about one-third of the total installed capacity in 2019 [4].

With the increasing solar market, the prices of solar panels continue to decline. Since solar panels contribute to half of the project price in any installed solar project, the LCOE also declines [2]. The continuously declining trend in the prices of solar PV modules since 2010 proves to be a prime driver, and this trend still continues. For crystalline modules, the average yearly price reduction between 2019 and 2020 is 5%–15%. The module price variation is dependent on the type of solar PV module and the technology embedded in it. Based on these factors, the modules sold at a price as low as US\$ 0.19/W to as high as US\$ 0.40/W for high efficiency, all black, and bifacial modules.

Because of their potentially increased power output per watt and narrowing cost, the bifacial solar PV module capture the greater market share and customer attention. In 2019, the market share for these bifacial solar PV modules was about 8%, which grew to about 17%–28% in 2020 [11]. Figure 4.4 shows the total installed solar PV system cost and their weighted averages for utility-scale systems from 2010 to 2020.

The continuous reduction in the total installation costs of the PV is related to different factors. Some of them include the optimization of manufacturing processes, reduced labor costs, and enhanced module efficiency. These technological advancements have led to an increased market

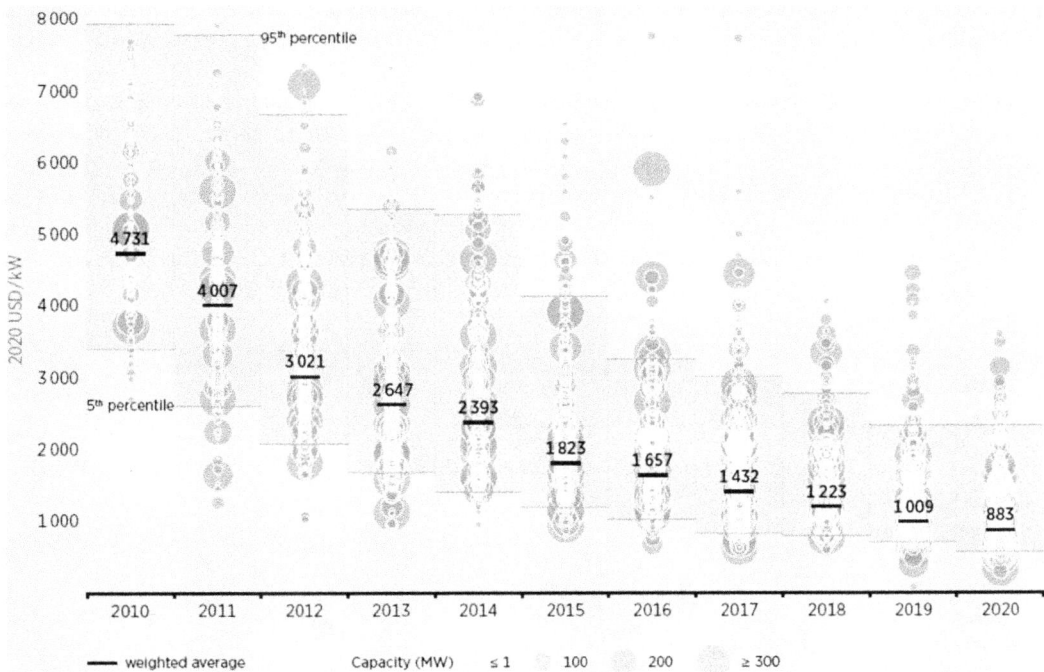

**FIGURE 4.4** Total installed PV system cost and weighted averages for utility-scale systems, 2010–2020. (From Ref. [5]. With permission.)

share of PV. There is a significant reduction in the total installed costs in historical PV markets, including China, India, Japan, the Republic of Korea, the United States, and Germany. In short, for the PV market development consideration, it will play a key role in future net zero greenhouse gas (GHG) emissions renewable energy systems until 2050 [12].

### 4.2.2 CONCENTRATING SOLAR POWER

Concentrating solar power (CSP) systems are installed in the areas with high direct irradiance. The sun's rays are concentrated using mirrors to create heat, which is then transferred through a heat transfer medium (normally molten salt or thermal oil). By this method, steam is generated and a thermodynamic cycle (steam Rankine cycle) is used to generate the electricity. The CSP systems are normally classified into line concentrating or point concentrating systems according to the mechanism by which solar collectors concentrate the solar irradiations. Nowadays, the existing system uses the linear concentrating type of arrangement called parabolic trough collectors (PTCs), and they are normally comprised of a holding structure with focused curved mirrors, a receiver tube, and a pylon foundation. Power towers or solar towers (STs) are the most widely deployed point concentrating CSP technology. In STs, the collectors are called heliostats and are arranged in a semi-circular pattern around a large central receiver tower. The heliostats are controlled individually to track the sun, equipped with dual axes to optimize the solar radiation concentration on the central receiver. The receiver absorbs the heat from the heat transfer oil or molten salt and generates electricity by the same thermodynamic cycle. STs can achieve very high solar concentration factors as compared to the PTCs and, therefore, can operate at very high temperatures.

In the early years of the CSP plant development, they were considered uneconomic and unwarranted because of their additional thermal energy storage issues. Since 2015, this problem has been solved and there has been hardly any project that is built without thermal energy storage. The average thermal energy storage capacity for PTC plants increased from 3.3 hours between 2010 and 2014 to 5.7 hours between 2015 and 2019, which is an increment of almost three-quarters. Similarly, for STs, the average thermal storage capacity has increased by 54% from 5 hours in 2010–2014 to 7.7 hours in 2015–2019.

The total installed costs for both PTCs and STs are controlled by the components that make up the whole solar field. In PTC systems, the components account for 39% of the total installed costs, whereas in STs, the share of the solar field components is at a lower percentage of 28% as shown in Table 4.1.

The LCOE for CSP plants fell significantly between 2010 and 2020 due to the reduction in the total installed costs, operations, and maintenance costs, increasing capacity factors, and declining financing costs. Between 2010 and 2020, the global weighted-average LCOE of the newly installed CSP plants fell by 68% from 0.34 US$/kWh in 2010 to 0.108 US$/kWh in 2020.

**TABLE 4.1**
**CSP Total Installed Costs by Components, 2020**

| | Parabolic Trough Collector (PTC) (%) | Solar Tower (ST) (%) |
|---|---|---|
| Owner's cost | 8 | 9 |
| Indirect EPC cost | 12 | 15 |
| Thermal storage | 17 | 12 |
| Power block | 18 | 16 |
| Tower | 0 | 2 |
| Receiver | 6 | 18 |
| Solar field | 39 | 28 |

*Source:* From Ref. [5]. With permission.

This is due to 50% reduction in the global weighted-average of newly installed CSP projects, a 41% increase in the capacity factors (from 30% to 42%), and a strong decline in the operations and maintenance cost [5].

## 4.2.3 TECHNOLOGICAL DEVELOPMENT

The sun is an inexhaustible source of energy and consequential various technologies are under development for harnessing solar energy. Researchers anticipate that the sun can suffice world energy demand in a sustainable manner and decarbonize the power industry if the technologies for its exploitation are readily available [13]. The advancement in solar energy technologies is a key step toward green and sustainable environment. However, various critical problems, such as efficiency, the inadequate balance of system performance, and economic merits hinder the development of the solar system. The process of harnessing of solar energy is categorized as active and passive technologies.

The passive solar energy technologies are simple since they do not use mechanical or electrical equipment, but are; however, inefficient. The work on the principle of accumulation of heat for primarily space heating purposes. The application of space heating lies in winter; however, there is an enormous amount of solar heat potential during the summer seasons. This has paved the way for active solar energy technologies such as PVs or CSP system. The active technologies convert solar energy into heat or electricity using mechanical and electrical equipment [14].

Recently, the PVs technology has become a highly desirable option for solar energy systems and has experienced significant research growth. PVs convert solar energy directly into electricity using a voltage difference created through a p-n junction. Figures 4.5 and 4.6 show a structure of simple PV solar cells and the compilation of cells into a module and solar panel, respectively.

The main hurdle is the efficiency of solar cells, and this has generated tremendous interest in researchers. The development of solar cells is summarized in four generations depending on the timeline and the material. The first-generation solar cell is categorized based on single and multi-crystalline silicon and is still dominating the market. These cells offer commercial efficiencies of up to 20%. In response to the high manufacturing costs of the first-generation cells, second-generation cells were developed. These are based on silicon or other metals, such as silver, indium, or tellurium; however, to reduce the price, the thickness of the film was brought down to the nanometer scale. The third and the fourth-generation solar cells are in the laboratory development stage, and they involve the use of quantum dots, organic solar cells, photochemical sells, and conjectural generation consisting of composite materials. The limiting efficiency of solar cells using any semiconductor is 33.7%; however, the stacking of solar cells can increase the efficiency by up to 50%, which is a key research area in the domain of solar energy technologies.

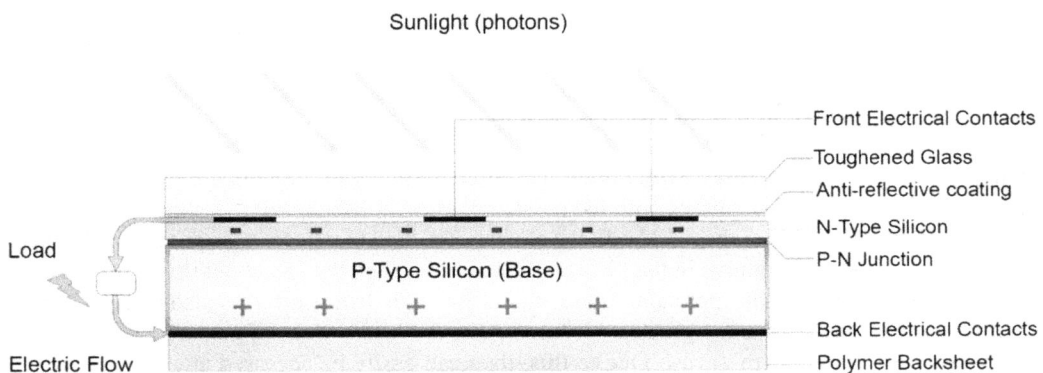

**FIGURE 4.5**   Schematics of a typical solar photovoltaic cell.

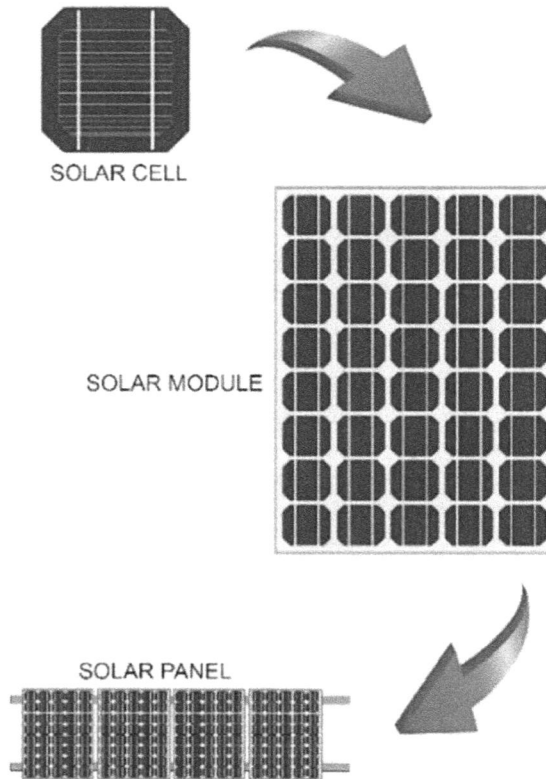

**FIGURE 4.6**   Arrangement of solar photovoltaic cells in solar panel.

The high initial cost of a solar cell system is still the major obstacle to its development as evident from Figure 4.1. Besides, the efficiency of domestic panels is still lower in the range of 10%–20% [15]. Moreover, solar cell technologies depend on rare and semi-rare materials, therefore, the advancement in the recycling of the spent metals is also underway. A solar cell is an efficient sustainable development source and is a key energy source for the processes in photosynthetic organisms. The researchers have mapped the direction and flow of solar energy in such organisms, which could be the most efficient way of harnessing solar energy. The researchers from Graphene Flagship improved the durability of perovskite solar cells by adding $MoS_2$ flakes as an active buffer layer [16]. Hence, the research work is going on in fabricating and innovating solar cells to improve their performance and durability, and, consequentially reduce the cost of solar cell technologies.

## 4.3   GEOTHERMAL ENERGY TRANSITION

Geothermal energy has been used as an important source of energy throughout human history and its use has gone through certain phases with the passage of time. It is believed that this form of energy existed since the formation of the world billions of years ago, and therefore, it is considered the oldest form of energy production source in the history of mankind. When the fact about this form of energy is revealed that it is renewable, potential to use, and utilize it in different forms also increases [17].

This source of energy has the greatest resource potential in the areas of active geothermal sites near the surface of the Earth's crust. Due to this, they can easily be accessed at a lower cost. The power generation from geothermal energy is somewhat unique and different in nature as compared to other renewable power generation technologies. Even though geothermal is a mature and commercially proven technology having a low cost of power generation, still research and development are required for more innovative, low-cost drilling techniques. This will lower the development cost and utilize the full potential of geothermal resources by making them even more economical from energy generation aspect. If we look back before 20th century, the only predominant use of geothermal energy was personal such as bathing [18]. The concept of using this energy for electricity production begins in the 20th century, and the first geothermal energy-based power plant was installed in Larderello region of Italy in 1904 and started producing electricity with a capacity of 250kW in 1913 [17]. The geothermal power plant has made a great contribution to electricity production not only for the industrial sector but also for the development of the integrated systems that produce electricity in general. Until 1942, the total electricity generation from geothermal power plants was around 128 GW. Later on, small-scale geothermal plants were installed in various regions of the world for fulfilling the electricity needs, and for the deeper exploration of geothermal energy production started. Many countries recorded their development in power production from geothermal power plants especially between 1950 and 1960, including New Zealand, the USA, and Mexico [19].

By the 21st century, geothermal energy became much more mature and exhibited a successful behavior in electricity generation. According to the report published by International Renewable Energy Agency (IRENA), the installed capacity and annual electricity generation by geothermal energy systems between 1995 and 2019 are shown in Figure 4.7. Considerably less geothermal installed capacity had been added in 2020 [20]. Turkey increased its geothermal installed capacity by 99 MW and some small expansions installed in New Zealand, the USA, and Mexico [20].

The IRENA study evidenced that the electricity production from geothermal energy has been accelerated in the last 3 years. By the end of 2020, geothermal power plants accounted for 0.5% of the total installed renewable power generation capacity globally, with a total installed capacity of 14 GW. These numbers are 41% higher than the installed capacity in 2010.

With rising commodity and drilling costs between 2000 and 2009, the total installed costs for geothermal power plants rises by 60%–70% in that period [21]. The total installation costs for geothermal power plants by project, technology, and capacity from 2007 to 2021 are shown in Figure 4.8. Based on the data available in IRENA power generation costs 2020, the installed costs for geothermal after 2010 declined within the range of 2,000–7,000 US$/kW, although there were a number of project outliers. However, since 2015, the installation costs have shown an increasing trend followed by an interruption in 2019 when the installation cost slightly dropped down.

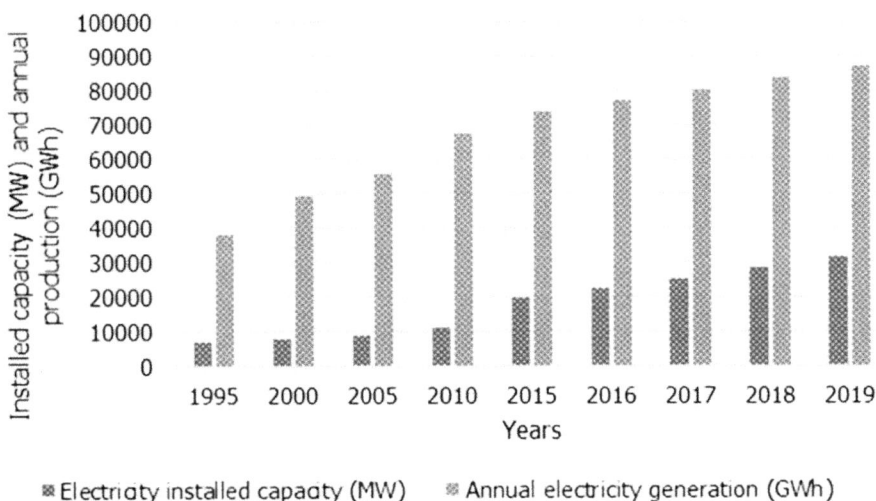

**FIGURE 4.7** Global installed capacity and annual electricity generation by geothermal energy. (From Ref. [17]. With permission.)

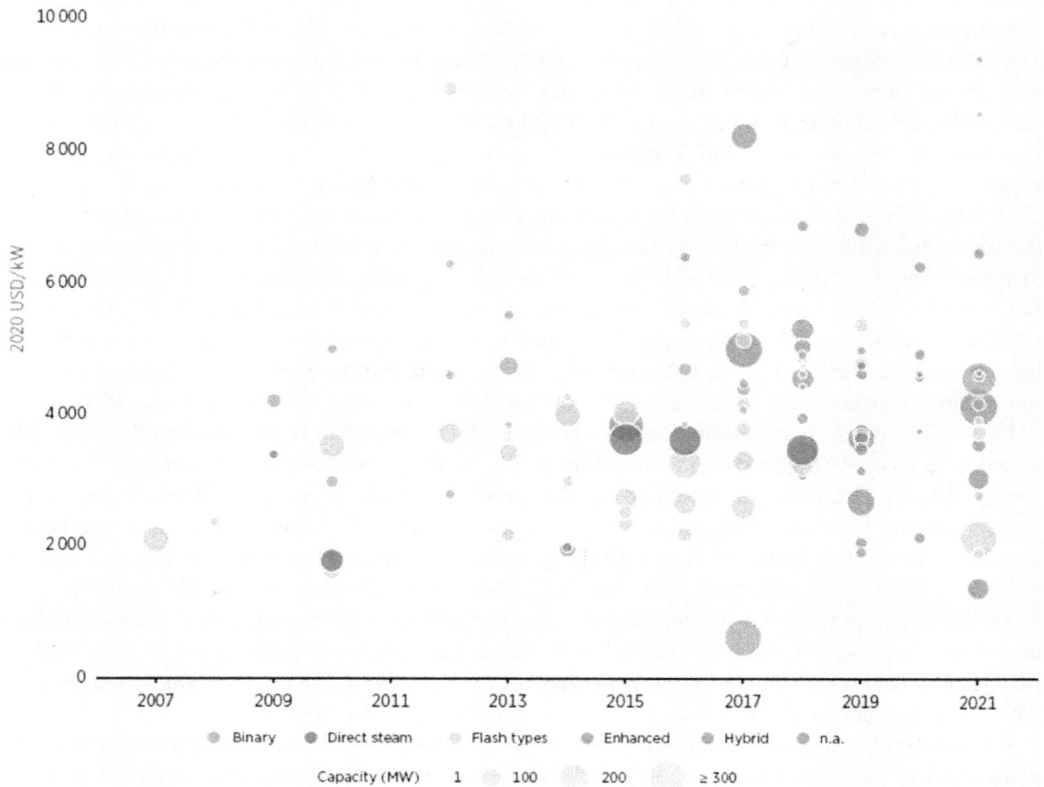

**FIGURE 4.8** Total Installed costs by project, technology, and capacity for geothermal power, 2007–2021. (From Ref. [5]. With permission.)

As mentioned previously, the total installation cost, weighted-average cost of capital, economic lifetime, and operations and maintenance costs of a geothermal power plant determine its LCOE. As compared to solar and wind technologies, geothermal power plant requires continuous optimization throughout its lifetime with proper, sophisticated management of the reservoir and the production wells to ensure the pure production of energy output to meet the needs. The LCOE varied as low as 0.04 US$/kWh for existing geothermal projects to as high as 0.17 US$/kWh for small geothermal installed projects in remote areas [5].

For the geothermal projects commissioned in 2010, the global weighted-average LCOE increased from 0.05 US$/kWh to 0.07 US$/kWh in 2020. There are annual variations in the capacity factor values of newly commissioned projects, which is less significant as compared to some other technologies. Due to this reason, the LCOE values tend to follow the trends in the total installed costs. The LCOE values remain stable at around 0.07 US$/kWh for the period 2019–2021. This depends on whether the projects meet their commissioning goals or not, particularly for the larger projects in pipeline.

Geothermal energy is one of the most abundant existing renewable sources of energy without showing any intermittence as in solar, wind, and biomass applications. Geothermal-based integrated power plants have great potential for multi-generation, which will help to increase thermodynamic and thermos-economic performances, reduces cost and environmental effects, and increase sustainability. Finally, improvement in the geothermal power plants-based multi-generation systems would help us to reduce the GHG emissions and other disastrous environmental effects and improves the environmental sustainability.

### 4.3.1 TECHNOLOGY DEVELOPMENT

The geothermal resources are in the crust of Earth and are in the temperature range of 20°C–220°C. Most of the geothermal energy (68%) lies in the low temperature of around 130°C while the remaining is above 130°C. The low-temperature geothermal heat source is used for direct use such as bathing, swimming, space heating and cooling, and agriculture while the high temperature is used for electricity power generation [22].

The direct use of geothermal energy is simple and mature where the major cost of the systems stems from the cost of well. Water or brine solution is used to extract the geothermal energy through an injection and the production well normally ranges from 500 to 2,700 m [23]. For the direct use application, the key advancement area is the research, development, and innovation in the drilling techniques. The full potential of geothermal energy can be attained if the cost of drilling is reduced.

Furthermore, advanced simulation methodologies for geothermal reservoirs are needed to increase the exploitation of geothermal energy. The high temperature and enthalpy of geothermal resources are used to produce power through a steam, organic Rankine cycle, or a flash cycle [24]. The schematics of simple flash cycles are shown in Figure 4.9 where the hot geothermal fluid is separated into vapor and liquid in a flashing device. The vapors later produce power while the liquid is fed back to the geothermal reservoirs using the injection well. The simple steam cycle is limited to the relative high reservoir temperatures and less exhaustion of the heat source is achieved.

The alternative schemes such as the organic Rankine cycle (as shown in Figure 4.10) have been investigated for power generation from a relatively low-temperature geothermal source.

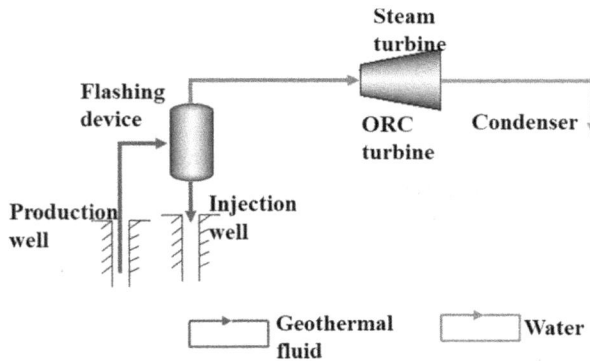

**FIGURE 4.9**  Direct geothermal steam cycle.

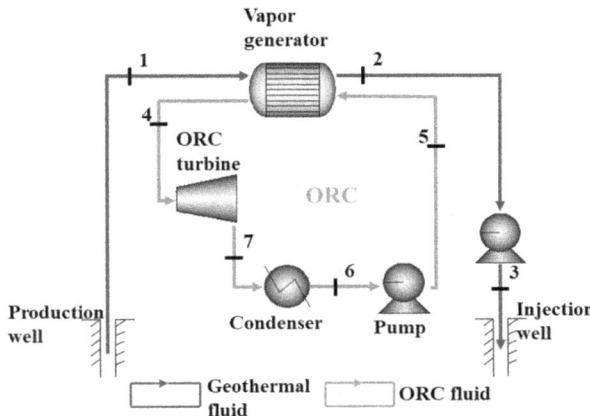

**FIGURE 4.10**  Simple binary geothermal organic Rankine cycle.

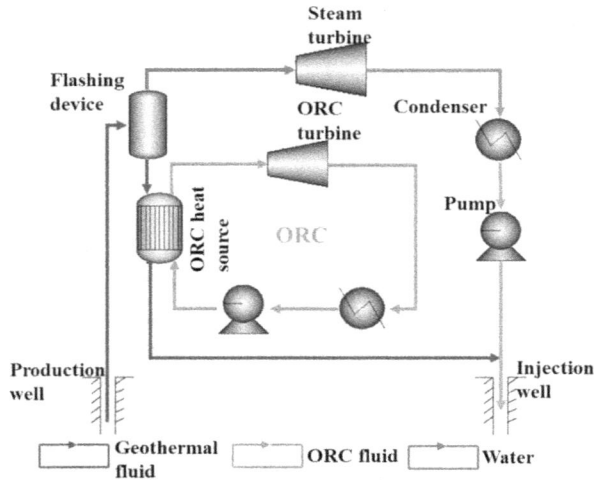

**FIGURE 4.11**   Schematics of a flash-binary ORC system.

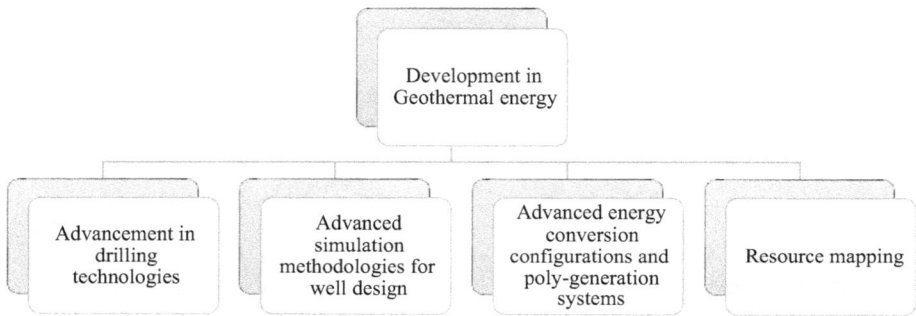

**FIGURE 4.12**   Key development domains in geothermal energy.

The above schematics are simple; however, they result in incomplete exhaustion of the geothermal heat source. For better performance, advanced schematics have been examined, which can fully exhaust the heat source temperatures. One advanced configuration is shown in Figure 4.11, which integrates the above two configurations. In this configuration, the liquid after the flashing device, which is still at high temperatures is used as the heat source for the ORC. This way the temperature at the injection well can be reduced further and more power can be obtained.

In addition to the technology development and the configuration optimization, which involves strengthening the evaluation mechanism and the development of key equipment, it is also imperative to develop a policy framework by legislating the laws and incentivizing the development of geothermal systems. Figure 4.12 summarized the key advancement areas for the development of geothermal energy systems. The geothermal locations are limited and are distributed; therefore, it is also essential to quantify the geothermal potential of various locations through resource mapping.

## 4.4   BIOENERGY TRANSITION

With rapid industrialization and population growth, the energy demand continues to increase. The excessive and aggregating use of conventional fossil fuels, fluctuating fuel prices, and increasing GHG emissions triggers the world to shift from the fossil fuels to biofuels [25]. A variety of feedstocks and their quality are the main source of power generation from bioenergy. A variety of different combustion technologies can be used for power generation from bioenergy. These technologies are mature, commercially available with long track records, and a wide range of suppliers. However, less mature but innovative system technologies are also present that includes the atmospheric biomass gasification and pyrolysis. Although these technologies are being tried out commercially for power generation, they are still at the research and developmental stage. The mature bioenergy conversion technologies include direct combustion in boilers, low percentage cofiring, anaerobic digestion, municipal solid waste incineration, landfill gas, and combined heat and power (CHP). Feedstock type, the conversion process, and power generation technology are the main factors that are considered very important for power generation from bioenergy. For the economic success of the biomass project, feedstock availability is one of the main elements.

Unlike solar and wind, the economics of biomass power generation is dependent on the availability of the feedstock supply that is predictable, sourced sustainably, has low cost, and of adequate quality over the long term. Biomass feedstocks are very heterogeneous as their chemical composition is highly dependent on the plant species. There is a high variation in the cost of feedstock price per unit of energy. Physical properties of the feedstock are as important as the cost, as different feedstock has a different density, particle size, combustion capacity, ash content, and moisture. These factors also affect the transportation, pretreatment, storage costs, and appropriateness of the different conversion technologies. Different regions of the world have different costs for power generation from biomass. The total cost of biomass power generation contains both the technology component and the local cost component. The project investment cost tends to be lower in the countries with emerging economies as compared to the OECD countries. The reason is that the emerging economies are often benefited from the low labor and commodity costs.

Figure 4.13 represents the total installed cost of biomass-based power generation projects for a variety of different feedstock types from 2000 to 2020. It can be seen that the deployment pattern of the feedstock varies by country/region, it can be seen that Europe and North America have higher total installed costs across the feedstock whereas it is lower in Asia and South America. This is due

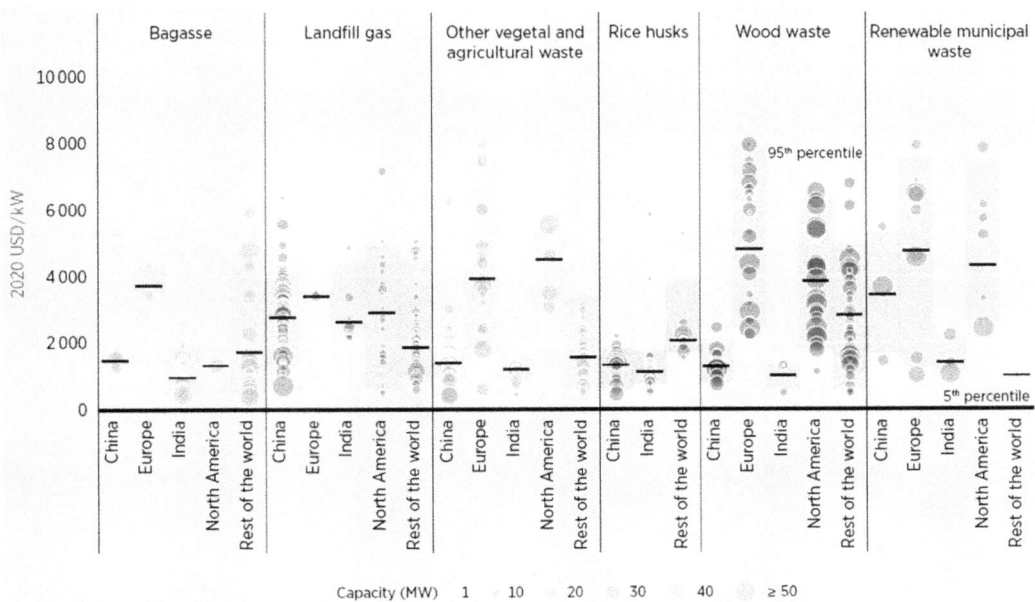

**FIGURE 4.13** Total installed cost of biomass projects by selected feedstock and country/region, 2010–2020. (From Ref. [5]. With permission.)

to the reason that the installed bioenergy projects in OECD countries are often based on wood or use combustible renewable municipal or industrial waste.

For China, between 2000 and 2020, the total installed cost of the project across all the feedstock ranges from a low value of 634 US$/kW for rice husk to as high as 5,304 US$/kW for renewable municipal waste. In India, this range is as low as 514 US$/kW for bagasse projects to a high value of 4,356 US$/kW for landfill gas projects.

If we analyze the bioenergy markets, then overall, it accounts for an estimated 11.6% (44 EJ) of total final energy consumption in 2019 [26]. Globally there was an increase of 5.8% in power generation from bioenergy from 137 GW in 2019 to 145 GW in 2020. China had the largest power generation capacity by the end of 2020 followed by the USA, Brazil, Germany, India, the United Kingdom, Sweden, and Japan. There was an increase of 6.4% in total bioelectricity generation from 566 TWh in 2019 to 602 TWh in 2020 with China as a leading bioelectricity producer [27], as shown in Figure 4.14.

According to China's 13th 5-year plan (2016–2020), the electricity generation from bioenergy increased by 26% from 17.8 GW in 2019 to 22.5 GW in 2020 [3]. Seventy-seven new projects with a combined capacity of 1.7 GW were approved in 20 different Chinese provinces. The United States had the second largest national bio-power capacity and generation in 2020. However, the electricity generation from bioenergy fell from 2.5% to 62 TWh continuing the trend of recent years [26]. Brazil, being the third largest bioelectricity producer globally, generates most of its electricity based on sugarcane bagasse [3,27]. In the EU, the bio-power generation increased by 4% in 2020 to 205 TWh and the bio-power capacity also jumps to 4% to 48 GW in the same year. Germany, being the region's best bioelectricity producer, increased its capacity to 400 MW in 2020 and generation jumps to 0.8% to 51 TWh [28].

Similarly, in the United Kingdom, the bioenergy-based power plants increase in capacity from 135 MW to 8 GW, and there is a rise in generation by 5.5% with the increase in large scale pellet fired generation, biogas, and municipal solid waste plants [29].

While bioenergy can directly replace the conventional fossil fuel technology for heating, transport, and electricity generation, biomass-based power plants could play a pivotal role in sustaining the global economy and making the environment even more sustainable. This would lower the GHG emissions and reduce the carbon footprints by minimizing the use of fossil fuel-based feedstock such as plastics and by replacing energy-intensive materials such as steel, concrete, and agriculture-based materials. Figure 4.15 gives an overview of the estimated shares of bioenergy in total final energy consumption—overall and by the end-use sector.

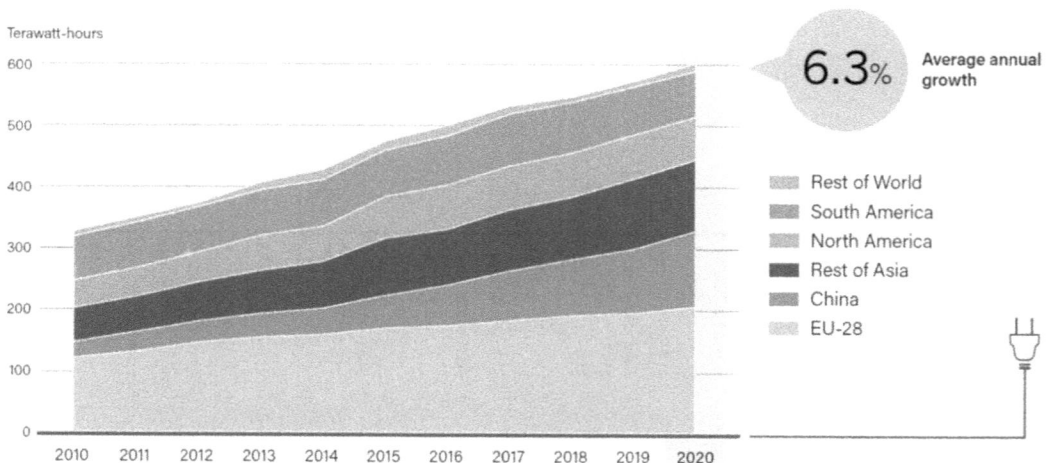

**FIGURE 4.14** Global bioelectricity generation by region, 2010–2020. (From Ref. [26]. With permission.)

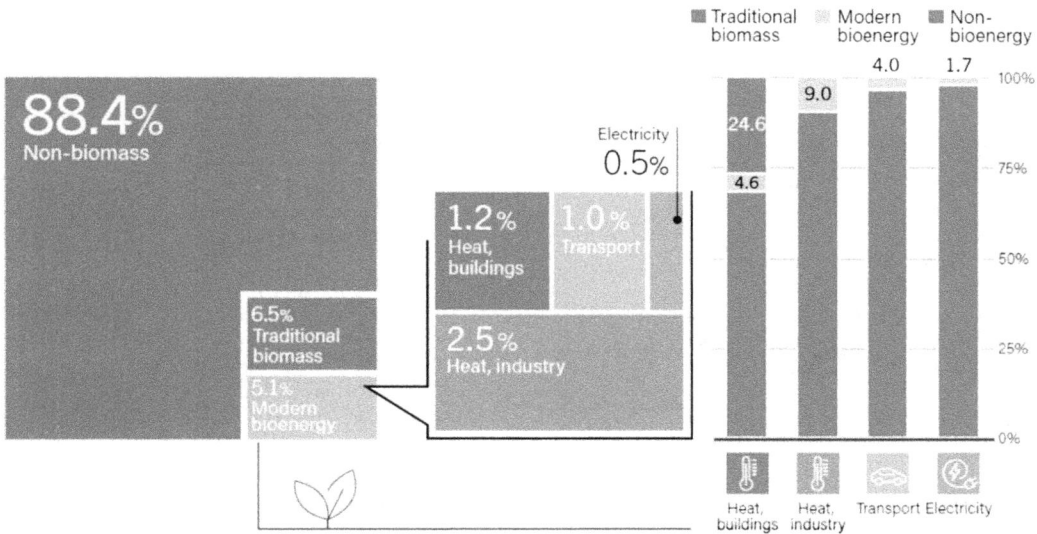

**FIGURE 4.15** Estimated shares of bioenergy in total final energy consumption, overall and by end-use, 2019. (From Ref. [27]. With permission.)

### 4.4.1 TECHNOLOGY DEVELOPMENT

The biomass or the organic matter present in nature or produced because of human activities can be used through its conversion to produce electricity, for CHP applications, transportation fuels, and to produce valuable chemicals for manufacturing industry. The conventional use of biomass is for heating, which is inefficient and harms the environment. The more sustainable conversion of biomass can be achieved through processes like gasification, pyrolysis, fermentation, anaerobic digestion, liquefaction, cracking, and so on. The energy conversion cycle from biomass is summarized in Figure 4.16.

The feedstock of biomass has the advantage of versatility and flexibility, and it ranges from aquatic, terrestrial, industrial, and agricultural waste. The biomass feedstock has different compositions and characteristics, which can hinder the performance of the conversion system. Therefore, depending on the nature of feedstock and the product, the biomass undergoes different conversion processes. Bioenergy systems are flexible, can be stored more easily compared to other renewable systems, and are better suited for small- to medium-scale power generation systems [30].

Nonetheless, the flexibility of biomass feedstock and the conversion processes has a downside and an inappropriate choice of conversion process for a given feedstock can limit the efficiency. The whole portfolio of biomass energy systems involves feedstock, technology, and the product.

The development of bioenergy systems involves understanding and addressing the interfaces between the feedstock to the product [31]. The biomass feedstock undergoes different pretreatments such as thermal, mechanical, or chemical to improve the performance of the system with the versatile feedstock. The pretreatments of feedstock can result in the improvement in performance. Ordonez-Frías et al. [32] showed that the palletization of the feedstock and the anaerobic digestion are economically more feasible for oil palm residues. Furthermore, in the premise of bioenergy, the focus has been shifted to grid expansion and integration. The development of new infrastructure for biomass energy systems incurs substantial upfront costs and is, therefore, economically unviable. The researcher has investigated the use of existing infrastructures of fossil-fueled plants for the biomass. Ozonah et al. [33] applied artificial neutral network systems to optimize the operational parameter during the cofiring of coal and biomass blend and showed the viability of using existing

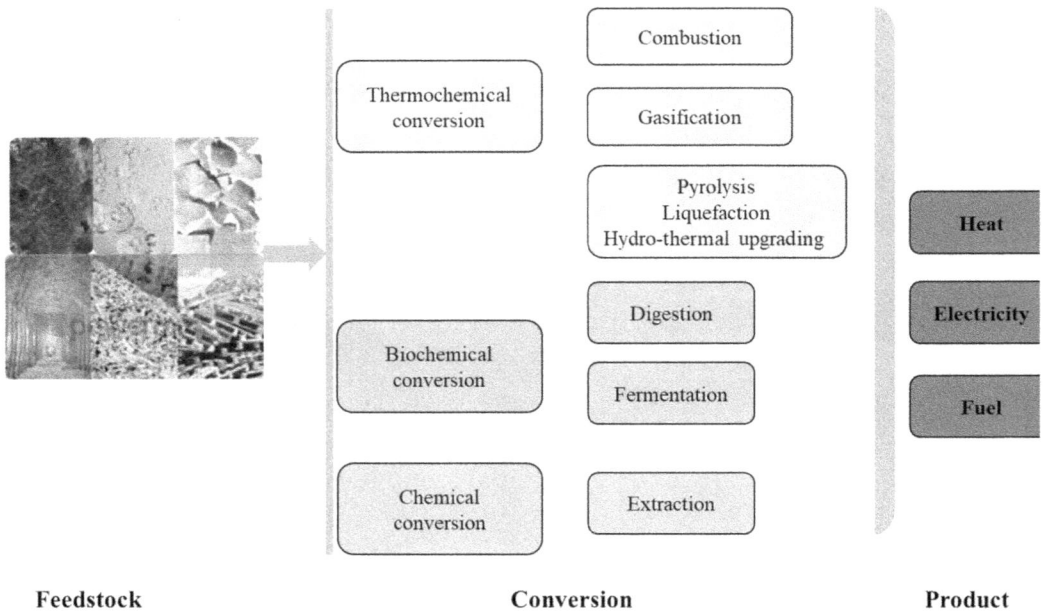

FIGURE 4.16    Biomass energy conversion.

infrastructure for the biomass energy systems. Biomasses have multiple applications from power production to the production of transport fuels and other chemicals that have industrial applications as well. The production of hydrogen through biomass is also a technology development area, which is widely investigated to sustain the upcoming hydrogen economy.

Therefore, the advanced conversion technologies of biomass into hydrogen and transport fuel have also been rigorously investigated. Technologies, such as cell immobilization, nanotechnology, mathematical optimization tools, and technologies for biogas upgrading using renewable $H_2$ are comprehensively investigated [34]. However, the cost of implementation of these technologies is yet to be understood for their industrial penetration. The advancement and development in bioenergy technologies is making way for diverse applications and the optimization through pretreatment is necessary to make the technology competitive to its counter technologies. An advanced exergoenvironmental method is applied to the bioenergy systems to ascertain their energy and economic merits. The analysis can offer detailed information from the economic and environmental perspective and thus can indicate potential measures to enhance the performance of the system. Aghbashlo et al. [35] recently examined the merits and demerits of exergoenvironmental analysis, and presented how it can be used to optimize the interfaces between the bio energy conversion chains and improved the efficiency and the capacity factor of the bioenergy systems. The supply of biomass is not ubiquitous in nature and is unevenly distributed where the countryside tends to have more resources than the city side. Thus, the site selection before the bioenergy system development is critical and there is a pressing appeal to develop methodologies that can do a preliminary energy and cost analysis of a biomass system at a given location. The summary of advancement areas in the biomass energy are given in Figure 4.17.

## 4.5   WIND ENERGY TRANSITION

Renewable power costs continue to decline in 2020 following and maintaining the trends from the past decade. Mature technologies such as bio-power, solar, and hydropower are typically low-cost renewable power generation sources and are competitive in the regions where available resources

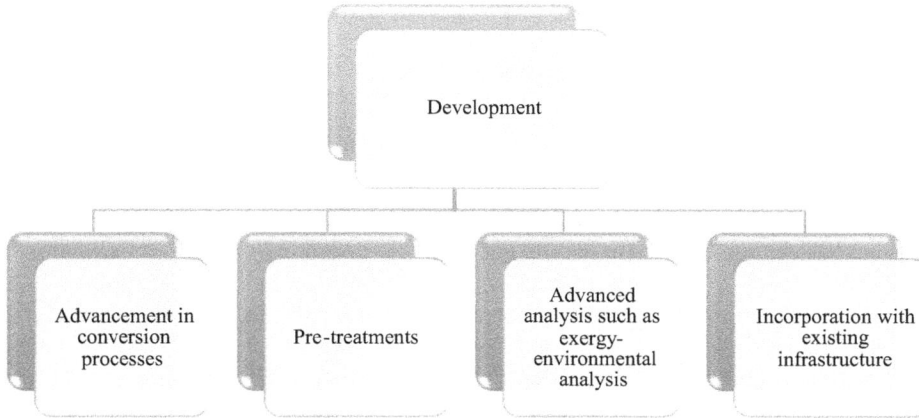

**FIGURE 4.17** Key development domains in bioenergy sector.

exist. However, the decade was more notable due to the paid improvements in the technology and competitiveness of solar and wind power. The global market analysis of 2020 showed that an estimate of 93 GW of wind power capacity was installed in that year. This capacity includes around 87 GW of onshore wind capacity and 6 GW offshore wind capacity [36]. This was the highest ever recorded wind energy installation, which was 45% higher as compared to the previous high in 2015 (63.8 GW), and a 53% increase was also been recorded as compared to the 2019 installations [37]. China has shown explosive growth in new wind energy installations and continues to take the lead in the global development of wind power with its share of the global market increasing by 8.5% last year. Due to the pandemic-related restriction in 2020, supply chains got disrupted, causing most of the wind energy workforce unavailable. This resulted in the postponed or canceled auctions, delayed investments, forced delays, or project cancelations in many countries, particularly in the onshore sector [26,37]. Even with the challenges related to the pandemic and issues of global health, the total wind energy installation by 2020 was 743 GW (707.4 GW onshore and 35.6 GW offshore), which was 14% more as compared to 2019, as shown in Figure 4.18.

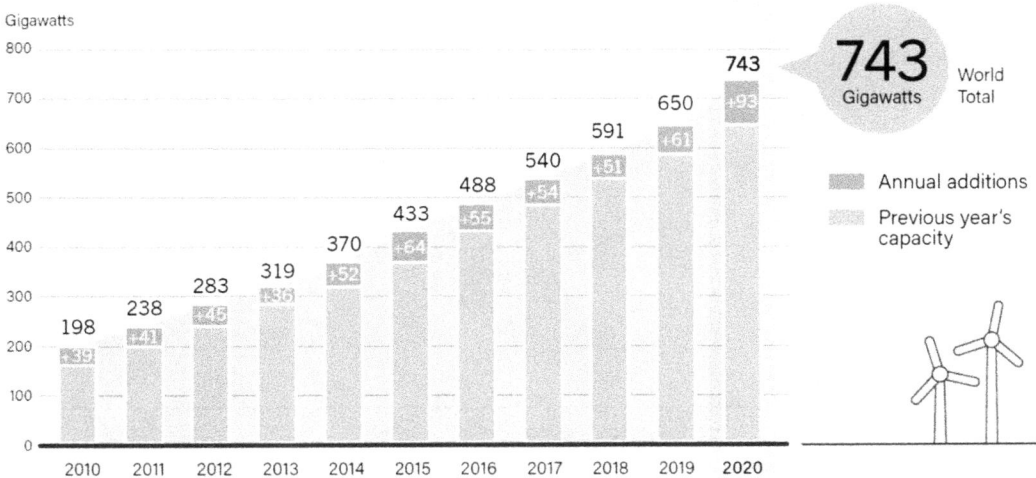

**FIGURE 4.18** Global capacity and annual addition in wind power, 2010–2020. (From Ref. [37]. With permission.)

### 4.5.1 Onshore Wind Energy

There is a significant advancement in onshore wind energy technology over the last decade. The technological improvements in making the larger and more reliable turbines with higher hub heights and larger rotor diameters increase the capacity factor. In addition to these technological developments, total installed costs, operation, and maintenance cost, and LCOE continuously declines as a result of increased competitiveness and maturity in the sector.

In 2020, the onshore wind energy deployment was the second highest to be installed after solar PV and considerably higher than that in 2019 due to the surge in projects in China. Nowadays, mostly installed onshore wind turbines are the horizontal axis type predominantly using three blades and upwind blades configuration. Wind turbines contribute the major share to the total installed cost of the wind project. Wind turbines accounts for about 64%–84% of the total installation costs of an onshore wind turbine project [38]. The total installed cost of the onshore wind projects fell by 74% between 1983 and 2020. In 1983, the total installed cost was 5.241 US$/kW, whereas, in 2020, it fell to 1,355 US$/kW [5], as shown in Figure 4.19. Between 2010 and 2020, the total installed cost of the onshore wind fell by 31% from 1,975 to 1,355 US$/kW with 10% decline in successive years from 2010 to 2020. Globally, the average total installation cost of wind projects fell by 9% for every doubling in the onshore wind capacity. In addition to the total installed costs of the onshore wind, the capacity factor is also very important. It represents the annual energy output from a wind farm as a percentage of the farm's maximum output.

The global weightage average capacity factor of the onshore wind rise by 81% between 1983 and 2020. A similar trend has been observed in the last 10 years from 2010 to 2020. There was almost one-third increase in the capacity factor from 27% in 2010 to 36% in 2020, whereas the capacity factor remained at 36% between 2019 and 2020. China's higher share in the global deployment of wind energy had a significant impact on the capacity factor. Similarly, the LCOE declined by 87% from 0.311 US$/kWh in 1983 to 0.041 US$/kWh in 2019. Consequently, wind energy competes with hydropower as the most competitive renewable technology without financial support.

### 4.5.2 Offshore Wind Energy

In 2010, offshore wind technology was relatively a developing technology, which matured rapidly with advancements and improvements in wind technology, and ever since offshore technology has also been improved. There is an increase of 11 folds in the global offshore wind installation capacity

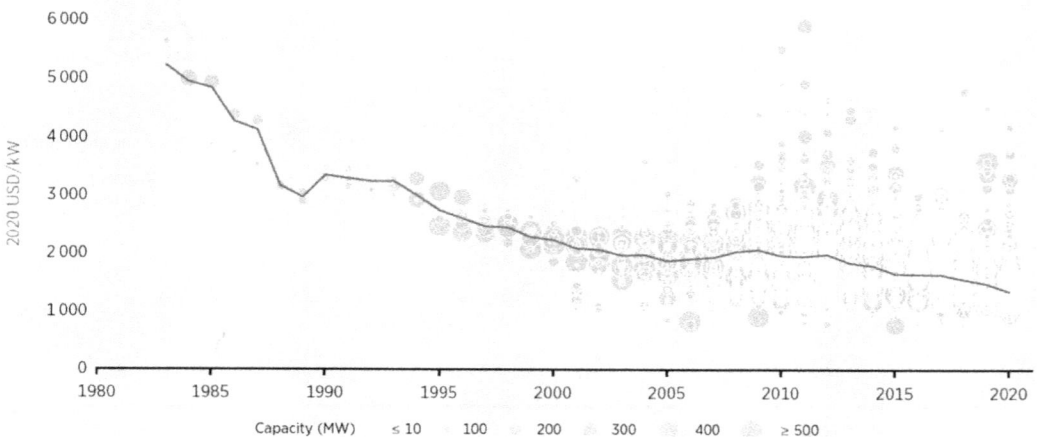

**FIGURE 4.19** Total installed costs and global weighted-average of onshore wind projects, 1983–2020. (From Ref. [5]. With permission.)

between 2010 and 2020 from 3.1 to 34.4 GW [4]. Offshore wind energy makes up just under 5% of global onshore and offshore wind energy deployment. The plans for future targets and deployment for offshore wind energy have been expanding with decrease in cost and maturity in the technology. There is an annual capacity addition of over 5 GW from 2017 to 2020.

Unlike the onshore wind energy projects, offshore wind farms must compete in terms of operation and maintenance costs in challenging and harsh environments, which make these projects expensive with significantly longer lead times. In recent times, with an increase in offshore wind farm deployment, cost reductions have been unlocked. Compared to onshore wind energy, offshore wind farms have higher installed costs but these farms are advantageous in terms of economies of scale. Meanwhile, their capacity factor is higher and have stable energy output (due to reduced wind shear and turbulence and higher average wind speeds).

According to the IRENA, Renewables Power Generation cost report 2021 [5], there is an increase of 121% in the deployment capacity of the offshore wind energy from 136 to 301 MW and there are still some ongoing offshore wind energy projects having capacity increasing 1 GW. From 2000 to 2008, the global weighted-average total installed cost for offshore wind farms increased from 2,592 to 5,500 US$/kW and fluctuates around 5,000 US$/kW within the period 2008–2015 as the offshore wind farm projects move away from the shores to the deeper waters. The cost declined after 2015 to 3,185 US$/kW in 2020.

At the end of 2020, 18 countries had offshore wind energy capacity plants in operation, unchanged from 2019 [39]. The United Kingdom has the leading share of the total capacity of 10.4 GW followed by China (10 GW), Germany (7.7 GW), the Netherlands (2.6 GW), Belgium (2.3 GW), and Denmark (1.7 GW) in 2020 [37]. A total of 70% of the global offshore wind power capacity has been installed in Europe (the number declined from 75% in 2019 and 78% in 2018). Asia has the rest of the 30% of the offshore wind energy capacity with China leading the number [40], as shown in Figure 4.20. Looking toward the total installed cost trend for offshore wind capacity by country, it is important to understand how the cost structures are evolving. China, being the largest to deploy the wind energy capacity globally, experienced a decline in the total installed cost by 32% between 2010 and 2020, falling from 4,476 to 2,968 US$/kW. Globally, the total installed cost of the offshore wind energy capacity fell by 40% from 5,323 to 3,185 US$/kW between 2015 and 2020.

The capacity factors for offshore wind farms are very wide due to the meteorological difference among different farm sites. The global capacity factor for newly installed and commissioned offshore wind farms grew from 38% to 40% between 2010 and 2020.

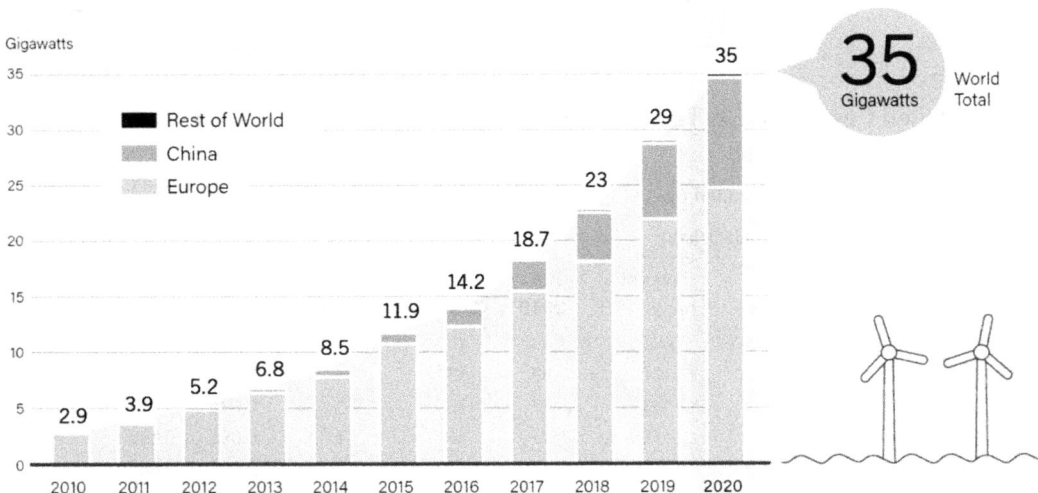

**FIGURE 4.20** Global offshore wind power capacity by region, 2010–2020. (From Ref. [26]. With permission.)

However, the capacity factor range of newly commissioned projects was between 33% and 47%. From 2010 to 2020, the improvements in the global capacity factors in the countries with offshore wind installation show that China has made a remarkable improvement with a 23% increase over the period. Table 4.2 shows the average capacity factors for offshore wind energy projects in six dominant regions of the world.

Similarly, the LCOE of offshore wind fell by 48% from 0.162 to 0.084 US$/kW in the span from 2010 to 2020. The detailed analysis of the LCOE for offshore and onshore wind energy can be studied from various published reports in 2021 [3,5,26,41]. Table 4.3 shows the summarized values for the total installed cost, capacity factors and LCOE for solar PV, geothermal, bioenergy, and wind energy.

**TABLE 4.2**

**Capacity Factors for Offshore Wind Projects in Six Countries, 2010–2020**

| | 2010 (%) | 2020 (%) | Percentage Change, 2010–2020 (%) |
|---|---|---|---|
| Belgium | 38 | 41 | ⬆ 8 |
| China | 30 | 37 | ⬆ 23 |
| Denmark[a] | 44 | 50 | ⬆ 14 |
| Germany | 46 | 45 | ⬇ 2 |
| Japan[a] | 28 | 30 | ⬆ 7 |
| Netherlands[b] | 48 | 47 | ⬇ 2 |
| United Kingdom | 36 | 38 | ⬆ 6 |

*Source:* From Ref. [5]. With permission.

[a] Countries with data only for projects commissioned in 2019.

[b] The Netherlands had no commissioned project in 2010, so data for projects commissioned in 2015 are shown.

**TABLE 4.3**

**Total Installed Cost, Capacity Factors and Levelized Cost of Electricity Trends by Technology, 2010 and 2020**

| | Total Installed Cost (2020 USD/kW) | | | Capacity Factor (%) | | | Levelized Cost of Electricity (2020 USD/kW) | | |
|---|---|---|---|---|---|---|---|---|---|
| | 2010 | 2020 | Percent Change (%) | 2010 | 2020 | Percent Change (%) | 2010 | 2020 | Percent Change (%) |
| Solar PV | 4,731 | 883 | −81 | 14 | 16 | 17 | 0.381 | 0.57 | −85 |
| Geothermal | 2,620 | 4,468 | 71 | 87 | 83 | −5 | 0.049 | 0.071 | 45 |
| Bioenergy | 2,619 | 2,543 | −3 | 72 | 70 | −2 | 0.076 | 0.076 | 0 |
| Onshore wind | 1,971 | 1,355 | −31 | 27 | 36 | 31 | 0.089 | 0.039 | −56 |
| Offshore wind | 4,706 | 3,185 | −32 | 38 | 40 | 6 | 0.162 | 0.084 | −48 |

*Source:* From Ref. [3]. With permission.

### 4.5.3 TECHNOLOGY DEVELOPMENT

The simplified wind energy system is illustrated in Figure 4.21. In a wind energy system, maneuvering the variation in wind energy with altitude as well as during the day is a key to the efficient and reliable operation [42]. The power produced by the wind turbine is proportional to the cube of the wind speed, and therefore, it is essential to control the rotational speed of the turbine at the wind gust to avoid the failure of the turbine. Thus, various methods have been researched such as stall control, active stall control, and pitch control to avoid the turbine breakdown in wind gust conditions [43].

The main challenge associated with a wind turbine is the fluctuation in the wind speed. Therefore, variable speed turbine has been developed to work around this issue [42,43]. These turbines store the varying wind speed as the rotation energy by adjusting the turbine speed. The conventional method of controlling the varying speed is by adding a variable resistance in the induction motor as presented in Figure 4.22. This allows to retain the variation in power within 2%–4%. Another advanced method is using a wound rotor induction generator known as doubly fed induction generator (DFIG) where a partial power converter via slip ring is attached to the rotor. In the case of high wind speed, the generator is operating at high speeds, and the power is delivered through both the rotor and the stator [42].

Figure 4.23 shows a detailed description of the wind turbine and its components. In the wind energy conversion system, the turbines are the huge components and can account for

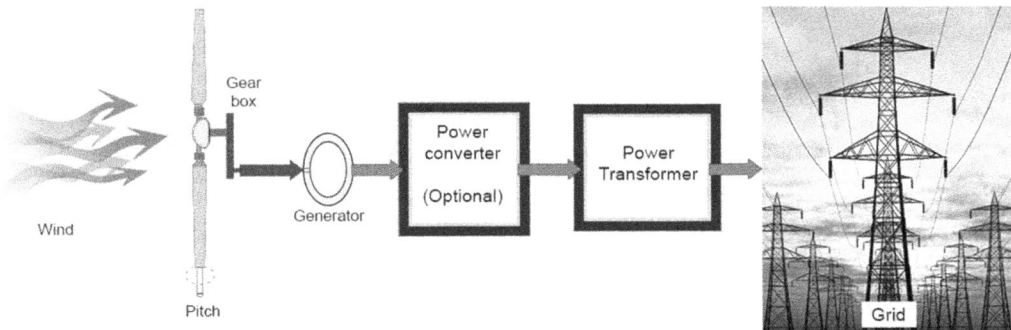

**FIGURE 4.21** Schematic and typical components of wind energy systems.

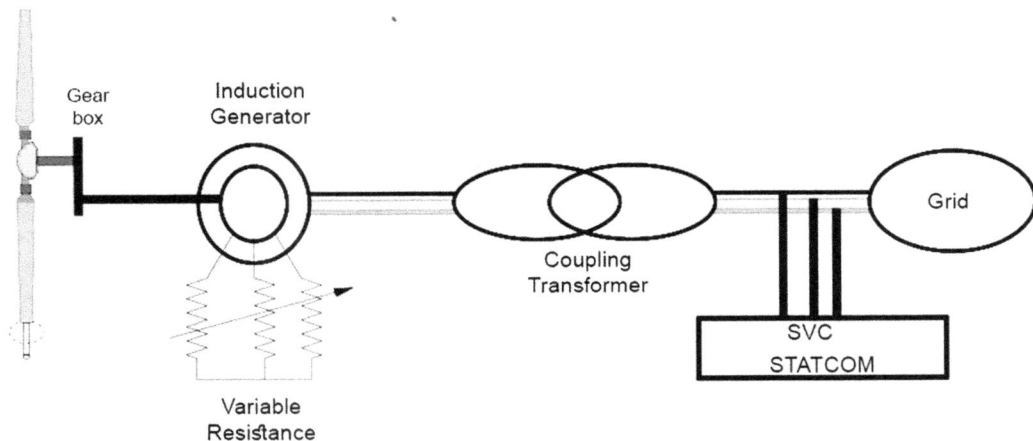

**FIGURE 4.22** Wind energy system with variable wind resistance.

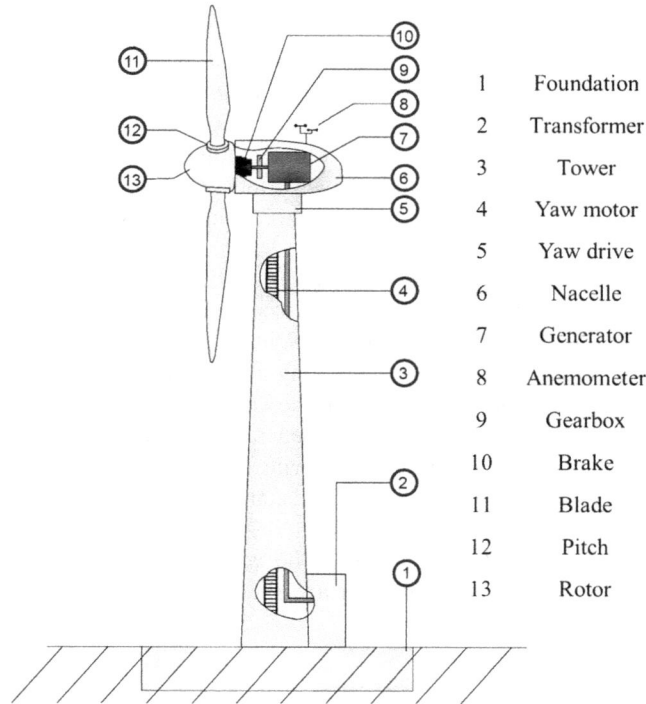

| | |
|---|---|
| 1 | Foundation |
| 2 | Transformer |
| 3 | Tower |
| 4 | Yaw motor |
| 5 | Yaw drive |
| 6 | Nacelle |
| 7 | Generator |
| 8 | Anemometer |
| 9 | Gearbox |
| 10 | Brake |
| 11 | Blade |
| 12 | Pitch |
| 13 | Rotor |

**FIGURE 4.23**   The schematic and components of a typical wind tower.

64%–84% of the total installed cost of an onshore wind turbine [8]. The latest SG 5.8-170 model by SIEMENS Gamesa has a rotor diameter of 155 m, and a tower height of up to 165 m with a total swept area of 18,868 m$^2$ for a nominal power of 5.8 MW. The normal wind turbines are the horizontal axis predominantly using pitch and variable control. Wind turbines with larger rotor diameters can increase power generation at marginal wind speeds. In addition, as the altitude increases, the wind speed increases, and as the power produced is proportional to the cube of wind speed, therefore, the net power increases with the height of the tower as well. The higher turbines can reduce the total installment cost and can increase the capacity factor of a wind turbine as well. Since the last decade, the evolution of wind turbines has seen large rotor diameters and tower heights. However, the larger turbine size means more production, transportation, and deployment cost.

Therefore, for increasing the share of wind energy, it is essential to develop technologies and advanced modeling techniques to reduce the turbine cost and ensure the structural integrity of the turbine. In recent years, despite the increase in rotor height and capacities, the decrease in turbine costs has been seen due to increase in competition among the manufacturers. Furthermore, the renewable energy system ambitions set by many countries has helped grow the supply chain and mature the industry, and so, results in reduction in the cost of wind energy [44].

The prospects of offshore wind turbines have also considerably increased recently [45]. The key development area in offshore wind turbines, besides the turbine development, is the installation and assembly of the system. The development of the installation vessels for a wind turbine is underway that can handle the various required support structures. The foundation technology of an offshore wind turbine is another key research area that needs to be reliable and friendly for installation as well as to aquatic life. Finally, an advanced economic analysis is needed to find cost-effective solutions for the deployment of offshore wind turbines.

## 4.6  RENEWABLE ENERGY DISTRIBUTION SYSTEMS

The renewable energy systems can be designed as standalone systems or as utility-interactive systems, as shown in Figure 4.24. For a standalone system, the involvement of a battery bank generates another research area for renewable systems. The performance limitation and short lifetime of battery banks have considerable room for improvement and development. For the interactive system, the research areas are focused on the inverter design and control methodologies using advanced control logic and components.

## 4.7  RENEWABLE ENERGY ADAPTATION

As discussed in previous sections, the transition of the energy sector toward renewable sources is eminent. However, the key driver for increase in adoption of any technology is due to technology development and the policies and frameworks developed by the governments [46]. A study of the history of energy transitions, that is, transition from wood to coal, from coal to oil and now from fossil fuels to renewable energy, will reveal that the key source of these transitions is particularly the latter two, is due to the government policies. May it be the industrial revolution, depleting resources or the decarbonization drive to mitigate the climate change, energy transition from one source to another is promoted and supported by the government. The present transition to renewable energy

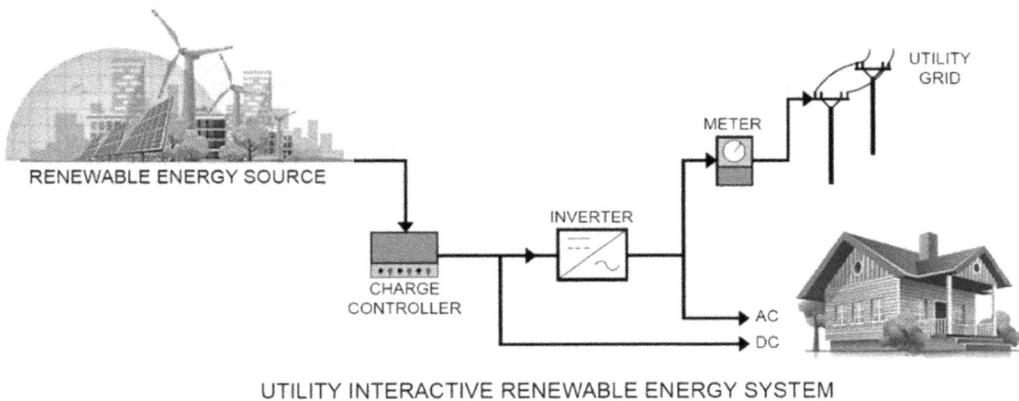

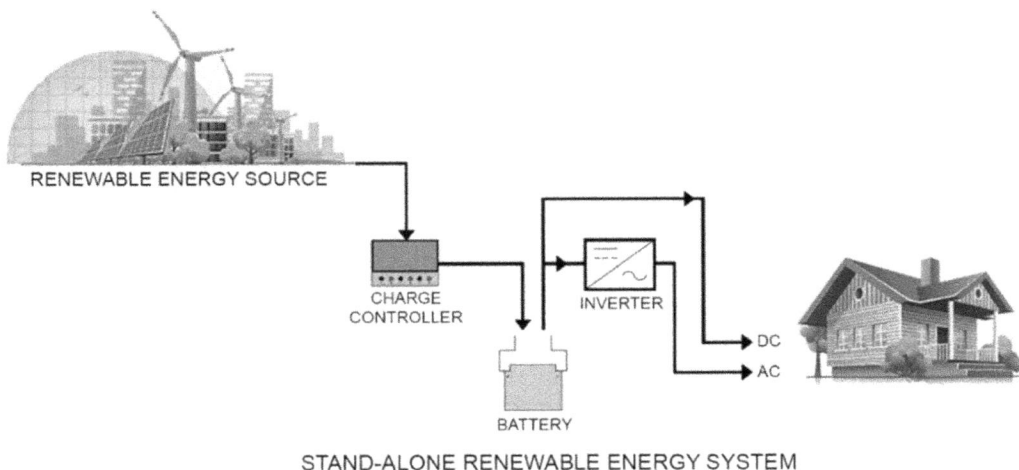

UTILITY INTERACTIVE RENEWABLE ENERGY SYSTEM

STAND-ALONE RENEWABLE ENERGY SYSTEM

**FIGURE 4.24**  Interactive and standalone renewable energy systems.

sources is to tackle the challenges in mitigation of climate change and cope with the depleting resources of fossil fuels. Key sectors that are the biggest contributors of $CO_2$ with high consumption of fossil fuels can be divided into three categories, that is, power sector (electricity), transportation sector, and cooling and heating sectors. The added advantages of adopting to renewable energy include increased energy security, enhanced system resilience, wider energy access, and monetary benefits for prosumers and utilities.

The major energy source utilized for the heating sector is gas. This sector is pervasive in all areas, that is, domestic, commercial, and industry and bulks of energy resources are used for this. Similarly, for the transportation sector, petroleum products are the major sources of energy. The electricity/power sector is mainly dependent on fossil fuels with a major transition toward renewable energy. Although each sector has its own key energy source, the carbon emissions in all sectors are compelling the sectors to change the primary energy sources and move toward low carbon emissions with increased penetration of renewable energy. This is mainly achieved by developing the low carbon technologies such as the use of electric heat pumps for heating to reduce the emissions in the heating and cooling sector. Similarly, there must be a policy shift for new transportation resources ranging from cars to trains, to use electricity as the primary energy source. Power generation is commonly considered to be the sector, which produces most of the emissions and is the key to minimization of emissions as all other sectors are becoming dependent on electricity/power.

As a result of many policy decisions such as the Paris Agreement 2015 [47] or the European Green Deal [48], increased penetration of renewable energy is expected to be expedited. This is mainly due to the policies adopted by different countries. It can be seen from Figure 4.25 [49] that a large number of countries have adopted many policies to reduce carbon emissions from 2014 to 2016 in all three sectors. It can be seen from the figure that most of the countries are focused on regulation of power sector to minimize carbon emissions. The transportation sector is also getting attention; however, to develop a stronger electrification network for transportation, de-carbonization of the power sector is the prerequisite. The cooling and heating sector is not gaining as much attention in policies because the focus of heating in the recent past has been more toward district heating to improve efficiency [50].

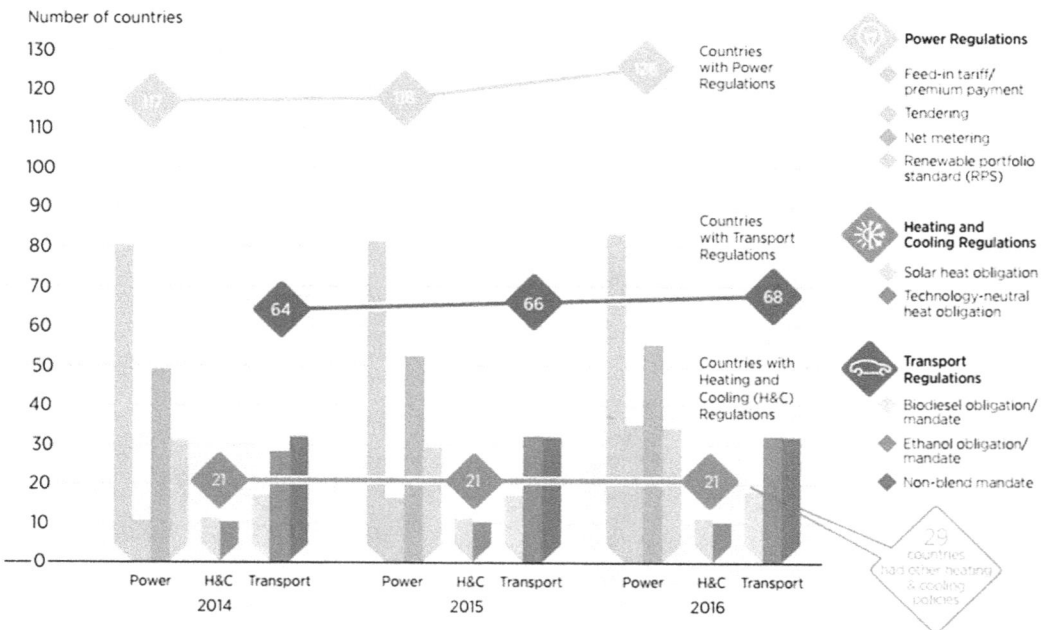

**FIGURE 4.25** Countries offering incentives and mandates for adoption of renewable energy.

Another interesting aspect of the heating and cooling sector is its impact on the power network. Utilities are already facing major technological challenges to ensure reliable and secure supply with increasing number of electric vehicles. As initial steps to tackle these challenges, the policies are designed to increase penetration of renewable energy by introducing Feed-in-tariff/premium payments, net metering, and opening the markets for competitive bidding/auctions [46]. On the other side, governments are adopting policies such as ban on diesel cars in the future and providing incentives for the purchase of electric vehicles.

Adoption of renewable energy and future plans can be clearly visualized from the energy flow chart of any country. For example, UK energy flow chart for 2020 is given in Figure 4.26a. From the figure, it can be clearly seen that the transport sector uses the most energy followed by the domestic consumers. Petroleum and gas are major energy sources, which are mainly used in transport and domestic sectors respectively. Gas is also used in electricity generation; however, bioenergy,

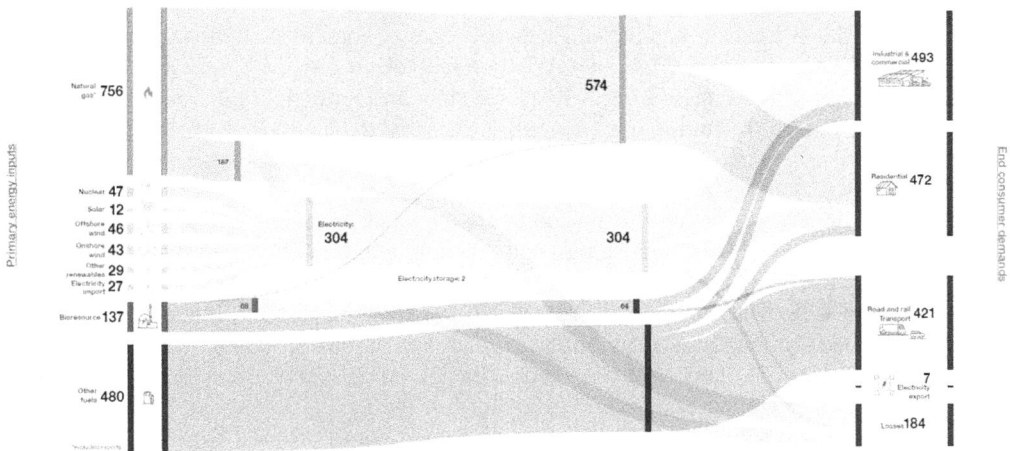

(a)

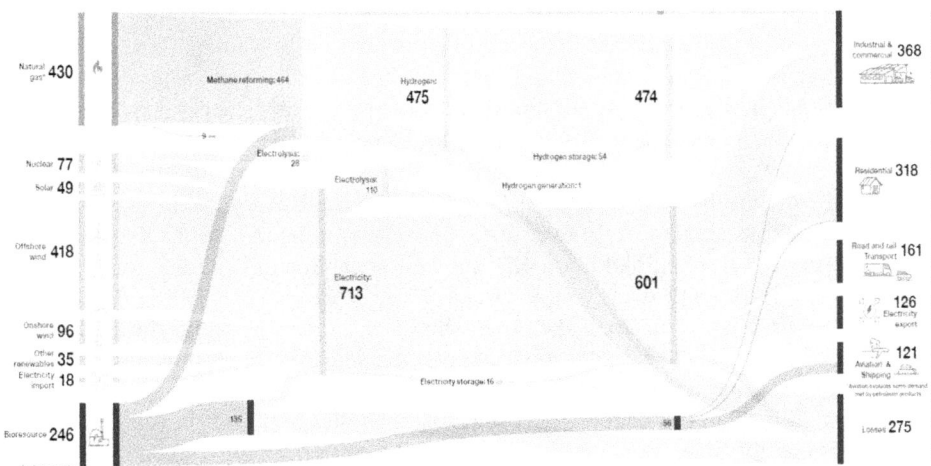

(b)

FIGURE 4.26   UK energy flow chart comparison. (a) 2020: energy demand and supply (TWh); (b) Projected energy flow chart for UK 2050 with system transformation. (From Ref. [51]. With permission.)

renewable energy resources, and coal are also used. Increasing penetration of renewable energy in the future is envisioned by the UK government.

This is eminent from the report by the National Grid [51] that the system transformation scenarios give a variation of energy supplied by hydrogen, however, apart from a small contribution of gas, almost all energy is produced by renewable energy sources. From Figure 4.26b [51], it is evident that wind energy is more suitable for the United Kingdom as the primary source of energy, followed by nuclear and then PV. The technological challenges involved in developing such a system with low penetration of conventional fossil fuel-based power plants will require significant input from hydrogen and other technologies to store energy to provide the necessary support to maintain the frequency of power system.

The European Union (EU) [46] is one of the major contributors to carbon emissions. Analysis of the energy flow in the EU (given in Figure 4.27a) shows that a significant amount of energy in the EU is used in transportation. The bulk of the energy is produced using nonrenewable energy sources with high use of oil, coal, and natural gas. Bull of electricity is generated from coal, however, after natural gas, hydroelectric power generation takes the third place.

Interestingly, the EU uses a number of renewable energy sources including, geothermal, solar thermal, solar PV, nuclear, and wind energy to produce electricity. The Sankey diagrams show projection of energy flow for the EU in 2050 as can show in Figure 4.27b. It can be observed that as opposed to the UK, in the future, the use of oil is expected to be prevalent in the transport sector with dominant use of gas in the manufacturing industry and buildings for space heating. However, overall electricity generation is expected to be more than double that in 2018. Due to the larger area and diverse climate, PV generation is expected to grow on a large scale alongside wind generation. Hydrogen is expected to be used in most the sectors to provide reliable energy sources.

The United States uses almost one-third of its present energy resources to generate energy [53]. From Figure 4.28, like the EU and the UK, bulk of energy is used in the transportation sector. Within the last 2 years, solar generation has increased up to 30% and wind generation has increased up to 20% [53]. This shows an increasing trend in renewable energy penetration in the energy sector. The United States was a signatory of the Paris Agreement; however, later on, with the change of government, the United States decided not to proceed with the terms agreed. Once again, the government has shown its commitment to tackle challenge of global warming and reducing carbon emissions.

From the above, the energy flows in different countries with sectoral usage show the transformation of many sectors and a major increase in the use of electricity as the source of energy. A glimpse into the future shows that most of the energy resources will be directly used for power generation.

According to a study by Larsson [54], to achieve the global renewable energy road map (Remap) prepared by IRENA, the renewable energy footprint for some countries will require to be almost 60% of their total generation capacity. For instance, China could increase the share of renewable energy in its energy use from 7% in 2015 to 67% in 2050. In the EU, the renewable energy penetration can increase from about 17% to over 70%. Similarly, India and the United States could see shares increase to two-thirds [54]. Figure 4.29 shows comparison of different countries, use of electricity, and share of renewable energy in power.

Overview of energy flow of different countries shows an increasing trend toward renewable energy. Studies show that to achieve the global targets of combating climate change, the global energy transition of energy toward renewable is eminent. However, this brings a number of challenges, which are mainly technical challenges. The primary challenge till recent years was the high prices of renewable energy generation, but the gradual decline in prices of wind and solar PV generation shows positive signs for adoption of renewable energy. On the other side, although renewable energy is distributed in nature and does not need to be centralized, this brings a number of challenges. This requires detailed monitoring at the consumer level to tackle the technical challenges. Installation of monitoring equipment at the distribution level, which was often considered suitable for medium and high voltage lines. Moreover, the consumers will require to be proactive in energy

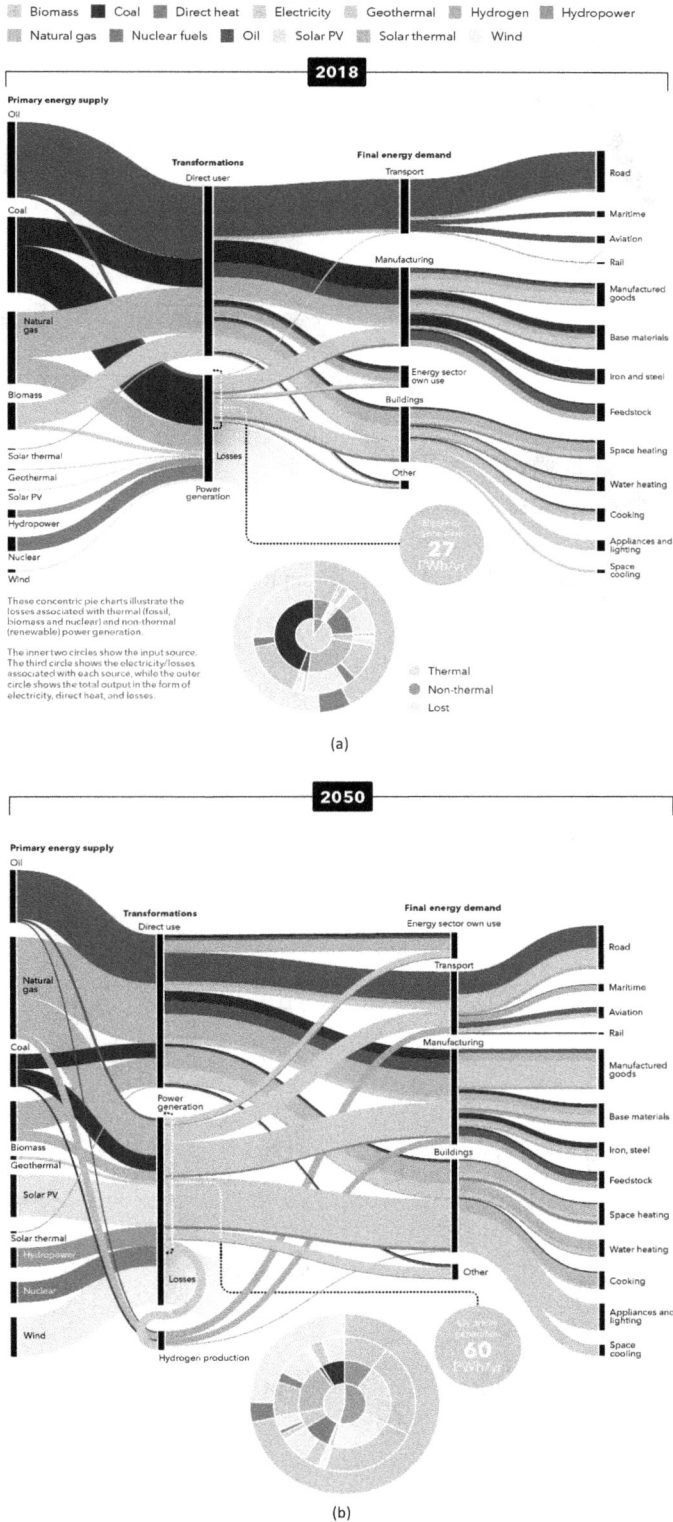

**FIGURE 4.27**  Energy flow chart for EU: (a) 2018 (b) 2050. (From Ref. [52]. With permission.)

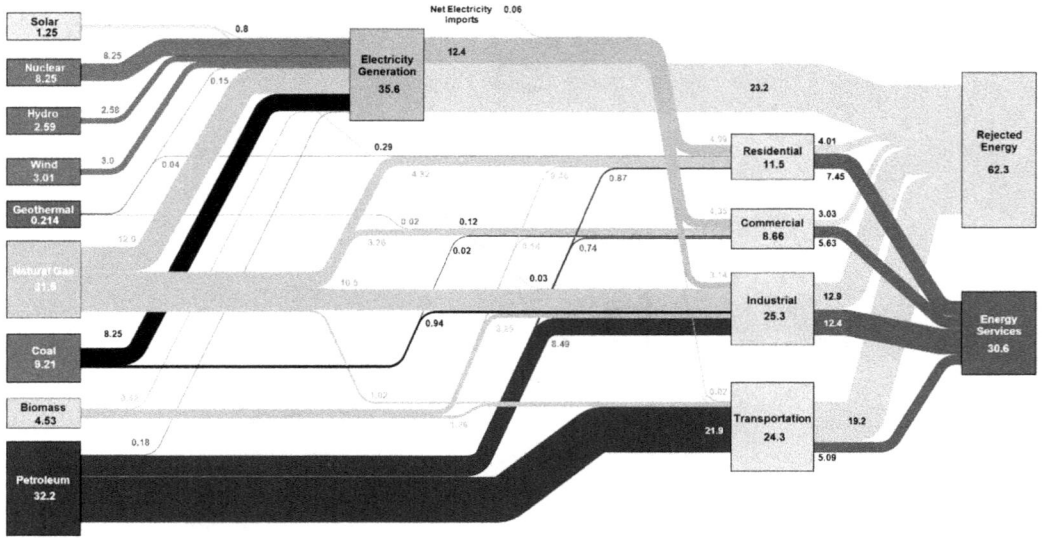

**FIGURE 4.28**  Energy flow chart of the USA, 2020. (From Ref. [53]. With permission.)

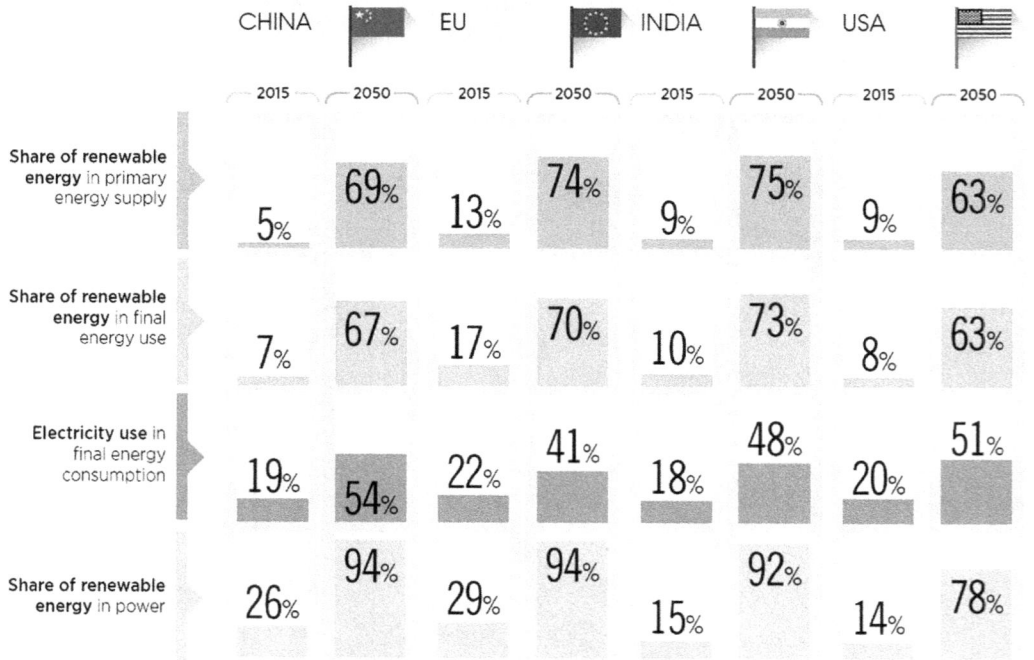

| | CHINA | | EU | | INDIA | | USA | |
|---|---|---|---|---|---|---|---|---|
| | 2015 | 2050 | 2015 | 2050 | 2015 | 2050 | 2015 | 2050 |
| Share of renewable energy in primary energy supply | 5% | 69% | 13% | 74% | 9% | 75% | 9% | 63% |
| Share of renewable energy in final energy use | 7% | 67% | 17% | 70% | 10% | 73% | 8% | 63% |
| Electricity use in final energy consumption | 19% | 54% | 22% | 41% | 18% | 48% | 20% | 51% |
| Share of renewable energy in power | 26% | 94% | 29% | 94% | 15% | 92% | 14% | 78% |

**FIGURE 4.29**  Key indicator for energy transition according to RENA Remap case. (From Ref. [54]. With permission.)

management as with the increased share of renewable energy, the conventional load-led generation paradigm will change into generation-led load [55].

This necessitates consumer participation in energy management, and consequently, a number of changes at the consumer end, including automation and smart devices will be required. New load and generation modeling techniques will be required [56] and the future power grids will face new planning and operational challenges [57].

Key challenges faced by the grid will include grid stability ranging from voltage stability to frequency stability. The intermittency of renewable energy sources will pose challenges that require novel solutions. Problems from generation of surplus energy to insufficient energy will need to be tackled with sufficient storage but in a cost-effective manner. More research on alternative technologies such as hydrogen production, fuel cells, grid-level battery storage, and super capacitors will be required to maintain the system security and reliability. Despite these challenges, the stability of the power system will eventually bring new concepts such as inter-continental interconnectors, which will enable the EU to use the abundant solar potential of African deserts such as the Sahara and coastal wind energy in Europe can be used in different countries. The future of energy can see a successful transition to renewable sources only with global efforts such as the Paris Agreement.

## 4.8   CONCLUSION

This chapter discusses different energy sources and technological advancements in the use of these energy sources. A number of aspects are identified that drive the transition of energy, and the most important one is identified to be the mitigation of climate change. However, decreasing the cost of renewable energy technologies has been identified to be another key element resulting in expediting the transition. The future scenarios of energy for different regions of countries are discussed and it is concluded that the future of the energy tends to be high renewable penetration. New technologies, such as hydrogen generation, are expected to mitigate the intermittency of renewable generation and provide the much needed energy storage solutions. However, the transition of energy will highly depend on both sides, that is, the generation side and the consumption side. Thus, the future of the energy system will be decided by the technologies not only at the generation side but the technological advancements at the consumer side will dictate the future of energy generation.

## REFERENCES

1. Petit V. The energy transition: An overview of the true challenge of the 21st century. 2017. Doi: 10.1007/978-3-319-50292-2.
2. Anderson BP. Review: Renewable energy; A primer for the twenty-first century. 2020;1. Doi: 10.5070/g314344210.
3. IEA (2021), Global Energy Review 2021, IEA, Paris. https://www.iea.org/reports/global-energy-review-2021
4. International Renewable Energy Agency, IRENA. World energy transitions outlook. 2022, 1.5 Pathway, 2022.
5. International Renewable Energy Agency, IRENA. Renewable power generation costs in 2020. 2020.
6. Bhattacharya M, Paramati SR, Ozturk I, Bhattacharya S. The effect of renewable energy consumption on economic growth: Evidence from top 38 countries. *Appl Energy* 2016;162:733–41.
7. International Renewable Energy Agency. Renewable power generation costs in 2019. 2020.
8. Vinet L, Zhedanov A. A "missing" family of classical orthogonal polynomials. 2011;44. Doi: 10.1088/1751-8113/44/8/085201.
9. Moriarty P, Honnery D. Can renewable energy power the future? *Energy Policy* 2016. Doi: 10.1016/j.enpol.2016.02.051.
10. O'Shaughnessy E, Cruce JR, Xu K. Too much of a good thing? Global trends in the curtailment of solar PV. *Sol Energy* 2020;208:1068–77. Doi: 10.1016/j.solener.2020.08.075.
11. Trube J, Fischer M, Erfert G, Li CC, Ni P, Woodhouse M, Li P, Metz A, Saha I, Chen R, Wang Q. International technology roadmap for photovoltaic (ITRPV). *VDMA Photovoltaic Equipment* 2018;24:77–97.

12. Breyer C, Bogdanov D, Aghahosseini A, Gulagi A, Child M, Oyewo AS, et al. Solar photovoltaics demand for the global energy transition in the power sector. *Prog Photovoltaics Res Appl* 2018;26:505–23. Doi: 10.1002/pip.2950.

13. Kabir E, Kumar P, Kumar S, Adelodun AA, Kim KH. Solar energy: Potential and future prospects. *Renew Sustain Energy Rev* 2018;82:894–900. Doi: 10.1016/j.rser.2017.09.094.

14. Breeze P. *Power Generation Technologies*. Newnes; 2019 Feb 21.

15. Sarwar S, Lee Ms, Park S, Dao TT, Ullah A, Hong S, et al. Transformation of a liquid electrolyte to a gel inside dye sensitized solar cells for better stability and performance. *Thin Solid Films* 2020;704:138024. Doi: 10.1016/j.tsf.2020.138024.

16. Solar cells: Few-layer MoS$_2$ flakes as active buffer layer for stable perovskite solar cells. *Adv Energy Mater* 2016;6. Doi: 10.1002/aenm.201670095.

17. Dincer I. Geothermal energy systems. 2021. Doi: 10.1016/c2019-0-01786-4.

18. Toth A, Bobok E. Flow and heat transfer in geothermal systems: Basic equations for describing and modeling geothermal phenomena and technologies. *Flow Heat Transf Geotherm Syst Basic Equations Descr Model Geotherm Phenom Technol* 2016:1–382.

19. Kutscher CF. The status and future of geothermal electric power. *Am Sol Energy Soc Conf* 2000:9.

20. International Renewable Energy Agency (IRENA). Renewable capacity statistics. 2021.

21. IPCC. Renewable energy sources and climate change mitigation. n.d. https://www.ipcc.ch/report/renewable-energy-sources-and-climate-change-mitigation/ (accessed July 24, 2021).

22. Lebbihiat N, Atia A, Arıcı M, Meneceur N. Geothermal energy use in Algeria: A review on the current status compared to the worldwide, utilization opportunities and countermeasures. *J Clean Prod* 2021:302. Doi: 10.1016/j.jclepro.2021.126950.

23. Hou J, Cao M, Liu P. Development and utilization of geothermal energy in China: Current practices and future strategies. *Renew Energy* 2018;125:401–12. Doi: 10.1016/j.renene.2018.02.115.

24. Kurnia JC, Putra ZA, Muraza O, Ghoreishi-Madiseh SA, Sasmito AP. Numerical evaluation, process design and techno-economic analysis of geothermal energy extraction from abandoned oil wells in Malaysia. *Renew Energy* 2021;175:868–79.

25. Sarangi PK, Nanda S, Mohanty P. Recent advancements in biofuels and bioenergy utilization. 2018. Doi: 10.1007/978-981-13-1307-3.

26. Murdock HE, Gibb D, André T, Sawin JL, Brown A, Appavou F, Ellis G, Epp B, Guerra F, Joubert F, Kamara R. Renewables 2020-Global Status Report.

27. Cozzi L, Gould T, Bouckart S, Crow D, Kim TY, Mcglade C, Olejarnik P, Wanner B, Wetzel D. World Energy Outlook 2020. vol. 2020:1–461.

28. Information portal for renewable energies - time series for renewable energies. n.d. https://www.erneuerbare-energien.de/EE/Navigation/DE/Service/Erneuerbare_Energien_in_Zahlen/Zeitreihen/zeitreihen.html (accessed July 25, 2021).

29. Department for Business, Energy and Industrial Strategy. UK energy trends: Section 1- total energy. 2022.

30. Röder M, Mohr A, Liu Y. Sustainable bioenergy solutions to enable development in low- and middle-income countries beyond technology and energy access. *Biomass Bioenergy* 2020:143. Doi: 10.1016/j.biombioe.2020.105876.

31. Alhazmi H, Loy ACM. A review on environmental assessment of conversion of agriculture waste to bio-energy via different thermochemical routes: Current and future trends. *Bioresour Technol Rep* 2021;14:100682. Doi: 10.1016/j.biteb.2021.100682.

32. Srinivas S. *Bioenergy Systems for Sustainable Energy Access*. vol. 4. Elsevier; 2017. Doi: 10.1016/B978-0-12-409548-9.10149-6.

33. Ozonoh M, Oboirien BO, Daramola MO. Optimization of process variables during torrefaction of coal/biomass/waste tyre blends: Application of artificial neural network & response surface methodology. *Biomass Bioenergy* 2020:143:105808.

34. Karthikeya K, Sarma MK, Ramkumar N, Subudhi S. Exploring optimal strategies for aquatic macrophyte pre-treatment: Sustainable feedstock for biohydrogen production. *Biomass Bioenergy* 2020;140:105678.

35. Aghbashlo M, Khounani Z, Hosseinzadeh-Bandbafha H, Gupta VK, Amiri H, Lam SS, et al. Exergoenvironmental analysis of bioenergy systems: A comprehensive review. *Renew Sustain Energy Rev* 2021;149:111399.

36. International Renewable Energy Agency. Renewable capacity statistics 2016. 2016.

37. Council GW. GWEC Global Wind Report 2019. Global Wind Energy Council: Bonn, Germany. 2017.

38. Roubanis N, Dahlstrom C, Noizette P. Renewable energy statistics. *Statistics in Focus–Eurostat*. 2010 Nov;56:1–8.

39. Ramirez L, Fraile D, Brindley G. Offshore wind in Europe: Key trends and statistics 2019.

40. Soares-Ramos EP, de Oliveira-Assis L, Sarrias-Mena R, Fernández-Ramírez LM. Current status and future trends of offshore wind power in Europe. *Energy*. 2020 Jul 1;202:117787.

41. International Renewable Energy Agency (IRENA). Renewable Capacity Statistics 2020. 2020.

42. Yahyaoui I. Advances in renewable energies and power technologies. 2018;2. Doi: 10.1016/c2016-0-04919-7.

43. Sadorsky P. Wind energy for sustainable development: Driving factors and future outlook. *J Clean Prod* 2021;289:125779. Doi: 10.1016/j.jclepro.2020.125779.

44. Hernandez-Estrada E, Lastres-Danguillecourt O, Robles-Ocampo JB, Lopez-Lopez A, Sevilla-Camacho PY, Perez-Sariñana BY, et al. Considerations for the structural analysis and design of wind turbine towers: A review. *Renew Sustain Energy Rev* 2020:110447.

45. Jiang Z. Installation of offshore wind turbines: A technical review. *Renew Sustain Energy Rev* 2021;139:110576.

46. Lu Y, Khan ZA, Alvarez-Alvarado MS, Zhang Y, Huang Z, Imran M. A critical review of sustainable energy policies for the promotion of renewable energy sources. *Sustainability* 2020;12:5078.

47. Agreement P. Paris agreement. In Report of the Conference of the Parties to the United Nations Framework Convention on Climate Change (21st Session, 2015: Paris). Retrieved December 2015 Dec (Vol. 4, p. 2017). HeinOnline.

48. Siddi, M. The European Green Deal: Assessing its current state and future implementation, Finnish Institute of International Affairs. FIIA working paper, 2020/114, 2020. https://www. fiia. fi/en/publication/the-european-green-deal. (accessed November 2021).

49. Murdock HE, Collier U, Adib R, Hawila D, Bianco E, Muller S, Ferroukhi R, Renner M, Nagpal D, Lins C, Frankl P. Renewable energy policies in a time of transition. 2018.

50. Ma Z, Knotzer A, Billanes JD, Jørgensen BN. A literature review of energy flexibility in district heating with a survey of the stakeholders' participation. *Renew Sustain Energy Rev* 2020;123:109750.

51. National Grid E. Future energy scenarios. 2021.

52. DNV GL. Energy transition outlook 2020- A global and regional forecast to 2050. DNV GL Energy Transit Outlook 2020:306.

53. Laboratory LLN. *Estimated U.S. Energy Consumption in 2020*. Lawrence Livermore Natl Lab 2020. https://flowcharts.llnl.gov/commodities/energy.

54. Larsson M. *Global Energy Transformation*. 2018. Doi: 10.1057/9780230244092.

55. Khan ZA, Jayaweera D, Gunduz H. Smart meter data taxonomy for demand side management in smart grids. *2016 Int. Conf. Probabilistic Methods Appl. to Power Syst., IEEE* 2016, pp. 1–8.

56. Gunduz H, Khan ZA, Altamimi A, Jayaweera D. An innovative methodology for load and generation modelling in a reliability assessment with PV and smart meter readings. *IEEE Power Energy Soc Gen Meet IEEE* 2018, pp. 1–5.

57. Khan ZA, Jayaweera D. Planning and operational challenges in a smart grid. *Smart Power Syst Renew Energy Syst Integr* Springer; 2016, pp. 153–77.

# 5 A Global Hydropower Generation, Potentials, and Externalities

*Maksud Bekchanov*
University of Hamburg

## CONTENTS

## 5.1   INTRODUCTION

Population growth and economic development trigger demand for energy consumption. Despite increasing coverage of population with electricity access, about 800 million people in Sub-Saharan Africa and South Asia still live without connection to electricity grids (as of 2020; IEA 2020). Depletion of fossil fuel stocks and increasing costs of these resources over time exacerbate the challenge of closing the energy supply and demand gap. Unequal distribution of fossil energy resources across the countries also leads to political conflicts which may intensify under increasing energy scarcity (Asif and Muneer, 2007). Since the extraction of fossil fuel resources is a major source of greenhouse gas (GHG) emissions, climate change policies and regulations restrict the exploitation of fossil fuel resources (IPCC, 2014). Consequently, renewable energy sources are gaining traction as an alternative option to meet the increasing energy demand and the reducing atmospheric GHG concentration.

Hydropower is the largest source of renewable energy production at present. Hydropower generation accounts for 16% of total electricity supply and 72% of total renewable energy production globally (IEA, 2016). It is the key source of energy contributing to more than 95% of national energy supply in countries such as Norway, Iceland, Nepal, and Tajikistan. In addition to electricity supply, hydropower systems provide clean water for irrigation, residential areas, and industrial sites. Water reservoir banks are also used for developing catering and recreational businesses. Hydropower projects improve water regulation capacity and thus may increase the society's resilience against extreme events such as drought and floods. The hydropower sector has been also promoted as a clean energy option due to its lower carbon emission intensity compared to fossil fuel-based energy production. Thus, additional efforts are recommended to expand the sector in clean energy transition policy

DOI: 10.1201/9781003315353-7

roadmaps. For instance, International Renewable Energy Agency (IRENA) shows the worthiness of allocating investments of about US$ 1.7 trillion (by 2050) in installing additional 850 GW hydropower capacity for enhancing the Paris Agreement goals (IRENA, 2020).

Given the recently renewed interest in hydropower projects, this study presents the development of the hydropower sector over the 20th century highlighting the current status and potential for further developments across the world. As there are intense debates between environmentalists and development practitioners over the usefulness of the large dams, this study also summarizes key arguments by opponents and proponents of the large dams. Through the synthesis of these various viewpoints, important recommendations are derived for more sustainable and inclusive development of hydropower developments.

## 5.2  HYDROPOWER GENERATION OPTIONS

Hydropower generation is based on converting the kinetic energy of falling water flow (from higher to lower head) into electrical energy (Breeze, 2019). The kinetic energy due to the gravity of the falling water rotates the turbine which is connected to the alternator to generate electricity. In most cases, a reservoir or dam is constructed to maintain kinetic energy from water. Although the construction of dam and hydropower generation facility comes at substantial costs, especially for large dams, operating costs are very low making the overall costs affordable to users and hydropower projects attractive to investors.

Three types of hydropower generation plants are differentiated from each other: impoundment, run-of-river, and pumped storage. Impoundment facilities to generate hydropower are more common and accompanied with dam or reservoir construction. Run-of-river facilities may not require dam construction and utilize the natural decline of river bed elevation. A pumped storage facility works like a giant battery using solar or wind energy for pumping water from a reservoir at a lower elevation to a reservoir at a higher elevation when electricity demand is low and releasing water from the upper reservoir to generate hydropower when demand is high.

Hydropower plants also vary according to their size from small projects suitable for a single household or community (village) to large projects to supply industries or urban areas with electricity. Usually, four categories of hydropower plants are differentiated according to their size: micro, mini, small, and large hydropower plants (Breeze, 2019). The hydropower plants with a capacity of 1–100 kW are categorized as micro hydropower plants, 100 kW–1 MW as mini-hydropower plants, and from 1 to 10 MW as small hydropower plants. Upper limit of the capacity of small hydropower plants can be up to 30 MW in some countries. Hydropower plants with a capacity higher than 10 MW (or 30 MW in some countries) are considered large plants.

## 5.3  GLOBAL HYDROPOWER CAPACITY DEVELOPMENT OVER TIME

Hydropower generation technologies emerged by the end of the 19th century. The first hydropower plant that served commercial customers started operating in Wisconsin, USA, in 1882 (IHA, 2020a). Following this project, hundreds of hydropower plants were launched within a decade in the United States. The hydropower generation technology was also spreading to other parts of the world at that time. The first three-phase hydroelectricity generation system was built in Germany in 1891. The first publicly owned hydropower plant was constructed in Australia in 1895. Hydropower technology was also brought to China in 1905.

Hydropower generation technology design was improved and the technology was spread rapidly over the 20th century (Figure 5.1). Especially, rapid population growth and economic development in the postwar period triggered the development of hydropower projects throughout Western Europe, the former Soviet Union, North America, Australia, and Japan (Biswas and Tortajada, 2001). Hydropower was considered a low-cost energy supply option to meet the growing needs of energy-intensive industries such as aluminum smelting and steelworks. Many developing countries in Asia and Africa are also in need of urgent development processes after gaining their independencies.

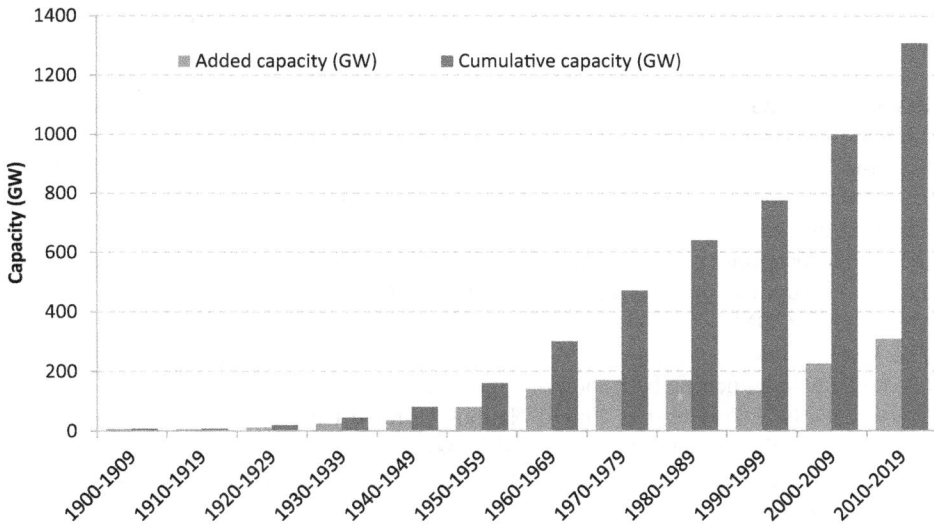

**FIGURE 5.1**    Hydropower capacity development over time. (See IHA, 2020a.)

Thus, dam construction became a symbol of national development and pride in countries such as India, Egypt, Zambia, and Ghana at that time.

By 1975, the developed countries in North America and Western Europe completed their large dam development programs while exploiting the most economically efficient sites (Biswas and Tortajada, 2001). In the post-1975 period, the constructions of large water infrastructures were pending or rarely occurred in the developing countries (mentioned above) due to various financial and political reasons. Instead, the hydropower development focus shifted to other developing countries such as Brazil, China, Indonesia, Malaysia, and Turkey. Major hydropower system developments especially took place in China and Brazil over the last decades of the 20th century. Three Gorges Dam with a capacity of 22,500 MW was constructed in China. The Itaipu Dam with a capacity of 12,600 MW was launched in 1984 on the territory shared between Brazil and Paraguay. Its capacity expanded to 14,000 GW over time, being the second largest after China's Three Gorges Dam.

Concerns over the social and environmental impacts of large dams became common in the late 1980s. Political conflicts between the great powers (the United States and the former Soviet Union) also played an important role in supporting or criticizing the large dam developments in the developing world (Biswas and Tortajada, 2001). Consequently, financial resources were limited for hydropower developments causing the stagnation of hydropower capacity development. Credit and other forms of financial support from international financial organizations including the World Bank ceased in the late 1990s, freezing the hydropower developments in the developing countries.

With the increased environmental and social impact consciousness toward the end of the 20th century, the role of hydropower for economic development was reassessed. A landmark report of 2000 by the World Commission on Dams pointed out environmental and social impact externalities of the hydropower projects and initiated a change in planning and development while focusing on sustainability and social impact aspects of hydropower projects (WCD, 2000a). The International Hydropower Association (IHA) also developed Sustainability Guidelines in 2004. This work was a basis for the development of the Hydropower Sustainability Assessment Protocol (HSAP) that highlighted a need for considering the interests of all stakeholders impacted and assessing the feasibility of the projects for their entire lifecycle (IHA, 2010).

Hydropower projects gained renewed momentum since the beginning of the 21st century (Figure 5.1). New projects were launched across Asia and South America. Global hydropower generation capacity consequently increased by more than 500 GW during the period from 2000 to 2019.

Multiple factors played an important role in increased interest and investment in the development of hydropower generation capacity (IHA, 2020a):

- Brazil and China needed a low-cost and sustainable energy source for their national development that was met through the development of hydropower projects;
- South-to-South investment and trade boom became essential for financing and enhancing technology transfers for hydropower sector developments;
- The role of hydropower developments in achieving internationally agreed commitments including Sustainable Development Goals (SDGs) and Climate Targets was recognized, and the hydropower projects attracted international funds through the Clean Development Mechanism (CDM);
- The World Bank also strengthened financial and technical support and facilitated private investments in launching hydropower projects;
- More recently, following the United Nations Climate Change Conference in 2015 (21st Conference of the Parties (COP21) in Paris), many countries endorsed "Intended Nationally Determined Contributions (INDCs)" that consider the expansion of hydropower developments as an option to cope with climate change.

## 5.4  HYDROPOWER GENERATION: CURRENT STATUS

Hydropower installed capacity increased from 160 to 1,308 GW (as of 2019) at a global scale since 1960 (Figure 5.2). The hydropower sector produces 4,306 TWh of electricity annually, contributing to 16% of the overall electricity supply. Almost 40% of the hydropower production capacities are located in East Asia and the Pacific region. Despite the considerable potential, the hydropower sector is the least developed across Africa and accounts for only 3% of global installed capacity. Underdevelopment of the economies in the region may constrain investment in hydropower developments. The regional landscape may also cause difficulties in developing potential hydropower projects requiring higher investments per hydropower generation capacity than in other parts of the world.

Among the world countries, China leads in hydropower generation, owning hydropower plants with a total installed capacity of 356.4 GW (Figure 5.3). Development policies to improve water, food, energy, and income security largely enhanced hydropower developments across the country. Next in the rank are Brazil, Canada, and the USA each having capacity of about 100 GW.

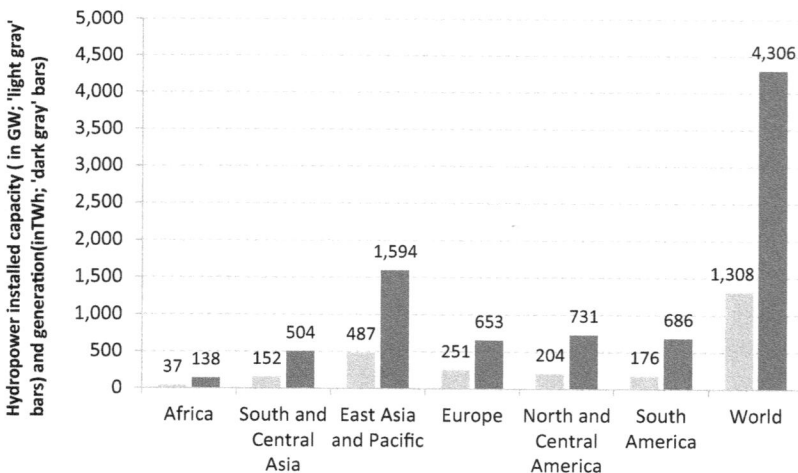

**FIGURE 5.2**  Hydropower installed capacity and generation across the world in 2019. (From IHA, *Hydropower Status Report: Sector Trends and Insights*, 2020b. With permission.)

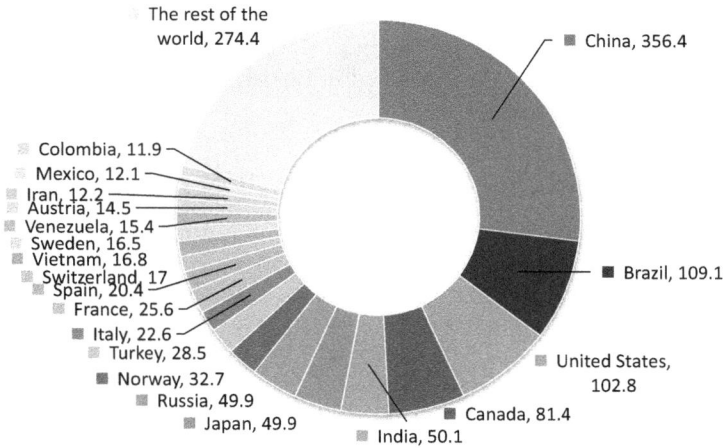

**FIGURE 5.3** Hydropower generation capacity (GW) by top 20 countries (in 2019). (From IHA, *Hydropower Status Report: Sector Trends and Insights*, 2020b. With permission.)

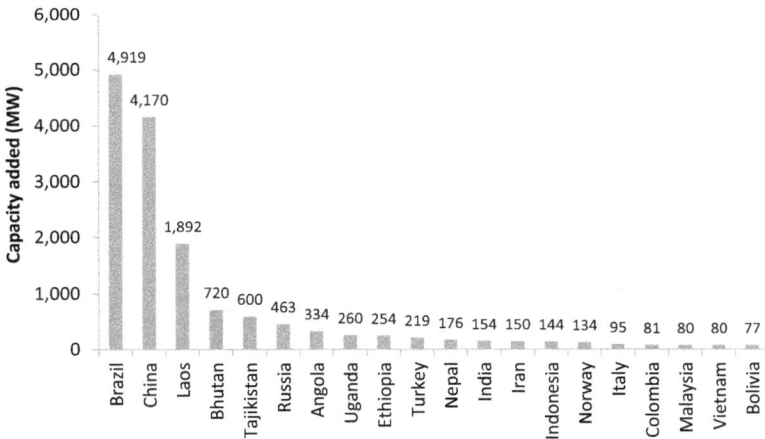

**FIGURE 5.4** Capacity added by top 20 countries (in 2019). (From IHA, *Hydropower Status Report: Sector Trends and Insights*, 2020b. With permission.)

Then comes India, Japan, and Russia having capacities of about 50 GW individually. These seven countries mentioned above account for more than 60% of global hydropower generation capacity together. Thus, given the heterogeneous environmental conditions and economic development levels, hydropower capacity is heterogeneously developed in the world and the substantial hydropower generation capacity is located in the territories of a countable number of countries. Most of these countries also own large terrestrial areas except Japan which has a large hydropower generation capacity despite its much smaller territorial size compared to the remaining six countries (e.g., China, Brazil, the USA, Canada, India, and Russia).

Recent increases in hydropower capacity mostly occurred in the countries of South America, East Asia, and South Asia. Global hydropower capacity increased from 1,292 to 1,308 GW in the last years (between 2018 and 2019). This global expansion of the hydropower capacity required annual investments of US$50 billion (as of 2018). Additional installed capacities were reported for 50 countries (IHA, 2020b). The highest hydropower capacity increases were observed in Brazil, China, and Laos (4.6, 4.2, and 1.9 GW, respectively; Figure 5.4). Major hydropower projects that enhanced the power generation capacity include the Belo Monte project in Brazil (11,233 MW), Xayaburi in Laos (1,285 MW), and Wunonglong in China (990 MW). Considerable increases in hydropower installed capacity were also observed in Bhutan, Tajikistan, Russia, Angola, Uganda, Ethiopia, and Turkey during the last years.

## 5.5 POTENTIALS OF HYDROPOWER GENERATION ACROSS THE WORLD

Although hydropower generation potential was almost fully utilized in some parts of the world such as North America and Western Europe, large potential has still remained unutilized in vast areas in the southern hemisphere. More than 3,700 dams of medium and large sizes were planned or under development in the last decade (Zarfl et al., 2015). These plants can double the global hydropower generation capacity once they are launched. Given the increased policy and financial support, previously ignored hydropower plans can get the attraction. The theoretical potential of global hydropower generation varies from 30,700 to 126,700 TWh across the studies (Gernaat et al., 2017). Technically feasible full potential of hydropower generation that comes at a cost lesser than US$ 0.50 per kWh is assessed to be 13,270 TWh (Figure 5.5), of which 5,670 TWh of electricity can be generated without causing considerable damage to ecosystems. Especially, river basins in the Asia-Pacific, South America, and Africa regions have large potential for hydropower developments. In South and East Asia, the Ganges-Brahmaputra and Yangtze River basins can host most of these projects. In South America, major hydropower developments can occur in the Amazon and La Plata river basins. Rapid development of hydropower systems is expected across Africa since the technical and economic potentials have been largely underexploited in the region despite a high energy demand for human needs and industrial development. In Europe, further development of hydropower systems will mainly take place in the Balkan region.

Since electricity from fossil fuel-based energy sources costs about US$ 0.05–0.10/kWh, hydropower projects that can supply electricity at a cost less than US$ 0.10/kWh are financially feasible. As estimated, the potential of global hydropower production that can be supplied with costs less than US$ 0.10/kWh is 5,700 TWh, of which 3,290 TWh is ecologically acceptable (Figure 5.6). These potentials are mainly located in the Asia Pacific, South America, and Africa. The Asia-Pacific region accounts for 37% of this potential, South America accounts for 28%, and Africa accounts for 25%. Further development of hydropower projects that cost less than US$ 0.10/kWh is quite limited in scope in Europe and North America. Most of the potential hydropower projects in these regions are

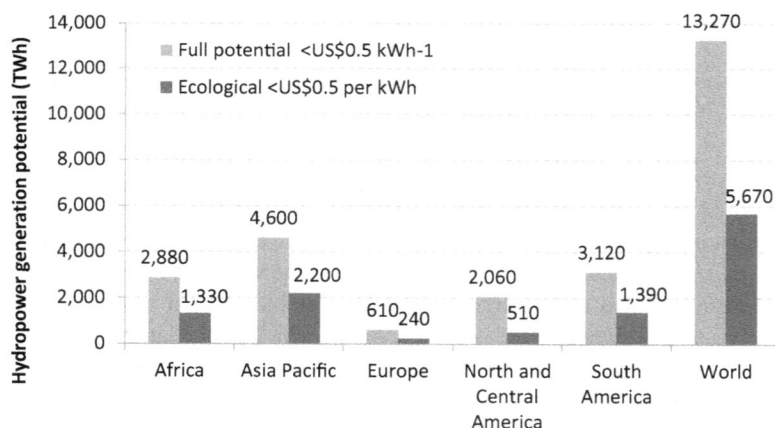

**FIGURE 5.5** Technical potential (with costs less than US$ 0.5 per kWh) of hydropower generation across the world. (From Gernaat, D.E.H.J., Bogaart, P.W., van Vuuren, D.P., Biemans, H., Niessink, R., High-resolution assessment of global technical and economic hydropower potential, *Nature Energy*, 2, 821–828. 2017. With permission.)

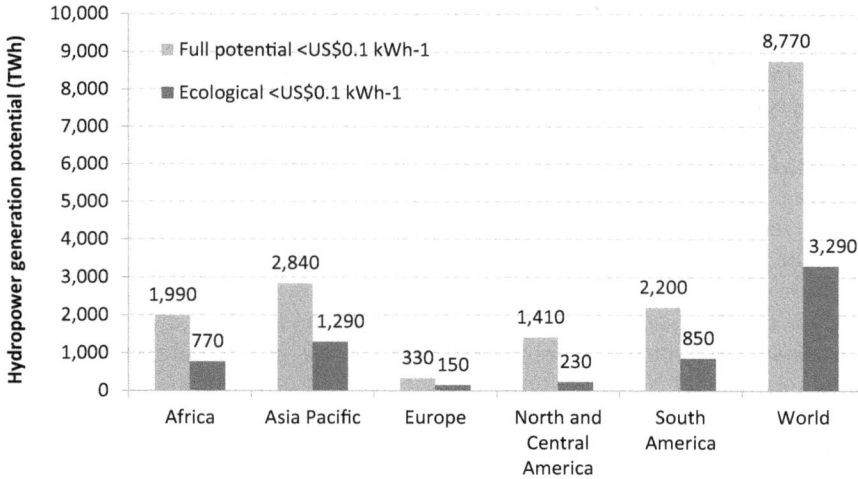

**FIGURE 5.6** Economic potential (with costs less than US$ 0.1 per kWh) of hydropower generation across the world. (From Gernaat, D.E.H.J., Bogaart, P.W., van Vuuren, D.P., Biemans, H., Niessink, R., High-resolution assessment of global technical and economic hydropower potential, *Nature Energy*, 2, 821–828. 2017. With permission.)

also in conflict with ensuring environmental security. Detailed maps on the locations of the current and potential dams being constructed or planned also confirm the findings presented in this section (Lehner et al., 2011; Mulligan et al., 2020; http://globaldamwatch.org/map/).

## 5.6 ECONOMIC AND INCOME DISTRIBUTION IMPACTS

Despite the considerable energy supply enhancement potential of hydropower dams and renewed policy interest in their further development, opinions on their socioeconomic benefits vary. Economy-wide modeling-based studies reported positive economic gains and income distribution advantages of the hydropower dams (Bhatia et al., 2008). Using the cases of Bhakra (in India), Aswan (in Egypt), and Sobradinho Dams (in Brazil), Bhatia and coauthors (2008) reported considerable indirect income generation effects in addition to direct positive economic benefits. Especially, in the case of the Bakra dam, the poor people did not get much directly from the change but they became major beneficiaries due to increased employment opportunities. The outcomes of this study, however, did not become popular as the findings are based on the examples of a few selected dams.

In contrast, several studies showed that economic impacts from dam developments can be disproportional bringing additional gains to a group residing in one part of a river basin yet reducing incomes for the people in the remaining parts of the basin (Duflo and Pande, 2007; Bird and Wallace, 2001). Large dam constructions have direct and indirect consequences due to the reallocation and displacement of local communities and the loss of access to natural assets (Richter et al., 2010). In Africa, the development of dams correlated with increased risks of malaria infection among people living in close vicinity to the reservoirs (Kibret et al., 2015). When hydropower projects take place in transboundary river basins, potential for conflicts increases among the riparian governments (Bekchanov et al., 2015). Dams, therefore, intensify the complexity of water, energy, and food nexus because of the multiple interests and social backgrounds of the riparian users and stakeholders (Vörösmarty et al., 2010).

Multipurpose hydropower projects may support irrigation sometimes but go into conflict with food production systems often. Global assessment of the trade-offs between hydropower and irrigation developments reveals that 54% of global installed hydropower capacity competed with irrigation and only 8% of the hydropower capacity complements irrigation (Zeng et al., 2017). Conflicting interests dominate in the river basins located in the Central parts of the USA, Northern Europe, North and Eastern Africa, Central Asia, and South and Southeast Asia. Hydropower and irrigation are mostly complementary in the Yellow and Yangtze River basins of East Asia, the East and West Coasts of the USA, Canada, Russia, and the majority of the river basins in Southeast Asia. Thus, policies that support hydropower development to mitigate climate change should additionally consider the food security impacts.

It was found that there is no correlation between planned hydropower capacity development and the rate of public access to electricity per country (Zarfl et al., 2015). In India, the technically low potential of hydropower developments in large parts of the country impedes achieving the goal of full electricity access for every citizen. In countries like Brazil and Congo, despite the large potential for hydropower development, improved electricity access may require additional investments in developing the national electricity grid. Yet, expansion of the hydropower systems by private investors with the aim of enhancing industrial development or earning revenues from the energy exports go in conflict with improving public access to electricity. For instance, countries such as Kenya and Tanzania could already provide electricity access for the whole population in case the currently developed hydropower sources were not used for meeting industrial needs such as mining operations but allocated for reducing energy poverty. When hydropower projects were planned to enhance energy-intensive industries such as aluminum melting, their potential for job creation and income equality is also very limited as investigated in the case of Brazil (Fearnside, 2016).

Evidence-based analysis of the costs of the large dam projects also reveals that actual construction costs are too high and hardly bear positive economic gains (Ansar et al., 2014). As estimated, actual cost overruns were on average 96% of the initially estimated costs of hydropower project developments. In addition, large dams took longer time for construction than the initially scheduled time. The actual construction schedule was on average 44% higher than the initially estimated time requirement. Hydropower dams produce only 80% of the targeted output in the first year after the launch (WCD, 2000a). These factors together may greatly reduce the economic efficiency of the large dams even without considering their environmental and social externalities (Ansar et al., 2014).

Last but not least, expected global warming may alter water availability and water demands in many river basins and consequently reduce the economic feasibility of the hydropower projects. Impacts of climate change are more pronounced in developing countries where a majority of future dams are expected to be built (Moran et al., 2018). Expected water scarcity due to climate change may reduce water inflow into reservoirs and reduce stored water consequently reducing the projected production efficiency and hydropower supply. Increased frequencies of extreme events such as flooding and landslides induced by climate change may accelerate sedimentation of the reservoirs further decreasing their efficiency. Heavy rains combined with storms may also increase the risk of failure of dams greatly threatening ecosystems and human lives.

## 5.7  ENVIRONMENTAL EXTERNALITIES

Opinions on the environmental effects of hydropower constructions are also heterogeneous among researchers, practitioners, investors and policy makers. Hydropower developments were often advocated by the proponents for their low carbon gas emissions in contrast to fossil fuel-based energy production with an enormous carbon footprint. As reported, the carbon footprint of hydropower is almost 30–100 times lower than that of fossil fuel-based production (Azevedo, 2011; IPCC, 2014). Given these low carbon reductions, many governments considered hydropower developments in their INDCs to enhance Paris Agreement for keeping global temperature rise below 2°C (Zarfl et al., 2015). Yet, some studies reported enormous greenhouse gas emissions following the dam constructions due to degradation of accumulated organic substances under anoxic conditions (Gibson et al., 2017). This issue is especially serious in tropical regions where most of the future dams are planned to be constructed (Fearnside, 2015).

Large dams have extensive impact on biodiversity since freshwaters are one of the most diverse ecosystems across the world (Zarfl et al., 2019). While water bodies cover only 2.3% of terrestrial ecosystems, they host about 9.5% of known animal species (Reid et al., 2018). Dam constructions change the duration, volume, temperature, turbidity, and chemistry of flow, consequently influencing aquatic plants and organisms and causing irreversible loss of species and ecosystems (McCartney, 2009; Bird and Wallace, 2001). Flooding large areas of forests for constructing large dams leads to biodiversity loss and reduces the natural capacity for carbon sequestration.

Development of dams has a substantial impact on freely flowing rivers causing fragmentation and preventing the free movement of aquatic organisms (Zarfl et al., 2019). Thus, dams put enormous pressure on freshwater megafauna causing the decline of their population. Examples are the declines of the species of the dolphins in the Indus River, paddlefish and sturgeons in the Yangtze River, and the beluga in the Russian rivers (He et al., 2019). New dam developments will globally defragment 25 large rivers out of 120 freely flowing rivers (Zarfl et al., 2019). Future developments of hydropower dams mainly occur in the most ecologically sensitive regions of the world, including the Amazon, Mekong, and Congo River basins. These basins together account for almost 20% of the global diversity of fish species.

Due to impoundment, globally 25%–30% of sediment fluxes transported in river systems are trapped in reservoirs (Vörösmarty et al., 2003). Sedimentation occurs due to the lowered velocity of river flow and increased deposition of weathered rocks and organic and chemical matters in the reservoirs (McCartney, 2009). Deposited sediment fluxes are accumulated over time consequently reducing the useful storage volume of the reservoirs. As estimated, reservoirs across the world lose 0.5%–1% of their storage volume annually because of sediment deposition (Mahmood, 1987). Since the sediments such as freshening soils are also trapped in upstream reservoirs, the reduced sediment load downstream of the dam has adverse effects on vegetation and ecology. The reduced sediment load leads to erosion of the downstream river shores and degradation of coastal deltas and flood plains. Removing sediments from the reservoir is quite a costly process and greatly reduces hydropower generation benefits.

## 5.8 HYDROPOWER POLITICS

Development of large dams has substantial impacts on national water, food, and energy security and may change power balances in transboundary basins. This often leads to interstate conflicts, unless mutually acceptable solutions and agreements are reached. Conflicts often occur over the use of dams due to energy security ambitions of one riparian country and water and food security concerns of other riparian countries (Bhaduri and Bekchanov, 2017). Intense debates over the development of water infrastructure were observed, for instance, in the Aral Sea and Nile River basins (Bekchanov, 2014; Paulos, 2019). The territory of these river basins is shared by several countries, and unilateral actions of upstream countries over using water resources or changing the water flow regimes impact on downstream water availability. Due to the interdependence of the countries and stakeholders in a river basin, unilateral actions of changing any component in the system may resonate with conflicts, unless mutual cooperation and effective coordination of the actions are foreseen.

In the Aral Sea basin, major water storage facilities are located in upstream countries which are interested in increased hydropower generation benefits while large areas of irrigation are located in downstream countries which are interested in reliable water supply from upstream dams for water and food security (Bekchanov et al., 2015). Here, energy demand is highest in the winter season due to increasing electricity needs for heating houses while irrigation water demand is highest in the summer vegetation season. Hydroinfrastructural systems were designed in the Soviet period to store water in upstream reservoirs in winter and to release the stored water to meet irrigation demands in summer (Bekchanov et al. 2018; Schrader et al., 2019). In turn, downstream countries supplied the upstream countries with oil and gas at a cheaper price during winter. Yet, the aftermath of the independence (since 1991), this collaboration scheme did not continue further. Due to the disintegration of the economic and financial cooperation and frequent failure in paying for the deliv-ered energy, gas, and oil supply from downstream countries ceased. In turn, upstream countries saw hydropower generation as an option to meet winter energy demand by increasing water releases from the reservoirs in winter thus causing undesired floods downstream and leaving less water for summer vegetation. This interstate conflict over meeting energy and irrigation demands was further intensified with the initiatives of the upstream countries to build additional water reservoirs for improved hydropower generation for industrial development. Only after political regime changes (since 2016) in Uzbekistan, the largest country in the region in terms of population and irrigated lands, further steps were taken to resolve the conflicts and find out effective solutions that meet the mutual interests of all riparian countries. At present, the cooperation is continuing over sharing the common water resources, regulating water releases from the reservoirs, developing hydrotechnical infrastructure, and joining the efforts to eliminate further environmental degradation in the basin.

Conflicts over water sharing and further development of the hydropower facilities are also tense in the Nile river basin due to heterogeneous distribution of water resources in the region (Farah and Opanga, 2016; Paulos, 2019). The territory of the basin is shared by 11 riparian states including Egypt, Sudan, South Sudan, Ethiopia, Eritrea, Kenya, Uganda, Tanzania, Burundi, Rwanda, and the Democratic Republic of Congo (Khalifa et al., 2022). Egypt is the most powerful both economically and militarily in the region, yet largely relies on water supplies from upstream regions. Upstream countries such as Ethiopia are also strengthening their economy and developing national infrastruc-ture, implying increased demands for water and energy. Ethiopia has control of over 85% of water resources entering Egypt.

The Nile has been governed according to the treaties signed in 1929 and 1959. These treaties prioritize water use rights to Egypt. In addition, Egypt was given the sole authority to develop any hydrotechnical infrastructure without seeking the consent of the remaining riparian countries and the power to veto any infrastructure establishments that are in conflict with their interests. On the other hand, upstream countries are interested in increased uses of water for improving their economic welfare and seek options for more equitable uses of water in the basin. Particularly, the development of the Renaissance Dam by Ethiopia (upstream country) raised the concerns of down-stream countries as they expected reduced water availability downstream due to new developments. The downstream countries were reluctant to the development of the Renaissance dam because they expected that this development gives control over the Nile water resources to Ethiopia. Some stud-ies, in contrast, try to justify the usefulness of the upstream dam to downstream countries point-ing out the surplus amount of hydroelectricity that can be sold to Egypt and the reduced sediment transportation to downstream reaches, which may decrease the efficiency of downstream dams (e.g., Aswan) otherwise (Lewis, 2013). Increased involvement of China in development projects in the region has been largely influencing hydropower politics and governance. Particularly, China's involvement has improved the financial feasibility and investment availability for hydropower proj-ects which were earlier imped due to poor economic conditions and insufficient state savings.

## 5.9 INTEGRATIVE ASSESSMENT OF DAM CONSTRUCTION

In addition to the income generation and energy production effects of the dams, their environmental impacts, social acceptability, and economic externalities are recommended to be assessed at the initial, mid-term, and final stages of the project developments to prevent cost overruns, various conflicts, and irreversible environmental degradation (WCD, 2000b). World Commission on Dams developed seven strategic priorities to ensure the sustainability and inclusiveness of large water and energy infrastructure development plans and processes:

- gaining public acceptance;
- comprehensive options assessment;
- addressing existing dams;

- sustaining rivers and livelihoods;
- recognizing entitlements and sharing the benefits;
- ensuring compliance;
- sharing rivers for peace, development, and security.

The criteria of "gaining public acceptance" is met through a thorough analysis of the stakeholders, negotiated decision-making processes, and reaching free and informed consent of the stakeholder for the changes. "Comprehensive assessment of the options" includes preliminary assessments of environmental, economic, social, health, income distribution, greenhouse gas emission impacts, and risks related to the dam development. "Addressing existing dams" considers socioeconomic and environmental impacts and improved management of reservoir operations. The strategy of "sustaining rivers and livelihoods" ensures sufficient flows to environmental sites and maintains a productive fishery. In case dam developments require settlements and cause externalities to some stakeholders, adequate compensations and benefit-sharing options should be also provided. The strategic priority of "ensuring compliance" requires minimum or zero deviations from the actual plans, agreements, and fairness principles in the project by involving independent expert panels in assessing the project plans, trust funds, and performance bonds for monitoring the process. When large dams are built in shared river basins, new developments should not start or intensify conflicts but ensure peace, development, and security.

Brown and colleagues (2009) developed a framework for a comparative analysis of the costs and benefits of new dam developments while considering all biophysical, socioeconomic, and geopolitical aspects. The framework integrates both quantitative indicators based on the technical measurement and qualitative judgments based on expert opinions. Biophysical indicators include water retention time in reservoir, natural value, biodiversity impact, greenhouse emissions, the length of downstream river left dry, reservoir induced seismicity, and surface area. Socioeconomic impact indicators comprise the impact on social networks (Buckner scale), the number of sites with cultural significance, associated health incidents, income generation prospects (irrigation, hydropower, recreation, industry), relocation costs, and hedonic value of the landscape. Geopolitical impact assessment considers the number of people affected downstream, downstream irrigation, storage capacity of downstream dams, number of political boundaries, number of intergovernmental agreements and institutions related to shared waterways, democratization index, historical stability and tensions, and governance quality. Although the multicriteria assessment framework may ignore some important indicators or overestimate the weights for some other values, it clearly shows that in addition to economic gains sustainability and inclusiveness aspects are also important for evaluating the feasibility of large dams.

## 5.10  CONCLUSION

This study presented the historical developments, current status, and future potential and prospects of the hydropower sector while comparing the various opinions on the performance of the hydropower projects as stated by both opponents and proponents. As shown above, the hydropower sector emerged at the beginning of the 20th century and expanded rapidly during the 1950s and 1960s, especially in the developed part of the world. After a stagnation phase in the last decades of the 20th century, hydropower projects gained renewed attraction by investors. Especially, China, Brazil, and India have hosted the world's largest hydropower projects. These countries together with other countries in the Asia Pacific and South America have still substantial potential for further expansion of the hydropower sector. The hydropower sector was underdeveloped across Africa currently, but the potential for future hydropower projects is also large in this region. Overall, the hydropower sector produces 4,306 TWh of electricity globally at present; the economically feasible potential for further development of hydropower electricity is 8,770 TWh, of which 3,200 TWh is also producible without considerable harm to the environment.

In addition, hydropower projects became attractive from the climate policy perspective and many countries considered them in their climate action policies because of lower GHG emissions compared to fossil fuel-based energy generation. Especially, small-scale hydropower projects are commonly agreed upon as effective options to improve energy, water, and food security at the community level. Yet, intense debates occur over the usefulness of large dams. Particularly, loss of biodiversity and rare species, third-party impacts, conflicts over the regulating reservoir water releases, social unrest, and enormous costs of resettlements are key concerns associated with large-scale hydropower projects. These social and environmental externalities reduce the financial attractiveness of large-scale hydropower projects. Therefore, a comprehensive economic, environmental, and social assessment of new developments of the large dams is recommended while involving all affected stakeholders in decision-making processes and following the guidelines of ensuring security, sustainability, fairness, and inclusiveness.

## REFERENCES

Ansar A., Flyvberg B., Budzier A., Lunn D. (2014) Should we build more large dams? The actual costs of hydropower megaproject development. *Energy Policy*, 69:43–66.

Asif M., Muneer T. (2007) Energy supply, its demand and security issues for developed and emerging economies. *Renewable and Sustainable Energy Reviews*, 11:1388–1413.

Azevedo A. (2011) Re´plica da Associac¸a˜o Brasileira do Alumı´nio (ABAL) a` revista Polı´tica Ambiental n 7. http://www.conservacao.org/publicacoes/files/politicaambiental7_replica.pdf.

Bekchanov M., Djanibekov N., Lamers J.P.A. (2018) Water in Greater Central Asia: A Cross-cutting issue. In: Squires V., Qi L. (eds.), *Sustainable Land Management in Greater Central Asia*. UK: Routledge, 211–236.

Bekchanov M., Ringler C., Bhaduri A., Jeuland M. (2015) How would the Rogun Dam affect water and energy scarcity in Central Asia?, *Water International*, 40(5–6):856–876, Doi: 10.1080/02508060.2015.1051788.

Bekchanov M. (2014) Efficient Water Allocation and Water Conservation Policy Modeling in the Aral Sea Basin. Doctoral thesis at Faculty of Agriculture. Bonn: University of Bonn.

Bhaduri A., Bekchanov M. (2017) Exploring benefits and scope of cooperation in transboundary water sharing in the Amu Darya Basin. In: Dinar A., Tsur Y. (eds.), Management of Transboundary Water Resources Under Scarcity: A Multidisciplinary Approach. New Jersey: World Scientific, 35–63.

Bhatia R., Cestti R., Scatasta M., Malik R.P.S. (2008) *Indirect Economic Impacts of Dams: Case Studies from India, Egypt and Brazil*. Washington, DC: World Bank.

Bird J., Wallace P. (2001) Dams and development – an insight to the report of the world commission on dams. *Irrigation and Drainage*, 50:53–64.

Biswas A.K., Tortajada C. (2001) Development and large dams: A global perspective, *International Journal of Water Resources Development*, 17(1):9–21, Doi: 10.1080/07900620120025024.

Breeze P. 2019. Chapter 8 – Hydropower. In: Breeze P. (ed.), *Power Generation Technologies* (Third Edition), Oxford and Boston: Newnes, 173–201.

Brown P.H., Tullos D., Tilt B., Magee D., Wolf A.T. (2009) Modeling the costs and benefits of dam construction from a multidisciplinary perspective, *Journal of Environmental Management*, 90(Supplement 3):S303–S311, Doi: 10.1016/j.jenvman.2008.07.025.

Duflo E., Pande R. (2007) Dams. *Quarterly Journal of Economics*, 122:601–646.

Farah I., Opanga V. (2016). Hydro-politics of the Nile: The role of South Sudan. *Development*, 59:308–313, Doi: 10.1057/s41301-017-0117-6.

Fearnside P.M. (2015) Emissions from tropical hydropower and the IPCC. *Environmental Science & Policy*. 50:225–239, Doi: 10.1016/j.envsci.2015.03.002.

Fearnside P.M. (2016) Environmental and social impacts of hydroelectric dams in Brazilian Amazonia: Implications for the aluminum industry. *World Development*, 77:48–65, Doi: 10.1016/j.worlddev.2015.08.015.

Gernaat D.E.H.J., Bogaart P.W., van Vuuren D.P., Biemans H., Niessink R. (2017) High-resolution assessment of global technical and economic hydropower potential. *Nature Energy*, 2:821–828.

Gibson L., Wilman E.N., Laurance W.F. (2017) How green is 'green' energy? *Trends in Ecology & Evolution*, 32:922–935.

He F., Zarfl C., Bremerich V., David J.N., Hogan Z., Kalinkat G., Tockner K., Jähnig S.C. (2019) The global decline of freshwater megafauna. *Global Change Biology*, 25:3883–3892.

International Energy Agency (IEA). (2016) *World Energy Outlook*. IEA: Paris. Available online at: https://www.iea.org/reports/world-energy-outlook-2015.

International Energy Agency (IEA). (2020) *SDG7: Data and Projections.* Available online at: https://www.iea.org/reports/sdg7-data-and-projections/access-to-electricity.

International Hydropower Association (IHA). (2010) *Hydropower Sustainability Assessment Protocol.* Available online at: http://www.hydrosustainability.org.

International Hydropower Association (IHA). (2020a) *A Brief History of Hydropower.* Available online at: https://www.hydropower.org/iha/discover-history-of-hydropower.

International Hydropower Association (IHA). (2020b) *Hydropower Status Report: Sector Trends and Insights.* Available online at: https://www.hydropower.org/publications/2020-hydropower-status-report.

Intergovernmental Panel on Climate Change (IPCC). (2014) Climate change 2014: Synthesis report. *Contribution of Working Groups I, II and III to the Fifth Assessment Report of the Intergovernmental Panel on Climate Change* Core Writing Team, R.K. Pachauri and L.A. Meyer (eds.) Geneva: IPCC, 151 p.

International Renewable Energy Agency (IRENA). (2020) *Global Renewables Outlook: Energy transformation 2050* (Edition: 2020). Abu Dhabi: IRENA.

Khalifa M., Bekchanov M., Osman-Elasha B. (2022) Nationally determined contributions to foster water-energy-food-environmental security through transboundary cooperation in the Nile Basin. In: Asif M. (ed.), *Handbook of Energy and Environmental Security*, Oxford: Elsevier, 429–452.

Kibret S., Lautze J., McCartney M., Wilson G.G., Nhamo L. (2015) Malaria impact of large dams in sub-Saharan Africa: maps, estimates and predictions. *Malaria Journal*, 14:339. Doi: 10.1186/s12936-015-0873-2.

Lehner B., Reidy Liermann C., Revenga C., Vörösmarty C., Fekete B., Crouzet P., Döll P., Endejan M., Frenken K., Magome J., Nilsson C., Robertson J.C., Rodel R., Sindorf N., Wisser D. (2011) High-resolution mapping of the world's reservoirs and dams for sustainable river-flow management. *Frontiers in Ecology and the Environment*, 9(9):494–502.

Lewis M.W. (2013) Egyptian protests, Ethiopian dams and the hydro-politics of the Nile basin. *GeoCurrents*, 4 June.

Mahmood K. (1987) *Reservoir Sedimentation—Impact, Extent and Mitigation.* World Bank Technical Paper No. 71. Washington, DC: World Bank.

McCartney M. (2009) Living with dams: Managing the environmental impacts, *Water Policy*, 11:121–139. Doi: 10.2166/wp.2009.108.

Moran E.F., Lopez M.C., Moore N., Müller N., Hyndman D.W. (2018) Sustainable hydropower in the 21st century. *Proceedings of the National Academy of Sciences of the United States of America*, 115:11891–11898.

Mulligan M., van Soesbergen A., Sáenz L. (2020) GOODD, a global dataset of more than 38,000 georeferenced dams. *Scientific Data*, 7:31. Doi: 10.1038/s41597-020-0362-5.

Paulos H.B. (2019) *The Water-Energy-Food Nexus in the Eastern Nile Basin: Transboundary Interlinkages, Climate Change and Scope for Cooperation.* Doctoral Thesis at Faculty of Agriculture. Bonn: University of Bonn.

Reid A.J., Carlson A.K., Creed I.F., Eliason E.J., Gell P.A., Johnson P.T.J., Kidd K.A., MacCormack T.J., Olden J.D., Ormerod S.J., et al. (2018) Emerging threats and persistent conservation challenges for freshwater biodiversity. *Biological Reviews*, 94:849–873.

Richter B.D., Postel S., Revenga C., Lehner B., Churchill A. (2010) Lost in development's shadow: The downstream human consequences of dams. *Water Alternatives*, 3:14–42.

Schrader F., Kamolidinov A., Bekchanov M., Laldjebaev M., and Tsani S. (2019). Hydropower. In: Xenarios S. et al. (eds.), *The Aral Sea Basin: Water for Sustainable Development in Central Asia*, London and New York: Routledge, 52–66.

Vörösmarty C.J., McIntyre P.B., Gessner M.O., Dudgeon D., Prusevich A., Green P., Glidden S., Bunn S., Sullivan C.A., Liermann C.R., Davies P.M. (2010) Global threats to human water security and river biodiversity. *Nature* 467:555–561.

Vörösmarty C.J., Meybeck M., Fekete B., Sharma K., Green P., Syvitski J.P.M. (2003) Anthropogenic sediment retention: Major global impact from registered river impoundments. *Global and Planetary Change* 39(1):169–190.

World Commission on Dams (WCD). (2000a) *Cross-Check Survey: Final Report.* Cape Town, South Africa: University of Cape Town.

World Commission on Dams (WCD). (2000b) *Dams and Development: A New Framework for Decision Making.* London: Earthscan.

Zarfl C., Berlekamp J., He F., Jähnig S.C., Darwall W., Tockner K. (2019) Future large hydropower dams impact global freshwater megafauna. *Scientific Reports*, 9(1):18531. Doi: 10.1038/s41598-019-54980-8.

Zarfl C., Lumsdon A.E., Berlekamp J., Tydecks L., Tockner, K. (2015) A global boom in hydropower dam construction. *Aquatic Sciences*, 77:161–170.

Zeng R., Cai X., Ringler C., Zhu T. (2017) Hydropower versus irrigation — an analysis of global patterns. *Environmental Research Letters*, 12:034006.S.

# 6 A Multigenerational Solar Energy-Driven System for a Residential Building

*Khaled H.M. Al-Hamed and Ibrahim Dincer*
Ontario Tech. University

## CONTENTS

## 6.1 INTRODUCTION

The need of using renewable energy sources to replace fossil fuel-based power plants is growing as global catastrophes are taking place, such as wildfires, droughts, and floods. Renewable energy sources, namely, solar, geothermal, wind, tidal waves, and others, must be implemented in all the economic sectors of each country. One of these economic sectors is the residential sector. This sector in particular has its unique set of challenges [1]. For instance, a residential building that is occupied by people requires not only electricity but also different types of energy in the form of heat supplied to the buildings, namely heating, cooling, and hot water. Another struggle with implementing renewables for this sector is that the energy demands of this sector vary according to a pattern different than the pattern of energy production by these environment-/weather-dependent renewable energy sources. This difference between the pattern of energy production by renewables and the pattern of energy consumption by the residential sector causes what is known as the energy mismatch problem between supply and demand. A third obstacle in the face of the utilization of renewable energy sources is the inefficiency of the energy systems that convert the energy received by these sources to useful outputs used by the residents of buildings. Addressing these three challenges of implementing renewable energy sources in the residential economic sector will bring the world closer to a more sustainable and clean future.

To resolve the first challenge of providing various types of energy to residential buildings, a number of concepts have been proposed in the literature. One such concept is multigeneration. This concept can be defined as the use of an energy source or multiple sources to produce more than a single useful output through the use of thermal processes in a system. There are a few classifications that belong to this concept depending on the number of outputs produced by the energy-driven

DOI: 10.1201/9781003315353-8

system. A system is classified as cogeneration if it produces two outputs, such as power and heating or power and cooling [2]. Another classification is trigeneration, which is when an energy system produces three useful outputs. Some possible combinations of outputs are power-heating-fuel production, power-hydrogen production-desalination [3], power-heating-hot-water production, and power-heating-drying. The third classification which is the most recent one is multigeneration. In this relatively new classification, an energy-driven system produces more than three useful outputs. A group of researchers have studied multigeneration systems that generate different combinations of useful outputs for residential buildings [4–6]. Examples of these combinations include power-heating-cooling-hot-water production, power-drying-hydrogen production-desalination [7], power-cooling-drying-fuel production-hot water, and many other combinations that can be easily found in the literature of multigeneration-integrated systems [8].

Secondly, the mismatch between energy supply and demand is a problem that can be addressed appropriately using energy storage methods. There are many strategies and technologies that have been developed and proposed in the literature and industry regarding resolving this problem. One of the most common strategies already in place is storing the excess electricity generated in the power grid network and then redirecting it to places that suffer from electricity shortages. Another strategy is the storage of excess energy from renewables in the form of chemical energy, that is, alternative fuels, such as hydrogen, ammonia, and methanol. When the demand profile is higher than the supply, these fuels are injected back into the energy system to provide the residential building with sufficient energy [9]. The main issue with this kind of energy storage method is the relatively high costs involved in manufacturing, installing, and operating the devices for fuel production and consumption. An alternative cheaper option is the use of thermal energy storage units [10]. These units use heat transfer fluids which are cheap and long-lasting, and can handle a wide range of temperature levels. They can be easily integrated into almost any renewable energy system without considerable modifications to the original system [11]. For example, a thermal energy storage unit is typically integrated with concentrated solar collectors which store the excess thermal energy at high-to-intermediate temperature levels in the absence of sunlight [12]. Another way of integrating thermal energy storage units is with geothermal power plants that need to operate under steady-state conditions [13]. When the supply of this plant is excessive to the demand by the residential buildings, the thermal energy is stored in these units, and when the supply is coming short, the thermal energy storage units can assist the power plant in increasing production. This control method can aid in reducing the capital cost of installing new geothermal power plants in remote areas that are growing in population which results in a growing demand for energy. Furthermore, they can store waste heat from industrial processes to regulate the use of this heat for multigeneration purposes. The waste heat can also be at low temperatures, as from photovoltaic (PV) solar cells, and then it is upgraded using a high-temperature heat pump to reach a high enough temperature level for storage and later extraction by a Rankine cycle [14].

To address the third challenge of the low performance of these renewable energy systems compared to the conventional ones, it is important to find analysis tools that can assist engineers and researchers identify the weaknesses in the energy systems and suggest some techniques to improve their performance. The most common tool for analyzing energy systems is the energy analysis method. This is an industrial standard and it has been used for many decades. This method can help engineers predict the expected power, heating, and other outputs of the energy system when it operates under certain conditions. In addition, it provides means of comparing possible energy systems for a specific application. In addition, the results of this analysis act as inputs to the economic analysis which provides more insights that are necessary when comparing suitable energy systems [15]. Another valuable tool in the analysis of energy systems for residential applications is the exergy analysis. This type of analysis stems from the second law of thermodynamics, which help energy engineers design systems according to the combination of energy quantity and quality, unlike the energy analysis which only accounts for the quantity of energy [16]. Putting the quality of energy into consideration helps engineers better manage the limited energy sources in a more efficient way by matching the supply and demand of the same type of energy. To illustrate this, if the energy source has a combination of electrical energy and thermal energy, as in a PV/thermal solar cell, then the exergy analysis guides us to direct the electrical energy to meet the electricity demands of the residential building and use the thermal energy for multigeneration purposes, such as space heating, cooling, and hot-water production. Consequently, the energy system can meet all the different types of energy demands by the residential building and operate at a higher energy and exergy efficiency without wasting high-grade energy sources on low-grade energy demands.

The objectives of this particular study are listed as follows:

- To develop an integrated solar energy-driven system for the multigeneration of power, heating, cooling, and hot-water production for a residential building.
- To analyze the integrated system using energy and exergy analyses and evaluate the overall energy and exergy efficiencies of the system.
- To use exergy destruction rates as a gauging method for the weak components of the integrated system.
- To conduct parametric studies on the integrated system to see how it behaves under varying conditions.

This chapter started with a brief introduction to renewable energy technologies and their challenges and solutions. The next section describes the integrated multigeneration system considered here. After that, a thermodynamic model based on energy and exergy analyses is developed to evaluate the system. The fourth section presents and discusses the results of this analysis in two subsections, which are a base case of a residential building and parametric studies. The last section summarizes the study and mentions the main findings of the integrated system analysis.

## 6.2 SYSTEM DESCRIPTION

In Figure 6.1, a multigeneration system producing power, heat, cooling, and hot water is shown schematically. This section describes the operation of the integrated system. The first step in this system is pumping water using pump 1 to the solar collectors at high pressure. This water stream combines with some water stream leaving the thermal energy storage unit at valve 2. The combined water stream enters the solar collectors and receives a lot of thermal energy from the sun to produce steam at high pressure, namely state 2. Then, the steam is split into two streams: One is to supply thermal energy to be stored in the thermal energy storage unit through heat exchanger 2 (HX2), while the second stream is expanded by the steam turbine (ST) to produce mechanical power which is then converted to electrical power using an electric generator. Part of this electric power is supplied back to the integrated system to drive all the six pumps in the system. After the steam is expanded by the ST, it becomes an intermediate-temperature-saturated water mixture. To recover some of the waste heat for the purposes of multigeneration, this water mixture passes through the generator of the absorption chiller unit and through heat exchanger 1 (HX1) to supply heat to the building.

The absorption chiller unit operates using an ammonia–water mixture. The unit starts by absorbing heat from the saturated water mixture at state 5 to separate the ammonia from the water. Ammonia becomes a vapor and leaves the generator at state 13 to enter the condenser. The water leaves the generator at state 14 which contains some thermal energy that can be recovered using the regenerator. The condenser loses heat to the freshwater stream leaving the reverse osmosis module to provide hot water to the building. After the ammonia is condensed, it is expanded by a valve to reduce both pressure and temperature to lower levels. This ammonia stream is then evaporated at low pressure in the evaporator to provide the building with the required cooling effect. In the absorber, the ammonia recombines with the expanded water leaving the regenerator and expansion valve 2. As it recombines, it releases heat into the environment. The ammonia–water mixture is again pumped to high pressure and recovers heat in the regenerator from the water leaving the generator. This completes the cycle of the absorption chiller unit.

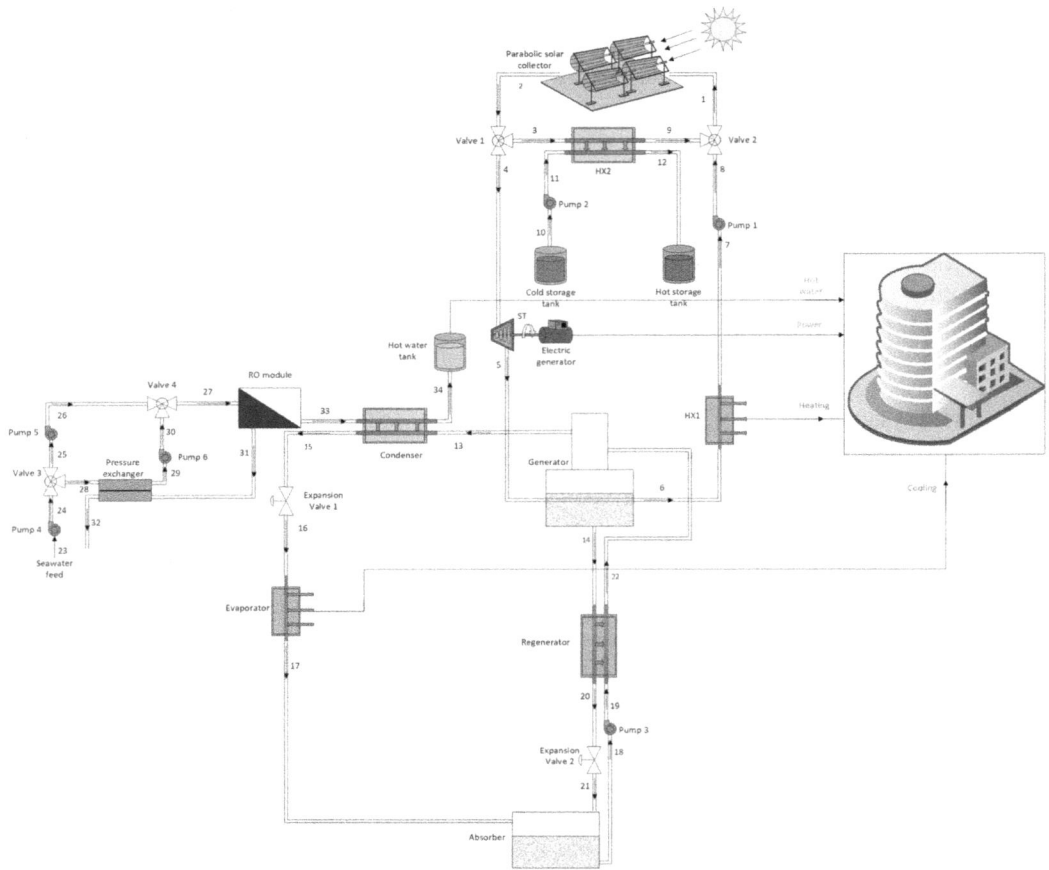

**FIGURE 6.1**    A schematic drawing of the solar energy-driven multigeneration system.

The reverse osmosis (RO) unit starts by receiving a filtered seawater feed from a source. Pump 4 increases the pressure of this seawater slightly and then the sweater stream gets separated into two streams. The first one goes directly to the reverse osmosis module for desalination after it passes through pump 5 which increases the seawater pressure to the required amount. The second stream recovers some pressure from the brine leaving the reverse osmosis module as it goes through the pressure exchanger. The seawater stream is now at some intermediate pressure and needs to be pressurized even further to the required amount using pump 6. Pressure recovering from the brine stream reduces the electric demands on pump 5 which are normally high. At the reverse osmosis module, freshwater is produced, and a brine stream leaves the module and enters the pressure exchanger before it is released back to the environment. The freshwater is heated to a suitable storage temperature, around 80°C, using the waste heat recovered from the condenser of the absorption chiller unit. Hot freshwater is discharged to the building on demand.

## 6.3    ANALYSIS

In this section, the modeling of the multigeneration system described in the previous section is introduced. The model developed here is based on the four balance equations that describe the mass, energy, entropy, and exergy flows and transfers across each component of the integrated system. These balance equations are based on the conservation laws of mass and energy, and on the second law of thermodynamics which is stated in two forms, namely entropy and exergy forms. In the entropy form, entropy generation is present in every component that undergoes a thermal process. An equivalent statement can be written in terms of exergy, which is that exergy is destroyed when a thermal process takes place in each component. Furthermore, a software package of Engineering Equation Solver (EES) [17] is used for analyses and calculations.

The thermodynamic model developed for the integrated system is constructed by applying the four balance equations to each component present in the system. The main assumptions introduced in this model can be listed as follows:

- All components, apart from the storage tanks, operate under steady-state conditions.
- The specific kinetic and potential energy differences between the inlet and outlet of the components are assumed too small compared to the thermal energy terms.
- The pressure losses through heat exchangers, connecting pipes, the solar collector, generator, separating valves, and absorber are considered small and can be neglected in this analysis.
- The separating valves do not affect the thermodynamic properties of the fluids passing through them and they only change the mass flow rates.
- The isentropic efficiency of the steam turbine is taken to be 85%, and the electric generator has a mechanical-to-electric conversion efficiency of 95%.
- The pressure exchanger is assumed to have an efficiency of 96%.
- All heat exchangers, which exchange heat between fluids within the integrated system, are assumed to not lose or gain heat to or from the environment.

Table 6.1 shows the four balance equations written for each component of the integrated system. To compliment the balance equations and to find some of the quantities related to the integrated system, specific models are introduced next for some of the components of the integrated system. To start with, the solar collectors receive heat from the sun according to the following relation

$$\dot{Q}_{solar} = \dot{S}_{solar} A_{solar}, \tag{6.1}$$

where $\dot{S}_{solar}$ is the solar irradiance hitting the collectors in kW/m$^2$ and $A_{solar}$ is the area of the solar collectors in m$^2$. According to Zarza et al. [18], the solar collectors area can be calculated using the parameters of the modules that compose the solar collectors total area. To find the solar collector area, the following empirical expression is used:

$$A_{solar} = nL(w - D_o), \tag{6.2}$$

where $n$ is the entire number of modules, $L$ is the module length, $w$ is the module width, and $D_o$ is the outer diameter of the module pipe. From the experimental design proposed and tested by Zarza et al. [18], the dimensions of each module are 12.27 m for the module length, 5.76 m for the module width, and 0.07 m for the outer diameter of the module pipe. The ratio of thermal energy stored by redirecting some steam into the hot storage tank is identified by the given expression:

$$\varepsilon = \frac{\dot{m}_3}{\dot{m}_2}. \tag{6.3}$$

To model the reverse osmosis desalination cycle, a pressure-based model was developed by Mistry et al. [19]. The seawater feed is assumed to enter the cycle at atmospheric pressure and temperature and with a salinity of 35 g/kg. The pressure necessary at the inlet of the RO module is found by knowing the brine osmotic pressure leaving the module ($P_{Osmotic,brine}$), the pinch pressure ($P_{pinch,RO}$), and the pressure loss across the module (($P_{loss,RO}$). This is expressed mathematically as

**TABLE 6.1**

**Mass and energy balance equations for the components of the integrated system**

| Component | Mass balance | Energy balance |
|---|---|---|
| Pump 1 | $\dot{m}_7 = \dot{m}_8$ | $\dot{m}_7\,h_7 + \dot{W}_{Pump1} = \dot{m}_8\,h_8$ |
| Concentrated solar collectors | $\dot{m}_1 = \dot{m}_2$ | $\dot{m}_1\,h_1 + \dot{Q}_{solar} = \dot{m}_2\,h_2$ |
| Pump 2 | $\dot{m}_{10} = \dot{m}_{11}$ | $\dot{m}_{10}\,h_{10} + \dot{W}_{Pump2} = \dot{m}_{11}\,h_{11}$ |
| Heat exchanger 2 (HX2) | $\dot{m}_3 = \dot{m}_9$, $\dot{m}_{11} = \dot{m}_{12}$ | $\dot{m}_3\,h_3 + \dot{m}_{11}\,h_{11} = \dot{m}_9\,h_9 + \dot{m}_{12}\,h_{12}$ |
| Steam turbine (ST) | $\dot{m}_4 = \dot{m}_5$ | $\dot{m}_4\,h_4 = \dot{m}_5\,h_5 + \dot{W}_{ST}$ |
| Generator | $\dot{m}_5 = \dot{m}_6$, $\dot{m}_{22} = \dot{m}_{13} + \dot{m}_{14}$ | $\dot{m}_5\,h_5 + \dot{m}_{22}\,h_{22} = \dot{m}_6\,h_6 + \dot{m}_{13}\,h_{13} + \dot{m}_{14}\,h_{14}$ |
| Condenser | $\dot{m}_{13} = \dot{m}_{15}$, $\dot{m}_{33} = \dot{m}_{34}$ | $\dot{m}_{13}\,h_{13} + \dot{m}_{33}\,h_{33} = \dot{m}_{15}\,h_{15} + \dot{m}_{34}\,h_{34}$ |
| Expansion valve 1 | $\dot{m}_{15} = \dot{m}_{16}$ | $\dot{m}_{15}\,h_{15} = \dot{m}_{16}\,h_{16}$ |
| Evaporator | $\dot{m}_{16} = \dot{m}_{17}$ | $\dot{m}_{16}\,h_{16} + \dot{Q}_{cooling} = \dot{m}_{17}\,h_{17}$ |
| Absorber | $\dot{m}_{17} + \dot{m}_{21} = \dot{m}_{18}$ | $\dot{m}_{17}\,h_{17} + \dot{m}_{21}\,h_{21} = \dot{m}_{18}\,h_{18} + \dot{Q}_{abs}$ |
| Pump 3 | $\dot{m}_{18} = \dot{m}_{19}$ | $\dot{m}_{18}\,h_{18} + \dot{W}_{Pump3} = \dot{m}_{19}\,h_{19}$ |
| Expansion valve 2 | $\dot{m}_{20} = \dot{m}_{21}$ | $\dot{m}_{20}\,h_{20} = \dot{m}_{21}\,h_{21}$ |
| Regenerator | $\dot{m}_{14} = \dot{m}_{20}$, $\dot{m}_{19} = \dot{m}_{22}$ | $\dot{m}_{14}\,h_{14} + \dot{m}_{19}\,h_{19} = \dot{m}_{20}\,h_{20} + \dot{m}_{22}\,h_{22}$ |
| Heat exchanger 1 (HX1) | $\dot{m}_6 = \dot{m}_7$ | $\dot{m}_6\,h_6 = \dot{m}_7\,h_7 + \dot{Q}_{HX1}$ |
| Pump 4 | $\dot{m}_{23} = \dot{m}_{24}$ | $\dot{m}_{23}\,h_{23} + \dot{W}_{Pump4} = \dot{m}_{24}\,h_{24}$ |
| Pump 5 | $\dot{m}_{25} = \dot{m}_{26}$ | $\dot{m}_{25}\,h_{25} + \dot{W}_{Pump5} = \dot{m}_{26}\,h_{26}$ |
| Pump 6 | $\dot{m}_{29} = \dot{m}_{30}$ | $\dot{m}_{29}\,h_{29} + \dot{W}_{Pump6} = \dot{m}_{30}\,h_{30}$ |
| RO module | $\dot{m}_{27} = \dot{m}_{31} + \dot{m}_{33}$ | $\dot{m}_{27}\,h_{27} = \dot{m}_{31}\,h_{31} + \dot{m}_{33}\,h_{33}$ |
| Pressure exchanger | $\dot{m}_{28} = \dot{m}_{29}$, $\dot{m}_{31} = \dot{m}_{32}$ | $\dot{m}_{28}\,h_{28} + \dot{m}_{31}\,h_{31} = \dot{m}_{29}\,h_{29} + \dot{m}_{32}\,h_{32}$ |

$$P_{27} = P_{Osmotic,brine} + P_{pinch,RO} + P_{loss,RO}. \tag{6.4}$$

In the current RO module, it is assumed to have a pinch pressure of 1010 kPa, and the pressure loss is 202 kPa as given by Nayar et al. [20]. The osmotic pressure of the brine can be found by knowing the salinity and temperature of the brine leaving the RO module. These two values are evaluated using the thermodynamic model and the thermophysical properties for seawater given in references [21,22]. In addition, it is assumed that the recovery rate of the RO module is 40% of pure freshwater (salinity is zero). The salinity for the inlet and outlets of the RO module follows the conservation law of mass for salts in the water, which is given as follows:

$$\dot{m}_{27}S_{27} = \dot{m}_{31}S_{31} + \dot{m}_{33}S_{33}, \tag{6.5}$$

where $S$ is the salinity in g/kg. The input power of all the three pumps in this cycle is evaluated using this expression

$$\dot{W}_{Pump} = \frac{\dot{m}_{inlet}v_{inlet}\left(P_{outlet} - P_{inlet}\right)}{\eta_p}, \tag{6.6}$$

where $\eta_p$ is the pump efficiency and it is assumed to be 85%. To conduct the second law analysis on these three pumps, the entropy generation expression for compression is used here, which was given by Mistry et al. [19]:

$$\dot{S}_{gen,Pump} = \frac{\dot{m}_{inlet} v_{inlet} \left(P_{outlet} - P_{inlet}\right)}{T_{inlet}} \left(\frac{1}{\eta_p} - 1\right). \tag{6.7}$$

The last component in this cycle is the pressure exchanger. The pressure recovered by this device is evaluated using this expression:

$$P_{29} = P_{28} + \eta_c \eta_e \left(\frac{v_{31}}{v_{28}}\right)(P_{31} - P_o), \tag{6.8}$$

where $\eta_c$ and $\eta_e$ are the compression and expansion efficiencies, respectively, and they both are assumed to be 98% [23]. $v$ is the specific volume in m³/kg, and $P_o$ is the atmospheric pressure in kPa. This expression is only valid when the mass flow rates of the brine and the recovering seawater stream are equal. Therefore, at Valve 3, the mass flow rate split is 40% toward Pump 5, and 60% (equal to the brine leaving the RO module) heads toward the pressure exchanger. To calculate the entropy generation rate of the pressure exchanger, the below expression is used:

$$\dot{S}_{gen,PX} = \frac{\dot{m}_{28} v_{28} \left(P_{29} - P_{28}\right)}{T_{28}} \left(\frac{1}{\eta_c} - 1\right) + \frac{\dot{m}_{31} v_{31} \left(P_{31} - P_{32}\right)}{T_{31}} \left(1 - \eta_e\right). \tag{6.9}$$

Lastly, it is desirable to introduce some overall performance indicators to evaluate the integrated multigeneration system. The first parameter is the overall energy efficiency of the integrated system:

$$\eta_{en,overall} = \frac{\dot{W}_{net} + \dot{Q}_{cooling} + \dot{Q}_{heating} + \dot{m}_{12}(h_{12} - h_{11}) + (\dot{m}_{34}h_{34} - \dot{m}_{23}h_{23})}{\dot{Q}_{solar}}, \tag{6.10}$$

whereas the overall exergy efficiency is

$$\eta_{ex,overall} = \frac{\dot{W}_{net} + \dot{Q}_{cooling}\left(\frac{T_o}{T_{cooling}} - 1\right) + \dot{Q}_{heating}\left(1 - \frac{T_o}{T_{heating}}\right) + \dot{m}_{12}(ex_{12} - ex_{11}) + (\dot{m}_{34}ex_{34} - \dot{m}_{23}ex_{23})}{\dot{Q}_{solar}\left(1 - \frac{T_o}{T_{solar}}\right)}. \tag{6.11}$$

The net power is calculated by considering the electric generator output minus the power inputs of all the six pumps in the integrated system (Table 6.2).

## 6.4 RESULTS AND DISCUSSION

Now, we present the results of the thermodynamic analysis of the integrated system for the chosen building. After that, parametric studies of the integrated system are performed to see how this system behaves under varying conditions.

### 6.4.1 BASE CASE

For this base case, we are selecting a residential building located in Ottawa, Ontario that has power, cooling, heating, and hot-water demands of 50 kW, 8.7 kW, 13.8 kW, and 0.03 kg/s. These building demands are adopted from Seyam et al. [24]. Input values to the integrated system are mentioned first.

## TABLE 6.2
### Entropy and exergy balance equations for the components of the integrated system

| Component | Entropy balance | Exergy balance |
|---|---|---|
| Pump 1 | $\dot{m}_7 \, s_7 + \dot{S}_{gen, Pump1} = \dot{m}_8 \, s_8$ | $\dot{m}_7 \, ex_7 + \dot{W}_{Pump1} = \dot{m}_8 \, ex_8 + \dot{Ex}_{dest, Pump1}$ |
| Concentrated solar collectors | $\dot{m}_1 \, s_1 + \dfrac{\dot{Q}_{solar}}{T_{solar}} + \dot{S}_{gen, solar} = \dot{m}_2 \, s_2$ | $\dot{m}_1 \, ex_1 + \dot{Q}_{solar}\left(1 - \dfrac{T_o}{T_{solar}}\right) = \dot{m}_2 \, ex_2 + \dot{Ex}_{dest, solar}$ |
| Pump 2 | $\dot{m}_{10} \, s_{10} + \dot{S}_{gen, Pump2} = \dot{m}_{11} \, s_{11}$ | $\dot{m}_{10} \, ex_{10} + \dot{W}_{Pump2} = \dot{m}_{11} \, ex_{11} + \dot{Ex}_{dest, Pump2}$ |
| Heat exchanger 2 (HX2) | $\dot{m}_3 \, s_3 + \dot{m}_{11} \, s_{11} + \dot{S}_{gen, HX2} = \dot{m}_9 \, s_9 + \dot{m}_{12} \, s_{12}$ | $\dot{m}_3 \, ex_3 + \dot{m}_{11} \, ex_{11} = \dot{m}_9 \, ex_9 + \dot{m}_{12} \, ex_{12} + \dot{Ex}_{dest, HX2}$ |
| Steam turbine (ST) | $\dot{m}_4 \, s_4 + \dot{S}_{gen, ST} = \dot{m}_5 \, s_5$ | $\dot{m}_4 \, ex_4 = \dot{m}_5 \, ex_5 + \dot{W}_{ST} + \dot{Ex}_{dest, ST}$ |
| Generator | $\dot{m}_5 \, s_5 + \dot{m}_{22} \, s_{22} + \dot{S}_{gen, generator} = \dot{m}_6 \, s_6 + \dot{m}_{13} \, s_{13} + \dot{m}_{14} \, s_{14}$ | $\dot{m}_5 \, ex_5 + \dot{m}_{22} \, ex_{22} = \dot{m}_6 \, ex_6 + \dot{m}_{13} \, ex_{13} + \dot{m}_{14} \, ex_{14} + \dot{Ex}_{dest, generator}$ |
| Condenser | $\dot{m}_{13} \, s_{13} + \dot{m}_{33} \, s_{33} + \dot{S}_{gen, cond} = \dot{m}_{15} \, s_{15} + \dot{m}_{34} \, s_{34}$ | $\dot{m}_{13} \, ex_{13} + \dot{m}_{33} \, ex_{33} = \dot{m}_{15} \, ex_{15} + \dot{m}_{34} \, ex_{34} + \dot{Ex}_{dest, cond}$ |
| Expansion valve 1 | $\dot{m}_{15} \, s_{15} + \dot{S}_{gen, exp1} = \dot{m}_{16} \, s_{16}$ | $\dot{m}_{15} \, ex_{15} = \dot{m}_{16} \, ex_{16} + \dot{Ex}_{dest, exp1}$ |
| Evaporator | $\dot{m}_{16} \, s_{16} + \dfrac{\dot{Q}_{cooling}}{T_{cooling}} + \dot{S}_{gen, evap} = \dot{m}_{17} \, s_{17}$ | $\dot{m}_{16} \, ex_{16} + \dot{Q}_{cooling}\left(1 - \dfrac{T_o}{T_{cooling}}\right) = \dot{m}_{17} \, ex_{17} + \dot{Ex}_{dest, evap}$ |
| Absorber | $\dot{m}_{17} \, s_{17} + \dot{m}_{21} \, s_{21} + \dot{S}_{gen, abs} = \dot{m}_{18} \, s_{18} + \dfrac{\dot{Q}_{abs}}{T_o}$ | $\dot{m}_{17} \, ex_{17} + \dot{m}_{21} \, ex_{21} = \dot{m}_{18} \, ex_{18} + \dot{Ex}_{dest, abs}$ |
| Pump 3 | $\dot{m}_{18} \, s_{18} + \dot{S}_{gen, Pump3} = \dot{m}_{19} \, s_{19}$ | $\dot{m}_{18} \, ex_{18} + \dot{W}_{Pump3} = \dot{m}_{19} \, ex_{19} + \dot{Ex}_{dest, Pump3}$ |
| Expansion valve 2 | $\dot{m}_{20} \, s_{20} + \dot{S}_{gen, exp2} = \dot{m}_{21} \, s_{21}$ | $\dot{m}_{20} \, ex_{20} = \dot{m}_{21} \, ex_{21} + \dot{Ex}_{dest, exp2}$ |
| Regenerator | $\dot{m}_{14} \, s_{14} + \dot{m}_{19} \, s_{19} + \dot{S}_{gen, Regenerator} = \dot{m}_{20} \, s_{20} + \dot{m}_{22} \, s_{22}$ | $\dot{m}_{14} \, ex_{14} + \dot{m}_{19} \, ex_{19} = \dot{m}_{20} \, ex_{20} + \dot{m}_{22} \, ex_{22} + \dot{Ex}_{dest, Regenerator}$ |
| Heat exchanger 1 (HX1) | $\dot{m}_6 \, s_6 + \dot{S}_{gen, HX1} = \dot{m}_7 \, s_7 + \dfrac{\dot{Q}_{HX1}}{T_o}$ | $\dot{m}_6 \, ex_6 = \dot{m}_7 \, ex_7 + \dot{Ex}_{dest, HX1}$ |
| Pump 4 | $\dot{m}_{23} \, s_{23} + \dot{S}_{gen, Pump4} = \dot{m}_{24} \, s_{24}$ | $\dot{m}_{23} \, ex_{23} + \dot{W}_{Pump4} = \dot{m}_{24} \, ex_{24} + \dot{Ex}_{dest, Pump4}$ |
| Pump 5 | $\dot{m}_{25} \, s_{25} + \dot{S}_{gen, Pump5} = \dot{m}_{26} \, s_{26}$ | $\dot{m}_{25} \, ex_{25} + \dot{W}_{Pump5} = \dot{m}_{26} \, ex_{26} + \dot{Ex}_{dest, Pump5}$ |
| Pump 6 | $\dot{m}_{29} \, s_{29} + \dot{S}_{gen, Pump6} = \dot{m}_{30} \, s_{30}$ | $\dot{m}_{29} \, ex_{29} + \dot{W}_{Pump6} = \dot{m}_{30} \, ex_{30} + \dot{Ex}_{dest, Pump6}$ |
| RO module | $\dot{m}_{27} \, s_{27} + \dot{S}_{gen, RO} = \dot{m}_{31} \, s_{31} + \dot{m}_{33} \, s_{33}$ | $\dot{m}_{27} \, ex_{27} = \dot{m}_{31} \, ex_{31} + \dot{m}_{33} \, ex_{33} + \dot{Ex}_{dest, RO}$ |
| Pressure exchanger | $\dot{m}_{28} \, s_{28} + \dot{m}_{31} \, s_{31} + \dot{S}_{gen, PX} = \dot{m}_{29} \, s_{29} + \dot{m}_{32} \, s_{32}$ | $\dot{m}_{28} \, ex_{28} + \dot{m}_{31} \, ex_{31} = \dot{m}_{29} \, ex_{29} + \dot{m}_{32} \, ex_{32} + \dot{Ex}_{dest, PX}$ |

In this case, we are considering 11 modules connected together to form the solar collectors of the integrated system. The solar irradiance is taken as 400 W/m², the thermal energy storage heat transfer fluid is chosen to be Dowtherm A, the ambient pressure and temperature are 101 kPa and 301 K, respectively, and the solar temperature is 5770 K. After inputting these values and the energy demands into the thermodynamic model of the integrated system, the following results are given.

Table 6.3 presents the thermodynamic properties of the states presented in the integrated system. The first thing to be noticed is some of the steam generated by the solar collector is stored in the thermal energy storage unit due to the low-power demands by the building. The ratio $\varepsilon$ is 0.158. Since the mass flow rate going through the solar collector is relatively small and the solar irradiance is moderate, the steam temperature leaving the solar collectors is high at a value of 770 K. The freshwater production in this case is double the amount needed by the building which is at 0.06011 kg/s. This cannot be controlled directly by changing the parameters of the integrated system easily due to the integration of the condenser of the absorption chiller unit and the reverse osmosis water heating part as it will be discussed later in a parametric study. The excess hot-water production is stored in the tank for later use. The overall energy and exergy efficiencies of the integrated system at this point of operation are 39.2% and 19.9%, respectively. The low overall exergy efficiency is mainly due to the low grade of the outputs, namely cooling, heating, and hot water. The overall energy efficiency is higher because some of the thermal energy received by the solar collector is stored in the thermal energy storage unit to be used when the sun is absent.

In Figure 6.2, the power output of the steam turbine and power inputs of the six pumps. In addition, it presents the heat transfer rates of some major components of the integrated system. Firstly, we look at the power values of the pumps. Pumps 1 and 5 are the two highest powering consuming pumps at values of 0.242 and 0.386 kW, respectively. These are a magnitude higher than the other pumps mainly because they increase the pressure of the fluids passing through them by significant values. For example, Pump 1 increases the pressure of water from a value of 120 kPa to a value of 3,200 kPa, while pump 2 increases the pressure from 101 to 110 kPa, and this pump consumes 0.0452 kW only. What is important is that the ST electric power output, 50.7 kW, can easily meet the consumption of all the six pumps without any tremendous sacrifice in power production to the building. Next, the heat transfer rate received by the solar collector is 307.2 kW. Since a portion of this heat is transferred to the thermal energy storage unit and converted to mechanical power by the ST, the generator receives a small amount of heat, 11.3 kW, compared to the amount received by the solar collectors. In addition, the condenser provides an adequate amount of heat, which is 8.85 kW, to raise the temperature of freshwater coming out of the reverse osmosis cycle to an acceptable temperature and with sufficient production for the building.

In Figure 6.3, the exergy analysis of the integrated system results in showing the weak spots in the integrated system which are also the spots for possible improvement. The figure shows that the top four exergy destruction rates occur at the parabolic solar collectors, HX1, ST, and HX2, in descending order. Their values are 171.8, 37.5, 7.26, and 6.76 kW, respectively. The main reason behind the high-exergy destruction rate in HX1 is the relative heat losses since most of the heat generated is lost to the environment and not used by the building. Looking at the components of the reverse osmosis cycle, it is noticed that the highest exergy destruction rate happens at the RO module followed by pump 5. Their values are 0.572, and 0.0578 kW, respectively. The other components are considerably more exergetically efficient.

## 6.4.2 Parametric Studies

Next, it is of interest to vary some of the operating parameters of the present integrated system to observe its performance and other important factors. The first parameter varied is the solar irradiance hitting the solar collector. Figure 6.4 presents the how the overall energy and exergy efficiencies are affected by this parameter. As the solar irradiance increases from 400 to 600 W/m², the overall energy efficiency increases significantly and in a nonlinear manner from 39.2% to 59.5%.

TABLE 6.3

Thermodynamic states of the integrated system

| State # | Fluid | Mass flow rate (kg/s) | Temperature (K) | Pressure (kPa) | Specific Enthalpy (kJ/kg) | Specific entropy (kJ/kg/K) | Specific exergy (kJ/kg) | Ammonia/ water mass fraction |
|---|---|---|---|---|---|---|---|---|
| 1 | Water | 0.09289 | 306.1 | 3,200 | 140.8 | 0.4758 | 3.285 | - |
| 2 | Water | 0.09289 | 770 | 3,200 | 3,448 | 7.195 | 1,288 | - |
| 3 | Water | 0.0147 | 770 | 3,200 | 3,448 | 7.195 | 1,288 | - |
| 4 | Water | 0.07819 | 770 | 3,200 | 3,448 | 7.195 | 1,288 | - |
| 5 | Water | 0.07819 | 417.9 | 120 | 2,765 | 7.503 | 511.9 | - |
| 6 | Water | 0.07819 | 377.9 | 120 | 2,620 | 7.131 | 479.3 | - |
| 7 | Water | 0.07819 | 306 | 120 | 137.8 | 0.4758 | 0.1907 | - |
| 8 | Water | 0.07819 | 306.1 | 3,200 | 140.8 | 0.4758 | 3.285 | - |
| 9 | Water | 0.0147 | 306.1 | 3,200 | 140.8 | 0.4758 | 3.285 | - |
| 10 | Dowtherm A | 0.1231 | 301 | 101 | 21.03 | 0.06705 | 0 | - |
| 11 | Dowtherm A | 0.1231 | 301 | 110 | 21.04 | 0.06705 | 0.008545 | - |
| 12 | Dowtherm A | 0.1231 | 510 | 110 | 416 | 1.052 | 98.55 | - |
| 13 | Ammonia/water | 0.007053 | 354.5 | 1,520 | 1434 | 4.598 | 385.8 | 0.9861 |
| 14 | Ammonia/water | 0.03535 | 354.5 | 1,520 | 129.3 | 1.003 | 46.62 | 0.4979 |
| 15 | Ammonia/water | 0.007053 | 312.8 | 1,520 | 179.1 | 0.6953 | 306.3 | 0.9861 |
| 16 | Ammonia/water | 0.007053 | 278.7 | 518 | 179.1 | 0.6965 | 305.9 | 0.9861 |
| 17 | Ammonia/water | 0.007053 | 278.7 | 518 | 1413 | 5.027 | 236 | 0.9861 |
| 18 | Ammonia/water | 0.0424 | 302.9 | 518 | −97.84 | 0.2914 | 53.29 | 0.5791 |
| 19 | Ammonia/water | 0.0424 | 303 | 1,520 | −96.57 | 0.2914 | 54.56 | 0.5791 |
| 20 | Ammonia/water | 0.03535 | 308.2 | 1,520 | −81.79 | 0.3656 | 27.48 | 0.4979 |
| 21 | Ammonia/water | 0.03535 | 308.4 | 518 | −81.79 | 0.3696 | 26.29 | 0.4979 |
| 22 | Ammonia/water | 0.0424 | 341.2 | 1,520 | 79.42 | 0.8382 | 65.96 | 0.5791 |
| 23 | Seawater (35 g/kg salinity) | 0.1503 | 301 | 101 | 112.6 | 0.3934 | 0 | - |
| 24 | Seawater (35 g/kg salinity) | 0.1503 | 302 | 202 | 112.7 | 0.3935 | 0.09898 | - |
| 25 | Seawater (35 g/kg salinity) | 0.06011 | 302 | 202 | 112.7 | 0.3935 | 0.09898 | - |
| 26 | Seawater (35 g/kg salinity) | 0.06011 | 303 | 5,778 | 119.1 | 0.3967 | 5.567 | - |
| 27 | Seawater (35 g/kg salinity) | 0.1503 | 303 | 5,778 | 119.1 | 0.3967 | 5.567 | - |
| 28 | Seawater (35 g/kg salinity) | 0.09016 | 302 | 202 | 112.7 | 0.3935 | 0.09898 | - |
| 29 | Seawater (35 g/kg salinity) | 0.09016 | 302 | 5,353 | 118.6 | 0.3964 | 5.15 | - |
| 30 | Seawater (35 g/kg salinity) | 0.09016 | 303 | 5778 | 119.1 | 0.3967 | 5.567 | - |
| 31 | Brine (58.3 g/kg salinity) | 0.09016 | 303 | 5,576 | 115.6 | 0.3943 | 2.781 | - |
| 32 | Brine (58.3 g/kg salinity) | 0.09016 | 301 | 101 | 109.7 | 0.392 | −2.477 | - |
| 33 | Freshwater | 0.06011 | 303 | 101 | 124.4 | 0.4319 | 0.2238 | - |
| 34 | Freshwater | 0.06011 | 338 | 101 | 271.6 | 0.8918 | 8.999 | - |

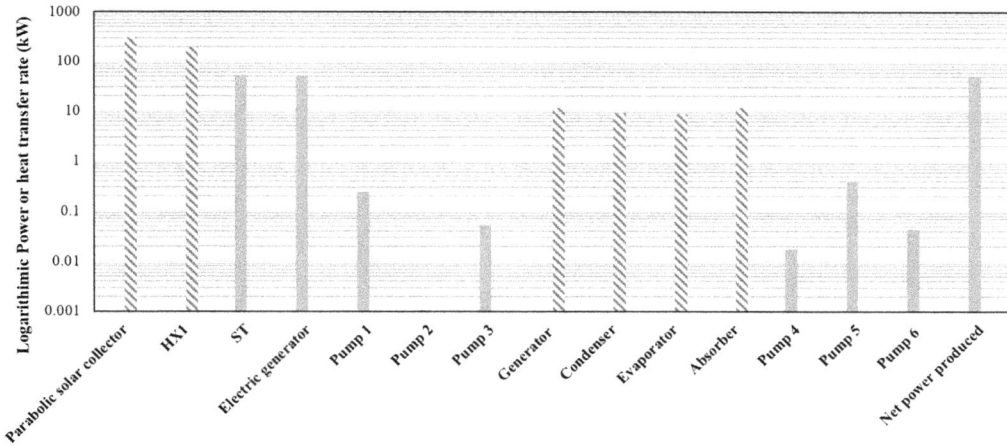

**FIGURE 6.2** Power or heat transfer rates of the components of the present integrated system. The logarithmic scale is used for a better display of values. Solid gray is for the power values and gray diagonal lines are for the heat transfer rates.

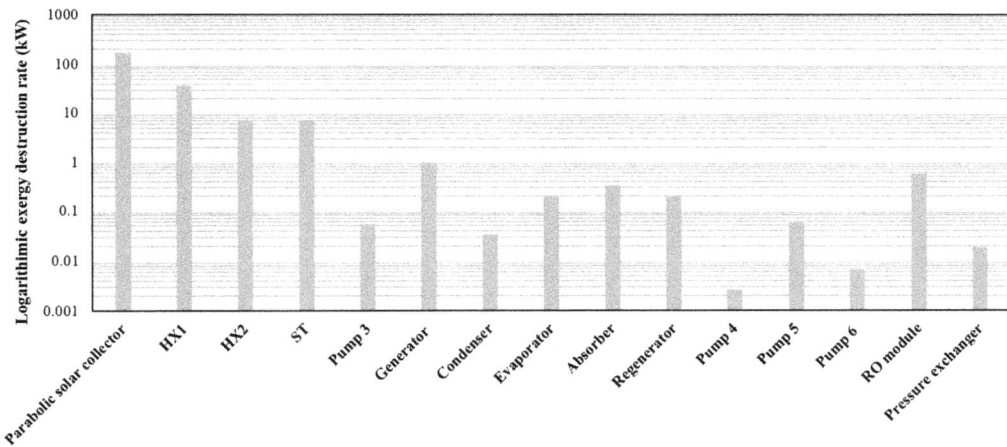

**FIGURE 6.3** Exergy destruction rates of the components of the present integrated system. The logarithmic scale is used for a better display of values.

The overall exergy efficiency of the integrated system only increases slightly by a difference of almost 1.5%. Another observation from the figure is that the gap between the energy and exergy efficiencies is widening. This is mainly due to the fact that the additional thermal energy received by the solar collector is stored thermally, which is a low-grade energy output, and no power production increase is present. The exergy efficiency is mostly affected by the power production of the integrated system, not the thermal outputs, namely cooling, heating, thermal energy storage, and hot-water production.

Figure 6.5 shows how solar irradiance affects the ratio of steam generated is redirected to the thermal energy storage unit and the exergy destruction rate of HX2. The curves explain the efficiencies behaviors outlined above. As more steam is redirected to the thermal energy storage unit, more thermal energy is stored efficiently and only a relatively ratio of heat is lost. This increase in ratio directly affects the overall energy efficiency and not the overall exergy efficiency, since it is

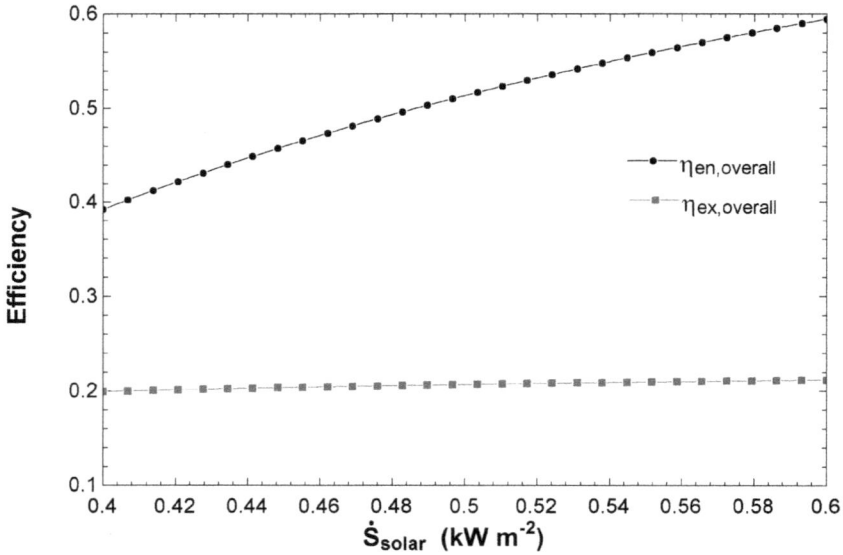

**FIGURE 6.4**  Influences of solar irradiance on overall energy and exergy efficiencies.

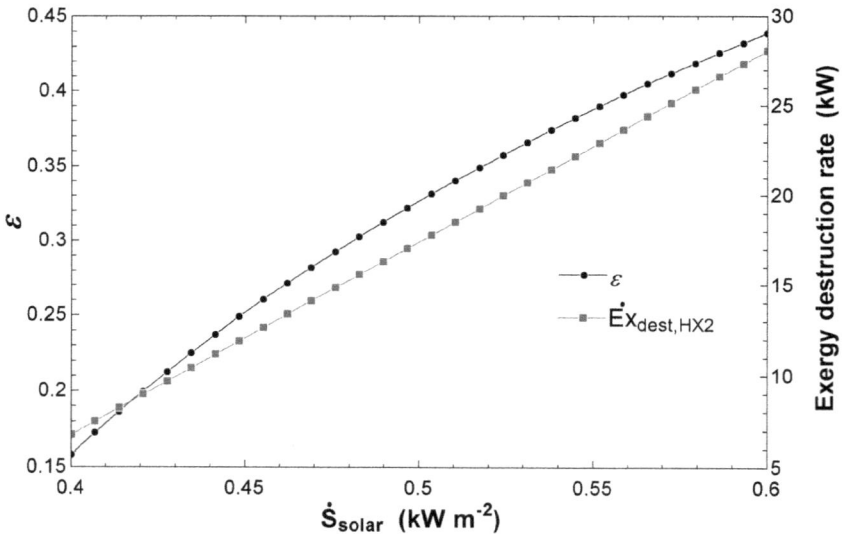

**FIGURE 6.5**  Influences of solar irradiance on $\varepsilon$ of the thermal energy storage unit and exergy destruction rate of HX2.

a quantity of energy increase and not a quality of energy. The quality of energy outputs is still low since no additional power is produced. It is clearly noticed that the behavior of the $\varepsilon$ ratio and the one for the overall energy efficiency are similar. Another obstacle that prevents the increase of overall exergy efficiency is the increase in the exergy destruction rate of HX2. As the solar irradiance increases, this exergy destruction rate increases noticeably from 6.76 to 28.1 kW.

Another parameter related to the solar collector is investigated next. Figure 6.6 shows how the incremental addition of modules to the solar collector changes the performance of the integrated system energetically and exergetically. Similar to the previous parametric study, increasing the

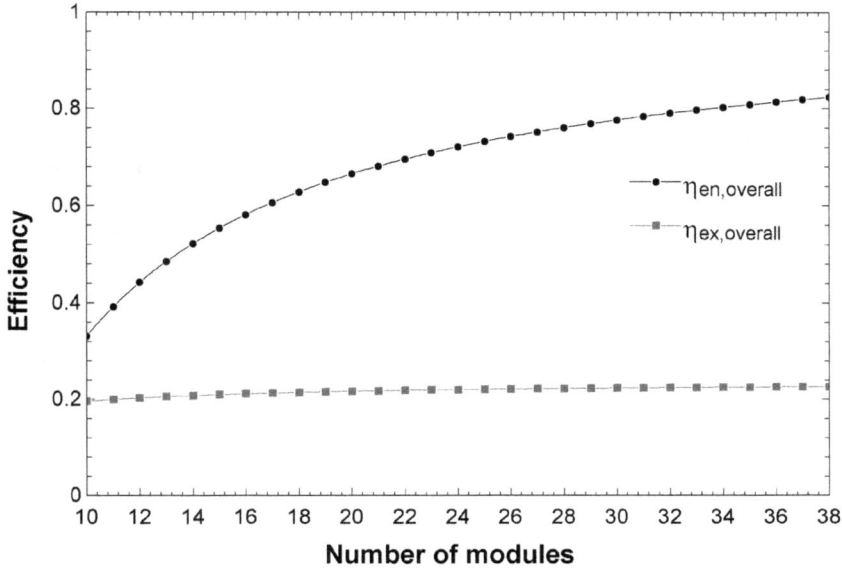

**FIGURE 6.6** Influences of the number of modules on overall energy and exergy efficiencies.

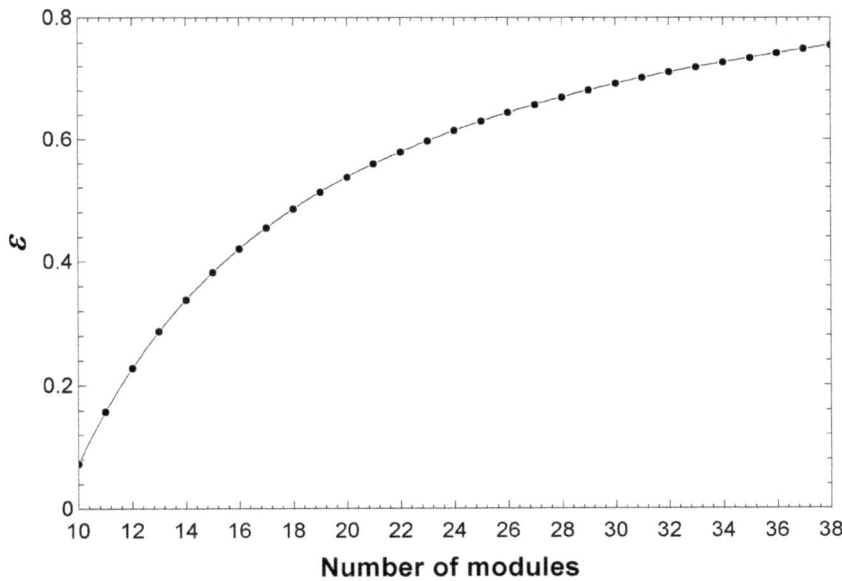

**FIGURE 6.7** Influences of the number of modules on $\varepsilon$ of the thermal energy storage unit.

number of modules makes the solar collector receive more thermal energy from the sunrays. This additional thermal energy is primarily stored in the thermal energy storage unit and this increases the overall energy efficiency from 33.2% to 82.9% as the number of modules goes from 10 to 39.

There is a small enhancement in the exergetic performance of the integrated system, but it is not as great as the energetic performance. The reason for this is due to the increase in the amount of steam redirected to the thermal energy storage unit which does not provide an addition to the quality of energy outputs of the integrated system, such as power production. This increase in redirected steam is evident in Figure 6.7.

Figure 6.8 presents the last parameter considered in this section. The parameter is the cooling load demanded by the residential building. Since the integrated system connects the solar collector-ST cycle to the reverse osmosis cycle through the absorption chiller, it is important to understand how changes in the absorption chiller affect both of these two connected cycles. Firstly, as the cooling load increases, the overall energy and exergy efficiencies have opposite trends. The energy efficiency increases with increasing cooling load from 37.4% to 40.5%, while the exergy efficiency decreases from 20.1% to 19.9% as the cooling load goes from zero to 15 kW provided to the building. This drop in exergy efficiency is marginal and it is because more low-grade energy output, namely cooling, is produced compared to a high-grade energy output, namely power. The energy efficiency grows since this indicator does not account for the quality of the energy outputs, only the quantity.

The increase in cooling load also influences the amount of thermal energy stored, as well as the hot-water production rate. This is shown in Figure 6.9. For the same solar energy received by the solar collector, more energy needs to be provided to the ST and the bottoming absorption chiller to produce more cooling. This means that less steam is redirected to the thermal energy storage unit as clearly shown in this figure. The ratio decreases from 0.167 to 0.152. In contrast, the mass flow rate of the hot water produced by the reverse osmosis cycle is increased with increasing the cooling load. This is because there is more heat available by the condenser to heat larger quantities of ambient temperature freshwater leaving the desalination cycle.

The entanglement of outputs of the integrated system is one of the main problems with multi-generation-integrated systems in general. If the energy and hot-water demands are to be met appropriately and without any excess production, the production of a single, provided to the building, needs to be independent from the production of another output of the integrated system. A complex control system is required to be placed for such a system.

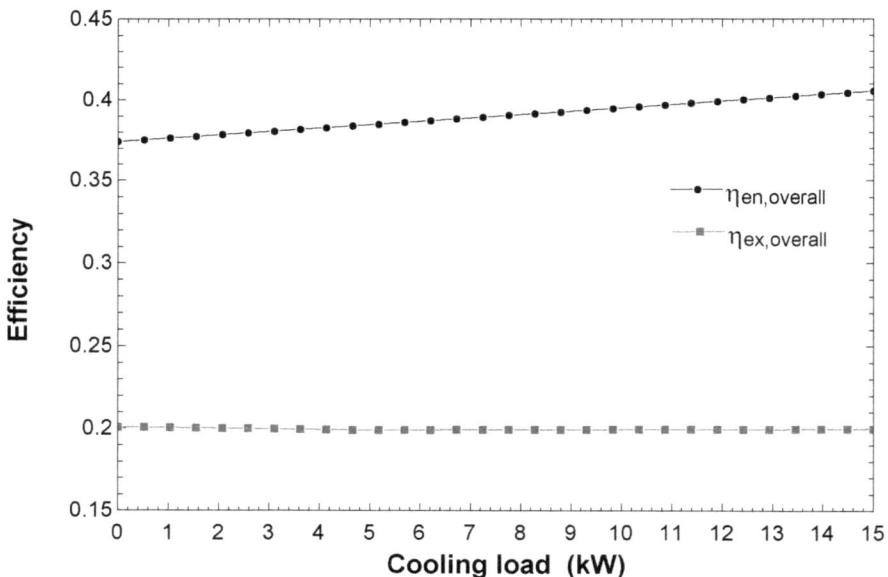

**FIGURE 6.8**   Influences of cooling load on overall energy and exergy efficiencies.

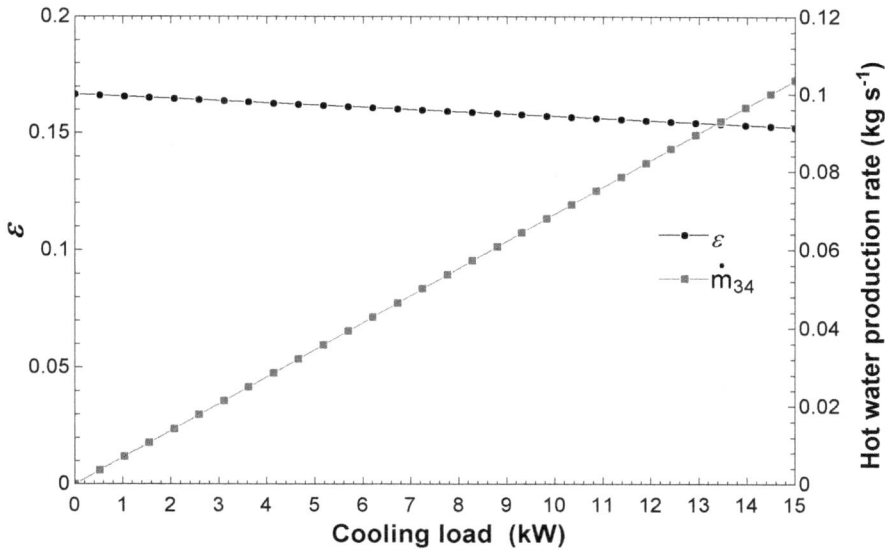

**FIGURE 6.9** Influences of cooling load on $\varepsilon$ of the thermal energy storage unit and hot-water production rate.

## 6.5 CONCLUSIONS

Power, cooling, heating, and hot water are provided to a residential building by the utilization of an integrated system that uses solar energy as a renewable energy source. This integrated system has been described and analyzed thermodynamically. A base case for a residential building located in Ottawa, Ontario is selected to operate this integrated system and evaluate its performance energetically and exergetically. The main outcomes are the evaluation of the overall energy and exergy efficiencies that are calculated to be 39.2% and 19.9%, respectively. In addition, the identification of the weakest spots using the exergy analysis of the integrated system has been conducted. The parabolic solar collectors, HX1, ST, and HX2, are identified to be the least-efficient components exergetically. Their exergy destruction rates are 171.8, 37.5, 7.26, and 6.76 kW, respectively.

In addition, three parametric studies of the integrated system have been performed. The three varied parameters are solar irradiance, the number of modules in the solar collector, and the cooling load of the residential building. The main conclusion from the first two parametric studies is that as more solar energy is received by the integrated system, the overall energy efficiency is improved tremendously from 33.2% to 82.9%, while the overall exergy efficiency is not affected by much. The results of the third parametric study show two main things. Increasing the cooling load from 0 to 15 kW enhances the overall energy efficiency by around 3% and the hot-water production. The increase in hot-water production along with the increase in cooling provided to the building is caused by the integration of two cycles, namely the absorption chiller and the reverse osmosis cycles.

## NOMENCLATURE

$A_{solar}$     solar collector area (m²)
$D_o$     module pipe outer pipe diameter (m)

| | |
|---|---|
| $ex$ | specific exergy (kJ/kg) |
| $Ex$ | exergy rate (kW) |
| $h$ | specific enthalpy (kJ/kg) |
| $L$ | module length (m) |
| $\dot{m}$ | mass flow rate (kg/s) |
| $n$ | total number of modules in the solar collector |
| $P$ | pressure (kPa) |
| $\dot{Q}$ | heat transfer rate (kW) |
| $s$ | specific entropy (kJ/kg/K) |
| $S$ | salinity (g/kg) |
| $\dot{S}$ | entropy rate (kW/K) |
| $\dot{S}_{solar}$ | solar irradiance (kW/m$^2$) |
| $T$ | temperature (K) |
| $v$ | specific volume (m$^3$/kg) |
| $w$ | module width (m) |
| $\dot{W}$ | power (kW) |

## GREEK LETTERS

| | |
|---|---|
| $\varepsilon$ | steam redirected ratio as per Equation (6.3) |
| $\eta$ | efficiency |

## SUBSCRIPTS

| | |
|---|---|
| abs | absorber |
| c | compression |
| cond | condenser |
| dest | destruction |
| e | expansion |
| en | energy |
| evap | evaporator |
| ex | exergy |
| exp1,2 | expansion valve 1,2 |
| gen | generation |
| HX1,2 | heat exchanger 1,2 |
| net | overall power of the integrated system |
| o | reference conditions |
| overall | overall integrated system |
| p | pump |
| PX | pressure exchanger |
| RO | reverse osmosis |
| ST | steam turbine |

## REFERENCES

1. Abu-Rayash A, Dincer I. A Sustainable trigeneration system for residential applications. *J Energy Resour Technol* 2020;143:1–18. Doi:10.1115/1.4047599.
2. Al-Hamed KHM, Dincer I, Rosen MA. Investigation of elastocaloric cooling option in a solar energy-driven system. *Int J Refrig* 2020. Doi:10.1016/j.ijrefrig.2020.07.015.

3. Safari F, Dincer I. Development and analysis of a novel biomass-based integrated system for multigeneration with hydrogen production. *Int J Hydrogen Energy* 2019;44:3511–26. Doi:10.1016/j.ijhydene.2018.12.101.

4. Karaca AE, Dincer I. A new integrated solar energy based system for residential houses. *Energy Convers Manag* 2020;221:113112. Doi:10.1016/j.enconman.2020.113112.

5. Panchal S, Dincer I, Agelin-Chaab M. Analysis and evaluation of a new renewable energy based integrated system for residential applications. *Energy Build* 2016;128:900–10. Doi:10.1016/j.enbuild.2016.07.038.

6. Sorgulu F, Dincer I. A renewable source based hydrogen energy system for residential applications. *Int J Hydrogen Energy* 2018;43:5842–51. Doi:10.1016/j.ijhydene.2017.10.101.

7. Ishaq H, Dincer I, Naterer GF. New trigeneration system integrated with desalination and industrial waste heat recovery for hydrogen production. *Appl Therm Eng* 2018;142:767–78. Doi:10.1016/j.applthermaleng.2018.07.019.

8. Dincer I, Bicer Y. *Integrated Energy Systems for Multigeneration*. Elsevier; 2019.

9. Khalid F, Dincer I, Rosen MA. Techno-economic assessment of a solar-geothermal multigeneration system for buildings. *Int J Hydrogen Energy* 2017;42:21454–62. Doi:10.1016/j.ijhydene.2017.03.185.

10. Erdemir D, Dincer I. Potential use of thermal energy storage for shifting cooling and heating load to off-peak load: A case study for residential building in Canada. *Energy Storage* 2020;2. Doi:10.1002/est2.125.

11. Dincer I, Rosen MA. *Thermal Energy Storage: Systems and Applications*. John Wiley & Sons; 2002.

12. Bolt AA, Dincer I, Agelin-Chaab M. An integrated heat recovery and storage system for a residential building. *Energy Storage* 2019;1. Doi:10.1002/est2.82.

13. Al-Hamed KHM, Dincer I. Investigation of a concentrated solar-geothermal integrated system with a combined ejector-absorption refrigeration cycle for a small community. *Int J Refrig* 2019;106:407–26. Doi:10.1016/j.ijrefrig.2019.06.026.

14. Al-Hamed KHM, Dincer I. Development of a solar farm powered quadric system with thermal energy storage option. *Energy Convers Manag* 2020;217:112981. Doi:10.1016/j.enconman.2020.112981.

15. Abu-Rayash A, Dincer I. Techno-economic evaluation of a residential roof-mounted solar system and its power generation: A case study in Canada. *Int J Glob Warm* 2019;19:202–19. Doi:10.1504/IJGW.2019.101782.

16. Dincer I, Rosen MA. *Exergy: Energy, Environment and Sustainable Development*. Newnes; 2012.

17. EES: Engineering Equation Solver. F-Chart Software : Engineering Software. 2012. http://fchart.com/ees/ (accessed June 17, 2019).

18. Zarza E, Rojas ME, González L, Caballero JM, Rueda F. INDITEP: The first pre-commercial DSG solar power plant. *Sol Energy* 2006;80:1270–6. Doi:10.1016/j.solener.2005.04.019.

19. Mistry KH, McGovern RK, Thiel GP, Summers EK, Zubair SM, Lienhard JH. Entropy generation analysis of desalination technologies. *Entropy* 2011;13:1829–64. Doi:10.3390/e13101829.

20. Nayar KG, Fernandes J, McGovern RK, Dominguez K, Al-Anzi B, Lienhard JH. Costs and energy needs of RO-ED hybrid systems for zero brine discharge seawater desalination. *Int Desalin Assoc World Congr Desalin Water Reuse* 2017.

21. Nayar KG, Sharqawy MH, Banchik LD, Lienhard JH. Thermophysical properties of seawater: A review and new correlations that include pressure dependence. *Desalination* 2016;390:1–24. Doi:10.1016/j.desal.2016.02.024.

22. Sharqawy MH, Lienhard V JH, Zubair SM. Thermophysical properties of seawater: A review of existing correlations and data. *Desalin Water Treat* 2010;16:354–80. Doi:10.5004/dwt.2010.1079.

23. Mistry KH, Lienhard V. JH. An economics-based second law efficiency. *Entropy* 2013;15:2736–65. Doi:10.3390/e15072736.

24. Seyam S, Dincer I, Agelin-Chaab M. Thermodynamic analysis of a hybrid energy system using geothermal and solar energy sources with thermal storage in a residential building. *Energy Storage* 2020;2. Doi:10.1002/est2.103.

# 7 Toward Building Sector Energy Transition

*Niccolò Aste, Claudio Del Pero, and Fabrizio Leonforte*
Politecnico di Milano

## CONTENTS

DOI: 10.1201/9781003315353-9

## 7.1  BUILDINGS, ENERGY, AND EMISSIONS

Buildings consume energy mainly to maintain indoor comfort conditions. This is referred to as operational consumption. In addition, we have to take into account the consumption due to the construction/decommission phases, related to the sector industries, the transport of materials and components, their installation, and, at the end of their life cycle, their disposal. Each of these consumption items is associated with a specific level of atmospheric pollutant emissions, according to quantities and profiles that vary by building type and age, use, country, and climate.

Focusing in depth on buildings in a book about energy and environmental impacts may seem misleading. Through this chapter, readers will have a clear idea of what energy transition means and what technologies, tools, and strategies are needed to pursue the evolution that must necessarily characterize the 21st century. But perhaps it is not so clear that the main causes of the problems we face in preserving the health of the planet lie right in the building sector. As we will see below, the construction and operation (not to mention decommissioning and disposal) of buildings, infrastructure, and cities accounts for most of the planet's energy consumption and therefore, the related emissions and impacts. It is, therefore, not secondary to analyze the role and weight of the world's building stock and industry in the current critical situation. It is absolutely strategic to understand what tools and methods are available to make it a key player in the energy transition framework.

In this section, therefore, the complex relationship between buildings, energy, and emissions will be analyzed with reference to global and national data and different types, uses, and age groups.

Subsequently, in Section 7.2, a number of strategies and solutions are outlined, which are considered to be the most feasible and effective in resolving the critical issues identified above.

As it will be seen, the main areas of intervention concern the design process, construction technologies, technical installations, and smart grid interface. Furthermore, in Section 7.3.3, some fields of action, on which it would be appropriate to work in order to implement the presented strategies and solutions, are suggested. Finally, the conclusions summarize the general picture of the buildings sector and present possible medium-term results for the energy transition.

### 7.1.1  ENERGY

#### 7.1.1.1  The Current Global Situation

The buildings sector accounted for the largest share of both global final energy use (35%) and energy-related $CO_2$ emissions (38%) in 2019 (IEA 2020a) (Figure 7.1), in a slightly lower proportion than the previous year (36% and 39% (IEA 2019a)) due to increase of transport and nonbuilding industry. Building construction industry[1] in 2019, in fact, registered a reduction of its proportion of final energy[2] and carbon emissions with respect to previous year since other sectors increased more, while direct and indirect emissions[3] remained stable.

The energy use trend in the buildings sector has increased constantly since the industrial revolution, achieving an annual average global growth rate of about 1.1% in the last two decades. The growth of the floor area—driven principally by an expanding population, increasing purchasing power in emerging economies, and growing commercial activity—is a critical driver of rising energy demand in the buildings sector. Buildings' floor space in the overall buildings sector has increased by around 65% since 2000, reaching 245 billion m$^2$ in 2019 (IEA 2019b), pushing building energy demand and global emissions (Figure 7.2).

---

[1] Buildings construction industry is the portion (estimated) of overall industry devoted to manufacturing building construction materials such as steel, cement and glass.

[2] Final energy consumption covers all energy supplied to the final consumer for all energy uses. It is usually disaggregated into the final end-use sectors: industry, transport, households, services and agriculture.

[3] Direct emissions are those emitted from buildings. Indirect emissions are emissions from power generation for electricity and commercial heat.

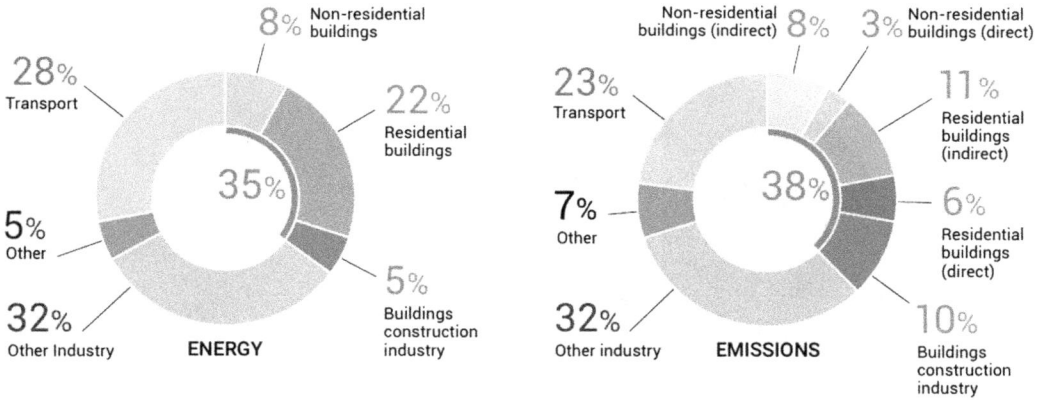

**FIGURE 7.1** Global share of buildings and construction final energy and emissions, 2019. (Adapted from IEA, 2020.)

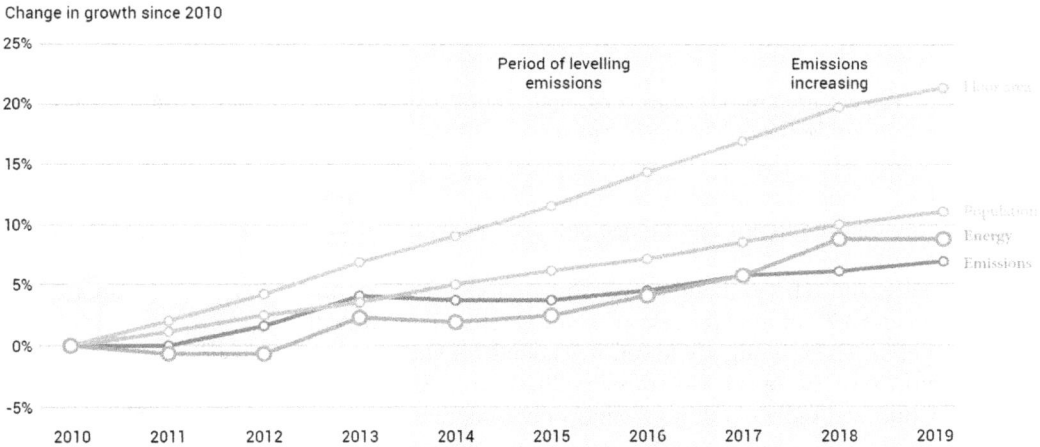

**FIGURE 7.2** Change in global drivers of trends in buildings energy and emissions 2010–2019. (Adapted from IEA, 2020.)

For the first time since 2010, the building's energy consumption growth slowed down despite the floor area increasing, thanks to the improvement of energy intensity of floor space,[4] while $CO_2$ emissions reached their highest level ever (IEA 2020).

Some countries should be considered pivotal to have a better understanding of the international trends in building energy consumption. The United States (US), European Union (EU), China, Russia, and India account for about 60% of global energy consumption (Figure 7.3) (IEA 2019c). In more detail, the US and China are the countries with the largest building energy consumption, with a rate of 16% each in 2019; the EU followed behind, taking up 12% in the same year; India and Russia are the fourth and fifth largest energy-consuming countries in the world, with a consumption of 8% and 7%, respectively. For such reason, in the present work, such countries are considered representative of the worldwide energy trend.

[4] Energy intensity of floor space: energy consumption per square meter of floor space (kWh/m2).

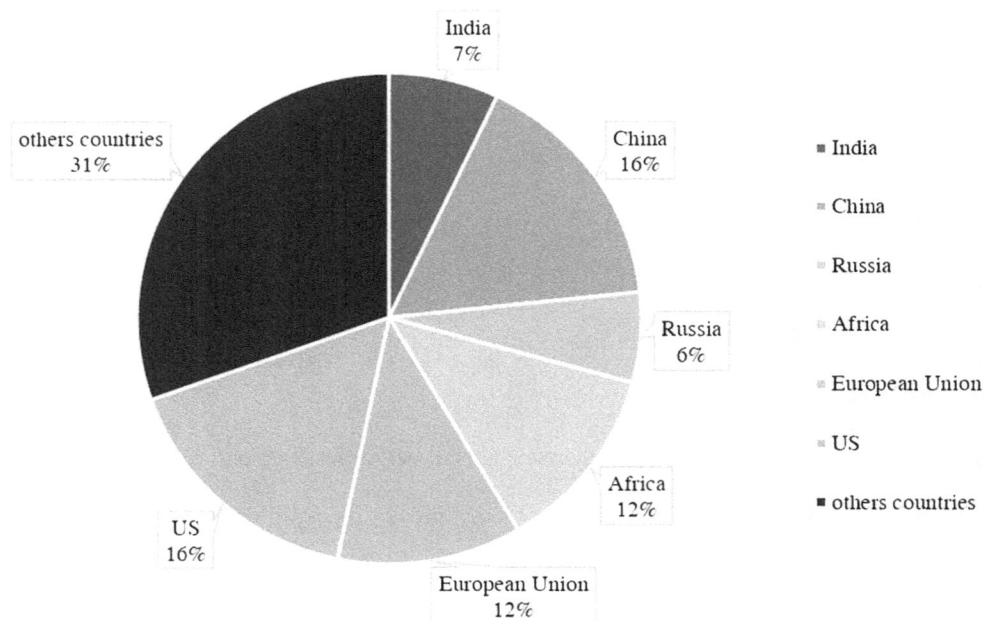

**FIGURE 7.3** Percentage of world energy consumption in 2019 for different countries. (Data elaborated from IEA, *Perspectives for the Clean Energy Transition 2019*, 2019c.)

Of course, Africa also deserves to be mentioned; in fact, it accounts for about 12% of the overall energy consumption in buildings, and such a share will grow in the next year (European Commission 2020).

As evidence, the African building construction sector accounts for 57% of total final energy consumption and 32% of total process-related $CO_2$ emissions (IEA 2020).

In Europe, the share is slightly lower: it accounts for about 40% in terms of energy and 36% in terms of emissions (European Commission 2020). In ASEAN (Association of South-East Asian Nations),[5] China and India's buildings' energy consumption accounted for 26% of total final energy consumption and 24% of total process and energy-related $CO_2$ (IEA 2020). Similar values, equal to 24% and 21%, respectively, for energy use and $CO_2$ emissions are achieved in Central and South America, while in North America, the share is about 19% and 16%, respectively, for energy and $CO_2$ emissions (EIA 2021).

In Australia, buildings consume 19% of the total energy and are responsible for a similar share of greenhouse gas emissions (23%) (Australian Academy of Science 2016).

### 7.1.1.2 Energy Consumption by Age Class

To better understand the building consumption phenomena over the years and countries, the main features of the worldwide building stock are outlined hereafter.

Currently, the overall building stock accounts for about 245 billion m², with an average age between 12 and 15 years for most emerging economies and 30–40 years for advanced economies. Considering that about 70 billion m² (IEA 2013) will be demolished in the next 30 years, more than half of the existing stock is likely to be in use in 2050 (Figure 7.4). In fact, the average lifetime of a building varies from 30 to 50 years for commercial typology to 70–100 years for modern residential constructions and 150 years or more for historic buildings. Of course, low-quality constructions can reduce the lifetime of buildings to 30 years or less, especially in rapidly emerging economies IEA 2019).

---

[5] Brunei, Cambodia, Indonesia, Laos, Malaysia, Myanmar, Philippines, Singapore, Thailand, and Vietnam.

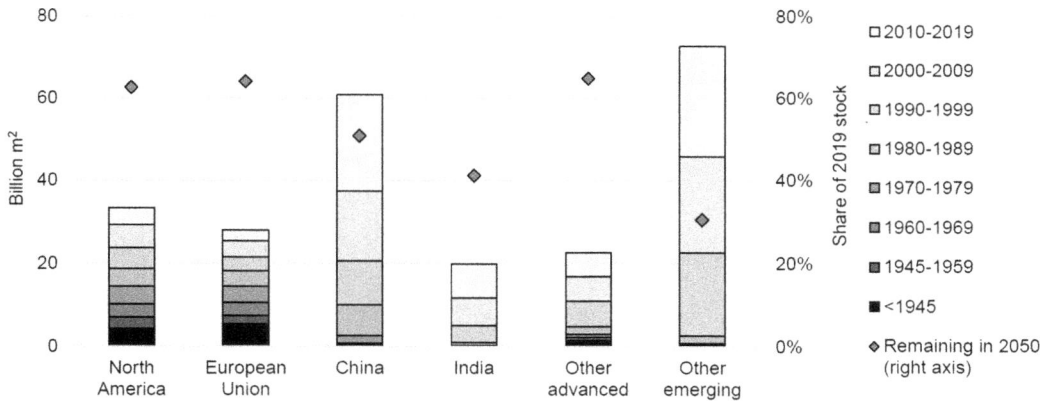

**FIGURE 7.4** Building stock by year of construction. (Adapted from IEA, *Energy Technology Perspectives 2020*, 2020b, IEA, *Transition to Sustainable Buildings. Strategies and Opportunities to 2050*, 2013.)

Such features (the age and the quality of the construction) greatly affect the heating and cooling demand. Buildings constructed before 1960, for example, can require three times (or more) as much heat as those built in accordance with current building codes. In general, building codes tend to be updated over time toward higher energy efficiency and lower final consumption. As a consequence, since 2000, the energy requirements of new buildings have reduced by around 20% globally and by more than 30% in the US and the EU (IEA 2019d). However, currently, around 60% of the global building stock in use today was raised when there were no code requirements regarding energy performance, and this rises to 85% or more in most emerging economies (IEA 2020b).

As shown in Figure 7.4, in fact, emerging countries such as China, India, and Africa have experienced a huge increase in building construction in the last decades in comparison to Europe where more than 50% of the buildings were built before 1960. Nevertheless, most emerging countries have not mandatory building codes able to limit energy consumption (Global Alliance for Buildings and Construction 2019, WRI 2016). Of course, the buildings' energy consumption is affected by the construction year but also by the climate in which they are located.

### 7.1.1.3 Energy Consumption by Region and Building Use

An overview of the final energy use per square meter by world region and building type, according to the input data collected from different sources, is shown in Figure 7.5. Because the sources of buildings' energy vary greatly, for example, significant amounts of coal and biomass are burned on-site in China and India and a much higher share of electricity is used in other countries, the following data are reported in terms of primary energy use.

At the EU level, the average annual specific consumption per square meter for all types of buildings is around 180 kWh/m$^2$ (European Commission 2020). Of course, it differs among different climates and countries: from 55 kWh/m² in Malta and 70 kWh/m² for Portugal or Cyprus to 300 kWh/m² in Romania (or 285 kWh/m² in Latvia and Estonia), which is significantly higher than the EU average. However, even for countries with a similar climate, significant discrepancies exist (e.g., 200 kWh/m² in Sweden, 18% lower than Finland).

More in detail, considering the residential sector at the EU level, it is possible to notice that the single-family buildings are generally characterized by consumption of 15% higher than multifamily ones. The opposite proportion can be registered in North and South America, while in Asia, the two considered buildings' typologies are characterized by almost the same consumption.

Non-residential buildings are on average 40% more energy-intensive than residential buildings (250 kWh/m$^2$ compared to 180 kWh/m$^2$). As for the residential buildings, energy consumption per square meter in services is heterogeneous. Italy, Malta, and Estonia use by far the largest

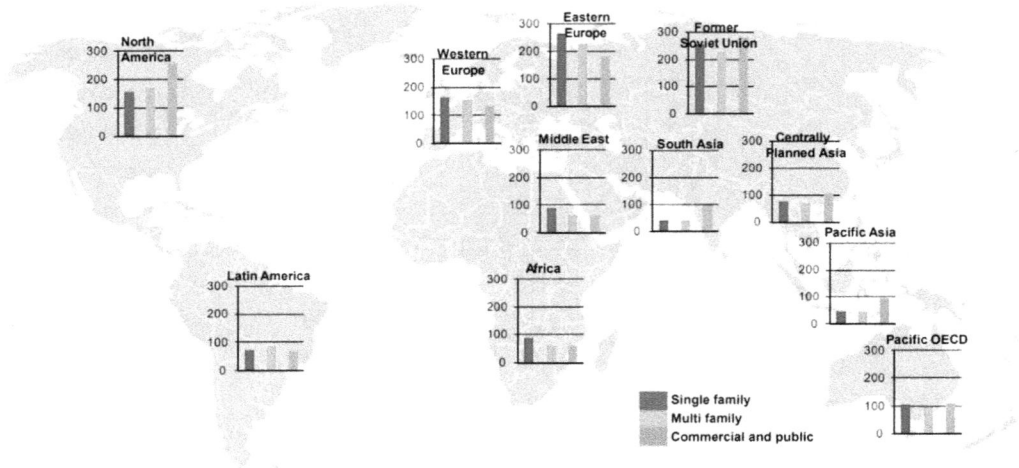

**FIGURE 7.5** Final heating and cooling specific energy consumption by region and building type in 2005 [kWh/m² per year]. (Adapted from, Ürge-Vorsatz, D, Eyre, N., Graham, P., Harvey, D., Hertwich, E., Jiang, Y., Jochem, E., Chapter 10- energy end-use: Buildings. In *Global Energy Assessment Writing Team, Global Energy Assessment: Toward a Sustainable Future* (pp. 649–760), Cambridge University Press, 2012, Doi: 10.1017/CBO9780511793677.016.)

amount of energy per m² (more than 1.5 times higher than the EU average). For the other European countries, energy consumption per square meter is much more homogeneous: most countries use between 200 and 300 kWh/m². A similar value can be considered in Russia, in which, due to the climatic condition and the features of the old building stock, the average energy consumption is about 295 kWh/m² (Sirviö and Illikainen 2015).

North America is generally characterized by high differences between residential and services buildings: the energy consumption of the latter, in fact, is about 280 kWh/m²y, almost twice the residential one. A similar swing can be noticed in China, where the consumption of service and residential buildings are, respectively, 200 and 80 kWh/m²y (Zhang et al. 2020). In Australia, the average building consumption is slightly lower (95 kWh/m²y) than the above-mentioned countries. Finally, in Africa, even if the average building energy consumption is quite similar to the last one, about 80% of such amount comes from traditional biomass for which reliable data does not exist (Kalua 2020, EIA 2018). It must be also noted that in many countries within African regions, building energy consumption information is hardly ever readily available.

### 7.1.1.4 Energy Consumption by End-Use

Today, space heating is the largest energy end-use in the buildings sector, consuming more than 11,000 TWh annually on a global basis. It represents more than a third of the energy demand in buildings, equal to about 36,000 TWh/y (IEA 2020), three-quarters of which is consumed in the US, the EU, China, and the Russian Federation (Russia) (IEA 2019c) (Figure 7.6).

Water heating consumed 6745 TWh in 2017, while in 2000, it was 5582 TWh (IEA 2019c). Excluding the traditional use of solid biomass in developing countries, sanitary hot water production in buildings is also dominated by fossil fuel use and low-efficiency electrical technologies. Sales of solar thermal technologies, led by growth in China, and energy-efficient heat pump water heaters, particularly in Japan, the US, and the EU, have progressively been taken up since 2010 but were not enough to offset rapidly rising energy service demand. Cooking globally consumes about as much energy as water heating but relies much less on fossil fuels. Around two-thirds of estimated energy consumption for cooking is from inefficient and traditional use of solid biomass, with nearly

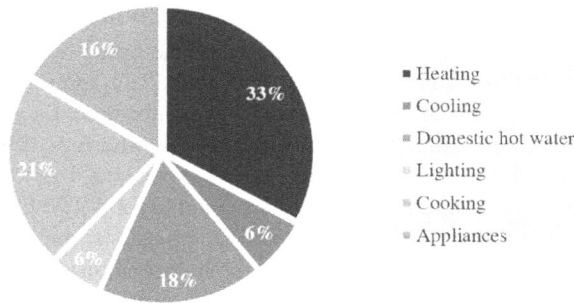

**FIGURE 7.6** Share of energy by end-use in 2017. (Data elaborated from IEA, *Perspectives for the Clean Energy Transition 2019*, 2019c.)

a quarter of households worldwide depending on this basic source for daily cooking needs. Coal, oil, and natural gas account for about 27% of energy use for cooking, while a very small share of renewables (e.g., modern biomass cookstoves) and electricity—despite being used as a principal energy source for cooking by roughly 20% of households—accounts for 6% of energy consumption for cooking purposes. Global electricity use for lighting and appliances grew by an average of 2.2% per year between 2000 and 2017, twice as fast as overall buildings sector energy demand, and accounted for 5000 TWh in 2017 (IEA 2019c). This was driven primarily by emerging economies, where household appliances ownership rates increased substantially over the period. For instance, ownership of major appliances (e.g., refrigerators, clothes washers, and televisions) has doubled in China and Indonesia, and more than tripled in India since 2000. Energy use from other electrical plug-loads, such as smaller electronics and connected devices, also have increased significantly since 2000, a trend that was common in both advanced and emerging economies. By contrast, energy efficiency improvements, notably the mass deployment of light-emitting diode (LED) technologies in recent years, helped to reduce the average energy intensity of lighting per square meter by more than 15% since 2010, even with rising demand for lighting services (IEA 2019c).

Space cooling, which today only accounts for 6% of energy use in the buildings sector, is by far the fastest-growing end-use worldwide, as more and more air conditioners (ACs) are acquired and used. According to the IPCC (Intergovernmental Panel on Climate Change), the global air conditioning energy demand must be considered one of the main issues to address in the next future (Birmingham Energy Institute 2015). In fact, it will grow 33-folds from 300 TWh/y in 2000 to more than 10,000 TWh/y in 2100, with most of the growth in developing economies. In fact, currently, in much of the developing world, the take-up of air conditioning is still very low, but this cannot last as incomes start to rise. In China, for example, less than 1% of urban households owned an air conditioner in 1990, but by 2003 that number had soared to 62%. The same process is now starting in India, where the number of room air conditioners rose from 2 million in 2006 to 5 million by 2011 and is forecast to reach 200 million by 2030 (IEA 2020b).

### 7.1.2 Emissions

Buildings' energy consumption provides some insight into the impact of the sector on global greenhouse gas (GHG) emissions, but a more in-depth analysis of the causes of emissions is needed to better understand the phenomenon. Building emissions can be divided into direct, indirect, and embodied emissions. Direct emissions are those caused by fossil fuels exploited directly in-site, for heating and cooking, while indirect emissions are those caused by the production of electricity and district heat for building use. Both refer to emissions related to building operations while emissions due to construction and demolition are defined as "embodied emissions" and refer to the emissions of industry and transportation sectors, related to the construction of the building.

### 7.1.2.1  General Framework on Direct, Indirect, and Embodied Emissions

Global energy-related emissions from the buildings sector increased by 25% over the 2000–2017 period and accounted for 28% of overall global energy-related $CO_2$ emissions in 2019 (IEA 2020). Buildings' direct emissions from coal, oil, and natural gas combustion, for space and water heating (80%) and cooking (16%) (IEA 2020b), have increased only slightly since 2000, reaching 3 gigatons of $CO_2$ (Gt $CO_2$) in 2019, 24% of total building-related $CO_2$ emissions. Indirect emissions from electricity and district heating consumption in buildings accounted for 6.9 Gt $CO_2$ in 2019 (Figure 7.7), corresponding to more than 55% of building-related $CO_2$ emissions, meaning that large efforts of decarbonization of the buildings sector must go through power and district heating sectors. Overall, residential buildings account for 80% of floor area, over 70% of energy demand, and around 60% of $CO_2$ emissions related to the buildings sector operations (Figure 7.8).

### 7.1.2.2  Direct Emissions

The buildings sector emission increase is due to higher activity levels in regions where electricity remains carbon-intensive, resulting in growing indirect emissions (i.e. electricity), while the use of coal, oil, and natural gas for heating and cooking maintains a steady level of direct emissions (IEA 2020).

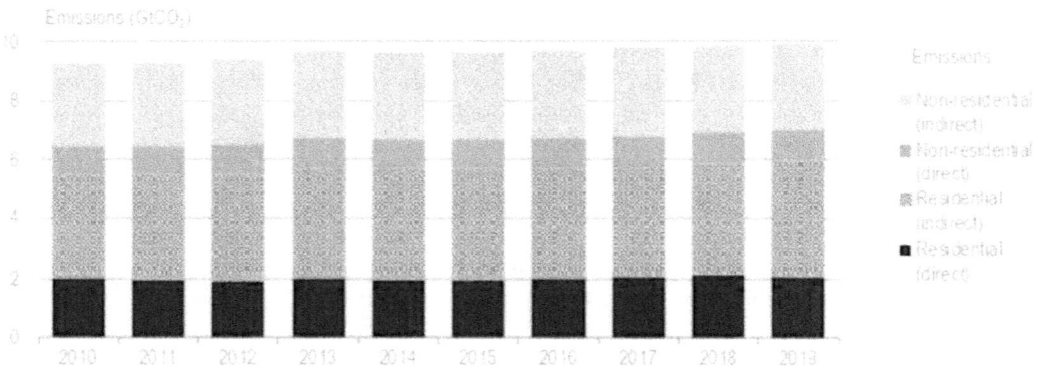

**FIGURE 7.7**  Buildings sector's global energy-related emissions (GtCO₂) by building type and indicator, 2010–2019. (Adapted from IEA, 2020.)

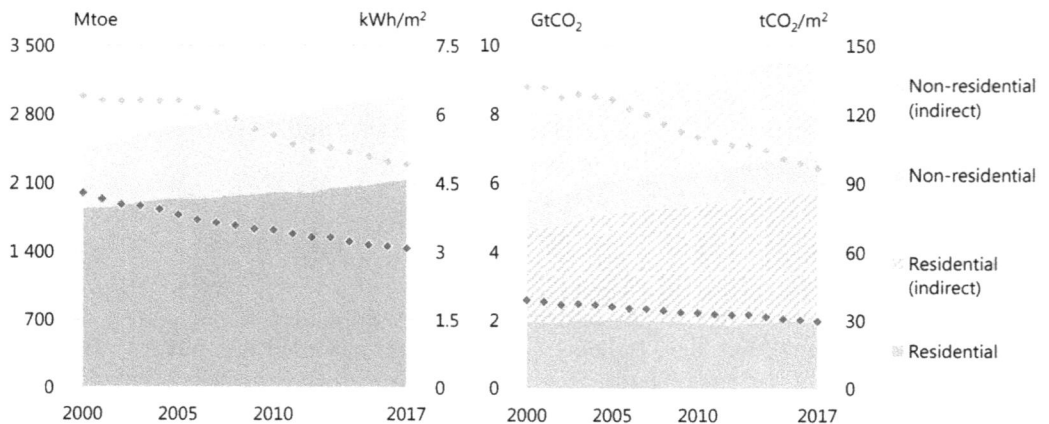

**FIGURE 7.8**  Global buildings sector energy use, energy intensity, and $CO_2$ emissions intensity by subsector, 2000–2017. (Adapted from IEA, *Perspectives for the Clean Energy Transition 2019*, 2019c.)

In terms of energy intensity[6] and carbon intensity,[7] both residential and non-residential buildings experienced a decrease over the past 20 years. The construction of new buildings with lower carbon emissions mainly contributed to reducing the overall energy and carbon impact of the construction sector, even if the largest fraction of existent buildings is characterized by poor performances.

The energy conversion devices that lead to direct emissions in the buildings sector (e.g. natural gas combustion for space and water heating) have a short lifetime (about 15 years) compared with energy power systems (indirect emissions) and industrial assets (embodied emissions). However, it should be noted that the buildings in which they are installed will condition energy consumption and subsequent emissions of the sector for decades (IEA 2020b).

### 7.1.2.3 Indirect Emissions

Unlike direct emissions from fossil fuels, indirect emissions from buildings rose more than 35% between 2000 and 2017, as electricity demand in buildings soared. The buildings sector now represents more than 55% of global final electricity consumption, and the continuous demand for air conditioning and electrical devices in buildings is placing an increasing need to reduce carbon emissions in the power sector. In fact, in the last 20 years, electricity demand in buildings globally grew more than five times faster than improvements in power sector carbon intensity.[8] As reported above, the principal source of energy demand and related $CO_2$ emissions in buildings is due to the residential subsector, which represents about 80% of global floor area, although accounts for nearly 70% of energy consumption in buildings and around 60% of buildings-related $CO_2$ emissions (IEA 2019c). This is due to a few notable differences between residential and non-residential (e.g. commercial, public, and other services) buildings. Non-residential buildings consume 46% of electricity use in the sector and consequently cause around 43% of indirect $CO_2$ emissions (IEA 2019c). In addition, a substantial share (30%, or 7734 TWh/y) of residential energy use is from the traditional use of solid biomass in developing countries. That biomass is generally considered carbon neutral, although it is extremely energy-intensive and causes hazardous indoor and local air pollution. It can also contribute to deforestation and land erosion (Hussaini et al. 2018). The important role of traditional use of solid biomass for meeting energy needs in buildings in emerging economies is evident when looking at $CO_2$ emissions across advanced and emerging economies (Figure 7.9).

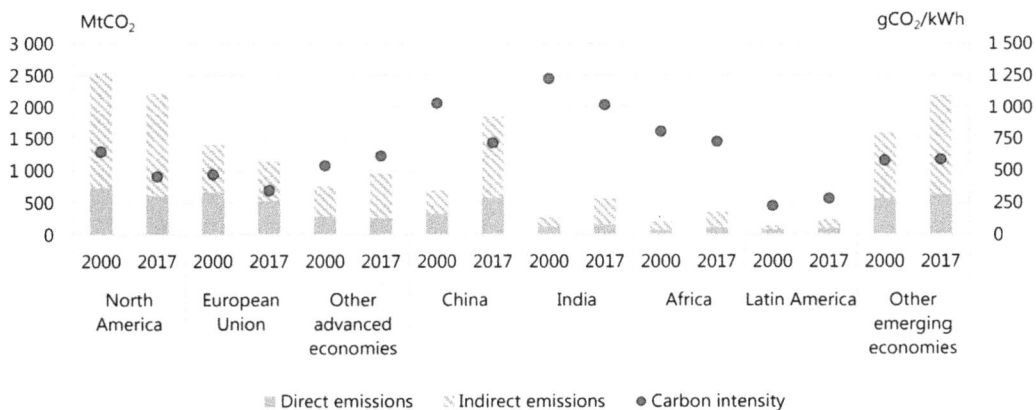

**FIGURE 7.9** Buildings-related $CO_2$ emissions and power sector carbon intensity by region, 2000–2017. (Adapted from IEA, *Perspectives for the Clean Energy Transition 2019*, 2019c.)

[6] Building energy intensity: total final energy use per floor area (kWh/m²).
[7] Building carbon intensity: tonnes $CO_2$ emissions per floor area (tCO₂/m²).
[8] Power sector carbon intensity: tons of $CO_2$ emissions per kilowatt-hour of electricity consumption (tCO₂/kWh).

In 2017, 45% of buildings' energy-related $CO_2$ emissions were in advanced economies (two-thirds if China is included). When indirect emissions from electricity and heat consumption are excluded (to remove the effect of power sector carbon intensities across countries), advanced economies represented half of the buildings-related emissions in 2017, or nearly 70% if China is included. That share is changing, however, as the growing demand for modern energy services (notably electricity) in emerging economies is driving up energy-related emissions. In places like India, $CO_2$ emissions related to buildings more than doubled between 2000 and 2017, while indirect emissions nearly tripled, despite falling carbon intensities of electricity generation. By contrast, indirect emissions from electricity and commercial heat use in North America and the EU fell, thanks to low-carbon electricity generation and relatively stable electricity demand in recent years and despite increasing electrification of buildings.

### 7.1.2.4 Embodied Emissions

In addition to fossil fuels combustion and emissions related to upstream power generation, the buildings sector also represents a considerable portion of emissions from material consumption, as it accounts for 50% of cement and 30% of steel demand (IEA 2020b). The production, transportation, and use of all construction materials for buildings resulted in energy and process $CO_2$ emissions of 3.5 Gt in 2019 or 10% of all energy sector emissions. When added to emissions from energy consumption, this means buildings and construction accounted for around 38% of global $CO_2$ emissions in 2019. Global demand for these bulk materials for construction and renovation has been rising rapidly in recent decades, in parallel with economic and demographic growth. Despite improvements in manufacturing processes, the emissions related to construction materials are therefore rising, especially in emerging economies (Toosi et al. 2020). Curbing this rise through material efficiency measures could make a major contribution to wider efforts to cut emissions in hard-to-abate materials sectors. Among the different construction materials used in the buildings sector, steel and cement are those with the higher impact (IEA 2020b). In particular, worldwide, the sector consumes over 2 Gt of cement and 0.5 Gt of steel, which corresponds to around 50% and 30% of total cement and steel demand, respectively. Glass accounts for most of the rest, in the form of flat glass for windows and glass fibers for insulation, alongside aluminum, plastics and other insulation materials (e.g. rock fibers) (IEA 2020b).

### 7.1.2.5 Emissions per Capita

Up to this point, the data presented refer to building quantities or individual subsectors. In addition, it may be interesting to analyze aggregations with respect to users or inhabitants.

It has to be noticed that, even though metropolitan areas or cities occupy just less than 5% of the earth's land area, they consume more than 75% of the natural resources used by humans and emit 60%–80% of the GHGs (Musango 2017).

According to a study carried out on 79 cities by Cities Climate Leadership Group (C40) (C40 team 2018) in partnership with the University of Leeds (United Kingdom), the University of New South Wales (Australia), and Arup, individual consumption-based GHG emissions per city vary widely from 1.8 to 25.9 $tCO_2$/capita, with an average value of 10.7 $tCO_2$/capita. There is a significant regional variation: most of the analyzed cities in South and West Asia, Africa, and Southeast Asia have individual GHG emissions below 5 $tCO_2$/capita. The median for cities in Latin America and East Asia lies between 5 and 10 $tCO_2$/capita, while cities in Europe, North America, and Oceania have the highest per capita emissions, between 10 and 25 $tCO_2$/capita.

Utilities and housing (rent, maintenance, and repair of water, electricity, gas, and other fuels), capital (Business investment in physical assets such as infrastructure, construction, and machinery), transportation (public and private), food supply. This consists of the categories food and non-alcoholic beverages (93%), alcoholic beverages and tobacco (7%), and government services generally contributing most to consumption-based GHG emissions, although with significant regional variation. For example, on a relative basis, transportation (private and public) emissions are highest for cities in Latin America and capital is most significant for cities in East and Southeast Asia, while emissions from food are largest for cities in South and West Asia. The above five categories make up over 70% of total consumption-based GHG emissions. Clothing (including footwear), furnishings/ household equipment, and restaurants, hotels, recreation, and culture, make up a further 7% and 6% of consumption-based GHG emissions, respectively.

## 7.2 STRATEGIES AND SOLUTIONS FOR THE BUILDINGS SECTOR

Once the situation has been ascertained and the role of the construction sector in the severe global energy and environmental crisis clarified, it seems appropriate to identify most promising strategies and solutions to enable the energy transition.

### 7.2.1 ELECTRIFICATION OF THE BUILDINGS SECTOR

Making the buildings sector more efficient (and clean) means drastically reducing consumption, but obviously, it is not possible to reduce it to zero. In view of the energy transition, then, particular attention must be paid to the energy carriers used. As we have seen, nowadays, the majority of these are fossil fuels, and it is crucial to replace them with the only one that is currently able to fully exploit the potential of renewable energy sources. Thus, the energy transition in the buildings sector means the first and foremost electrification.

This process is already underway: around 68% of the buildings' expected energy demand in 2050 will be covered by electricity (IRENA 2019). Almost the total amount of such electricity must be provided by on-site or nearby renewables, which must be necessarily integrated into the building/ district design. In parallel, precise rules and guidelines on the impact in terms of EE of different technologies/materials must be disseminated to support designers, builders, and decision-makers.

Different approaches must be set for different countries. Electrification of buildings implies the availability of reliable grids. Establishing a robust framework in middle- and low-income countries is fundamental to fulfilling the objective.

Of course, the process will require a substantial and rapid shift toward investment in clean energy technologies. In 2017, the share of overall clean energy investment, covering both low-carbon energy supply and energy efficiency, was only 32%, compared to 51% for fossil fuel investment. The share of clean energy in global energy investments would need to increase substantially to achieve the goal— more than doubling by the mid-2030s and reaching around three-quarters of investments by 2050. By contrast, the share of fossil fuel investment falls to just over 10% by 2050. The share of networks in total investment remains relatively stable at around 15% in 2050 (compared to 17% in 2015) (IEA 2019e).

Efforts and support for renewables must therefore be intensified, but switching to a greener vector is not enough. It must be ensured that the opportunities offered by electricity are fully and efficiently exploited, for example through the options described below.

### 7.2.2 TOWARD A SUSTAINABLE ARCHITECTURE

It is quite clear that the sector as a whole is far from being sustainable, even though current strategies and policies are pushing toward so-called sustainable architecture.

According to the current definition, the aim of sustainable architecture is to limit the environmental impact of the built environment, setting as design goals the containment of pollutants, energy efficiency, improvement of health, comfort and quality of use for inhabitants/users, achievable through the application of appropriate methods, technologies, and materials.

The final result should be the buildings sector capable of satisfying functional needs in the best possible way, taking into account environmental issues and natural resources from the earliest stages of the design process, without causing damage or discomfort to the ecosystem and dialoguing harmoniously with the context.

Thus, if we want to enable the energy transition, sustainable architecture becomes one of the main tools. A rather simple deduction, but not easy to apply. Going into more detail, it would be necessary to transform buildings from inefficient energy consumers into zero-energy, or zero-carbon, structures. This means highly energy-efficient buildings with all remaining operational energy (OE) used from renewable energy, preferably by means of on-site but also off-site production, to annually achieve net-zero carbon emissions in the operational phase. Obviously, this is neither an easy nor quick process, so in the mid-term, it may be sufficient to aim for low-energy buildings or nZEB (nearly zero-energy buildings).

Furthermore, the EE and emissions caused by the production and disposal processes must also be reduced to an absolute minimum.

This new paradigm must be applied not only to new buildings, which will account for about 210 billion $m^2$ by 2050 (the majority of which will be built in low- and middle-income countries), but also to 78 billion m⁻ of existing buildings which must be subjected to renovation (currently with at an average rate of about 1.5%) (IEA 2021). This will bring about multiple co-benefits to society, including lower energy costs, more comfortable and healthy internal conditions, energy security, and resilience against climate change effects.

Benefits and performances associated with sustainable buildings are well-documented and demonstrate how remarkable results can be achieved not only in new construction but also in retrofit operations.

Just to mention a virtuous example, the European Performance of Buildings Directive requires all new buildings to be nZEB by 2020, including existing buildings undergoing major renovations. However, it isn't still possible to properly compare the ambition level of national nZEB definitions due to different indicators, calculation methodologies, applied primary energy factors, system boundaries, etc. In Europe, according to the ZEBRA database (ZEBRA 2021), the primary energy consumption due to heating cooling and lighting for nZEB residential buildings is on average about 55 kWh/$m^2$y, with higher values in cold climates (up to 100 kWh/$m^2$y) and lower in warm countries such as Italy (with an average consumption of 15 kWh/$m^2$y), while for non-residential buildings, it varies between 125 and 55 kWh/$m^2$y.

In China, although some provinces and cities, such as Beijing, Hebei, Shandong, Jiangsu, Hubei, and Fujian, adopted measures to promote the development of nZEB in accordance with local conditions, at present it is difficult to directly move from the current building code to the ZEB (zero-energy buildings, or also NZEB: net-zero energy buildings) level (Liu et al. 2019) thus, reliable acquired data are not available. However, in some studies, typical buildings for the main cities were simulated. The results show that the primary energy demand for heating, cooling, and lighting can be about 110 and 60 kWh/$m^2$y, respectively, for residential and for public buildings (Yang et al. 2019).

According to New Buildings Institute, which monitors the pioneering zero-energy buildings in North America, the average consumption from all fuels (grid-delivered and on-site-generated electricity, natural gas, district energy, and delivered fuels) is about 10 kWh/$m^2$y for residentials and 20 kWh/$m^2$y for public ones (New Building Institute 2020, Ürge et al. 2012).

Very little information and even not updates about high-performance buildings in Russia can be found in literature. According to World Bank (2014), in 2010, the heating energy intensity for new, multifamily high rise buildings in Russia is about 77 kWh/$m^2$y, while no data have been provided for other types of constructions.

Besides, regarding the opportunities offered by building rehabilitation, it should be noted that the bulk of the current global building stock was built without significant energy performance requirements and for that reason offers a high potential for energy-saving measures. Thus, energy renovation is a key element to shifting to a low-carbon building stock, especially in Europe which hosts the greatest amount of existing buildings to be retrofitted. More in detail in the EU, on average, the primary energy saving due to a deep renovation of existing buildings is about 66% both for residential and non-residential buildings, which corresponds to an energy saving of about 122 and 167 kWh/$m^2$y, respectively (European Commission 2019). Of course, such value varies greatly in EU:

in Italy, the potential energy saving for residential building stock is about 90 kW/m²y, and in Spain and Portugal are, respectively, 70 and 45 kWh/m²y. Higher energy-saving potential can be achieved in Germany and France, with values up to 160 kWh/m²y. For non-residential building stock, a similar trend can be identified.

In the US, even if a deep retrofit of existing buildings allows an energy saving of about 45% (IEA 2017), demolition and reconstruction operations are generally preferred unless buildings are historical or with significant value.

Similarly, in China, the retrofit of existing buildings can reduce the energy consumption on average by about 100 kWh/m²y, which corresponds to an energy saving of about 50% (He et al. 2021).

In Russia, the retrofit of existing building stock can yield energy intensities of roughly 150 kWh/m²y, again about 50% lower than the initial value (Word Bank 2014).

In Africa, since the population and, thus, the housing demand is exponentially growing, the number of constructions is not enough to meet the demand. For example, in Nairobi, one of the biggest cities in East Africa, despite the increased levels of urbanization, only 35,000 new homes are built against a demand of 120,000 housing units per year (Van Noppen 2013). The result of this mismatch has led to increase of slums and informal settlements resulting that 60% of the city's population living there (UN-Habitat 2006), characterized by not monitored energy consumption. In such respect, the retrofit of existing buildings cannot be considered a valuable strategy for the African continent, while it is necessary to boost the high-quality construction also able of coping with global warming and cooling demand.

The analysis of the potential of low-energy buildings would not be complete without an in-depth look at EE, which can even be a major factor in the overall balance.

Within conventional buildings' life cycle, operational and EE account for about 90%–70% and 10%–30%, respectively (Ingrao et al. 2018). However, the proportion of buildings' embodied impacts (e.g. EE and carbon) increases considerably along with energy efficiency. This is particularly true in low-energy buildings and even more in zero-energy buildings.

According to Sartori and Hestnes (2007), which collected the data on 60 conventional and low-energy buildings, EE accounts for a share between 2% and 38% and 9% and 46% of total energy use in traditional and low-energy buildings, respectively. Similar results have been acquired by Chastas et al. (2016) who compared the EE and the OE of 90 residential buildings characterized by different energy labels. According to the results, the authors concluded that the EE contribution ranges between 6% and 20% in traditional buildings, 11%–33% in passive buildings, 26%–57% in low-energy buildings, and 74%–100% in net-zero energy buildings.

Of course, it should be noted that the contribution of EE share is also climate dependent (Nebel et al. 2008): in heating-dominated regions, generally characterized by high OE, the EE represents a smaller percentage of the life cycle energy use (Ibn et al. 2013).

Plank (2008), for instance, shows that embodied emissions for a building in heating-dominated regions account for 10% of life cycle emissions. Koezjakov et al. (2018) investigated the relative significance of EE in the Dutch context for different residential archetypes and under various improved efficiency scenarios. They concluded that EE represents 10%–12% of life cycle energy use in conventional homes and 31%–46% in advanced new homes that use higher levels of insulation. They also predict that in 2050, as compared with 2015, the share of EE would increase by 26% (assuming a renovation rate of 1.4% for existing buildings) or 35% (assuming a renovation rate of 1.9%), while OE would decrease by 19% or 46%, respectively (Figure 7.10).

### 7.2.3 FROM SMART BUILDING TO SMART CITY

In the Information and Communication Technologies (ICT) era, the natural evolution of the sustainable building is the smart building, which enhances its potential and performance.

Smart buildings are buildings that integrate and account for intelligence, enterprise, control, and materials/construction as an entire building system, with adaptability, not reactivity, at its core, in

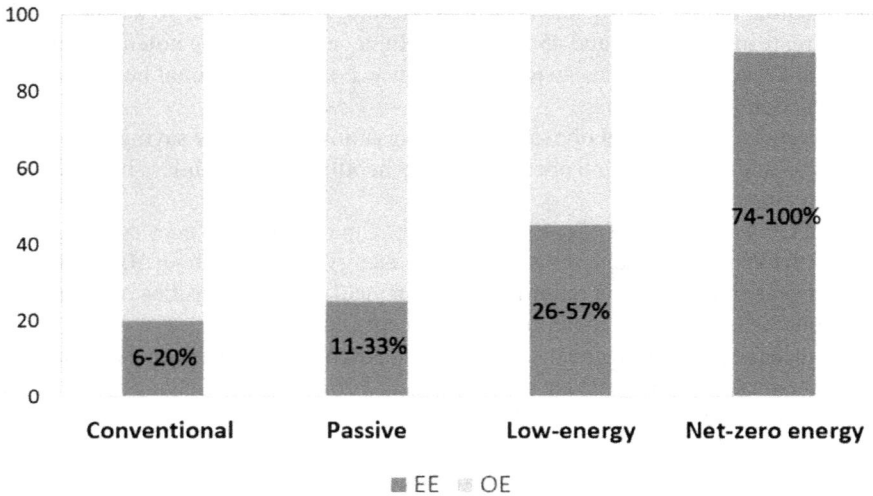

**FIGURE 7.10** Significance of EE and OE in the life cycle energy use of residential buildings. (Drawn based on the data presented by Chastas, P., Theodosiou, T., Bikas, D., Embodied energy in residential buildings-towards the nearly zero-energy building: A literature review. *Building and Environment*, 105, 267–282, 2016.)

order to meet the drivers for building progression: energy and efficiency, longevity, and comfort/satisfaction (Buckman et al. 2014, Al Dakheel et al. 2020).

Basically, through appropriate automation and control systems, it is possible to optimize the operation of the building, making it more efficient, especially from an energy point of view. Smart buildings interact with users, process behavioral, meteorological, operational, and economic data and exchange energy and information flows with the outside world. They are one of the key nodes in a distributed generation system based on renewables, where buildings become both producers and consumers of energy (prosumers) and interact effectively with each other, according to the specific conditions and needs.

In addition to their specific efficiency increase, the strength of smart buildings lies mainly in their interactivity, which is in turn enabled by their inclusion in the so-called smart grids. These are electricity grids equipped with smart sensors and devices that collect information on energy demand and supply, both in real-time and over the long term, optimizing energy distribution.

The energy produced by the sun, wind, biomass, etc. can be consumed directly on-site, or be fed into the grid meeting the needs of other users, who in turn will supply their surpluses when occurring, minimizing waste or the storage burden.

To make the energy transition in the buildings sector possible and effective by making full use of renewables, the organic system buildings-network-cities is thus an essential step, without which the epoch-making change would not be possible.

Around the world, smart grid technology is growing steadily; between 2017 and 2023, the global market is expected to triple in size reaching some 61 billion $. The key regions incorporating smart grid technology include North America, Europe, and the Asia Pacific. The latter will see the fastest growth and is expected to become the largest market for smart grid technologies, which include electrical supply grids that use connectivity, demand response, and renewable energy sources to increase energy efficiency (Figure 7.11).

According to the JRC database (Gangale 2017), in the EU, about 950 projects related to smart buildings and smart grids have been developed. About 57% of them are research & development (R&D) projects, while the remaining part is demonstration projects. The high number of R&D projects suggests that, even if some smart grid solutions are getting close to the commercialization phase,

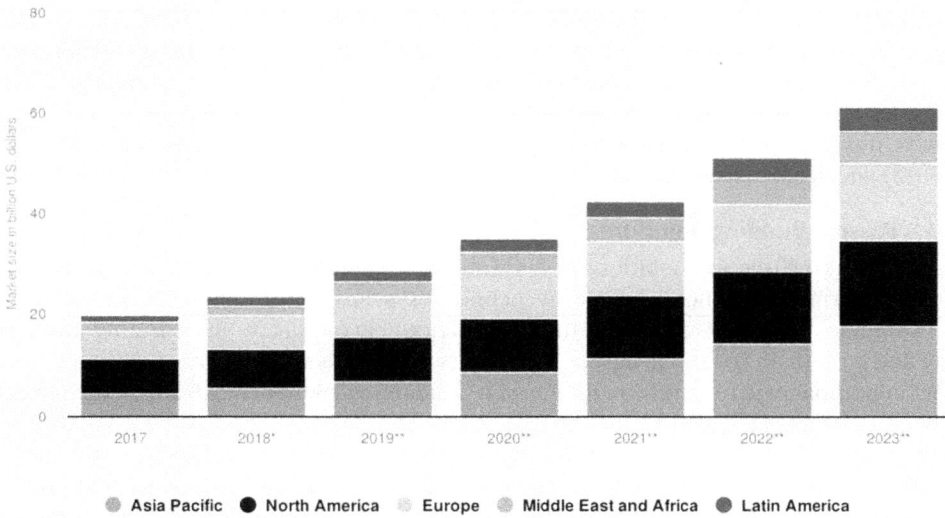

**FIGURE 7.11**  Market value of smart grids worldwide from 2017 to 2023, by region. (Adapted from Statista Market value of smart grids worldwide from 2017 to 2023, by region, 2021, https://www.statista.com/statistics/246154/global-smart-grid-market-size-by-region/ (Accessed September 29, 2021).)

R&D efforts are still required in many fields to investigate new options and features as well as their integration and interoperability within the grid.

### 7.2.4 SUSTAINABLE DESIGN APPROACH

Using a combination of available technologies and approaches, sustainable (or even better zero-energy) buildings can be already widely achieved. The best-practice approach, using energy as efficiently as possible and meeting the energy needs by renewable energy supplies, can be adopted and applied in new and existing buildings across countries with diverse applications. This will result in different combinations of solutions that are appropriate for different specific contexts, while enabling buildings that are fit for purpose, future-proofed, and resilient to climate change effects.

Nevertheless, in spite of the abundance of established or innovative solutions and practices, their systemic and synergistic integration to obtain optimal results is still difficult to fully reach.

In this sense, the planning, estimation, and design phases are crucial, but it is very difficult to establish precise rules to be followed, valid in all cases. However, it is still possible to outline basic indications, to be declined specifically on a case-by-case basis.

First of all, it is essential to adopt an integrated project approach, in which the various professional skills dialogue and interact in order to find synergistic solutions. Buildings are complex organisms and must necessarily be designed according to a complex and multidisciplinary logic.

Secondly, the Life Cycle Analysis (LCA) should always be taken into account. LCA allows the quantification of potential environmental and human health impacts associated with a good or service, starting from resource consumption (energy included) and associated emissions. It takes into account the entire life cycle of the analyzed system, from the acquisition of raw materials to end-of-life disposal, including manufacturing, distribution, and operation phases. Having this kind of data available, it would be paradoxical to design a high-efficiency building, the construction of which costs more energy than the energy consumed during its whole use. This statement is not self-evident, since many high-performance technological components have considerable energy content.

Furthermore, if energy performance and efficiency are prioritized, it is necessary to use design tools capable of assessing in detail the repercussions of each material, technological or installation choice, in terms of consumption and emissions. Accurate and reliable energy simulation tools are commonly available on the market, but they must be used in a competent and well-balanced way to avoid misleading predictions and decisions.

Finally, it is possible to operate according to some priorities (De Groote and Fabbri 2019, Butera et al. 2015) which are considered valid with due adaptation to each different context, as follows.

### 7.2.4.1 Passive Building Solutions

Generally speaking, passive solutions in buildings are technologies that are able to provide the required performance without the use of technical systems. All the passive technologies and design strategies aimed to reduce the thermal energy demand due to the building envelope for heating and cooling in buildings have to be put into practice before any other measure, ensuring comfort conditions with the minimum possible amount of energy. In detail, in countries with large heating loads, advanced insulation and proper architectural choices are pivotal to reduce thermal losses. In hot climates, where cooling loads are set to increase substantially, the reduction of the loads can be done through many low-cost and local components, such as cool roofs, shading systems, and high thermal inertia materials (e.g. stone or bricks). Heat flows can be better managed, preferably through natural ventilation or with mechanical ventilation systems that manage airflow and reduce unnecessary cooling energy demand. In mixed climates with both space heating and cooling loads, the multiplicity of seasonal constraints requires solutions to address both. For example, low-emissivity windows can reflect solar radiation during the summer to minimize heat gain as well as reflect radiative heat from the inside during winter to minimize heat loss.

### 7.2.4.2 High-Efficiency Technical and Control Systems

Once the building's energy demand for heating, cooling, and lighting has been minimized thanks to the envelope's configuration, there is still a not negligible amount of energy to be supplied through the technical systems to meet the internal comfort needs. Of course, these plants must be highly efficient and compatible with Renewable Energy Sources (RES) exploitation. A good example of this are heat pumps, which can use photovoltaic and geothermal energy at the same time, with increasingly high coefficients of performance over time.

In order to maximize their potential, these installations must be coupled with the introduction of control/management systems, able to optimize the operational strategy of the building. This can be as simple as programming gradual power ramp-up before building occupants wake in the morning, or more complex, by using artificial intelligence tools to properly manage load profiles. A sample of advanced HAVAC system is shown hereafter (Figures 7.12 and 7.13).

### 7.2.4.3 Renewable Energy Production and Self-Consumption

In a zero-energy building, RES systems are installed to completely cover or also exceed the total energy demand with the minimum exchange of energy with the grid (thus stimulating energy management, storage, and exchange at the district level). This is the most ambitious objective, but, for the time being, it may also be sufficient to cover a significant part of the consumption, as in the case of nZEB, where self-production averages around 50%.

### 7.2.5 Strategic Technologies

In the previous section, methods and strategies to be applied to enable the energy transition and efficiency in the buildings sector have been explained. As a corollary to these guidelines, a number of technological solutions (IREA 2019, Bean et al. 2017) strategic to enable their implementation, are set out below.

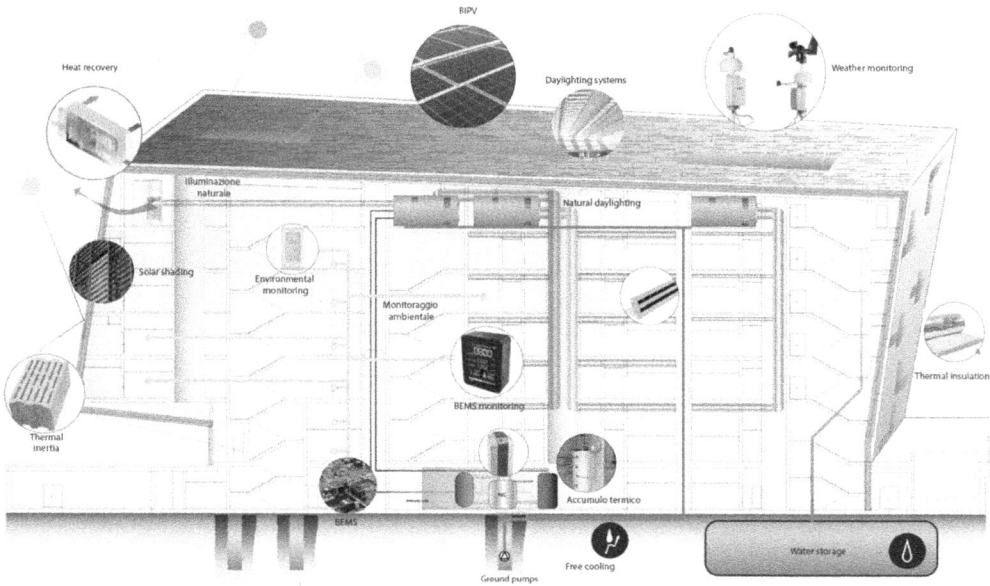

**FIGURE 7.12** Advanced HVAC system and control.

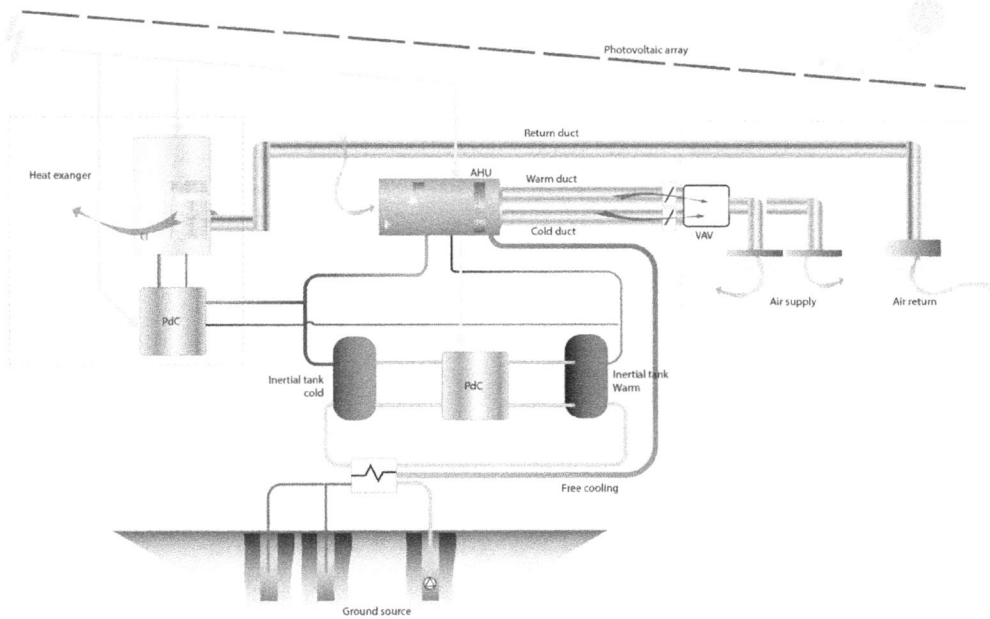

**FIGURE 7.13** Advanced HVAC concept.

### 7.2.5.1 Sustainable Envelope Technologies

A material or component is defined as sustainable if it substantially reduces impacts and energy consumption. As already introduced, the energy transition of the buildings sector must not only address consumption in the operational phase, but its entire life cycle starting from the construction. In this sense, efficient, low-energy, and possibly recyclable/reusable materials and components should be used, such as high-performance selective glazing (Aste et al. 2018), and opaque materials with high thermal insulation and inertia (Aste et al. 2015), solar reflective materials, etc. As buildings become more efficient and the grid decarbonizes, embodied carbon (that is the carbon associated with materials extraction, transport, and manufacture) increases in significance; for such reason, just low-carbon materials should be used, thus preferring local/natural materials and—more in general—those with the lowest EE (Butera et al. 2015).

### 7.2.5.2 Building Integrated RES

As consumption nodes in a distributed generation perspective, buildings must produce as much energy as possible to meet their own needs. This is possible through the integration of renewable energy systems, mainly solar, wind, biomass, geothermal, and hydrothermal, depending on the availability of the specific context.

In general, the most widespread source that can be easily exploited in urban areas is solar energy, which can be converted into electricity (photovoltaics, PV), heat (solar thermal), or both electricity and thermal energy (less diffused hybrid photovoltaic/thermal technology, PVT).

At present and with respect to future projections, PV represents the most effective technology, as they directly produce electricity, can be applied in any context, can be easily integrated into roofs and facades, and are economically competitive (several countries have already reached or exceeded the grid parity) (Figure 7.14).

### 7.2.5.3 Heat Pumps

Besides the fact that they run mainly on electricity and are therefore compatible with most RES, heat pumps can be considered the most versatile and high-efficiency solution for space heating, cooling, and Domestic Hot Water (DHW) preparation. In fact, a heat pump has the potential to deliver significantly more energy than that is used to drive it, by exchanging thermal energy with the air, ground, water bodies, and sources of waste heat. Electrically driven heat pumps currently available on the market can easily deliver three to five units of energy for every one unit of electricity consumed. Heat pumps can be synergistically integrated with PV ad PVT technologies, obtaining solar-assisted heat pump systems. Heat pumps size ranges from small capacities (less than 1 kW, for a single-room application) to several MW (to serve large buildings/districts).

**FIGURE 7.14** BIPV integration example. (From Aste et al., Energy retrofit of commercial buildings: A case study on ERGO building. *International Conference on Energy Efficiency in Commercial Buildings IEECB'12*, 392–401, 2012. With permission; From Aste et al., nZEB: bridging the gap between design forecast and actual performance data. *Energy and Built Environment*, 2020. With permission.)

### 7.2.5.4   Building Automation

Home automation has many applications, but in the energy field, it essentially consists of EMS (energy management systems) with demand-side management capabilities. Building automation and controls coupled with smart metering allow buildings to react to the occupants' needs (user response) and to external signals (grid response); the energy use of the building should be continuously optimized, by ensuring that it is used only when and where necessary and that all technical building systems are properly integrated. Moreover, all consumers should be allowed to feed into the grid the electricity they generate but do not use and/or participate in demand-response activities. This will also enable synergies between smart buildings and electric vehicles. Cloud-based EMS must be preferred since they ease the application of district/city-scale management strategies, coordinating multiple buildings.

### 7.2.5.5   Energy Storage

To strengthen the energy transition of the buildings sector, most of the self-produced energy must be consumed on-site. This means consumption at the same time as production (preferable option) or deferred through storage systems. In this sense, energy storage possibilities should be encouraged in buildings and districts. Where possible, for example in urban areas, clustering buildings within local districts with common energy storage could enhance smart grid functionality and reduce operating costs. The market uptake of energy storage has so far been limited due to high prices (especially for electrical storage); however, economies of scale are leading to significantly reduced costs.

It should also be noted that, since the main final energy use in buildings is for thermal purposes, sensible thermal energy storages are particularly interesting when coupled with heat pumps and solar technologies, as they allow for efficient and low-cost accumulation of already converted electricity.

### 7.2.5.6   Energy Networks

Thermal networks and advanced electrical grids for smart districts/cities will have a key role in the energy transition. In fact, they enable the interconnections of distributed energy resources (renewables, combined heat, power generators, etc.), storage systems, and loads (electrical and thermal), balancing supply and demand locally, creating a multi-energy system (MES).

Such MESs increase the overall efficiency of a district or a group of buildings, by allowing waste heat recovery for heating, cooling, and domestic hot water applications, and optimal local dispatching of both thermal energy and electricity.

Innovative thermal and electrical MES must be designed with the specific aim to integrate multiple renewable energy sources (i.e. PV and geothermal energy) and energy storage technologies (i.e. sensible thermal storage, batteries, and hydrogen), and different types of thermal and electrical loads (space heating, space cooling, domestic hot water preparation, appliances, electric vehicles, etc.).

Such MES can represent a key solution not only for new urbanizations but also for the energy retrofit of existing buildings (e.g. by distributing groundwater to existing buildings, thus enabling the substitution of fossil fuels boilers with high-efficiency water-to-water heat pumps).

In parallel, there is a huge need to efficiently manage the local energy distribution, avoiding or minimizing the overload of the electrical system, with considerable advantages, such as:

*   significant reduction of the impact on the grid of the electric renewable energy sources;
*   minimization of the peak electricity demand due to the electrification of both the building and the transportation sectors.

## 7.3   ACTIONS

In the previous section, the role of the world's building stock within the global energy scenario has been clarified and the most promising solutions and strategies to enable the energy transition in this area have been identified. The opportunities available are therefore manifold, but a number of concrete actions need to be taken for them to be implemented. In this sense, the main areas of intervention are outlined below.

### 7.3.1  RESEARCH, DEVELOPMENT, AND APPLICATION

The buildings sector must be renewed, but above all, it must be innovated. Recent developments have shown how technologies and concepts that were only experimental until recently, can provide concrete and effective solutions. Consider the parable of PV, which until the turn of the century was an interesting but too expensive technology, and is now one of the leading solutions for the energy transition. The worlds of research and industry must therefore continue along the path of innovation, studying and developing solutions capable of exploiting technological progress and offering solutions that are always up-to-date and adapted to the specific moment.

Efforts must then be intensified so that these solutions find the correct application and are compatible with the specific needs and requirements of the sector. Innovation is integrated into buildings, not superimposed.

### 7.3.2  CULTURAL SWITCH

The construction sector has always been characterized by certain inertia toward change, in favor of maintaining established practices and techniques. Moreover, the logic of the market is generally speculative in the short term and hardly oriented to collective sustainability. Finally, it should be remembered that architecture is more and more perceived as an expressive art, in which form prevails over function and functionality.

It is therefore clear that a change in building and architectural culture is needed, as has happened in other sectors. If it is the market that strongly directs choices, it is necessary to operate on the market and especially on end users/buyers, through marketing operations capable of promoting energy efficiency and sustainability and depreciating old energy-consuming traditions.

No one wants to be treated as "Mr Climate-Dirty building" (Tapper 2019)…

### 7.3.3  CAPACITY BUILDING

Closely related to the previous one is the issue of training and updating professionals, technicians, and stakeholders in general. Constructing buildings according to innovative concepts requires new knowledge and new skills, which cannot be acquired roughly for the sole purpose of surviving on the market. All too often, riding on the coattails of fashion, mediocre architecture is passed off as sustainable in order to increase its value.

It is, therefore, necessary to strengthen the training and awareness of the sector operators, so that designers, technicians, laborers, etc. acquire the necessary skills to create new generation buildings that can effectively guarantee the promise of energy-saving and efficiency.

### 7.3.4  POLICY

Climate change, environmental crisis, and energy transition are all issues that concern the whole community. Then it is those who govern the community who must take responsibility for it. As has been the case for several decades, economic, building, and energy policies must aim to support sustainability and energy transition. However, in order to achieve tangible results in the short term, the interrelationships between the various sectors and the coordination between different countries must be intensified so as to act according to a global vision.

For example, legislative frameworks must be capable of progressively making measures to enable energy transition compulsory, but must also assume responsibility for regulating the sector appropriately (permits, authorizations, concessions, etc.) to prevent bureaucratic barriers from slowing down change. In the building and energy sectors, there are many actors with different and often conflicting interests. Politicians must strive to find the right balance so that they converge in a single direction: that of the energy transition.

### 7.3.5 INCENTIVES AND FINANCING

In general, within the buildings sector, energy efficiency, sustainability, and renewables are not clearly perceived as cost-effective. Actually, we know that this is not the case, but the benefits of the related measures are realized more slowly than in the rather quick time frame of the building market.

Even if not immediate, the advantages are real, including strictly economic ones, and above all, they benefit the whole community. However, action must be taken on individuals, combining obligations with measures to encourage virtuous choices. Actions such as tax relief, low-interest loans, and feed-in tariffs have already proved their worth, but have also proved to be fragmented and uncoordinated. In this sense, medium- and long-term promotion and incentive strategies, appropriately planned and supported by governments, would be a fundamental tool.

Another way forward could be energy efficiency finance. Investments in energy efficiency in buildings can have returns of up to 20% per year. However, the distributed and fragmented framework, the uncertainties related to usage patterns and occupants, as well as the lack of really specialized operators, relegate the related economic operations to a secondary level.

Compared to traditional speculative finance (which by the way is not extraneous to the global crisis), it would be appropriate to strengthen this new branch, in which the profits of individual investors would be combined with considerable benefits on a global scale.

## 7.4 CONCLUSIONS

Operational energy demand of the buildings sector currently accounts for the largest share of both global final energy use (35%) and energy-related $CO_2$ emissions (38%). The total world primary energy demand of existing buildings (245 billion m²), equal to 36,000 TWh/y, is subjected to a slow decrease thanks to energy retrofit and renovation interventions. In particular, considering the different energy-saving potentials mentioned in Section 7.2, it is possible to precautionary assume that the world average saving on retrofitted buildings from now until 2050 will be 1/3 of their energy demand. Since the average yearly retrofit rate of existing constructions is 1.5%, and taking into account also the quota of demolished buildings (around 2.3 billion m² per year), the expected total primary energy saving related to the exiting stock will reach 14,000 TWh/y in 2050. However, by 2050, the built floor area worldwide is expected to increase by 210 billion m² (+85% in respect to the current stock), most of which (three-quarters) in emerging markets and developing economies. With a conservative assumption that new constructions worldwide will have an energy demand that is 50% of the current mean value (with lower percentages in countries where nZEB regulations are in force but higher values in emerging economies), the impact in 2050 of all new buildings could be roughly estimated equal to 15,000 TWh/y, which has the same order of magnitude of the before-mentioned total saving.

In summary, according to the scenario envisaged, the total world primary energy demand of the buildings sector in 2050 could remain unchanged from the current situation. Savings are due to increased efficiency and renewable, in line with the energy transition.

Although the operational phase of buildings has the greatest impact on energy demand and related emissions (at present around two-thirds of the total), it must be noted that also the remaining one-third resulting from EE must be accounted for. This quota is also particularly relevant since is entirely consumed at the stage of material production, transportation, and construction. For reference purpose only, assuming an average EE of a new building equal to 1500 kWh/m², the constructions being built in 2050 (around 7 billion m²) will require an additional 10,500 TWh. Since such a quota is related to industry and transportation sectors, a radical change in manufacturing processes (e.g. strongly increasing RES exploitation) and selection of construction materials (e.g. promoting bio-based, recycled, and/or local materials to abate the embodied fraction) is needed.

In parallel, sustainable architecture, energy-efficient technical solutions but also focused policies and incentives can be considered key drivers both to increase the yearly retrofit rate of existing buildings and minimizing the operating energy demand in the sector.

## REFERENCES

Al Dakheel, J., Del Pero, C., Aste, N., & Leonforte, F., 2020. Smart buildings features and key performance indicators: A review. *Sustainable Cities and Society*, 61, 102328.

Aste, N., Adhikari, R. S., Buzzetti, M., Del Pero, C., Huerto-Cardenas, H. E., Leonforte, F., & Miglioli, A., 2020. nZEB: bridging the gap between design forecast and actual performance data. *Energy and Built Environment*, pp. 16–29.

Aste, N., Adhikari, R. S., Del Pero, C., & Tagliabue, L. C., 2012. Energy retrofit of commercial buildings: A case study on ERGO building. In *International Conference on Energy Efficiency in Commercial Buildings IEECB'12*, Frankfurt, Germany (pp. 392–401).

Aste, N., Buzzetti, M., Del Pero, C., & Leonforte, F., 2018. Glazing's techno-economic performance: A comparison of window features in office buildings in different climates. *Energy and Buildings*, 159, 123–135.

Aste, N., Leonforte, F., Manfren, M., & Mazzon, M., 2015. Thermal inertia and energy efficiency–Parametric simulation assessment on a calibrated case study. *Applied Energy*, 145, 111–123.

Australian Government Department Of Defence, 2016. *Energy for Australia in the 21st Century: The central Role of Electricity*, Camberra.

Birmingham Energy Institute, 2015, *Doing Cold Smarter, University of* Birmingham, *Edgbaston, Birmingham*, United Kingdom.

Bean, F., De Groote, M., & Volt, J., 2017. *Opening the Door to Smart Buildings*. Buildings Performance Institute Europe (BPIE), Brussels.

Buckman, A. H., Mayfield, M., & Beck, S. B., 2014. What is a smart building? *Smart and Sustainable Built Environment*, 3(2), 92–109.

Butera, F., Adhikari, R. S., & Aste, N., 2015. Sustainable building design for tropical climates. UNON, Publishing Services Section, Nairobi.

Chastas, P., Theodosiou, T., & Bikas, D., 2016. Embodied energy in residential buildings-towards the nearly zero energy building: A literature review. *Building and Environment*, 105, 267–282.

C40 team, 2018. *Consumption-Based GHG Emissions of C40 Cities.* https://cdn.locomotive.works/sites/5ab410c8a2f42204838f797e/content_entry5ab410fb74c4833febe6c81a/5ad4c0c274c4837def5d3b91/files/C40_GHGE-Report_040518.pdf?1540555698

De Groote, M., & Fabbri, M., 2019. *Smart Buildings in a Decarbonised Energy System.* Buildings Performance Institute Europe (BPIE). https://www.bpie.eu/wp-content/uploads/2016/11/BPIE-10-principles-final.pdf

Gangale, F., Vasiljevska, J., Covrig, C., Mengolini, A., & Fulli, G., 2017. Smart grid projects outlook 2017: Publications Office of the European Union, JRC106796, https://publications.jrc.ec.europa.eu/repository/handle/JRC106796.

Global Alliance for Buildings and Construction, International Energy Agency and the United Nations Environment Programme,2019. *2019 Global Status Report for Buildings and Construction*: Towards a zero-emission, efficient and resilient buildings and construction sector. https://iea.blob.core.windows.net/assets/3da9daf9-ef75-4a37-b3da-a09224e299dc/2019_Global_Status_Report_for_Buildings_and_Construction.pdf

EIA, 2021. *Annual Energy Review.* https://www.eia.gov/totalenergy/data/annual/ (Accessed September 29, 2021)

EIA, 2018. *Energy Implications of Higher Economic Growth in Africa.* U.S. Department of Energy Washington.

European Commission, 2019, *Comprehensive Study of building Energy Renovation Activities and the Uptake of Nearly Zero-Energy Buildings in the EU.* Ipsos Belgium, Brussels.

European Commission, Department of Energy, 2020. *In Focus: Energy Efficiency in Buildings.* European Commission, Brussels.

He, Q., Hossain, M. U., Ng, S. T., Skitmore, M., & Augenbroe, G., 2021. A cost-effective building retrofit decision-making model–Example of China's temperate and mixed climate zones. *Journal of Cleaner Production*, 280, 124370.

Hussaini, M., Hamza, D., & Usman, M. 2018. Assessment of consumption rate of solid biomass fuels and the consequent environmental impact in Maiduguri metropolis. *Open Journal of Air Pollution*, 7, 34–47.

Ibn-Mohammed, T., Greenough, R., Taylor, S., Ozawa-Meida, L., & Acquaye, A., 2013. Operational vs. embodied emissions in buildings—A review of current trends. *Energy and Buildings*, 66, 232–245.

IEA, 2013. *Transition to Sustainable Buildings. Strategies and Opportunities to 2050.* IEA Publications, 9 rue de la Fédération, 75739 PARIS CEDEX 15.

IEA, 2017. *Deep Energy Retrofit – Case Studies Business and Technical Concepts for Deep Energy Retrofit of Public Buildings. Energy in Buildings and Communities Programme. Annex 61, Subtask A.* New Buildings Institute (NBI).

IEA, 2019a. *Global Status Report for Buildings and Construction 2019.*

IEA, 2019b. *Material Efficiency in Clean Energy Transitions.* IEA Publications.

IEA, 2019c. *Perspectives for the Clean Energy Transition 2019.* IEA Publications.

IEA, 2019d. *Material Efficiency in Clean Energy Transitions.* IEA Publications.

IEA, 2019e. *Perspectives for the Clean Energy Transition. The Critical Role of Buildings.* IEA Publications.

IEA, 2020a. *Global Status Report for Buildings and Construction 2020.* Nairobi.

IEA, 2020b. *Energy Technology Perspectives 2020.* IEA Publications.

IEA, 2021. *Building Envelopes*, online: https://www.iea.org/reports/building-envelopes (Accessed September 29, 2021).

Ingrao, C., Messineo, A., Beltramo, R., Yigitcanlar, T., & Ioppolo, G., 2018. How can life cycle thinking support sustainability of buildings? Investigating life cycle assessment applications for energy efficiency and environmental performance. *Journal of Cleaner Production*, 201, 556–569. Doi: 10.1016/j.jclepro.2018.08.080.

International Renewable Energy Agency, 2019. *Global Energy Transformation. A roadmap to 2050.* International Renewable Energy Agency, Abu Dhabi.

Kalua, A., 2020. Urban residential building energy consumption by end-use in Malawi. *Buildings*, 10(2), 31.

Koezjakov, A., Urge-Vorsatz, D., Crijns-Graus, W., & Van den Broek, M., 2018. The relationship between operational energy demand and embodied energy in Dutch residential buildings. *Energy and Buildings*, 165, 233–245.

Liu, Z., Zhou, Q., Tian, Z., He, B. J., & Jin, G., 2019. A comprehensive analysis on definitions, development, and policies of nearly zero energy buildings in China. *Renewable and Sustainable Energy Reviews*, 114, 109314.

Musango, J. K., Currie, P., & Robinson, B., 2017. *Urban Metabolism for Resource Efficient Cities: From Theory to Implementation.* UN Environment: Paris.

Nebel, B., Alcorn, A., & Wittstock, B., 2008. *Life Cycle Assessment: Adopting and Adapting Overseas LCA Data and Methodologies for Building Materials in New Zealand.* Ministry of Agriculture and Forestry.

New Buildings Institute, 2020. *2020 Getting to Zero Buildings List.* New Buildings Institute, Portland.

Plank, R., 2008. The principles of sustainable construction, *IES Journal Part A* 1 (4) 301–307.

Sartori, I., & Hestnes, A.G., 2007. Energy use in the life cycle of conventional and low-energy buildings: A review article. *Energy and Buildings*, 39(3), 249–257.

Sirviö, A., & Illikainen, K., 2015. *Sustainable Buildings for the High North. Energy Performance, Technologies and Challenges of New Buildings in Russia and Scandinavia.* http://www.oamk.fi/epooki/2015/sustainable-buildings-high-north-energy-performance-technologies-and-challenges-new-buildings-russia-and-scandinavia/

Statista, Market value of smart grids worldwide from 2017 to 2023, by region. https://www.statista.com/statistics/246154/global-smart-grid-market-size-by-region/ (Accessed September 29, 2021).

Tapper, J., 2019, 28 July. Experts call for ban on glass skyscrapers to save energy in climate crisis. *The Guardian*, https://www.theguardian.com/environment/2019/jul/28/ban-all-glass-skscrapers-to-save-energy-in-climate-crisis.

Toosi, H. A., Lavagna, M., Leonforte, F., Del Pero, C., & Aste, N., 2020. Life cycle sustainability assessment in building energy retrofitting; A review. *Sustainable Cities and Society*, 60, 102248.

UN-Habitat, 2006, *Nairobi Urban Sector Profile.*

Ürge-Vorsatz, D., Eyre, N., Graham, P., Harvey, D., Hertwich, E., Jiang, Y., Jochem, E., 2012. Chapter 10- energy end-use: Buildings. In *Global Energy Assessment Writing Team, Global Energy Assessment: Toward a Sustainable Future* (pp. 649–760), Cambridge University Press, Doi: 10.1017/CBO9780511793677.016.

Van Noppen, A, 2013. *The ABC's of Affordable Housing in Kenya.* https://acumen.org/wp-content/uploads/2013/03/ABCs-of-Affordable-Housing-in-Kenya.pdf

WRI Ross Centre for Sustainable Cities, World Resources Institute (WRI), 2016. *Accelerating Building Efficiency – Eight Actions for Urban Leaders*, https://publications.wri.org/buildingefficiency/.

World Bank, 2014. *Energy Efficiency in Russia: Untapped Reserves.* https://documents1.worldbank.org/curated/en/573971468107682519/pdf/469360WP0Box331C10EE1in1Russia1engl.pdf

Yang, X., Zhang, S., & Xu, W., 2019. Impact of zero energy buildings on medium-to-long term building energy consumption in China. *Energy Policy*, 129, 574–586.

ZEBRA, 2020. Data tool. https://zebra-monitoring.enerdata.net/nzeb-activities/panel-distribution.html (Accessed September 29, 2021).

Zhang, S., Xu, W., Wang, K., Feng, W., Athienitis, A., Hua, G., … Lyu, Y., 2020. Scenarios of energy reduction potential of zero energy building promotion in the Asia-Pacific region to year 2050. *Energy*, 213, 118792.

IRENA (2019), Global energy transformation: The REmap transition pathway (Background report to 2019 edition), International Renewable Energy Agency, Abu Dhabi.

# 8 Electric Vehicles and Future of Transport Sector

*Hüseyin Turan Arat*
Sinop University

*Mustafa Kaan Baltacioğlu and Cağlar Conker*
İskenderun Technical University

## CONTENTS

## 8.1 BASIC DESCRIPTIONS, HISTORY, AND DEFINITION OF ELECTRIC VEHICLE

Everywhere in our lives, we sincerely use tools and apparatuses that we take for granted and which we often do not think of, but which "make people's lives easier" sincerely as a source of power; electricity. Undoubtedly, it is one of the most important discoveries in the world.

On the other hand, the age of "transportation by machines" started with the discovery of steam engines. Fossil resources were limited at that time, and phase 3 of the industrial revolution was trying to obtain asphalt roads everywhere and put a lot of vehicles on them. Numerically, the "transport sector" is responsible for 28% of the world's emissions nowadays [1].

The state-of-the-art vehicle has just reached its autonomous features this year (detailed information will be given under the next headings). Travel needs power in order for this completely user-friendly vehicle technology to perform its most important task. It needs energy to drive and generate power. In new vehicles, this power is provided by an electrical storage device. It is called "battery." The power in this battery then uses the electric-powered motor to turn the wheels and generate "motion." This system, which we call the electric motor (EM), was invented in the early 1900s. Figure 8.1 depicts a toy-to-reality electric vehicle (EV). Similarly, the battery in a remote-control toy car you played with when you were young runs the toy by transferring its energy to small gears or a dynamo. It can also be controlled manually with the signal receiver. When its size and carrying capacity are increased, this time it turns into a battery-powered car. This time, the battery capacity, the load it will carry and the electric motor capacity increase. Finally, it turns into a vehicle that enables travel safely for a child and a family.

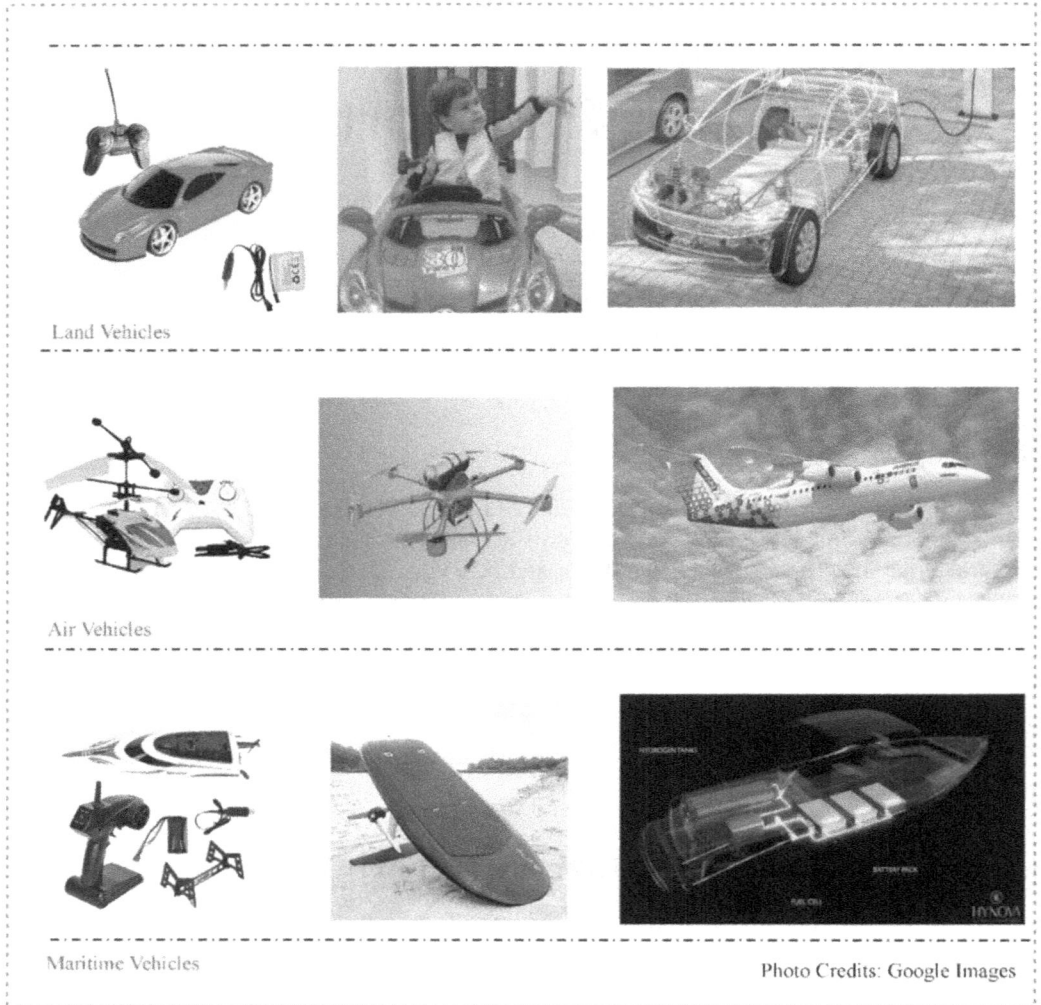

**FIGURE 8.1** Toy to vehicle of EVs.

Each phenomenon has to live its own time in order to be realized. The first discovery of EVs was also within such a period of time. While it has been 200 years for EVs to start mass production, the first studies were undoubtedly realized with the discovery of the EM. When Michael Faraday first created "motion" from a moving wire and magnet, the studies gained momentum in this direction. Anyos Jendlik is one of the important names in EM studies. Thomas Davenport obtained the first EM patent in 1830–1834. The "propulsion apparatus" of the EV was found, but this discovery was not used in EVs at that time. Because there was a need for an energy storage system to "run" this engine. When the dates show the year 1859, Gaston Plante found the lead-acid battery [2]. With this invention, both the energy provider, storing and the energy transmitter have been introduced.

At that time, studies were continuing all over the world, but the state of technology and access to information was not "one click away" as it is today. Prof. Sibrandus Stratingh and his assistant Becker, in 1834, with a total weight of 3 kg; developed a vehicle that could move for 20 minutes and had a carrying capacity of 1.5 kg. Later, with the development of EMs and batteries, interest in EVs suddenly increased. Especially between the years 1860 and 1908, EVs would be preferred because of the vibration, noise, and prices of internal combustion engines. Thomas Parker in 1884, Henry G. Morris in 1894 (electrobat), and Willam Morrision in 1895 (with his six-seater EV) are among the names that signed this sector [2].

In addition to these developments, again between the years 1870 and 1910, the process of "combining different components and upgrading the system." that is, providing movement by using the internal combustion engine and EM together, is performed and named "hybrid electric vehicle" (HEV) by William H. Phaton and Ferdinand Porshe [2].

While turning the energy of vehicles into motion with electric propulsion, the trains used in railway work with electricity became quite popular, and even while academic and engineering studies were continuing without slowing down, the "silence period" for EVs began in 1908. Henry Ford launched the Model T into mass production; it disrupted the supply and demand balances and switched to a system that would completely change the automotive market. Industry 2.0 forced EVs to move to a silent period. Undoubtedly, the most important reason for this has been financial concerns. Until both the second industrial period and the oil crisis broke out in this period between 1910 and 1970, vehicle prices were almost three times higher than their equivalent internal combustion vehicles (circa; 650 $ for ICE vehicle and 1750 $ for EV).

The break of silence during this period was triggered by the oil crisis, the Clean Air act, and space exploration necessitating electrical and battery systems. If we define the years 1970–1997–2008 as the millennium years for EVs, it will not be wrong. Because in 1970 and 1971, the Clean Air act and NASA Luna Rover emerged, respectively, and later, automotive companies started both hybrid and battery EV (BEV) prototypes. In 1997, the first mass-produced HEV, the Toyota Prius, took its place on the road. Significant gains have been achieved from HEVs, which act as a technological bridge between EVs and internal combustion engines. When the calendars showed 2008, the Tesla Roadstar with IM (induction motor) drive mechanism, equipped with lithium-ion batteries which have a 350 km driving range and 200 km speed, was launched.

Ever since this period, the EV sector, which has increased its momentum with both the Paris agreement and the Euro emission standards, shapes the vehicle market of the future with an unstoppable state and potential. The most striking example is that while there were 30 companies in the world established in 2008 regarding EVs and apparatus, nowadays there are 400+ companies only in China, according to the Center for Strategic and International Studies [3]. Again, according to Global EV Outlook 2020 [4], while the global home car stock was 0.2 million in 2010, it was 7.2 million vehicles in 2019.

After these historical developments, the definition of the basic question "What is an EV?" can be expressed in the simplest terms, "an electric vehicle is an electric-powered car." If it will be expressed in more technical detail, it is defined as "the general name of the vehicles that provide movement through the wheels of the rotational movement obtained by the evolution of the electrical energy taken from the batteries and running the EM." When talking about an EV, it should first show the structural design of this vehicle. Although we will go into much more detail in the next chapters, the main components of an average EV and their comparison with other vehicles are shown in Figure 8.2 [5]. Usually, it consists of a motor, battery, management systems control unit, charging systems, and reducer (for things like a transmission). Although these are the main descriptions of the systems, more detailed information will be given in the next sections.

The first thing customers feel when they get into an EV is the silence of the vehicle. Driving dynamism is highly facilitated electronically. It has become quite preferable for user convenience. Especially compared to the early years of EVs, the "short-range" issue is also being solved with each new generation produced.

The most important advantages of an EV will be that it will be environmentally friendly, silent, can be recharged from home, and use relatively less energy because it contains fewer mechanical parts, have fewer maintenance costs, and provide convenience in urban use, etc. Along with these, one of the important disadvantages that should be known is that the charging duration is still not as short as conventional vehicles (<7 minutes), the battery temperature system should be controlled

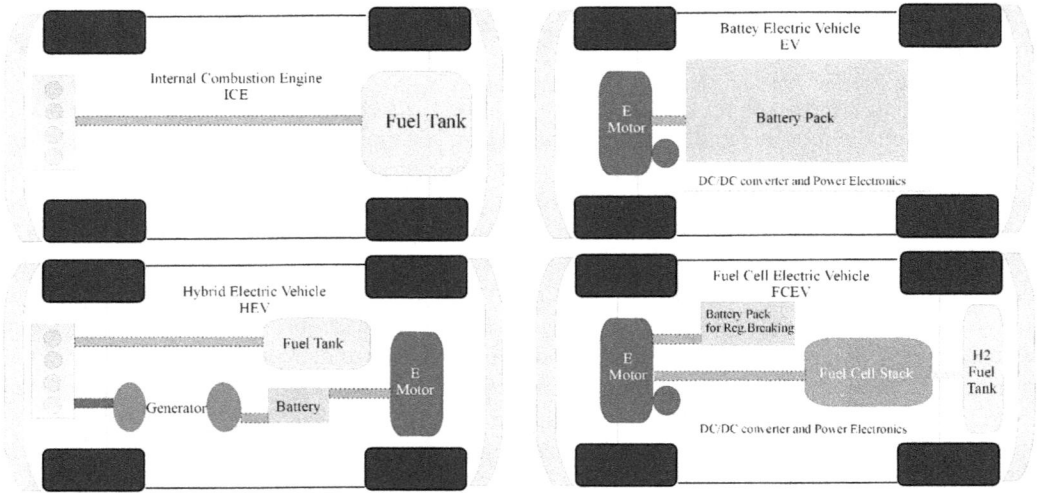

**FIGURE 8.2** Comparative components details of vehicles. (Adopted from Ref. [5].)

well, and it is not completely compared under variable weather and road conditions. Also, as an emission reducer, in this book the authors split EVs into two colors: Dark gray EVs and Light gray EVs.

EVs (BEV, FCEV (Fuel Cell EV), FCHEV (Fuel Cell Hybrid EV)) do not generate emissions while driving, like HEVs and ICEs. Because EVs do not need fossil fuel combustion to convert chemical energy into work. But it is important through that which power plants obtain the electrical energy to EVs for they fill in their batteries before driving. Dark gray EVs take the electricity from the main lines (grid) that are assumed to be provided by fossil-based power plants. Although this issue is not part of the travel, it causes emissions in electrical power generation. Light gray EVs are defined as EVs that provide the electrical energy to be used from alternative and clean energy sources.

## 8.2   STATE OF ART ON EVs TECHNOLOGY AND MERITS & DEMERITS OF EVs

As an engineering approach, if things are efficient, powerful, durable, and cheap, then they are more preferable than others. From this point of view, it seemed much more reasonable to light up the room with a bulb instead of using a kerosene lamp. So why did the "age of silence" arise for the EVs in the previous episode? Yes, the answer is obvious; capital. Like all systems, as technology develops and demand increases, the price of the product produced becomes more affordable due to overproduction. If we assume that they have almost the same service life and initial investment costs, only power and efficiency remain reasons for preference. Figure 8.3 shows the WTW efficiency analysis picture taken from another study by the authors [6].

Well to Wheel analysis offers you an efficiency comparison from the generation of energy to the consumption of it. In EVs, unlike ICEs, there is not a huge amount of heat energy loss. Therefore, it can be said that EVs are more efficient compared to these vehicles. (This comparison is the result of a direct energy efficiency comparison. The production and cost of the energy source are excluded from the evaluation.)

The evolution of common transport engineering devices is from air to land and from land to sea. To make it more understandable, new technology is both costly and valuable in terms of usage for early adopters. That's why technologies with high added values are used in future-oriented products as the technology of the future. For example, hydrogen fuel was used in space exploration for the first time. It is the same for electrical propulsion. First of all, products are evaluated in the fields

## Well to wheel (WTW) Efficiencies

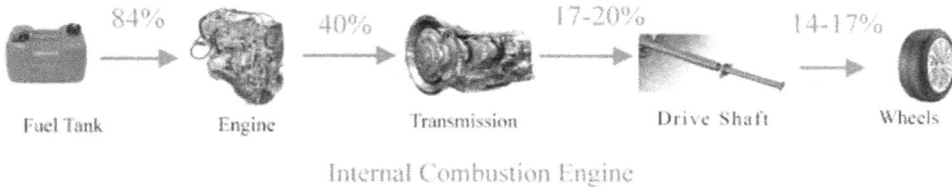

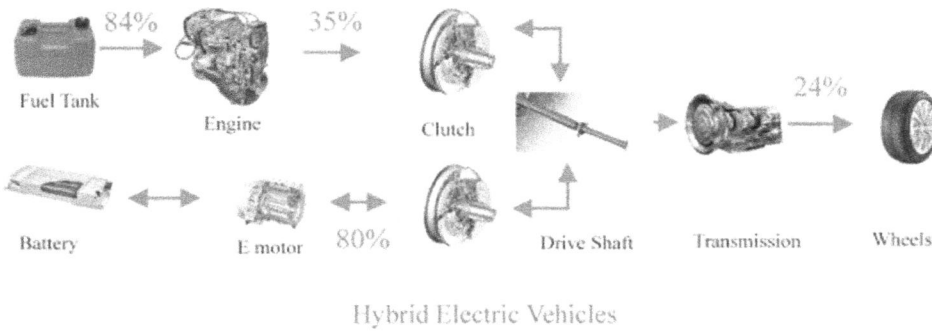

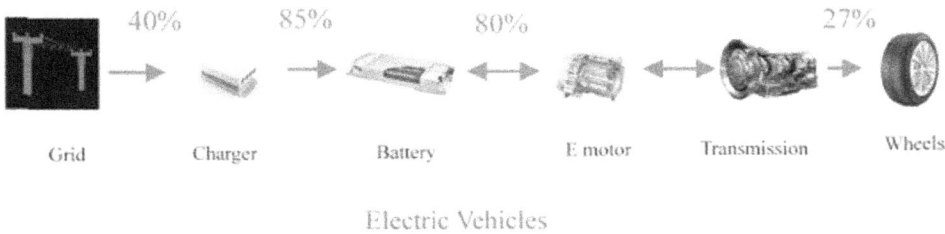

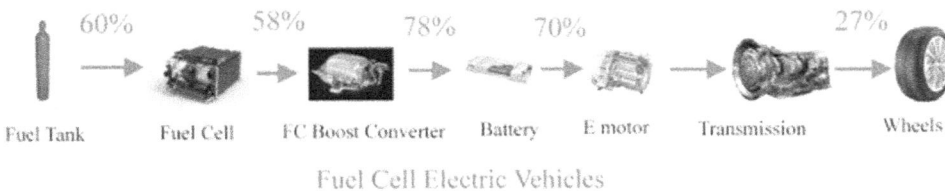

**FIGURE 8.3** WTW analyses. (From Ref. [6]. With permission.)

of space and aviation. Then, after the price breakdown, they are applied to land vehicles. Since 1970, most automotive companies have announced "hydrogen car prototypes." EVs have disrupted their quiet ages in these years. One of the biggest problems with general conventional tools is the depletion of fossil resources. Increasing energy demand, increasing population, and the number of vehicles resulted in much more fossil fuel usage. So, extra emission releases and climate change problems have occurred.

Especially after these two transportation sectors (air and land), the process comes to marine vehicles. Because marine vehicles require more and sustainable power than others. In order to obtain this power, it is necessary to choose the fuel according to its high energy requirement, or it is necessary to increase the energy source. To meet this demand, there are various systems that offer a diverse spectrum of energy, which include large diesel engines or nuclear reactors for submarines.

EVs are used in all vehicle technologies. However, this process may seem like there is confusion this time. Because of the Paris meeting declaration, a 2° mandatory average climate temperature reduction and emission reduction situations, researchers and manufacturers are pushed to find new and different solutions. Although the gas turbines and jet engines used in the aviation industry provide the required power and speed, they are responsible for 10% of total world fossil fuel usage and 7% of total released world emissions. It is very nice to circumnavigate the world in less than 24 hours, but the increasing number of flights shows us that we should not ignore what is happening in the sky.

It is already known that 37% of the aircraft to be produced by 2050 will be fully electric or hybrid-powered. Stakeholders and market leaders with a large budget are getting ready for this transition. One of the most famous one, named E-fanX, is manufactured by Rolls-Royce, Siemens and Airbus companies' cooperation. In this vehicle, the aircraft is hybridized with three normal fans and a 2 MW electric fan. The system is prepared according to the serial configuration (which will be explained in detail in the next section). The first real flights of this "bird" are performed [7].

This situation in aviation is one of the first signs that small prototypes will turn to mass usage. On land, the system is developing much faster than expected. The "Musk effect" has increased globalization in EVs and made EVs become an investment and production tool rather than a dream. EVs will be one of the most important historical milestones of the next 20 years, with technology that develops in every aspect and renews itself every day. A general list of advantages and disadvantages of HEVs, ICEs, and EVs is shown in Table 8.1 [5].

Researchers are still working hard to remove demerits. Battery management systems are being studied, which are directly related to the travel range. One of the most important foreseeable future

## TABLE 8.1
### The Bridge Role of HEVs Respect to ICE and EV

| Types of Vehicle | ICE | HEV | EV |
|---|---|---|---|
| *Advantages* | • Range<br>• Lower cost and maintenance<br>• Basic coding<br>• Known proven technology<br>• Widespread use (land, air, sea app.)<br>• Refueling time | • Two or more energy source<br>• Range<br>• Generated technology<br>• Regular costs<br>• Good wheel to wheel efficiency<br>• Regenerative breaking<br>• Multi usage driving | • Zero emission vehicles<br>• More efficient motors<br>• Lighter weights<br>• Newly and generative technology<br>• Hydrogen usage (FCEV) |
| *Disadvantages* | • Limited oil reserves<br>• Emissions<br>• 35%–40% efficiency<br>• Complicated and heavy engines<br>• More energy destruction (chem. →rotating→mech.) | • Complicated algorithm and coding<br>• lower but still emissions<br>• Heavyweight vehicles<br>• Challenges in production<br>• Higher cost in maintenance | • Limited range<br>• Higher cost and maintenance<br>• Complicated and difficult coding<br>• Infrastructure and electrification problems<br>• Battery charging repeatability and battery life |

*Source:* From Ref. [5]. With permission.

concerns here is the battery recycling problem. Everything has a lifetime as well as a life in its batteries. The State-of-Charge (SOC) and State-of-Health (SOH) topics will be conveyed to the readers in detail in other sections.

## 8.3 EV TYPES AND TOPOLOGIES: HEVs, BEVs, AND FCEVs

EVs are vehicles that use the electric propulsion system. They are listed and named as HEV (*Hybrid Electric Vehicle*), PHEV (*Plug-in Hybrid EV*), BEV (*Battery EV*), FCEV (*Fuel Cell EV*) and FCHEV (*Fuel Cell Hybrid EV*).

HEV vehicles are in the category of "hybrid" vehicles that have both internal combustion and electric engines. The topology of these vehicles is examined under four main titles: series, parallel, power split, and complex. Before moving on to these details; determining the definition of hybridization rate is important. These are micro, mild, full hybrid, and plug-in HEV. World Hybrid & Electric Vehicles mentions for 2021 can be listed as the micro and mild hybrid will be 20.1 million, full and plug-in hybrid sales will be 4.1 million and EVs will be nearly 1 million. EMs propulsion was added to the vehicles to achieve this goal. Small EMs are used for start/stop systems, while medium EMs for serial/parallel HEVs and large EMs for complex HEV systems. These functionalities lead EMs to be indispensable for all transportation sectors. The importance staging is determined by due to hybridization ratios [5].

Here, the main information and key points will be given about EV topologies which have several detailed types of research conducted by engineers and scientists. Parallel HEV, the internal combustion engine, and EM are engines with a propulsion system that can work both separately and together. The first-generation HEVs were generally produced with this topology. The structure that enables the use of both power supplies simultaneously and separately is named, in its simplest term, a planetary gear. This device, which gathers the ICE and EM output energy on the same common shaft, is integrated with the necessary electronic software to enable the use of the desired engine and power.

The most important detail in the series HEV system is that the internal combustion engine only feeds the EM via the generator. In other words, only EM gives the required power to the wheels. ICE gives the necessary power to EM. It is used in systems that require more power and less complex coding compared to the parallel configuration similar to "power generator" usage when the electricity cuts out in your home. ICE operates and rotates the connected shaft, which is connected to the EM motor. Electric energy turns to kinetic energy and motion.

Due to instant need, the series & parallel and complex HEV topologies can be used to obtain the serial or/and parallel system. In the complex hybrid vehicle, instead of the generator used in the serial & parallel systems, there is an electric machine that operates as a generator when necessary. In other words, when great power is required, the EM and ICE run the wheels together when the quiet and clean operation is required, only the EM drives. In this case, the battery can be continuously filled with the electrical energy produced by the internal combustion engine. In addition to these, the electric machine between the two systems can be operated as a motor and can act as a starter for the ICE. Plug-in HEV (PHEV) has the capability to be charged from a station (Alternative Current/Direct Current, AC/DC) which HEVs do not. Plug-in HEVs, in addition to those mentioned above, can also be charged from the mains, like BEVs.

One of the important parameters that it cannot pass without mentioning in HEVs is the "regenerative braking system." This system uses the energy we lose as heat when we brake in conventional vehicles; it is used to fill the battery. Rotational kinetic energy coming from the axles is converted into electrical energy when the EM is used as a generator, and this energy is stored in the battery. If an EM connected to the axle does not drive the axle, which is not in generator mode, then it creates a resistance (torque in the opposite direction) for the axle. This resistor provides braking [8].

BEV is one of the most popular and probably the first models to come to mind when the EV is mentioned. Those batteries are charged via plug and do not contain moving parts that need lubrication.

**TABLE 8.2**
**HEV Design Equations and Formulas**

| Eq. No. | Description | Formula |
|---|---|---|
| 1 | Vehicle acceleration | $dV_v/dt = \left( \sum F_t - \sum F_{tr} \right) \big/ \delta M_v$ |
| 2 | Vehicle torque | $T_v = \left( f_R M_v g \cos\varnothing + M_v g \sin\varnothing + \frac{1}{2} C_D \rho A_v V_v^2 \right) r_w$ |
| 4 | EM power | $P_m = (T_m \omega_m)/\eta_m$ |
| 5 | Battery power | $P_{\text{batt}} = V_{oc} I_{\text{batt}} - I_{\text{batt}}^2 R_{\text{bat}}$ |
| 6 | Battery current | $I_{\text{batt}} = \dfrac{V_{oc} - \sqrt{V_{oc}^2 - 4 R_{\text{batt}} * P_{\text{batt}}}}{2 R_{\text{batt}}}$ |
| 7 | Battery SOC | $SOC(k+1) = SOC(k) - \left[ (I_{\text{batt}} * \Delta t)/Q_c \right]$ |

*Source:* From Ref. [5]. With permission.

Subscripts: $F_t$ is traction effort; $F_{tr}$ is total resistance; $\delta$ is a mass factor; $V_v$, $M_v$, $A_v$ are the speed, mass, and frontal area of the vehicle; $r_w$ is the wheel radius; $f_R$ is the rolling resistance (friction) coefficient; $C_D$ and $\rho$, are the drag coefficient and air density; $\varnothing$ is the road angle and g is gravitational acceleration; $m_{ve0}$ *is* nominal vehicle mass; $m_{ce}$ is combustion engine; $m_{ba}$ is battery system; $m_{em}$ is electric machine masses; $T_m$ is the motor torque; $\omega_m$ is the angular speed of the motor; $\eta_m$ is the efficiency of the motor; $Q_c$ is battery capacity; $I_{\text{batt}}$ Battery current; SOC State-of-Charge of battery; (*k*) is SOC discrete-time index; *R*, resistance; *V*, Voltage.

Although the logic of the process is relatively easy, the control system is a bit complicated when it comes to moving a vehicle of a certain mass. BEVs have some variations and include an AC/DC converter and/or a DC/DC converter according to the battery and EMs in their entire system. Generally, the most popular EVs are in this class. FCEVs are EVs that contain a fuel cell (proton exchange membrane fuel cell, PEMFC). Protons and electrons are separated with the help of membranes and electrons generate electricity to be consumed by an EM. Oxygen from the air and the neutral hydrogen bring out the water as the only emission. Although there are many types used in the industry, the most suitable and portable fuel cell is PEM (Proton Exchange Membrane) type fuel cell. This type of fuel cell is preferable due to the suitable operating temperature (20°–90°) [6,9]. In order to reduce the cost of FCs, research and developments still keep going about catalyst and membrane materials.

Table 8.2 contains some basic formulas required for a HEV to be able to move. More examples and mathematical models are available in Ref. [10].

## 8.4   HEART, BRAIN, AND MUSCLE: BATTERIES, MANAGEMENT SYSTEM, AND ELECTRIC MOTOR

One of the main rules emerging in engineering research for years is the fact that we need energy to get power. Based on our adherence to the laws of thermodynamics, it can be easily said that it has been stored this energy that has been obtained. Important headlines about all types of energy; production, storage, distribution, and consumption. The storage of electrical energy is also carried out by batteries, which we see as the heart of EVs. When compared to energy storage capacities with respect to discharge, they can be listed as flywheels, batteries, supercapacitors, compressed air, and fuel cells are energy storage devices with high capacities, respectively. Although there are many types of batteries, they can be listed for vehicle applications. They are lithium-ion, nickel-metal hydride, lead-acid, and ultracapacitors.

In its simplest form, a battery is an electrochemical device that converts electrical energy into potential energy when the battery is charging and potential energy into electrical energy while it is discharged. Cells with a simple system consisting of two electrodes (negative and positive) and one electrolyte come together to form the battery. These energy storage devices have basic productional features such as:

- State of Charge (SOC),
- State of Health (SOH),
- State of Function (SOF),
- State of Energy (SOE),
- Lifetime and capacity,
- Cycle (one charge and discharge),
- Safety,
- Power/weight ratio.

For example, the general life of EV batteries on the market is 8–10 years, or up to 100–120 k mi. The mention of these terms here is to be remembered for further reminder. Mathematical formulas, comparisons between each other, and chemical reaction equations can be found both in open sources and in books [11–13].

The variety and efficiency of batteries used in EVs have created a considerable market. Its technology is improving, and the costs of existing ones are decreasing. Due to the increasing demand in recent years for EVs, interest in batteries, which are the heart of these vehicles, continues to be sought from both industry and academia. The best example is that the unit energy/price currency ($/kWh) obtained from batteries has decreased by 80% in the last 5 years. According to the BloombergNEF Electric Vehicle Outlook Report 2020, this value is expected to be a bit more than $100 in the next 5 years. To make an immediate comparison, it can be reached from WOS (web of science) search section and type "battery." As of March 12th, 2021, there will be 281,784 documents. 118.578 of this belongs to the last 5 years (2021–2017) [14].

Most of the studies carried out in recent years are the cycles that are directly related to the life of the battery and the "fast charging and slow discharge" principle studies, which have a great effect on battery efficiency. A lot of work is being done on the battery charging time, which can be counted among the three main disadvantages of EVs (range, battery, and cost). Attempts have been made to charge from home, main grid charging, onboard charging, and even to charge when the vehicle is in motion.

As in human beings, batteries, which are the heart of the EV, should also be taken care of. To do this, battery management systems (BMS) are being developed. Detailed information about this subject, which can be written as a book on its own, can be found in [15].

The "motion" devices of EVs are EMs, which act as muscles of the vehicle. It can be said that these devices, which appear in the literature as electric traction motors, EM, or direct motors, are machines that convert electrical energy into mechanical motion energy in the simplest terms. They convert the electromagnetic effect into rotational motion by means of a rotor and stator. The types that can be used in EVs are shown in Figure 8.4 [16].

In a previous study [16], the authors quoted the desired characteristics of EMs used in EVs as follows:

- sufficient torque and power generation,
- wide speed range for all type conditions between the start-stop period,
- offering high-efficiency use for the whole driving and regenerative braking,
- instant response to torque demand,
- high stability and robustness,
- acceptable cost.

The most important arguments to be considered here are the efficiency of the EM and the power curves of the EM. EMs generally work with an efficiency of more than 85%. Although the use of

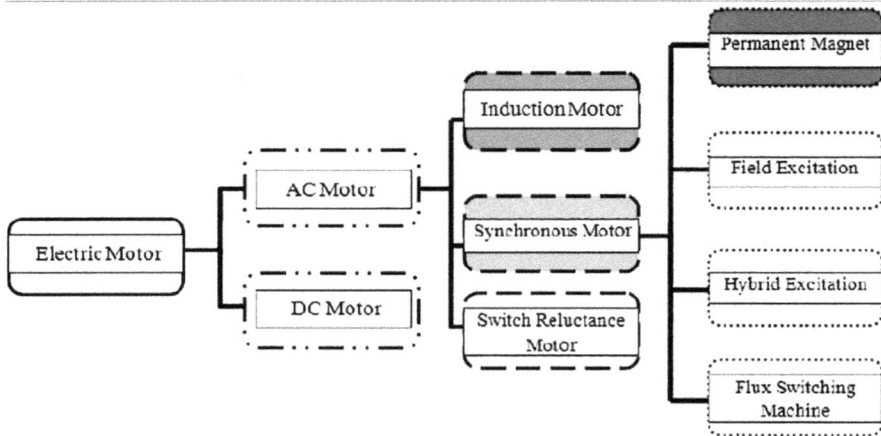

**FIGURE 8.4** Typical EMs which are widely used in EVs. (From Ref. [16]. With permission.)

different EM motors is entirely at the initiative of the manufacturers, two concepts emerge when comparing the most obvious ones (PM-Permanent magnet and AM or IM-Asecronys or IM). One is the price; the other is the loss. In addition to the high power and efficiency required especially in EMs, EM speed is also a key point factor that will cause significant variations.

PM EMs are generally used in HEVs. In addition, in regenerative braking, these motors are selected from EMs that act as motors/generators. In order to perform this task, they need various converters. In order to optimize these actuators (AC/DC or DC/DC converters) in terms of both control and minimum usage, some corporations uses IM EMs in its vehicles. Asecron motor can continue its generator function with reverse motion when high magnetism occurs. Although IM offers more opportunities in terms of both price and usability; PM motors are used more widely, especially in HEVs and FCEVs. "Thermal" effects, which are an important point in both batteries and EMs, are a crucial issue that needs to be controlled and regulated. Air gap magnetic flux for EMs and battery thermal stress are being investigated deeply by scientists and manufacturers. It shows that thermal system management is an important issue.

The word "manageable" should be able to be made in an EV without human interaction (driver). Computers have made our work much easier since they sped up technology very much and took all our transaction work to a digital system. It is called an ECU and it stands for an electronic control unit. This unit is the center where the systems community that helps you drive the way you want to drive by waiting for your command is managed.

EVMS (electric vehicle management system) is basically located on three legs: Thermal MS, Battery MS, and Energy MS. As we mentioned before, energy transformations always reveal heat energy. This is the law of nature and thermodynamics. The heat energy generated in the battery needs to be regulated and cooled. In addition, the high electromagnetic oscillation that occurs in EMs necessitates temperature control in motion. So, we need to control thermal stress and fluctuation. In addition to all these, we should definitely not ignore the "energy optimization," which affects the "range" and all other important parameters. Driving the vehicle efficiently and consciously will positively affect the life and performance of both components and the vehicle.

The heart and muscle duo (battery and motor) has a direct relationship with material technology. Batteries to be produced with materials and conductors that will increase the SOC and SOH of the battery, extend its lifecycle and increase its efficiency are going to be more preferable. In EMs, it is necessary to develop materials that will realize these "motion actors," which already have high efficiency, with less "loss." Thermal, mechanical, and electrical losses in magnetized MS and IMs should be eliminated or minimized. Even with an insensitive calculation, it can be said that 40% of the EVs cost come from batteries and 20% from EMs. At the end of the day, 60% is a crucial rate in determining the market price of an EV.

## 8.5 CONTROL STRATEGIES OF EVs AND AUTONOMOUS (SELF-DRIVING) VEHICLES

Nowadays, with the increasing dependence on fossil fuels and the environmental pollution caused by these fuels, the search for new energy sources has come to the fore again. In this context, manufacturers develop technologies that will enable the use of different energy sources, such as hybrid, electric and hydrogen vehicles. In addition, goals such as reducing traffic accidents by preventing human-induced errors, minimizing traffic congestion, and reducing transportation costs have highlighted the work for autonomous (self-driving) vehicles.

Vehicle structures, which previously required complete driver control, have gained driver support systems by increasing their control performance with technological developments. In this state, the system has become safer with the anti-lock braking system (ABS), electronic brake-force distribution system (EBD), and systems that protect the vehicle direction balance Electronic Stability Program (ESP). All of the subsystems listed above have their own control systems and can help the driver maintain control of the vehicle and prevent some accidents that may occur due to the driver's mistakes. Despite these developments, driver errors continue to be one of the biggest factors in the occurrence of traffic accidents. For this reason, the automotive industry plans to focus on autonomous vehicles to solve existing vehicle, traffic, and driver problems. Considering that some companies are already testing their autonomous vehicles in the traffic environment, it seems likely that significant developments will be achieved in the near future. Autonomous vehicles are expected to be a solution to many problems encountered today. With autonomous vehicles, it will be possible to prevent fatal accidents due to human error, to prevent traffic congestion by removing traffic lights and using intersections more effectively, and to reduce vehicle-related air pollution by reducing the time spent in traffic. It is expected that autonomous vehicles will provide the solution to many other problems besides the problems listed above.

Figure 8.5 presents the result of a function-based conceptual design study of an autonomous vehicle system. This block diagram shows the functions of the autonomous system to be developed. These functions are as follows: briefly detection; autonomous consultant control function where perceived information is evaluated; communication function; sub-control functions of the system; and system behavior intervention functions to be used for routing to the system. In Figure 8.5, the features that should be added within the scope of the transformation of automobiles to autonomous vehicles are shown, taking into account the system architecture. In this architecture, the autonomous consultant control system is located at the top. This system should be a system that combines the information from the sensors with the information received from the communication system and can make decisions that are close to human logic by evaluating them with its artificial intelligence support. In this case, the decisions made are transferred to the lower level modular sub-control systems, and with this system, brake, speed, and direction controls are made within the scope of controlling the vehicle.

These systems perform their functions with the relevant motion generators. For this purpose, there is a need for a control system built into the system for braking, adjusting the speed, and controlling the direction of the vehicle. In case of an error or an abnormal situation in the subsystems within the building, it may be possible to bypass the subsystems. In such a case, the autonomous advisor control system will be able to directly control the relevant additional motion generators to intervene in the vehicle. In order for the decisions to be made in the system to be carried out in the best way possible, the sensor system must collect all the necessary information, process it properly, and transmit the results to the autonomous consultant control system. Some basic sensors to be included in such a sensor system are listed in Figure 8.6 [17].

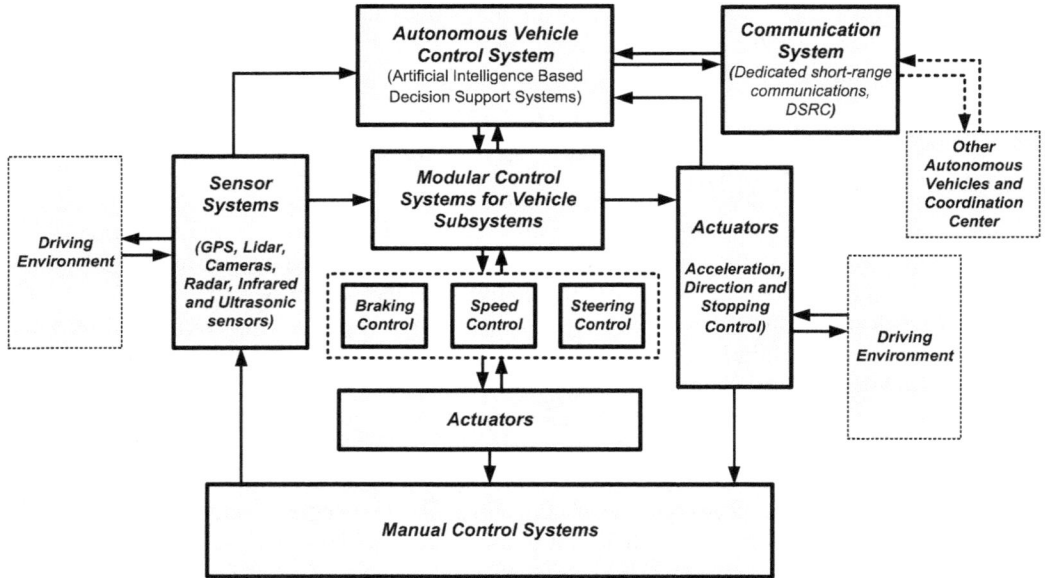

**FIGURE 8.5**   Function-based conceptual design of autonomous (self-driving) vehicle system. (From Ref. [17]. With permission.)

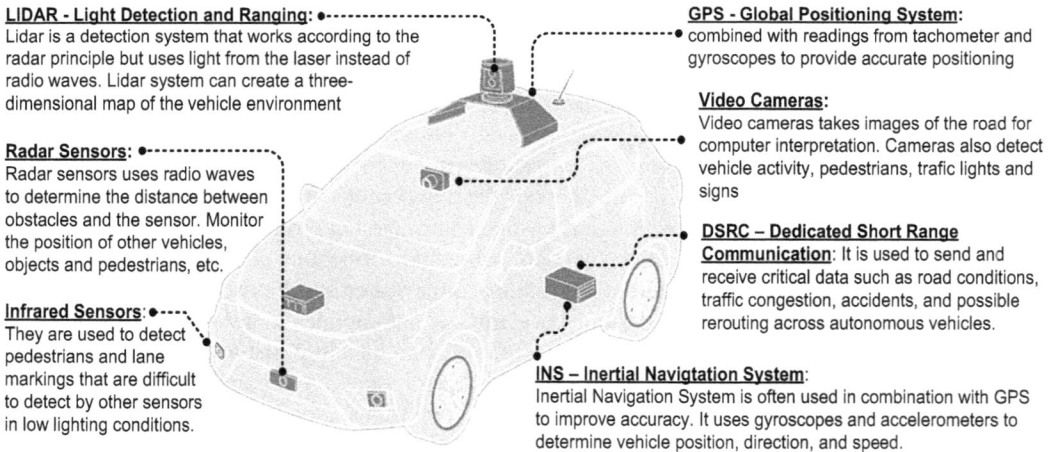

**LIDAR - Light Detection and Ranging:**
Lidar is a detection system that works according to the radar principle but uses light from the laser instead of radio waves. Lidar system can create a three-dimensional map of the vehicle environment

**Radar Sensors:**
Radar sensors uses radio waves to determine the distance between obstacles and the sensor. Monitor the position of other vehicles, objects and pedestrians, etc.

**Infrared Sensors:**
They are used to detect pedestrians and lane markings that are difficult to detect by other sensors in low lighting conditions.

**GPS - Global Positioning System:**
combined with readings from tachometer and gyroscopes to provide accurate positioning

**Video Cameras:**
Video cameras takes images of the road for computer interpretation. Cameras also detect vehicle activity, pedestrians, trafic lights and signs

**DSRC – Dedicated Short Range Communication:** It is used to send and receive critical data such as road conditions, traffic congestion, accidents, and possible rerouting across autonomous vehicles.

**INS – Inertial Navigation System:**
Inertial Navigation System is often used in combination with GPS to improve accuracy. It uses gyroscopes and accelerometers to determine vehicle position, direction, and speed.

**FIGURE 8.6**   Sensor system and components for an autonomous vehicle.

Autonomous cars combine various sensors to detect their surroundings, including radar, laser light, ultrasonic sensor, a global positioning system (GPS), inertial measurement unit (IMU), and computer vision and evaluate the results in a way that makes sense. The data obtained from the sensors are analyzed and the control of the vehicle (acceleration, braking, and steering) is carried out without the need for any human intervention in line with the decisions taken. Advanced vehicle control systems monitor the relevant units by constantly checking the road condition, environmental conditions, pedestrian and other vehicle status, traffic signs, and real-time location of the vehicle.

In Figure 8.6, information about sensors and other system elements that will take place in an autonomous vehicle is presented. When the properties of the sensors in the figure are examined, it is seen that these sensors complement each other's properties. In this state, the system is actually trying to get a complete picture of the environmental situation. It will not make much sense to transfer

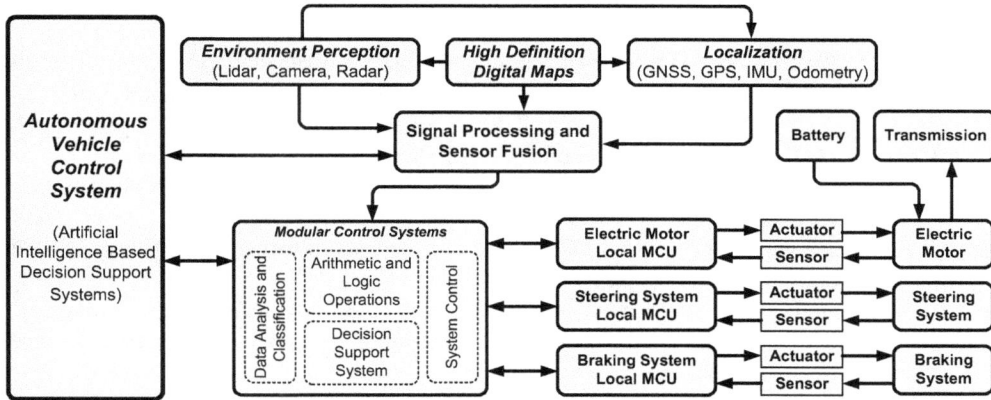

**FIGURE 8.7** Autonomous vehicle modular sub-control system. (From Ref. [17]. With permission.)

all the information collected by all the sensors to the system as it is. It is necessary to process this information and to make a preliminary evaluation, albeit to a certain extent. In this way, the autonomous system will be able to make faster decisions by reducing the load. In addition, it is necessary to review the sensing information obtained in the signal processing and evaluation parts of the subsystems within the sensor system and to perform accuracy analyzes and present these results to the system with the reliability coefficient together with the measurement information. In this way, the autonomous consultant control system will be able to make its decisions on a more solid basis.

One of the most technical parts of the autonomous system to be developed is the modular sub-control systems in the building and the parts that control the braking, speed, and direction of the vehicle have been controlled by these systems. This part of the system is shown in Figure 8.7 in more detail than the general form given in Figure 8.5. Figure 8.7 shows the basic sensors, control units, and main equipment needed by an autonomous vehicle. As can be seen from the figure, the data obtained from many sensors are collected in the intermediate signal processing and evaluation unit, transformed into an appropriate form, and transferred to the electronic control system.

The electronic control system is responsible for the identification and classification of the incoming data, the execution of the necessary arithmetic-logical operations, and the control of the subsystems in line with the decisions taken. The decision taken by the electronic control system is transferred to the microcontroller of the relevant hardware, and the microcontroller system performs the desired function by operating the necessary actuators. The fulfillment of the desired task status of the sub-function is monitored by the relevant sensors, obtained by the microcontroller, and transferred to the electronic control system in an appropriate form. As a result of the harmonious operation of all subsystems, the control system carries out the acceleration, braking, and steering control of the vehicle, depending on environmental conditions [17].

## 8.6  GENERAL CONCLUSION AND FURTHER RECOMMENDATIONS

The authors share basic and identified information with the reader about the EV sector, which continues to rise and shine successfully as one of the sectors least affected by the COVID-19 pandemic.

As a general conclusion and as a future recommendation, the authors list the most important parts below:

1. Electric power is used in transportation equipment that counts as space/air/land/sea/bicycle/all kinds of vehicles.
2. The increasing population and the need for energy show us how important electrical energy is and should be evolved in this direction.

3. According to the easy, simple, and traditional counterparts, a safety vehicle family started to form.

4. In the last 20 years, there has been no other transport vehicle that has grown on itself increasingly every day. And each type of EV is environmentally more advantageous.

5. Considering the "supply-demand balance" of the "energy" sector in the near future, which will cease to be fossil-based, it will be in need of alternative energy sources.

6. Autonomous vehicles will be an important part of our lives. They will provide convenience and be more profitable than other markets in terms of price/performance.

7. Although the Paris Agreement requirement cannot reach full target in 2030 and 2050, it will make great improvements with EVs.

8. Situations that are still inconvenient, such as material technology, new engine studies, the use of magnetic fields, mobile and on-road charge technology, and the elimination of range problems, will disappear in the short term.

9. FCEV (especially in heavy vehicles, aviation, and sea vehicles) will increase its dominance.

10. EVs and electricity prices will decrease in countries with strong electrical infrastructure or open to development.

11. The maintenance costs of EVs will be relatively lower than those of other vehicles.

12. While battery technology is developing fastly; on the other hand; battery recycling systems need to be developed quickly.

13. Energy only changes shape. The EV market will continue to multiply its capacity with energy storage methods that will be generated with new technologies via decreasing the unit cost.

14. EVs are superior to other vehicles both environmentally and efficiently, and are preferred for their silent operation, low maintenance costs, and user-friendly interfaces.

15. The EV market, which is thought to have a place in the transport market ten times as much as it is in 2050, will spread to all transport areas.

16. Engineering departments will be opened in universities for EVs (i.e., electric vehicle engineering).

17. Driving EVs with electricity throughout from alternative energy sources will enable us to take steps toward a sustainable system.

18. HEVs have served as perfect bridges between ICEs and EVs. It has benefited from both its experience and its technological infrastructure. Although the use of HEVs will continue in vehicles requiring high power, it will not be able to prevent EVs from taking over, especially in city tours and light-duty vehicles.

19. FCEVs will maintain their place in the sector with the increase of hydrogen filling stations, the change of fuel cell catalyst material, the increase in safe driving and the breakthroughs made by the manufacturers in this direction.

20. All EVs must have a worldwide standard.

The authors hope that they were able to make a presentation that would enable everyone to accumulate knowledge of EVs.

## AUTHOR CONTRIBUTIONS

Both authors participated equally in the preparation of this article. All authors have read and agreed to the published version of the manuscript.

## FUNDING

This research received no external funding.

## CONFLICTS OF INTEREST

The authors declare no conflict of interest.

## REFERENCES

1. Tanç, B., Arat, H. T., Conker, Ç., Baltacioğlu, E., & Aydin, K. Energy distribution analyses of an additional traction battery on hydrogen fuel cell hybrid electric vehicle. *International Journal of Hydrogen Energy*, 45(49), 26344–26356, (2020).
2. https://interestingengineering.com/a-brief-history-and-evolution-of-electric-cars (Accessed on 15 March 21).
3. Ladislaw, Z., Tsafos, G.-S., Carey, L., Nakano, & Chase. *Industrial Policy, Trade, and Clean Energy Supply Chains*, A Report of the CSIS Energy Security and Climate Change Program & BloombergNEF; CSIS, (2021).
4. IEA. Global EV outlook 2020, IEA, Paris. https://www.iea.org/reports/global-ev-outlook-2020 (Accessed on 15 March 2021), (2020).
5. Arat, H. T. Simulation of diesel hybrid electric vehicle containing hydrogen enriched CI engine. *International Journal of Hydrogen Energy*, 44(20), 10139–10146, (2019).
6. Tanç, B., Arat, H. T., Baltacıoğlu, E., & Aydın, K. Overview of the next quarter century vision of hydrogen fuel cell electric vehicles. *International Journal of Hydrogen Energy*, 44(20), 10120–10128, (2019).
7. https://www.airbus.com/innovation/zero-emission/electric-flight/e-fan-x.html (Accessed on 15 March 21).
8. Altindemir, E. *Hibrid Elektrikli Taşıtlarda Rejeneratif Frenleme*. Thesis (M.Sc.), Institute of Science and Technology, İstanbul Technical University, (2008).
9. İnci, M., Büyük, M., Demir, M. H., & İlbey, G. A review and research on fuel cell electric vehicles: Topologies, power electronic converters, energy management methods, technical challenges, marketing and future aspects. *Renewable and Sustainable Energy Reviews*, 137, 110648, (2021).
10. Lui, W. *Hybrid Electric Vehicle System Modelling and Control*. Wiley, Online ISBN 9781119278924, Doi: 10.1002/9781119278924, (2017).
11. Emadi, A. *Advanced Electric Drive Vehicles*. CRC Press Taylor & Francis Group, ISBN 13: 978-1-4665-9770-9 (E-book), (2015).
12. Jiang, J., and Zhang, C. *Fundamentals and Applications of Lithium-Ion Batteries in Electric Drive Vehicles*. Wiley, Online ISBN 978-1-118-41478-1, (2015).
13. Mehrdad E., et al. *Modern Electric, Hybrid Electric, and Fuel Cell Vehicles: Fundamentals, Theory, and Design*. CRC Press LLC, ISBN 0-8493-3154-4, (2005).
14. http://apps.webofknowledge.com/Search.do?product=WOS&SID=E2vOCTdhQS3uFpbB2K8&search_mode=GeneralSearch&prID=ddfe8fc5-5454-486a-b600-16169d00bd71 (Accessed on 15 March 21).
15. Xiong, R. *Battery Management Algorithm for Electric Vehicles*. Springer, ISBN 978-981-15-0248-4 (eBook) Doi: 10.1007/978-981-15-0248-4, (2020).
16. Arat, H. Numerical comparison of different electric motors (IM and PM) effects on a hybrid electric vehicle. *Avrupa Bilim ve Teknoloji Dergisi*, 14, 378–387, Doi: 10.31590/ejosat.494127, (2018).
17. Conker, Ç. & Yavuz, H. *Mekatronik Mühendisliğine Giriş, Bölüm 16: Mekatronik Yaklaşım ile Mühendislik Problemlerinin Analizi*. İstanbul: Papatya Yayıncılık, pp. 571–610. ISBN 978-605-9594-52-3, (2018).

# 9 Hydrogen and Fuel Cells

*Saeed-ur-Rehman*
Korea Institute of Energy Research

*Hafiz Ahmad Ishfaq, Zubair Masaud,
Muhammad Haseeb Hassan, and Hafiz Ali Muhammad*
Korea Institute of Energy Research
University of Science and Technology (UST)

*Muhammad Zubair Khan*
Pak-Austria Fachhochschule: Institute of
Applied Sciences and Technology

## CONTENTS

DOI: 10.1201/9781003315353-11

## 9.1 INTRODUCTION

Consumption of energy has become a daily norm in our life, and people's reliance on energy for daily activities is progressively growing as well. Meanwhile, the global energy demand is also rapidly increasing due to a noticeable increase in the human population around the globe. Various statistics predict that by the year 2040, the demand for energy will exhibit a sharp rise of 25% (Pudasainee, Kurian, and Gupta 2020). At present, nations all around the world primarily depend on fossil fuels as a source of energy, including coal, petroleum oil, and gas. It is reported that 84% of the energy generated in the world is from fossil fuels (Iordache, Gheorghe, and Iordache 2013; Nejat Veziroglu 2012; Veziroğlu and Şahin 2008; Rusman and Dahari 2016; Sun et al. 2018). Fossil fuels are geographically distributed around the world in limited amounts. It is estimated that the supply of coal reserves will run out in the next 150 years, natural gas in 60 years, and petroleum reserves in the upcoming 40 years (Midilli et al. 2005). In addition, the use of fossil fuels has resulted in catastrophic issues of environmental pollution and global warming. This is due to the fact that the burning of fossil fuels for energy generates harmful gases such as $NO_x$, $SO_x$, $CO_x$, and microscopic particulate matter (Comar and Nelson 1975; Zecca and Chiari 2010; Barbir, Veziroğlu, and Plass 1990). Living in such an environment not only reduces the quality of life but could also result in the ultimate demise of human living standards. This makes humanity's reliance on fossil fuels questionable considering the negative impact of burning fossil fuels.

To solve the global energy crisis, scientists and researchers around the globe have come to terms that there is a need for alternative fuels and energy conversion technologies to produce a clean and efficient supply of energy, which has resulted in the idea of the hydrogen economy. This concept dates to be as old as two centuries and it was introduced by a British scientist J.B.S. Haldane while the term "hydrogen economy" was introduced by John Bockris in the 1970s. A hydrogen economy depends on

**TABLE 9.1**

**Comparing the Lower and Higher Heating Value of Hydrogen as Compared to Other Fuels**

| Fuel | Energy Content (MJ/kg) | |
|---|---|---|
| | Lower Heating Value | Higher Heating Value |
| Gaseous hydrogen | 119.96 | 141.88 |
| Liquid hydrogen | 120.04 | 141.77 |
| Natural gas | 47.13 | 52.21 |
| Liquefied natural gas (LNG) | 48.62 | 55.19 |
| Still gas (in refineries) | 46.89 | 50.94 |
| Crude oil | 42.68 | 45.53 |
| Liquefied petroleum gas (LPG) | 46.60 | 50.14 |
| Conventional gasoline | 43.44 | 46.52 |
| Reformulated or low-sulfur gasoline (RFG) | 42.35 | 45.42 |
| Conventional diesel | 42.78 | 45.76 |
| Low-sulfur diesel | 42.60 | 45.56 |
| Coal (wet basis) | 22.73 | 23.96 |
| Bituminous coal (wet basis) | 26.12 | 27.26 |
| Coking coal (wet basis) | 28.60 | 29.86 |
| Methanol | 20.09 | 22.88 |
| Ethanol | 26.95 | 29.84 |

*Source:* Reprinted from Abe, J.O., A.P.I. Popoola, E. Ajenifuja, O.M. Popoola, "Science direct hydrogen energy, economy and storage: Review and recommendation." *International Journal of Hydrogen Energy*, 44(29), 15072–86, 2019. Doi: 10.1016/j.ijhydene.2019.04.068. With permission.

hydrogen as a fuel, a mode of energy storage, and its transport. Hydrogen is the first element on the periodic table and the lightest element present. Hydrogen has the highest known calorific value for any fuel as highlighted in Table 9.1. It can be seen that the calorific values of hydrogen are astonishingly higher when compared with fossil fuels such as petroleum and natural gas (Abe et al. 2019). More importantly, hydrogen is the cleanest fuel because its burning produces only water.

Although hydrogen is found abundantly on our planet, it is present almost only in a bonded state with other elements such as water, carbohydrates, and hydrocarbons. Therefore, using hydrogen as a fuel is not a simple task, as it is first required to be freed from the bonded state, which demands a certain amount of energy. Once pure hydrogen is obtained, it is required to be stored, transported, and converted to energy on demand. This is where new energy technologies need to play a key role to produce, store and convert the economically feasible, environmentally friendly, and energy-efficient fuel in the form of hydrogen. Among the various new energy conversion technologies, this chapter is focused on the fuel cells for efficient utilization and generation of hydrogen fuel. A fuel cell is a highly efficient electrochemical device, which converts hydrogen directly into electricity through electrochemical oxidation while generating pure water as its byproduct. On the other hand, a reversible fuel cell splits water into hydrogen and oxygen (fuel cell electrolysis) while using electricity from an external source such as a renewable energy source. Fuel cells present a highly efficient and clean path for generating electricity and hydrogen and, therefore, have the potential to become a primary component of the hydrogen economy (Figure 9.1).

This chapter is divided into five sections. Section 9.2 discusses hydrogen production methods specifically fuel cell electrolysis and the major industries producing hydrogen by electrolysis. Section 9.3 explains the key challenges associated with hydrogen storage along with a brief discussion on the hydrogen storage methods. Section 9.4 details various fuel cell technologies. Section 9.5 highlights applications of fuel cell technology. Section 9.6 deals with the challenges regarding clean

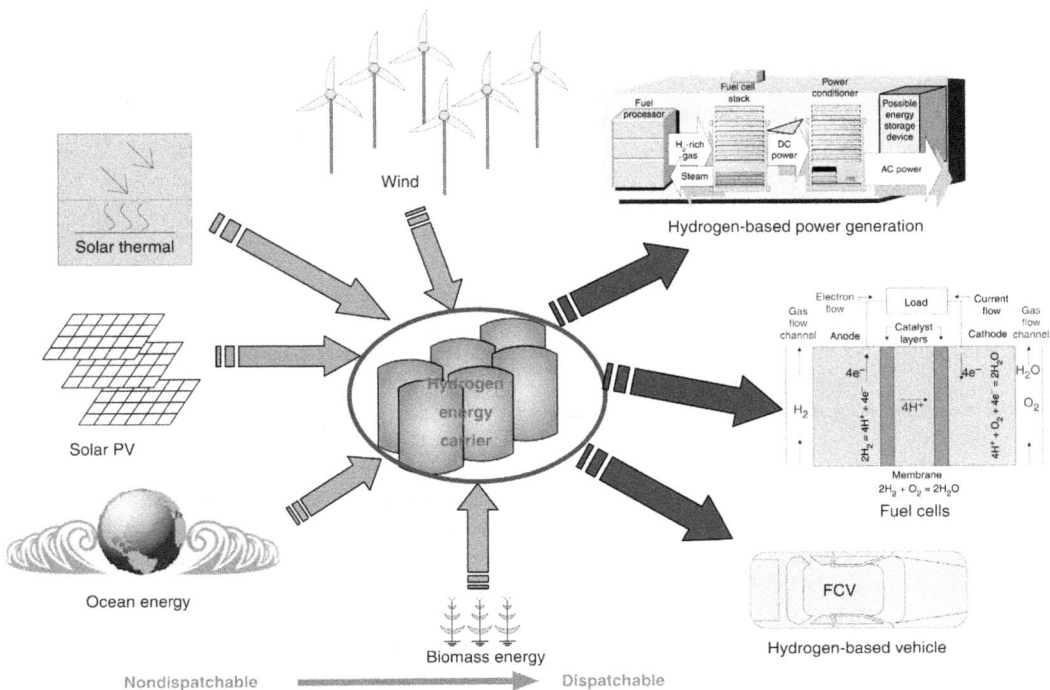

**FIGURE 9.1** The proposed outlook of the hydrogen economy. (Reprinted from Hashem Nehrir, M., C. Wang, 2015, Fuel cells. In *Electric Renewable Energy Systems*, Elsevier Inc, Doi: 10.1016/B978-0-12-804448-3.00006-2. With permission.)

and efficient energy generation via fuel cells. In the end, section 9.7 addresses the market and policy trends of hydrogen and fuel cells.

## 9.2  HYDROGEN PRODUCTION BY FUEL CELL ELECTROLYSIS

The hydrogen production processes can be classified into two broad categories, which include production from fossil fuels and renewable resources. Hydrogen is produced from fossil fuels by hydrocarbon reforming and pyrolysis techniques that are not environmentally friendly due to the emission of $SO_x$, $NO_x$, and $CO_x$. On the other hand, hydrogen is produced by biomass processing and water splitting, which are considered environmentally friendly processes due to the renewable nature of the feedstock. This chapter is focused on the production of hydrogen by water splitting through electrolysis and particularly through fuel cell electrolysis. Various methods for hydrogen production along with their advantages, disadvantages, and efficiency are summarized in Table 9.2.

Electrolysis is one of the most capable processes for producing hydrogen because it uses renewable water as a reactant and produces hydrogen along with pure industrial-grade oxygen as a byproduct. Electrolysis is currently being used for producing pure hydrogen for electronics, pharmaceutical, food, and other industries and it is regarded as the potential method to produce hydrogen fuel. Electrolysis exploits the fact that a water molecule consists of two atoms of hydrogen, bonded with an atom of oxygen. Hence, by providing external energy in the form of electrical current, the water molecule splits into its constituent elements releasing hydrogen and oxygen. The primary advantage of using electrolysis is that it is a clean and green process that produces hydrogen without the evolution of any harmful gases,

---

### TABLE 9.2
### Various Hydrogen Production Methods and Their Advantages, Disadvantages, and Efficiency

| Hydrogen Production Method | Advantages | Disadvantages | Efficiency |
|---|---|---|---|
| Steam reforming | Mature technology | Generates $CO_2$, CO, unstable $H_2$ supply | 74–85 |
| Partial oxidation | Developed technology | Produced petroleum coke and other heavy oils with $H_2$ production | 60–75 |
| Auto thermal reforming | Existing infrastructure and already established | Generates $CO_2$, utilization of fossil fuels | 60–75 |
| Bio photolysis | Requires mild working conditions, $CO_2$ consumption with the production of $O_2$ | Use of expensive materials, lower production of $H_2$, requirement of sunlight | 10–11 |
| Dark fermentation | No need of sunlight, simple method, no $O_2$ limitation, $CO_2$-neutral | Low $H_2$ yields, large volume of reactor, elimination of fatty acids | 60–80 |
| Photo fermentation | $CO_2$ neutral, recycling of wastewater | Low efficiency, required sunlight, large volume of reactor, low yield of $H_2$, $O_2$-sensitivity | 0.1 |
| Gasification | Inexpensive feedstock, widely used, neutral $CO_2$ | Fluctuating $H_2$ yield, seasonal availability, and formation of tar | 30–40 |
| Pyrolysis | Inexpensive feedstock, widely used, neutral $CO_2$ | Fluctuating $H_2$ yield, seasonal availability, and formation of tar | 35–50 |
| Thermolysis | $O_2$ byproduct, clean and sustainable | Corrosion problems, high capital costs, Use of toxic elements | 20–45 |
| Photolysis | Zero emissions, $O_2$ byproduct, availability of feedstock | Need sunlight, low efficiency, noneffective photocatalytic material | 0.06 |
| Electrolysis | Green and developed technology, $O_2$ byproduct, existing infrastructure | $H_2$ storage and transportation issues | 60–80 |

which may hinder the balance of the environment. Various electrolysis methods based on the difference in operating conditions, or the ionic agent are alkaline water electrolysis cell (AEC), polymer electrolyte membrane electrolysis cell (PEMEC), and high-temperature electrolysis using solid oxide electrolysis cell (SOEC) or proton-conducting electrolysis cell (PCEC).

## 9.2.1 Alkaline Water Electrolysis Cell

Alkaline water electrolysis is a well-established technology for hydrogen production that is operating worldwide up to megawatt scale commercial plants. AEC consists of two compartments containing alkaline solution consisting of KOH and NaOH that are separated by an asbestos diaphragm. Ni-based electrodes are dipped into the solution, which produced hydrogen and hydroxyl ($OH^-$) ions at the cathode when a DC electric current is applied. Hydroxyl ions then move toward the anode through the diaphragm under the influence of DC electric current and produce oxygen and water molecules at the anode. The diaphragm separates the two chambers and also the produced hydrogen and oxygen gases. Alkaline water electrolysis operates are low temperatures from 30°C to 80°C (Figure 9.2).

However, it has the disadvantages of low current densities (<400 mA/cm$^2$), low energy efficiency, and low operating pressure. The current research trend in alkaline water electrolysis is to replace the asbestos diaphragm with a polymer anion exchange membrane for obtaining better performance.

## 9.2.2 Polymer Electrolyte Membrane Electrolysis Cell

PEMEC technology is based on the PEM fuel cell technology and a typical PEMEC consists of a membrane electrode assembly (MEA), which contains the electrodes, polymer electrolyte membrane, bipolar plate, and a gas diffusion layer. In PEMEC, water is supplied to the anode where it is converted to oxygen and protons ($H^+$) and electrons. Protons then travel through the polymer electrolyte membrane and convert into hydrogen after combining with the electron at the cathode.

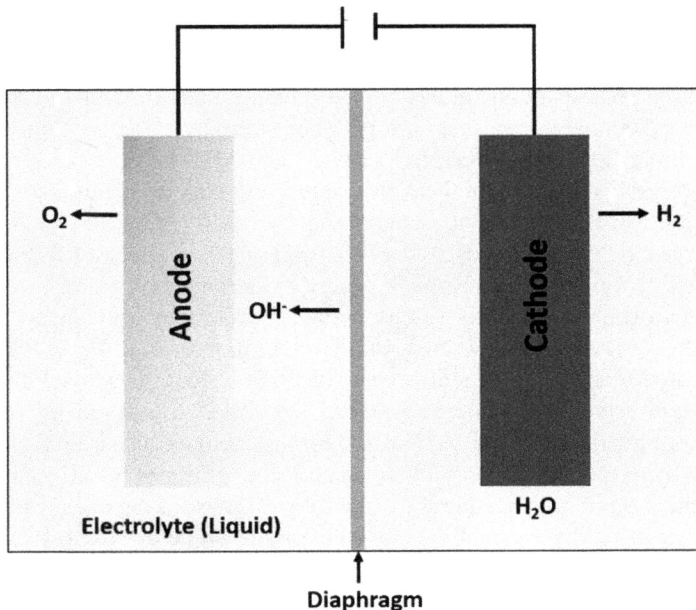

**FIGURE 9.2** Schematic illustrating the working principle of an alkaline water electrolysis cell (AEC).

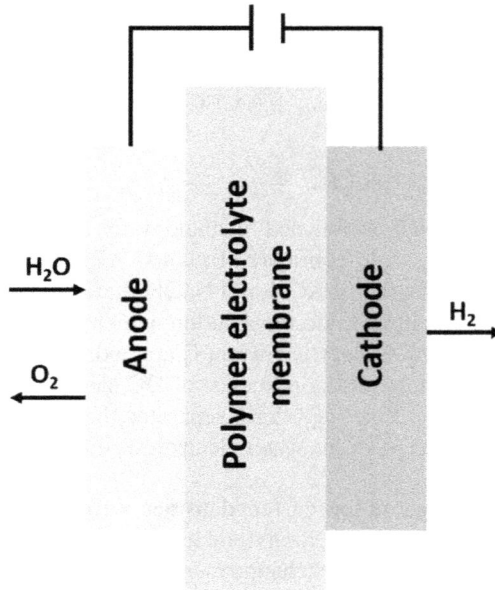

**FIGURE 9.3**   Illustration of a typical membrane electrode assembly (MEA) in the PEMEC.

PEMEC has a high capital cost due to the high material costs of the polymer membranes (perfluorinated polymers) and precious-metal-based (Pt, Ir, Ru) catalysts (Figure 9.3).

However, PEMFC systems provide the advantages of better efficiency, higher current densities, and compact design as compared to the AEC systems. PEMECs also give the advantage of low-temperature operation (20°C–80°C). The recent research trend in PEMECs is to reduce the capital cost while maintaining the high efficiency.

### 9.2.3   HIGH-TEMPERATURE ELECTROLYSIS USING SOEC OR PCEC

High-temperature electrolysis of steam has gained increased attention due to the better thermodynamics at elevated temperatures, that is, a better performance can be obtained by using low electrical energy input, low electrocatalytic activity, and low electrolyte resistance. The high-temperature electrolysis is founded on the solid oxide ion conductor technology that is applied in solid oxide fuel cells (SOFCs) and proton-conducting fuel cells (PCFCs). A SOEC typically uses a zirconia-based ceramic membrane capable of transporting oxide ions ($O^{2-}$). Steam is converted to hydrogen and oxide ions at the Ni cathode. These oxide ions then travel through the ceramic zirconia membrane toward the perovskite-based anode where they are converted to oxygen. On the other hand, the working principle of a PCEC is similar to that of the PEMEC, however, conducted at elevated temperatures using steam. In PCEC, steam is provided at the perovskite anode where it is converted to oxygen and protons ($H^+$), these protons then travel to the Ni cathode through the barium zirconate ($BaZrO_3$)-based proton-conducting ceramic electrolyte and are converted to pure hydrogen. A comparison between the working of SOEC and PCEC is shown in Figure 9.4. In the case of high-temperature electrolysis, most of the energy needed to split the water molecule is acquired by thermal means, which reduces the electrical energy requirement, thus, improving the overall process efficiency. This also provides the advantages of reduced capital and running costs and can also provide waste heat recovery. High-temperature electrolysis is current in research and development (R&D) phase and needs to be developed into a stable technology for commercialization.

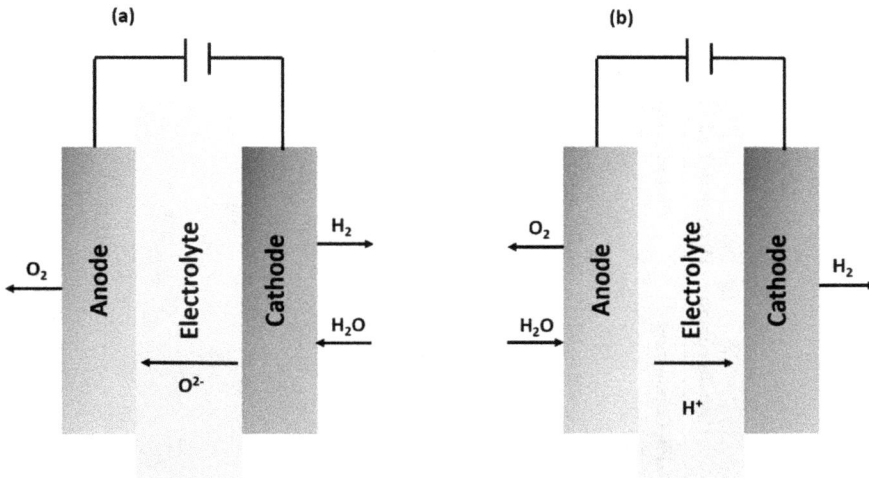

**FIGURE 9.4**  Comparison of working of a typical (a) SOEC and (b) PCEC.

Operating principle, materials, and working parameters of various electrolysis technologies that are discussed above are summarized in Table 9.3. Major companies that are playing key roles in the R&D of water electrolysis and are also developing electrolysis systems for commercial purposes are listed in Table 9.4.

## 9.3  HYDROGEN STORAGE AND TRANSPORTATION

In the preceding section of this chapter, the production of $H_2$ has been discussed. For hydrogen to be used as a commercial fuel, it requires excellent storage and transportation from the production location to the consumers. Therefore, it is compulsory to ensure that the materials for hydrogen storage and transportation are reliable and cost-effective. Hydrogen storage and transportation is a highly challenging area in the establishment of the hydrogen economy. Following are some of the major challenges faced in storing $H_2$ (Mori and Hirose 2009; Schüth 2009).

- $H_2$ has a low density, which limits the amount of storage, consequently, a tank with a large volume and weight is needed.
- $H_2$ can be explosive when it comes in contact with atmospheric oxygen.
- When compared with other fuels such as petroleum, the cost of hydrogen storage is too high.
- The refueling for $H_2$ fuels is slow. According to the US department of energy (DOE), the refueling time should be under 3 minutes.
- The lifetime and durability of the $H_2$ storage system are not up to the mark.

Various applications demand different requirements in terms of hydrogen storage, volumetric capacity, gravimetric capacity, and cost. In the case of stationary applications, the weight and the volume factors of the $H_2$ storage are not that prominent. However, in mobile applications, special considerations are needed for storage design because a mobile vehicle can manage a limited amount of weight and volume. Therefore, a balance is needed while designing the storage system of $H_2$ for mobile applications. Conventionally, hydrogen is stored as compressed or liquefied gas and these are regarded as the state-of-the-art hydrogen storage technologies. However, more compact solutions are needed for the portable and mobile applications and the solid-state hydrogen storage materials are being developed as promising and highly efficient hydrogen storage media.

**TABLE 9.3**

**Comparison of Various Electrolysis Technologies and Their Key Characteristics**

| Characteristic | | Alkaline Water Electrolysis Cell | Polymer Electrolyte Membrane Electrolysis Cell | High-Temperature Electrolysis Cell | |
|---|---|---|---|---|---|
| | | | | Oxide Conducting | Proton Conducting |
| Chemical reaction | Anode | $2OH^- \rightarrow 1/2O_2 + H_2O + 2e^-$ | $2H_2O \rightarrow 4H^+ + 4e^- + O_2$ | $2O^{2-} \rightarrow 2e^- + O_2$ | $H_2O \rightarrow 2H^+ + 1/2O_2 + 2e^-$ |
| | Cathode | $2H_2O + 2e^- \rightarrow H_2 + 2OH^-$ | $4H^+ + 4e^- \rightarrow 2H_2$ | $4H_2O + 4e^- \rightarrow 2H_2 + 2O^{2-}$ | $2H^+ + 2e^- \rightarrow H_2$ |
| | Overall | $H_2O \rightarrow H_2 + 1/2O_2$ | $H_2O \rightarrow H_2 + 1/2O_2$ | $H_2O \rightarrow H_2 + 1/2O_2$ | $H_2O \rightarrow H_2 + 1/2O_2$ |
| Materials | Anode | Mostly Ni materials | Pt | Perovskite oxides | Perovskite oxides |
| | Cathode | Pt and Ni-based alloys | Pt | NiO | NiO |
| | Electrolyte | KOH, NaOH, and $H_2SO_4$ | Nafion | Zirconia or Ceria based electrolytes | Barium zirconate-based electrolytes |
| Operation temperature (°C) | | 20–80 | 20–200 | 500–1,000 | 500–800 |
| Efficiency (%) | | 60–70 | 65–82 | Up to 100 | Up to 100 |
| Charge carrier | | OH- | $H^+$ | $O^{2-}$ | $H^+$ |
| Electrolyte | | Liquid | Solid | Solid | Solid |
| Cost | | Low | High | High | High |
| Durability | | High | Low | Low | Low |
| System lifetime (year) | | 20–30 | 10–20 | - | - |
| Status | | Commercialized | Early commercialized | R&D | R&D |

**TABLE 9.4**

**Companies Manufacturing the Electrolyzers with Their Important Characteristics**

| Company | Electrolyzer Type | Location | Efficiency (%) | Op. T. (°C) | Main Characteristic | Ref. |
|---|---|---|---|---|---|---|
| Giner, Inc. | PEMEC | USA | 70–90 | <45 | Power density > 1 kw/h | Hamdan (2011) |
| McPhy energy | AEC | France | 65–70 | 60–80 | Lifespan >80,000 hours | President and Energy (2019) |
| Hydrogenics (Cummins, Inc.) | Mainly PEMEC | Canada | – | 70–90 | Developed world's largest PEM electrolyzer with air-liquid company | Cummins (2021) |
| Green hydrogen systems | AEC | Denmark | 85 | – | Produces >99.98% pure $H_2$ | "Green Hydrogen Systems Electrolyzers" (n.d.) |
| Next hydrogen | AEC | Canada | – | 10–35 | Purity >99.999% | NextHydrogen (2021) |
| ITM power | PEMEC | British | Above 77 | 50–90 | Largest electrolyzer factory in the world | BvDEP (2019) |

## 9.3.1 COMPRESSED GAS STORAGE

Compressed gas is the most common method of hydrogen storage. In this mode of storage, challenges are faced due to the low density of hydrogen gas. For laboratory and industrial applications, the hydrogen gas is compressed to a maximum of 20 MPa; however, the hydrogen tanks for mobile applications can store gas from 35 to 70 MPa. Compressed gas is an efficient way of storing hydrogen because the volumetric density of hydrogen can be increased by increasing the pressure of the gas. Most commonly, metallic containers are used for industrial applications with pressure from 20 to 30 MPa but these containers can contain only up to 1 wt. % of hydrogen (Type 1). For producing lightweight cylinders, part of the cylinder is replaced with fiber resin composite (Type 2). Metallic part is further reduced by using carbon fiber embedded polymer matrix with a thin metallic liner (Type 3). Complete polymer cylinders are also suggested for lightweight applications (Type 4). As there is always a risk of gas leakage when stored at such high pressures, for compressed hydrogen storage, the cylinder material must have a high tensile strength, low density, and it should not be permeable to hydrogen.

## 9.3.2 LIQUEFIED HYDROGEN STORAGE

Liquefaction of hydrogen (cryogenic hydrogen) at a temperature of 20 K is an advanced and also a conventional method of its storage. This is advantageous because conversion to the liquid form increases the density and the storage efficiency of the hydrogen storage. Hydrogen liquefaction is an energy-intensive and time-consuming process and about 40% of energy contents are lost for liquid hydrogen as compared to the 10% loss for compressed hydrogen. To store the liquid hydrogen, 20 K temperature is needed to be maintained, which requires efficiently insulated containers. For that purpose, the liquid $H_2$ tanks are designed with multilayers of metals, with the inner layer being insulated with aluminum foils. Moreover, glass wool is also installed between the inner and outer layers. In addition, the space between the inner and outer layers is also vacuumed. However, the volumetric energy density of such liquid hydrogen is still lower compared to alternative options of gasoline and diesel (Chen and Jiang 2016). That is why the research has now recently shifted to cryo-compressed hydrogen, which includes the benefits of both cryogenic and compressed hydrogen. Cryo-compressed containers are required to store hydrogen at cryogenic temperature (20 K) and a pressure of 30 MPa. Limitation of the liquefied hydrogen storage is high energy requirements, cost of the pressure vessel, and the production rate of the hydrogen.

### 9.3.3 SOLID-STATE HYDROGEN STORAGE

Compressed and liquefied hydrogen storage lack the necessary gravimetric and volumetric capacity. For achieving high-density hydrogen storage, the solid-state storage systems are being developed in which hydrogen is absorbed into solid-state materials and then removed for use by applying some external form of energy such as heat. The hydrogen can be absorbed physically or chemically and this is known as physical sorption (physisorption) and chemical sorption (chemisorption).

In the physisorption, the hydrogen is adsorbed on the high surface area solids and that interaction is developed through the weak Van der Waal forces of attraction. These weak Van der Waal forces are particularly useful in this case because they can be easily broken down with minimal thermal stimulation. The weak bonds can be established with zeolites and carbon-based materials such as activated carbon, fullerenes, carbon fibers, carbon nanotubes, covalent organic framework (COF), metal-organic framework (MOF), and polymer intrinsic microporosity (PIMs). Although this kind of system is associated with fast kinetics, the practical applications are still limited because extremely low temperatures are required for high hydrogen storage capacity. In the chemisorption, the hydrogen is chemically bonded with solids, for example, metal hydrides. This involves the processes of hydrogenation and dehydrogenation. The hydrides are classified as classical metal hydrides, chemical hydrides, and light metal complex hydrides. The complex hydrides require complex hydrogenation and dehydrogenation processes, and a lack of reversibility reaction. The chemical hydrides have many advantages such as the easy release of hydrogen and the compose of lighter elements, but they also face the typical problems related to complex hydrides. For this purpose, metal hydrides hold the most promise among the hydrides. The metal hydrides are associated with significant advantages of high reversibility of hydrogenation and dehydrogenation, a higher degree of safety, low-pressure requirements, high energy density, and little external energy requirements, among others. Figure 9.5

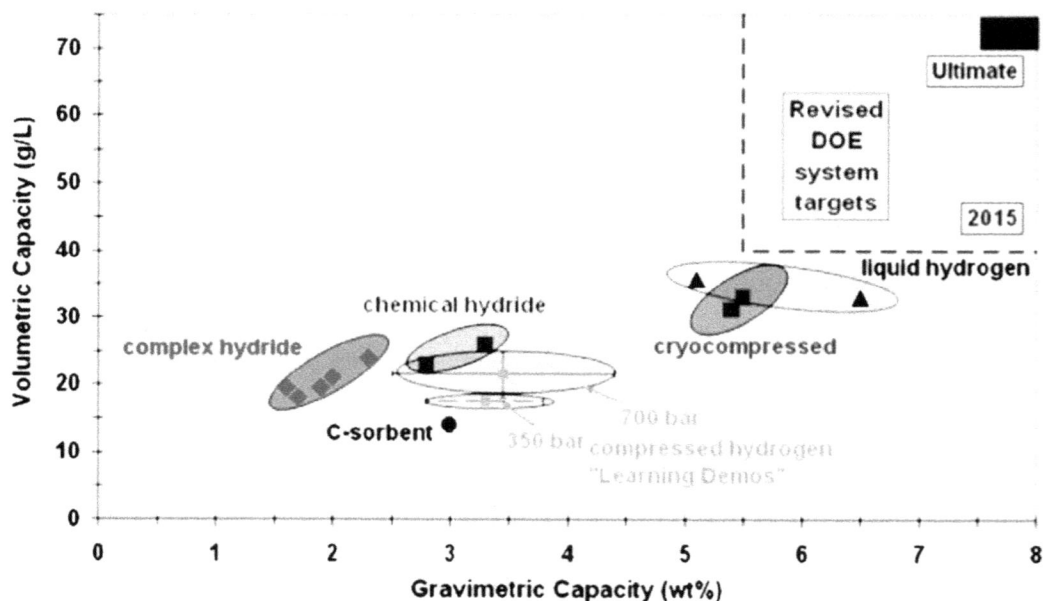

**FIGURE 9.5** The volumetric and gravimetric capacity of various hydrogen storage technologies published by US DOE. (Reproduced from Durbin, D.J., "Review of hydrogen storage techniques for on board vehicle applications." 1–23, Doi: 10.1016/j.ijhydene.2013.07.058. With permission.)

displays various $H_2$ storage system's gravimetric and volumetric capacities. According to this data, no system has yet achieved the required gravimetric and volumetric capacity requirements set by the DOE; therefore, extensive research work is still required in this field to make hydrogen storage more efficient and feasible.

## 9.4 FUEL CELL TECHNOLOGIES

Fuel cells are mainly open-thermodynamic systems, and they typically operate based on the electrochemical reactions where they utilize reactants from the outer source. Fuel cells are promising substitutes to traditional methods of electricity production mainly for small-scale applications. Hydrogen fuels have considerable chemical energy as compared to typically used materials in the battery; therefore, they are extensively utilized for various energy-related applications. A Ragone plot is displayed in Figure 9.6, which shows that fuel cells have the highest energy density than the other various types of energy devices.

Fuel cells have quite easy design and consistent operation. In addition, the use of hydrogen as fuel makes them the greenest and environmentally friendly energy system. Presently, fuel cells have both small- and large-scale applications, including combined heat and power (CHP) systems, electric vehicles, and even equipment required for military communication (Xu, Kong, and Wen 2004; Larrosa-Guerrero et al. 2010). Besides these benefits, there are some drawbacks to employing fuel cells. For instance, the durability of the fuel cells reduces by the degradation and poisoning of fuel cell components due to the presence of impurities in the gas stream. The less power density per volume and low accessibility are other major questions that fuel cell technology must overcome (Dogdibegovic et al. 2021; Alias et al. 2020). Though great development is still to be witnessed, optimistic growth is seen in recent years (Wang et al. 2021).

Fuel cells produce power and heat employing electrochemical response, which is the reverse of electrolysis reaction. It occurs among hydrogen and oxygen to generate the water (Brandon and Kurban 2017). Fuel cells have various designs, but they all are working on this basic principle to generate electricity. The major difference in the various designs of the fuel cell is the difference in the characteristics of electrolytes used (Cook 2002). Figure 9.6 shows the basic working principle of all fuel cell types with their general trends in efficiency, temperature, fabrication, and materials cost. Eq. (9.1) shows the main electrochemical reaction of the fuel cell.

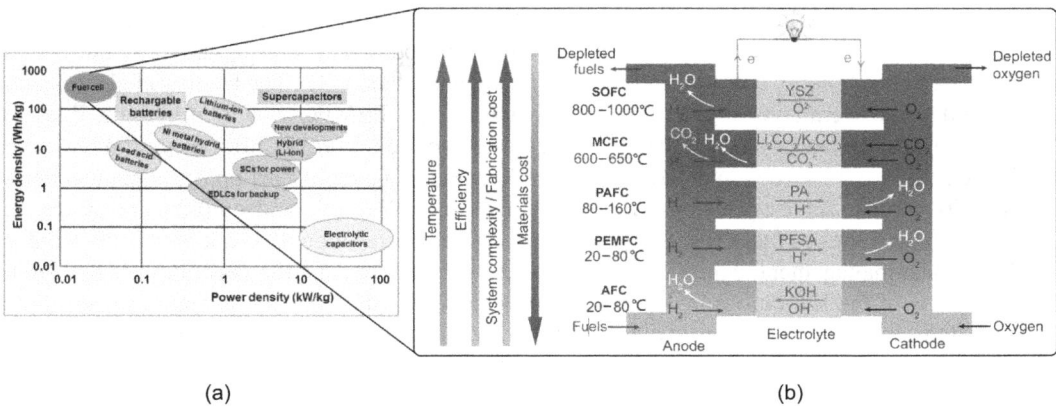

(a)                                    (b)

**FIGURE 9.6** (a) Ragone plot and (b) working principle of various fuel cell technologies and their general trend in the relationship between materials cost, fabrication cost, system complexity, efficiency, and operating temperature. (From Wang, S., S.P. Jiang, "Prospects of fuel cell technologies." *National Science Review*, 4(2), 2017, Doi: 10.1093/nsr/nww099. With permission; From Muthukumar, M., N. Rengarajan, B. Velliyangiri, M.A. Omprakas, C.B. Rohit, U.K. Raja, "The development of fuel cell electric vehicles – A review." *Materials Today: Proceedings*, 2020, Doi: 10.1016/j.matpr.2020.03.679. With permission.)

$$2H_2(g) + O_2(g) \rightarrow 2H_2O + \text{energy} \tag{9.1}$$

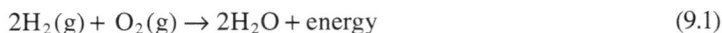

A typical fuel cell consists of four major components: anode, electrolyte, cathode, and interconnect. At the anode (also called fuel electrode) side of the fuel cell, hydrogen fuel is converted into $H^+$ ions and electrons, while at the cathode side (also called air electrode), oxygen is reduced to $O^{2-}$ ions, which react with produced $H^+$ on the anode side to complete the reaction by generating water. Based on the type of electrolyte used, either generated $H^+$ or $O^{2-}$ ions are transferred through an ion-conducting electrolyte while the electrons travel through the outer circuit (the interconnect) to produce the electric power. The electrochemical reactions occurring at the fuel and air electrodes of fuel cells are shown in Eqs. (9.2) and (9.3).

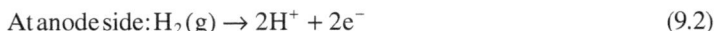

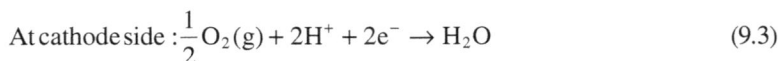

$$\text{At anode side}: H_2(g) \rightarrow 2H^+ + 2e^- \tag{9.2}$$

$$\text{At cathode side}: \frac{1}{2}O_2(g) + 2H^+ + 2e^- \rightarrow H_2O \tag{9.3}$$

Fuel cells are classified into six main groups according to the type of fuel and electrolyte used:

### 9.4.1 PROTON EXCHANGE MEMBRANE FUEL CELLS

The proton exchange membrane fuel cell (PEMFC) consists of a solid polymer electrolyte/membrane that permits the flow of $H^+$ ions from the fuel to the air electrode. The PEMFCs operate at a temperature and pressure range of 70°C–80°C and 1–2 bar, respectively (Ijaodola et al. 2018). The cell potential of PEMFC stacks is around 1.1 V for one cell and rises in proportion with the number of cells in the stack.

The electrochemical reaction for PEMFC is the same as described in Eqs. (9.1–9.3). In PEMFCs, electrons transfer from the anode side through an external circuit to reach the cathode. At the same time, the $H^+$ ions move across the electrolyte toward the cathode. The $H^+$ ions are combined with oxide ions on the cathode side to complete the reaction. The fuel and air electrode of PEMFCs are made up of graphite bipolar plates and these plates have flow channels, which direct the reactant flow into the cell. Consequently, the design and geometry of the bipolar plate and its channel have a substantial effect on cell performance. The water and heat management in the cell also depends on the geometry of bipolar plates. The designs of PEMFCs differ depending on the required output voltage of the cell. The final design of PEMFCs must be inexpensive, simple to fabricate, and able to compare with other energy conversion devices (Liu, Tu, and Chan 2021). The loading of electrocatalyst is vital as the overall cost of the PEMFCs depends on it. A PEMFC usually utilizes the Pt as an electrocatalyst to enhance the chemical reaction, which is quite expensive. Still, the optimization in the Pt catalyst layer is needed to reduce the cost of PEMFCs. The rapid start-up, excellent mechanical stability, a broad scale of electric power produced from mW to kW, and simple scale-up are the benefits of these types of fuel cells. Sluggish oxygen reduction reaction (ORR) kinetics, inadequate water and heat management, contamination of the electrocatalysts, and the need for pure hydrogen fuel are some of the limitations of PEMFCs. Irrespective, PEMFC has the potential to substitute gasoline and conventional engines in automobile and aeronautics applications (Baroutaji et al. 2021; Moreno et al. 2015).

### 9.4.2 ALKALINE FUEL CELL

Alkaline fuel cell (AFC) is another type of fuel cell that utilizes the solution of alkaline NaOH or KOH as an electrolyte material. KOH is widely used as an AFC electrolyte material due to its superior conductivity as compared to the other alkaline-based solutions (McLean et al. 2002). AFCs are typically operated at lower temperatures in the range of 23°C–70°C. AFCs have several advantages of high efficiency, rapid start-up, higher activity toward the ORR, and best heat management. In AFCs, Pt electrocatalyst can be replaced by Ni and its alloy with the other transition metals at the fuel electrode making this fuel cell technology inexpensive. As compared to PEMFCs, AFCs have better tolerance toward CO poisoning to a certain level due to their higher electrocatalytic activity. The biggest challenge in AFCs is to enhance their resistance toward $CO_2$, which is the main reaction byproduct of hydrocarbon fuels. This $CO_2$ is consumed by the electrolyte to form the carbonate salt, which deteriorates the ionic conductivity of the AFC electrolyte and decreases its performance and efficiency (Ran et al. 2018). The electrochemical reactions of AFCs are described in Eqs. (9.4) and (9.5):

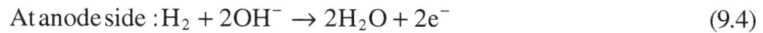

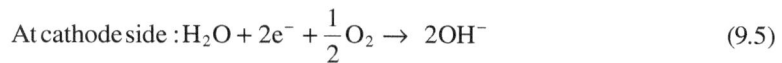

$$\text{At anode side} : H_2 + 2OH^- \rightarrow 2H_2O + 2e^- \tag{9.4}$$

$$\text{At cathode side} : H_2O + 2e^- + \frac{1}{2}O_2 \rightarrow 2OH^- \tag{9.5}$$

### 9.4.3 DIRECT ALCOHOL FUEL CELLS

Direct alcohol fuel cells (DAFCs) have operating temperatures less than 100°C and are mostly used for portable-type energy applications where the output power requirement is less than 250 W. Methanol and ethanol are used as fuels in DAFCs. The catalysts layer consists of Pt and Ru where Ru enhances the tolerance of Pt toward the CO poisoning. DAFCs have many benefits, such as short start-up time, consumption of waste resources as fuels, superior energy density, ease of fuel transportation, and cost-effectiveness (Abdelkareem, Morohashi, and Nakagawa 2007). The main obstacle encountered in DAFCs is the crossover of fuel since fuel goes from the fuel to the air electrode because of the variation in concentration, which reduces the overall electrochemical performance and exterminates the cathode. The alcohols are extremely combustible and produce some toxicity especially if methanol is used as fuel. Moreover, the catalyst layer used in DAFCs has Pt and Ru metals, which are precious and expensive metals (Basri et al. 2010). The overall electrochemical reactions of DAFCs are as follows:

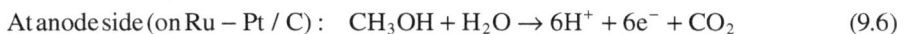

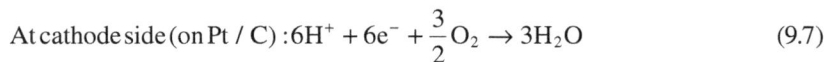

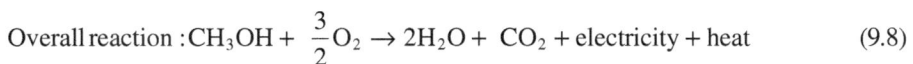

$$\text{At anode side (on Ru} - \text{Pt / C)}: \quad CH_3OH + H_2O \rightarrow 6H^+ + 6e^- + CO_2 \tag{9.6}$$

$$\text{At cathode side (on Pt / C)} : 6H^+ + 6e^- + \frac{3}{2}O_2 \rightarrow 3H_2O \tag{9.7}$$

$$\text{Overall reaction} : CH_3OH + \frac{3}{2}O_2 \rightarrow 2H_2O + CO_2 + \text{electricity} + \text{heat} \tag{9.8}$$

### 9.4.4 PHOSPHORIC ACID FUEL CELLS

Phosphoric acid fuel cell (PAFC) utilizes phosphoric acid $H_3PO_4$ as an electrolyte hence named as PAFCs. This fuel cell typically operates between 150°C and 230°C, with an optimal temperature of 180°C. The greater resistance to CO poisoning and smaller Pt catalyst loading as compared to PEMFCs make the PAFC attractive to AFCs. In addition, the waste heat can be utilized effectively in PAFCs. Due to their higher operating temperatures, PAFCs can be used in the CHP applications (Ito 2017). Major drawbacks of PAFCs are the great price, prolonged start-up time, and low ionic conductivity. Because it is operated at high temperatures than PEMFCs and AFCs, the materials for this type of fuel cell are limited. The thermal expansion coefficient (TEC) requirements further limit the materials to certain choices. The electrochemical reactions in PAFCs are like PEMFCs (Stonehart and Wheeler 2006).

## 9.4.5 Molten Carbonate Fuel Cells

Molten carbonate fuel cells (MCFCs) are also operated at relatively high temperatures than PAFCs in the range of 550°C–700°C. They utilize salt of molten carbonate mainly lithium and potassium carbonates, as an electrolyte material that is mixed in a porous and chemically inert matrix of $\beta$-$Al_2O_3$ solid electrolyte, which is called a *base*. A variety of fuels are used for MCFCs, containing hydrogen with oxygen as oxidants. High efficiency is the biggest advantage of MCFCs. $CO_2$ can also be used as an oxidant in MCFCs indicating that they contribute toward the carbon capture and storage (CCS) issue (Spinelli et al. 2018). Moreover, noble metals like Pt and Ru are not used as electrocatalysts in these fuel cells because of their high operating temperature, which makes this technology cost-effective. Degradation of cell components at high temperatures, the difficulty in handling the molten carbonate liquid, and slow start-up time are the major drawbacks of MCFCs (Kulkarni and Giddey 2012). The two electrochemical oxidation reactions take place at an anode and are summarized as follows:

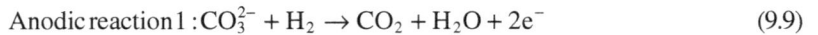

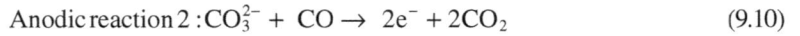

$$\text{Anodic reaction 1}: CO_3^{2-} + H_2 \rightarrow CO_2 + H_2O + 2e^- \tag{9.9}$$

$$\text{Anodic reaction 2}: CO_3^{2-} + CO \rightarrow 2e^- + 2CO_2 \tag{9.10}$$

The reduction reaction takes place at the cathode side and generates the carbonate ions $\left(CO_3^{2-}\right)$ from the carbon dioxide and oxygen reaction as indicated in Eq. (9.9). The generated carbonate ions travel to the anode side via molten carbonate electrolyte.

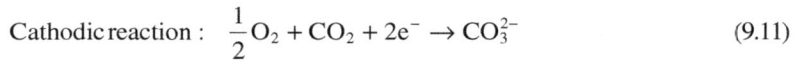

$$\text{Cathodic reaction}: \quad \frac{1}{2}O_2 + CO_2 + 2e^- \rightarrow CO_3^{2-} \tag{9.11}$$

## 9.4.6 Solid Oxide Fuel Cells

SOFCs are the high-temperature fuel cells that typically operate at higher temperatures in the range of 600°C–800°C than all the categories of fuel cells. The SOFC electrolyte is made up of ceramic material typically yttria- or scandia-stabilized zirconia, which has high ionic conductivity and excellent thermal and chemical stability. Some requirements must be considered for the cathode materials for SOFC, such as good thermal stability, mixed ionic and electronic conductivity (MIEC), and catalytic activity for ORR. The oldest cathode material that fulfills all these requirements is lanthanum strontium manganite (LSM). But MIEC materials such as Sr- and Fe-doped lanthanum cobaltite (LSCF), Sr-doped lanthanum cobaltite (LSC), and Sr-doped samarium cobaltite (SSC) are widely used SOFC cathode materials these days. The anode is made up of NiO-YSZ, which boosts the kinetics of $H_2$ oxidation (Hussain and Yangping 2020; Tsipis and Kharton 2008). Oxygen is reduced at the cathode at 800°C, whereas fuel ($H_2$) oxidation takes place at the anode side. The anode must be porous to transport the generated products because of fuel oxidation away from the electrolyte/anode interfaces. The electrochemical reactions that occur at SOFC electrodes are as follows:

$$\text{At anode side}: H_2 + O^{2-} \rightarrow H_2O + 2e^- \tag{9.12}$$

$$\text{At cathode side}: 2e^- + \frac{1}{2}O_2 \rightarrow O^{2-} \tag{9.13}$$

The benefits of SOFC are enormous; hence, they are widely used in several applications. The efficiency of SOFCs is high up to 80% due to its effective utilization of waste heat to generate more electric power by running the gas turbines. These fuel cells are performed well in the absence of noble metal catalysts such as Pt making them affordable that run up to 80,000 hours under fuel cell conditions (Steele and Heinzel 2001). SOFCs have excellent fuel diversification as compared to other categories of fuel cells. Conversely, SOFC systems have slow start-up and cooling downtimes. Due to its higher operation temperature, only selected materials are used due to the chemical and mechanical compatibility problems (Li, Wang, and Liu 2020). Several authors have provided viable solutions to alleviate the operation temperature of SOFC and declared that if effective and sustainable countermeasures are developed, SOFC is an efficient way for energy production to new production.

### 9.4.7 PROTON CERAMICS FUEL CELLS

The high-temperature of SOFCs is the biggest hindrance to their commercialization. Therefore, there is an urgent requirement for the development of low-temperature SOFCs (LT-SOFCs). In this respect, PCFCs are recently developed as promising LT-SOFCs since they require low activation energy for the transport of protons through the proton-conducting electrolyte such as $BaZr_{0.4}Ce_{0.4}Y_{0.1}Yb_{0.1}O_{3-\delta}$ (BZCYYb). The $H^+$ ions are the charge carries in this technology as compared to the $O^{2-}$ ions in SOFCs. These protons are transferred from anode to the cathode via the proton-conducting electrolyte where these protons react with the oxides ions to generate water at the cathode. The generation of water at cathode is the major benefit of this technology, which retards the fuel dilution and hence enhances the fuel utilization and efficiency of PCFCs. The huge over potentials of cathodes and the poor chemical stability and sinterability of the proton-conducting oxides are the major barriers to this technology (Kim et al. 2019). The fuel cell reactions of PCFCs are summarized as follows:

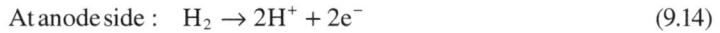

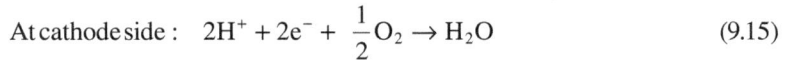

$$At\,anode\,side: \quad H_2 \rightarrow 2H^+ + 2e^- \tag{9.14}$$

$$At\,cathode\,side: \quad 2H^+ + 2e^- + \frac{1}{2}O_2 \rightarrow H_2O \tag{9.15}$$

The highest efficiency of fuel cells has fascinated several researchers; thus, they have developed a broad range of fuel cell types. Therefore, there is a pressing appeal to compare the various aspects of fuel cells, which emphasizes their merits and demerits so that they can be used in different applications according to the requirements. Table 9.5 summarizes the operational and technical characteristics of various fuel cell technologies.

## 9.5  GLOBAL APPLICATIONS OF HYDROGEN AND FUEL CELLS

With numerous categories of available fuel cell technologies, there is a need to describe, which technology meets the requirements of a certain application. As described, fuel cells can deliver a broad range of electric energy, that is, 1–10 MW; therefore, they can effectively use in nearly any application that requires electrical energy. They have the potential to use in low-range power devices and even electronic equipment such as cellular phones. Intermediate scale power applications contain fuel cell cars, household appliances, and public transport. Lastly, in the huge range (1–10 MW) of power applications, fuel cells are mainly employed in distributed power systems. Table 9.6 summarizes the main applications along with the pros and cons of different fuel cell technologies. As mentioned above, all types of fuel cells have many applications, as shown in Figure 9.7, and these applications can be broadly classified as (Mekhilef, Saidur, and Safari 2012):

1. Applications related to zero emissions such as the urban areas, automobiles, and various industrial facilities, which have stringent emission standards.
2. Applications which require high power such as telecommunication and data processing.
3. Applications related to the utilization of biological waste gases such as waste treatment factories.

**TABLE 9.5**

**Summary of Operational and Technical Characteristics of Various Fuel Cell Technologies**

| | Fuel cell type | PEMFC | AFC | DAFC | PAFC | MCFC | SOFC |
|---|---|---|---|---|---|---|---|
| Operational characteristics | Electrolyte/ membrane | Polymer-Nafion | Alkaline KOH | Polymer-Nafion | Phosphoric acid | Molten carbonate | Yttria-/Scandia-stabilized Zirconia |
| | Fuel | Pure $H_2$ | Pure $H_2$ | $CH_3OH$ | Pure $H_2$ | $H_2$, CO, $CH_4$, others | $H_2$, CO, $CH_4$, others |
| | Oxidant | $O_2$ in air | $O_2$ in air | $O_2$ in air | $O_2$ in air | $O_2$ in air | $O_2$ in air |
| | Anode reaction | $H_2 + 2OH^- \rightarrow 2H_2O + 2e^-$ | $H_2 + 2OH^- \rightarrow 2H_2O + 2e^-$ | $CH_3OH + H_2O \rightarrow 6H^+ + 6e^- + CO_2$ | $H_2 + 2OH^- \rightarrow 2H_2O + 2e^-$ | $CO_3^{2-} + H_2 \rightarrow CO_2 + H_2O + 2e^-$ | $H_2 + O^{2-} \rightarrow H_2O + 2e^-$ |
| | Cathode reaction | $H_2O + 2e^- + 1/2O_2 \rightarrow 2OH^-$ | $2H_2O + 4e^- + O_2 \rightarrow 4OH^-$ | $6H^+ + 6e^- + 3/2O_2 \rightarrow 3H_2O$ | $H_2O + 2e^- + 1/2O_2 \rightarrow 2OH^-$ | $1/2O_2 + CO_2 + 2e^- \rightarrow CO_3^{2-}$ | $2e^- + 1/2O_2 \rightarrow O^{2-}$ |
| | Cogeneration | No | No | No | Yes | Yes | Yes |
| Technical characteristics | Operation temperature | 70°C–80°C | 25°C–70°C | >60°C | 180°C | 550°C–700°C | 700°C–900°C |
| | Range of power | 2 W–500kW | 20 W–500kW | 100 mW–1kW | 50 kW–1 MW | <1 kW–1 MW | 5 kW–3 MW |
| | Cell electrical efficiency | 50%–70% | 60%–70% | 20%–30% | 55% | 55% | 60%–65% |
| | Stack electrical efficiency | 30%–50% | 62% | 10%–25% | 40% | 45%–55% | 55%–60% |

*Source:* From Abdelkareem, M.A., K. Elsaid, T. Wilberforce, M. Kamil, E.T. Sayed, A. Olabi, "Environmental aspects of fuel cells: A review." *Science of the Total Environment*, 752, 2021, Doi: 10.1016/j.scitotenv.2020.141803. With permission.

**TABLE 9.6**

**Advantages, Disadvantages, and Applications of Different Fuel Cell Technologies**

| Fuel cell types | Advantages | Disadvantages | Applications |
|---|---|---|---|
| PEMFC | Wide power range, quick start-up, high power density | Slow ORR, water management issues, CO poisoning | Portable power, backup power, transportation |
| AFC | Inexpensive, rapid start-up, simple heat management, high activity | Low tolerance to $CO_2$, Requires pure oxygen | Military, submarines, aviation, backup power |
| DAFC | No $CO_2$ emissions, availability of fuel, Use of cheap fuel, quick start-up | Crossover of fuel, Use of expensive Pt & Ru catalysts, Cathode poisoning, Toxic fuel | Electronic devices |
| PAFC | High tolerance to $CO_2$, inexpensive due to lower Pt usage | Slow start-up, Intolerant to CO, Low power density, have limited materials | Distributed generation |
| MCFC | High efficiency, supports internal reforming, fuel diversification, can be used with gas turbines | Poor long-term stability, long start-up time, Use of selected materials, cathode poisoning | Large, distributed generation, auxiliary power, electric utility |
| SOFC | | | |

*Source:* From Abdelkareem, M.A., K. Elsaid, T. Wilberforce, M. Kamil, E.T. Sayed, A. Olabi, "Environmental aspects of fuel cells: A review." *Science of the Total Environment*, 752, 2021, Doi: 10.1016/j.scitotenv.2020.141803. With permission; From Mekhilef, S., R. Saidur, A. Safari, "Comparative study of different fuel cell technologies." *Renewable and Sustainable Energy Reviews*, 2012, Doi: 10.1016/j.rser.2011.09.020. With permission.

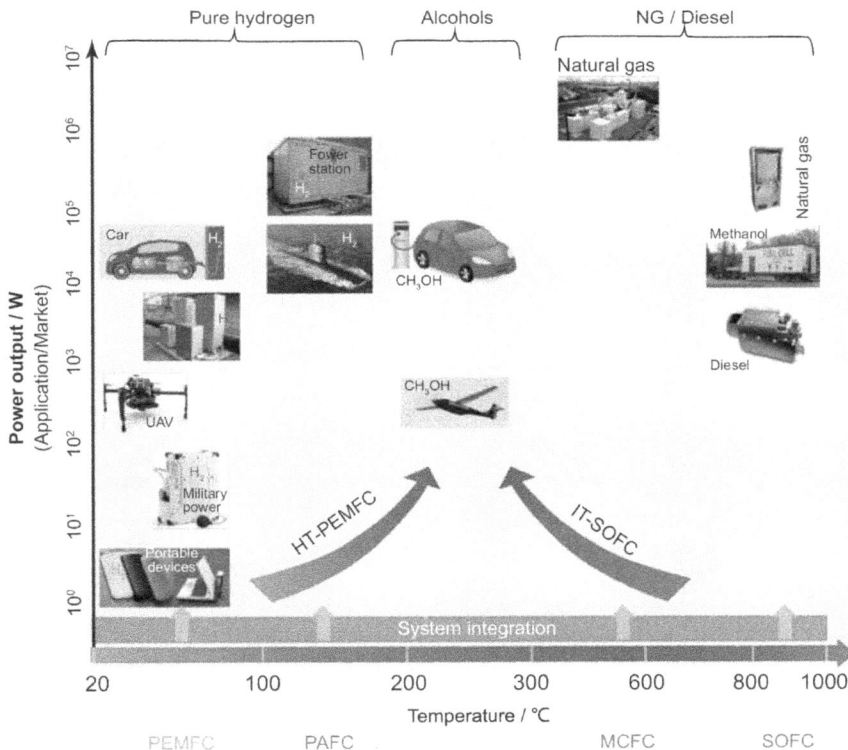

**FIGURE 9.7** Main applications and power ranges of PEMFCs, PAFCs, MCFCs, and SOFCs as a function of their operating temperature. The arrows indicate the R&D trend in the development of HT-PEMFCs and IT-SOFCs. (Reproduced from Wang, S., S.P. Jiang, "Prospects of fuel cell technologies." *National Science Review*, 4(2), 2017, Doi: 10.1093/nsr/nww099. With permission.)

Among these fuel cell applications, the transportation sector has major applications of fuel cells and is described in detail in this section. The suitability of hydrogen fuel cells differs between transport modes and has been used in a variety of transport sectors, including land, air, and sea along with passengers and freight. In 2015, the target value of the share of renewable energy in UK transport was around 10% (Committee on Climate Change 2017). This 5.8% reduction in the target value poses serious concerns among the UK government to take stronger actions. Hydrogen is one of three main alternatives for zero $CO_2$ transport along with biofuels and electric vehicles (EVs). Hydrogen prevents the usage of land and air quality effects of biofuels, and the narrow range and prolonged recharging times of EVs. Due to particulates, $SO_x$, and $NO_x$ emissions, Europe had faced above a half million premature deaths in a year (Ferreira et al. 2017). In addition, around 92% of the overall population is subjected to the bad quality of air that goes beyond the World Health Organization (WHO) limits (World Health Organization 2016). Due to such reasons, the majority of the cities have declared prohibitions on all diesel-driven trucks and cars by 2025, and especially France and UK have announced the ban on all ICEs by 2040 (Staffell et al. 2019). Therefore, hydrogen vehicles are the urgent priority since they have the potential to tackle climate change by improving the quality of air. The hydrogen and fuel cells have many road transport applications, including fuel cell electric vehicles (FCEVs), hydrogen powertrains, fuel cell buses, trucks, and motorbikes.

### 9.5.1 Fuel Cell Electric Vehicles

FCEVs are typically constructed in a distinct configuration of frequently used ICEs. In gasoline or diesel-driven vehicles, the energy generated from the hydrogen fuel is transferred from the engine to the vehicle's wheels through mechanical way; however, the way of transmission of energy is by the electrical way in FCEVs. Generally, the PEMFC stack is used as an electrical source for FCEVs because of its high power density, minimum operation temperature, rapid start-up, and lightweight. The various car manufacturing companies, which are doing R&D on FCEVs are listed in Table 9.7, which suggests that most of the companies use PEMFCs in FCEVs. The FCEVs can compete with the conventional internal combustion engines (ICEs), plug-in hybrid vehicles (PHEVs), and battery electric vehicles (BEVs) in terms of:

1. **Lifetime:** Batteries have typically high charging/discharging rates, and their lifetime is severely influenced by overcharging and local climate. This could be depicted by the <5 years lifetime of batteries when Tesla had expected such batteries to last for about 10–15 years

**TABLE 9.7**
**Various Car Companies Undertaking the Expansion of FCEVs**

| Company | Fuel used | Fuel cell | System Type |
|---|---|---|---|
| Ford | Hydrogen/methanol | Direct or indirect | Only fuel cell |
| General motors | Hydrogen/methanol | Direct or indirect | Fuel-cell-battery hybrid |
| Nissan | Methanol | Indirect | Fuel-cell-battery hybrid |
| Mazda | Hydrogen | Direct or indirect | Fuel-cell with ultra-capacitor hybrid |
| Honda | Hydrogen/methanol | Direct or indirect | Fuel-cell with ultra-capacitor hybrid |
| ZeTech | Hydrogen | Direct | Fuel-cell-battery hybrid |
| Daimler-Chrysler | Hydrogen/methanol | Direct or indirect | Fuel cell, Fuel-cell-battery hybrid |
| Renault | Hydrogen | Direct | Hybrid of fuel-cell-battery |
| Toyota | Methanol | Direct or indirect | Hybrid of fuel-cell-battery |

*Source:* From McNicol, B.D., D.A.J. Rand, K.R. Williams, "Fuel cells for road transportation purposes - yes or no?" *Journal of Power Sources*, 100(1–2), 2001, Doi: 10.1016/S0378-7753(01)00882-5. With permission.

(Staffell et al. 2019). Conversely, hydrogen tanks have quick fill-up and high discharging without compromising the lifetime.

2. **Capital cost:** The FCEVs normally have high operating and capital costs as compared to BEVs these days: around \$60–75 k for the Hyundai ix35 or Toyota-Mirai versus \$24–30 k for the Nissan Leaf or Renault Zoe (Staffell et al. 2019). But FCEVs are considered a cost-effective technology in the future when their manufacturing demands increase and ultimately end up as cheaper alternatives (Brandon and Kurban 2017).

3. **Refueling and range time:** BEVs have shorter driving ranges and longer refueling times than FCEVs. These ranges of FCEVs are also like the conventional vehicles, that is, ca. 3 minutes and 500 miles (Pollet, Staffell, and Shang 2012).

4. **Emissions:** FCEVs are known as low-carbon technology from the production point of view and have zero emissions from the usage point of view. BEVs also have negligible emissions whereas ICEs have reduced decarbonization potential. Although adding biofuels with diesel and petrol can alleviate the $CO_2$ emissions, but the quality of air remains the biggest challenge in the case of ICEs.

5. **Requirements of infrastructure:** The EV chargers can serve a smaller number of vehicles than the hydrogen fueling stations in the case of FCEVs. Nevertheless, the hydrogen refuelers are much expensive as compared to the electric charging posts, but the costs of hydrogen refuelers are expected to drop by around two-thirds once their demand is increased (International Energy Agency 2015).

6. **User's experience:** FCEVs are much quieter, have fewer vibrations, and have no issues with gear shifting and hence they offer a much smoother experience than ICEs.

## 9.5.2 Fuel Cell Buses

The fuel cell buses are becoming mature due to the enormous attention they are receiving around the globe. The TCO, abbreviated as the total cost of ownership, of fuel cell buses, has 10%–20% larger than diesel by 2030. Europe has faced considerable early deployment in fuel-cell-hydrogen buses with around 7 million km of operating experience. China had ordered around 300 fuel cell buses for Foshan City and hence has the world's biggest bus market. The reliability of such buses is depicted from their long operating hours of around 18,000 hours in London. In addition, in California, ten fuel cell buses have passed the 12,000 hours of operation and one of them has reached the 22,400 hours of operation time, which is very much close to the target of 25,000 hours operation time set by the US DOE. Availability of fuel cell buses exceeds 90% in Europe compared to the 85% target, with 95% availability of hydrogen refueling station.

## 9.5.3 Fuel Cell Trucks

Trucks have the significant potential to adopt the fuel cells. However, higher durability is required for trucks due to their high traveling distance, with one program aiming at a 50,000-hour lifetime of the stack. Low fuel and high efficiency costs are also important. Toyota and Kenworth are considering the massive production of fuel cell trucks, and Nikola in the United States is also creating a long-distance heavy goods vehicle by exploiting the liquefied hydrogen. However, fuel cell trucks have lower acceptance than fuel cell buses because of the heavy goods vehicle market, which is extremely cost-sensitive with the negligible support from government or intervention. With the ban on diesel trucks in some major cities, the interest in fuel cell trucks is rising day by day.

## 9.5.4 Trains and Ships

Hydrogen trains are the effective choices on the roads, which are comparatively hard to electrify economically because of the large route length or limited space in the urban regions. In Germany, the testing of the fuel cell trains, which have roof-mounted tanks of hydrogen has been begun. Alstom has announced its plans to convert the electric and diesel trains to hydrogen in the United Kingdom to eliminate the requirement for line electrification to meet the targets of the government of removing all the diesel trains by 2040 (Gerlici et al. 2018). The hydrogen power trains are 50% more costly than the diesel-powered trains but the economic feasibility depends on the lower fuel cost and one study has concluded that from the TCO viewpoint, FCEV trains are competing the diesel-driven trains from an economic perspective. The maritime applications hold significant potential for hydrogen utilization. In addition, with the growth of urban ports and the Baltic Sea (the controlled emissions zones), hydrogen has gained huge attention in marine applications.

### 9.5.5　Airplanes

Aeronautics is the toughest area to decarbonize and decreasing the emissions from jet propulsion has noticed a small improvement. In 2016, the International Civil Aviation Organization approved to limit emission in air applications at 2020 levels, but mainly by offsetting the carbon instead of low-emission fuels (Lyle 2018). For this purpose, some ideas of hybrid electric are being examined, though the reduction in the emission has been seen. Biofuels can be the fine alternatives because of their superior power density to batteries or hydrogen, but they are not emission-free and may remain expensive with their low accessibility. The hydrogen needs to liquefy to be used as a propulsion fuel so that it can supply according to the desired range. Since the fuel cells do not have the ability to take off, the combustion turbines are therefore needed. Moreover, the use of hydrogen in jet applications produce water vapors causing the radiative forcing and contributes toward the global warming. This poses a serious question on the climate benefits of using hydrogen for aviation. The deployment of hydrogen is, therefore, not considered feasible before 2050 with an exemption of low-flying jet applications (Staffell et al. 2019). Hence, there is more room for R&D on opening the doors of hydrogen in low-emission aviation applications.

### 9.5.6　Stationary Applications of Fuel Cells

SOFCs cover a major part of stationary applications of fuel cells and hence will be discussed in this section. In stack system applications, in 1989 Japan started the first research on SOFC stack system by fabricating the first 100 W SOFC stack. The power of the SOFC stacks was increased to kilowatts with the development of planar SOFCs with the more focus on the enhancing the durability of these stacks. For this purpose, The NTEL, National Energy Technology Laboratory, in the United States collaborated with PNNL, Pacific Northwest National Laboratory, to develop the highly efficient, cost-effective, and fuel diversified SOFC stack of 10 kW power. These stacks are subsequently used in the hybrid power generation systems, which comes in the main stationary applications of SOFCs. The optimization of SOFC exhausted heat led to the development of hybrid power generations. In this regard, the Siemens Westinghouse made the hybrid project with the focus on the fabrication of tubular SOFCs stacks for the efficient power generation system. According to this project, the SOFC system was combined with turbines to utilize its byproduct heat. This augmented idea enhances the entire efficiency of the system and reduces the electricity cost without the use of extra fuel. With the expanding of this idea, several industries such as General Electric and Rolls Royce also established SOFC stack, which was combined with the gas turbines (GT) for the generation system of hybrid power. This idea of hybrid system was also projected to the utilization of other thermal generators such as steam turbines (ST) and Stirling heat engines (SHE). The several hybrid systems of SOFC-GT-ST, SOFC-ST, and SOFC-SHE are therefore studied and developed as the efficient power generators for the hybrid power generation. In further maturation of power generation system, the waste heat recovery from SOFC was not only limited to the heat generators but also expanded to the idea of the satisfying the heat demand. For this purpose, the integrated SOFC system for CHP was developed as an efficient system because of the absence of separate cooling and heating systems.

This system was first commercialized for the application of domestic power generation. This was demonstrated by various companies such as Ceres Power in UK, Hexis in Switzerland, and Ceramic Fuel Cells Ltd in Australia. After the development of CHP systems, the SOFC-based CCHP (combined cooling, heating, and power) came into the existence. A SOFC-CCHP system has the water and space heating as well as cooling requirements, which enhance the overall efficiency of the system more than that of the typical CHP system. However, the transition of such integrated systems from the laboratory and simulation scale research to the commercialization level must overcome the several environmental, economic, technical, and political barriers.

## 9.6 FUEL CELL CHALLENGES

Fuel cells have the highest efficiency, reliability, and zero emissions. Then what are the reasons that restrict them to make it a mature technology? The answer is the various technical barriers that caused the hindrance of fuel cell deployment since their invention. These technical factors include the issues of their high costs especially the expensive materials and the poor long-term durability of the fuel cells due to the degradation and contamination of their electrode materials pose serious challenges to their commercialization. These major technical barriers and challenges are discussed in this section.

### 9.6.1 Cost Challenges

The manufacturing cost of the fuel cells contains the equipment capital, materials, labor, component fabrication, design, and assembly, which are essential in the overall manufacturing of fuel cells and stacks. Since, the major applications of fuel cells include the PEMFCs, so the cost challenges related to them are discussed only in this section. According to the recent estimates, the cost of an 80-kW fuel cell stack for automobile applications costs around $73/kW. In the other study, Ahluwalia, Wang, and Peng (2015) produced an 80 $kW_{net}$ Argonne PEM fuel cell stack and Yang (2013) examined the costs of this PEM fuel cell stack and indicated that the entire cost of this stack is about $30 k/W. The PEM electrodes (anode, cathode, and catalyst layer) itself represented a major part of the whole cost of the stack which is about 51%. The stack assembly constitutes about 7% of the overall cost. The rest of the other elements (bipolar plates and seals) cover the remaining percentage of the cost, as shown in Figure 9.8a. On the other hand, in the case of a fuel cell stack system of 80 $kW_{net}$ the cost would be increased to $59 k/W. The 50% of this cost is represented by

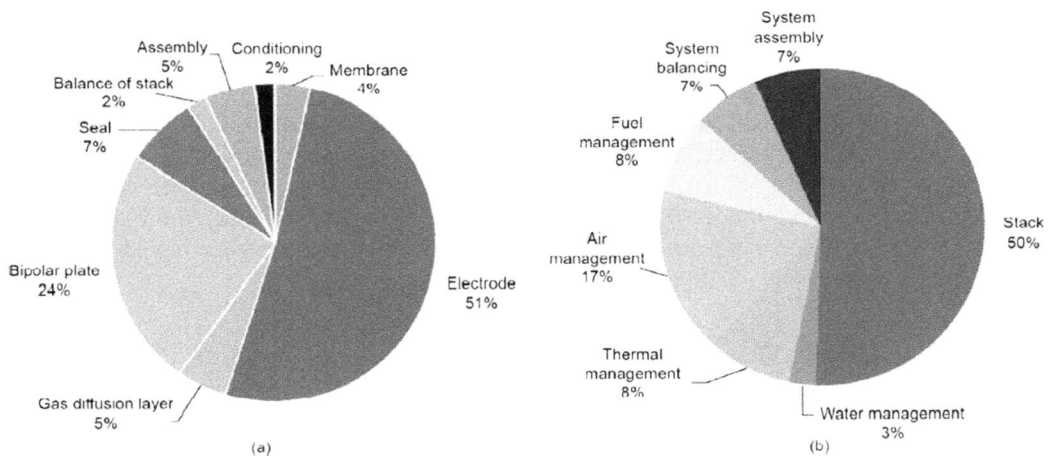

FIGURE 9.8 A pie chart shows the cost division of various components of an 80 $kW_{net}$ PEMFC for (a) stack; (b) stack system. (Adapted from Wang, J., H. Wang, Y. Fan, "Techno-economic challenges of fuel cell commercialization." *Engineering*, 2018, Doi: 10.1016/j.eng.2018.05.007.)

**TABLE 9.8**

**The Cost Comparison between FCEVs, PHEVs, and IECs**

| Type of Vehicle | Engine Type | Efficiency (%) | Power (hp) | Cost ($) | | | | | |
|---|---|---|---|---|---|---|---|---|---|
| | | | | Body | Propulsion System | Transmission | Fuel Tank | Others | Overall |
| Fuel cell vehicle | FCEVs | 60–70 | 112 | 7,902.5 | 13,964 | 316.1 | 750 | 1,422.45 | 24,355.05 |
| Honda Civic Hybrid | PHEVs | 10–16 | 110 | 120,025 | 7,215 | 2,405 | 240.5 | 2,164.5 | 132,050 |
| Honda Civic Sedan | ICEs | 10–16 | 140 | 7,902.5 | 4,741.5 | 1,580.5 | 158.05 | 1,422.45 | 15,805 |

*Source:* From Wang, J., H. Wang, and Y. Fan, "Techno-economic challenges of fuel cell commercialization." *Engineering*, 2018, Doi: 10.1016/j.eng.2018.05.007. With permission; From Elnozahy, A., A.K.A. Rahman, H.H Ali, "A cost comparison between fuel cell, hybrid and conventional vehicles." 2014. With permission.

its stack with the fuel, air, and thermal management are the other factors that should be considered they enhance the overall cost of fuel cell stack system shown in Figure 9.8b. In comparison with the ICEs entrenched technology, the cost of the fuel cells must be comparable without compromising on their benefits that they provide. For example, Elnozahy Rahman, and Ali (2014) have compared the costs of the FCEVs with the IECs and PHEVs, and their analysis, as shown in Table 9.8, had indicated that FCEVs cost higher (~$25 k) than the IECs (~$15 k) and PHEVs (~$24 k). Though the entire fuel cell cost is greater as compared to IECs (Table 9.3), the overall spread is equivalent to IECs due to the superior efficiency of fuel cells. Consequently, the fuel cell cost would not be the major factor in end-user acceptance. Nevertheless, the cost of maintenance and repair is essential for the end-user acceptance and stack service, which is almost neglected in considering the overall cost of the fuel cell stack. The failure of any fuel cell component due to the long-term operation and cyclic conditions would result in the cell failure and hence the stack also. Generally, the disassembly of the entire stack is required to replace the failed part of the stack and this adds the 100% cost of the balancing and conditioning of the stack and stack system (Wang 2015; Wang and Jiang 2017). As represented in Figure 9.3, such stack conditioning can represent around 22% of the entire cost of the stack system. This means that the 22% cost of assembly, stack balancing, and conditioning are added up for every repair since these are the essential treatments for every fuel cell stack system. Therefore, the cost of the whole stack is increased for every single failure of the fuel cell's components because the maintenance and repair cost can exceed 60% of the entire cost of the fuel cell system. The failure of fuel cell components mainly occurs due to the degradation and poisoning of the fuel cell components, which affects the durability and reliability of the fuel cell stack system. Thus, it can be concluded that durability and reliability are an effective strategy for significantly decreasing the cost of the entire fuel cell system and enhancing the acceptance of end-user of the system (Wang, Wang, and Fan 2018).

## 9.6.2 Reliability and Durability Challenges

As mentioned above, the impact of repair and maintenance on the entire expense of the fuel cell system is high due to the unexpected failures which cause the reliability and durability issue of hydrogen fuel cells, and these are the other biggest challenges which hinder them to become a mature technology. The PEMFCs are mainly used for automobile applications where they are operated at a severe and wide range of cyclic and operating conditions. The PEMFCs stack is rapidly cycled between low and high-temperature, gas compositions, voltages, and humidity which results in the chemical and physical failure of the cell's components (Burlatsky et al. 2019; Martin, Kopasz, and McMurphy 2010). For example, high voltages and loading cycles, which are typical operating conditions of FCEVs, enhance the degradation of Pt electrocatalysts and mechanical stresses at membranes, respectively, results in the catastrophic failure of cell and hence the durability of the stacks are reduced (Liu et al. 2020). The degradation of the quick start-up and shut down is the biggest issue in PEM-FCEVs. Under such harsh conditions, the local overpotentials can exceed the 1.5 V at which the carbon support in PEMFCs corrodes easily. The catalyst degradation of PEMFCs could be mitigated using Pt-alloys rather than Pt metals that improved its durability and activity under severe operating and cyclic conditions. In SOFCs, the Ni-coarsening and Ni-reoxidation of NiO-YSZ anode under accelerated operating conditions results in the reduction of its electrocatalytic activity and delamination and cracking of the anode layer (Hagen et al. 2006). The cation segregation at cathode/electrolyte interface of SOFCs under high current density causes the formation of insulating phases, which results in the spallation of the cathode layer (Khan et al. 2021; Li et al. 2017). The careful selection of materials and essential treatments are required to prevent the formation of secondary phases during SOFC operation.

## 9.7  MARKET AND POLICY TRENDS ABOUT HYDROGEN AND FUEL CELLS

As described above, the main applications of hydrogen fuel cells are in the transport sector, which mainly depends on the future targets and policy framework (McNicol, Rand, and Williams 2001). These policies, as well as investments substantially, differ from country to country. The current and future national targets for hydrogen and fuel cell technology uptake in six prominent countries are summarized in Tables 9.9 and 9.10 compares the level of funding provided by these countries to raise the interest in hydrogen and fuel cells. These Tables (9.9 and 9.10) indicate the level of interest taken by the top six leading countries in making such policies.

The policy support for the deployment of hydrogen and fuel cell technologies is guided by different public priorities including climate change, air pollution, energy safety, economic development, and affordability (Staffell and Dodds 2017). For instance, the need for air quality improvement due to transportation is the main driving force that runs the US policy implying that there are no national objectives for utilizing the fuel cells in the fixed applications. In Japan, energy security is the main

**TABLE 9.9**
**A Summary of Uptake Goals of Hydrogen and Fuels in Various Countries**

| Country | Fuel Cell Cars | | | Hydrogen Refueling Stations | | |
|---|---|---|---|---|---|---|
| | 2020 | 2025 | 2030 | 2020 | 2025 | 2030 |
| Japan | 40,000 | 200,000 | 800,000 | 160 | 320 | 900 |
| China | 3,000 (only Shanghai) | 50,000 | 1 M | 100 | 1,000 | |
| South Korea | 10,000 | 100,000 | 630,000 | 100 | 210 | 520 |
| US | 0 | 3.3M | | 100 (California) | | |
| UK | 100% ZEV[a] by 2040 | | | 30 | 150 | |
| Germany | 100% ZEV[a] by 2040 | | | 400 | | |

*Source:* From Tlili, O., C. Mansilla, D. Frimat, Y. Perez, "Hydrogen market penetration feasibility assessment: Mobility and natural gas markets in the US, Europe, China and Japan." *International Journal of Hydrogen Energy* 44(31), 2019, Doi: 10.1016/j.ijhydene.2019. With permission; From Staffell, I., D. Scamman, A.V. Abad, P. Balcombe, P.E. Dodds, P. Ekins, N. Shah, K.R. Ward, "The role of hydrogen and fuel cells in the global energy system." *Energy and Environmental Science*, 2019, Doi: 10.1039/c8ee01157e. With permission.

[a]  Zero emission vehicle.

**TABLE 9.10**

**A Summary of Support Presented by Different Countries for Hydrogen and Fuel Cells in 2017**

| Country | Fuel Cell Vehicles | Hydrogen Refueling |
| --- | --- | --- |
| Japan | $147 m | $61 m |
| China | $1700/kW (around $5,700 per vehicle) | $1.1 m per unit |
| South Korea | $5.5 m (around $31,000 per vehicle) | |
| US | $13,000 per vehicle | 30% of cost (~ $30,000) |
| UK | $3.3 m (~60% of refueling cost) | |
| Germany | $4,000 per vehicle | $466 m |

*Source:* From Tlili, O., C. Mansilla, D. Frimat, Y. Perez, "Hydrogen market penetration feasibility assessment: Mobility and natural gas markets in the US, Europe, China and Japan." *International Journal of Hydrogen Energy* 44(31), 2019, Doi: 10.1016/j. ijhydene.2019. With permission; From Staffell, I., D. Scamman, A.V. Abad, P. Balcombe, P.E. Dodds, P. Ekins, N. Shah, K.R. Ward, "The role of hydrogen and fuel cells in the global energy system." *Energy and Environmental Science*, 2019, Doi: 10.1039/c8ee01157e. With permission.

reason to promote hydrogen to support the national industries and lessen the environmental impacts by developing the three stages to make Japan a hydrogen society by (i) encouraging FCEVs, and hydrogen production, (ii) integrating and developing the supply chains of hydrogen by 2030 into the energy system, and (iii) establishing the supply of zero-carbon hydrogen by 2040. Similarly, China has made a policy to reduce the issues of urban air quality by boosting economic growth via fabricating hydrogen fuel cells as a part of the MC2025 (Made in China 2025) policy. Even the France and UK made the policy to terminate the sale of diesel and petrol-based cars from 2040 and the Netherlands from 2030. Norway also plans to end the sale of diesel cars from 2025 and replace them with hydrogen passenger cars. Despite all of this, further policies, development, research projects, and funding are needed to widen the awareness of hydrogen and fuel cell technology to increase the acceptance of the public. A stable framework of policy and obvious long-term vision are the most important things from society's perspectives. The GDP per capita is another crucial factor that is directly related to the market uptake of hydrogen and fuel cells. However, in FCEVs area, the United States led with 2,750 sold FCEVs by 2017, which is more than the combined sold figure of FCEVs in Japan and Europe. The Fuel Cell Vehicle Technology Roadmap 2016 (FCVTR-2016) of China gained the target of 5,000 FCEVs in 2020 with the more FCEVs (in millions shown in Table 9.5) in 2030 (Tlili et al. 2019; Staffell et al. 2019). The EU has anticipated around 50,000 fuel cell systems by 2020, but in the "ene.project" they installed only 1,046 systems and the rest of the systems will be installed by 2021 indicating the rise in the market size of hydrogen and fuel cells globally.

## 9.8   CONCLUSION

Energy production, storage, and delivery technologies remain in a ceaseless state of development for the supply of sustainable, robust, and clean energy. Therefore, energy is perhaps the hot topic in the 21st century. With the fast exhaustion of fossil fuels and with progressive contamination of atmosphere caused by huge consumption of fossil fuels, there is an appeal to utilize energy and to look for inexhaustible and clean fuel sources that can substitute fossil fuels to empower the practical improvement of both the economy and society. Hydrogen as a fuel has the potential to replace fossil fuels as it is recognized as zero-carbon emission fuel or green fuel; therefore, it neither contributes to pollution nor global warming. In this chapter, the detailed discussion of the production of hydrogen by electrolysis method concludes that it is the most efficient, economic, and green way to generate hydrogen as compared to the other conventional production methods of hydrogen. The storage of hydrogen is a crucial issue after production and various methods of hydrogen storage have been discussed in this chapter. Hydrogen and fuel cells are not synonymous; they can either utilize separately or in combination depending on the type of fuel cell technology. The high- and low-temperature fuel cells have their own benefits and drawbacks and can be used in a variety of applications depending on the requirements. FCEVs, hydrogen powertrains, fuel cell buses, trucks, and motorbikes are the various non-stationary applications of hydrogen and fuel cells. However, SOFCs cover a major part of stationary applications, e.g., stacks, GT, and other hybrid power systems of hydrogen and fuel cells. However, the high cost and durability challenges of hydrogen and fuel cells are the crucial issues that hinder them to become mature technology. The commercialization of hydrogen and fuel cells requires a well-defined and focused energy policy, which is also one of the greatest challenges for hydrogen and fuel cells. The review of market and policy trends about hydrogen and fuel cells concludes that Japan, Korea, China, the USA, and few European countries are enhancing the market value of hydrogen and fuel cells by investing more and more in this technology. Energy generation through hydrogen fuel cells will be a hot topic in the future and the commercialization of fuel cells will not take a long time as the research on them is boosting day by day.

## REFERENCES

Abdelkareem, M.A., K. Elsaid, T. Wilberforce, M. Kamil, E.T. Sayed, and A. Olabi. 2021. "Environmental aspects of fuel cells: A review." *Science of the Total Environment* 752. Doi: 10.1016/j.scitotenv.2020.141803.

Abdelkareem, M.A., N. Morohashi, and N. Nakagawa. 2007. "Factors affecting methanol transport in a passive DMFC employing a porous carbon plate." *Journal of Power Sources* 172(2). Doi: 10.1016/j.jpowsour.2007.05.015.

Abe, J.O., A.P.I. Popoola, E. Ajenifuja, and O.M. Popoola. 2019. "Sciencedirect hydrogen energy, economy and storage : Review and recommendation." *International Journal of Hydrogen Energy* 44(29): 15072–86. Doi: 10.1016/j.ijhydene.2019.04.068.

Ahluwalia, R.K., X. Wang, and J.-k. Peng. 2014. "Fuel cells systems analysis" *US Department of Energy Hydrogen and Fuel Cells Program 2011 Annual Merit Review and Peer Evaluation Meeting.* Washington, DC: US Department of Energy.

Alias, M.S., S.K. Kamarudin, A.M. Zainoodin, and M.S. Masdar. 2020. "Active direct methanol fuel cell: An overview." *International Journal of Hydrogen Energy.* Doi: 10.1016/j.ijhydene.2020.04.202.

Barbir, F., T.N. Veziroğlu, and H.J. Plass. 1990. "Environmental damage due to fossil fuels use." *International Journal of Hydrogen Energy* 15(10): 739–49. Doi: 10.1016/0360-3199(90)90005-J.

Baroutaji, A., A. Arjunan, M. Ramadan, J. Robinson, A. Alaswad, M.A. Abdelkareem, and A.G. Olabi. 2021. "Advancements and prospects of thermal management and waste heat recovery of PEMFC." *International Journal of Thermofluids* 9. Doi: 10.1016/j.ijft.2021.100064.

Basri, S., S.K. Kamarudin, W.R.W. Daud, and Z. Yaakub. 2010. "Nanocatalyst for direct methanol fuel cell (DMFC)." *International Journal of Hydrogen Energy.* Doi: 10.1016/j.ijhydene.2010.05.111.

Brandon, N.P., and Z. Kurban. 2017. "Clean energy and the hydrogen economy." *Philosophical Transactions of the Royal Society A: Mathematical, Physical and Engineering Sciences* 375(2098). Doi: 10.1098/rsta.2016.0400.

Burlatsky, S.F., V. Atrazhev, N. Cipollini, D. Condit, and N. Erikhman. 2019. "Aspects of PEMFC degradation." *ECS Transactions* 1(8). Doi: 10.1149/1.2214557.

BvDEP. 2019. "ITM power PLC." *Fame* 49(February): 30–80.

Chen, K., and S.P. Jiang. 2016. "Review—materials degradation of solid oxide electrolysis cells." *Journal of The Electrochemical Society* 163(11): F3070–83. Doi: 10.1149/2.0101611jes.

Comar, C. L., and N. Nelson. 1975. "Health effects of fossil fuel combustion products: Report of a workshop." *Environmental Health Perspectives* 12(December): 149–70. Doi: 10.1289/ehp.7512149.

Committee on Climate Change. 2017. "2017 Report to parliament - meeting carbon budgets: Closing the policy gap." *Microbiology and Molecular Biology Reviews : MMBR* (68) (June): 203.

Cook, B. 2002. "Introduction to fuel cells and hydrogen technology." *Engineering Science and Education Journal* 11(6). Doi: 10.1049/esej:20020601.

Cummins. 2021. "Hydrogenics." https://www.cummins.com/.

Dogdibegovic, E., Y. Cheng, F. Shen, R. Wang, B. Hu, and M.C. Tucker. 2021. "Scaleup and manufacturability of symmetric-structured metal-supported solid oxide fuel cells." *Journal of Power Sources* 489. Doi: 10.1016/j.jpowsour.2020.229439.

Durbin, D.J. 2013. "Review of hydrogen storage techniques for on board vehicle applications," 1–23. Doi: 10.1016/j.ijhydene.2013.07.058.

Elnozahy, A., A.K.A. Rahman, and H.H Ali. 2014. "A cost comparison between fuel cell, hybrid and conventional vehicles." *In Proceedings of the 16th International Middle-east Power Systems Conference—MEPCON*, Cairo, Egypt, 23–25.

Ferreira, J., J. Leitão, A. Monteiro, M. Lopes, and A.I. Miranda. 2017. "National emission ceilings in Portugal—trends, compliance and projections." *Air Quality, Atmosphere and Health* 10 9). Doi: 10.1007/s11869-017-0496-6.

Gerlici, J., M. Gorbunov, K. Kravchenko, O. Prosvirova, T. Lack, and V. Hauser. 2018. "Assessment of innovative methods of the rolling stock brake system efficiency increasing." *Manufacturing Technology* 18 (1). Doi: 10.21062/ujep/49.2018/a/1213-2489/MT/18/1/35.

Hagen, A., R. Barfod, P.V. Hendriksen, Y.-L. Liu, and S. Ramousse. 2006. "Degradation of anode supported SOFCs as a function of temperature and current load." *Journal of The Electrochemical Society* 153(6). Doi: 10.1149/1.2193400.

Hamdan, M. 2011. "*PEM Electrolyzer Incorporating an Advanced Low-Cost Membrane. No. DOE/GO/18065-22*" Giner, Inc./Giner Electrochemical Systems, LLC, Newton, MA.

Hashem Nehrir, M., and C. Wang. 2015. Fuel cells. In *Electric Renewable Energy Systems*. Elsevier Inc. Doi: 10.1016/B978-0-12-804448-3.00006-2.

Hussain, S., and L. Yangping. 2020. "Review of solid oxide fuel cell materials: Cathode, anode, and electrolyte." *Energy Transitions* 4 (2). Doi: 10.1007/s41825-020-00029-8.

Ijaodola, O., E. Ogungbemi, F.N. Khatib, T. Wilberforce, M. Ramadan, Z.E. Hassan, J. Thompson, and A.G. Olabi. 2018. "Evaluating the effect of metal bipolar plate coating on the performance of proton exchange membrane fuel cells." *Energies* 11 (11). Doi: 10.3390/en11113203.

International Energy Agency. 2015. "Technology roadmap: Hydrogen and fuel cells." International Energy Agency (IEA): Paris, France.

Iordache, I., A.V. Gheorghe, and M. Iordache. 2013. "Towards a hydrogen economy in Romania: Statistics, technical and scientific general aspects." *International Journal of Hydrogen Energy* 38(28): 12231–40. Doi: 10.1016/j.ijhydene.2013.07.034.

Ito, H. 2017. "Economic and environmental assessment of phosphoric acid fuel cell-based combined heat and power system for an apartment complex." *International Journal of Hydrogen Energy* 42 (23). Doi: 10.1016/j.ijhydene.2017.05.038.

Khan, M.Z., R.H. Song, M.T. Mehran, S.B. Lee, and T.H. Lim. 2021. "Controlling cation migration and interdiffusion across cathode/interlayer/electrolyte interfaces of solid oxide fuel cells: A review." *Ceramics International*. Doi: 10.1016/j.ceramint.2020.11.002.

Kim, J., S. Sengodan, S. Kim, O. Kwon, Y. Bu, and G. Kim. 2019. "Proton conducting oxides: A review of materials and applications for renewable energy conversion and storage." *Renewable and Sustainable Energy Reviews*. Doi: 10.1016/j.rser.2019.04.042.

Kulkarni, A., and S. Giddey. 2012. "Materials issues and recent developments in molten carbonate fuel cells." *Journal of Solid State Electrochemistry*. Doi: 10.1007/s10008-012-1771-y.

Larrosa-Guerrero, A., K. Scott, I. M. Head, F. Mateo, A. Ginesta, and C. Godinez. 2010. "Effect of temperature on the performance of microbial fuel cells." *Fuel* 89 (12). Doi: 10.1016/j.fuel.2010.06.025.

Li, W., Y. Wang, and W. Liu. 2020. "A review of solid oxide fuel cell application." In *IOP Conference Series: Earth and Environmental Science*, Vol. 619. Doi: 10.1088/1755-1315/619/1/012012.

Li, Y., W. Zhang, Y. Zheng, J. Chen, B. Yu, Y. Chen, and M. Liu. 2017. "Controlling cation segregation in perovskite-based electrodes for high electro-catalytic activity and durability." *Chemical Society Reviews*. Doi: 10.1039/c7cs00120g.

Liu, H., J. Chen, D. Hissel, J. Lu, M. Hou, and Z. Shao. 2020. "Prognostics methods and degradation indexes of proton exchange membrane fuel cells: A review." *Renewable and Sustainable Energy Reviews* 123. Doi: 10.1016/j.rser.2020.109721.

Liu, Y., Z. Tu, and S.H. Chan. 2021. "Applications of ejectors in proton exchange membrane fuel cells: A review." *Fuel Processing Technology*. Doi: 10.1016/j.fuproc.2020.106683.

Lyle, C. 2018. "Beyond the Icao's Corsia: Towards a more climatically effective strategy for mitigation of civil-aviation emissions." *Climate Law*. Doi: 10.1163/18786561–00801004.

Martin, K.E., J.P. Kopasz, and K.W. McMurphy. 2010. "Status of fuel cells and the challenges facing fuel cell technology today." In *ACS Symposium Series*. Vol. 1040. Doi: 10.1021/bk-2010-1040.ch001.

McLean, G.F., T. Niet, S. Prince-Richard, and N. Djilali. 2002. "An assessment of alkaline fuel cell technology." *International Journal of Hydrogen Energy* 27(5). Doi: 10.1016/S0360-3199(01)00181-1.

McNicol, B.D., D.A.J. Rand, and K.R. Williams. 2001. "Fuel cells for road transportation purposes - yes or no?" *Journal of Power Sources* 100(1–2). Doi: 10.1016/S0378-7753(01)00882-5.

Mekhilef, S., R. Saidur, and A. Safari. 2012. "Comparative study of different fuel cell technologies." *Renewable and Sustainable Energy Reviews*. Doi: 10.1016/j.rser.2011.09.020.

Midilli, A., M. Ay, I. Dincer, and M. A. Rosen. 2005. "On hydrogen and hydrogen energy strategies I : Current status and needs." *Renewable and Sustainable Energy Reviews* 9(3): 255–71. Doi: 10.1016/j.rser.2004.05.003.

Moreno, G.N., M.C. Molina, D. Gervasio, and J.F.P. Robles. 2015. "Approaches to polymer electrolyte membrane fuel cells (PEMFCs) and their cost." *Renewable and Sustainable Energy Reviews*. Doi: 10.1016/j.rser.2015.07.157.

Mori, D., and K. Hirose. 2009. "Recent challenges of hydrogen storage technologies for fuel cell vehicles." *International Journal of Hydrogen Energy* 34(10): 4569–74. Doi: 10.1016/j.ijhydene.2008.07.115.

Muthukumar, M., N. Rengarajan, B. Velliyangiri, M.A. Omprakas, C.B. Rohit, and U. Kartheek Raja. 2020. "The development of fuel cell electric vehicles – A review." *Materials Today: Proceedings*. Doi: 10.1016/j.matpr.2020.03.679.

Nejat Veziroglu, T. 2012. "Conversion to hydrogen economy." *Energy Procedia* 29: 654–56. Doi: 10.1016/j.egypro.2012.09.075.

NextHydrogen. 2021. "Next hydrogen." https://nexthydrogen.com/.

Pollet, B.G., I. Staffell, and J.L. Shang. 2012. "Current status of hybrid, battery and fuel cell electric vehicles: From electrochemistry to market prospects." *Electrochimica Acta*. Doi: 10.1016/j.electacta.2012.03.172.

President, Executive Vice, and Renewable Energy. 2019. "Edf's vision and ambition on hydrogen Mitei's" The MIT Energy Initiative's 2019 Spring Symposium.

Pudasainee, D., V. Kurian, and R. Gupta. 2020. Coal: Past, present, and future sustainable use. In *Future Energy: Improved, Sustainable and Clean Options for Our Planet*. Elsevier Ltd. Doi: 10.1016/B978-0-08-102886-5.00002-5.

Ran, J., L. Ding, C. Chu, X. Liang, T. Pan, D. Yu, and T. Xu. 2018. "Highly conductive and stabilized side-chain-type anion exchange membranes: Ideal alternatives for alkaline fuel cell applications." *Journal of Materials Chemistry A* 6(35). Doi: 10.1039/c8ta05876h.

Rusman, N.A.A., and M. Dahari. 2016. "A review on the current progress of metal hydrides material for solid-state hydrogen storage applications." *International Journal of Hydrogen Energy* 41(28): 12108–26. Doi: 10.1016/j.ijhydene.2016.05.244.

Schüth, F. 2009. "Challenges in hydrogen storage." *European Physical Journal: Special Topics* 176(1): 155–66. Doi: 10.1140/epjst/e2009-01155-x.

Spinelli, M., S. Campanari, S. Consonni, M.C. Romano, T. Kreutz, H. Ghezel-Ayagh, and S. Jolly. 2018. "Molten carbonate fuel cells for retrofitting postcombustion $CO_2$ capture in coal and natural gas power plants." *Journal of Electrochemical Energy Conversion and Storage* 15(3). Doi: 10.1115/1.4038601.

Staffell, I., and P.E. Dodds. 2017. "The role of hydrogen and fuel cells in future energy systems." H2FC SUPERGEN, London, UK.

Staffell, I., D. Scamman, A.V. Abad, P. Balcombe, P.E. Dodds, P. Ekins, N. Shah, and K.R. Ward. 2019. "The role of hydrogen and fuel cells in the global energy system." *Energy and Environmental Science*. Doi: 10.1039/c8ee01157e.

Steele, B.C.H., and A. Heinzel. 2001. "Materials for fuel-cell technologies." *Nature*. Doi: 10.1038/35104620.

Stonehart, P., and D. Wheeler. 2006. "Phosphoric acid fuel cells (PAFCs) for utilities: Electrocatalyst crystallite design, carbon support, and matrix materials challenges." In *Modern Aspects of Electrochemistry*. Doi: 10.1007/0-387-25838-8_4.

Sun, Y., C. Shen, Q. Lai, W. Liu, D.W. Wang, and K.F. Aguey-Zinsou. 2018. "Tailoring magnesium based materials for hydrogen storage through synthesis: Current state of the art." *Energy Storage Materials* 10(January 2017): 168–98. Doi: 10.1016/j.ensm.2017.01.010.

Tlili, O., C. Mansilla, D. Frimat, and Y. Perez. 2019. "Hydrogen market penetration feasibility assessment: Mobility and natural gas markets in the US, Europe, China and Japan." *International Journal of Hydrogen Energy* 44(31). Doi: 10.1016/j.ijhydene.2019.04.226.

Tsipis, E.V., and V.V. Kharton. 2008. "Electrode materials and reaction mechanisms in solid oxide fuel cells: A brief review." *Journal of Solid State Electrochemistry* 12(11). Doi: 10.1007/s10008-008-0611-6.

Veziroğlu, T.N., and S. Şahin. 2008. "21st century's energy: Hydrogen energy system." *Energy Conversion and Management* 49(7): 1820–31. Doi: 10.1016/j.enconman.2007.08.015.

Wang, J. 2015. "Barriers of scaling-up fuel cells: Cost, durability and reliability." *Energy* 80. Doi: 10.1016/j.energy.2014.12.007.

Wang, Y., F. Chu, J. Zeng, Q. Wang, T. Naren, Y. Li, Y. Cheng, Y. Lei, and F. Wu. 2021. "Single atom catalysts for fuel cells and rechargeable batteries: Principles, advances, and opportunities." *ACS Nano.* Doi: 10.1021/acsnano.0c08652.

Wang, S., and S.P. Jiang. 2017. "Prospects of fuel cell technologies." *National Science Review* 4(2). Doi: 10.1093/nsr/nww099.

Wang, J., H. Wang, and Y. Fan. 2018. "Techno-economic challenges of fuel cell commercialization." *Engineering.* Doi: 10.1016/j.eng.2018.05.007.

World Health Organization. 2016. "Ambient air pollution: A global assessment of exposure and burden of disease." *Clean Air Journal* 26 (2). Doi: 10.17159/2410-972x/2016/v26n2a4.

Xu, H., L. Kong, and X. Wen. 2004. "Fuel cell power system and high power DC-DC converter." *IEEE Transactions on Power Electronics* 19(5). Doi: 10.1109/TPEL.2004.833440.

Yang, Y. 2013. "PEM fuel cell system manufacturing cost analysis for automotive applications." Austin Power Engineering LLC: Wellesley, MA, USA.

Zecca, A., and L. Chiari. 2010. "Fossil-fuel constraints on global warming." *Energy Policy* 38(1): 1–3. Doi: 10.1016/j.enpol.2009.06.068.

# 10 Energy Storage Systems

*Ghulam M. Mustafa*
University of Education Lahore

*Ghulam Ali*
National University of Sciences and Technology (NUST)

## CONTENTS

DOI: 10.1201/9781003315353-12

## 10.1   HISTORY OF ENERGY STORAGE

The ever-increasing usage of portable electronics and the development of the economy have attracted the interest of industrial developers and researchers to store energy in any form. Energy storage is essential to meet these challenges faced by economies across the globe. There are different types of energy storage systems as they can be stored in many forms like chemical, thermal, electrical, or electrochemical. The heating of geezers or air-conditioned cooling is produced using electricity through thermal energy. Electrical energy is being harvested from natural resources like solar and wind energy. The energy can be stored in chemical form through fossil fuels. The chemical and electrical energies are interconvertible in electrochemical energy storage systems for a specific period. The primary energy storage devices include batteries and supercapacitors [1]. To accomplish these energy needs, it is much needed to design such energy storage devices (ESDs) that have various aspects listed as: (i) their performance should be high, (ii) they must be light in weight and smaller in size, and (iii) they must be economical and versatile. To design ESDs, various inorganic, organic, and composites have been investigated extensively. Among these, transition metal oxides (TMOs), conducting polymers, and carbonaceous materials like carbon nanotubes (CNTs) and graphene remained at the top of the table. The fabrication of these materials with nanoscale dimensions paved the way for the technological revolution by exhibiting newly designed possibilities of electronic and electrochemical devices [2].

A panoramic view of the historical evolution of ESDs shows that the fabrication of ESDs started in the mid-19th century, when British scientist William Grove designed a prototype of a fuel cell in 1839 that can store a large amount of energy. Using Grove's approach for the generation of electricity, in the 1960s, the National Aeronautics and Space Administration (NASA) designed the first commercial fuel cell. To increase its ability to produce energy, many modifications were made. The other energy storage devices like batteries and capacitors were designed later by the researchers to store electrical energy. In 1859, a scientist named Gaston Plante manufactured the first lead-acid battery. Another scientist Emile Alphonse Faure firstly invented the sticky plates and then designed the first lead-acid battery at a commercial scale in 1880. To increase the storage capacity of the battery and to store charges, sulfuric acid and paste of lead powder were used to coat the lead plates. Using this process, various attempts have been made by the researchers to increase the storage capacity. Afterward, using Faure's technique, Sellon repeated the process to design a battery. They noticed an improved storage capacity compared to Faure's battery when the same paste was used on a perforated plate. An American oil company named Standard Oil of Ohio (SOHIO) developed the first-ever electronic double-layer capacitor in 1961. After that, the developers and researchers tend their interest in these devices and started working to improve their outputs. The energy storage applications got a new revolution with the fabrication of solar cell devices because they were designed to remarkably increase the field of energy storage [3]. Various types of materials including metal oxides and carbon-based materials along with hybrids and composites were then used by the researchers to increase the stability and conversion efficiency of these cells.

## 10.2   ENERGY STORAGE SYSTEMS

Over the centuries, a continuous improvement of human living standards demands high energy consumption. To meet these growing energy needs, it is much needed to manage additional energy sources. Different research groups have tried to manufacture such systems where greater efficiency can be achieved with minimal energy losses. Fossil fuels were used to provide energy in the form of heat, but their major drawback includes carbon dioxide emission, which limits their applications. Besides, the burning of fossil fuels releases some pollutants into the environment that indirectly harms human beings and other creatures on the planet. Therefore, we need to look forward to some alternate sources of energy. One of the alternate sources of energy is the nuclear power plant, which does not produce carbon dioxide and other pollutants like fossil fuels; however, their use involves some security concerns. Since the raw material required for this system is not economical, they are not suitable for domestic or commercial use [4]. The only solution to environmental problems such as global warming followed by air pollution is to use renewable energy sources and energy waste recovery systems. Renewable energy sources are wind, bioenergy, geothermal, solar, and hydropower. As renewable sources have significant fluctuations and intermittent nature, they cannot stand alone in power plants. It is much needed to design energy storage systems that can combine renewable energy sources to cope with the energy demands [5].

Despite the storage of excess energy, the energy storage systems have some other benefits, as discussed here:

- The emission of carbon dioxide and fuel consumption is reduced by utilizing renewable energy sources in energy storage systems.
- Because of the intermittent nature of smoothing renewable energy fluctuations, the balance between energy supply and demand is maintained.
- The issue of load shedding can be reduced in this way, and a huge capacity of storage can be attained.
- In the remote areas that are suffering the electricity deficit, sustainable and convenient energy storage systems are being used due to their flexibility.

To manage the intermittent nature and fluctuations of renewable sources, flexibility is provided by the energy storage system (ESS) to the systems. From another point of view, it will be worthy to say that the balance between energy supply and demand is maintained by ESS [6]. There are various beneficial aspects of the storage systems that are: (i) they can increase the grid stability and performance of the system; (ii) they can enhance the penetration of renewable energy sources; and (iii) the use of fossil fuel energy resources can be reduced, and (iv) the emission of pollutants to the environment can be controlled. Typically, three major energy storage systems are electrical, mechanical, and thermal energy storage systems. The electrical energy storage systems mostly consist of direct electrical or electrochemical energy storage, whereas the mechanical storage system is composed of flywheels, pumped hydro energy storage, compressed air energy storage, and gravity. In contrast, thermal energy is categorized into sensible and latent energy storage systems [7,8].

Recently, Behabtu et al. [9] presented an inclusive review on energy storage technologies (ESTs) and highlighted their potential role in the evolution of renewable energy sources, which comprises plenty of information. They established a comparison among different technologies based on technical, economic, and environmental perspectives. The key factors to probe the technical performance of any energy storage technology (EST) include power density, energy density, self-discharge, discharge time, response time, lifetime, efficiency, and maturity level of technology. Contrarily, the economic criteria are based on the capital cost (including power and energy) invested to generate 1 kWh of energy from these technologies. In addition, they also analyzed the environmental impact of different ESTs quantitatively and presented it very well. Maximum values of power range, energy density, power density, and round-trip efficiencies are presented in Table 10.1, while the estimated minimum capital cost for these technologies is presented in Table 10.2 [9].

## 10.3  THERMAL ENERGY STORAGE

Early human beings used thermal energy to cook their food. Similarly, human bodies need an ambient temperature for their survival. Apart from the other thermal energy sources, the sun is considered a major contributor of heat to the earth. It is much needed to maintain the thermal energy so that mechanical work, heat, and electricity can be generated [10,11]. Thermal energy can be attained by either natural ways or fuel sources, and some of the energy is lost. Utilizing fuels burning, energy needs are being fulfilled, but the drawback of burning these fuels is that they emit carbon dioxide gas that further affects the atmosphere (global warming). Owing to this reason, researchers have

**TABLE 10.1**

**Technical Characteristics of Pumped Hydro Energy Storage (PHES), Flywheel Energy Storage (FES), Compressed Air Energy Storage (CAES), Lead-Acid (Pb-A), Nickel-Cadmium (Ni-Cd), Sodium Sulfur (Na-S), Lithium-Ion (Li-ion), Vanadium Redox Battery (VRB) and Supercapacitor Energy Storage (SCES)**

| Energy Storage Technologies | Power Range (MW) | Energy Density (Wh/l) | Power Density (W/l) | Round-Trip Efficiency (%) |
|---|---|---|---|---|
| PHES | 10–5,000 | 0.5–1.5 | 0.5–1.5 | 65–87 |
| FES | 0–0.25 | 20–80 | 1,000–2,000 | 90–95 |
| CAES | 5–1,000 | 3–6 | 0.5–2 | 50–89 |
| Pb-A | 0–40 | 50–80 | 10–400 | 70–90 |
| Ni-Cd | 0–40 | 60–150 | 150–300 | 85–90 |
| Na-S | 0.05–34 | 150–250 | 130–230 | 85–90 |
| Li-ion | 0–100 | 200–500 | 500–2,000 | 90–97 |
| VRB | 0.3-3 | 20–70 | 0.5-2 | 85–90 |
| SCES | 0-0.3 | 2.5–15 | 500–5,000 | 95–98 |

**TABLE 10.2**

**Estimated Minimum Capital Cost and Environmental Impact of Pumped Hydro Energy Storage (PHES), Flywheel Energy Storage (FES), Compressed Air Energy Storage (CAES), Lead-Acid (Pb-A), Nickel-Cadmium (Ni-Cd), Sodium Sulfur (Na-S), Lithium-Ion (Li-ion), Vanadium Redox Battery (VRB) and Supercapacitor Energy Storage (SCES)**

| Energy Storage Technologies | Total Capital Cost | | Environmental Impact |
|---|---|---|---|
| | Power Cost $/kW | Energy Cost $/kWh | |
| PHES | 500–2,000 | 5–100 | High |
| FES | 250–350 | 500–1,000 | Very low |
| CAES | 400–800 | 2–50 | Medium |
| Pb-A | 200–300 | 120–150 | High |
| Ni-Cd | 500–1,500 | 800–1,500 | High |
| Na-S | 1,000–3,000 | 300–500 | High |
| Li-ion | 900–4,000 | 300–1,300 | Medium |
| VRB | 600–1,500 | 150–1,000 | Medium |
| SCES | 100–450 | 300–2,000 | Very low |

been trying to utilize renewable energy sources to generate and conserve energy because fossil fuel resources are limited. Energy conservation can be obtained through thermal energy storage [12]. A detailed discussion of heat sources is given below.

### 10.3.1 SOLAR THERMAL ENERGY

Earth receives solar radiation from sun-generated at the core of the sun due to the nuclear fusion reactions. It will be worth worthy to fulfill the world's total energy need if solar energy is completely harnessed. Concentrated solar power plants (CSP) or photovoltaic (PV) cells use solar radiation to generate electricity. The other way to use solar radiation by solar thermal appliances is their

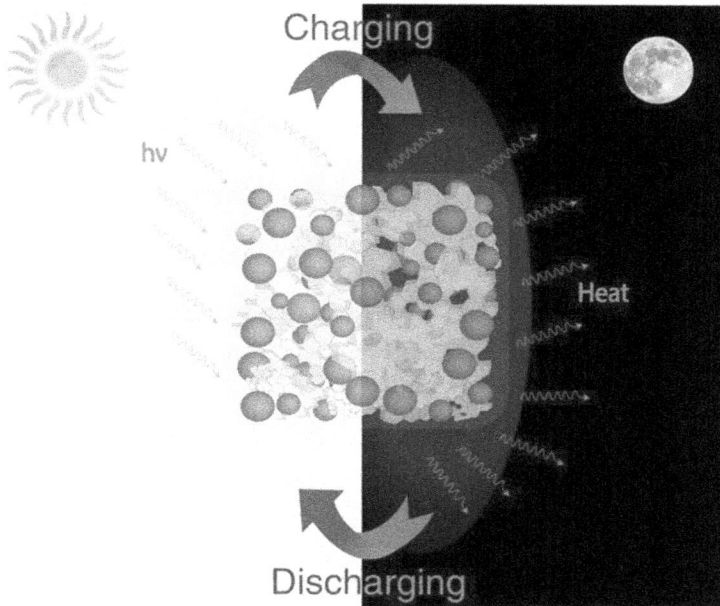

**FIGURE 10.1**  Schematic representation of solar thermal energy harvesting mechanism. (Adapted from Ref. [14].)

utilization in a hot water supply or absorption refrigeration and space heating [13]. The energy storage systems such as battery-based electricity and thermal energy storage systems are required during peak consumption hours. The comparison between these two systems is given here:

- A round-trip efficiency of 50%–100% in TES systems can be obtained where source energy form was stored as low-grade thermal energy. The thermal energy storage mechanism is most appropriate for thermal electricity generation plants, including nuclear reactors or CSP.
- A round-trip efficiency of 80%–100% in battery storage is achieved where source energy form was stored as high-grade electrical energy.
- The TES had improved cycle life and performance as compared to the battery technologies.
- There are limited toxicities in materials used as TES.

During the daytime, TES systems store solar thermal energy. During the night hours, this stored energy is used for electricity generation, and the graphical representation of such a process is shown in Figure 10.1 [14].

## 10.3.2  GEOTHERMAL ENERGY

Earth is one the best example of the TES system itself, as it is composed of heat in its layers. Below the earth's crust, such as the mantle along with outer and inner cores, the heat is stored that can be utilized. Inside the earth's crust, the underground water columns are heated by the heat emanating from the mantle. After an extended shutdown period, the geothermal energy resources can be fully recovered to their pre-exploitation states and so are considered long-term energy sources. The geothermal fluid is present with higher temperatures up to 180°C at lower depths, near the earth's plate tectonics. The geothermal plants are formed at this place owing to the complexity of drilling at high depths. The steam created with a temperature >150°C can generate electricity using geothermal fluids [15]. Solar thermal energy can only be used during the daytime, but geothermal energy can be used any time of the day depending upon the energy needs because of its availability during the whole day.

### 10.3.3 THERMAL ENERGY STORAGE SYSTEMS

In TES systems, the excess heat is stored in the form of thermal energy to be used later. There are two basic categories of thermal energy storage. The first one is sensible energy storage, whereas the other one is latent energy storage. In heat-sensitive energy storage systems, when a material is heated up, its temperature rises, so the heat gain is sensible. As the temperature rises, the internal energy of that system increases. However, in latent energy storage, the system deals with latent heat, which means there is a phase change experienced by the substance when the heat is given to the system or extracted from the system. Typically, a phase change is a change from solid to liquid phase or liquid to the solid phase. In this process, the state of the matter changes from solid to liquid, but the temperature during these phase changes remains the same. In sensible heat storage, we use high thermal mass, density, or specific heat. That means the amount of energy we stored is $Q = mC_p\Delta T = \rho V C_p \Delta T$, where m is the mass, $C_p$ is the specific heat, $V$ is the volume, $\rho$ is the density, and $T$ is the temperature. For a given $\Delta T$, $Q$ increases as density and specific capacity increase. For maximum energy storage, the product of $\rho C_p$ must be maximum. On the other hand, in latent heat storage, we use the heat of fusion such as ice storage, phase change materials, and molten salts. The phase change materials are organic alkanes and hydrated salts [16].

## 10.4 MECHANICAL ENERGY STORAGE

Due to the various interesting aspects such as cost, environmental impact, and sustainability, mechanical energy storage (MES) systems have attracted the interest of researchers toward their importance. The categories of such systems comprised flywheel, pumped hydro, and compressed air energy storage systems. The main problem is finding a suitable system among these based on load nature, available space, and source of energy. The common benefits among these MESSs include no environmental effects and a relatively fast response. As these energy storage systems create fewer pollutants either in construction levels or in operational levels, they play a vital role in enhancing air quality [17]. The detail of each category of MESSs is given herein.

### 10.4.1 FLYWHEEL ENERGY STORAGE

The energy can be stored in the flywheel energy storage (FES) system using the rotation of mass for the short term in the form of kinetic energy. By considering the cost-effective parameter and their fast response, the flywheels are suitable energy storage systems. The major use of flywheels is the regeneration of braking power in trains, cars, and locomotives. To give power for acceleration, the focus should be on storing the braking energy lost. Some parameters should be considered first for storing the energy, that is: (i) the peak power should be decreased, (ii) efficiency must be improved, and (iii) the fuel consumption and emissions should be reduced. The use of material depends on the flywheel's shape factor (K) that is proportional to the specific energy stored per unit mass. The energy stored by the flywheel is given by the expression $E = \frac{1}{2} I\omega^2$, where $\omega$ and $I$ are rotational speed and moment of inertia, respectively. To calculate the ratio between specific energy stored and unit mass, the expression $E/m = K \sigma_{max}/\rho$ is used. In this expression, $E$, $\sigma_{max}$, $\rho$, and $K$ are stored energy, maximum stress, the density of the flywheel, and shape factor, respectively. Because of the intermittency nature of the flywheels, they can be combined with renewable sources to improve the lifetime of batteries rather than their use in energy storage systems.

## 10.4.2  Pumped Hydro Energy Storage

Owing to the flexibility, low maintenance cost, and long life cycle, the pumped hydro energy storage (PHES) is considered one of the best mechanical energy storage systems. There are three basic parts of this system: the pumping system, the upper reservoir, and the hydro turbine. In the presence of excess energy, the water can be utilized again when required, as caused by its pumping from lower to lower reservoir. Depending on the potential gravitational energy, the positive pressure difference is provided by the upper container concerning the lower container. Therefore, by using hydro turbines, the power can be produced through potential gravitational energy.

## 10.4.3  Compressed Air Energy Storage

To store the excess energy underground, compressed air energy storage (CAES) is used. For the generation of electricity, the compressed air should be exposed to heat addition before entering into the expander. Some amount of the compressed air can pass through a natural gas turbine that will generate electricity, and the remaining can be utilized to heat the compressed air flow. There is no high maintenance needed for the compressed air energy storage system as it is an eco-friendly energy storage system. The porous rocks, as well as leached out the salt dome and mired hard rock storage facility, are various categories of underground air storage. For commercial and industrial utilizations, underground air storage is being used [18].

## 10.5  ELECTROMAGNETIC ENERGY STORAGE

Electricity storage in the form of electrical energy is problematic because of many factors, such as high device losses and poor efficiencies. Therefore, to be stored in a secure, healthy, accessible, and environmentally sustainable manner, electrical energy has to be turned into another form of energy. Electrical energy is converted into electromagnetic energy in electromagnetic energy storage devices by various technologies, such as condensers and superconducting electromagnets. The purpose of the report by Wang et al. [19] is to create a novel phase-shift nanocomposite for effective conversion and storage of electromagnetic and solar energy. By using $PEG/SiO_2$ as the form-stable phase-shift material and well-dispersed $Fe_3O_4$-functionalized graphene nanosheets as the energy converter, the multifunctional nanocomposites are formulated. The magneto-thermal effect of $Fe_3O_4$ nanoparticles and the light graphene harvesting properties in $Fe_3O_4$-GNS allowed for effective magnetic or light-to-heat conversion under alternating magnetic fields or solar illumination. Due to PEG's melting phase transition, the heat produced during the energy conversion process is deposited in the form-stable phase change material. Besides, nanocomposites exhibit high energy storage capability with excellent thermal stability and reversibility of the phase change material. There are two electrical conductors in capacitors, which are isolated by a dielectric material from one another. There is an accumulation of electrical charges on the side of the applied current as an electrical current is added to the conductors. As a consequence, between the conductor plates, energy is contained within the electrical field. There is a greater energy density for electrochemical capacitors than for other capacitors. At or above 600 V, certain double-layer capacitors have a voltage rating. A hybridized electromagnetic-turboelectric Nanogenerator for combined scavenging of wind energy, with an electromagnetic generator and a turboelectric Nanogenerator. Under a loading resistance of 10 M, the TENG can supply maximum output power of about 1.7 mW, whereas the EMG can supply maximum output power of about 2.5 mW under a loading resistance of 1 kΩ [20]. To store the energy generated from the TENG in a capacitor, a power management circuit has been developed, resulting in an improved energy storage performance of up to 112% compared to the conventional rectifier. By adjusting the inductance and flipping the on/off period, the charging rate of the capacitor was optimized using the TENG. The hybridized Nanogenerator has a greater capacity to sustainably control temperature sensors relative to the individual TENG or EMG, exhibiting the future applications of hybridized. Using the hybridized Nanogenerator, a temperature sensor can be sustainably powered, exhibiting the viability of the hybridized Nanogenerator as a practical power source for the realization of self-powered sensor systems. Using several gear modules and an array of disc Hal Bach magnets, we proposed a non-resonant finger-triggered electromagnetic energy harvester that can produce considerable voltage and power under low input-frequency vibrations. Using a multiplication gear module, the suggested EMEH transforms the applied low frequencies into higher frequencies and forwards the linear excitation by gear sections to a rotational module. The multiplier gear module was used to improve the magnet's rotation speed. In the energy generation portion, an array of disc Hal Bach magnets were used for concentrating the magnetic flux toward the coils. An open-circuit voltage of 1.39 V was produced by the proposed energy harvester at an average power of 7.68 mW, with an optimum load at an input frequency of 3 Hz. The conservation and dissipation processes of electromagnetic energy in nanostructures are based on both the properties of the material and the geometry. Using rigorous coupled-wave analysis, the distributions of local energy density and power dissipation in nano gratings were investigated. The improvement of absorption was seen to be followed by the improvement of energy storage for both the material at the resonance of its dielectric function defined by the classical Lorentz oscillator and the nanostructures at the resonance caused by its geometric arrangement. The presence of a strong local electric field at the geometry-induced resonance in nano gratings is directly related to the overall storage of electrical energy. Local energy storage and dissipation research will also help obtain a greater understanding of global energy storage and dissipation in the photovoltaic and heat transfer applications of nanostructures.

## 10.6   ELECTROCHEMICAL ENERGY STORAGE

Nonconventional energy devices have been recently designed due to the increasing need for electricity and the reduction of fossil fuels. The most substantial energy storage devices are batteries and supercapacitors. The following subunits explained them very well.

### 10.6.1   SUPERCAPACITORS

In contrast with the other energy storage devices, supercapacitors are most worthwhile and rechargeable. Their power density is high, and they have a fast charge/discharge capacity. Due to these aspects, they are much-demanding energy storage devices. They have a resemblance with solid electrode batteries, but their power density is much higher than batteries. They need low maintenance, and their life cycle is comparatively long. The traditional supercapacitor is made up of two electrodes, an electrolyte, and a separator that separates them, as shown in Figure 10.2.

There are various parameters, including electrodes and separator along with electrolyte and current collector that are responsible for the performance of a supercapacitor. They can attain power density up to >10 W/kg and their cycle life is a hundred times longer than batteries. They are charged or discharged within seconds, and their self-discharging is also low which affects their performances [3]. Through the capacitance of the supercapacitors, the amount of stored charge can be measured from the following Eq. (10.1), while Eqs. (10.2) and (10.3) can be used to measure the energy density and maximum power, respectively.

$$C = I/m(dV/dt) \tag{10.1}$$

$$\text{Energy density} = \frac{1}{2}CV^2 \tag{10.2}$$

$$\text{Maximum power} = Vi^2/4m\,R_s \tag{10.3}$$

Herein, $m$, $I$, and $dV/dt$ are the mass, current (measured in amperes), and the rate of change of voltage (measured in V/s), respectively. The term $V$ in the second equation is the voltage ($V$), and $V_i$ and $R_s$ in

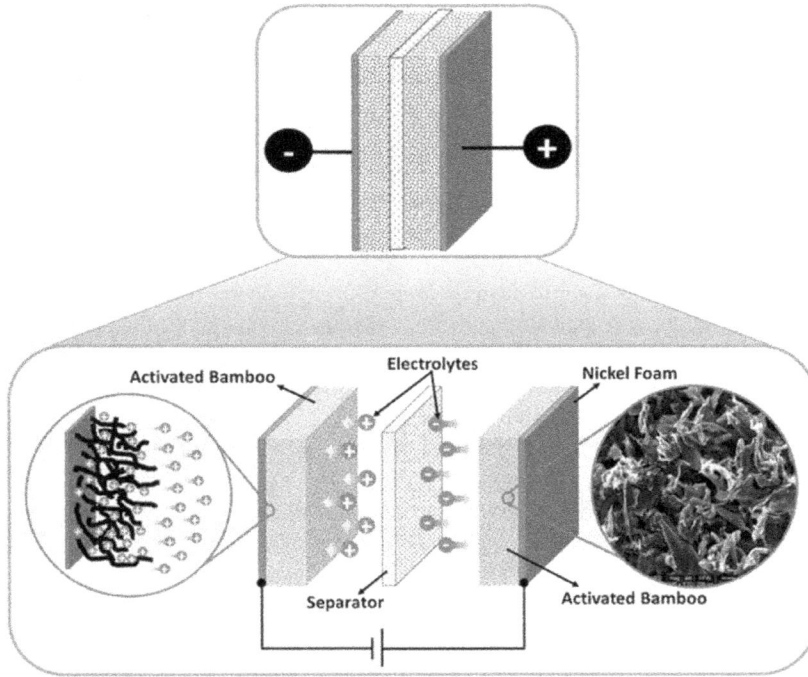

**FIGURE 10.2** Schematic representation of supercapacitor with its individual components. (Adapted from Ref. [22].)

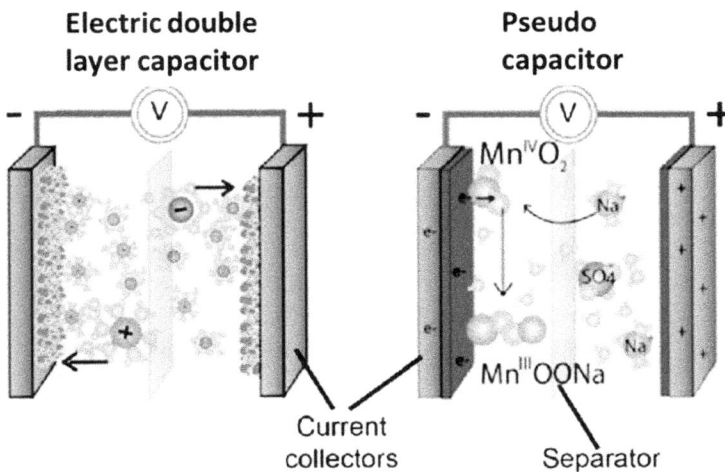

**FIGURE 10.3** Schematics of electric double-layer capacitor and pseudo-capacitor. (Adapted from Ref. [23].)

the third equation are initial voltage and equivalent series resistance, respectively. Depending on the charge storage mechanism, the supercapacitors can be categorized into two forms. The first category is an electrochemical double-layer capacitor (EDLC) with an almost equal value of specific capacitance for electrodes as a parallel plate capacitor, and the second category is a pseudocapacitor (PC). The charge is stored at the electrolyte/electrode interface in EDLCs. In comparison to EDLCs, PC stores charge using faradaic reactions [22]. The schematics of EDLCs and PCs are shown in Figure 10.3.

In supercapacitors, two highly conductive electrodes are separated by an electrolyte. The mobile ionic species are present in an electrolyte. When a voltage is applied to the conventional capacitor, the charges are stored at the surface of the electrodes. In supercapacitors, the adsorption of anions and cations balanced the charges on the electrodes. The highly reversible ion adsorption mechanism is responsible for the excellent performances of the supercapacitors. Batteries and supercapacitors are two well-known electrochemical energy storage devices. Various benefits of supercapacitors make them superior over batteries, and some are listed here. Batteries are unsafe to use at the industrial level, and their lifetime is also short, whereas supercapacitors are secure to be used and they last longer. They are replacing the conventional capacitors and batteries as emerging devices for storing and converting energy. Due to various facts like high power density, the greater capability of charge/discharge, long cycle stability, and great efficiencies, they exhibit many advantages and potentials. Currently, supercapacitors are being used in various fields like aerospace, military fields, energy automotive, portable electronic devices, etc. In EDLCs, the electrode materials are made up of porous carbons with a high accessible surface that acts as an electric sponge. During charging, the adsorption of ions occurs, and during discharge, these adsorbed ions are released. Researchers are busy finding materials that can store charges by other physical mechanisms. In PCs, on the oxide-based material surface, redox reactions occur. The concept of charge storage in PCs is now improving due to the nanostructured oxides and hybrid carbon/oxide electrodes. In literature, we have seen that researchers are either focusing on electrolytes or investigating the characteristics of the electrode materials [24].

### 10.6.1.1   Electrochemical Double-Layer Capacitors

At the interface of electrolyte and electrode, electrostatic attractions form the oppositely charged layers. These reactions occur between the charges (at the surface of the electrode) and the ions (in an electrolyte). In this way, the charges are stored in supercapacitor electrodes. In EDLCs, large ion adsorption is produced due to the porous carbon used as electrode materials providing high surface area. Helmholtz proposed a model in 1853 to calculate the double-layer capacitance that is:

$$C = \varepsilon_o \varepsilon_r \quad A/d \tag{10.4}$$

Here, $\varepsilon_o$, $\varepsilon_r$, $A$, and $d$ are the vacuum permittivity, electrolyte relative permittivity, electrode surface area, and the effective thickness of the double layer, respectively. The electrochemical response of the materials can be investigated through different techniques like small-angle X-ray or neutron scatterings. The nanoscale graphitic units make up the nanoporous carbons. The nanoporous carbons have pores with ranges between micropores to mesopores as they did not show three-dimensional long-range ordering like nanotubes or slit pores. To design effective supercapacitors, reliable and precise characterizations are required to check the electrochemical performance of the carbons at the porous electrolyte/carbon interface [25]. Theoretical models are used to determine the pore size distributions of porous specimens and specific surface area from gas adsorption methods.

### 10.6.1.2   Charge Storage Mechanism

In EDLCs, the charge is stored by the adsorption of ions at the surface of the electrodes. Over a century, the theoretical electrochemistry based on the Gouy–Chapman–Stern theory predicted that the polarization of the electrolyte balanced the electrode charge near an extended surface. There are two important parameters upon which the Debye length depends; (i) solvent permittivity and (ii) electrolyte concentration. Generally, the Debye length is between 1 and 10 nm range where the decay of the electrostatic potential and ionic charge distribution happens. The comparison of electrochemical interfaces between the solvent-based electrolyte and pure ionic liquids has been recently claimed by the surface force apparatus experiments. From the Gouy–Chapman–Stern theory, it can be predicted that the application of potential increased the capacitance in the planar graphitic electrodes. The nature of the electrolyte depends upon the differential capacitances having different shapes, like bell or camel shape, as explained by the Mean-field theories. In supercapacitors, these capacitance values are too low. For extreme confinement, where the diameter of the ions is equivalent to the pore sizes, the imbalance amount of counter-ions and co-ions inside the pores compensated for the electronic charge inside the electrode. One or more mechanisms are noticed as a function of voltage in a combined system of electrodes and electrolytes (choice of a solvent and nature of the ions). At low voltage, the ion swapping can be explained by either infrared spectroscopy or molecular dynamics and nuclear magnetic resonance (NMR). At high voltage or for large ion sizes, the adsorption of the counter-ions can also be seen while de-sorption of co-ions is less frequently noticed. Recently, an NMR study on a solvent-based electrolyte revealed that electrode polarization is an important factor to describe the charge storage mechanism. For negative polarization, the counter-ion adsorption was noticed, whereas, for positive polarization, the ion swapping was seen. The excess ionic charge is attained by various factors such as mobilities and relative sizes of the co-ions or counter-ions, as well as ion reordering and kinetic phenomena.

### 10.6.1.3 Dynamics of Charging and Discharging

The supercapacitors possess high power densities as compared to the batteries because they need a few seconds to charge or discharge. It is required for a supercapacitor to have high capacitance without reducing actual power density. Generally, it is noticed that when microporous carbons are used for supercapacitors, the dynamics of liquids remarkably slow down. Luckily, such results were not attained when the experiment was done using carbide-derived carbons (CDCs). When the electrolytes were designed by dissolving the organic ions into acetonitrile, the room temperature measured values of resistivity lie in the range of 50–200 $\Omega$-cm. To observe the dynamical response of the supercapacitors, their fact charging delays the application of the in situ methods. Mostly, the charging time of supercapacitors is shorter than the time required to record a spectrum. Though small-angle X-ray spectroscopy or infrared spectroscopy can be used to investigate the charge/discharge cycles of supercapacitors with a scan rate as low as 5 mV/s. Molecular simulations have been used to study the microscopic origin of fast charging. Particularly, the routes of the molecules are defined by the molecular dynamics that can further be used to extract transport characteristics, including diffusion coefficients [26].

### 10.6.1.4 Oxide-Based Supercapacitors

Some materials were recognized in the 1970s that experience redox reactions near or at their surfaces. The electrochemical characteristics of such materials were similar to supercapacitors where the energy can be stored at the electrolyte/electrode interface forming the electrical double layer. Due to these redox reactions, such pseudocapacitive materials possess greater charge storage than EDLCs. Therefore, the pseudocapacitors have attracted the interest of the researchers toward them due to the long life cycle, the power density of EDLCs, and the high energy density of typical battery electrode materials [27].

### 10.6.1.5 The Mechanism of Pseudo-Capacitors

The oxides of manganese and ruthenium are the most studied pseudocapacitors. Particularly, the pseudocapacitive response is observed in such materials that exhibit high charge/discharge rates and high energy densities representing electrochemical signatures. To depict the pseudocapacitance, a system was designed using niobium pentoxide. For lithium, 70% of its theoretical capacity having a value of 125 mAh/g, was stored in just a minute. However, the charge stored through redox reactions took place not only at the surface but also in the bulk of the material. The semi-infinite diffusion did not control the charge storage mechanism, that is, intercalation pseudocapacitance. The fast kinetic behavior was noticed in double-layer capacitors instead of the redox reactions for charge storage. The electrochemical pseudocapacitive behaviors are related to some important parameters listed here. The potential linearly depends upon the state of charge, and redox peaks are observed at small voltages as well as the charge storage capacity is independent of rate. The pseudocapacitive materials possess all these aspects.

### 10.6.1.6   Pseudo-intercalation Reactions

On intercalation, there is no phase transformation in the structure of pseudocapacitive materials. The synthesis of nanoscale material had a great impact on the pseudocapacitive materials as it is noticed that if the crystallite is lesser than 20 nm, the electrochemical properties tend to become more capacitor-like. Due to the suppression of a phase transition or a large number of surface sites for charge storage, the potential dependence on the state of charge becomes linear. The reduced diffusion distance provides a route toward greater power density; hence, the smaller crystallite sizes influenced the kinetics. The pseudocapacitance can either be extrinsic or intrinsic to a material. The shape or size of the particle had no influence on the electrochemical signatures of double-layer capacitors in an intrinsic pseudocapacitor. Comparatively, the only nanoparticle that represents pseudocapacitor properties in an extrinsic pseudocapacitor, and these materials, when present in bulk form, did not represent a pseudocapacitive response. A wide range of sulfides, nitrides, and transition metal oxides are being used by the researchers to design such materials that exhibit the battery-like high energy density and capacitor-like high power density.

### 10.6.1.7   Hybrid Supercapacitors

It is the need of the hour to manufacture composite hybrid materials that possess increased energy density along with high power density. The next-generation hybrid supercapacitors are created using these fast discharge/charge materials. The high-rated $Li_4Ti_5O_{12}$ (LTO) hybrid supercapacitors, lithium-ion insertion material, have been used that have a discharge/charge rate of 30C with a lithium capacity of 20 mAh/g. The oxides of manganese and ruthenium work in an aqueous electrolyte, whereas the potential window is also limited to about 1 V. On the other hand, LTO works in non-aqueous electrolytes operated at high voltage. During the lithiation, the LTO represents phase transformation as well as a voltage plateau; thus, it cannot be considered a pseudocapacitive material. During the de-intercalation reaction, the LTO supports rapid $Li^+$ ion transport and exhibits zero strain. Despite that, the LTO can be used in a hybrid supercapacitor as an electrode. Such devices combine a capacitive electrode with a faradaic battery. Using a capacitive porous-carbon positive electrode and Li-ion battery negative graphite electrode, the first generation of commercial hybrid supercapacitor (Li-ion capacitor) was designed. Such types of materials like LTO could improve the power capability and energy density of the devices [28,29].

## 10.6.2   Batteries

Batteries energy storage systems (BESS) are one of the extensively used electrochemical energy storage devices where the electrical energy is stored as chemical energy. There are two electrodes in batteries, namely the cathode and anode. The solid or liquid electrolyte just permits the movement of ions and opposes the electrons. The electrolyte is an aqueous solution containing sulfuric acid in the lead-ion battery. A separator between the two electrodes prevents the touching of these electrodes. Regardless of the charging or discharging of the system, the conventional electropositive electrode is termed an anode. The applications of batteries are countless, starting from button cells (used in watches) to megawatt load-leveling systems. The Volta's cell, invented in 1800, was the first battery that was designed using the brine solution (electrolyte), cardboard (separator), and alternating discs of copper and zinc (electrodes). After that, in 1836, Daniel's cell was invented, having two electrolytes. Later, in 1866, the Leclanche cell was designed using carbon and zinc anodes. Before 1949, no dry cell was fabricated where the cathode was made up of manganese oxide, the anode of zinc, and electrolytes of alkaline. All these discussed cells were non-rechargeable as they were the primary cells. The secondary or rechargeable batteries were manufactured using lead-acid in 1859, then nickel-cadmium (Ni-Cd) batteries came in 1899. Later, in the mid-1980s, a nickel-metal hybrid (NiMH) was designed, and lastly, the lithium-ion batteries in 1977 were made. The usages of Ni-Cd batteries in high-power consumer devices are still countless, including their applications in electric razors and aero engines along with gardening tools. Due to the poisonous nature of Cd, these batteries are being replaced with lithium-ion batteries for consumer applications [30]. Different types of batteries and their working mechanism are discussed below.

### 10.6.2.1  Lead Batteries

Globally, rechargeable lead batteries are extensively used due to their energy production in MWh, safety, and reliability. They have been used in transportation, mostly in trucks and cars. From the domestic to the commercial level, they are being used in telecommunication and uninterrupted power supply and secure power and energy storage. The dominancy of lead batteries in energy storage applications is due to their recyclability. At the end of life, these batteries are collected to be recycled, and 90% of the material can be recovered through recycling. New lead batteries are designed with recovered lead through recycling. Long cycle life is required for a battery to be used in energy storage systems, and the battery must avoid sulfation. To achieve 90% of energy efficiency, two parameters play a vital role in battery management and charging. In domestic and commercial places, lead batteries can be used for load management and frequency regulations. Despite the recycling parameter, the other important factor of lead batteries is their availability at cheap rates.

### 10.6.2.2  Working Mechanism of Lead Batteries

The materials used in lead-acid batteries are (i) lead peroxide at the positive electrode, (ii) sponge lead at the negative electrode, and (iii) an electrolyte of diluted sulfuric acid. The water to acid ratio is kept at 3:1. The lead-acid storage battery is formed by dipping the lead peroxide plate and sponge plate in dilute sulfuric acid. A load is connected externally between these plates. In diluted sulfuric acid, the molecules of acid split into positive hydrogen ions and negative sulfate ions. The hydrogen ions on reaching the lead peroxide plate, receive an electron from it and become hydrogen atom, which again attacks lead peroxide, and form lead oxide and water.

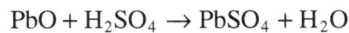

$$PbO_2 + 2H \rightarrow PbO + H_2O$$

$$PbO + H_2SO_4 \rightarrow PbSO_4 + H_2O$$

$$PbO_2 + H_2SO_4 + 2H \rightarrow PbSO_4 + 2H_2O$$

The formation of water and lead sulfate was caused by the reactions occurring between sulfuric acid and lead oxide. Within the solution, the negatively charged ions of sulfate are free to move while a few of them approach the lead plate. Radical sulfates are formed when these sulfate ions release an electron that further generates lead sulfate by reacting with pure lead. There is an imbalance between the two plates of the battery. At the lead peroxide plate, the electrons are taken by the hydrogen ions, whereas the electrons are given by the sulfate ions at the lead plate. To maintain the imbalance of electrons, a current will flow through the external load between the plates. This whole process is termed as discharge process of lead-acid batteries.

The charging process will start by disconnecting the load and connecting the dc source. The negative and positive terminals of the source are connected with the lead peroxide-covered lead plate and lead sulfate-covered lead peroxide plate, respectively. Even the density of sulfuric acid is reduced during discharge, but it still exists in the solution. As the hydrogen ions are positive, they make transitions toward the negative terminal. The hydrogen atoms are formed when hydrogen ions take an electron from the negative terminal.

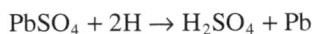

$$PbSO_4 + 2H \rightarrow H_2SO_4 + Pb$$

These hydrogen atoms then attack lead sulfate and form lead and sulfuric acid. The negative sulfate ions move toward the electrode connected with the positive terminal of the dc source, where they will give up their extra electrons and become radical sulfate.

$$PbSO_4 + 2H_2 + SO_4 \rightarrow PbO_2 + 2H_2SO_4$$

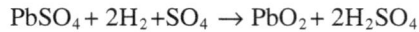

This radical sulfate reacts with lead sulfate of anode and forms lead peroxide and sulfuric acid.

### 10.6.2.3 Vanadium Redox Flow Batteries

A type of redox flow battery, vanadium redox battery (VRB), uses vanadium redox couples at both electrodes to store energy where $V^{4+}/V^{5+}$ can be used at positive half-cell and $V^{2+}/V^{3+}$ at negative half-cell. An electrolyte of sulfuric acid was used where these chemicals were dissolved [31]. For a specific application, the energy and power can be optimized as they do not depend upon each other in VRB. The VRB was firstly designed in the late 1980s.

### 10.6.2.4 Working Mechanism of Vanadium Flow Redox Batteries

For the discharge process, in the negative half-cell, the $V^{2+}$ releases an electron to become oxidized to $V^{3+}$. In the external circuit, this released electron does work. From the external circuit, the acceptance of an electron by $V^{5+}$ occurs to form $V^{4+}$ in the positive half-cell. To maintain charge neutrality, the exchange of hydrogen ions occurs between the two half cells. The half cells are separated by the permeable polymer membrane where hydrogen ions can be diffused. Across the membrane, both the water and the charged vanadium species can be diffused. The balance between the negative and positive sides of the system is not required in a vanadium-only system. The presence of vanadium in the form of oxy-cations is observed in the positive half-cell. The temperature between 10°C and 40°C is the most suitable range for VRB. The reported DC-DC efficiency of up to 80% is noticed for such batteries. In contrast, with other RFBs, the cell voltages and energy and power densities are higher for VRB [32]. These batteries can be used in steel manufacturing.

### 10.6.2.5 Nickel-Cadmium (Ni-Cd) Batteries

For sustainable use in electrode technologies and packaging, a traditional battery, nickel-cadmium, has been used since the 1910s. The installation of such batteries is quite easy, and they also provide consistent services and long life.

### 10.6.2.6 Working Mechanism of Ni-Cd Batteries

These batteries are still in use nowadays, where the pocket plate technology was used in early Ni-Cd cells that are now replaced with foam plates or fiber plates. The electrode design of cells having fiber and pocket plates is almost similar to that of the Cd negative and Ni positive. On the other hand, plastic-bonded negatives have been used in foam and sintered positives. The water replenishment is needed to compensate for all the industrial Ni-Cd designs where gases must be dissipated that are formed due to overcharge. Changing levels of recombination are allowed by the separator designs. Concerning the electrolyte, some maintenance-free operations are designed for off-grid renewable energy and telecom systems. These batteries have been used for energy storage systems and stabilizing wind-energy systems.

### 10.6.2.7 Sodium Sulfur (NaS) Batteries

In the 1960s, Ford Motor Company designed the first sodium sulfur battery for stationary systems. These can be operated even at a high-temperature range between 300°C and 350°C. These batteries can be used for energy as their efficiency is up to 90%.

### 10.6.2.8 Working Mechanism of Sodium Sulfur Batteries

In NaS batteries, the active material at the positive electrode is molten sulfur, and at the negative electrode, molten sodium is used. These electrodes are parted away through a separator, or an electrolyte made up of sodium alumina (solid ceramic). Only positively charged sodium ions can pass through the ceramic. In the discharge process, one electron from an individual sodium atom is released from sodium metal, forming the sodium ion which further makes transportation through the electrolyte toward the positive electrode side. Owing to this transfer of an electron, the poly-sulfide is formed when the electrons are taken up by the molten sulfur. For balancing the electron charge flow, the positively charged sodium ions move toward the positive electrode side. The reverse process is seen during charging. To assist the process, a temperature greater than 300°C is needed for the battery. The efficiency of such batteries is up to 89% [33].

### 10.6.2.9 Lithium-Ion Batteries

Lithium-ion (Li-ion) batteries had an important role in different fields of life, such as in electron-ics and communication or computation as well as entertainment and information. The element Li is the simplest metal in the periodic table that has some interesting parameters, including light-weight and high energy density and operating voltage. From a battery application point of view, it can be used in both non-rechargeable and rechargeable batteries. In the 1960s, the primary bat-teries were designed where Li reacts with the cathodes, including Al, I, or $MnO_2$. In rechargeable batteries, a molten form of lithium and sulfur was used as electrode materials. The electrolyte made up of molten salt was used to separate these electrodes. The operating temperature of such batteries was up to 450°C. Whittingham et al. proposed an idea in the 1970s to develop batteries that can be operated at room temperature where the electrochemical intercalation was observed when guest molecules interacted with layered hosts. The chemical reactions occurring even at high temperatures controlled the chemical intercalations. There were few structural changes in the electron density of the host, but other electronic factors were tuned according to the require-ment. These electronic factors were conductivity or superconducting transition temperatures and the density of electronic states. The reaction barriers were overcome by applying an electric field in electrochemical intercalations. Both the intercalations and deintercalation unwrapped the new perspective for battery systems that caused released energy and reversible storage [34]. The work-ing principles of Li-ion batteries have been represented by the schematic diagram of research work by Iqbal et al. [3] (Figure 10.4).

The fascinating features of energy storage systems were as follows: (i) they have high compo-sitional ranges, (ii) they contain large energies, (iii) their reversibility was facile, (iv) their crystal structure was stable, and (v) their electronic structures were predictable. For rechargeable batter-ies, the electrochemical intercalations were described by Whittingham, where the energy can be stored or released in energy storage systems. Originally, the Li batteries working on the intercala-tion process were made having components such as (i) an anode of pure metal lithium, (ii) a cath-ode of layered transition metal composition, and (iii) an organic electrolyte. There were two main purposes of this electrolyte. First, it transported the Li ions, and second, it opposes the flow of elec-trons between a cathode and an anode. As there were no inactive atoms at the anode, the storage capacity was not diluted. The operating voltages were offered by the organic electrolytes where the oxidation or reduction of electrolytes was not possible. When these batteries were used at a commercial level, the dendrites grew across the electrolyte or lost electrochemical and mechani-cal contact with the anode due to repeated cycling, hence, causing the flames. To overcome this flammability, pure lithium was switched with Li-Al alloy, but the cycle life was also shortened. In contrast with the intercalations, the energy density was much higher for pure metal anodes. The energy density was diluted in intercalated systems due to the inert material in the anodes. Different anodes were designed with lithium for commercial use, but in 1991, Sony developed Li-ion bat-teries by replacing metallic lithium with lithium at the anode and using carbon as a host structure. The other method to improve the performance of batteries was the use of an efficient cathode mate-rial consisting of layered transition metal oxides so that high energy density and high voltage can be achieved. A well-suited electrolyte was also needed to have a wide electrochemical window to operate high voltages of the metal-oxide cathode. For the Li-ion batteries, the basic problem was to find a suitable anode that can replace Li metal because of the dendrite challenges. As the Li ions can be accepted by the carbon materials by intercalation, researchers turned their interest toward carbon materials as intercalation anodes [35].

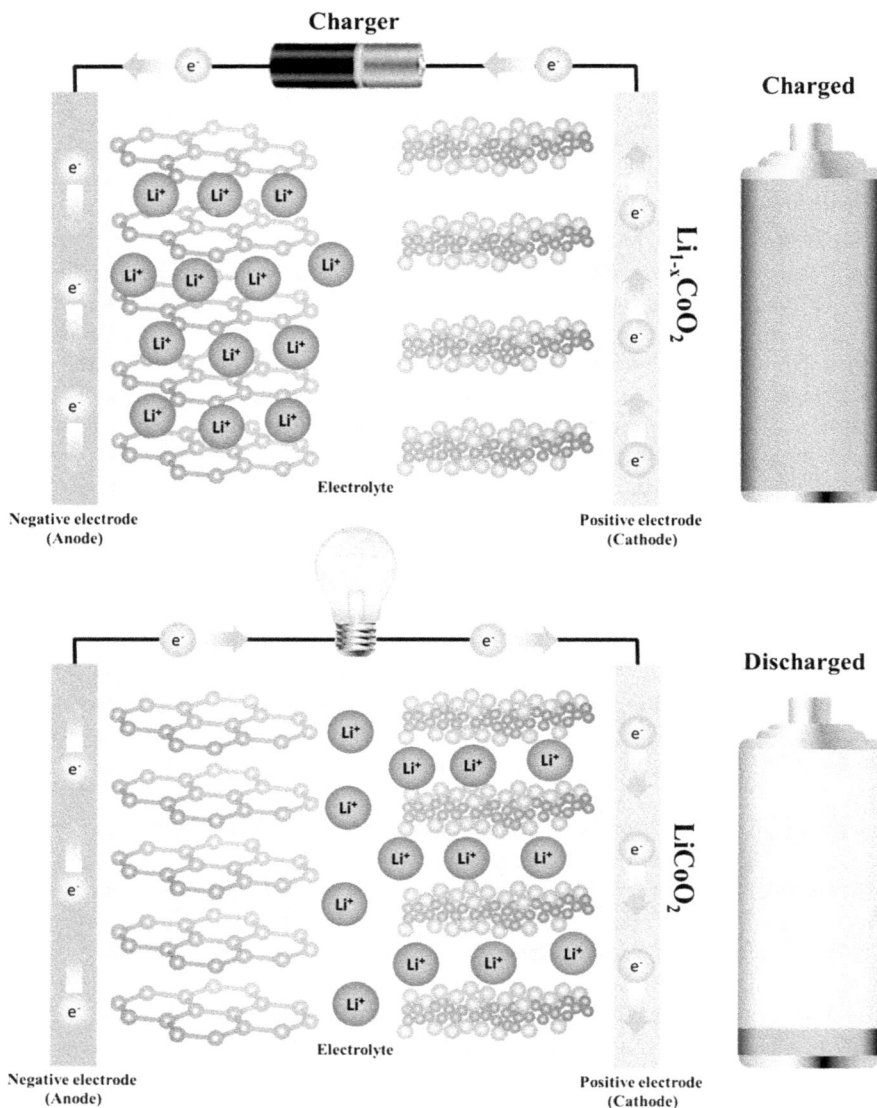

**FIGURE 10.4**  Schematic representation for working principles of lithium-ion batteries.

First of all, the Li-ion batteries were developed by Sony, where soft carbon was used as the anode and propylene carbonate as the electrolyte. There were no detrimental side reactions, and an energy density of 80 Whkg$^{-1}$ was obtained. Then, the replacement of soft carbon with hard carbon was done to improve Li insertion and voltage capacity. With this replacement, the energy density increased up to 120 kWh/kg. Lastly, graphite was used in a place of hard carbons as it is noticed that most cell phones need almost 3 V to operate. An ethylene carbonate can be used to overcome the challenge of detrimental side reactions with propylene carbonates. The nanocomposites are being designed using nickel, manganese, or cobalt at M place in the $Li_xMO_2$ formula. These composites are available at low cost, and their stability against oxidation of electrolytes is greater, and also their energy densities are high. It is seen that the gravimetric energy density of $LiNiO_2$ is 20% more than the $LiCoO_2$. To prevent the removal of all the lithium, lengthen the cycle life and increase the discharge density, a minor addition of aluminum was used in $Li_x(Ni_{0.8}Co_{0.15}Al_{0.05})O_2$. To improve the performance of

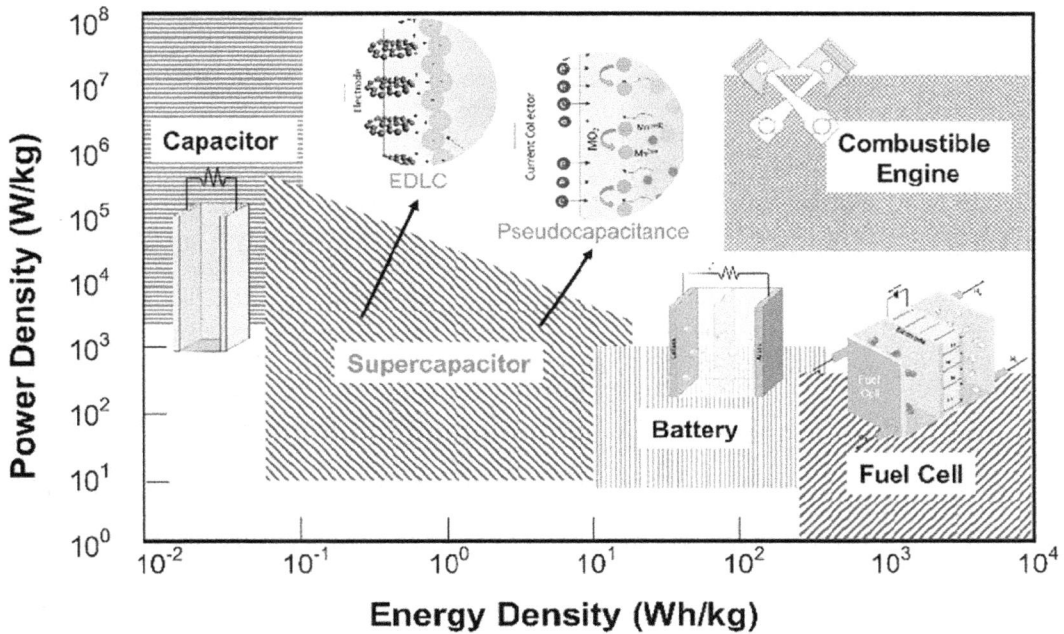

**FIGURE 10.5**  Ragone plot for several types of electrochemical energy storage devices. (Adapted from Ref. [37].)

the batteries, different cathode materials are still being investigated [36]. The energy and power densities of different electrochemical energy storage devices are shown in Figure 10.5.

## 10.7  NEXT-GENERATION BATTERIES TECHNOLOGY

For grid energy storage and vehicle electrification, the future of portable energy sources is provided by next-generation batteries. The cell-level energy of typical Li-ion batteries can be increased twice by using the Li metal batteries that are among the best next-generation battery technologies. The rapid cell death of Li metal is mainly caused by the breakdown of electrolytes followed by the increased internal resistance and dendrite formation. This effect must be overcome using stripping and plating of Li metal where interfacial reactions occur. The mass transport of ions may influence the plating of Li metal within the electrolyte. To fulfill the energy needs with the progress of modern society, it is necessary to design substitutive energy storage systems. From this point of view, the main contestants are electrochemical systems, as Li-ion batteries are being used in almost every field of life.

Sodium-ion batteries (SIBs) have recently emerged as a strong alternate for the use of energy storage systems (ESS) with the advantages of low cost and abundant resources of sodium [38]. However, SIBs face the problem of a larger ionic radius of 0.97 and a higher redox potential of $-2.71$ V (vs. standard hydrogen electrode), which results in sluggish kinetics and low energy density [39]. But SIBs have made significant progress in recent years and emerged with high-performance cathode and anode materials [40].

Various research groups have explored next-generation battery technologies using different materials. Some latest investigated materials for next-generation battery technologies are summarized here. Xu et al. [41] turned their focus toward potassium ion batteries (KIBs) and explored their importance over sodium and lithium-ion batteries. The comparison between Na-ion batteries to Li-ion counterparts has been investigated by Abraham [42]. The KIBs possess some interesting parameters due to which they can sometimes surpass Li-ion batteries. These parameters are: (i) they have superior ionic mobility; (ii) their operating voltages are high; (iii) they are economical; and (iv) their resources are abundant and readily available. It is also noted that the anode materials used in Li-ion batteries may show excellent properties but their response to KIBs will be different. To increase the capacity, there are two approaches for KIBs. The first one is to find a novel intercalated anode material, and the second one is to improve the material capacity by coating or doping carbon-based materials. In $KO_2$ batteries, the reversibility of the battery system was ensured by the final discharge product $KO_2$. That product was thermodynamically and kinetically stable in contrast with $NaO_2$ and $LiO_2$ batteries. To overcome the issues of energy scarcity and to solve global warming, Jiao et al. [43] used $CO_2$ to manufacture Li-$CO_2$ batteries. These batteries were designed to utilize and capture $CO_2$, but some issues still exist. The drawbacks of such batteries include: (i) the bandgap of $Li_2CO_3$ is wide that it may cause high overpotential; (ii) the cyclic performance is poor; (iii) the rate capacity is weak; and (iv) the discharge capacity is low. Thus, it is worthy to say that the slow reaction kinetics of $CO_2$ gas caused these issues. For the promotion of electrochemical reactions of $CO_2$, it is needed to develop stable electrolytes and effective cathode catalysts.

Dorfler et al. [44] examined the working of Li-S batteries for the electrification progress within the vehicle sectors. Some important parameters of such sectors were kept in mind: electric vertical take-off and landing along with high-altitude pseudosatellite and long endurance and also large commercial vehicles (buses and trucks). Concerning the system costs, power, and also the cycle life, the needs for the performance of the battery system change. The common requirement for all these applications is a high gravimetric energy density up to 400 Wh/kg. It may enable the vehicles to get longer mission duration and lighter weight, enhanced payload, and extended range. They discussed the performance advantages and limitations, modeling, and the latest progress of the Li-S batteries. Oesch et al. [45] studied the integrated next-generation technologies to generate, store and distribute the electrical energy for small spacecraft according to their requirements. Different material designs and the engineering of next-generation flow-battery technologies were studied by Park et al. [46]. Newly designed hybrid energy conversion and storage systems were developed using a flowing electrolyte because of their cost-effectiveness. The potential opportunities were employed for the advanced electrical energy storage technologies by flow-based lithium-ion and organic redox-active materials and photoelectrochemical batteries and metal-air cells.

## 10.8   CONCLUSION

This chapter provides an overview of the brief history, current development, and future perspectives of energy storage media. Here, we systematically described the role and capacity of different energy storage systems to realize their importance for advanced technological evolution. A comprehensive comparison based on the technical, economic, and environmental impact is established, providing a panoramic view of whole energy storage systems. Among these systems, batteries and super-capacitors are emerging candidates due to their superior salient features, for instance, small size, high energy and power density, lighter weight, and long cyclic life. In the present scenario, there is an urgent need to energy storage media that can have high storage capacity with minimum losses. Therefore, there are plenty of rooms at the bottom to be explored, which could serve to enhance the energy storage capabilities of existing energy storage devices.

## REFERENCES

1. T.F. Yi, H.M.K. Sari, X. Li, F. Wang, Y.R. Zhu, J. Hu, … X. Li, *Nano Energy*. (2021), p. 105955.
2. C.A. Zhou, Z.J. Yao, X.H. Xia, X.L. Wang, C.D. Gu, and J.P. Tu, *Mater. Today Nano*, 11 (2020) p. 100085.
3. S. Iqbal, H. Khatoon, A.H. Pandit, and S. Ahmad, *Mater. Sci. Energy Technol.* 2(3) (2019), pp. 417–428.
4. E. Morofsky, *Thermal Energy Storage for Sustainable Energy Consumption*, (2007), pp. 3–22. Springer, Netherlands.

5. H. Ibrahim, A. Ilinca and J. Perron, *Renew. Sust. Energ. Rev.* 12(5) (2008), pp. 1221–1250.

6. A.G. Olabi, *Energy*, 136 (2017), pp. 1–6.

7. Y. Balali and S. Stegen, *Renew. Sust. Energ. Rev.* 135 (2021), p. 110185.

8. J. Zheng and L.A. Archer, *Sci. Adv.* 7(2) (2021), p. eabe0219.

9. H.A. Behabtu, M. Messagie, T. Coosemans, M. Berecibar, K. Anlay Fante, A.A. Kebede and J.V. Mierlo, *Sustainability* 12(24) (2020), p. 10511.

10. G. Alva, Y. Lin and G. Fang, *Energy* 144 (2018), pp. 341–378.

11. L.F. Cabeza, *Advances in Thermal Energy Storage Systems* (2015), (pp. 37–54). Woodhead Publishing, United Kingdom.

12. F.S. Barnes and J.G. Levine, *Large Energy Storage Systems Handbook.* (2011) CRC Press, United States.

13. A. Anastasovski, *J. Clean Prod.* 278 (2021), p. 123476.

14. M. Mehrali, E. Johan, M. Shahi and A. Mahmoudi, *Chem. Eng. J.* 405 (2021), p. 126624.

15. H. Aydin and S. Merey, *Renew. Energ.* 164 (2021), pp. 1076–1088.

16. F. Fornarelli, S.M. Camporeale and B. Fortunato, *Heat Transfer Eng.* 42(1) (2021), pp. 1–22.

17. A.H. Alami, *Mechanical Energy Storage for Renewable and Sustainable Energy Resources*, Springer, United States 2020.

18. P.M. Congedo, C. Baglivo and L. Carrieri, *Energ. Buildings* 224 (2020), p. 110234.

19. W. Wang, M.M. Umair, J. Qiu, X. Fan, Z. Cui, Y. Yao and B. Tang, *Energ. Convers. Manage.* 196 (2019), pp. 1299–1305.

20. X. Wang and Y. Yang, *Nano Energy.* 42 (2017), p. 36–41.

21. C. Zequine, C.K. Ranaweera, Z. Wang, S. Singh, P. Tripathi, O.N. Srivastava … R.K. Gupta, *Sci. Rep.* 6(1) (2016), pp. 1–10.

22. K.T. Kubra, A. Javaid, R. Sharif, G. Ali, F. Iqbal, A. Salman, F. Shaheen, A. Butt and F.J. Iftikhar, *J. Mater. Sci: Mater. El.* 31(15) (2020), pp. 12455–12466.

23. U. Gulzar, S. Goriparti, E. Miele, T. Li, G. Maidecchi, A. Toma … R.P. Zaccaria, *J. Mater. Chem. A* 4(43) (2016), pp. 16771–16800.

24. M. Salanne, B. Rotenberg, K. Naoi, K. Kaneko, P.L. Taberna, C.P. Grey … P. Simon, *Nat. Energy* 1(6) (2016), pp. 1–10.

25. R. Ahmad, N. Iqbal, M.M. Baig, T. Noor, G. Ali and I.H. Gul, *Electrochim. Acta* 364 (2020), p. 137147.

26. S. Dörfler, S. Walus, L. Locke, A. Fotouhi, D.J. Auger, N. Shateri … S. Kaskel, *Energy Technol.* 9(1) (2021), p. 2000694.

27. K.T. Kubra, R. Sharif, B. Patil, A. Javaid, S. Shahzadi, A. Salman, S. Siddique and G. Ali, *J. Alloy. Compd.* 815 (2020), p. 152104.

28. H. Zhou, L. Zhang, D. Zhang, S. Chen, P.R. Coxon, X. He … S. Ding, *Sci. Rep.* 6(1) (2016), pp. 1–11.

29. A. Vlad, N. Singh, J. Rolland, S. Melinte, P.M. Ajayan and J.F. Gohy, *Sci. Rep.* 4 (2014), p. 4315.

30. G. Karkera, M.A. Reddy and M. Fichtner, *J. Power Sources* 481 (2021), p. 228877.

31. S. Mehboob, G. Ali, S. Abbas, K.Y. Chung and H.Y. Ha, *J. Ind. Eng. Chem.* 80 (2019) pp. 450–460.

32. S. Mehboob, G. Ali, H.J. Shin, J. Hwang, S. Abbas, K.Y. Chung and H.Y. Ha, *Appl. Energ.* 229 (2018), pp. 910–921.

33. I. Hasa, S. Mariyappan, D. Saurel, P. Adelhelm, A.Y. Koposov, C. Masquelier … M. Casas-Cabanas, *J. Power Sources* 482 (2021), p. 228872.

34. A. El-Kharbachi, O. Zavorotynska, M. Latroche, F. Cuevas, V. Yartys and M. Fichtner, *J. Alloy. Compd.* 817 (2020), p. 153261.

35. K.M. Abraham, *ACS Energy Lett.* 5(11) (2020), pp. 3544–3547.

36. G. Crabtree, E. Kocs and L. Trahey, *MRS Bull.* 40(12) (2015), pp. 1067–1078.

37. F.A. Permatasari, M.A. Irham, S.Z. Bisri and F. Iskandar, *Nanomaterials* 11(1) (2021), p. 91.

38. G. Ali, J.H. Lee, S.H. Oh, H.G. Jung and K.Y. Chung, Nano Energy 42 (2017), pp. 106–114.

39. D.S. Bhange, G. Ali, J.Y. Kim, K.Y. Chung and K.W. Nam, *J. Power Sources* 366 (2017), pp. 115–122.

40. R. Ata Ur, G. Ali, A. Badshah, K.Y. Chung, K.W. Nam, M. Jawad, … S.M. Abbas, *Nanoscale* 9(28) (2017), pp. 9859–9871.

41. J. Xu, S. Dou, X. Cui, W. Liu, Z. Zhang, Y. Deng, … Y. Chen, *Energy Storage Mater.* 34 (2020), pp. 85–106.

42. K.M. Abraham, *ACS Energy Lett.* 5(11) (2020), pp. 3544–3547.

43. Y. Jiao, J. Qin, H.M.K. Sari, D. Li, X. Li and X. Sun, *Energy Storage Mater.* (2020), pp. 1–94.

44. S. Dörfler, S. Walus, J. Locke, A. Fotouhi, D.J. Auger, N. Shateri, … S. Kaskel, *Energy Technol.* 9(1) (2021), p. 2000694.

45. C. Oesch, C. Pearson, W.H. Francis and R. Lidvall, In *AIAA Scitech 2021 Forum* (2021), p. 0803.

46. M. Park, J. Ryu, W. Wang and J. Cho, *Nat. Rev. Mater.* 2(1) (2016), pp. 1–18.

# 11 Lithium-Based Battery Systems
*Technological and Environmental Challenges and Opportunities*

*C. M. Costa*
University of Minho

*J. C. Barbosa*
University of Minho
University of Trás-os-Montes e Alto Douro

*R. Gonçalves and S. Ferdov*
University of Minho

*S. Lanceros-Mendez*
BCMaterials, Basque Center for Materials,
Applications and Nanostructures
Ikerbasque, Basque Foundation for Science

## CONTENTS

DOI: 10.1201/9781003315353-13

## 11.1  INTRODUCTION

Energy consumption and management have represented, since the First Industrial Revolution in the eighteenth century, a central role in the development of new technologies and massification of mechanization production, allowing new and affordable products. Initially obtained from coal and then from petrol, increasing energy used for industrial processes and mobility lead to the increase of $CO_2$ gas emissions to the atmosphere. With the technological evolution and environmental concerns, these non-renewable sources of energy based on fossil fuels were revised, and new renewable sources were developed and implemented such as hydroelectric, solar, and wind (Qadir et al. 2021). The fast increase of $CO_2$ gas emissions in the last years leads to an increase in the earth's temperature, originating environmental consequences including extreme weather conditions, glaciers melting, and health diseases (Ahmed et al. 2020). The European Commission (EC) faced with all these issues by establishing as key targets until 2030 the decrease in at least 40% the greenhouse gas emissions, reaching a 32% share of renewable energy and a 32.5% improvement in energy efficiency (European Union 2021a). In addition, in 2015, the United Nations (UN) adopted 17 global objectives focusing on social, economic, and environmental issues to accomplish by 2030. Within the 17 global goals, objectives as "7- Affordable and clean energy" and "13- Climate change" are related to the sustainability of the consumed energy and its relationship with the environment and atmosphere (United Nations 2021). Furthermore, the European Commission (EC) goal is a net-zero greenhouse gas emissions by 2050, where, in 2030, the reduction of the greenhouse gas emissions was revised to be at least 55% compared with the levels measured in 1990 (European Union 2021b).

The development of the new energy scenario is focused on three technological modifications: (i) energy savings on the demand side; (ii) efficiency progress during its production; and (iii) substitute fossil fuels with renewable energy sources. Clean energy is thus nowadays a very important topic that should be centered not only in its production step but also in its storage, allowing a clean and affordable distribution, as shown in Figure 11.1. To initiate this development strategy, the first challenge is to extend the renewable energy in the supply system, where each country should obtain its own energy from renewable sources such as hydropower, wind, or solar, among others (Lund 2007).

# Towards a Clean Energy Scenario

**FIGURE 11.1**  Technological modifications for a clean energy scenario.

## 11.2   ENERGY STORAGE SYSTEMS

With the increase in the energy obtained from renewable energy sources and their implementation across different countries around the world, the efficient storage of this energy is essential to guarantee affordability, accessibility, and reliability of the system. To ensure this goal, different storage systems have been used by exploring the conversion of the generated energy into chemical and potential energy, as the most implemented processes. Thus, different energy storage systems including electrochemical (supercapacitors, hydrogen, and batteries), electrical (magnetic energy storage), mechanical (pumping, compressed air, and flywheel), and thermal energy storage (ceramic thermal storage and thermal fluid storage) have been implemented in different applications (Cheng, Sami, and Wu 2017). Different efficiencies, power, power and energy density, and response time have been obtained across all these systems. Supercapacitors are the most efficient system with values higher than 95%, but, on the other hand, show lower power (< 1 MW) and energy densities (1 Wh–1 kWh) compared to some other systems. Pumped hydro shows the highest energy density (400 MWh–20 GWh) and power (100 MW–2 GW) but has lower efficiency (70%–80%) and response time (12 minutes) (Abdi et al. 2017). Battery energy storage systems, on the other hand, present a good compromise between these properties with efficiencies of 60%–80%, power that can range up to 100 MW, and energy density of 1 kWh–200 MWh with response times in the order of seconds. Furthermore, batteries are safe systems capable of distributing energy to devices from small sizes such as laptops and smartphones to large-scale applications such as electric vehicles and power plants.

### 11.2.1   BATTERY SYSTEMS

Batteries are the most used energy storage system because of their high capacity, versatility, low cost, and good efficiency compared to the other storage systems (Komarnicki, Lombardi, and Styczynski 2017). Within electrochemical systems, different types can be found including supercapacitors, solid-state batteries, fuel cells, flow batteries, and metal-air batteries, among others. Cell types based on nickel-metal hydride, alkaline (zinc-manganese dioxide), silver oxide, sodium-ion, and, the most used, lithium-ion batteries, are examples of the various possibilities for energy storage (Abdi et al. 2017). In these cells, the chemical energy is converted to electrical energy and vice versa. This principle is driven by redox reactions that in 1800 were implemented by Alessandro Volta in the development of the first voltaic system (Erenoğlu, Erdinç, and Taşcıkaraoğlu 2019). Since that year, developments have been made to improve this promising system. Efforts in addressing the issues of electrolyte leakage and improvement in the durability of the battery were made by Daniell and Leclanché (Sarakonsri and Kumar 2010). After the first functional rechargeable batteries in 1859, different systems were used and nowadays lithium-ion batteries are the most used battery system (Job 2020).

Battery systems are typically composed of two electrodes (cathode and anode) separated by a membrane soaked in an electrolyte solution (see a typical example in Figure 11.2a, for example). Each electrode is composed of three different materials: (i) active material, responsible for the battery capacity that should be capable of donating (cathode) and storing (anode) the ionic species of the battery system; (ii) conductive material, allowing to improve the electrical conductivity within the electrode; and (iii) polymer binder, responsible for the structural integrity of the electrode components (Liu, Neale, and Cao 2016, Lee 2019). The membrane has the main function of physically separating both electrodes, avoiding short circuits, while allowing the ionic specie passage through electrolyte within its structure during the battery operation. The membrane typically presents a porous structure with the electrolyte embedded within this structure, to allow the ionic conductivity of the system (Costa et al. 2019). During battery operation, electrons from the electrodes move along an external circuit through a current collector. A copper metal current collector is located at the anode electrode, and an aluminum metal current collector is located at the cathode electrode. These two metals' current collectors with high electric conductivity show different redox potentials, ensuring battery operation (Gonçalves et al. 2021).

During battery operation, the ionic species and the electrons move from the cathode to the anode during the charge and to the opposite direction (from anode to cathode) during the discharge. The applied voltage to the battery system during the charging process is responsible to promote sufficient energy to break the bonds between the ionic species and the active material promoting the ion storage in the anode electrode. During the discharge process, a spontaneous reaction occurs to lower the energy state, promoting the ion to return to the cathode, releasing energy. This released energy will then be consumed by the connected device (Gonçalves et al. 2019).

## 11.3 LITHIUM-BASED BATTERY SYSTEMS

Lithium-ion batteries (LIBs) were introduced by Sony in the portable device market and nowadays are widely used for various applications from smartphones to electric vehicles. This is based on their excellent properties such as being cheap and light, showing higher energy and power density, lower auto-discharge, no memory effect, and a higher number of charge/discharge cycles, among others, when compared to other battery systems such as Ni-Cd, nickel-cadmium; NiMH, or nickel-metal hydride. There are several types of rechargeable lithium-based battery systems available for use. LIBs show the highest energy density (both gravimetric and volumetric) and theoretical specific energy of 387 Wh/kg (for graphite/$LiCoO_2$). Nevertheless, their practical energy is around 200 Wh/kg (Nitta et al. 2015, Huang et al. 2021, Xie and Lu 2020).

Figure 11.2 shows the schematic representation of different types of LIBs based on the materials that constitute the battery. The most widely used are those that are identified as lithium-polymer batteries.

### 11.3.1 LITHIUM-BASED BATTERY TYPES

#### 11.3.1.1 Lithium-Polymer Batteries

In lithium-polymer batteries, the separator is in the form of porous or gel-like electrolyte with the main function, together with the needed ionic conductivity, of improving mechanical flexibility and robustness. This component is the main difference between this and the rest of the Li-ion-based

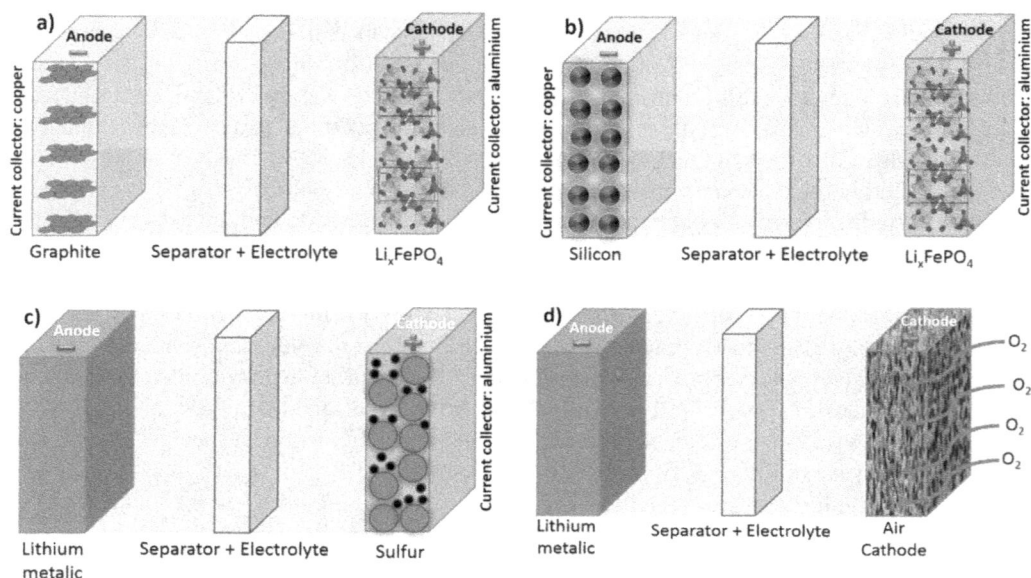

**FIGURE 11.2** Schematic representation of different lithium-ion battery types identified as: (a) Lithium-polymer, (b) lithium-silicon, (c) lithium-sulfur, and (d) lithium-air types.

batteries. In relation to the materials for the electrodes, the anode is basically composed of graphite and the cathode can be composed of different materials, including lithium cobalt oxide ($LiCoO_2$), lithium manganese dioxide ($LiMnO_2$), lithium iron phosphate ($LiFePO_4$), lithium nickel cobalt manganese oxide ($LiNiCoMnO_2$), lithium nickel manganese spinel ($LiNi_{0.5}Mn_{1.5}O_4$), or lithium nickel cobalt aluminum oxide ($LiNiCoAlO_2$), among others (Nitta et al. 2015).

### 11.3.1.2  Lithium-Silicon Batteries

An interesting type of LIBs are the batteries where the anode material is based on silicon, replacing graphite, and the cathodes are similar to the ones for lithium-polymer batteries (Zuo et al. 2017). Silicon is a promising material for anode due to its high gravimetric and volumetric capacity of 4,200 mAh/g and 9,785 mAh/cm$^3$, respectively, and for being very abundant in the earth's crust and nontoxic (Yang, Yuan, et al. 2020). Its main disadvantages for its use in LIB applications are its low ionic and electrical conductivity and high volume variation during the charge/discharge processes (Yang, Yuan, et al. 2020). Considering its high energy density and to improve its conductivity, the strategy is to dope silicon with carbon materials, metals, or metal oxides, among others. In addition, to reduce volume variation, the possibility of reducing particle size to the nanometer scale to accommodate stress during battery operation has been explored (Feng et al. 2018). This possibility has—as a negative aspect—the fact of being expensive and consequently increasing the cost of the batteries, making their commercialization difficult (Zuo et al. 2017). The massive commercialization of LIBs with silicon as anode will be reached when the problems of this material are solved. Further, strategies related to improving polymer binders, conductive additives, and electrolytes for this specific anode material have to be explored.

### 11.3.1.3  Lithium-Sulfur Batteries

Sulfur is used as cathode material in lithium-sulfur (Li-S) batteries based on its natural abundance and being nontoxic, but mainly for its higher theoretical specific capacity (1,675 mAh/g) that is 3–5 folds higher when compared to conventional lithium-ion batteries (Zhao et al. 2018). Annually, several millions of tons of sulfur are separated from crude oil refineries and stored on open grounds without utilization, the implementation of Li-S batteries being an opportunity for the valorization of this element (Ma et al. 2015).

Li-S batteries are composed of a metallic Li anode, which leads to safety concerns, but the main challenge lies in the sulfur cathode used in this battery. Sulfur is an insulating material (electrical conductivity: $1 \times 10^{-15}$ S/m), and this material undergoes 80% volume change during charge/discharge behavior. Further, sulfur is not compatible with carbonate-based solvents commonly used in LIBs (Rana et al. 2019) and dissolves in most solvents compatible with the Li anode (Zhao et al. 2018). Presently, ether-based solvents are employed for this battery system, where sulfur has moderate solubility (Li et al. 2019). An assembled Li-S cell is charged and ready to be discharged once after its assembly. During the discharge process, sulfur undergoes reduction in several stages. Sulfur has over 30 solid allotropes, and the most common form is the cycle octasulfur ($S_8$), which is the thermodynamically stable form followed by the cyclic $S_{12}$ allotrope. During the discharge process, sulfur forms reduction products such as $Li_2S_8$, $Li_2S_6$, $Li_2S_4$, $Li_2S_2$, and $Li_2S$ (Zhao et al. 2018, Li et al. 2019). $Li_2S$ is the final discharged product, which is also an insulating material. On recharge, $Li_2S$ gradually oxidizes back to form $S_8$ (Li et al. 2018). The reduction products of sulfur, such as $Li_2S_8$, $Li_2S_6$, and $Li_2S_4$ are soluble in the liquid electrolyte employed ($LiSO_3CF_3$ and $LiN(SO_2CF_3)_2$ in dimethyl ether (DME) and 1,3-dioxolane (DOL)), while the other products ($Li_2S_2$ and $Li_2S$) are insoluble in this electrolyte (Zhang 2013). Further, the high-order polysulfide reduction products dissolved in the electrolyte cross the separator and get reduced at the Li anode, reducing either as a passivating $Li_2S$ deposit on the Li anode or as a low-order polysulfide, which again diffuses back to the cathode, where it gets reoxidized to high-order polysulfides. In this way, the dissolved polysulfides shuttle between anode and cathode, undergoing reduction and oxidation, respectively. This polysulfide shuttle phenomenon causes the cell to undergo a kind of self-discharge. The use of dry solid electrolytes in Li-S cells is considered to be the ultimate approach to prevent sulfur and poly-sulfide dissolution into the electrolyte. In addition, solid polymer electrolytes (SPEs) are considered to suppress the dendritic growth of lithium and enhance safety.

### 11.3.1.4 Lithium-Air Batteries

Another LIBs type with a high theoretical capacity (3,505 Wh/kg) are lithium-air batteries, which are composed of a lithium-metal anode, a separator with an electrolyte solution, and a carbon-supported air electrode where oxygen from the atmosphere dissolves in the electrolyte within the pores (Jung et al. 2012).

The oxygen reduction process leads to the formation of lithium peroxide ($Li_2O_2$, $2Li + O_2 + 2e^- \leftrightarrow Li_2O_2$), and the process is reversed on charging (Bruce, Hardwick, and Abraham 2011).

This type of battery is of low cost because the cathode material is mainly carbon and the substrate that supports the carbon materials must be porous to allow the formation of $Li_2O_2$(Tan et al. 2017). Pore size has been shown to affect battery performance, and pores in the range from 10 to 200 nm seem to be ideal for improved performance (Bruce, Hardwick, and Abraham 2011). This battery type should operate at low overpotentials (below 1 V) to maximize energy density, and $Li_2O_2$ film growth depends on the electrolyte solution (Aurbach et al. 2016). Considering this fact to obtain more stable chemical and electrochemical electrolytes, the solvent used in the electrolytes for this battery type is ether-based or dimethyl sulfoxide (DMSO)-based, among others (Balaish, Kraytsberg, and Ein-Eli 2014). In this battery type, more efforts are needed to understand the mechanisms behind the operation of Li-air cells at different rates as well as decomposition reactions of the electrolyte solution. It is also important to address the physicochemical properties related to the electrochemical behavior of the electrodes and the electrolyte to provide pathways for future commercialization in areas such as electric vehicles with improved autonomy.

### 11.3.2 COMPARATIVE ANALYSES OF THE DIFFERENT BATTERY TYPES

Different lithium-ion battery types have been introduced, where each type has advantages and disadvantages, as indicated in Table 11.1. The most used type is the lithium-polymer battery, taking

### TABLE 11.1
### Comparative Analysis of Different LIBs Types

| Battery Types | Advantages | Disadvantages |
|---|---|---|
| Lithium-polymer | • Lightweight,<br>• Small internal resistance,<br>• Good discharge characteristics, and<br>• Improved safety from explosion. | • Slightly expensive comparative to other LIBs, and<br>• Less recharge life about 300–400 cycles. |
| Lithium-silicon | • Higher gravimetric capacity and volumetric capacity,<br>• Low cost,<br>• Environment friendly, and<br>• Nontoxic. | • Drastic volume expansion,<br>• Low electrode structure integrity, and<br>• Electronic isolation. |
| Lithium-sulfur | • Higher theoretical specific capacity (1,675 mAh/g), and<br>• Theoretical energy density (2,600 Wh/kg). | • Poor electrical and ionic conductivity,<br>• High volume expansion/contraction,<br>• Dissolution of lithium polysulfide, and<br>• Slightly soluble in many polar electrolyte solvents. |
| Lithium-air | • High theoretical energy density (3,505 Wh/kg)<br>• Low cost | • Stability of the electrolyte,<br>• $O_2$ solubility and diffusivity, and<br>• Membrane allowing $O_2$ to pass and blocking $CO_2$ |

into account that its specific practical capacity approaches the theoretical capacity when compared to other battery types.

In addition, lithium-sulfur and lithium-air types show larger theoretical but low practical capacity due to drawbacks that need to be resolved in the coming years. Among those two battery types, the batteries that are closer to commercialization are lithium-sulfur with the implementation of solid polymer electrolytes (SPEs) to prevent sulfur and polysulfide dissolution into the electrolyte and to suppress dendritic growth of lithium, allowing enhanced safety. These Li-S cells could be applied to the next generation of electric vehicles to increase their autonomy.

## 11.4 LITHIUM SOURCES, EXTRACTION, AND PRICE

Lithium is considered an essential material for the energy transition to a more sustainable model, where the burn of fossil fuels is being replaced by the use of renewable energies combined with efficient storage systems (Richardson 2013). This leads to a growing demand for lithium raw materials worldwide, increasing their market share related to battery production. This trend is particularly significant in recent years, as it took just about 3 years to almost double the market share of lithium used for battery production, from 35% to 65% (Martin et al. 2017, Tadesse et al. 2019). This increase in demand for lithium minerals leads to an increase in their cost as a direct consequence. For example, according to BMO Capital Markets, the lithium carbonate cost, which is the main source of lithium materials, increased by 61% in just 1 year (2017) due to its skyrocketing demand. Apart from this, the policies applied by several countries to incentive the expansion of electric mobility also result in an increase in lithium demand, affecting, even more, their costs (Sterba et al. 2019). Additionally, lithium is a highly reactive element that does not occur naturally in metallic form. Instead, it is concentrated in water solutions or immobilized in various minerals unevenly distributed in the earth's crust. According to a 2021 United States Geological Survey (USGS) report, the overall identified lithium resources in the world are 86 Mt, of which 58% are in South America (Bolivia, 21 Mt; Argentina, 19.3 Mt; Chile, 9.6 Mt), and the rest is scattered in Australia (6.4 Mt), China (5.1 Mt), Congo (3 Mt), Canada (2.9 Mt), Germany (2.7 Mt), and other countries. However, due to a variety of economic and geopolitical reasons, the majority of lithium mine production is currently concentrated in Australia (49%), Chile (22%), China (17%), and Argentina (7.6%) (2021) and relies mainly on brine and pegmatite deposits. Thus, lithium materials are usually explored from brines or hard rocks, as shown in Figure 11.3.

Waters with high concentrations of dissolved lithium, calcium, magnesium, sodium, potassium, boron, and other salts are called *brine deposits* (Talens Peiró, Villalba Méndez, and Ayres 2013, Gruber et al. 2011). Typically, brines occupy the pores of rocks formed by evaporation in confined basins of arid regions. The most common lithium brine deposits are the playas (salt flats, salt pans, or salars), representing dry lakes in desert areas (Talens Peiró, Villalba Méndez, and Ayres 2013, Baker 2021). Lithium in salar deposits accumulates by drainage from nearby rocks exposed to erosion and geothermal activity. The extraction of metal relies on pumping the brine to the surface and then concentrating it by evaporation in a chain of shallow ponds. The duration of evaporation varies between 12 and 18 months and depends on a combination of factors such as seasonal variations in solar radiation, humidity, winds, precipitation, and the Mg/Li ratio in the brine (high magnesium content reduces the evaporation rate) (Cabello 2021). After reaching a lithium concentration of 5.5%–6.5% (An et al. 2012), the solutions are further processed in a plant to produce lithium carbonate, lithium hydroxide, or salts of boron, magnesium, potassium, and sodium (Talens Peiró, Villalba Méndez, and Ayres 2013). Among the operating deposits, the salars are among those with the highest capacity for large-scale and long-term lithium production. For example, the *Salar de Atacama* (Chile) is operational since 1984 and continues to be the richest commercial lithium brine deposit in the world (Cabello 2021). The brines with the largest resources of lithium are located in Uyuni, Bolivia (10.2 Mt), Atacama, Chile (6.3 Mt), Qaidam, China (2.02 Mt), Zabuye, China (1.53 Mt), and Rincon, Argentina (1.12 Mt) (Gruber et al. 2011). These types of deposits account for 66%–69% of the world's

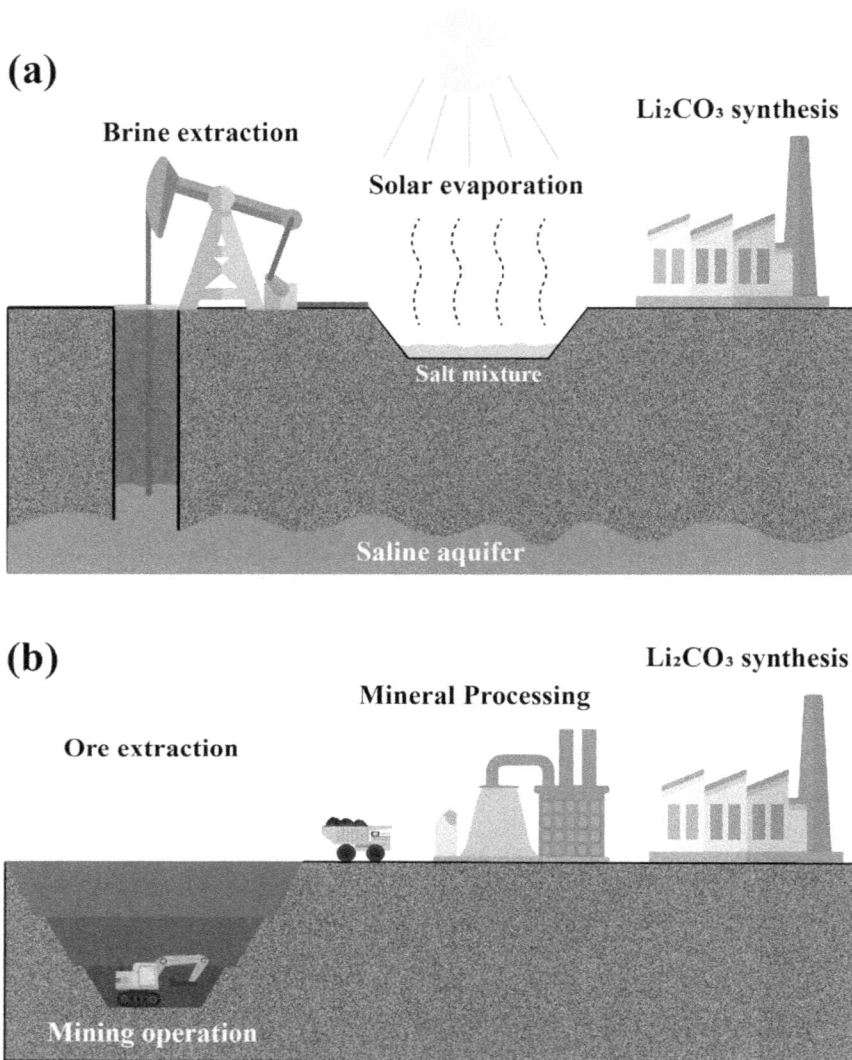

FIGURE 11.3   Lithium production methods in brines (a) and hard rocks (b).

lithium resources (Gruber et al. 2011, Christmann et al. 2015, Kesler et al. 2012). However, just three countries, Chile (60%), China (20%), and Argentina (14%), supply more than 90% of lithium from brine deposits (USGS 2021, Talens Peiró, Villalba Méndez, and Ayres 2013).

Other less common brine deposits are the *deep oil fields* such as the Smackover Formation (Arkansas, USA), with an average concentration of 0.015% (from 50 to 572 mg/L) and an estimated 750,000 tones of Li metal equivalent (Kumar et al. 2019 and Gruber et al. 2011). However, more than 75% of oil wells are deeper than 4,500 m (Dyman et al., 1990), which requires additional pumping costs that are profitable if the extraction is carried out together with the oil (Gruber et al. 2011). A feasible technology for lithium extraction from petroleum brines is membrane separation (Kumar et al. 2019).

*Geothermal brine deposits* are an emerging source of lithium. They are hot saline water solutions circulating through hard rock and accumulate substances (Li, Ca, and Na-K chlorides) leached from the rock (Kesler et al. 2012). These brines are typically pumped to the surface by a geothermal power plant, used to generate energy, and then returned to the underground reservoir via an injection well or discharged to a surface water body. Some plants have implemented technologies to extract valuable materials such as lithium, manganese, and zinc from brines prior to disposal to maximize efficiency and reduce waste. For example, 2 million tons of lithium resource with concentrations of 0.02% have been reported in the Salton Sea (California, USA) (Miles et al. 2009, Kesler et al. 2012). Extraction of lithium from geothermal brines includes precipitation, reverse osmosis, and ion exchange methods (Kalmykov et al. 2021) (Figure 11.3a).

Lithium in *seawater* is the most evenly distributed reserve. However, its concentration is quite low (0.17 ppm) compared to salars (1,000–3,000 ppm), and the magnesium–lithium ratio is high (Tahil 2007, Talens Peiró, Villalba Méndez, and Ayres 2013). Moreover, only about 20% of the lithium in seawater is recoverable by methods such as nanofiltration, ion exchange resins, solvent extraction, coprecipitation, and adsorption (mainly by manganese oxides), and electrochemical reactions (Gruber et al. 2011, Kim et al. 2015). The listed extraction approaches are still not cost-competitive at the current price of lithium.

*Pegmatite (hard rock) deposits* are intrusive igneous rocks formed in the last stage of the crystallization of magma. They consist of minerals rarely found in other types of rocks rich in lithium and other valuable elements such as tin, tantalum, niobium, beryllium, cesium, rare earth, and others (London 2018). Lithium in pegmatites accounts for up to 26% of the world's resources (Gruber et al. 2011). It concentrates in spodumene ($LiAlSi_2O_6$) among other silicate minerals such as petalite ($LiAlSi_4O_{10}$), lepidolite [$(KLi_2Al(Al,Si)_3O_{10}(F,OH)_2$)], and eucryptite ($LiAlSiO_4$) (Gruber et al. 2011). Pegmatite ores contain on an average 0.58%–1.59% of lithium, which is much higher than the richest salar (0.14%, Salar de Atacama, Chile) (Gruber et al. 2011). These deposits are usually smaller and have a shorter lifetime than the brines, but their more regular distribution in the earth's crust and independence on weather conditions ensures more secure supply lines to lithium end users (Figure 11.3b).

Unlike the brines, the lithium from pegmatite minerals can be extracted directly as LiOH (Fosu et al. 2020), which is a more valuable end product. However, the extraction is more energy-intensive due to drilling, crushing, and separation of hard rocks followed by high-temperature (1,100°C) conversion of the monoclinic α-form to the chemically more reactive tetragonal β-form of spodumene. Additional steps include acid ($H_2SO_4$) roasting, water leaching, filtration, purification (precipitation and/or ion exchange), electrolysis (when intended to produce LiOH), or concentration by ion exchange and precipitation (when intended to produce $Li_2CO_3$) that eventually leads to end products of lithium hydroxide or lithium carbonate (Rioyo, Tuset, and Grau 2020, Dessemond et al. 2019).

The largest lithium pegmatite deposit is at King's Mountain, North Carolina (USA), which is no longer mined, although large lithium reserves remain (5.9 Mt). Among the operating deposits, Greenbushes (Australia) is the largest with a resource of 560,000 tonnes of lithium ore with an average concentration of about 1.6% (Gruber et al. 2011) and supplying 85% of world lithium from pegmatites (Talens Peiró, Villalba Méndez, and Ayres 2013). Australia alone is responsible for nearly half of the world's lithium mine production, with 40,000 tons in 2020 (2021). Among the prospective deposits, Portugal hosts one of the most significant spodumene-bearing pegmatites in Western Europe, with an average grade of 1.06% and total resources of 270,000–285,900 tons of $Li_2O$ (U.S. Geological Survey 2021, Serra et al. 2020). Currently, Portugal is the largest lithium mine producer (900 tons for 2020) (2021) in Europe, and these resources back the plants of the European Union (EU) to decrease its dependence on imported lithium (86% of Li in the EU is from outside suppliers) (https://www.europarl.europa.eu/doceo/document/E-9-2019-003008_EN.html).

In *sedimentary deposits*, lithium is typically concentrated in layered silicates (clays and micas) and accounts for 8% of the world's resources (Vine 1975, Gruber et al. 2011). Sedimentary deposits are formed when lithium is leached from volcanic minerals into basins, where it is immobilized during mineral formation mainly by precipitation processes. Unlike the pegmatites, lithium in sedimentary rock minerals is weakly bonded to the crystal structure. Moreover, the sedimentary rocks are softer than the pegmatites and require less energy for mining. The lithium extraction methods include a combination of limestone–gypsum roasting, selective chlorination, roasting-water leaching, inverse Hofmann–Klemen process, and ion exchange (Crocker and Lien 1987, Jaynes and Bigham 1987). The concentration of lithium in deposits varies from 0.27% in Kings Valley (Nevada, USA) to 0.7% in Hector (California, USA), which is currently the largest sedimentary deposit with 2.0 Mt of lithium resource (Gruber et al. 2011). However, the complex extraction methodology (one of the big problems is the excess of leaching reagents necessary to extract the lithium from the sediment) and the lower amounts of lithium make this source of lithium less attractive than the pegmatites. An exceptional example among the sedimentary deposits is in the Jadar Valley (Serbia), where the rocks are composed mainly of the rare mineral jadarite ($LiNaB_3SiO_7(OH)$), which, unlike the typical lithium-bearing minerals, is a nesosilicate that offers approaches for lithium extraction similar to the pegmatite deposits. Ore reserves of 16.6 Mt at 1.81% $Li_2O$ and 13.4% $B_2O_3$ make the Jadar Valley one of the global lithium deposits whose exploration onset was intended for 2026 (https://www.riotinto.com/-/media/Content/Documents/Invest/Reserves-and-resources/RT-Jadar-Reserves-2020.pdf?rev=c31b0fd7047f46dcba39bc3dea49ccb6).

The price of lithium is one of the driving forces for the development of the lithium mining industry. The two main commercial products of lithium mining are lithium carbonate ($Li_2CO_3$) and lithium hydroxide (LiOH). Lithium hydroxide is generally more expensive (2021), primarily due to higher production costs. Another reason is that lithium hydroxide decomposes at a lower temperature, and it is preferred to produce more sustainable, durable, efficient, and safer batteries, which is one of the main reasons why electric vehicle manufacturers choose LiOH for their batteries (Takemura et al. 2005). Between 2002 and 2018, the increasing demand for LIBs has driven up the price by more than tenfolds, that is, from 1,590 to 16,500 US $ for a ton of lithium carbonate (Kundu et al. 2017). In response to the rising cost, between the tears 2012 and 2018, the global lithium supply increased by 140% (US Geological, Orienteering Survey, and US Geological Survey 2009). Currently, 71% of lithium is consumed in batteries (2021), which largely determines the price. According to S&P Global Platts Analytics (https://www.spglobal.com/en/research-insights/articles/lithium-supply-is-set-to-triple-by-2025-will-it-be-enough), more than 2 million electric vehicles were sold in 2018 alone, which coincided with the peak in lithium prices. In 2019, lithium carbonate prices fell sharply (12,700 US $) (https://www.statista.com/statistics/606350/battery-grade-lithium-carbonate-price/) as the market moved toward oversupply and electric vehicle growth slowed down. In early 2020, the economic impact of the global COVID-19 pandemic was another major factor in the decline in lithium demand (2021) with a further drop in the price to 8,000 US $ (https://www.statista.com/statistics/606350/battery-grade-lithium-carbonate-price/). Nevertheless, the production of lithium is expected to almost triple to more than 1.5 million tons by 2025 (https://www.spglobal.com/en/research-insights/articles/lithium-supply-is-set-to-triple-by-2025-will-it-be-enough).

These lithium reserves are not equally distributed around the world, being more concentrated in South America and Australia, as previously mentioned, but these numbers can change quickly, as the interest in lithium materials led to a growing number of studies and prospection activities in several countries, to find new sources and, consequently, reduce the price. This trend is particularly observed in Europe, where the current explorations are residual. Viability studies and preliminary activities are being carried out in Portugal, Germany, Sweden, and Serbia, and several other countries are following this trend, as presented in Figure 11.4 (Gourcerol et al. 2019).

## 11.5 RECYCLING OF LITHIUM-ION BATTERIES

The strong growth of the exploration of the lithium reserves in recent years calls for different solutions to warrant a sustainable industry and reduce the environmental impact. Some studies point out that the lithium demand in the incoming years can grow to values between 74% and 248% of the actual reserves (Pehlken, Albach, and Vogt 2017). This means that lithium extraction cannot be enough by itself to meet the demand requirements. The rising price, the amount of used LIBs, their relatively short lifetime (5–10 years), and the decarbonization plans of many countries have increased the interest in lithium recycling technologies. The cathode of a standard lithium-ion

**FIGURE 11.4** Lithium exploration licensed areas in 2019.

battery (31% of the battery) consists of 7% Li, 1.5% Al, 49% Ni, and 9% Co. The rest of the battery consists of 17% Cu, 22% graphite, 8% Al, 3% separator plastic, 4% carbon black and binder, and 15% electrolyte solution (https://cen.acs.org/materials/energy-storage/time-serious-recycling-lithium/97/i28), providing good opportunities for sustainable recycling.

In this context, efforts to establish efficient and viable recycling strategies are being carried out, both at laboratory and industry scales, which will allow reusing the spent lithium resources, reducing the pressure on natural reserves (Costa et al. 2021). This approach also allows the reuse of other used materials, such as cobalt or nickel, that due to their scarcity are expensive to extract and process. Before recycling, large batteries need to be discharged, the electrolyte removed, and disassembled. For electric vehicles, disassembly can be challenging because batteries from different manufacturers have different sizes, weights, pack configurations, locations, and the presence of various hazardous materials, including nanoparticles (Harper et al. 2019). In addition, physical disassembly involves a combination of mechanical, magnetic (cast steel and fractions with different magnetic susceptibility), flotation (separation based on hydrophilicity; e.g., graphite is hydrophobic and $LiCoO_2$ is hydrophilic), solvent extraction (recovery of $LiPF_6$ from polypropylene carbonate via supercritical $CO_2$ extraction) (Grützke et al. 2015), and gravity separation methods. Thus, the most valuable components (electrodes, electrolytes, aluminum, and copper foil) are separated for further recycling during this process.

Thus, the development of recycling strategies can be advantageous both environmentally and economically, as the reduction of the need to extract new materials will predictably cause a drop in the battery prices, allowing an even higher establishment of these devices in a wide range of applications, particularly with the expansion of the electric vehicle market observed in recent years (Mo and Jeon 2018). The market context exposed before also contributes to the growing interest in the recycling of LIBs, as the rising costs of raw materials encourage the manufacturers to develop cheaper solutions, making recycling a more competitive option (Busch, Dawson, and Roelich 2017). Several governments also incentive the application of recycling LIBs, such as the European Union, which set the recycling efficiency target of 50% in the short term.

Different approaches can be taken when it comes to battery recycling strategies, which are schematized in Figure 11.5. These approaches allow to reintroduce the recycled materials at different steps of the value chain, reducing both the extraction of new raw materials and the disposal of spent materials, in a circular economy context.

When the second use of the battery is no longer possible, direct recycling methods should be the next taken approach.

*Direct recycling* involves the regeneration of spent battery components. Attractive features of this technology are the higher quality of the recovered material and the low amount of waste compared to pyrometallurgical and hydrometallurgical methods. However, complexity and production cost are still significant drawbacks (Harper et al. 2019). For example, the physical separation of cathode and anode can easily lead to contamination with Al or Cu (present in electrode foils), which strongly reduces battery performance. Typically, in spent LIBs, part of the lithium in the cathode is lost either at the anode or in the electrolyte during the charging and discharging processes. The recovery of these losses relies on a lithiation process that restores the lithium defects in the crystal structure of the cathode (Yang, Lu, et al. 2020). However, the efficiency of the process depends on the crystal structure and chemistry of the cathode. For example, layered lithium cobalt oxide cathodes can experience a recovery of up to 70% of their original value, while this value can be much lower for other types of cathodes (Gaines 2018).

*Bioleaching* is an approach adapted from the mining industry and relies on redoxolysis, acidolysis, and complexolysis processes during the biological activity of bacteria and fungi (Roy, Cao, and Madhavi 2021). The general mechanism involves the biogenic synthesis of sulfuric acid ($H_2SO_4$), promoting the dissolution of metals in LIBs. Subsequently, the elements are recovered through a series of precipitation reactions by varying the pH. Some of the advantages of bioleaching are the lower

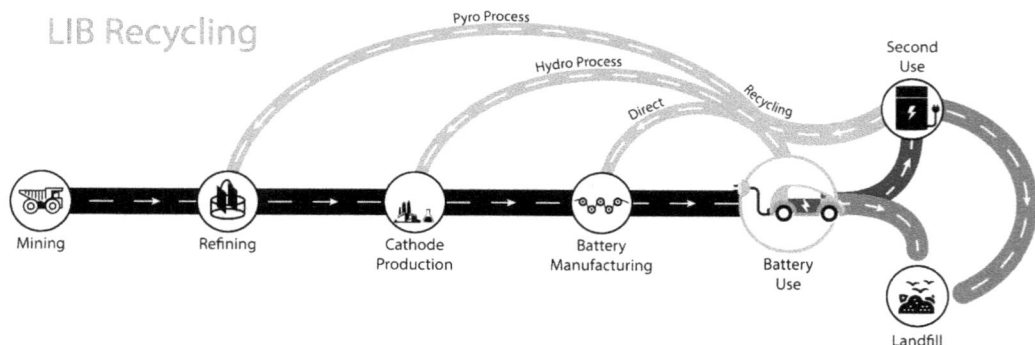

**FIGURE 11.5** Lithium-ion battery recycling pathways and strategies. (From, Costa, C.M., J.C. Barbosa, R. Gonçalves, H. Castro, F.J. Del Campo, S. Lanceros-Méndez, "Recycling and environmental issues of lithium-ion batteries: Advances, challenges and opportunities." *Energy Storage Materials*, 37, 433–465, 2021, Doi: 10.1016/j.enset al. 2021. With permission.)

amounts of harmful gases, low operating costs, high efficiency at low concentrations, and expected high metal recovery (100% Cu, 100% Li, 77% Mn, 75% Al, 64% Co, and 54% Ni) (Bahaloo-Horeh and Mousavi 2017). However, the kinetics of the extraction process is slow. In addition, some components of the batteries (electrolytes and binders) are toxic and hinder the development of microorganisms (Roy, Cao, and Madhavi 2021), and the technology needs to be significantly improved.

These methods also include physical and non-destructive processes, such as gravity or magnetic separation, that allow the recovery of the battery components without any chemical changes. The recovered components, usually active materials, can further be used in the assembly of new batteries (Shi, Chen, and Chen 2018).

Finally, a last and deepest step in battery recycling includes the chemical processes that reduce the battery components to the simpler forms of metals and materials that can be used later to resynthesize new components. This means that these processes are the most complex and energy-demanding and should only be applied when the other described options are no longer available. The most common approaches in these processes are the use of high temperatures (pyrometallurgical), where the materials are reduced to their metallic form, allowing for their posterior used as alloys, and acid leaching (hydrometallurgical), which originates solutions with metal ions that can be later recovered by precipitation or solvent evaporation (Lien 2018).

The most common strategies include second use of the batteries, direct recycling, and hydro- and pyrometallurgical processes. Their extraction from metal oxides is usually achieved by *pyrometallurgical* processing. In this method, the batteries are treated at high temperatures, leading to the carbonization of the organic components (plastics and electrolytes) and the formation of Co, Cu, Fe, and Ni alloys. The remaining slag contains Al, Mg, and Li and is further processed by $H_2SO_4$/$H_2O_2$ solutions (Harper et al. 2019). However, the high-temperature treatment is associated with the release of toxic gases from the burnet organics that require further remediation. To lower the temperature of metal extraction, there are reports on the addition of salt cosolvents, which can also be used to obtain water-soluble salts of the elements of interest (Fan et al. 2020). The pyrometallurgical technology is readily available, has low complexity, low production cost, and does not require pre-sorting of batteries. However, the low quality of the recovered material, high energy consumption, high investment cost, and waste generation are serious drawbacks (Harper et al. 2019). Moreover, all organic components (anode, separator, and electrolyte) of the battery are lost.

*Hydrometallurgical* processing involves leaching acids and reducing agents to obtain a solution of the elements. After extraction, the elements are recovered by pH-dependent precipitation reactions. The technology is more complex and costly than pyrometallurgical processing (Harper et al. 2019). It also produces large amounts of acidic waste and hazardous gases ($Cl_2$, $SO_3$, and $NO_x$) (Roy, Cao, and Madhavi 2021). However, it allows higher quantity (more than 90% leaching yield of Li, Ni, and Co) (Meshram, Pandey, and Mankhand 2015) and better quality of recovered metals. In addition, the method is energy-efficient and, unlike pyrometallurgical processing, allows almost direct recovery of Li along with other valuable elements such as Co, Ni, and Cu (Harper et al. 2019).

Depending on the objective of the recycling process, each path has different advantages and disadvantages. Direct recycling methods are usually more environmentally friendly, as they rely almost exclusively on physical processes, without the need of using hazardous materials.

However, recycling is frequently not efficient, and this method does not allow the processing of different kinds of materials at once. On the other hand, pyrometallurgical processes have a high recycling efficiency, with large recovery rates for different materials, depending just on temperature variation. However, the high temperatures needed mean that these processes are energy-demanding. Finally, hydrometallurgical processes are the most efficient in the recovery of spent materials, as well as the most unspecific, which implies that a larger number of different materials can be recovered in the same process. Nevertheless, the use of acids and toxic reagents makes them the most harmful of the processes for the environment, and the complexity of the method also makes them more costly than other processes (Costa et al. 2021).

However, lithium-ion battery recycling is not yet well established worldwide. There are only a couple of industries worldwide devoted to this field, with Umicore being the leader nowadays, with

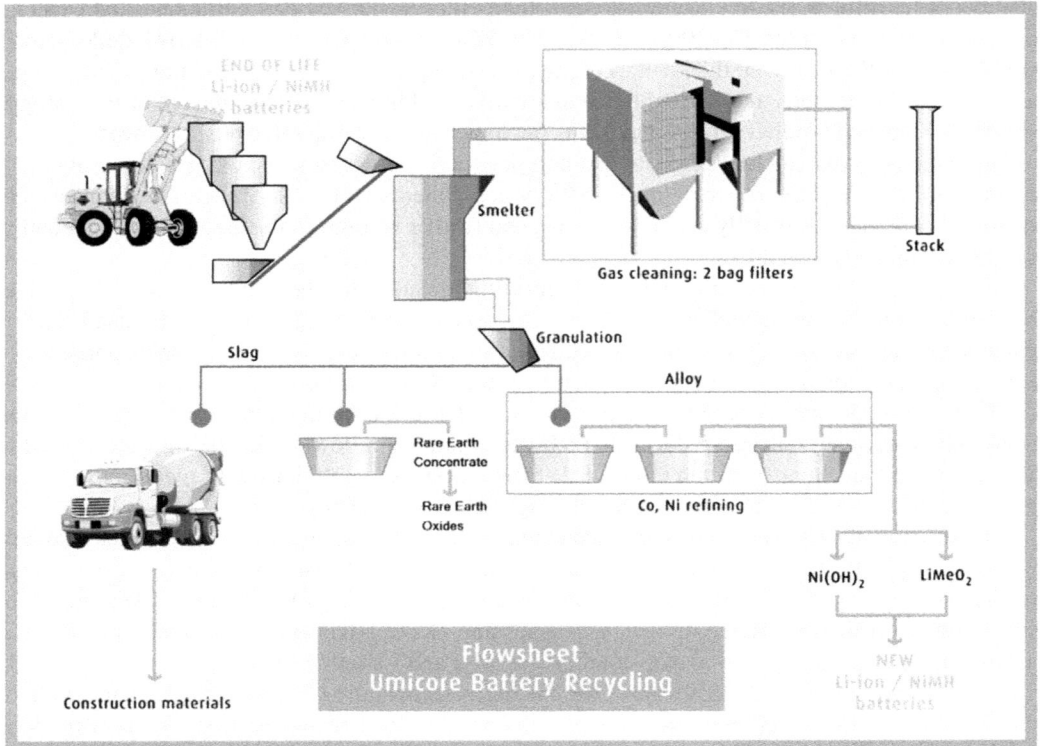

**FIGURE 11.6** Umicore process flow for battery recycling. (From, Tytgat, J, "The recycling efficiency of Li-ion EV batteries according to the European Commission Regulation, and the relation with the End-of-Life Vehicles Directive recycling rate." *World Electric Vehicle Journal*, 6 (4), 2013, Doi: 10.3390/wevj6041039. With permission.)

its process flow achieving a recycling efficiency of about 60% (Tytgat 2013). The Umicore recycling flowchart is schematized in Figure 11.6. Other companies developed their own recycling flows with different methods and distinct efficiencies. The great variety of used active materials and battery architectures leads to a high diversity of battery types, meaning that the implementation of a standard recycling process that can be effective to all types of batteries is difficult to achieve at the present moment. Standardization of the lithium-ion battery structure and materials would be an important step to facilitate the recycling process, similarly to the case of lead acid batteries, in which recycling rates can achieve 99% in several countries. These batteries possess the same main characteristics everywhere, which allowed the development of a unique and effective recycling process (Gaines 2014).

## 11.6   ENVIRONMENTAL ISSUES RELATED TO LITHIUM BATTERIES

While batteries can reduce the use of fossil fuels, the environmental problems of manufacturing LIBs begin with mining. Brine and hard rock mining contribute to global warming as extraction technologies are energy-intensive and rely on fossil fuels. For example, extracting 1 ton of lithium carbonate from spodumene requires 24,000 L of water, 1.66 GJ of gas, and 1.01 GJ of electricity (Talens Peiró, Villalba Méndez, and Ayres 2013). By comparison, brine extraction requires up to 500,000 gallons (almost 2 million L) of water and 12.53–18.80 GJ/$m^2$ of solar radiation per ton of lithium carbonate (Sohn et al. 2009). Although it is not drinking water, the extraction of saline water affects groundwater in salars, and many mining areas are already experiencing water shortages (Angulakshmi and Stephan 2014). For example, in the Salar de Atacama (Chile), mining processes consume 65% of the water, leading to groundwater depletion, soil contamination, and other problems (Angulakshmi and Stephan 2014, Tahil 2007). In addition, nearly 80% of the brine is extracted within 10 km from a coastline, usually discharging untreated wastewater brine directly into the marine environment. The main risks to marine life and marine ecosystems are that brine significantly increases the salinity of the receiving seawater and pollutes the seas with the toxic chemicals used in the desalination process (Jones et al. 2019).

The environmental problems associated with the exploration of hard rock deposits are less addressed in the literature, and current concerns relate mainly to the potential contamination of drinking water by beryllium- and fluorine-rich pegmatites (Bradley, McCauley, and Stillings 2017). However, considering that hundreds of tons of rock are excavated and crushed during mining operations, waste rock can be expected to accumulate and increase the amount of dust and particles in the air. Mining operations are also water-dependent, and depletion of ground and surface water resources at mining sites should be carefully evaluated. Other conceivable problems include the removal of large horizons of fertile soil, erosion, and leaching of toxic and radioactive elements from the waste rock, and the unknown impacts on biodiversity around the mining site.

Another emerging problem is the increasing number of electric vehicles expected to generate 2 million tons of used LIBs per year by 2030 (https://cen.acs.org/materials/energy-storage/time-serious-recycling-lithium/97/i28). However, the disposal and recycling of batteries are still far behind the expected production rates. According to Eurostat, in 2018, less than half of the batteries are collected for recycling in the European Union (https://ec.europa.eu/eurostat/web/products-eurostat-news/-/ddn-20200723-1), which is highly insufficient for sustainable reuse of lithium metal.

The recycling of LIBs is a useful way to mitigate the environmental impacts caused by their production and use. These impacts need to be identified and mitigated, and different strategies can be approached to fulfill this objective.

Depending on the extraction method, the production of LIBs origins all kinds of impacts associated with the mining activities. The visual impacts, production of waste waters, and ecosystem destruction are the most common and relevant, and can be significantly reduced by developing suitable recycling processes. Recycling will directly decrease the amount of raw materials and, consequently, reduce the extraction impacts. The processing of the materials and synthesis of the battery components have impacts essentially upon the use of thermal processes and toxic reagents, such as acids and organic solvents. These issues can be mitigated by developing environmentally friendly methods and using natural materials to replace the synthetic ones used nowadays (Wanger 2011).

During the use phase, the most important issues are associated with the risks of explosion and electrolyte leakages in the LIBs. These limitations occur due to the used materials which are liquid and easily flammable, increasing the risks. The development of solid electrolytes is an important step toward the mitigation of these problems, as they combine the functions of separator and electrolyte, allowing the production of fully solid batteries with reduced risks, and preventing the need to apply strong encapsulations to the device (Barbosa et al. 2021). Also, the commonly used materials such as chromium, cobalt, or nickel are related to high levels of toxicity and are found in high concentrations in LIBs, representing significant risks for human health if they are not well cased in the battery structure (Kang, Chen, and Ogunseitan 2013).

Finally, the end of life of a battery brings the need to understand what to do with the spent materials. The disposal of these materials without further processing caused significant negative impacts at different levels on the disposal place, due to the time it takes to naturally degrade those spent materials. Thus, as stated before, is important to develop efficient recycling methods for all battery components to reduce the disposal rate and extraction of raw materials (Dunn et al. 2012).

## 11.7 LIFECYCLE ASSESSMENT AND ECONOMIC ANALYSIS

Lifecycle assessment (LCA) is an interesting tool to characterize all the lifecycle phases of a given material or product. LCA can evaluate potential hazards and impacts of the different stages, from the production to the end of life of the product (Zhao and You 2019). When correctly applied, LCA allows to find the most impacting activities associated with the lifecycle, meaning that it would be needed to mitigate them by finding better alternatives. However, LCA is difficult to apply accurately, because frequently there are not enough data, meaning that several assumptions have to be made, making the LCA less precise (Zackrisson, Avellán, and Orlenius 2010).

In the particular case of LIBs, LCA is very useful to identify where the main economic and environmental issues are in the value chain and to understand to what extent is recycling more advantageous than producing new devices. The most environmentally impacting activities are associated with the extraction, processing, and production of the cells during the production phase. In a scenario with high recycling rates, these impacts can be reduced from 10% to 30%, due to the reduction in the dependence on natural resources (Zackrisson et al. 2016). Furthermore, the predicted increase in the extraction costs associated with the depletion of natural resources will add an economical advantage to recycling (Wanger 2011). The study of other scenarios, which can include different production methods or the use of more sustainable materials can also give valuable information regarding the economic and environmental consequences of each action.

### 11.7.1 Market Developments and Trends

With the increasing demand for LIBs for electronic devices and for the growing electric mobility industry, a significant increase in the global LIB market is expected in the next years. This growth is already noticeable in the last 5 years, mainly driven by the increase of the electric vehicle fleet worldwide as it is shown in Figure 11.7 (Blomgren 2016).

However, the steady growth of the LIB market was affected by the recent events associated with the global Covid-19 pandemics. According to the MarketstandMarkets, this was caused due to the disruption of the supply chain in the Asia-Pacific region and the lack of raw materials that are mainly produced in this region. The speed of recovery of this market is dependent on the government's incentives for the electric vehicle purchase and on the stabilization of the supply chain for the raw materials (Wen et al. 2021).

Despite the fact that electric mobility is taking a significant share of the LIB market, other application areas must be also taken into account. With the growing number of interconnected devices,

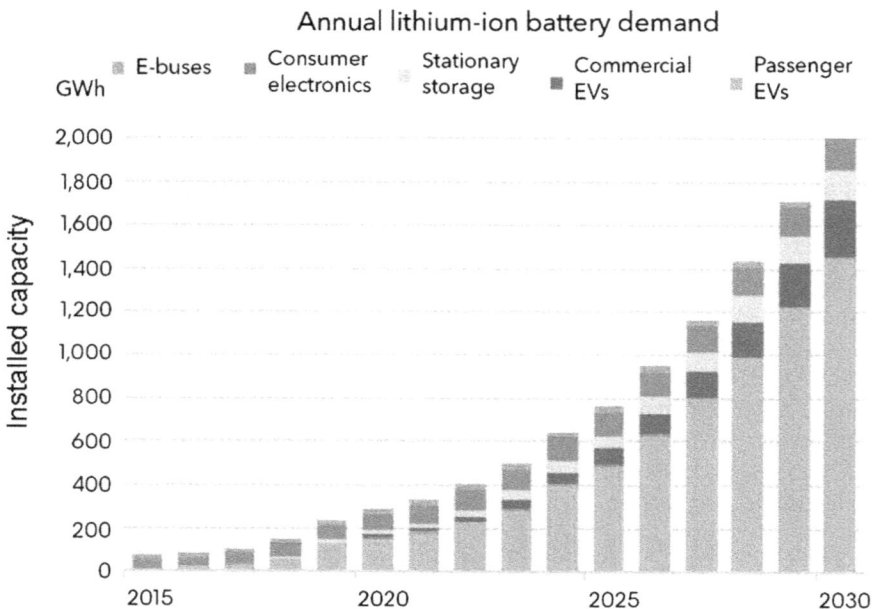

**FIGURE 11.7** Predicted evolution of the LIB demand until 2030 in the different implementation areas.

and the emerging of the industry 4.0 and the Internet of Things concepts, the need for smaller energy storage systems will also strongly grow. Even though the LIBs used for these applications are small, their high numbers make them a major player in the battery market (Pillot 2016).

All of these factors associated with the aforementioned recycling issues will expectedly create a strong market niche, which will be able to create value in all the phases of the LIB lifecycle. As this market is still in its early stages of development, the right decisions toward a circular economy model, where batteries are reused as many times as possible, and then recycled to minimize the extraction of more raw materials, can make it turn into an example of profitability integrated with sustainability.

## 11.8   CONCLUSION AND FUTURE PERSPECTIVES

Nowadays, considering the increasing mobility of the general population and goods, as well as the industrial processes and electricity production that are increasingly relying on renewable energy sources, lithium-ion batteries are being increasingly used as energy storage systems and will play a key role in the transition to avoid the environmental impacts of the fossil fuel era. In this chapter, the different types of lithium-ion batteries are presented, as well as the main advantages and disadvantages of each one. Given that they are implemented in a wide range of applications, such as laptops, smartphones, electric vehicles, or even home storage systems, a heavy reliance on this technology will lead to a growing demand for lithium and other materials needed to manufacture batteries. Lithium-ion sources, extraction, price, and recycling of lithium-ion batteries have been also addressed for a transition to a more sustainable economy model, based on circular economy and renewable energy sources. The different recycling methods and strategies have been presented, showing that there is a need to standardize battery materials and devices as well as to further incentive battery recycling to reduce environmental issues and promote an improved sustainable approach in this field.

## ACKNOWLEDGMENTS

The authors thank FCT (Fundação para a Ciência e Tecnologia) for financial support under the framework of Strategic Funding UID/CTM/50025/2021, UIDB/04650/2020 and support from FEDER funds through the COMPETE 2020 Programme (projects PTDC/FIS-MAC/28157/2017). Grants SFRH/BD/140842/2018 (J.C.B.) and contracts under the Stimulus of Scientific Employment, Individual Support CEECIND/00833/2017 (R.G.) and 2020.04028 CEECIND (C.M.C.) are also acknowledged as well as financial support from the Basque Government under the ELKARTEK program and the BIDEKO project, funded by MCIN/AEI, NextGenerationEU, PRTR.

## REFERENCES

Abdi, H., B. Mohammadi-ivatloo, S. Javadi, A.R. Khodaei, and E. Dehnavi. 2017. "Chapter 7- energy storage systems." In *Distributed Generation Systems*, edited by G. B. Gharehpetian and S. Mohammad Mousavi Agah, 333–368, Oxford: Elsevier.

Ahmed, R., G. Liu, B. Yousaf, Q. Abbas, H. Ullah, and M.U. Ali. 2020. "Recent advances in carbon-based renewable adsorbent for selective carbon dioxide capture and separation-A review." *Journal of Cleaner Production* 242:118409. Doi: 10.1016/j.jclepro.2019.118409.

An, J.W., D.J. Kang, K.T. Tran, M.J. Kim, T. Lim, and T. Tran. 2012. "Recovery of lithium from Uyuni salar brine." *Hydrometallurgy* 117–118:64–70. Doi: 10.1016/j.hydromet.2012.02.008.

Angulakshmi, N., and A. Manuel Stephan. 2014. "Electrospun trilayer polymeric membranes as separator for lithium–ion batteries." *Electrochimica Acta* 127:167–172. Doi: 10.1016/j.electacta.2014.01.162.

Aurbach, D., B.D. McCloskey, L.F. Nazar, and P.G. Bruce. 2016. "Advances in understanding mechanisms underpinning lithium–air batteries." *Nature Energy* 1 (9):16128. Doi: 10.1038/nenergy.2016.128.

Bahaloo-Horeh, N., and S.M. Mousavi. 2017. "Enhanced recovery of valuable metals from spent lithium-ion batteries through optimization of organic acids produced by Aspergillus Niger." *Waste Management* 60:666–679. Doi: 10.1016/j.wasman.2016.10.034.

Baker, V.R. 2021 "Playa" In Encyclopedia / Physical Geography of Land, *Encyclopædia Britannica*.

Balaish, M., A. Kraytsberg, and Y. Ein-Eli. 2014. "A critical review on lithium–air battery electrolytes." *Physical Chemistry Chemical Physics* 16 (7):2801–2822. Doi: 10.1039/C3CP54165G.

Barbosa, J.C., R. Gonçalves, C.M. Costa, V.d.Z. Bermudez, A. Fidalgo-Marijuan, Q. Zhang, and S. Lanceros-Méndez. 2021. "Metal–organic frameworks and zeolite materials as active fillers for lithium-ion battery solid polymer electrolytes." *Materials Advances* 2 (12):3790–3805. Doi: 10.1039/D1MA00244A.

Blomgren, G.E. 2016. "The development and future of lithium ion batteries." *Journal of The Electrochemical Society* 164 (1):A5019-A5025. Doi: 10.1149/2.0251701jes.

Bradley, D.C., A.D. McCauley, and L.L. Stillings. 2017. Mineral-deposit model for lithium-cesium-tantalum pegmatites. In *Scientific Investigations Report*, Reston, VA.

Bruce, P.G., L.J. Hardwick, and K.M. Abraham. 2011. "Lithium-air and lithium-sulfur batteries." *MRS Bulletin* 36 (7):506–512. Doi: 10.1557/mrs.2011.157.

Busch, J., D. Dawson, and K. Roelich. 2017. "Closing the low-carbon material loop using a dynamic whole system approach." *Journal of Cleaner Production* 149:751–761. Doi: 10.1016/j.jclepro.2017.02.166.

Cabello, J. 2021. "Lithium brine production, reserves, resources and exploration in Chile: An updated review." *Ore Geology Reviews* 128:103883. Doi: 10.1016/j.oregeorev.2020.103883.

Cheng, M., S.S. Sami, and J. Wu. 2017. "Benefits of using virtual energy storage system for power system frequency response." *Applied Energy* 194:376–385. Doi: 10.1016/j.apenergy.2016.06.113.

Christmann, P., E. Gloaguen, J.-F. Labbé, J. Melleton, and P. Piantone. 2015. "Chapter 1- Global lithium resources and sustainability issues." In *Lithium Process Chemistry*, edited by A. Chagnes and J. Światowska, 1–40. Amsterdam: Elsevier.

Costa, C.M., J.C. Barbosa, R. Gonçalves, H. Castro, F.J. Del Campo, and S. Lanceros-Méndez. 2021. "Recycling and environmental issues of lithium-ion batteries: Advances, challenges and opportunities." *Energy Storage Materials* 37:433–465 Doi: 10.1016/j.ensm.2021.02.032.

Costa, C.M., Y.-H. Lee, J.-H. Kim, S.-Y. Lee, and S. Lanceros-Méndez. 2019. "Recent advances on separator membranes for lithium-ion battery applications: From porous membranes to solid electrolytes." *Energy Storage Materials* 22:346–375 Doi: 10.1016/j.ensm.2019.07.024.

Crocker, L., and R.H. Lien. 1987. *Lithium and Its Recovery from Low-Grade Nevada Clays Bulletin*, U.S. Department of Energy Office of Scientific and Technical Information, Utah: United States.

Dessemond, C., F. Lajoie-Leroux, G. Soucy, N. Laroche, and J.-F. Magnan. 2019. "Spodumene: The lithium market, resources and processes." *Minerals* 9 (6). Doi: 10.3390/min9060334.

Dunn, J.B., L. Gaines, J. Sullivan, and M.Q. Wang. 2012. "Impact of recycling on cradle-to-gate energy consumption and greenhouse gas emissions of automotive lithium-ion batteries." *Environmental Science & Technology* 46 (22):12704–12710. Doi: 10.1021/es302420z.

Dyman, T.S., D.T. Nielson, R.C. Obuch, J.K. Baird, and R.A. Wise. 1990. Summary of deep oil and gas wells and reservoirs in the U.S. In *Geological Survey: United States Department of Interior*, Virginia: United States.

Erenoğlu, A.K., O.Erdinç, and A. Taşcıkaraoğlu. 2019. "Chapter 1- History of electricity." In *Pathways to a Smarter Power System*, edited by A. Taşcıkaraoğlu and O.Erdinç, 1–27. Academic Press, London, United Kingdom.

European Union. 2021a. "2030 climate & energy framework." https://ec.europa.eu/clima/policies/strategies/2030_en.

European Union. 2021b. "European climate law." https://ec.europa.eu/clima/policies/eu-climate-action/law_en.

Fan, E., L. Li, Z. Wang, J. Lin, Y. Huang, Y. Yao, R. Chen, and F. Wu. 2020. "Sustainable recycling technology for Li-ion batteries and beyond: Challenges and future prospects." *Chemical Reviews* 120 (14):7020–7063. doi: 10.1021/acs.chemrev.9b00535.

Feng, K., M. Li, W. Liu, A.G. Kashkooli, X. Xiao, M. Cai, and Z. Chen. 2018. "Silicon-based anodes for lithium-ion batteries: From fundamentals to practical applications." *Small* 14 (8):1702737. Doi: 10.1002/smll.201702737.

Fosu, A.Y., N. Kanari, J. Vaughan, and A. Chagnes. 2020. "Literature review and thermodynamic modelling of roasting processes for lithium extraction from spodumene." *Metals* 10 (10). Doi: 10.3390/met10101312.

Gaines, L. 2014. "The future of automotive lithium-ion battery recycling: Charting a sustainable course." *Sustainable Materials and Technologies* 1–2:2–7. Doi: 10.1016/j.susmat.2014.10.001.

Gaines, L. 2018. "Lithium-ion battery recycling processes: Research towards a sustainable course." *Sustainable Materials and Technologies* 17:e00068. Doi: 10.1016/j.susmat.2018.e00068.

Gonçalves, R., P. Dias, L. Hilliou, P. Costa, M.M. Silva, C.M. Costa, S. Corona-Galván, and S. Lanceros-Méndez. 2021. "Optimized printed cathode electrodes for high performance batteries." 9 (1):2000805. Doi: 10.1002/ente.202000805.

Gonçalves, R., T. Marques-Almeida, D. Miranda, M.M. Silva, V.F. Cardoso, C.M. Costa, and S. Lanceros-Méndez. 2019. "Enhanced performance of fluorinated separator membranes for lithium ion batteries through surface micropatterning." *Energy Storage Materials* 21:124–135 Doi: 10.1016/j.ensm.2019.05.044.

Gourcerol, B., E. Gloaguen, J. Melleton, J. Tuduri, and X. Galiegue. 2019. "Re-assessing the European lithium resource potential – A review of hard-rock resources and metallogeny." *Ore Geology Reviews* 109:494–519 Doi: 10.1016/j.oregeorev.2019.04.015.

Gruber, P.W., P.A. Medina, G.A. Keoleian, S.E. Kesler, M.P. Everson, and T.J. Wallington. 2011. "Global lithium availability." *Journal of Industrial Ecology* 15 (5):760–775. Doi: 10.1111/j.1530-9290.2011.00359.x.

Grützke, M., X. Mönnighoff, F. Horstmann, V. Kraft, M. Winter, and S. Nowak. 2015. "Extraction of lithium-ion battery electrolytes with liquid and supercritical carbon dioxide and additional solvents." *RSC Advances* 5 (54):43209–43217. Doi: 10.1039/C5RA04451K.

Harper, G., R. Sommerville, E. Kendrick, L. Driscoll, P. Slater, R. Stolkin, A. Walton, P. Christensen, O. Heidrich, S. Lambert, A. Abbott, K. Ryder, L. Gaines, and P. Anderson. 2019. "Recycling lithium-ion batteries from electric vehicles." *Nature* 575 (7781):75–86. Doi: 10.1038/s41586-019-1682-5.

Huang, W., X. Feng, X. Han, W. Zhang, and F. Jiang. 2021. "Questions and answers relating to lithium-ion battery safety issues." *Cell Reports Physical Science* 2 (1):100285. Doi: 10.1016/j.xcrp.2020.100285.

Huang, T., M.-C. Long, G. Wu, Y.-Z. Wang, and X.-L. Wang. 2019. "Poly(ionic liquid)-based hybrid hierarchical free-standing electrolytes with enhanced ion transport and fire retardancy towards long-cycle-life and safe lithium batteries." *ChemElectroChem* 6 (14):3674–3683. Doi: 10.1002/celc.201900686.

Jaynes, W.F., and J.M. Bigham. 1987. "Charge reduction, octahedral charge, and lithium retention in heated, Li-saturated smectites." *Clays and Clay Minerals* 35 (6):440–448. Doi: 10.1346/CCMN.1987.0350604.

Job, R. 2020. *Electrochemical Energy Storage.* De Gruyter, Berlin/Boston.

Jones, E., M. Qadir, M.T.H. van Vliet, V. Smakhtin, and S.-m. Kang. 2019. "The state of desalination and brine production: A global outlook." *Science of the Total Environment* 657:1343–1356. Doi: 10.1016/j.scitotenv.2018.12.076.

Jung, H.-G., J. Hassoun, J.-B. Park, Y.-K. Sun, and B. Scrosati. 2012. "An improved high-performance lithium–air battery." *Nature Chemistry* 4 (7):579–585. Doi: 10.1038/nchem.1376.

Kalmykov, D., S. Makaev, G. Golubev, I. Eremeev, V. Vasilevsky, J. Song, T. He, and A. Volkov. 2021. "Operation of three-stage process of lithium recovery from geothermal brine: Simulation." *Membranes* 11 (3). Doi: 10.3390/membranes11030175.

Kang, D.H.P., M. Chen, and O.A. Ogunseitan. 2013. "Potential environmental and human health impacts of rechargeable lithium batteries in electronic waste." *Environmental Science & Technology* 47 (10):5495–5503. Doi: 10.1021/es400614y.

Kesler, S.E., P.W. Gruber, P.A. Medina, G.A. Keoleian, M.P. Everson, and T.J. Wallington. 2012. "Global lithium resources: Relative importance of pegmatite, brine and other deposits." *Ore Geology Reviews* 48:55–69. Doi: 10.1016/j.oregeorev.2012.05.006.

Kim, J.-S., Y.-H. Lee, S. Choi, J. Shin, H.-C. Dinh, and J.W. Choi. 2015. "An electrochemical cell for selective lithium capture from seawater." *Environmental Science & Technology* 49 (16):9415–9422. doi: 10.1021/acs.est.5b00032.

Komarnicki, P., P. Lombardi, and Z. Styczynski. 2017. "Electric Energy Storage System." In *Electric Energy Storage Systems: Flexibility Options for Smart Grids*, edited by P. Komarnicki, P. Lombardi and Z. Styczynski, 37–95. Berlin, Heidelberg: Springer Berlin Heidelberg.

Kumar, A., H. Fukuda, T. Alan Hatton, and J.H. Lienhard. 2019. "Lithium recovery from oil and gas produced water: A need for a growing energy industry." *ACS Energy Letters* 4 (6):1471–1474. Doi: 10.1021/acsenergylett.9b00779.

Kundu, M., C.M. Costa, J. Dias, A. Maceiras, J.L. Vilas, and S. Lanceros-Méndez. 2017. "On the relevance of the polar β-phase of poly(vinylidene fluoride) for high performance lithium-ion battery separators." *The Journal of Physical Chemistry C* 121 (47):26216–26225. Doi: 10.1021/acs.jpcc.7b09227.

Lee, Y.K. 2019. "The effect of active material, conductive additives, and binder in a cathode composite electrode on battery performance." 12 (4):658.

Li, T., X. Bai, U. Gulzar, Y.-J. Bai, C. Capiglia, W. Deng, X. Zhou, Z. Liu, Z. Feng, and R.P. Zaccaria. 2019. "A comprehensive understanding of lithium–sulfur battery technology." *Advanced Functional Materials* 29 (32):1901730. Doi: 10.1002/adfm.201901730.

Li, G., S. Wang, Y. Zhang, M. Li, Z. Chen, and J. Lu. 2018. "Revisiting the role of polysulfides in lithium–sulfur batteries." *Advanced Materials* 30 (22):1705590. Doi: 10.1002/adma.201705590.

Lien, L. 2018. "Recycling lithium batteries using membrane technologies." *ECS Meeting Abstracts* MA2018-01 611.

Liu, C., Z.G. Neale, and G. Cao. 2016. "Understanding electrochemical potentials of cathode materials in rechargeable batteries." *Materials Today* 19 (2):109–123. Doi: 10.1016/j.mattod.2015.10.009.

London, D. 2018. "Ore-forming processes within granitic pegmatites." *Ore Geology Reviews* 101:349–383. Doi: 10.1016/j.oregeorev.2018.04.020.

Lund, H. 2007. "Renewable energy strategies for sustainable development." *Energy* 32 (6):912–919. Doi: 10.1016/j.energy.2006.10.017.

Ma, L., K.E. Hendrickson, S. Wei, and L.A. Archer. 2015. "Nanomaterials: Science and applications in the lithium–sulfur battery." *Nano Today* 10 (3):315–338. Doi: 10.1016/j.nantod.2015.04.011.

Martin, G., L. Rentsch, M. Höck, and M. Bertau. 2017. "Lithium market research – global supply, future demand and price development." *Energy Storage Materials* 6:171–179. Doi: 10.1016/j.ensm.2016.11.004.

Meshram, P., B.D. Pandey, and T.R. Mankhand. 2015. "Hydrometallurgical processing of spent lithium ion batteries (LIBs) in the presence of a reducing agent with emphasis on kinetics of leaching." *Chemical Engineering Journal* 281:418–427. Doi: 10.1016/j.cej.2015.06.071.

Miles, A.K., M.A. Ricca, A. Meckstroth, and S.E. Spring. 2009. Salton sea ecosystem monitoring project. In *Geological Survey Open-File Report* 1276, 150 p.

Mo, J.Y., and W. Jeon. 2018. "The impact of electric vehicle demand and battery recycling on price dynamics of lithium-ion battery cathode materials: A vector error correction model (VECM) analysis." *Sustainability* 10 (8):2870.

Nitta, N., F. Wu, J.T. Lee, and G. Yushin. 2015. "Li-ion battery materials: Present and future." *Materials Today* 18 (5):252–264. Doi: 10.1016/j.mattod.2014.10.040.

Pehlken, A., S. Albach, and T. Vogt. 2017. "Is there a resource constraint related to lithium ion batteries in cars?" *The International Journal of Life Cycle Assessment* 22 (1):40–53.

Pillot, C. 2016. "The rechargeable battery market and main trends 2015–2025." In *Avicenne Energy, Presentation at 18th International Meeting on Lithium Batteries*, Chicago, IL.

Qadir, S.A., H. Al-Motairi, F. Tahir, and L. Al-Fagih. 2021. "Incentives and strategies for financing the renewable energy transition: A review." *Energy Reports* 7:3590–3606. Doi: 10.1016/j.egyr.2021.06.041.

Rana, M., S. Abdul Ahad, M. Li, B. Luo, L. Wang, I. Gentle, and R. Knibbe. 2019. "Review on areal capacities and long-term cycling performances of lithium sulfur battery at high sulfur loading." *Energy Storage Materials* 18:289–310. Doi: 10.1016/j.ensm.2018.12.024.

Richardson, D.B. 2013. "Electric vehicles and the electric grid: A review of modeling approaches, Impacts, and renewable energy integration." *Renewable and Sustainable Energy Reviews* 19:247–254. Doi: 10.1016/j.rser.2012.11.042.

Rioyo, J., S. Tuset, and R. Grau. 2020. "Lithium extraction from spodumene by the traditional sulfuric acid process: A review." *Mineral Processing and Extractive Metallurgy Review*:1–10. Doi: 10.1080/08827508.2020.1798234.

Roy, J.J., B. Cao, and S. Madhavi. 2021. "A review on the recycling of spent lithium-ion batteries (LIBs) by the bioleaching approach." *Chemosphere* 282:130944. Doi: 10.1016/j.chemosphere.2021.130944.

Sarakonsri, T., and R. Vasant Kumar. 2010. "Primary batteries." In *High Energy Density Lithium Batteries*, edited by Aifantis, K E., Hackney S. A., Kumar, R. V. 27–52. WILEY-VCH Verlag GmbH & Co. KGaA, Weinheim.

Serra, J.P., R.S. Pinto, J.C. Barbosa, D.M. Correia, R. Gonçalves, M.M. Silva, S. Lanceros-Mendez, and C.M. Costa. 2020. "Ionic liquid based Fluoropolymer solid electrolytes for Lithium-ion batteries." *Sustainable Materials and Technologies* 25:e00176. Doi: 10.1016/j.susmat.2020.e00176.

Shi, Y., G. Chen, and Z. Chen. 2018. "Effective regeneration of $LiCoO_2$ from spent lithium-ion batteries: a direct approach towards high-performance active particles." *Green Chemistry* 20 (4):851–862. Doi: 10.1039/C7GC02831H.

Sohn, J.-Y., J.S. Im, S.-J. Gwon, J.-H. Choi, J. Shin, and Y.-C. Nho. 2009. "Preparation and characterization of a PVDF-HFP/PEGDMA-coated PE separator for lithium-ion polymer battery by electron beam irradiation." *Radiation Physics and Chemistry* 78 (7):505–508. Doi: 10.1016/j.radphyschem.2009.03.035.

Sterba, J., A. Krzemień, P.R. Fernández, C.E. García-Miranda, and G.F. Valverde. 2019. "Lithium mining: Accelerating the transition to sustainable energy." *Resources Policy* 62:416–426. Doi: 10.1016/j.resourpol.2019.05.002.

Tadesse, B., F. Makuei, B. Albijanic, and L. Dyer. 2019. "The beneficiation of lithium minerals from hard rock ores: A review." *Minerals Engineering* 131:170–184. Doi: 10.1016/j.mineng.2018.11.023.

Tahil, W. 2007. *The Trouble with Lithium - Implications of Future PHEV Production for Lithium Demand*. Meridian International Research, accessed 02.07.2021

Takemura, D., S. Aihara, K. Hamano, M. Kise, T. Nishimura, H. Urushibata, and H. Yoshiyasu. 2005. "A powder particle size effect on ceramic powder based separator for lithium rechargeable battery." *Journal of Power Sources* 146 (1):779–783. Doi: 10.1016/j.jpowsour.2005.03.159.

Talens Peiró, L., G.V. Méndez, and R.U. Ayres. 2013. "Lithium: Sources, production, uses, and recovery outlook." *JOM* 65 (8):986–996. Doi: 10.1007/s11837-013-0666-4.

Tan, P., H.R. Jiang, X.B. Zhu, L. An, C.Y. Jung, M.C. Wu, L. Shi, W. Shyy, and T.S. Zhao. 2017. "Advances and challenges in lithium-air batteries." *Applied Energy* 204:780–806. Doi: 10.1016/j.apenergy.2017.07.054.

Tytgat, J. 2013. "The recycling efficiency of Li-ion EV batteries according to the European Commission Regulation, and the relation with the End-of-Life Vehicles Directive recycling rate." *World Electric Vehicle Journal* 6 (4). Doi: 10.3390/wevj6041039.

United Nations. 2021. "The 17 goals." https://sdgs.un.org/goals.

US Geological, Orienteering Survey, and US Geological Survey. 2009. *Mineral Commodity Summaries, 2009.* U.S. Government Printing Office, Virginia: United States.

USGS. 2021. Mineral commodity summaries 2021. In *Mineral Commodity Summaries*, Reston, VA.

Vine, J.D. 1975. "Lithium in sediments and brines--how, why and where to search." *Journal of Research of the U.S. Geological Survey* 3 (4):479–485.

Wanger, T.C. 2011. "The Lithium future—resources, recycling, and the environment." *Conservation Letters* 4 (3):202–206. Doi: 1111/j.1755-263X.2011.00166.x.

Wen, W., S. Yang, P. Zhou, and S. Z. Gao. 2021. "Impacts of COVID-19 on the electric vehicle industry: Evidence from China." *Renewable and Sustainable Energy Reviews* 144:111024. Doi: 10.1016/j.rser.2021.111024.

Xie, J., and Y.-C. Lu. 2020. "A retrospective on lithium-ion batteries." *Nature Communications* 11 (1):2499. Doi: 10.1038/s41467-020-16259-9.

Yang, T., Y. Lu, L. Li, D. Ge, H. Yang, W. Leng, H. Zhou, X. Han, N. Schmidt, M. Ellis, and Z. Li. 2020. "An effective relithiation process for recycling lithium-ion battery cathode materials." *Advanced Sustainable Systems* 4 (1):1900088. Doi: 10.1002/adsu.201900088.

Yang, Y., W. Yuan, W. Kang, Y. Ye, Q. Pan, X. Zhang, Y. Ke, C. Wang, Z. Qiu, and Y. Tang. 2020. "A review on silicon nanowire-based anodes for next-generation high-performance lithium-ion batteries from a material-based perspective." *Sustainable Energy & Fuels* 4 (4):1577–1594. Doi: 10.1039/C9SE01165J.

Zackrisson, M., L. Avellán, and J. Orlenius. 2010. "Life cycle assessment of lithium-ion batteries for plug-in hybrid electric vehicles – Critical issues." *Journal of Cleaner Production* 18 (15):1519–1529. Doi: 10.1016/j.jclepro.2010.06.004.

Zackrisson, M., K. Fransson, J. Hildenbrand, G. Lampic, and C. O'Dwyer. 2016. "Life cycle assessment of lithium-air battery cells." *Journal of Cleaner Production* 135:299–311. Doi: 10.1016/j.jclepro.2016.06.104.

Zhang, S.S. 2013. "Liquid electrolyte lithium/sulfur battery: Fundamental chemistry, problems, and solutions." *Journal of Power Sources* 231:153–162. Doi: 10.1016/j.jpowsour.2012.12.102.

Zhao, H., N. Deng, J. Yan, W. Kang, J. Ju, Y. Ruan, X. Wang, X. Zhuang, Q. Li, and B. Cheng. 2018. "A review on anode for lithium-sulfur batteries: Progress and prospects." *Chemical Engineering Journal* 347:343–365. Doi: 10.1016/j.cej.2018.04.112.

Zhao, S., and F. You. 2019. "Comparative life-cycle assessment of li-ion batteries through process-based and integrated hybrid approaches." *ACS Sustainable Chemistry & Engineering* 7 (5):5082–5094. Doi: 10.1021/acssuschemeng.8b05902.

Zuo, X., J. Zhu, P. Müller-Buschbaum, and Y.-J. Cheng. 2017. "Silicon based lithium-ion battery anodes: A chronicle perspective review." *Nano Energy* 31:113–143. Doi: 10.1016/j.nanoen.2016.11.013.

# 12 Demand-Side Management in Smart Grids

*Intisar Ali Sajjad*
University of Engineering & Technology, Taxila

*Haroon Farooq and Waqas Ali*
University of Engineering & Technology, Lahore

*Rehan Liaqat*
Government College University (GCU) Faisalabad

## CONTENTS

## 12.1 INTRODUCTION

In the traditional grid, variations in electrical demand are usually tackled by varying in the generation, whereas in the modern grid, devices at customers' locations also participate in grid management operations. Management of demand and generation occurs side by side, as shown in Figure 12.1.

The term demand-side management (DSM) was invented by Clark Gellings in 1984 at the US-based research institution, named Electric Power Research Institute (EPRI) [1]. DSM encompasses all the programs, technologies, and actions on the demand side of electricity metering that manage or lower electrical energy consumption, to cut down overall electricity system expenses

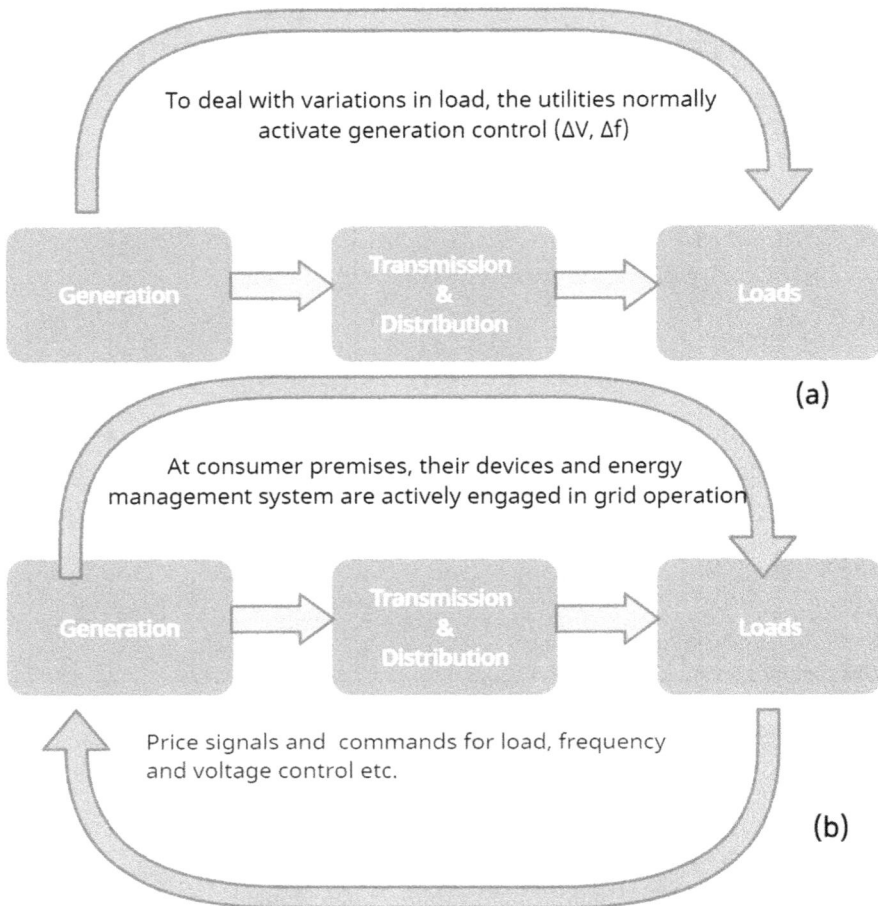

**FIGURE 12.1**  Control-centric comparison of (a) Traditional grid and (b) Smart grid.

or support in achieving the policy objectives related to power system supply–demand balance and environmental concerns [9].

Previously, DSM programs were confined to managing electrical demand only, but now, thermal energy demand is also incorporated through cogeneration and trigeneration technologies. Due to this difference, various definitions of DSM can be found in the literature with little modifications. Some researchers only include electrical energy management in DSM, while others consider other types of energy as well. The latter group considers DSM as a technique to reduce energy irrespective of the type at peak times, and some researchers extend the same definition by including the customers' behavior and response to time-varying prices.

Moreover, energy conservation and energy efficiency techniques are also included in some definitions, and others include microgeneration as well. However, the comprehensive definition given by Gellings and Chamberlin [2] states:

> DSM activities are those which involve actions on the demand side of the electric meter, either directly or indirectly stimulated by the utility. These activities include those commonly called load management, strategic conservation, electrification, strategic growth or deliberately increased market share.

DSM strategy is usually designed to motivate customers to vary their energy consumption patterns for multirange benefits for all the system participants. It includes all the monitoring, controlling, planning, and development strategies. Some load shape modifications desired by the operator through DSM programs are highlighted in Figure 12.2.

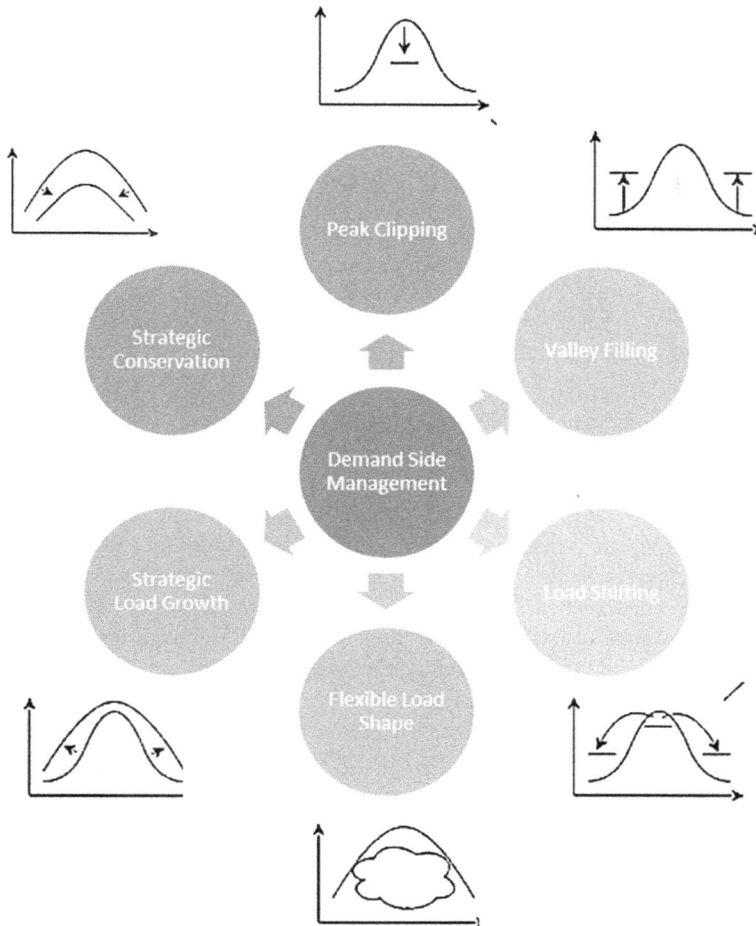

**FIGURE 12.2** Demand shaping techniques. (From Ref. [1]. With permission.)

## 12.2 NEED FOR DEMAND-SIDE MANAGEMENT

Urbanization has led the grid components to hit their designed limits. By employing DSM techniques, utilities can delay their investments with customers' cooperation as it is usually less expensive to control a load than to install a new power plant or energy storage system.

DSM promotes clean distributed energy resources which can be generated and consumed locally without excessively loading the transmission network. Now, the DSM is considered as a customer-driven program because all the benefits are not possible without a prompt response from the customers. Customers are inspired and engaged in these programs by providing incentives on their utility bills.

## 12.3 DSM COMPONENTS

The framework of DSM programs is designed to optimally manage system resources and customer loads with the participation of various system components. A brief description of some basic DSM components is as follows:

   a. **Local generators:** Usually, these are the distributed energy plants that generate electrical energy that can be consumed locally or exported to the grid.

**TABLE 12.1**
**Smart Grid Domains**

| | |
|---|---|
| Customers | Entities that take electrical energy for their own consumption |
| Market | A system for electricity purchasing and sale, based on pricing mechanisms |
| Service provider | An entity that provides electric services to retailers or end-use customers |
| Microgrid | Local small-scale power grid, including DERs and electric vehicles |
| Operation | Management of power production process and power market |
| Distribution | A process of electricity delivery to end-use customers |
| Transmission | Bulk transfer of electric energy, a process of electricity delivery |
| Generation | Production of bulk electric power, including storage and distributed sources |

b. **Smart devices:** These are the electrical loads having the capability of self-monitoring, data transmission, and can be controlled remotely by the customers through Web-based or mobile phone applications.

c. **Sensors:** Sensors measure different parameters such as temperature, light intensity, humidity, and persons' presence in a smart home and are helpful in monitoring and controlling of loads.

d. **Energy storage systems:** Energy storage systems provide the provision of storing energy during off-peak hours which can be later used during peak hours or contingencies.

e. **Energy management unit (EMU):** It is an information exchange unit for all the system elements and can be helpful in the management of system resources intelligently.

f. **Smart grid domains:** Various entities of the smart grid such as generation, distribution, transmission, market, operation, customers, and service providers are termed as domains. A brief description of these domains is given in Table 12.1.

## 12.4   TYPES OF DEMAND-SIDE MANAGEMENT

DSM can be classified into the following types:

a. Energy Efficiency (EE) and Energy Conservation (EC) Programs
b. Demand Response (DR) Programs
c. Energy Storage (ES) Programs

Some articles like [3] classify DSM into the first two categories, that is, (a) and (b), while others like [4] also include (c) as a third category.

### 12.4.1   ENERGY EFFICIENCY (EE) AND ENERGY CONSERVATION (EC) PROGRAMS

These programs include all permanent modifications in equipment design, material properties, process alterations, energy wastage minimization, energy recoveries, efficient usage of resources, and so on. Such modifications once employed start generating economic benefits instantly and keep on giving permanent savings of energy and emissions. Sometimes, energy conservation (EC) is also seen as a separate category of DSM which targets the users for behavioral changes to accomplish, process modifications, management and optimization of resources, energy recoveries, energy audits, efficient usage of energy, and so on.

### 12.4.2   DEMAND RESPONSE (DR) PROGRAMS

These are the programs in which customers modify their normal energy usage behavior by giving response to time-varying utility price signals in wholesale or retail electricity markets or in cases when system reliability is jeopardized. In DR programs, customers receive signals (price signals or system risk indication signals) from the utility and then they respond to these signals to modify their demand for due benefits. Such response is mostly given by special controllers called DR controllers which receive price signals from the utility and then control the customer appliances as per preferences set by them. A DR controller installed in a smart home is shown in Figure 12.3. DR programs comprise two different classes:

1. Incentive-based DR (IBDR) programs
2. Price-based DR (PBDR) programs

Their further classes are shown in Figure 12.4.

### 12.4.2.1 Incentive-Based DR Programs

These programs provide incentives to the end users for their load reduction when the grid reliability is at risk. In IBDR, there are following two ways to shift loads [5,6]:

1. The service provider directly controls the loads of customers as agreed by their contract.
2. The customers respond to different incentive measures/signals intended by the utility.

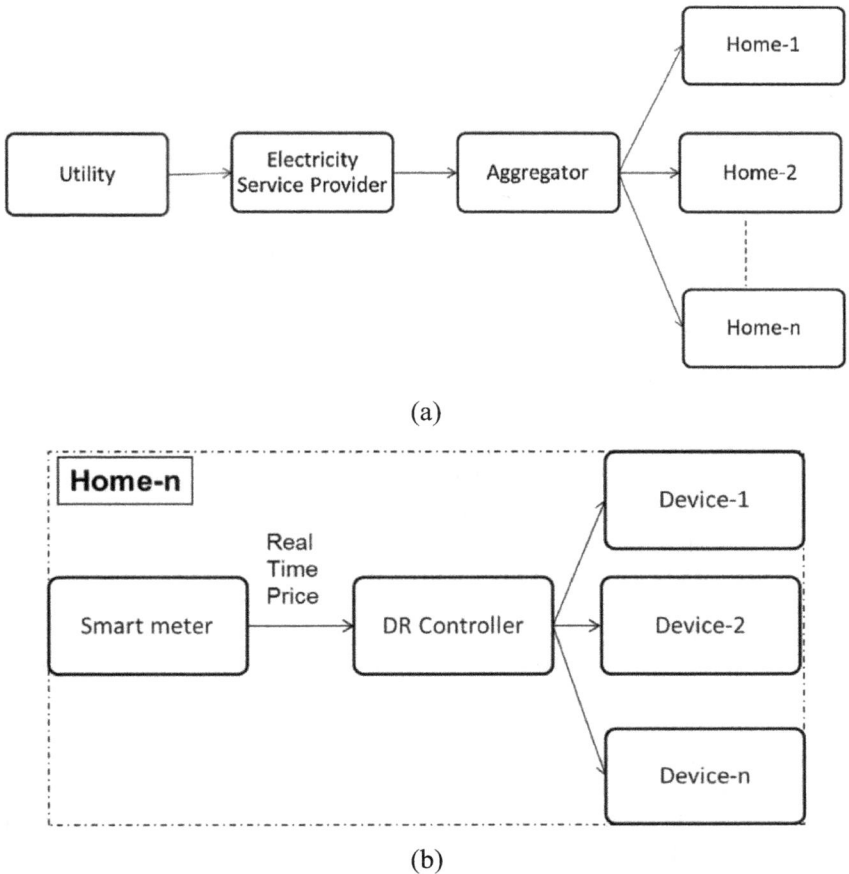

(a)

(b)

**FIGURE 12.3** (a) Smart homes in electricity market; (b) DR controller of smart home in smart grid environment.

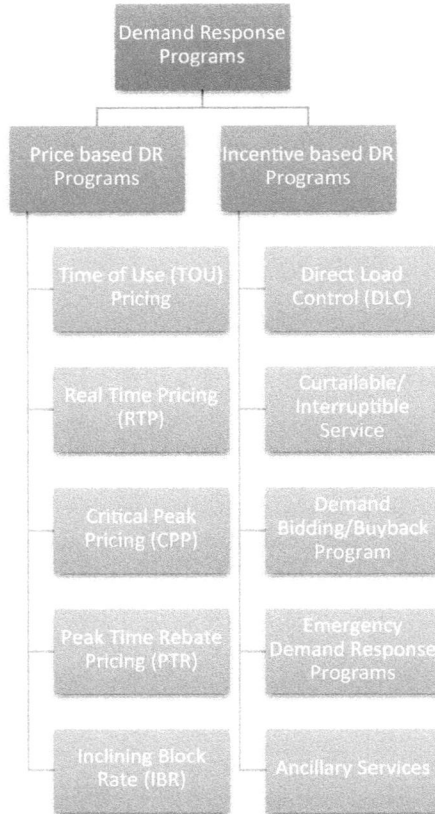

**FIGURE 12.4**  Classification of DR programs. (From Ref. [3]. With permission.)

These programs are more dominated by the supply side and are called dispatchable programs. Some examples of such programs are given in the following lines:

a. **Direct load control:** Generally, these programs are based on agreements between utility and customers, where the utilities are authorized to disconnect the specific customer loads remotely on a short notice to attain the desired demand reduction.

b. **Curtailable/interruptible rates:** In these programs, it is required for the participants to curtail/reduce their loads to a preagreed/predefined value; otherwise, fines would be charged for violating the agreement.

c. **Buyback/demand bidding programs:** In such programs, the stakeholders propose to reduce specific loads in the wholesale market. Costumers are penalized if they fail to reduce to the proposed amount of load.

d. **Emergency DR programs:** In case of some emergency, the participants are provided with monetary incentive for reducing their load during DR periods.

e. **Ancillary services market programs:** The participants submit bids for load reduction in the retail market. On the acceptance of their bids, they receive financial benefits for their standby loads as well as for curtailing loads as needed.

f. **Capacity market programs:** In these programs, a day-ahead notice of events is served to the participants. If agreed, the customers may receive payments for their load reduction or may bear penalties for failing to curtail the loads.

#### 12.4.2.2 Price-Based DR Programs

Price-based DR programs are based on changing the prices of electricity. In these programs, the participants change their energy utilization patterns to deal with the changing prices to benefit from the lower tariff intervals to avoid the higher tariff intervals. These programs are more dominated by the market/demand side. These are also termed as non-dispatchable DR resources. Some examples of PBDR programs are as follows:

a. **Time of use (TOU):** TOU tariffs impose higher electricity price during certain periods of time, so that the customers (re)arrange their loads to minimize energy costs. By linking TOU tariffs and DSM, the utilities significantly increase system security and reduce bulk energy sale–purchase costs.

b. **Critical peak pricing (CPP):** Generally, this pricing scheme is used during system contingencies, and it may be termed as a variant of the TOU tariff with lesser restrictions. These tariffs can be put on top of normal flat rate tariffs or TOU rate tariffs. The overall CPP time per annum is normally limited to few days or some hours.

c. **Real-time pricing (RTP):** This pricing scheme is considered as the most sophisticated type of PBDR. In this scheme, the customers are charged with varying prices that are normally communicated to them a day-ahead or a few hour-ahead.

The risk, reward, and uncertainty are correlated with different price-based programs, as shown in Figure 12.5. The circles in the figure show the uncertainty. More diameter of the circle represents more uncertainty and vice versa.

#### 12.4.2.3 Costs and Benefits Involved in DR Programs

Participants and program owners (utilities) are the active players in any DR program. To achieve benefits of DR programs, these players have to bear some costs. Brief details of these cost and different system benefits are listed in Tables 12.2 and 12.3 respectively.

### 12.4.3 ENERGY STORAGE (ES) PROGRAMS

The energy storage system plays an effective part in increasing system flexibility as it can be charged in renewable surplus time or off-peak time and can be discharged in peak time or gird interruptions. It can relieve the peak load through stored energy. It can also act as a static reserve and serve the functions of spinning reserves.

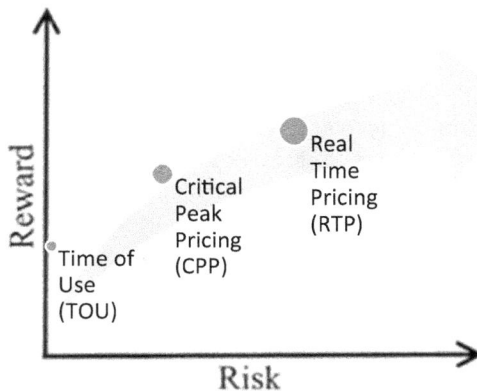

**FIGURE 12.5** Risk, reward, and uncertainty (circles) in price-based programs.

**TABLE 12.2**

**Initial and Running Costs Involved in DR Program**

| Market Player | Cost Type | Cost Head |
|---|---|---|
| Participant | Initial cost | • Enabling technology |
| | | • Response plan |
| | Running cost | • Inconvenience |
| | | • Lost business |
| | | • Rescheduling |
| | | • On-site generation |
| Program owner | Initial cost | • Metering and communication |
| | | • Billing system |
| | | • Customer education |
| | Running cost | • Administration |
| | | • Marketing |
| | | • Incentive payments |
| | | • Evaluation |

**TABLE 12.3**

**Benefits of DR Programs**

| Domain | Benefits |
|---|---|
| Participant | • Incentive payments |
| | • Bill savings |
| Market-wide | • Price reduction |
| | • Capacity increase |
| | • Customer education |
| | • Avoided/deferred infrastructure cost |
| Reliability | • Reduced outages |
| | • Customer participation |
| | • Diversified resources |
| Market performance | • Reduces market power |
| | • Options to customers |
| | • Reduces price volatility |

## 12.5   DSM AND POWER MARKET

DSM is a type of electricity management activity that entails a set of demand-side technology, strategies, and initiatives aimed at increasing energy efficiency, lowering costs, and lowering emissions [12]. There are three components to it: (i) demand response, (ii) energy storage management, and (iii) energy efficiency management. Energy efficiency management minimizes total energy usage by improving power facilities and enforcing regulatory guidelines. Through energy reserve, energy storage management may alleviate peak demand. DR focuses on shifting load using pricing schemes and other incentives. DSM can increase the efficiency of electricity products, optimize how they are used, save resources, protect the environment, and reduce total costs. Figure 12.6 depicts the DSM participants, benefits, policies, and classifications.

DR is a subset approach of DSM. As illustrated in Figure 12.4, it can be separated into IBDR and PBDR. Load shifting is done in two ways in IBDR: To begin with, the power sector has some direct control over customer load; secondly, customers' reactions to the power sector's incentive schemes [5,6].

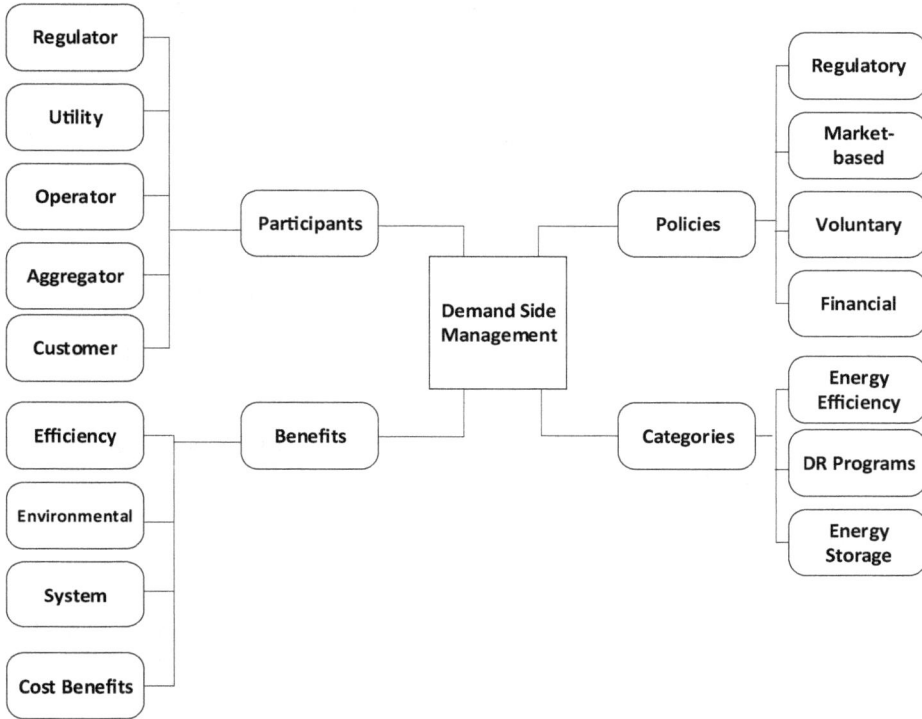

**FIGURE 12.6** Participants, benefits, policies and categories of DSM.

Direct load control (DLC), capacity market, demand bidding and buyback (DBB), interruptible load program (ILP), ancillary service market, and emergency DR are just a few of the IBDR concepts [6]. Normally, the power industry enters into contracts with its customers. Customers will receive a rebate if they follow the contract's terms, or they will face penalties if they do not. The backbone of DR is PBDR, which sets various rates for different hours to incentivize customers to reduce peak electricity consumption or transfer peak electricity consumption to off-peak periods. The rate is announced ahead of time or in real time.

The relationship between supply and demand can be realized through the implementation of DR programs. Customers can adjust their behavior voluntarily in response to grid conditions, and the supply side can use demand-side feedback to manage and schedule [7]. Not only may DR lower grid demand, but it can also move peak load to off-peak intervals, maximizing energy efficiency.

The price of electricity is typically determined in a smart grid system by the interaction and competition between different parties. The costs and benefits of every unit are major factors that drive the setting of energy prices for end users in a deregulated power market. Small users can only engage in the wholesale electricity market in a fully competitive power market.

Aggregator is an independent agent that offers a variety of novel services to small users, including management of the bills and home energy usage, electricity generation at home, and other such services. It provides an emerging service that assures that a group of residents and businesses save energy (costs) by purchasing and/or supplying energy in a flexible manner. The aggregator is also the entity responsible for ensuring that smart devices in a power system are connected and can respond flexibly to energy stock market pricing. The aggregator can act as a third party to aggregate the small users into a single buying agent to mediate with the stakeholders including electricity retailers, market operators, and transmission operators.

The aggregator is also the party that ensures that smart devices are connected in an energy system and can respond flexibly to electricity prices. An aggregator manages smart devices in the office or energy-intensive activities in the business sector. It calculates how much power usage can be increased or decreased in advance. This is what flexibility is all about: to keep the supply and demand for energy, and hence the power system in balance. The aggregator sells this flexibility on a separate flexibility market.

An aggregator controls smart device in-house or energy-using processes at companies. It predicts in advance how much power consumption can be increased or decreased. This is flexibility. The aggregator sells this flexibility on a separate flexibility market to keep the supply and demand for energy and thus the electricity grid in balance. The network operator can avoid peaks in the electrical network and minimize power outages by using the flexibility that the aggregator collects. In future, the aggregator is anticipated to play an increasingly prominent role. This is a win–win situation.

- The aggregator's flexibility services offer a more cost-effective option to grid reinforcement for addressing the energy transition's issues.
- Customers of energy pay less or receive more financial benefits in exchange for their unused energy.

The competitive structure of power market is shown in Figure 12.7.

There are certain intrinsic aspects to the electricity system. Power supply and demand, for example, must be balanced in real time. Large-scale power storage is currently impractical. Furthermore, the prices of producing power vary greatly depending on the sources and technology used.

Furthermore, users' electricity usage patterns change over time and are difficult to forecast. By influencing customer behavior, electricity pricing can help to reduce supply and demand mismatches. The electricity market can be classified into forward, day-ahead, intra-day, and balancing markets based on the competitive structure.

Figure 12.8 depicts the factors that influence DR decisions to encourage supply–demand balance.

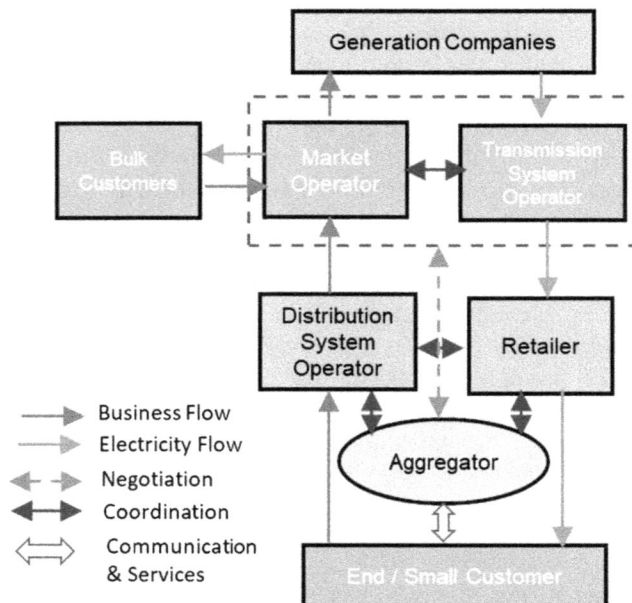

**FIGURE 12.7**　Competition in power market.

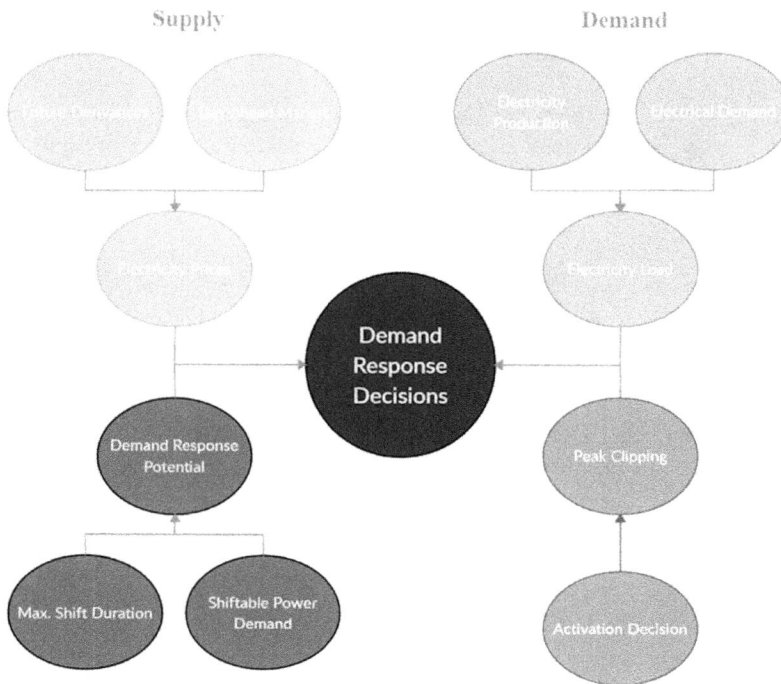

**FIGURE 12.8**  Factors influencing DR decisions.

## 12.6  DSM BARRIERS AND ENABLERS

There are different economic, technological, and social barriers in the successful implementation of DSM programs [8–10]. Classification and hierarchy of DR barriers are presented in Figure 12.9.

DR barrier classification theories consider the mutual interaction among various barrier classes among other factors. Economic and social (including behavioral and organizational) barriers are also considered vital as DR is eventually based upon the decisions made by individuals.

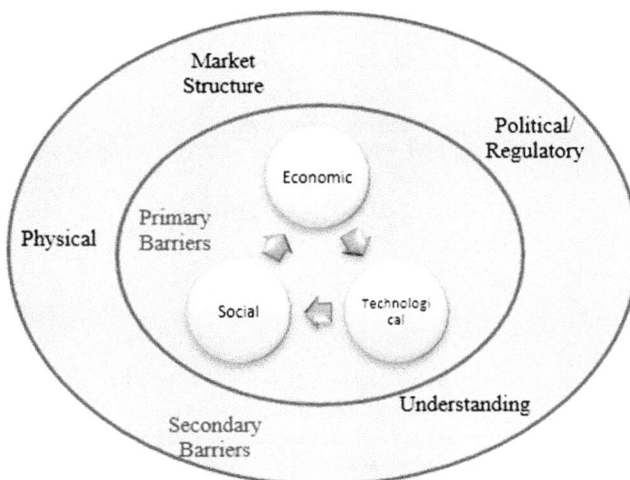

**FIGURE 12.9**  Classification and hierarchy of DR barriers.

Computation and communication after sensing are vital, for the implementation of smart grid and DR, which necessitates that a technological barrier class must be defined and classed as fundamental.

The remaining barriers, which are derived from fundamental socio-economic and technological facets, may be classified as secondary. The secondary barriers may be related to anthropogenic institutions (e.g., markets and regulation) or system feedbacks (e.g., electrical network constraints) [9]. Details of these barriers and corresponding enablers are summarized in Table 12.4.

**TABLE 12.4**
**Barriers and Corresponding Enablers of DR Programs**

| Barriers | | Enablers |
|---|---|---|
| Fundamental | Economic | |
| | **Market failures** | |
| | Imperfect information | • Improving the access to information by development of modified DR markets considering both buyers and sellers |
| | | • Enabling flexibility by developing parameters for the effective communication of user preferences |
| | Incomplete markets | • Pricing of externalities, such as emissions |
| | | • To eradicate "free riding" by adopting "DR exchanges" or similar schemes |
| | Imperfect competition | • To monitor the market power, particularly where only a small number of participants is in the market |
| | **Market barriers** | |
| | Access to capital | • Availability of government-supported loans on priority |
| | Uncertainty | • Contracts for difference |
| | Hidden costs | • Subsidize the cost of DR market |
| | System value | • Assessment techniques to evaluate the long-term benefits of DR |
| | Social | |
| | Organizational | |
| | Power | • Improve understanding of the value of DR among decision-makers |
| | Culture | • General education on DR and its benefits |
| | **Behavioral** | |
| | Form of information | • Careful design of user interfaces |
| | Credibility and trust | • Influx of new, third parties |
| | | • Transparency of data rights |
| | | • Enhancement of security of IT systems by deploying integrated design. |
| | | • Anonymity of Data. |
| | Values | • Advancement of DR organizations |
| | Inertia | • N/A |
| | Bounded rationality | • Automation |
| | **Technological** | |
| | *Sensing* | |
| | Metering | • Deployment of appropriate of metering infrastructure |
| | Energy service sensors | • Scrutinizing the energy services |
| | | • Ensuring DR schemes by induction of good requirement |
| | | • Compensation as per customer choices pertaining to energy services. |

*(Continued)*

**TABLE 12.4 (*Continued*)**
**Barriers and Corresponding Enablers of DR Programs**

| Barriers | Enablers |
|---|---|
| | **Computing** |
| Computing power | • Simple optimization algorithms |
| | • Distributing the computational load |
| | • Provision of necessary network resources |
| | **Communication** |
| Interoperability | • Open-ended technologies |
| | • Plug-in-based designs |
| | • Development of standards by establishing collaborations. |
| | • Development of common understanding among the industries. |
| Data security and privacy | • Enhancement of design to ensure security and privacy |
| | • Establish life-cycle management scheme for data |
| | • Enable instinctive classification for end-user configuration data |
| | • Adopt current state of the art |
| | **Technology standardization** |
| Multiple competing standards | • Consolidation of standards among energy, building and ICT sector alliances |
| Technology skills | |
| Insufficiently skilled workforce | • Outsourcing as an option |
| | • Address identified factors for retaining talent |
| | • Know the requisite skills mix |
| | • Target continual learners with tolerance of ambiguity |

**Secondary**

| Barriers | Enablers |
|---|---|
| | **Political/regulatory** |
| Taxes | • To allow the application of taxes, generation, storage, and consumption may be bifurcated |
| Energy markets regulation | • Systemic acceptance of the equivalence of demand and generation resources in energy markets |
| | • Improve cost reflectivity in energy markets. |
| End-user price regulation | • Allow end users to face full variation of wholesale energy-related markets |
| Unclear policy | • Forward standard guidelines on energy programs |
| Network operator regulation | • Giving priority to modernization and novel ideas |
| | **Market structures** |
| Baselines | • Approved standard procedures |
| Product definition | • Provide appropriate descriptions of various items in energy markets (bearing in mind the effect on transaction costs) |
| Complexity | • Decentralized optimization (system-of-systems approach) |
| | **Physical** |
| Distribution network constraints | • Real-time network pricing |
| Understanding | |
| Lack of understanding of DR | • To show the advantages of DR by deploying cost-benefit analysis |

## 12.7 DSM IN SMART BUILDINGS

Buildings are the major consumers of energy, as 40% of the global energy is used in buildings [11]. Buildings that provide productive and cost-effective environments through optimizing their elements such as structure, systems, services, management, and the interrelationships among the elements are called smart/intelligent buildings. Such buildings have mainly three following features [12]:

* Automatic control
* Learning abilities
* Occupancy trends incorporation

Demand-side management programs that consist of energy efficiency, conservation, and management programs can bring significant savings in energy and carbon footprints in the building sector. Buildings can also get benefits of demand-side management programs in smart grid environment by adding the following requirements:

* Incorporation of smart metering
* Demand response capabilities
* Distributed architecture
* Interoperability

Smart building research mainly has two domains: One domain is related to challenges in buildings' realization, and the second deals with users, user interactions among themselves and with enabling technologies of smart buildings.

The DSM framework in a smart building encompasses different entities of customer domain, smart grid domain, and communication infrastructure. All these entities are shown in Figure 12.10.

Understanding the energy demand profiles or load characterization plays a vital role in an effective DSM program as it will give useful information about customers' energy usage behavior.

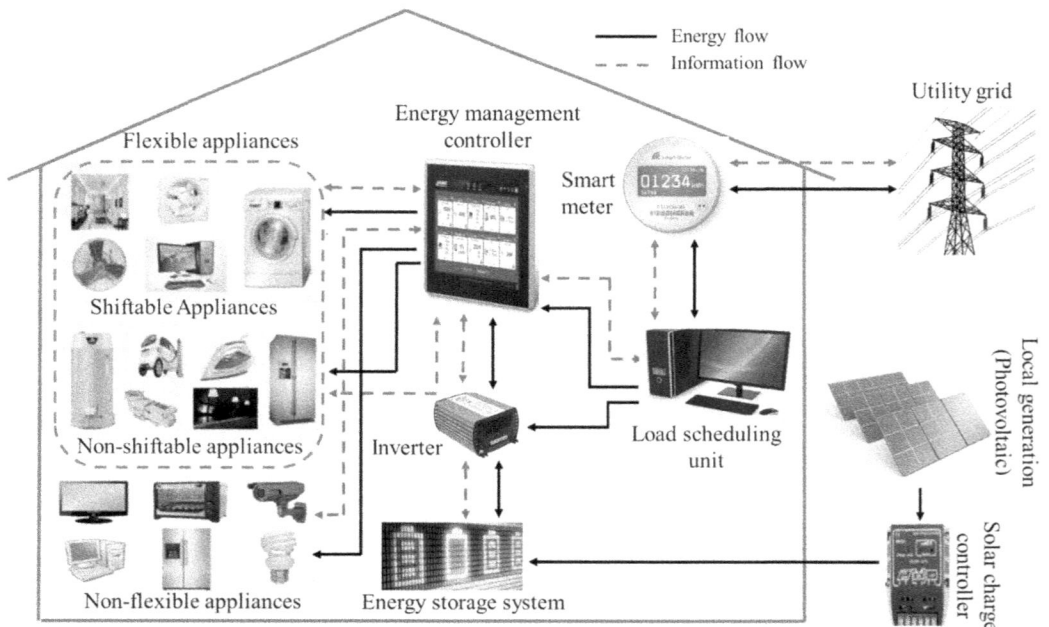

**FIGURE 12.10** Architecture of DSM framework.

Load characterization will also reveal the flexible intervals for DR programs. The next section presents a review on the demand characterization.

## 12.8 DEMAND CHARACTERIZATION

This "load follows supply" strategy necessitates a bidirectional communication infrastructure that allows information, including price signals, to flow freely between customers and utilities. The evolution of power systems foresees the optimization of electricity consumption by the dynamic management of controllable loads while maximizing the integration of renewable energy resources. Moreover, the likely transition toward dynamic tariffs and influx of information and communication technologies (ICT) will transform the passive end users into the active ones to ensure the utilization of all the available energy resources. Similarly, the deployment of distributed generation and storage (e.g., the electric vehicle) will transform the conventional consumers into the prosumers (simultaneously producer and consumer).

However, to make decisions on demand response measures, such as purchasing and selling of electricity, this active role necessitates near to real-time monitoring of electricity use, manageable demand, microgeneration, storage systems, and electricity prices. Due to all of these issues that must be addressed in order to achieve the overall optimal energy resource management, a system capable of automatically replicating customers' decisions while taking into account a variety of input signals and attempting to satisfy various constraints and preferences without compromising comfort or the quality of the energy service provided is required.

In order to design such a system, it is necessary to properly characterize customers' demands and loads in terms of normal appliance consumption, working cycles, technical limits, and controllability [13]. Although electricity usage is connected with uncertainty, the specific customers, for example, the residential sector, have a comparable portfolio of appliances, making it easy to design identical, though customizable, electricity management actions [14].

## 12.9 DEMAND FLEXIBILITY DEFINITIONS

Although it may appear that "flexibility" refers to the power system's ability to manage changes, it is not a unified concept with no widely agreed definition. There are several potential definitions available, some of which are included below.

The International Council on Large Electric Systems (French: Conseil International des Grands Réseaux Électriques, CIGRÉ) defined "flexibility" from the planner's perspective in 1995 through working group 37.10 as [15]: "the ability to adapt the planned development of the power system, quickly and at reasonable cost, to any change, foreseen or not, in the conditions which prevailed at the time it was planned."

In 2011, the International Energy Agency (IEA) described flexibility as [16]:

the extent to which a power system can modify electricity production or consumption in response to variability, expected or otherwise. In other words, it expresses the capability of a power system to maintain reliable supply in the face of rapid and large imbalances, whatever the cause.

In 2014, the European Union of the Electricity Industry (EURELECTRIC), as well as its affiliates and associates on various other continents, published a description of flexibility [17]. The Smart Grids Task Force of the European Commission adopted this description as the definition of flexibility [18]: "the modification of generation injection and/or consumption patterns in reaction to an external signal (price signal or activation) in order to provide a service within the energy system."

The Electric Power Research Institute (EPRI) defined flexibility in 2016 as [19]: "the ability to adapt to dynamic and changing conditions, for example, balancing supply and demand by the hour or minute, or deploying new generation and transmission resources over a period of years."

The Council of European Energy Regulators (CEER) held a public consultation in 2017 to establish flexibility definitions. The ETSO-E presented the definition in response to the consultation as [20]: "the active management of an asset that can impact system balance or grid power flows on a short-term basis, i.e. from day-ahead to real-time."

European DSOs also responded to the CEER consultation, citing [17] and [18] as sources for their proposed definition, which was published in [21] as: "the modification of generation injection and/or consumption patterns, on an individual or aggregated level, often in reaction to an external signal, in order to provide a service within the energy system or maintain stable grid operation."

CEER's final results, as well as the suggested definition of flexibility, were released in 2018 as [22]: "the capacity of the electricity system to respond to changes that may affect the balance of supply and demand at all times."

In 2018, IEA published a revised definition of flexibility as [23]: "all relevant characteristics of a power system that facilitates the reliable and cost-effective management of variability and uncertainty in both supply and demand."

The International Renewable Energy Agency (IRENA) defined flexibility in 2018 as [24]:

> the capability of a power system to cope with the variability and uncertainty that VRE (variable renewable energy) generation introduces into the system in different time scales, from the very short to the long term, avoiding curtailment of VRE and reliably supplying all the demanded energy to customers.

Many more definitions of flexibility are available, including ones from the academy, such as reference [25], which defines flexibility as "a system's ability to accommodate variability and uncertainty in the load-generation balance while maintaining satisfactory levels of performance for any time scale."

When examining these proposed definitions, it becomes clear that some confine the definition of flexibility to changes/alterations in supply and demand, while others do not. Furthermore, the wide range of interpretations of the proposed definitions leads to the broad statement that: Flexibility refers to the power system's ability to manage change.

In some ways, this remark demonstrates how the term "flexibility" is used to cover a wide range of requirements and facets of the power system. Therefore, the topic of flexibility is very complex to examine and necessitates differentiation to elaborate further.

## 12.10  CATEGORIES OF FLEXIBILITY

Flexibility is required not only for the balance of supply and demand but also to maintain voltages and the security of transfer capabilities [26]. As a result, flexibility solutions are explored for needs in both power system operation and planning, with timelines spanning from fractions of a second to years. Different categories related to flexibility as presented in Figure 12.11 are explained in the following lines.

### 12.10.1  FLEXIBILITY FOR POWER

Generation resources now also include the intermittent energy resources dependent upon weather conditions. Therefore, to ensure the stability in system frequency, it is required to maintain short-term balance between power demand and supply. It may be activated from a fraction of a second up to an hour.

### 12.10.2  FLEXIBILITY FOR ENERGY

Generation resources now have reduced amount of fuel storage-based energy. Therefore, to meet the wide range of demand scenarios, it is required to maintain medium- to long-term balance between power demand and supply. It may be activated from a few hours up to numerous years.

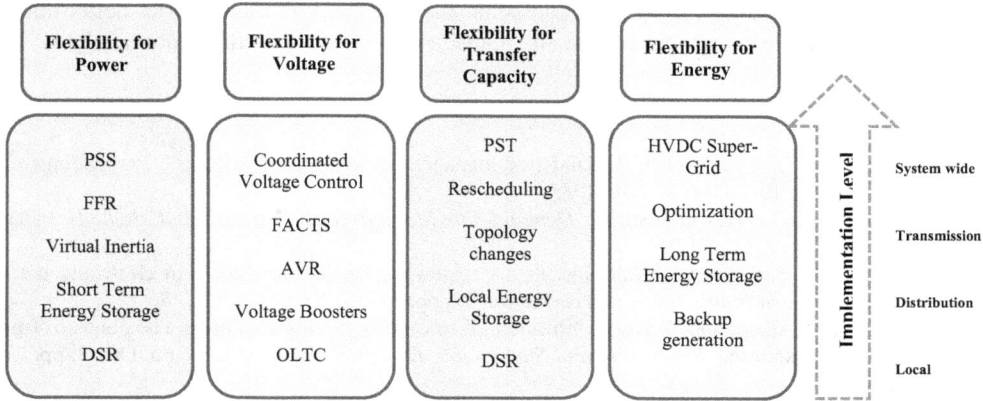

**FIGURE 12.11** Examples of flexibility solutions for each category, ranging from local to system-wide implementation. (From Ref. [26]. With permission.)

### 12.10.3 FLEXIBILITY FOR TRANSFER CAPACITY

The energy consumption has now increased with higher peak loads and corresponding supplies, which may result in increased congestion costs owing to bottlenecks caused by system constraints. Therefore, short- to medium-term capability is required to transfer power among the stakeholders. It may be activated from a few minutes up to numerous hours.

### 12.10.4 FLEXIBILITY FOR VOLTAGE

There has been increased influx of distributed generation resources in the distribution systems which has given rise several different operating condition and scenarios including bidirectional flow of power. Therefore, systems must have short-term capability of maintaining the bus voltages within standard limits. It may be activated from a few seconds up to several minutes.

## 12.11 CONCLUSIONS

The electric grid will be important to future deep decarbonization efforts, since it will be necessary to deliver clean and renewable energy to potential loads previously provided by fossil fuels, in addition to its current role. This progress will have to rely on loads playing a larger role in the future to make it technically and economically feasible.

Meanwhile, a slew of short- and medium-term disruptors will have an impact on the contribution of domestic loads to the grid. Just to name two:

- Heat pumps are gaining popularity as an alternative to resistive heating and could eventually replace natural gas furnaces. They do, however, have high peak loads during the coldest days of the winter that must be managed; integrated phase change thermal storage is being looked into as a solution.
- Load control algorithms have problem in managing thousands of loads to provide improved grid services while retaining user comfort. With its ability to handle enormous datasets, artificial intelligence may be the right tool for achieving advanced DSM control while taking individual user preferences into account.

In conclusion, to evolve clean electric smart grids will rely heavily on better load management of smart and connected devices. DSM has enormous, low-cost potential for grid services (e.g., load following, ramping, regulation, and peak shaving) that will help utilities better integrate high levels of renewable energy generation and manage future load growth economically.

## REFERENCES

1. C. W. Gellings, "The concept of demand-side management for electric utilities," *Proceedings of the IEEE,* vol. 73, no. 10, pp. 1468–1470, 1985.
2. C. W. Gellings and J. H. Chamberlin, *"Demand-side Management: Concepts and Methods.* Fairmont Press, 1993.
3. R. Sharifi, S. H. Fathi, and V. Vahidinasab, "A review on demand-side tools in electricity market," *Renewable and Sustainable Energy Reviews,* vol. 72, pp. 565–572, 05/01/2017.
4. Y. Changhui, M. Chen, and Z. Kaile, "Residential electricity pricing in China: The context of price-based demand response," *Renewable and Sustainable Energy Reviews,* vol. 81, no. Part 2, pp. 2870–2878, 2018.
5. P. Siano, "Demand response and smart grids—A survey," *Renewable and Sustainable Energy Reviews,* vol. 30, pp. 461–478, 02/01/2014.
6. J. Aghaei and M.-I. Alizadeh, "Demand response in smart electricity grids equipped with renewable energy sources: A review," *Renewable and Sustainable Energy Reviews,* vol. 18, pp. 64–72, 2013.
7. X. Zhang and F. Zhou, "Smart grid leads the journey to innovative smart home and energy consumption patterns," *Power System Protection and Control,* vol. 42, no. 5, pp. 59–67, 2014.
8. J. C. Cole, J. B. McDonald, X. Wen, and R. A. Kramer, "Marketing energy efficiency: perceived benefits and barriers to home energy efficiency," *Energy Efficiency,* pp. 1–14, 2018.
9. N. Good, K. A. Ellis, and P. Mancarella, "Review and classification of barriers and enablers of demand response in the smart grid," *Renewable and Sustainable Energy Reviews,* vol. 72, pp. 57–72, 2017.
10. S. Nolan and M. O'Malley, "Challenges and barriers to demand response deployment and evaluation," *Applied Energy,* vol. 152, pp. 1–10, 2015.
11. D. Kolokotsa, "The role of smart grids in the building sector," *Energy and Buildings,* vol. 116, pp. 703–708, 03/15/ 2016.
12. J. Martins et al., "Smart homes and smart buildings," In *2012 13th Biennial Baltic Electronics Conference,* Tallinn, Estonia, 2012, pp. 27–38: IEEE.
13. A. Soares, A. Gomes, and C. H. Antunes, "Domestic load characterization for demand-responsive energy management systems," In *2012 IEEE International Symposium on Sustainable Systems and Technology (ISSST),* Boston, MA, USA, 2012, pp. 1–6.
14. E. Carpaneto and G. Chicco, "Probabilistic characterisation of the aggregated residential load patterns," *IET Generation, Transmission & Distribution,* vol. 2, no. 3, pp. 373–382, 2008.
15. W. CIGRE, "Methods for planning under uncertainty: toward flexibility in power system development," *Electra,* no. 161, pp. 143–163, 1995.
16. O. Publishing and I. E. Agency, *Harnessing Variable Renewables: A Guide to the Balancing Challenge.* Organisation for Economic Co-operation and Development, Paris, France, 2011.
17. P. Mandatova and O. Mikhailova, "Flexibility and aggregation: Requirements for their interaction in the market," *Eurelectric: Brussels, Belgium,* 2014.
18. S. G. T. Force, "Regulatory recommendations for the deployment of flexibility," *EU SGTF-EG3 Report,* Brussels, Belgium, 2015.
19. EPRI, "Electric power system flexibility: Challenges and opportunities," Electric Power Research Institute (EPRI) Report, California, US, 2016.
20. ENTSO-E, "ENTSO-E response to the CEER public consultation guidelines of good practice for flexibility use at distribution level," [Online]. Available: https://docstore.entsoe.eu/Documents/Publications/Position%20papers%20and%20reports/170517_ENTSO-E_response_to_CEER_consultation_VF.pdf
21. "Flexibility in the energy transition – A Toolbox for Electricity DSOs," CEDEC, EDSO for smart grids, eurelectric, GEODE 2018, Available: https://www.edsoforsmartgrids.eu/flexibility-in-the-energy-transition-a-toolbox-forelectricity-dsos/.
22. "Flexibility use at distribution level - A CEER conclusions paper, Ref: C18-DS-42–04," Council of European Energy Regulators - Distribution Systems Working Group, Brussels, Belgium, "Flexibility use at distribution level—A CEER conclusions paper," July 17, 2018, Ref: C18-DS-42-04.
23. International Energy Agency (IEA), Paris, France, (2018). *Status of Power System Transformation 2018, Advanced Power Plant Flexibility.* [Online] Available: https://webstore.iea.org/status-of-power-system-transformation-2018.

24. E. Taibi et al., "Power system flexibility for the energy transition: Part 1, Overview for policy makers," International Renewable Energy Agency, Abu Dhabi, UAE, 2018.

25. H. Holttinen et al., "The flexibility workout: managing variable resources and assessing the need for power system modification," *IEEE Power and Energy Magazine,* vol. 11, no. 6, pp. 53–62, 2013.

26. A. Z. A. Emil Hillberg, B. Herndler, S. Wong, J. Pompee, S. L. O. Jean-Yves Bourmaud, G. Migliavacca, I. O. N. Kjetil Uhlen, H. Pihl, M. Norström, and J. R. R. G. B. E. Mattias Persson, "Flexibility needs in the future power system," 2019, Available: http://www.iea-isgan.org/ourwork/annex-6/.

# 13 Digitalization in the Energy Sector

*Muhammad Umer, Muhammad Abid, and Tahira Nazir*
COMSATS University Islamabad

*Zaineb Abid*
Cranfield University

## CONTENTS

## 13.1 INTRODUCTION

The word "digitization" refers to the translation of analog signals into digital ones and further involves the manipulation of the extracted data into information (Brennen & Kreiss, 2016). Across the world, businesses are undergoing digitalization by inducing digital technologies to increase the efficiency of the pre-existing business processes and attain higher profitability. These industries and businesses include but are not limited to health care, communication, transportation, financing, media, and tourism/hospitality. As result, a transformation has been observed across the business sector in terms of adding value to the prevailing market segments and at the same time identifying and exploiting the newer opportunities in the environment (Bloomberg, 2018; Varela, 2018). This unceasing shift of businesses toward digitalization has been an outcome of the ever-evolving technology that not only has extended the reliance of humans on computers, but also has increased the accessibility of the data to an extent that is currently beyond comprehension (Salminen, Ruohomaa, & Kantola, 2017; Sestino, Prete, Piper, & Guido, 2020).

Digital technologies everywhere are consistently shaping the way humans interact with entities in their environment. The scalability of digitalization on a global level can be understood from the fact that the magnitude of Internet traffic increased three-fold over the past 5 years, and approximately

90% of the data we currently have across various platforms were developed over the last 2 years only. Individuals in the today's world are more connected than ever, as more than half of the global population, that is, 3.5 billion, have access to the Internet that makes up 54% of the households. Similarly, there are more mobile subscriptions, that is, 8.3 billion, than the cumulative global population, that is., 7.9 billion. Moreover, the implementation of IoT (Internet of Things) has connected the utilities of individual users with larger communication networks, thus providing an enriching user experience in the form of personal health care, smart electricity grid, and surveillance, transportation, smart homes, etc. Currently, there are well over 31 billion devices smartly connected over numerous communication networks (Di Silvestre, Favuzza, Sanseverino, & Zizzo, 2018; IEA, 2017; Morley, Widdicks, & Hazas, 2018).

Considering this, the energy sector remains no exception in terms of adapting digitalization across various spectrums. Akin to improvement across various fields, digitalization has induced accessibility, productivity, safety, and sustainability to global energy deployments as well. However, the said progression has complicated the concerns and challenges associated with the security and privacy of its associated stakeholders (Judson, Soutar, & Mitchell, 2020; Küfeoglu, Liu, Anaya, & Pollitt, 2019). Taking into account the global energy demand and its respective production, there is a significant gap and energy deficit that needs to be filled, and for this purpose, there has been an ongoing global shift toward the adaptation of novel energy generation approaches that are bound to increase exponentially in the years to come. The solution to managing this ever-evolving situation lies with the effective and efficient implementation of digitalization practices in the energy sector. This is because the digitalization of the energy sector has, over the years, proved itself to usefully educate its relevant stakeholders in terms of investing in geographical regions with relatively more energy potential than the others. Also, it has allowed the relevant technical teams deployed on any given energy system to proactively predict faults within the systems and suggest optimal solutions (Küfeoglu et al., 2019; Lv, Fang, Yang, & Romero, 2020). Furthermore, the inclusion of AI (Artificial Intelligence) in the digitization of the energy sector has aided the energy management entities to precisely observe and predict the energy consumption trends, and subsequently allocate the accurate energy quota to the correct consumer at the right time, therefore reducing the overall cost associated with the said energy systems (Ahmad et al., 2021; Varela, 2018). The digitalization of the energy sector has also been the key enabler toward the observed transition to two major trends, including decarbonization and the decentralization of energy systems. Though there has been a major progression in the digitalization of the energy sector in the past decades, regardless of its global implementation is yet to be seen (Morley et al., 2018). Cumulatively, digitalization in the energy sector is set to enhance the sustainability of the said industry by assuring the intelligence, reliability, and efficiency of the deployed systems. Consequently, it is making the world a healthier place to live in, as well as empowering the suppliers and consumers with personalized control over energy utilization (Di Silvestre et al., 2018).

Considering the revolution that digitalization is bringing in the global energy sector, this chapter highlights its importance, overviews the current digitalization trends being followed by stakeholders around the world, and explains digitalization technologies being currently deployed to accomplish data-driven decision-making in alignment with their impact. It also addresses the concerns and challenges of digitalization and conclusively discusses the future of digitalization in the energy sector. This work will enable the researchers and practitioners to learn about the current dynamics of digitalization being followed across the globe and what needs to be done to ensure the sustainability of the energy sector.

## 13.2   DIGITALIZATION TRENDS

Digitalization enables unlocking sectorial opportunities by predicting, measuring, monitoring, and improving on basis of the gathered information from its technical deployments as well as its consumers (Varela, 2018).

To begin with, predicting involves maintaining equilibrium between supply and demand as a key task of renewable energy integration. This ought to be supported using the progressive and contemporary machine learning procedures and other relevant AI technologies, which support and simplify robust integration by providing reliable consumption trends, refining weather predictions, detecting potential risks proactively, and forecasting the performance of deployed technologies, and by proposing optimal solutions to manage the deployed infrastructure. This analytical approach can enable the energy systems to function rigorously and effectively (Ahmad et al., 2021; Sozontov, Ivanova, & Gibadullin, 2019).

Holistically, the functioning of the energy system can only be enhanced by acquainting it with digital measurements. However, the energy system offers innumerable quantifiable frameworks, which can be a breakthrough toward adapting the dynamic consumption trends and managing technical as well as non-technical aspects of the energy systems. To exemplify, in order to collect and send the domestic energy consumption data systematically to the energy provider, smart meters can be utilized to optimally record the energy supply and consumption (Le Ray, 2019; Varela, 2018).

Considering the aspect of monitoring, the energy sector focuses consistently on collecting the data from its environment, which further allows it to oversee and control all the network-related components in real time. Doing so potentially leads toward enhanced efficiency, optimization, and consumer choice (Klavsuts, Klavsuts, Rusina, & Khayrullina, 2019; Le Ray, 2019). Keeping in view the real-time monitoring, machine learning, concerning the field of security and maintenance, can collect and manipulate the live data during a component's lifetime including cutoffs and disconnections or system failures to augment the maintenance and to reorder schedules (Küfeoglu et al., 2019; Parida, Sjödin, & Reim, 2019; Varela, 2018).

Lastly, to improve, all three above-mentioned facets, that is, predicting, measuring, and monitoring, are the keys to enhance the power system. Digitalization in this regard enables the integration between the networks and its users and ultimately offers them immediate calculations and best prospects following through the tendency over a prolonged time with a high capacity or a huge velocity of statistics that can help in supervising the intricate calculations and augmenting the phases of decision-making. Digitalization, in general, aids in enhancing the gathering of data, overall analysis, forecasting, and improved measurement. This can ultimately lead to higher productivity, increased sustainability, improved safety, and cost-efficiency for both consumers and suppliers (Ali & Choi, 2020; Varela, 2018).

## 13.2.1 SUPPLY-SIDE POTENTIAL

The current digitalization trends have played a vital role in improving the output and reducing the expenditures regarding the procurements of the energy sector. New horizons could welcome the supply and demand sides by optimizing and automating the combination of innovative prospects. A few of the vital trends are discussed as follows (Specht & Madlener, 2019; Wu, Wu, Guerrero, & Vasquez, 2021):

- **Petroleum (coal, oil, and gas):** Compared to other sectors, a low level of digitalization has been seen in this sector globally as it is considered to be a highly hardware-intensive sector. However, the transformations can be observed in petroleum-based industries which have been digitalized by integrating their systems with the process automation, uniting sensors, remote control fiscal management, and prognostic maintenance to augment the functioning of assets and to reduce the general maintenance and production cost, which can easily be observed in new CTEP (combined thermal electric plant) (Arif & Al Senani, 2020; Mingaleva, Shironina, & Buzmakov, 2020).
- **Biomass, renewables, and waste energy:** Digitalization also improves the efficiency in terms of optimizing the entire system to an extent to understand the contemporary

trends, connectivity, information access, better analysis, and deployment of shareholders across the value chain. Considering the increased utilization of biomass, renewables, and wastes to produce energy, digitalization aims to set the bar high with every passing milestone, in terms of producing cleaner and greener energy. By the year 2030, digitalization is set to double the share of the aforementioned sources in the existing energy sector. This will further allow the stakeholders of the energy sector to battle climate changes across the globe in an effective manner (Lange, Pohl, & Santarius, 2020; Wu et al., 2021).

- **Power generation, transmission, and distribution:** The excessive consumption of renewable power sources can dissipate the power supply and counter the conventional business models of utility firms. Despite this, the process of digitalization, to decentralize the power system and augmenting a smooth inclusion of renewable power into the prevailing centralized power system, has been considered an essential component. In addition, it also offers a variety of business prospects for the industries by providing them sustainable, reliable, efficient, and customized power solutions (Chou & Yutami, 2014; Küfeoglu et al., 2019; Le Ray, 2019).

- **Hybrid energy supply:** Hybrid energy supply could serve as a solution to fulfill the need of bringing flexibility to the domain of renewable energy systems. The provision of manufacturing along with innumerable competitive power supplies, in particular solar thermal and biomass, PV and wind, PV and hydraulic, etc., accompanied by the storage system, could supply the requisite flexibility that would overcome the inconstant power production of renewable energies. However, merely by transforming the power source, several improvements could be observed in terms of maintaining the variability in prices and converting it into the cost-effective mode (Varela, 2018), considering which the flexibility in production could be enhanced. Also, the carbon footprint could be minimized. In this respect, a similar process has been observed in familiarizing the—demand-side flexibility, particularly, adjusting the time and the extent of consumption of energy to price and accessibility of power instead of the source of energy. The reliability of these structures could be gauged through the utility of digital tools and skills. Nowadays, new energy suppliers and producers, and traditional consumers are also participating in digital power sectors as a producer of energy by identifying themselves as "prosumers" (Botelho et al., 2021; Küfeoglu et al., 2019; Leal-Arcas, Lesniewska, & Proedrou, 2018). For instance, in UK and Germany, FIT (feed-in tariff) is fixed in its cost, and the prosumer obtains it for the renewable power supply which has not been consumed and is rather utilized in the grid. Such kinds of benefits allow the clients to remain as an energetic entity in the domain of the new power system, promising the swift evolution into the overall energy system and enabling the power grid to fulfill all the challenges of energy supply at a convenient and smooth pace of increased power demand. This can only be functional with smart and penetrating power devices (Meister, Ihle, Lehnhoff, & Uslar, 2018). In this way, the microgrids or distributed generation systems could generate the chance of energy trading not among individual microgrids rather multiple microgrids to optimize the utilization of renewable power sources. Also prosumers participate in boosting energy balance in the urban generation because they have been considered the competent stakeholders to the production facility in terms of either energy storage or selling the extra power to their local generation (Di Silvestre et al., 2018; Herenčić et al., 2019), whereas the digital technicalities could serve in stimulating the utility of waste heat from production to household nourishment or utilizing it for their purpose. Lastly, digitization also streamlines the autonomous implantation of business decisions among the multipurpose microgrids, enabling them to maintain a safe security system and the sanctity of the information (De Dutta & Prasad, 2020; Herenčić et al., 2019).

## 13.2.2 DEMAND-SIDE OPPORTUNITIES

Digitalization imparts considerable freedom to exploit the advantages in every power-consuming sector and resolve the constraints relating to the inadequate enforcement in these sectors which are discussed as follow:

- **Buildings:** Another influential trend of digitalization involves broadening the international housing application of IoT devices. One example is a smart regulator that could automate the customization of the heating and cooling of temperature. Furthermore, slim lights or other smart gadgets could maximize the consumption of the user's requirements. In this domain, the buildings' frameworks could also utilize energy savings by employing digital technicalities and equipment, which can ultimately disseminate accurate and timely information across the supply chain. For that purpose, it is called real-time construction management, enabling the collection of all evidence at the grassroots level and providing it a complete manifesto in terms of minimizing the cost, maximizing the revenues, and improving the productivity (Daniotti et al., 2020; Di Silvestre et al., 2018; Singh, 2019).
- **Industry:** Digitalization has had an impact on industries relating to the transformational process of the deliverance of their products. Globally, industries consume 38% of the total energy and release 24% of total carbon dioxide However, it has been ascertained that the up-gradation of digitalization contributes to power savings of almost 10%–20% on an average. Such a process could be optimized in addition to energy savings if buildings operationalized digitalization. However, distinct kinds of technicalities such as 3D printing, industrial robots, automation and advanced analytics, virtual reality, additive manufacturing, semantic technologies, and industrial IoT could become the exclusive platform for applying digitalization in different levels of industries (Di Silvestre et al., 2018; Varela, 2018; Wu et al., 2021).
- **Transport:** Digitalization has drastically transformed transportation as it has impacted all kinds of transportation including rails, roads, aviation, and maritime in terms of the functionality and processing of evidence and by bringing refinement in services, operations, and safety effectively. However, IoT could also boost the efficacy and effectiveness of transportation commodities from their beginning point to the point of consumption by combining it with highly intelligent sensors and also by maximizing the flow of traffic. In this way, IoT could become the reason to generate more potential on current frameworks of waterways, air, rail, and roads. Similarly, ITS (intelligent transport system) can condense the power usage by 25% through the modal shift, less travel, and decreasing per kilometer energy consumption (Creutzig et al., 2019; IEA, 2017; Noussan & Tagliapietra, 2020).

## 13.2.3 SYSTEM SUPPORTING OPPORTUNITIES

DTES (digital transformation of energy system) could likely mitigate the shortcomings at the grass-root level as well as promote innovative possibilities and induct the in-depth transfiguration in terms of allowing the entities, systems, and gadgets to connect and communicate. Here are some of its key features:

- **Creating trust and better connectivity and transparency:** The digitalization of the power system empowers its customers, dealers, and other partners to seize preferable connections to attain better aftermaths. Moreover, DPS (digitalization of power system) also inculcates satisfaction and builds trust among its various participants, which can easily entitle a competitive, more resilient, transparent, non-discriminatory, and open energy market, which will benefit both the society and the economy (Leal-Arcas et al., 2018;

Parida et al., 2019; Varela, 2018). This open data conception may authorize the entities to indulge in better decision-making for both businesses and policy-making phenomena by aiding the anticipation and inventiveness of the entire society.

- **Effective supply chain management:** Digitalization technicalities are likely to unravel the new horizons for business entities across the whole energy supply chain. However, the line of production of flexible renewable power supplies must be vigorous to influence its full potential, machine learning, advanced data analytics, and big data to the extent to optimize and manage various renewable power sources with high flexibility (Ahmad et al., 2021; De Dutta & Prasad, 2020; Küfeoglu et al., 2019). Similarly, digitalization is characterized to standardize the innumerable procedures and increase the swiftness as well as the cost across the whole power supply, by the provision of high visibility and innovations in the domain of the supply chain (Marmolejo-Saucedo & Hartmann, 2020; Muñoz-Villamizar, Solano, Quintero-Araujo, & Santos, 2019).
- **Innovative cost propositions:** Innovative cost or value propositions are also the aftermath of the application of digitalization in the corporate world. They empower the consumer to adhere to the monetary benefits and cash flows for the usefulness of companies by providing them the smart battery charging, automated demand response vehicle infrastructure, maximized efforts through sector coupling and vehicle to grid services, etc. However, consumers may utilize the optimized fiscal benefits on the platform of new cost and value propositions and could generate revenue streams for the well-being of the business (Morley et al., 2018; Varela, 2018).
- **Future forecasting and reliable outcome prediction:** The progression in digitalization like machine learning could prove to be the pillar of a reliable power structure. Such technicalities of digital mechanisms could unveil in prophesying the aftermaths of unseen data and have the potential to foresee and predict the forthcoming. However, the best prospects of consumer behavior, weather pattern, and demand data would have the capability to augment supply security and grid stability by maintaining the sustainability of the entire power system (Ahmad et al., 2021; Ali & Choi, 2020).

## 13.3  DIGITALIZATION TECHNOLOGIES AND THEIR IMPACT

The traditional practices being used in the energy sector to generate, manage, and disseminate electricity have been a leading reason for inefficiency and losses in terms of overall costs (Di Silvestre et al., 2018). To overcome the concurrent circumstances, the stakeholders associated with the energy sector are expected to figure out the optimal approaches to generate power, manage the accumulated energy, and evaluate consumption trends (Morley et al., 2018). The reliance on all these aspects is based upon a common denominator, that is, data. The effective and efficient collection and evaluation of the data can only enable the precise determination of how the energy systems are performing at a single point in time and further train the regarding systems to automatically control, predict, and respond to future situations. In terms of technological shift, the energy sector has been an early adopter of digital technologies (Küfeoglu et al., 2019). Dating back as far as the 1970s, the power utilities became the pioneers of digital adaptation by inducing the cutting-edge technologies present at that time to manage grid systems. Following them, the oil and gas sector as well divulged into digital technologies to make educated decisions regarding resource exploration, asset management, and dissemination of energy (IEA, 2017; Malecki & Moriset, 2007).

The adaptation of digitalization in the energy sector has been classified into three broader domains, that is, system balance, process optimization, and customer orientation: Firstly, the system balance comprises the processes that enable maintaining energy generation levels in alignment with the end-user demand. Secondly, process optimization aims to continually improve the internal process to enhance the efficiency and effectiveness of the overall systems. Thirdly, the customer orientation focuses on creating value addition for the consumer and generating increased revenues. The aforementioned three domains have been subcategorized into seven aspects that make up the latest technologies being used in the energy sector (Alekseev, Lobova, Bogoviz, & Ragulina, 2019; Parida et al., 2019).

To begin with, the system balance focuses on the effective and efficient management of the energy at both the power generation as well as consumer end by addressing smart grid and operation optimization, creating a smart market with integrated flexibility, and putting in place the anomaly identification systems (De Dutta & Prasad, 2020). As mentioned earlier, the availability of data in the energy sector is a grand challenge, and the situation worsens with most of the systems positioned as decentralized entities. This makes the availability of real-time data and its timely translation into usable information inconvenient, therefore resulting in substantial losses. To overcome the situation, the concept of smart grid systems comes into play (De Dutta & Prasad, 2020). As it suggests, the deployment of relevant hardware is coupled with supporting software that can sense and record the environmental data and transmit it to the control centers to take necessary actions. For example, the surge in user demand can load the transformers and can eventually cause its temperature to rise (Di Silvestre et al., 2018). With the necessary infrastructure in place, the rise in temperature levels can be easily detected and can enable the operations teams to balance the demand and supply levels. Apart from managing the grid systems smartly, the collected data can aid the stakeholders to optimize the energy systems by identifying the best-suited conditions upon which each of the technical components operates and then restructuring the overall system in alignment with the pre-processed information (Klavsuts et al., 2019; Lyons, 2019). Furthermore, the system balance attempts to shift the consumer demand in alignment with the production capacity, unlike the traditional approach where the production capacity was aligned with the consumer demand. This is where the implementation of smart meters comes into action. These devices enable consumers to monitor their usage patterns and adapt in real time (De Dutta & Prasad, 2020; Klavsuts et al., 2019). Moreover, the consumers are offered financial incentives to adopt the existing production patterns, thus averting the overburdening of the production infrastructure. In terms of carrying out financial transactions between the energy suppliers and consumers, the role of blockchain can play a pivotal role by carrying out the decentralized financial exchange between all the involved stakeholders. This brings financial flexibility to the system (De Dutta & Prasad, 2020; Moradi, Shahinzadeh, Nafisi, Gharehpetian, & Shaneh, 2019). Lastly, the balance of the system focuses on collecting the longitudinal data related to the environmental conditions an energy system is deployed into, the optimum performing conditions, varying consumer trends across the year, and more. Based upon that, anomalies existing in the regarding systems are identified and then later on rectified either by the system automatically or by the maintenance team manually (Ali & Choi, 2020; Svendsen, Tollefsen, Gjengedal, Goodwin, & Antonsen, 2018).

Secondly, the implementation of digitalization focuses on process optimization. Digitalization offers a broader spectrum of universal optimization solutions across various platforms, but a few remain attributed to the energy sector only (Alekseev et al., 2019; Küfeoglu et al., 2019). Considering this, the application of data analytics and machine learning can enable the stakeholders deployed at managing the identified anomalies in the energy system and translate the gathered data to proactively mitigate the potential problems that are probable to occur in the future. Moreover, the information gathered through the mediums of data analytics and machine learning can enable researchers and manufacturers to devise future systems that are more efficient and less prone to faults (Ahmad et al., 2021) and less downtime, and more productivity. This also gets proved to be useful in terms of deferred grid investments, declined energy consumption, and extended life of the energy system. Considering the application of digitalization, the administrative costs can be cut down by digitizing the paperwork using various utility software to effectively communicate with the internal as well as external stakeholders of a system, while the repetitive technical tasks can be automated using RPA (robotic process automation) (Anagnoste, 2018; Benndorf, Wystrcil, & Réhault, 2018). This leaves fewer chances for human error.

Finally, digitalization attempts to address customer orientation. Considering the past, the availability of power was considered a commodity by the users. However, the trend has shifted in the past few decades as consumers now focus on getting environment-friendly power (Lange et al., 2020; Moşteanu, Faccia, & Cavaliere, 2020). This has brought a significant shift toward renewable energy in the last decade. Apart from that, the customer orientation has evolved very much around the concept of transparency between the providers and consumers as well. The significant discrepancies between the estimated billing available to consumers at a single point in time during the month and the actual one issued by the supplier after the physical examination of the power meter gave rise to the frustration of customers (Meister et al., 2018; Varela, 2018). These billing devices over time proved themselves to be very useful in providing the consumers the very precise information about the power usage and the total amount of the bill. Also, the integration of a power meter with the smart-home-based devices allowed it to reflect the power consumption by each smart device individually, may it be one's refrigerator, microwave oven, or electrically powered car. This allowed the users not only to observe the billing information in real time but also helped them to redefine their living habits and consumption patterns. Apart from the consumer's side of smart meter application, the said deployment can also be used to gather data from users collectively and translate it into useful information for shaping the energy sector and making it more consumer friendly (De Dutta & Prasad, 2020; Parviainen, Tihinen, Kääriäinen, & Teppola, 2017; Wu et al., 2021).

## 13.4  CONCERNS AND CHALLENGES

### 13.4.1  FUNDAMENTAL PRIVACY AND SECURITY CHALLENGES AND OPPORTUNITIES

The energy sector is very sensitive and prone to be affected by any incident which could have detrimental and adverse effects on business, society, and critical setups. Historically, the referred personifications could cause an impulsive breakdown and may even bring certain civilizations to the brim of downfall (Onyeji, Bazilian, & Bronk, 2014; Venkatachary, Prasad, & Samikannu, 2017). However, the research has a consensus upon the reality that businesses tend to adopt the new digitalization solutions more quickly in the recent decades. Maintaining the maneuvering of certain legacy systems like SCADA (supervisory control and data acquisition) demonstrates the increased vulnerability to put a great burden upon the suppliers and end users of digital technology since the distinctive and unique mode of cyberthreats globally coerces the adopted systems to allow greater vulnerabilities, which signifies the intense need for personalized upgrading. However, an abundance of tracking sources, for example, license plate scanning, cell site location tracking, and data generated from wearable and connected vehicles and DNA repositories of people accumulate the substantial target for hackers. In this way, the today's industries are more jeopardized by an increased number of contemporary cyberthreats (Fortunato, 2020; Irmak & Erkek, 2018).

### 13.4.2  INDIVIDUAL CYBERSAFETY

It has been observed in the last few years that individual rights have become the main focus in terms of maintaining security, PII (personal identifiable information) ownership, privacy, and security system. It seems that this situation not only caters to the industry's expectations to evaluate, acquire knowledge, or correlate for resource planning and predictive maintenance, but there also exists a need to liquidate the databases by the provisions of utility services. Here the question arises if PII is being violated? The consequences to the individuals could be serious and undesirably can have an adverse life-changing impact. Such conception in most parts of the globe is overshadowed, particularly when the safety concerns of CNI (critical national infrastructure) systems are being violated in terms of societal and financial consequences (Le Ray, 2019; Rajavuori & Huhta, 2020).

### 13.4.3 NEED FOR PROACTIVE POLICIES ON CYBERSECURITY AT REGIONAL AND NATIONAL LEVELS

The threats related to distinct businesses and cultures are evident everywhere. Hence, considering the safety of all the related stakeholders, certain reliable protective strategies need to be incorporated to benefit them. Considering the devastating circumstances on a huge scale, innumerable infrastructures have been proposed to implement reliable defense mechanisms, and the assurance of the CNI is to be contemplated (Venkatachary, Prasad, & Samikannu, 2018).

### 13.4.4 CLOUD-BASED DIGITAL SOLUTION COULD INCREASE CYBERATTACK POTENTIALS

Contemporary software and hardware solutions, such as smart meters and IoT sensors which have been structured for the range of cloud-based facilities, tend to get attacked and increase the threats and risks to distinct businesses and cultures (Hossein Motlagh, Mohammadrezaei, Hunt, & Zakeri, 2020). An attack vector signifies itself as a transmitter or a medium, through which the invader could get a proscribed or outlawed accession of the network to transmit the baleful and malevolent consequences. Such attackers permit the vectors to install the malware and induce cyberattacks to seize the system vulnerabilities. Consequently, there is a dire need to implement adequate and secured measures for cyberprotection and to install reliable digital applications in businesses, offices, and homes (Andoni et al., 2019). Therefore, it is must to recognize that reliable data security systems are the key to the execution of digital solutions.

### 13.4.5 BENEFITS OF APPLYING CLOUD SOLUTIONS

Cloud services if connected with solutions rightly, not only foster the ICT (information and communications technology) development but also render the most resourceful scaling and considerably strengthen the security (Gouvea, Kapelianis, & Kassicieh, 2018). The services of cloud are resilient in terms of sharing of finances and entitle the individuals to utilize its services whenever and wherever they need. Moreover, it could be economically beneficial even at the time of any fluctuations in services or system load. The cloud solution also shares its flexible services upon a massive grade, and it permits to install the wide-ranging safety solutions at data centers that are most prominently operated by the providers (Hossein Motlagh et al., 2020; Mohammed, Ghareeb, Al-bayaty, & Aljawarneh, 2019).

### 13.4.6 USING AI TO INCREASE CYBERSECURITY

The need for AI systems is capturing attention and is essential due to the increased amount of digital data that needs to have a defense against the malefactors disguised in a multitude of digital signals. As AI can figure out the microscopic anomalies which couldn't be evident to anyone easily, it equips the fast-growing alleviations to have an impact against the attackers (Andoni et al., 2019; Hossein Motlagh et al., 2020).

### 13.4.7 STRENGTHENING END-USER RESPONSIBLE PRACTICE TO INCREASE CYBERSECURITY AND SAFETY

The indicators of CCP (cloud computing providers), undoubtedly, serve from the minimal of competent resources to employing the basic service paradigms at the greatest degree of compliance and safety. The data controller, however, has the responsibility to carry out the "last mile" of safety in terms of the configuration and the daily usage by their workers. For this purpose, some of the measures which are easily overlooked in terms of customers' reliance upon confidentiality, integrity, and availability, including the flexibility of technical tools, must not be considered trivial. Though the entire digital data security system becomes intricate in terms of its hybrid services, that is, maintaining individual security and modern data analysis, it turns out to be a challenge to infix any kind of analysis on the contemporary digital data set. Thus, two types of approaches relating the data protection are mentioned below (Küfeoglu et al., 2019; Mosteanu, 2020; Sozontov et al., 2019):

The privacy of the data could be secured to some extent by eliminating the identifiable data associated with an individual or a group. This can be accomplished by utilizing quasi-identifiers. A quasi-identifier is a piece of information that does not itself identify an individual but can do so when combined with other quasi-identifiers. The research on the algorithm indicating the greater level of confidentiality and safeguarding it against the personal level of identification due to the quasi-identifiers has been conducted globally. Practically, such research, when implemented on a daily usage, supposes a greater level of command in terms of choosing the validity of testing data, which could permit the data-based frameworks to function and add value to maintain the greater degree of privacy (Asghar, Dán, Miorandi, & Chlamtac, 2017; Khatoun & Zeadally, 2017).

Despite, the basic collection of its model which merely exhibits the information in groups is considered the simplest way to implement. However, mostly it has been taken as a safety tool to protect one's information. Such a combination might be observed as an ordinary mechanism in creating harmless outputs from data. However, it gives a wrong sense of safety. In this domain, a plethora of research indicates that by ignoring the fact either deliberately or unintentionally, it could be likely to set apart an individual by applying the various filters to the collection of datasets, which characterizes that only with true analysis, aggregate information could reveal the considerable personal information too (Khatoun & Zeadally, 2017; Molina-Solana, Ros, Ruiz, Gómez-Romero, & Martín-Bautista, 2017).

## 13.5   THE FUTURE OF DIGITALIZATION IN THE ENERGY SECTOR

Considering the prior discussion, it is very evident that the landscape of the energy sector is set to embrace digitalization more and more in the years to come. Regardless, the said sector faces numerous concerns and obstacles in the way to adapting digitalization to the full extent.

Overviewing the aspect of energy utilization, the industry across the globe consumes 38% of the total energy produced and accounts for 24% of global $CO_2$ emissions (IEA, 2017). With the inclining implementation of digitalization, the global industrial practices are expected to become more efficient, thus leading toward a substantial decline in both of the parameters significantly. Exclusively referring to the energy sector, digitalization is set to enhance the resource recovery process in the oil and gas sector by 5% (Lyons, 2019), therefore assuring sustainable levels of resource extraction and less resource wastage during the extraction process. As a result, digitalization is set to cut down the overall production costs of the oil and gas sector by 10%–20% using the data-enriched methods making precise estimations and effective decisions. Considering the coal sector, the implementation of digitalization apart from bringing efficiency and effectiveness to the existing processes will help in encountering the safety of the workers, which up until now has remained a compromised issue within the said sector (Kudryashova, Venger, & Zakharova, 2019; Nikitenko, Goosen, Malakhov, & Kizilov, 2021). Focusing onto the power industry, digitalization is set to enhance its stakeholder's experience in various ways that may include decreased capital expenditure (CAPEX) and operational expenditure (OPEX), enhanced power network efficiency, mitigated operational downtime, and extended overall lifespan of the infrastructure deployed to generate and transfer power (Di Silvestre et al., 2018; Loock, 2020). The magnitude of the benefit attributed to digitalization in the power sector can be gauged from the fact that the said adaptation will enable savings of up to $80 billion per annum throughout 2016–2040, that is, 5% of absolute power generation cost (IEA, 2017). Further, the data analytics will enable the reduction of overall operations and maintenance (O&M) costs by allowing the management to observe the operational patterns on a broader spectrum remotely, thus benefitting not only the producers but also the consumers, with an average benefit close to $20 billion per annum (Anagnoste, 2018). Also, the availability of digital data in the power sector will empower the engineers and other relevant stakeholders to design the networks with optimum power deliverance capacity and reduced network losses. This will be achieved by remotely monitoring the power centers and networks to observe the optimal conditions the existing systems are operating onto and then designing relatively much efficient systems (Cozzi & Franza, 2017; Parida et al., 2019). The available data will also guide the operational teams to proactively maintain the power systems and perform preventive maintenance routines to minimize the system downtime and avoid unplanned power outages. Cumulatively, the taken actions will extend the overall lifespan of the deployed systems by 5 years, reduce the projected investments in the power systems by $34 billion per annum, and cut down the network costs by $20 billion per annum (Lange et al., 2020).

## 13.6   CONCLUSION

Digitalization sets to break down the boundaries between energy systems deployed across the various sectors allowing more flexibility toward the integration of different systems with one another. Further, the said approach will enable pushing the accomplishment of achieving carbon neutrality to a greater extent (Varela, 2018). Keeping that in view, it is the responsibility of the relevant stakeholders (i.e., government, private firms) to realize the potential of implementing digitalization in the energy sector. Furthermore, the attributed stakeholders are required to devise effective strategies and put in place aligned policies and regulations to assure the effective transition toward a sustainable future of the energy sector. The energy sector leaders need to take cyberrisks into account when assessing the stability in their line of business. With digital assets deployed throughout every aspect of operations, companies that fail to maintain strong cybersecurity may face the risk of having their operations come to a sudden, expensive, and prolonged halt. While a home computer or a corporate server hit with a cyberattack can be halted, rebooted, and fixed without lasting damage to the equipment, operating technologies when get hit with a cyberattack may suffer physical damage, thus posing safety threats to personnel on-site and requiring extensive repairs or replacement before operations can resume (Cassotta & Sidortsov, 2019; Kucharska, 2020). Managing this risk will also be an integral part of any future energy business and a major consideration for investors.

## REFERENCES

Ahmad, T., Zhang, D., Huang, C., Zhang, H., Dai, N., Song, Y., & Chen, H. (2021). Artificial intelligence in sustainable energy industry: Status quo, challenges and opportunities. *Journal of Cleaner Production, 289*, 125834.

Alekseev, A. N., Lobova, S. V., Bogoviz, A. V., & Ragulina, Y. V. (2019). Digitalization of the Russian energy sector: State-of-the-art and potential for future research. *International Journal of Energy Economics and Policy, 9*(5), 274.

Ali, S. S., & Choi, B. J. (2020). State-of-the-art artificial intelligence techniques for distributed smart grids: A review. *Electronics, 9*(6), 1030.

Anagnoste, S. (2018). The road to intelligent automation in the energy sector. *Management Dynamics in the Knowledge Economy, 6*(3), 489–502.

Andoni, M., Robu, V., Flynn, D., Abram, S., Geach, D., Jenkins, D., ... Peacock, A. (2019). Blockchain technology in the energy sector: A systematic review of challenges and opportunities. *Renewable and Sustainable Energy Reviews, 100*, 143–174.

Arif, M., & Al Senani, A. M. (2020). Digitalization in oil and gas industry-a case study of a fully smart field in United Arab Emirates. *Paper Presented at the Abu Dhabi International Petroleum Exhibition & Conference*, United Arab Emirates.

Asghar, M. R., Dán, G., Miorandi, D., & Chlamtac, I. (2017). Smart meter data privacy: A survey. *IEEE Communications Surveys & Tutorials, 19*(4), 2820–2835.

Benndorf, G. A., Wystrcil, D., & Réhault, N. (2018). Energy performance optimization in buildings: A review on semantic interoperability, fault detection, and predictive control. *Applied Physics Reviews, 5*(4), 041501.

Bloomberg, J. (2018). Digitization, digitalization, and digital transformation: Confuse them at your peril. *Forbes. Retrieved on August, 28*, 2019.

Botelho, D., Dias, B., de Oliveira, L., Soares, T., Rezende, I., & Sousa, T. (2021). Innovative business models as drivers for prosumers integration-Enablers and barriers. *Renewable and Sustainable Energy Reviews, 144*, 111057.

Brennen, J. S., & Kreiss, D. (2016). Digitalization. *The International Encyclopedia of Communication Theory and Philosophy, 1*, 1–11.

Cassotta, S., & Sidortsov, R. (2019). Sustainable cybersecurity? Rethinking approaches to protecting energy infrastructure in the European High North. *Energy Research & Social Science, 51*, 129–133.

Chou, J.-S., & Yutami, I. G. A. N. (2014). Smart meter adoption and deployment strategy for residential buildings in Indonesia. *Applied Energy, 128*, 336–349. Doi:10.1016/j.apenergy.2014.04.083.

Cozzi, L., & Franza, V. (2017). Digitalization: A new era in energy: Extract form digitalization & energy report. *EEJ, 7*, 44.

Creutzig, F., Franzen, M., Moeckel, R., Heinrichs, D., Nagel, K., Nieland, S., & Weisz, H. (2019). Leveraging digitalization for sustainability in urban transport. *Global Sustainability, 2*, 1–6.

Daniotti, B., Pavan, A., Spagnolo, S. L., Caffi, V., Pasini, D., & Mirarchi, C. (2020). Evolution of the building sector due to digitalization In *BIM-Based Collaborative Building Process Management*, 1–16, New York, United States: Springer.

De Dutta, S., & Prasad, R. (2020). Digitalization of global cities and the smart grid. *Wireless Personal Communications, 113*, 1385–1395.

Di Silvestre, M. L., Favuzza, S., Sanseverino, E. R., & Zizzo, G. (2018). How decarbonization, digitalization and decentralization are changing key power infrastructures. *Renewable and Sustainable Energy Reviews, 93*, 483–498.

Fortunato, S. S. (2020). *Risk Management in ICS/SCADA Systems to Enhance Security within the Energy Sector*. Utica, NY: Utica College.

Gouvea, R., Kapelianis, D., & Kassicieh, S. (2018). Assessing the nexus of sustainability and information & communications technology. *Technological Forecasting and Social Change, 130*, 39–44.

Herenčić, L., Ilak, P., Rajšl, I., Zmijarević, Z., Cvitanović, M., Delimar, M., & Pećanac, B. (2019). Overview of the main challenges and threats for implementation of the advanced concept for decentralized trading in microgrids. *Paper Presented at the IEEE EUROCON 2019-18th International Conference on Smart Technologies*, Novi Sad, Serbia.

Hossein Motlagh, N., Mohammadrezaei, M., Hunt, J., & Zakeri, B. (2020). Internet of Things (IoT) and the energy sector. *Energies, 13*(2), 494.

IEA. (2017). *Digitalisation and EnergyIEA*. Paris: IEA.

Irmak, E., & Erkek, İ. (2018). An overview of cyber-attack vectors on SCADA systems. *Paper Presented at the 2018 6th International Symposium on Digital Forensic and Security (ISDFS)*, Antalya, Turkey.

Judson, E., Soutar, I., & Mitchell, C. (2020). Governance challenges emerging from energy digitalisation.

Khatoun, R., & Zeadally, S. (2017). Cybersecurity and privacy solutions in smart cities. *IEEE Communications Magazine, 55*(3), 51–59.

Klavsuts, I. L., Klavsuts, D. A., Rusina, A. G., & Khayrullina, M. V. (2019). Digitalization of resource consumption on the basis of innovative devices of demand side management in smart grids. *Paper Presented at the 2019 8th International Conference on Modern Power Systems (MPS)*, Cluj-Napoca, Romania.

Kucharska, A. (2020). Cybersecurity challenges in Poland in the face of energy transition. *Rocznik Instytutu Europy Środkowo-Wschodniej, 18*(1), 141–159.

Kudryashova, I., Venger, M., & Zakharova, N. (2019). Ecological aspects of coal mining enterprises' activity in conditions of digitalization. *Paper Presented at the E3S Web of Conferences*, Russia.

Küfeoglu, S., Liu, G., Anaya, K., & Pollitt, M. (2019). Digitalisation and New Business Models in Energy Sector. https://doi.org/10.17863/CAM.41226.

Lange, S., Pohl, J., & Santarius, T. (2020). Digitalization and energy consumption. Does ICT reduce energy demand? *Ecological economics, 176*, 106760.

Le Ray, G., Pinson, P., Bindner, H. W., Bessa, R. J. G. S. B., & Goude, Y. (2019). *On the Role of Smart Metering Data Analytics in the Energy Sector Digitization Process*. Technical university of Denmark, Ph. D Dissertation.

Leal-Arcas, R., Lesniewska, F., & Proedrou, F. (2018). Prosumers: new actors in EU energy security. In: Amtenbrink, F., Prévost, D., Wessel, R. (eds) *Netherlands Yearbook of International Law 2017* , vol. 48. T.M.C. Asser Press, The Hague. Doi: 10.1007/978-94-6265-243-9_5.

Loock, M. (2020). Unlocking the value of digitalization for the European energy transition: A typology of innovative business models. *Energy Research & Social Science, 69*, 101740.

Lv, Y., Fang, F., Yang, T., & Romero, C. E. (2020). An early fault detection method for induced draft fans based on MSET with informative memory matrix selection. *ISA Transactions, 102*, 325–334.

Lyons, L. (2019). *Digitalisation: Opportunities for Heating and Cooling*. Publications Office of the European Union, Luxembourg.

Malecki, E. J., & Moriset, B. (2007). *The Digital Economy: Business Organization, Production Processes and Regional Developments*. Routledge, New York, United States.

Marmolejo-Saucedo, J., & Hartmann, S. (2020). Trends in digitization of the supply chain: A brief literature review. *EAI Endorsed Transactions on Energy Web, 7*(29), 1–7.

Meister, J., Ihle, N., Lehnhoff, S., & Uslar, M. (2018). Smart grid digitalization in Germany by standardized advanced metering infrastructure and green button. In *Application of Smart Grid Technologies* (pp. 347–371): Elsevier, Amsterdam, Netherlands.

Mingaleva, Z., Shironina, E., & Buzmakov, D. (2020). Implementation of digitization and blockchain methods in the oil and gas sector. *Paper Presented at the International Conference on Integrated Science*, Kep, Cambodia.

Mohammed, T. A., Ghareeb, A., Al-bayaty, H., & Aljawarneh, S. (2019). Big data challenges and achievements: applications on smart cities and energy sector. *Paper Presented at the Proceedings of the Second International Conference on Data Science, E-Learning and Information Systems*, Dubai, United Arab Emirates.

Molina-Solana, M., Ros, M., Ruiz, M. D., Gómez-Romero, J., & Martín-Bautista, M. J. (2017). Data science for building energy management: A review. *Renewable and Sustainable Energy Reviews, 70*, 598–609.

Moradi, J., Shahinzadeh, H., Nafisi, H., Gharehpetian, G. B., & Shaneh, M. (2019). Blockchain, a sustainable solution for cybersecurity using cryptocurrency for financial transactions in smart grids. *Paper Presented at the 2019 24th Electrical Power Distribution Conference (EPDC)*, Khorramabad, Iran.

Morley, J., Widdicks, K., & Hazas, M. (2018). Digitalisation, energy and data demand: The impact of Internet traffic on overall and peak electricity consumption. *Energy Research & Social Science, 38*, 128–137.

Mosteanu, N. R. (2020). Artificial intelligence and cyber security–A shield against cyberattack as a risk business management tool–case of European countries. *Quality-Access to Success, 21*(175), 148–156.

Moşteanu, N. R., Faccia, A., & Cavaliere, L. P. L. (2020). Digitalization and green economy-changes of business perspectives. *Paper Presented at the Proceedings of the 2020 4th International Conference on Cloud and Big Data Computing*, Liverpool, United Kingdom.

Muñoz-Villamizar, A., Solano, E., Quintero-Araujo, C., & Santos, J. (2019). Sustainability and digitalization in supply chains: A bibliometric analysis. *Uncertain Supply Chain Management, 7*(4), 703–712.

Nikitenko, S., Goosen, E., Malakhov, Y. V., & Kizilov, S. (2021). Multifunctional safety system as a framework for the digital transformation of the coal mining industry. *Paper Presented at the IOP Conference Series: Earth and Environmental Science*, Yurga, Russian Federation.

Noussan, M., & Tagliapietra, S. (2020). The effect of digitalization in the energy consumption of passenger transport: An analysis of future scenarios for Europe. *Journal of Cleaner Production, 258*, 120926.

Onyeji, I., Bazilian, M., & Bronk, C. (2014). Cyber security and critical energy infrastructure. *The Electricity Journal, 27*(2), 52–60.

Parida, V., Sjödin, D., & Reim, W. (2019). *Reviewing Literature on Digitalization, Business Model Innovation, and Sustainable Industry: Past Achievements and Future Promises*. Multidisciplinary Digital Publishing Institute, Luleå, Sweden.

Parviainen, P., Tihinen, M., Kääriäinen, J., & Teppola, S. (2017). Tackling the digitalization challenge: how to benefit from digitalization in practice. *International Journal of Information Systems and Project Management, 5*(1), 63–77.

Rajavuori, M., & Huhta, K. (2020). Investment screening: Implications for the energy sector and energy security. *Energy policy, 144*, 111646.

Salminen, V., Ruohomaa, H., & Kantola, J. (2017). Digitalization and big data supporting responsible business co-evolution. In *Advances in Human Factors, Business Management, Training and Education* (pp. 1055–1067): Springer, Hämeenlinna, Finland.

Sestino, A., Prete, M. I., Piper, L., & Guido, G. (2020). Internet of Things and Big Data as enablers for business digitalization strategies. *Technovation*, 102173.

Singh, V. (2019). Digitalization, BIM ecosystem, and the future of built environment: How widely are we exploring the different possibilities? *Engineering, Construction and Architectural Management*.

Sozontov, A., Ivanova, M., & Gibadullin, A. (2019). Implementation of artificial intelligence in the electric power industry. *Paper Presented at the E3S Web of Conferences*, Irkutsk, Russia.

Specht, J. M., & Madlener, R. (2019). Energy Supplier 2.0: A conceptual business model for energy suppliers aggregating flexible distributed assets and policy issues raised. *Energy Policy, 135*, 110911.

Svendsen, A. B., Tollefsen, T., Gjengedal, T., Goodwin, M., & Antonsen, S. (2018). Digitalization of the power business: How to make this work? *Safety and Reliability–Safe Societies in a Changing World. Proceedings of ESREL 2018*, June 17–21, 2018, Trondheim, Norway.

Varela, I. (2018). Energy is essential, but utilities? Digitalization: What does it mean for the energy sector? In *Digital Marketplaces Unleashed* (pp. 829–838): Springer, Düsseldorf, Germany.

Venkatachary, S. K., Prasad, J., & Samikannu, R. (2017). Economic impacts of cyber security in energy sector: A review. *International Journal of Energy Economics and Policy, 7*(5), 250–262.

Venkatachary, S. K., Prasad, J., & Samikannu, R. (2018). Cybersecurity and cyber terrorism-in energy sector–a review. *Journal of Cyber Security Technology, 2*(3–4), 111–130.

Wu, Y., Wu, Y., Guerrero, J. M., & Vasquez, J. C. (2021). Digitalization and decentralization driving transactive energy Internet: Key technologies and infrastructures. *International Journal of Electrical Power & Energy Systems, 126*, 106593.

# 14 Application of Blockchain for Energy Transition Systems

*Gijs van Leeuwen*
Delft University of Technology

*Tarek AlSkaif and Bedir Tekinerdogan*
Wageningen University and Research

## CONTENTS

## 14.1 INTRODUCTION

In recent years, blockchain technology has been described as a promising and potentially disruptive technology that can create a whole new range of business models and use cases across many sectors of the economy [1], for example, in agriculture [2]. Another important domain is the energy sector, where blockchain can provide Information and Communication Technology (ICT) architecture that enables novel Energy Transition Systems (ETSs) that are associated with the trends of decentralization, decarbonization, and digitalization [3,4]. As a Distributed Ledger Technology (DLT), the disruptive potential of blockchain lies in its capacity to distribute the registration of transactive and administrative processes across a large number of actors in a secure and verifiable manner without requiring a central authority to be in control of this registration. This eliminates the need for a central controlling party on which all other actors must depend [1,5], reducing vulnerability against failures and cyber-physical attacks, increases transparency, and prevents potentially problematic misuse of market power by centralized actors. As a motivation for this book chapter, it will be described how the electricity system, in particular, is transforming in a fashion that makes it especially suitable for the application of blockchain-based systems.

The power grid and the surrounding system are expected to undergo numerous fundamental changes and transformations in the coming years [6]. It is expected that the amount of Distributed Energy Resources (DERs) in the system, including renewable energy generators, electric vehicle (EV) charging stations, and energy storage assets, will increase significantly, which will increase the complexity of coordination and control as well as the number and variety of participants in electricity markets [4,7,8]. In parallel, the grid and the surrounding system will be increasingly decentralized and thoroughly digitized. Rather than the current monolithic, hierarchical, centrally controlled grid, a modern grid would be smart, agile, and composed of different segments that may be islanded and operate independently. Local DERs and grid infrastructure are increasingly managed by smart energy management systems (EMSs), and numerous actors can trade energy through

local electricity markets (LEMs) [9]. In addition, larger grid assets such as solar farms, wind turbines, and industrial electric assets have to be integrated similarly. To enable such novel energy
systems, advanced ICT architecture and data management are required to house control algorithms
and connect the different actors and owners of energy assets.

Blockchain technology can facilitate such platform architecture and provides several benefits.
While it is possible for a single actor to control such systems in a centralized manner, this can bring
up multiple concerns of privacy, security, transparency, potential power abuse, and fees for participants. By using blockchain to facilitate novel ETSs, all stakeholders can access and verify all data
and processes on the network. Therefore, it is not required to trust a third party who is vulnerable
to system failures or attacks and can misuse their control over the network. Additionally, the digitalization and increase in overall complexity mean that the communication and data management
overhead increases significantly. Centralizing such vast amount of digital control could require very
advanced computational methods that may be infeasible and impractical [6]. For these reasons,
blockchain is considered a feasible solution for energy systems. The goal of this chapter is to provide
insights into the architectural components of blockchain, the potential applications of blockchain in
ETSs, and the future opportunities and challenges of blockchain in this domain.

In Section 14.2, we will give an overview of blockchain technology and describe a characterization framework that may be used to classify various purposes of blockchain in the energy system.
This characterization framework is constituted by various important components and properties
of a blockchain network. The first of these is the openness of network, where a distinction is made
between public, private, and consortium blockchains [10]. Another core component is the consensus
mechanism, which allows different self-interested actors on a blockchain network to agree on the
validity of transactions so that a central authority is not required to verify their contents. Finally, the
functionalities of smart contracts will be described as these greatly expand the range of potential
use cases for blockchain in the ETSs and power systems. In Section 14.3, we will describe several
proposed applications of blockchain in the energy and power sector by reviewing recent literature.
The most commonly proposed applications include peer-to-peer (P2P) trading and electric vehicle
(EV) charging. Other applications include the facilitation of microgrid networks, industrial compounds, and connecting grid operators and ancillary service providers [11]. The characterization
framework described earlier will be applied to the applications in the energy system in this section.
In Section 14.4, an outlook is given for the future development pathway of blockchain for energy
sector applications. Key strengths and challenges for the future potential of blockchain will be discussed and suggested research directions that should be pursued to further realize its potential. An
important barrier is the technological maturity, which involves aspects such as scalability, speed,
security, and user-friendliness [10]. Another challenge is the high energy usage, particularly for
PoW-based blockchain [12]. Finally, attention should also be given to societal, legal, and institutional barriers and implications [4,8,13,14]. A discussion and conclusion of the findings of this
chapter will be provided in Section 14.5.

## 14.2 FUNDAMENTALS OF BLOCKCHAIN

This section will introduce the fundamental aspects of blockchain technology, its valuable characteristics, and various important components and properties that make up a blockchain network.
These properties, being openness of the network, smart contract role, and consensus mechanism,
make up a characterization framework. This framework will be used in the next section to describe
applications in the energy system.

Fundamentally, the blockchain network is a collective digital database that enables a secure and
verifiable distributed or decentralized management of data among a group of users [15,16]. This
is the most important characteristic of a blockchain network that distinguishes it from a regular
centralized database. In a centralized database, all data is stored on one server that is operated by a
single controlling party, which means that the accuracy and security of this data are dependent on

the actions of this controlling party. If the data is not well protected from cyberattacks or if the controlling party has any interest in manipulating it, the contents of the database can be changed unbeknownst to other stakeholders. In contrast, on a blockchain network, all the data can be accessed by all nodes (i.e., users) on the network at all times. If any node changes the content of the database, every other node in the network can read that, and there is a guarantee that everybody has the same copy of the ledger. Thus, a distributed ledger cannot be tampered with as all nodes must approve of any changes to the content. Figure 14.1 represents the difference between a centralized system, where all participants rely on the central authority, and a decentralized system where all participants communicate to cross-validate the data. The changes are checked and verified by the rest of the network and written to the ledger using a consensus mechanism [17]. All changes to the database content, called transactions, are recorded chronologically, and grouped into blocks of data chained together. Understood in this way, a blockchain is an ever-expanding historical record of the evolution of the data contents. By accessing this historical record, any network user can reconstruct and verify the current status of the database. Because all users have the same historical record, it is impossible for any single user to manipulate the data.

This can be illustrated with the earliest and most well-known application of blockchain, which is cryptocurrency. For cryptocurrency, the blockchain database contains a registration of ownership of each unit of currency, and the historical record shows every occurrence where a unit of currency changed ownership between two network users. In this way, anyone with access to the blockchain can trace the history of cryptocurrency ownership all the way back to the creation, ensuring that the current owner is the rightful one. In this way, the need for a bank as a trusted intermediary is taken away. The function of the bank as a trusted intermediary is to ensure that transactions and registration of ownership of money are secure and verifiable. This can have disadvantages as it places a disproportionate amount of market power under the control of a small number of centralized entities. If the centralized entity or bank has other interests besides acting as a trusted intermediary, such as serving stakeholder interests, investing in the market, and realizing profits overall, this large centralized market power may be used to their advantage. Besides the risk of power misuse, the intermediary can also profit from transaction fees, and it becomes a prime target for any cyberattacks or other tampering attempts by malicious third parties.

The security and utility of a blockchain network is dependent on its openness to participation, of which there are various types and degrees. Typically, a distinction is made between permissionless or public networks on the one hand and permissioned networks on the other hand, which are further subdivided into private and consortium blockchain networks [18,19]. Public blockchains typically include cryptocurrencies and allow anyone to connect and participate on the network. Any person who wishes can access and write data on the blockchain, and the full anonymity of all participants

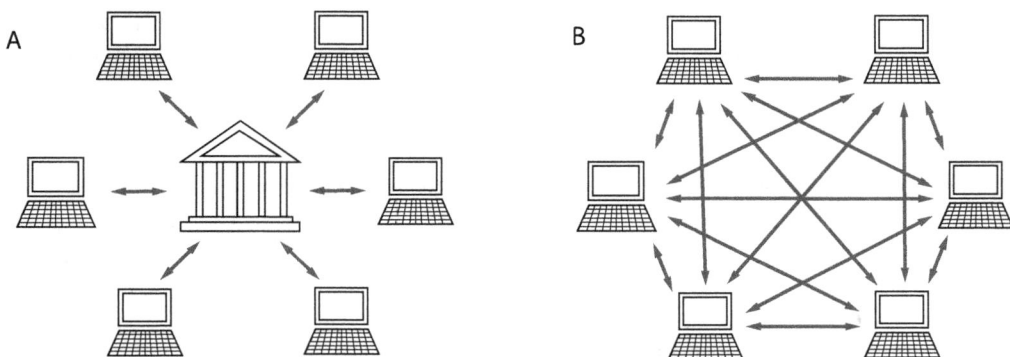

**FIGURE 14.1** This figure shows the difference between a centrally controlled database (a), where stakeholders must obtain information from an institutional authority that manages the data, and a distributed ledger or blockchain network (b), where every network participant has access to all the data.

is preserved. Furthermore, on a public blockchain network, every node can validate other transactions on the network, i.e., every node can participate in computing the consensus mechanism. For example, on the Bitcoin network, any node can participate in the mining process, which serves to validate new data blocks and create new bitcoins [19]. The main benefit of a public network is full decentralization and the highest level of security against tampering. The downside is the decreased computational efficiency as many nodes that participate in the validation process perform obsolete work (i.e., the validation process becomes more complex and computationally expensive as more nodes participate) [12].

In contrast, permissioned blockchains have clear rules and restrictions about who can read and write on the common database and who can validate transactions with the consensus mechanism [16,17]. Permissioned blockchains are thus more hierarchical and not fully decentralized, although they still use a distributed computation. Permissioned blockchains are overall more efficient and better suited for trusted, closed environments where the identity of at least some of the participants is known. Permissioned blockchains can be further subdivided in private and consortium blockchains. Private blockchains have strict access requirements where only known, invited users can join and participate, and the network as a whole is typically managed by a single party. On a consortium network, such access requirements can be more relaxed, and the network is more open, although this varies between networks. Several selected, trusted nodes can validate blocks, and ownership is often shared among several users rather than a single party.

Whereas many cryptocurrencies like Bitcoin operate on the first iteration of blockchain technology, where the database is used for ownership registration, the possibility to deploy smart contracts on the network has greatly expanded the blockchain's functionalities. This option came with the development of blockchain 2.0 technology in the Ethereum network [20]. Smart contracts can be considered stand-alone computer programs or software that reside on the blockchain network just like other nodes, with an address and public key. When certain predetermined conditions outlined in the contract are met, the contract will execute a piece of computer code that can make further changes to the database contents [16]. The conditions and code of the contract are visible to any user on the network. Smart contracts can be used to execute transactions or codify agreements or exchanges with more conditions than transfer of the currency ownership. Although this is a very useful property by itself, smart contracts also enable the execution of complex computer programs in a distributed way across the blockchain network, i.e., distributed computing.

The final important element of a blockchain network is the consensus mechanism, which are listed in Table 14.1. Consensus mechanisms ensure the validity of transactions and the formation of new blocks [15,17]. They can be computed either by a selected number of validator nodes (in permissioned networks) or by any node (in permissionless networks). Essentially, the validation of transactions and formation of blocks is an administrative task that must be executed by a node, or a group of nodes, within the network. The consensus mechanism ensures that a trustworthy user is selected to perform this action. There are many different algorithms that each use different methods

## TABLE 14.1
### Strengths and Weaknesses of the Different Consensus Mechanisms

| Type | Strengths | Weaknesses |
|------|-----------|------------|
| PoW | High security and privacy, tamper-resistant, high scalability | High energy consumption, high latency, low throughput |
| PoS | High transaction speed, energy efficient | Less security |
| BFT | Transaction speed, quick finality of validation | Less scalability and security |
| PoA | High computational efficiency, high security | Less privacy and scalability, relatively more centralized |

*BFT, Byzantium Fault Tolerance; PoA, Proof-of-Authority; PoS, Proof-of-Stake; PoW, Proof-of-Work.*

to achieve this goal, and they have to meet certain strict requirements to be fully functional. Various algorithms have different characteristics, advantages, and disadvantages in terms of scalability, efficiency, and security [17]. The operation and properties of a blockchain network are defined by the type of consensus mechanism used. The most commonly proposed consensus mechanisms include Proof-of-Work (PoW), Proof-of-Stake (PoS), Proof-of-Authority (PoA), and Byzantine Fault Tolerance (BFT) mechanisms.

The PoW mechanism was first used in Bitcoin and operates by making nodes compete to solve a cryptographic puzzle, which would ensure consensus [21]. The only way to solve this puzzle is by brute-force computational power, which is incentivized by rewarding the succeeding nodes with new bitcoins. The mechanism ensures that the administrative task of forming new blocks and validating transactions is carried out by nodes that put in a high effort. For this reason they are considered trustworthy. Because different nodes can solve the puzzle in parallel, multiple sidechains may be generated. It is then assumed that the longest chain is generated by the majority of the computational power, at which point this chain is accepted as the new final blockchain. This makes it impossible for attackers to get their manipulated side-chain accepted unless they control more than 51% of the computational power spent in achieving consensus, which is very difficult to achieve. Overall, PoW is well suited for public blockchains as it is highly scalable and secure. A major downside is that a huge amount of electricity is wasted in solving the cryptographic puzzles [12,22], and also the speed of processing the transactions is relatively low [15].

The PoS algorithm was developed to counter some of the disadvantages of the PoW algorithm [17]. Instead of relying on brute-force computational work, the PoS algorithm randomly selects a group of nodes to validate the network, which can be used on public and private networks. This is a lottery-type approach, where the chance of being selected increases with the amount of stake in the network—e.g., the amount of cryptocurrency held by a node. This greatly increases transaction speed and reduces energy consumption than PoW [19]. A downside is lower security compared to PoW, as attackers would have to control a majority of the total resources within the network rather than computational power. Since computational power requires hardware and software resources, PoS is considered less secure. Another commonly used set of consensus mechanisms is BFT algorithms, which are voting-based algorithms [17]. A verified and trustworthy set of validator nodes vote on the validity of transactions, and when enough votes are collected, the data is considered final. This makes BFT best suited to private networks where these validator nodes are known. The instant and straightforward finality provided by the BFT algorithm is a major advantage, while it suffers from relatively worse scalability, speed, and security. Finally, PoA is an algorithm similar to PoS in that validator nodes are randomly selected. Instead of selecting based on the size of their stake in the network, however, the PoA algorithm selects nodes based on the known validity of their identity or authority. This makes PoA-type algorithms especially suited for closed environments where privacy is a lower priority.

The three properties of the openness of a blockchain network, the smart contract role, and the consensus mechanism make up a characterization framework that we will use to describe applications within ETSs in the next section. Table 14.2 provides a general overview of this framework, the different properties, and the different subclasses of these properties.

### TABLE 14.2
### Characterization Framework for the Application of Blockchain Networks in ETSs

| Property | Subclasses |
| --- | --- |
| Network openness | Public, Consortium, Private |
| Smart contract role | Securing transactions, Distributed computing |
| Consensus mechanism | PoW, PoS, BFT, PoA |

## 14.3 APPLICATIONS IN THE ENERGY TRANSITION

In this section, we will describe in more detail the applications that blockchain has in the power grid, and surrounding system, as that is where most new applications and innovations have been proposed [13,23–26]. Blockchain networks may enable a range of novel transactive energy services in the power system. While the development of marketable blockchain applications is still at a relatively early stage, the amount of R&D on the blockchain is increasing as well as the number of piloting and experimentation projects [3,4,15]. These initiatives aim to explore and innovate new and creative solutions, many of which are associated with the trends described elsewhere in this book. Before describing several concrete applications from existing literature, we propose several relevant properties and types of ETSs. In conjunction with the blockchain characterization framework from Table 14.2, these properties will be used to classify the applications of blockchain within ETSs.

The use of blockchain in the energy system is connected with the emergence of smart power grid systems. The complexity of coordination in the power grid will increase significantly in the energy transition, with a higher geographical distribution of energy generators and more bidirectional electricity flows. Therefore, the grid will be increasingly equipped with smart sensing, control, and communication systems. Furthermore, the expected increase in prosumers and commercial and industrial aggregators of DERs would result in an increasing number of new proactive participants in the power system and electricity markets [7,27], which provides mutual services in many different forms. Existing actors such as energy service companies and grid operators may have to prepare for this digital transformation by innovating their business models and operations, and new actors could emerge. Another notable development is the fragmentation of the grid into smaller segments, such as microgrids. Microgrids are small grid segments that can function independently from the rest of the grid and can connect neighborhoods, communities, cities, organizations, or industrial compounds. By segmenting the grid into smaller parts in this way, there are several benefits concerning efficiency and resilience to blackouts or other system failures [6]. Microgrids that use EMSs or LEMs can be based on a blockchain network for its benefit.

A variety of actors can be connected on a blockchain network, including prosumers, industrial and commercial users, Transmission System Operators (TSOs), Distribution System Operators (DSOs), energy communities, energy service companies, Balancing Service Providers (BSPs), commercial aggregators, industries, and others. By being connected on the blockchain, such actors can cooperate and provide mutual flexibility services such as Demand Response (DR) [28,29], flexible EV charging [30], and P2P energy trading [26,27,31] in a secure and verifiable manner. We propose that such systems can be conveniently described as having different dimensions or layers, shown in Figure 14.2. First, there is the *physical infrastructure*, which includes grid infrastructure and tangible, physical energy assets such as generation units and storage systems. To guide and control energy flows and respect grid constraints and limitations within the physical infrastructure, there is a need for a *digital infrastructure*, where measurement, communication, and control systems reside, and information is collected, managed, and exchanged. The digital infrastructure resides above the layer of physical infrastructure and includes blockchain technology. It coordinates power flows in the physical layer and enables data-driven optimal decision support. Above the digital infrastructure, there is the *economic infrastructure*, where the different actors enter into agreements and cooperation on the conditions for the exchange of electricity or electricity services. These economic aspects and purposes guide the underlying cyber-physical systems and include market mechanisms, trading strategies, and exchange of services between different actors. Finally, we consider the fourth layer of *governance*. In this layer, agreements of governance between the different actors establish which actors operate which part of the infrastructure and what their mutual responsibilities are. It includes regulations, legal agreements, and different actors in the system and the relationships among them.

Blockchain is a technology that resides within the digital infrastructure and serves to interconnect mechanisms within the digital, economic, and physical infrastructure layers. Proposed blockchain applications typically prioritize solutions in either the physical or economic infrastructure. Therefore,

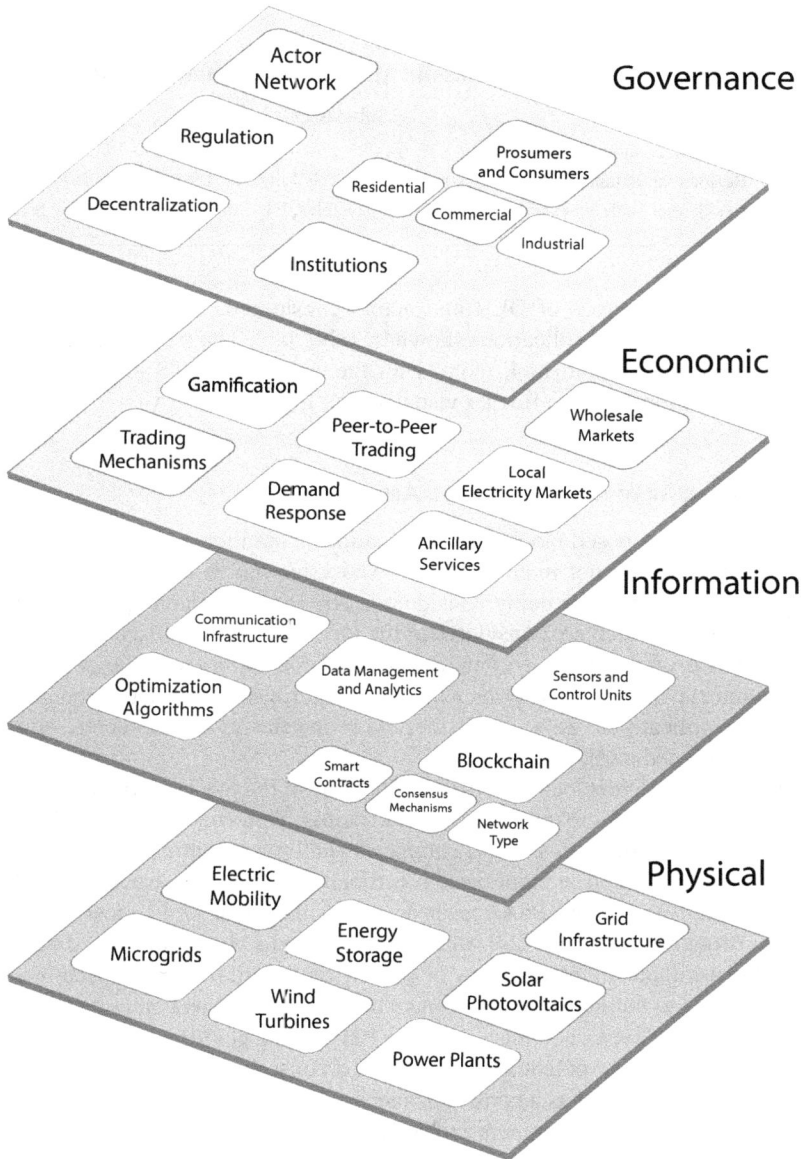

**FIGURE 14.2** This figure shows the different system layers that can typically be distinguished in power systems based on blockchain networks. Each layer shows a selection of concepts commonly encountered in this domain.

when describing proposed blockchain applications, we will distinguish between solutions focused on the physical and economic infrastructure in the following subsections. Section 14.3.1 describes applications where blockchain is mainly used for electricity trading, LEMs, and other primarily purposes in the economic infrastructure. In Section 14.3.2, we describe blockchain applications within the physical infrastructure that typically include technical optimization in EMSs. Implications of these findings for the governance layer will be described later in Section 14.5.

To describe the applications, we will use the characterization framework shown in Table 14.2. To further classify ETS applications, we will connect these aspects to the power grid system structure

**TABLE 14.3**

**Framework for ETS Properties for the Classification of Blockchain Applications**

| Property | Subclasses |
|---|---|
| Infrastructure layer | Physical, economic |
| Actor types | Residential, commercial and industrial prosumers, EV owners, TSOs, DSOs, BSPs, energy companies |
| Asset types | Residential-scale renewables, large-scale renewables, EVs, storage assets, electric heating units |

and components, including types of DER and actors considered. This results in a classification framework for blockchain ETS applications shown in Table 14.3. The studies referred to throughout these sections are shown in Table 14.4, along with the considered ETS properties. The different properties and interrelationships are further visualized in Figure 14.3.

## 14.3.1 Peer-to-Peer Trading and Market Applications

This section will review proposed blockchain applications from literature focused on the economic infrastructure. This review is not intended to be exhaustive but to indicate a variety of existing applications under development or being considered. The most common applications include P2P trading and EV charging, which are considered at the local scale of the distribution grid. At a larger scale, blockchain has been proposed to connect industrial consumers with the greater grid and utility actors or to connect different microgrids where each microgrid represents a separate node on the network. For these applications, we will consider the properties in the characterization framework of blockchain, as discussed earlier.

P2P trading between different local actors is perhaps one of the most commonly proposed applications of blockchain in the area of smart grids. Such trading platforms are especially well suited for using blockchain because of their sensitivity to tampering and smart contracts for settling transactions. When connected on a microgrid, the connection through physical infrastructure usually requires a permissioned blockchain and often a PoA consensus mechanism. One of the most well-known studies of this kind is the Brooklyn Microgrid [24], which implemented a blockchain-based microgrid energy market that functions without an intermediary or other trusted third party. The permissioned network uses PoA for consensus to facilitate the closed network and uses smart contracts to secure financial and energy transactions between prosumers. Another P2P trading platform based on a private network is proposed, which also has the potential to expand to a consortium network [10]. With the private network accessibility, PoA is used as a consensus mechanism as all participants are known. The network focuses on trading between prosumers and consumers that have access to solar PV, batteries, and smart meters. Smart contracts enable power and financial transactions.

Several other studies propose PoW-based blockchains, particularly for their benefits regarding security. Because PoW is well suited for public networks, it is also associated with increased privacy, although it does not strictly provide this when compared with PoS. In [5], a multi-signature blockchain based on PoW is proposed to focus on increasing cybersecurity and privacy in a trading network. Participants within the network, including prosumers and DSO, can negotiate energy prices anonymously, and smart contracts are used to settle transactions. The work in [18] uses a consortium blockchain for its benefits in the security domain. PoW is used as a consensus mechanism, and a general platform architecture is proposed for trading among many actors. The platform can be adopted for microgrids with prosumers and grid operators, as well as EV charging and vehicle-to-grid applications. Other studies propose a more general system architecture that allows for a variety of participants. For example, the system proposed in [32] can accommodate various end-users from residential, commercial, or industrial sectors. It focuses on demand-side management, DR, and P2P energy exchange and incorporates both trading and technical optimization. Smart contracts are

**TABLE 14.4**

**Several Reviewed Studies That Propose the Application of Blockchain-Based Solutions in the Energy Sector**

| Ref. | Focus | Domain | Blockchain Properties | | | Grid Properties | |
| --- | --- | --- | --- | --- | --- | --- | --- |
| | | | Network Type | Consensus | Smart Contract | Actors | Asset Types |
| [5] | Multi-signatures blockchain, privacy and cybersecurity | Economic | | PoW | Distributed contract for multi-signatures | Prosumer, DSO | |
| [10] | P2P trading | Economic | Permissioned | PoA | Power and financial transactions | Prosumers, consumers | Solar PV, battery, smart meter |
| [39] | EV trading and charging | Economic, physical | Consortium | | Transactions of EV charging | EV owners, new energy companies, government | EV charging stations |
| [35] | P2P trading | Economic | Consortium | | | Prosumers, households | Batteries, PV |
| [36] | P2P trading | Economic | Consortium | PBFT | P2P trading, gamification | Prosumers | |
| [30] | P2P trading | Economic | Consortium | PoW | Trading by charging and discharging between EVs | PHEV owners | EV charging stations |
| [18] | P2P trading, credit-based payment | Economic | Consortium | PoW | | Various | Various |
| [34] | Transactive energy, trading | Economic | Permissioned | | Securing optimal transactions between microgrids and local distribution grid | DSO, microgrid aggregators | Various |
| [32] | Demand-side management, P2P trading | Economic, EMS | | | Securing trading transactions between retailers and consumers | Residential, commercial, industrial, generation | Various |
| [40] | Security, EVs | Physical | Consortium | PoW/PoS | Validating transactions | EV owners | EV charging |
| [33] | Trading | Economic | | | | Prosumers | Various |
| [29] | Ancillary services TSO | Physical | Permissioned | PBFT | Validating transactions | TSO, DSO, distributed asset owner, aggregator, balance service provider | EV charging stations, distributed assets |
| [37] | Microgrid market, game theory | Economic | Consortium | Hyperledger | Securing transactions | DER owners | Various |
| [41] | EV charging | Physical, charging | Permissioned | BFT | Secure charging | Smart communities, EV owners | EVs |
| [28] | DR | Physical | Permissioned | PoS | Regulating energy flexibility services | DSO, prosumers, aggregators, retailers | Various |
| [31] | Decentralized optimization | Physical, trading | Permissioned | | Distributed optimization | Prosumers | EV, PV, battery |
| [24] | P2P trading | Economic | Permissioned | | Trading transactions | Prosumers, community members | Various, households |
| [38] | Decentralized optimization | Physical, economic | Permissioned | | Distributed optimization | Prosumer households | Various |

The general focus of each study is indicated and the domain of application (physical/economic). Furthermore, the blockchain properties are indicated as well as the different actors and assets that are connected to the system.

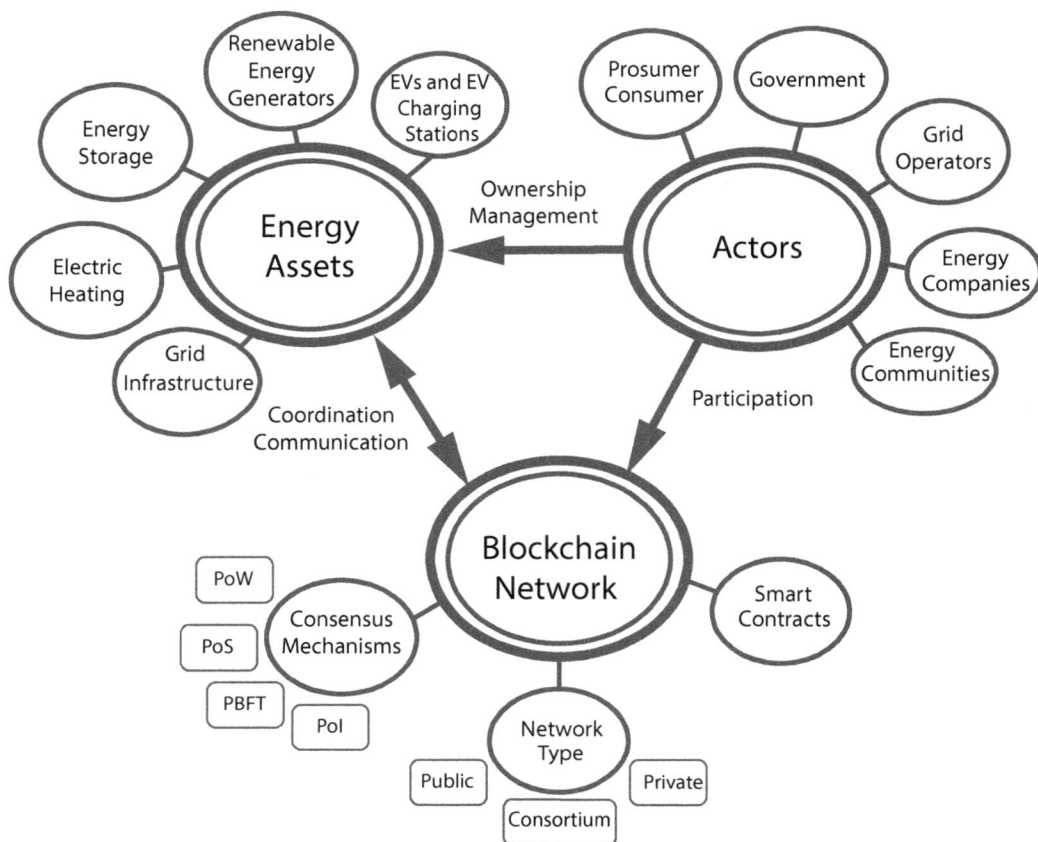

**FIGURE 14.3** This figure shows the interrelationships between the blockchain network, the participating actors, and the integrated physical assets in the network. The same subdivision is used as in the properties for reviewing existing studies.

used to validate transactions between the various actors, similarly as in [33], which propose distributed electricity trading system for P2P trading between prosumers that can facilitate a variety of DER. Smart contracts are used for securing trading transactions.

Although many articles focus on trading at the level of a single microgrid [34], they explore the use of blockchain for transactive energy services among different microgrids networked together. A larger scale is thus considered. In this way, the primary actors would not be individual households but rather DSOs, aggregators, BSPs, and providers of ancillary services that manage their local microgrid. The trading platform is based on a permissioned, private blockchain. Smart contracts regulate transactions between the microgrids on the network, and the local distribution grid that the DSO manages.

There are also P2P trading applications that are based on consortium blockchain networks. The work in [35] describes a consortium blockchain for prosumers energy trading which addresses the problem of privacy leakage without restricting the options for trading. Prosumers are assumed to have access to PV and batteries. Similarly, [36] provides a P2P trading system for prosumers and uses game theory for the pricing of energy. It is based on a consortium blockchain with the PBFT consensus mechanism and uses smart contracts to settle trading transactions and the game theoretic mechanism. Another article that uses game theory is [37], where a consortium blockchain facilitates a microgrid market, making use of Nash game equilibrium theory. Consensus is reached using Hyperledger technology [16], and smart contracts are used for transactions. In [26], two bilateral trading strategies for matching power supply and demand are proposed using a permissioned block-chain and smart contracts to settle the transactions.

## 14.3.2   ENERGY MANAGEMENT SYSTEMS AND TECHNICAL COORDINATION

As opposed to the application for trading and other economic purposes, blockchain can also be used for EMSs that enable the technical optimization of electricity flows throughout the physical grid. The work in [29] proposes a blockchain network that can facilitate the provision of ancillary services to the TSO by DER. It argues that a permissioned blockchain is most suited in this application due to the closed-off nature of the transmission grid and the existence of multiple actors (e.g., TSO, BSPs, aggregators, owners of DERs, and DSOs). The paper proposes using a PBFT consensus mechanism and smart contracts to validate transactions. Similar applications exist in the distribution grid. For example, [28] proposes the use of blockchain for facilitating DR and the management of flexible DER for prosumers and other stakeholders, including DSOs, retailers, and aggregators. The network is based on a PoS consensus mechanism, and smart contracts are used to provide flexibility services. There are also proposed solutions that integrate trading and technical optimization. A notable study is in [38], which utilizes smart contracts to perform distributed optimization. The work considers P2P trading to happen while ensuring that the operational constraints of the local grid infrastructure are respected, without the need for an aggregator or grid manager. This occurs on a permissioned network that can facilitate prosumer households with various DER. Similarly, it utilizes distributed optimization to integrate bilateral P2P energy trading and technical optimization of grid operation as an OPF problem. It provides a detailed description of the smart contract role. Prosumers on the microgrid network can deploy a variety of DER, such as EV, PV, and battery storage.

Another commonly proposed application is the charging of EVs and energy exchange between EVs. Such networks are typically based on consortium blockchain due to its flexibility for users to leave and enter the network. The work proposes a blockchain-based EV energy charging system that uses smart contracts to verify the exchange of energy in the charging process. The network provides a way for EV owners, energy companies, aggregators, and governments to cooperate in manag-ing such a network. In [30], a P2P trading platform for plug-in hybrid electric vehicles (PHEVs) is proposed and implemented? On a consortium blockchain, PoW is used as a consensus mechanism, and smart contracts validate charging and discharging transactions between EVs. Special atten-tion is given to security issues in EV charging by [40], who uses data coins and energy coins on a consortium blockchain. Consensus is reached by a PoW algorithm. The work in [41] employs a permissioned blockchain to facilitate an EV charging algorithm for EV owners, especially for use within smart communities. The platform uses smart contracts to verify charging transactions and the BFT consensus mechanism.

While there are various proposed applications of blockchain, the main fields of interest are P2P trading and LEMs, usage for EMSs and optimization, and smart, secure charging of EVs. In all of these areas, the use of blockchain would allow all participants or nodes on the network to cooper-ate and be connected without the requirement for a central authority to certify transactions, act as a market maker or coordinate the flow of electricity. Although there appears to be a preference for solutions at the local scale, there are also proposed blockchain applications that function on the larger scale of the transmission grid. In this way, proposed applications can connect a variety of actors, whether local actors such as prosumer households, communities, and small aggregators, or larger actors such as industrial consumers, microgrid managers, BSPs, DSOs, and TSOs.

It can be noticed from the surveyed articles that when there is a strong desire for high anonym-ity and security, PoW has been the consensus mechanism of preference, whereas PoA is suited for closed systems where participants or nodes are known, and some degree of centralization is acceptable. From this review, it appears that permissioned blockchains are more often proposed than public networks, likely because of the complex nature of the power grid and the physical infra-structure underlying the network. For EV charging, there appears to be a preference for consortium blockchains as these networks allow for more flexible connection and disconnection of new participants in the network (i.e., EV owners coming to charge their car). Blockchain, as a digital platform, can provide solutions for applications in both the economic and physical infrastructure of the smart grid, and there is potential for providing integrated solutions that combine both. However, there remains a challenge in the fourth layer of governance, which encompasses the underlying three.

## 14.4   FUTURE OUTLOOK AND POTENTIAL

To study the future strengths, benefits, opportunities, and challenges of blockchain in the energy sector, we will review several articles that take a broader perspective of blockchain technology and consider its expected potential and barriers for implementation at a societal level. As in Section 14.3, this review is not meant to be exhaustive but intended to indicate common findings. Key strengths include transparency, immutability, and security, whereas challenges pertain to the technological maturity of blockchain and broader societal aspects, including institutional, regulatory, and governmental challenges. From these findings, a discussion of the implications for governance aspects will be provided in the final Section 14.5.

One of the most comprehensive reviews from recent years is provided in [3], which looks at over 140 research projects across various domains in the energy sector. The study cites numerous challenges and barriers, including issues with scalability, speed, security, and high development costs of blockchain, all of which can be considered related to the technological immaturity of blockchain at present. Another important challenge lies in the regulatory and legal sphere and the lack of standardization and flexibility. Opportunities in applications include P2P trading, Internet-of-Things (IoT) applications, and EV charging. Key strengths of blockchain include transparency and tamper-proof transactions and the ability to empower consumers and small generators. The work in [13] looks at the impact of blockchain, particularly in energy market transformation, via a literature review and by conducting interviews with experts. Overall, the interviewed experts recognize the large and disruptive potential for blockchain, while the largest challenge lies in laws and regulations that are not favorable for blockchain and its novel applications and the technological maturity of blockchain. As far as future opportunities are concerned, the authors consider EV charging a strong short-term application with a broad impact, and P2P trading in microgrids to be the most important long-term application. The exact nature of blockchain implementation will depend on the innovation and development of business models within these domains.

A number of technical challenges and barriers for blockchain implementation and applications are addressed in [42]. These challenges include the throughput of transactions per minute, which is relatively slow for blockchains using PoW consensus, especially Bitcoin. Furthermore, high latency means that there is more time for attackers to tamper with transactions, and another challenge is that overall block size and bandwidth are limited. Other problems relate to security threats, wasting resources in PoW due to the mining process, and user-friendliness of blockchain. The study [23] considers in more detail the use of blockchain in power systems. It considers several applications, including electrical energy trading, renewable energy certification, and DR tracing. The applications can be divided into electricity markets (i.e., trading applications) and the provision of ancillary services. Key strengths identified include transparency, immutability, efficiency, security, and confidence in blockchain transactions. Challenges include scalability, speed, lack of flexibility, user-friendliness, and energy consumption.

A case study in Japan is used in [10] to identify challenges and opportunities in different dimensions. Key potential and strengths are recognized in the domain of distributed P2P trading, where security and transparency are increased and where prosumers are empowered. Challenges are identified in different dimensions, including technological, economic, social, environmental, and institutional. Technological challenges pertain to scalability, security, speed, and related problems. Here it should be clarified that, while security is generally considered a strength of blockchain, the security must still be proven in real-world applications, which is why some studies consider it still a challenge. Key challenges in the other dimensions include the ability to compete in the market and the need for behavioral change, stakeholder management, and public acceptance. Challenges are considered most significant in the institutional domain, where regulations surrounding privacy, licensing, and overall centralized decision-making, may inhibit the development and implementation of blockchain. In the end, the authors argue that there is a strong need and opportunity for the use of living labs, sandboxes, and other experimental sites where different stakeholders can collaborate holistically and pragmatically to innovate new solutions with blockchain.

Further attention to societal issues of institutions and governance is given in [7]. The authors focus on the potential impact of blockchain-based P2P trading and LEMs on the actor configuration within the Dutch electricity sector. From this study, implications can be drawn regarding broader trends of decentralization within the electricity sector and the relative influence of centralized and decentralized actors and authorities. They conclude that while blockchain may have a significant impact on the electricity sector, it may not be as disruptive and decentralizing as is sometimes suggested. Although new actors may emerge and fulfill novel functions that hitherto did not exist, there is also the risk that incumbent actors and central authorities will resist such changes and consolidate their position. Furthermore, the authors describe how new centralized functions may emerge taken up by incumbent or novel actors.

Finally, [8] looks at the shifting role of consumers and prosumers within the electricity market and considers how prosumers can be integrated into electricity markets from a legal perspective. Fundamentally, blockchain is considered a means of creating a trust for transactions between actors. The distinction between permissionless and permissioned blockchain is considered relevant when the degree of decentralization in the network of actors is studied. This is because permissioned blockchains still rely on a selected number of nodes to be trusted. Several important policy implications are identified, including the distribution of responsibilities in a decentralized system, the need to incentivize consumer or producer investment in flexibility services, and the need to strike a balance between empowerment and protection of end-users.

## 14.5   DISCUSSION AND CONCLUSION

This chapter started with a description of the distinctive qualities and properties of blockchain networks. By providing a means of distributed data management and computing, blockchain potentially removes the need for a trusted intermediary, reducing risks of tampering, cyber-physical attacks, and centralized power misuse in ETSs. As discussed, relevant properties of blockchain networks include the openness of the network, the role of smart contracts, and the type of consensus mechanism used. These properties have been defined as a characterization framework and combined with relevant properties for ETSs, including the infrastructural layers and types of assets and actors that are included. Using these properties, we have discussed several applications, where P2P trading, DR, and EV charging are among the foremost discussed. Overall, blockchain is often considered for trading and other economic purposes, such as LEM, and for enabling EMSs and microgrids.

From the studies reviewed, it becomes apparent that there is a comprehensive agreement and recognition of the strengths and benefits of blockchain. The transparency and immutability of a blockchain ledger are unique and highly beneficial properties, and while there remain barriers surrounding security, privacy, efficiency, and scalability, these problems may be mitigated as the technology matures over time. The societal and institutional questions are more complex, however. Although blockchain is sometimes hailed as a technology that can provide full decentralization, this is only true for certain types of networks, most notably public blockchains based on PoW consensus. In the power system, however, participants on LEMs or EMSs cannot enter or leave the connected blockchain network as easily because of the complexity and integrated connections in the underlying technological infrastructure. Depending on the type of platform, a physical connection through distribution or transmission lines is often required, meaning that the identity of participants would be known. This is per the observation that such networks often use PoA or similar consensus mechanisms. These mechanisms, as well as other permissioned networks, still rely on a certain degree of centralization as the ability to serve as a validator node is only provided to a small, select number of nodes.

Besides the limited number of validator nodes, permissioned blockchains that operate on a microgrid or other fixed grid infrastructure require that the party controlling the grid infrastructure also controls the blockchain network. While blockchain technology may enable decentralization in the information layer, it is important to consider to what extent decentralization is occurring in the other domains. Control of DER, distributed assets, solar farms, and other energy generation assets in the physical layer would have to be distributed using independent control stations for increased resilience and more convenient coordination. Market mechanisms within the economic layer would have to provide a fair and transparent means of exchange that rewards actors that contribute more to the operating of the system without providing unfair advantages. At the same time, such trading mechanisms must respect the boundaries and constraints set by the underlying physical infrastructure. In the information layer, blockchain can provide the architecture for digital overhead for all processes by providing distributed computation and data registration in a decentralized manner. However, although blockchain can register information transparently and securely, the data and information from the smart grid infrastructure and market mechanisms is highly complex and may be difficult to fully grasp for non-expert users. Thus, the true benefits of blockchain technology may only be realized when the different system layers are fully integrated and decentralized to the same degree and made insightful, understandable, and transparent in the blockchain ledger.

Finally, challenges may be most significant in the fourth layer of governance, encompassing all the underlying domains. This is particularly relevant for permissioned blockchain networks, which have restricted access conditions for participants and often use consensus mechanisms that limit rights for validating blocks and transactions to a selected number of users. For such networks, the conditions for access and validation rights must still be determined through negotiation between different actors. It is likely that in some cases, there will be the main actor that has control over the blockchain network: for microgrids and EV charging stations, this may be an aggregator, community organization, energy company, grid operator, or governmental body. Besides this controlling actor, there may be many other stakeholders, and whether or not these stakeholders' needs are taken into account is beyond the configuration of the blockchain network. This question of governance and politics should be carefully considered in future studies.

## REFERENCES

1. B. Bhushan, A. Khamparia, K. M. Sagayam, S. K. Sharma, M. A. Ahad, and N. C. Debnath, "Blockchain for smart cities: A review of architectures, integration trends and future research directions," *Sustainable Cities and Society*, vol. 61, no. March, p. 102360, 2020. [Online]. Available: Doi: 10.1016/j.scs.2020.102360.
2. G. A. Motta, B. Tekinerdogan, and I. N. Athanasiadis, "Blockchain applications in the agri-food domain: The first wave," *Frontiers in Blockchain*, vol. 3, no. February, pp. 1–13, 2020.
3. M. Andoni, V. Robu, D. Flynn, S. Abram, D. Geach, D. Jenkins, P. McCallum, and A. Peacock, "Blockchain technology in the energy sector: A systematic review of challenges and opportunities," *Renewable and Sustainable Energy Reviews*, vol. 100, no. February 2018, pp. 143–174, 2019. [Online]. Available: Doi: 10.1016/j.rser.2018.10.014.
4. A. Ahl, M. Yarime, K. Tanaka, and D. Sagawa, "Review of blockchain based distributed energy: Implications for institutional development," *Renewable and Sustainable Energy Reviews*, vol. 107, no. February, pp. 200–211, 2019. [Online]. Available: Doi: 10.1016/j.rser.2019.03.002.
5. N. Z. Aitzhan and D. Svetinovic, "Security and privacy in decentralized energy trading through multi-signatures, blockchain and anonymous messaging streams," *IEEE Transactions on Dependable and Secure Computing*, vol. 15, no. 5, pp. 840–852, 2018.
6. P. Fox-Penner, *Power After Carbon: Building a Clean, Resilient Grid*. Cambridge, MA: Harvard University Press, 2020, vol. 5, no. 1.

7. M. C. Buth, A. J. Wieczorek, and G. P. Verbong, "The promise of peer-to-peer trading? The potential impact of blockchain on the actor configuration in the Dutch electricity system," *Energy Research and Social Science*, vol. 53, no. February, pp. 194–205, 2019. [Online]. Available: Doi: 10.1016/j.erss.2019.02.021.

8. L. Diestelmeier, "Changing power: Shifting the role of electricity consumers with blockchain technology – Policy implications for EU electricity law," *Energy Policy*, vol. 128, no. April 2018, pp. 189–196, 2019. [Online]. Available: Doi: 10.1016/j.enpol.2018.12.065.

9. J. L. Crespo-Vazquez, T. AlSkaif, Á. M. Gonźalez-Rueda, and M. Gibescu, "A community-based energy market design using decentralized decision-making under uncertainty," *IEEE Transactions on Smart Grid*, vol. 12, pp. 1782–1793, 2020.

10. A. Ahl, M. Yarime, M. Goto, S. S. Chopra, N. M. Kumar, K. Tanaka, and D. Sagawa, "Exploring blockchain for the energy transition: Opportunities and challenges based on a case study in Japan," *Renewable and Sustainable Energy Reviews*, vol. 117, no. October 2019, p. 109488, 2020. [Online]. Available: Doi: 10.1016/j.rser.2019.109488.

11. M. F. Zia, M. Benbouzid, E. Elbouchikhi, S. M. Muyeen, K. Techato, and J. M. Guerrero, "Microgrid transactive energy: Review, architectures, distributed ledger technologies, and market analysis," *IEEE Access*, vol. 8, pp. 19 410–19 432, 2020.

12. J. Sedlmeir, H. U. Buhl, G. Fridgen, and R. Keller, "The energy consumption of blockchain technology: beyond myth," *Business & Information Systems Engineering*, vol. 62, no. 6, pp. 599–608, 2020.

13. V. Brilliantova and T. W. Thurner, "Blockchain and the future of energy," *Technology in Society*, vol. 57, no. July 2018, pp. 38–45, 2019. [Online]. Available: Doi: 10.1016/j.techsoc.2018.11.001.

14. L. de Almeida, N. Klausmann, H. van Soest, and V. Cappelli, "Peer-to-peer trading and energy community in the electricity market-analysing the literature on law and regulation and looking ahead to future challenges," *Robert Schuman Centre for Advanced Studies Research Paper No. RSCAS*, vol. 35, 2021. https://cadmus.eui.eu/bitstream/handle/1814/70457/RSC%202021_35rev.pdf?sequence=4&isAllowed=y)

15. J. Yli-Huumo, D. Ko, S. Park, and K. Smolander, "Where is current research on blockchain technology? - A systematic review," *PLoS One*, vol. 10, no. 11, p. e0163477, 2016.

16. Y. Wang, Z. Su, Q. Xu, T. Yang, and N. Zhang, "A novel charging scheme for electric vehicles with smart communities in vehicular networks," *IEEE Transactions on Vehicular Technology*, vol. 68, no. 9, pp. 8487–8501, 2019.

17. Z. Zheng, S. Xie, H. Dai, X. Chen, and H. Wang, "An overview of blockchain technology: Architecture, consensus, and future trends," *IEEE International Congress on Big Data (BigData Congress)*, Honolulu, pp. 557–564, 2017.

18. Z. Li, J. Kang, R. Yu, D. Ye, Q. Deng, and Y. Zhang, "Consortium blockchain for secure energy trading in industrial internet of things," *IEEE Transactions on Industrial Informatics*, vol. 14, no. 8, pp. 3690–3700, 2018.

19. X. Xu, I. Weber, M. Staples, L. Zhu, J. Bosch, L. Bass, C. Pautasso, and P. Rimba, "A taxonomy of blockchain-based systems for architecture design," *IEEE International Conference on Software Architecture*, Gothenburg, 2017.

20. G. Wood, "Ethereum: a secure decentralised generalised transaction ledger," Ethereum Project Yellow Paper, pp. 1–32, 2014. https://gavwood.com/paper.pdf

21. Y. Wang, W. Saad, Z. Han, H. V. Poor, and T. Başar, "A game-theoretic approach to energy trading in the smart grid," *IEEE Transactions on Smart Grid*, vol. 5, no. 3, pp. 1439–1450, 2014.

22. C. Stoll, L. Klaasen, and U. Gallersdörfer, "The carbon footprint of bitcoin," *Joule*, vol. 3, no. 7, pp. 1647–1661, 2019.

23. M. L. Di Silvestre, P. Gallo, J. M. Guerrero, R. Musca, E. Riva Sanseverino, G. Sciumè, J. C. Vásquez, and G. Zizzo, "Blockchain for power systems: Current trends and future applications," *Renewable and Sustainable Energy Reviews*, vol. 119, no. January 2019, p. 109585, 2020.

24. E. Mengelkamp, B. Notheisen, C. Beer, D. Dauer, and C. Weinhardt, "A blockchain-based smart grid: Towards sustainable local energy markets," *Computer Science - Research and Development*, vol. 33, no. 1–2, pp. 207–214, 2018.

25. A. Mosavi, M. Salimi, S. F. Ardabili, T. Rabczuk, S. Shamshirband, and A. R. Varkonyi-Koczy, "State of the art of machine learning models in energy systems, a systematic review," *Energies*, vol. 12, no. 7, p. 1301, 2019.

26. T. Alskaif, J. L. Crespo-Vazquez, M. Sekuloski, G. van Leeuwen, and J. P. Catalao, "Blockchain-based fully peer-to-peer energy trading strategies for residential energy systems," *IEEE Transactions on Industrial Informatics*, vol. 18, pp. 231–241, 2021.

27. T. Baroche, F. Moret, and P. Pinson, "Prosumer markets: A unified formulation," In *IEEE PowerTech Conference*, Milan, pp. 1–6, 2019.

28. C. Pop, T. Cioara, M. Antal, I. Anghel, I. Salomie, and M. Bertoncini, "Blockchain based decentralized management of demand response programs in smart energy grids," *Sensors (Switzerland)*, vol. 18, no. 1, p. 162, 2018.

29. T. AlSkaif, B. Holthuizen, W. Schram, I. Lampropoulos, and W. Van Sark, "A blockchain-based configuration for balancing the electricity grid with distributed assets," *World Electric Vehicle Journal*, vol. 11, no. 4, p. 62, 2020.

30. J. Kang, R. Yu, X. Huang, S. Maharjan, Y. Zhang, and E. Hossain, "Enabling localized peer-to-peer electricity trading among plug-in hybrid electric vehicles using consortium blockchains," *IEEE Transactions on Industrial Informatics*, vol. 13, no. 6, pp. 3154–3164, 2017.

31. G. van Leeuwen, T. AlSkaif, M. Gibescu, and W. van Sark, "An integrated blockchain-based energy management platform with bilateral trading for microgrid communities," *Applied Energy*, vol. 263, no. February, p. 114613, 2020. [Online]. Available: Doi: 10.1016/j.apenergy.2020.114613.

32. Y. Li, W. Yang, P. He, C. Chen, and X. Wang, "Design and management of a distributed hybrid energy system through smart contract and blockchain," *Applied Energy*, vol. 248, no. March, pp. 390–405, 2019. [Online]. Available: Doi: 10.1016/j.apenergy.2019.04.132.

33. F. Luo, Z. Y. Dong, G. Liang, J. Murata, and Z. Xu, "A distributed electricity trading system in active distribution networks based on multi-agent coalition and blockchain," *IEEE Transactions on Power Systems*, vol. 34, no. 5, pp. 4097–4108, 2019.

34. Z. Li, S. Bahramirad, A. Paaso, M. Yan, and M. Shahidehpour, "Blockchain for decentralized transactive energy management system in networked microgrids," *Electricity Journal*, vol. 32, no. 4, pp. 58–72, 2019. [Online]. Available: Doi: 10.1016/j.tej.2019.03.008.

35. K. Gai, Y. Wu, L. Zhu, M. Qiu, and M. Shen, "Privacy-preserving energy trading using consortium blockchain in smart grid," *IEEE Transactions on Industrial Informatics*, vol. 15, no. 6, pp. 3548–3558, 2019.

36. Y. Jiang, K. Zhou, X. Lu, and S. Yang, "Electricity trading pricing among prosumers with game theory-based model in energy blockchain environment," *Applied Energy*, vol. 271, no. January, p. 115239, 2020. [Online]. Available: Doi: 10.1016/j.apenergy.2020.115239.

37. W. Zhao, J. Lv, X. Yao, J. Zhao, Z. Jin, Y. Qiang, Z. Che, and C. Wei, "Consortium blockchain-based microgrid market transaction research," *Energies*, vol. 12, no. 20, 2019.

38. E. Munsing, J. Mather, and S. Moura, "Blockchains for decentralized optimization of energy resources in microgrid networks," In *1st Annual IEEE Conference on Control Technology and Applications, CCTA 2017*, Hawaii, vol. 2017, pp. 2164–2171, January, 2017.

39. Z. Fu, P. Dong, and Y. Ju, "An intelligent electric vehicle charging system for new energy companies based on consortium blockchain," *Journal of Cleaner Production*, vol. 261, p. 121219, 2020. [Online]. Available: Doi: 10.1016/j.jclepro.2020.121219.

40. H. Liu, Y. Zhang, and T. Yang, "Blockchain-enabled security in electric vehicles cloud and edge computing," *IEEE Network*, vol. 32, p. 78–83, 2018.

41. Z. Su, Y. Wang, Q. Xu, M. Fei, Y.-C. Tian, and N. Zhang, "A secure charging scheme for electric vehicles with smart communities in energy blockchain," *IEEE Internet of Things Journal*, vol. 6, no. 3, pp. 4601–4613, 2019.

42. F. Alam Khan, M. Asif, A. Ahmad, M. Alharbi, and H. Aljuaid, "Blockchain technology, improvement suggestions, security challenges on smart grid and its application in healthcare for sustainable development," *Sustainable Cities and Society*, vol. 55, no. January, p. 102018, 2020. [Online]. Available: Doi: 10.1016/j.scs.2020.102018.

# 15 Machine Learning Applications in the Petroleum Industry

*Ahmed Abdulhamid Mahmoud, Salaheldin Elkatatny, and Abdulazeez Abdulraheem*
King Fahd University of Petroleum & Minerals

## CONTENTS

## 15.1 INTRODUCTION

During the course of the last century, with the dominance of the "difficult-to-recover" hydrocarbon resources, the need for new operational approaches and business models for hydrocarbons exploration and production has increased to ensure the highest profitability from oil and gas production [1]. This is true for brownfields and greenfields, that is, well-developed and newly discovered fields, respectively.

Although most of the brownfields are large in their geometrical size and good in terms of their storage and transport properties of permeability and porosity, a small amount of the additional oil is recoverable by conventional and relatively inexpensive waterflooding. These fields produce with high water cut ratio [2]. Oil and gas companies usually spend extra money to perform different operations to maintain the desired production levels. These operations include: (i) well treatment such as matrix stimulation or hydraulic fracturing, (ii) extra drilling, and (iii) field-scale-enhanced oil recovery such as chemical or steam injection. In many cases, these operations are not enough for the invested money to be paid off; therefore, this leaves the well-developed fields in a slow process of dying.

On the other hand, development and maintenance of the greenfields is also difficult for many reasons: (i) low mobility of the oil through the reservoir rocks caused by the low permeability of the reservoir rocks or high oil viscosity, (ii) complexity of these reservoirs in terms of their geometry (e.g., the presence of the oil-saturated layer within faults and cracks), and (iii) presence of these reservoirs in places of harsh environmental conditions [3,4]. Due to these reasons, expensive technologies are required to develop these reservoirs and the profitability of producing oil from such reservoirs is questionable.

DOI: 10.1201/9781003315353-17

The procedures for handling the high uncertainties associated with high-value investments in the oil and gas upstream are performed manually by the decision makers and are based on expert knowledge. These procedures are time-consuming and their accuracy is dependent on the experience of the decision-maker. Consequently, it is highly desirable to speed up the hydrocarbon reservoir assessment processes, increase their efficiency, and make them more expert-independent.

The energy industry is flooded with huge chunks of data coming from different resources. Currently, the fourth generation of the industrial revolution is driving all industries toward making use of the latest technological developments for data digitization and the Internet of Things. Machine learning tools are considered the core driving force for the fourth generation of the industrial revolution. The oil and gas industry has also been undergoing digital transformation with the help of big data science and machine learning to provide data-driven solutions to enhance operations' performance, speed up the processes, boost efficiency, improve security, and lower the costs. All these have led to significant changes and transitions in the oil and gas industry.

This chapter contains four sections, viz., an introduction section, and three sections on various application areas. It covers the applications of machine learning techniques in different areas related to the petroleum industry such as the prediction and optimization of drillability of hydrocarbon wells, generating synthetic petrophysical logs for drilled wells, and optimization and forecasting of the oil and gas production. These cases represent important ones among many areas where machine learning is employed to enhance the efficiency of overall operations.

## 15.2  MACHINE LEARNING AND ITS APPLICATIONS IN THE UPSTREAM PETROLEUM INDUSTRY

Different machine learning models such as artificial neural networks (ANN), support vector machine (SVM), functional neural networks (FNN), and Random forests (RF) were applied extensively in different aspects of the energy sector.

ANN is the most commonly used machine learning tool in the energy sector. It was developed to work as a computing system that imitates the behaviors of the biological systems. ANN had been successfully applied in the energy sector for regression [5,6] and classification purposes [7]. Although it is available in different structures, the multilayered perceptron (MLP) that consists of a single input layer, one or many learning layers, and a single output layer is considered as the simplest ANN structure and is widely applied in the energy industry [8,9].

Another commonly used machine learning tool is the SVM which was developed earlier by Cortes and Vapnik [10], it was originally developed as a classification tool which was later been applied for regression problems, and it becomes suitable for solving nonlinear problems. In nature, the SVR model is comparable to the neural networks, and when the sigmoid kernel function is used with the SVR model, it becomes similar to the MLP model that consists of learning layers. Nowadays, SVR is extensively used in the energy sector for both the classification [11,12] and regression purposes [13–15].

FNN model was also used commonly in the energy sector [16,17]. Compared to the ANN, the neuron's function in the FNN is learned from the existing data, which means they are not constant. Therefore, the weights related to links are not needed, because the neuron functions include the effect of weights [18]. FNN contains an input layer, an output layer, and layers of computing units that are related to each other. In FNN, there are different arguments in neural functions instead of one argument such as in ANN [19].

RF is another machine learning tool that was proposed in 2001 by Breiman [20], and it consists of several decision trees; these work as a committee in determining overall ensembled model response. Recently, RF was also widely used in the energy sector and it showed high ability in classification [21,22] and optimization of different properties in this sector [23].

Machine learning models have been used widely to evaluate the different parameters and properties related to different aspects of upstream petroleum engineering such as reservoir characterization through improving the seismic image reconstruction [24], lithological facies prediction [25], prediction of the total organic carbon needed to evaluate the unconventional hydrocarbon resources [26–28], prediction of the irreducible water saturation [29], and reserve estimation [30,31].

For the borehole drilling process, the machine learning tools have been used to optimize drillability while drilling horizontally or vertically through different types of the formation [32–35], optimize drilling parameters to improve the drilling performance and efficiency [36], evaluate pore and fracture pressure [37–39], predict the equivalent circulation density [40,41], optimize drillbit selection [42], detect downhole torque in real time while drilling [43], and predict formation tops [7,44].

Several rock properties have also been successfully predicted using different machine learning techniques. These properties include porosity [45], permeability [46,47], static Young's modulus [17,48], static Poisson's ratio [49–51], and unconfined compressive strength [52].

Machine learning tools also showed high accuracy in estimating the drilling fluid rheological properties such as plastic viscosity and yield point for oil-based drilling fluid [53] and for different types of water-based drilling fluids such as spud mud [54,55], $CaCl_2$ brine-based mud [56], KCl brine-based mud [57], and invert emulsion-based mud [58]. Hydrocarbon fluid properties such as formation volume factor [59–61], bubble point pressure [62–64], and gas solubility [64] were also predicted with high accuracy using machine learning tools.

During hydrocarbon field development stage, machine learning tools have shown high efficiency for oil and gas production rate history matching [65], production forecasting [66,67], and prediction of condensate-to-gas ratio for retrograde gas condensate reservoirs [68]. Waterflooding performance [69], low-salinity flooding [70], chemical flooding [71], and $CO_2$ flooding [72] during the production maintenance stage have also been addressed with high accuracy in relevant predictions using machine learning methods.

## 15.3 DATA PREPARATION AND PREPROCESSING FOR MODELS TRAINING

As discussed earlier, the petroleum industry is flooded with a huge amount of data. Before feeding these data into the machine learning tools for predicting a required property, they must be preprocessed. The purpose of this stage is to achieve three roles: to remove all outliers, to eliminate all unrealistic values, and to clean the data by removing the noise of the input and output parameters.

The outliers are defined by many authors as all values of input parameters that are used for training the machine learning model or output parameter to be predicted by the model that falls outside a range of $\pm 3.0$ standard deviation [73].

The reports generated by the sensors involved in different operations related to the petroleum industry include many numerical values of several measured parameters. In some situations, these values may be unrealistic such as negative values, zeros, and constant large values. All rows of the data including these unrealistic values must be excluded from the input data to ensure that the machine learning model is trained using a valid database.

Figure 15.1 shows how the data are preprocessed for the removal of unrealistic values; the data of this figure are an examples of predicting the dynamic Young's modulus ($E_{dyn}$) from the surface-measurable drilling parameters such as the drillstring rotation (DSR), weight on bit (WOB), rate of penetration (ROP), torque, standpipe pressure (SPP), and the drilling fluid flowrate (Q). As shown in the figure, the data in rows 2 and 3 include many zero values indicating that these were not recorded properly during the drilling process. They are unrealistic or unrelated to the drilling phase because while drilling WOB and torque will always be present for a given ROP.

On the other hand, the data in row 700 show many 999 values, indicating a fault in the sensors while recording them. These values are not logical because an ROP of 999 ft/h is not physically possible. Therefore, the data in rows 2, 3, and 700 must be removed during the data preprocessing stage and they should not be considered for training the machine learning models.

### (a) Before Unrealistic Values Removal

| Index | Inputs | | | | | | Output |
| | WOB (klbf) | TORQUE (kft.lbf) | SPP (psi) | DSR (1/min) | ROP (ft/h) | Flow Rate (gpm) | $E_{dyn}$ (GPa) |
|---|---|---|---|---|---|---|---|
| 1 | 19.8 | 8.47 | 2394 | 108 | 35.2 | 798 | 30.1 |
| 2 | 0.0 | 0.00 | 2442 | 113 | 0.0 | 0 | 29.8 |
| 3 | -0.1 | 0.00 | 2464 | 115 | 0.0 | 0 | 29.7 |
| 4 | 20.1 | 8.15 | 2494 | 118 | 35.4 | 814 | 29.8 |
| 5 | 20.2 | 8.12 | 2501 | 119 | 35.3 | 816 | 29.8 |
| 6 | 20.3 | 8.07 | 2517 | 121 | 35.2 | 818 | 29.7 |
| ⋮ | ⋮ | ⋮ | ⋮ | ⋮ | ⋮ | ⋮ | ⋮ |
| 700 | 20.4 | 999 | 999 | 121 | 35.2 | 819 | 999 |
| 701 | 20.8 | 8.12 | 2538 | 126 | 36.4 | 822 | 29.6 |
| 702 | 20.9 | 8.13 | 2539 | 126 | 36.7 | 822 | 29.7 |
| 703 | 21.0 | 8.16 | 2559 | 128 | 37.6 | 823 | 29.8 |
| 704 | 21.3 | 8.19 | 2589 | 131 | 39.2 | 829 | 29.9 |
| 705 | 21.4 | 8.20 | 2600 | 132 | 39.8 | 831 | 29.9 |
| ⋮ | ⋮ | ⋮ | ⋮ | ⋮ | ⋮ | ⋮ | ⋮ |
| 998 | 22.1 | 8.24 | 2654 | 136 | 43.2 | 845 | 30.4 |
| 999 | 22.5 | 8.27 | 2642 | 134 | 43.6 | 848 | 29.9 |
| 1000 | 22.6 | 8.28 | 2640 | 134 | 44.0 | 848 | 29.7 |

### (b) After Unrealistic Values Removal

| Index | Inputs | | | | | | Output |
| | WOB (klbf) | TORQUE (kft.lbf) | SPP (psi) | DSR (1/min) | ROP (ft/h) | Flow Rate (gpm) | $E_{dyn}$ (GPa) |
|---|---|---|---|---|---|---|---|
| 1 | 19.8 | 8.47 | 2394 | 108 | 35.2 | 798 | 30.1 |
| 4 | 20.1 | 8.15 | 2494 | 118 | 35.4 | 814 | 29.8 |
| 5 | 20.2 | 8.12 | 2501 | 119 | 35.3 | 816 | 29.8 |
| 6 | 20.3 | 8.07 | 2517 | 121 | 35.2 | 818 | 29.7 |
| ⋮ | ⋮ | ⋮ | ⋮ | ⋮ | ⋮ | ⋮ | ⋮ |
| 701 | 20.8 | 8.12 | 2538 | 126 | 36.4 | 822 | 29.6 |
| 702 | 20.9 | 8.13 | 2539 | 126 | 36.7 | 822 | 29.7 |
| 703 | 21.0 | 8.16 | 2559 | 128 | 37.6 | 823 | 29.8 |
| 704 | 21.3 | 8.19 | 2589 | 131 | 39.2 | 829 | 29.9 |
| 705 | 21.4 | 8.20 | 2600 | 132 | 39.8 | 831 | 29.9 |
| ⋮ | ⋮ | ⋮ | ⋮ | ⋮ | ⋮ | ⋮ | ⋮ |
| 998 | 22.1 | 8.24 | 2654 | 136 | 43.2 | 845 | 30.4 |
| 999 | 22.5 | 8.27 | 2642 | 134 | 43.6 | 848 | 29.9 |
| 1000 | 22.6 | 8.28 | 2640 | 134 | 44.0 | 848 | 29.7 |

**FIGURE 15.1**  Preprocessing the data for unrealistic values removal. (a) The data before unrealistic values removal and (b) the data after unrealistic values removal.

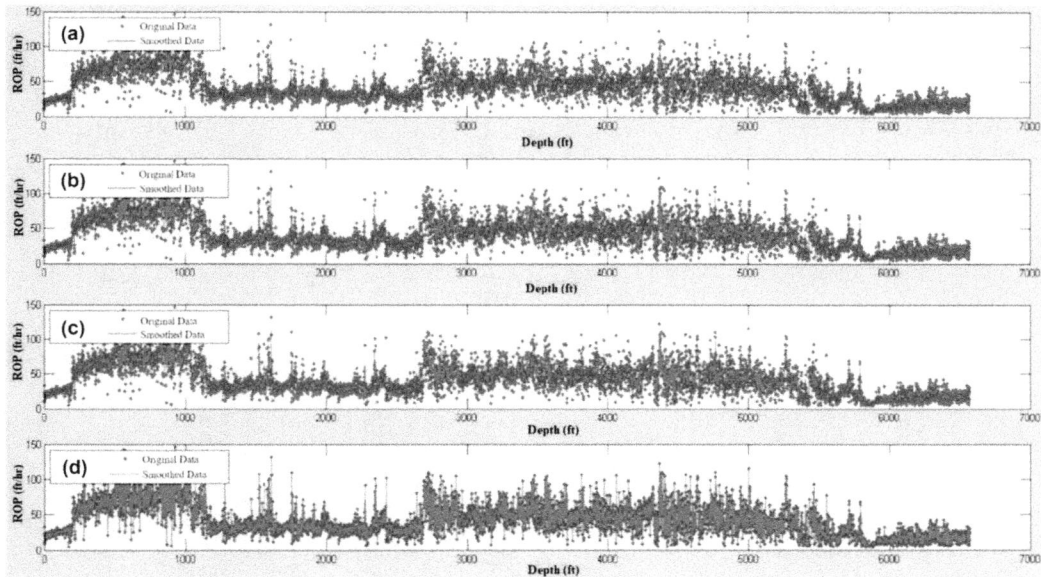

**FIGURE 15.2**   Comparison of the performance of MAT with spans of (a) 5, (b) 4, (c) 3, and (d) 2, in reducing the noise in the ROP for the training data. Light gray dots represent the original data and the dark gray curves denote the filtered data. (From Ref. [44]. With permission.)

The next stage of the data preprocessing is to remove the noise in the training data. The moving average technique (MAT) is commonly considered for data smoothing in the petroleum industry. Figure 15.2 compares the actual (light gray points) and the smoothed (dark gray curve) data for the ROP for different MAT spans of 5, 4, 3, and 2. The output showed that the use of MAT with a span of 5 (Figure 15.2a) is the best to smooth the ROP without significantly changing its structure compared to the use of 4, 3, and 2 spans (Figure 15.2b–d).

## 15.4   PREDICTION OF THE BOREHOLE DRILLABILITY

Optimizing borehole drillability is a very critical issue because of the high cost of the drilling operations. This section provides a summary of the most common applications of different machine learning models in estimating and optimizing the drillability of the hydrocarbon wells through the prediction of the rate of penetration (ROP). ROP is the rate at which the drillbit drills through the formation rocks and is measured as the number of feet drilled per hour of the drilling operation.

Bilgesu et al. [74] developed two models based on the artificial neural networks (ANN) to evaluate the ROP in nine formations. They optimized their first model to predict the ROP from the type of the formation, WOB, footage, DSR, drilled formation, drillbit features, mud circulation, and gross hours of drilling (GHD). For their second model, they did not include the bit features in the input dataset. As reported by them, both optimized ANN models accurately estimated the ROP.

Two ANN-based models for ROP estimation were suggested by Amar and Ibrahim [75] based on formation depth, pore pressure gradient, drillbit's tooth wear, WOB, equivalent circulation density (ECD), DSR, and Reynolds number function. The authors reported that their ANN-based models were more accurate than available correlations.

An ANN model was also optimized by Bataee and Mohseni [76] to estimate the ROP based on the bit diameter, depth, WOB, DSR, and mud weight. The data used in this work were collected from the Shadegan oil field. From this study, the authors were able to describe the effect of each input parameter on ROP prediction.

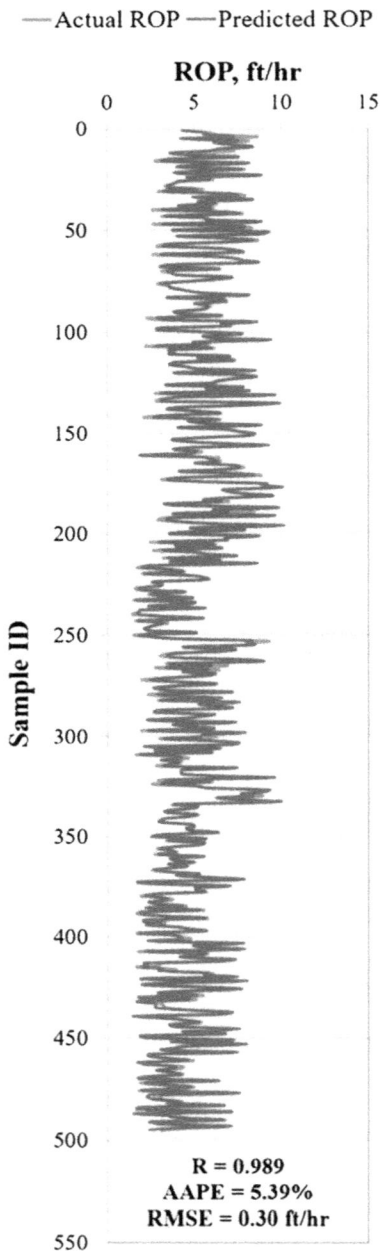

**FIGURE 15.3**  Comparison of the actual and ANN-based ROP (From Ref. [33]. With permission.)

In another study, Elkatatny [77] trained an ANN model on more than 3,333 data points of the drilling fluid flowrate, DSR, WOB, torque, SPP, the formation compressive strength (UCS), drilling fluid density, and plastic viscosity gathered from two wells to assess the ROP. The trained model predictions were validated using a third well yielding an AAPE of 4.0% only. After that, Al-AbdulJabbar et al. [78] developed another ANN model considering all the inputs considered by Elkatatny [77] except the drilling fluid plastic viscosity and density to allow for real-time estimation of the ROP while drilling. After training, this last model predicted the ROP while drilling with a high correlation coefficient (R) of greater than 0.94 for two different wells.

Ahmed et al. [79] developed a support vector regression (SVR) model to determine the ROP from ten parameters representing the drilling parameters as well as mud properties. This model assessed the ROP with an AAPE of only 2.83%. Al-AbdulJabbar et al. [32] optimized the ANN model design parameter to maximize its predictability for the ROP while horizontally drilling through carbonate formations by applying the algorithm of the self-adaptive differential evolution. This model was developed to assess the ROP from the formation petrophysical data coupled with the drilling parameters, and it assessed the ROP with an R of 0.96.

In another study, Al-AbdulJabbar et al. [33] optimized the ANN model for real-time estimation of the ROP for drilling a horizontal well in a natural gas-bearing sandstone formation. To allow for ROP prediction in real time, the ANN model was trained on real-time measurable parameters of Q, DSR, torque, WOB, and SSP. The optimized model was converted into empirical correlation which evaluated the ROP in real time with $R$ of 0.954, AAPE of 8.85%, and RMSE of 0.44 ft/h for the validation data.

An example of ROP prediction using an optimized ANN model is shown in Figure 15.3. This ROP profile was generated by Al-AbdulJabbar et al. [33] showing the possibility of real-time prediction of ROP based on the drilling data.

Recently, Alali et al. [80] proposed a two-phase-integrated and data-driven ROP optimization system. The authors developed a heat-map function for identifying the optimal ROP based on Q, DSR, and WOB. They also extracted an equation from their optimized ANN model for predicting the ROP in real time. However, they did not consider most of the parameters influencing the ROP such as the drilling torque and the SPP.

Table 15.1 summarizes some of the optimized machine learning models for ROP estimation. This table lists the optimized models, the inputs considered for training these models, the field or

## TABLE 15.1
## Summary of Some of the Optimized Machine Learning Models for ROP Prediction

| Authors | Inputs | Machine Learning Models | Field or Formation | Remarks |
|---|---|---|---|---|
| Bilgesu et al. [74] | Type of the formation, WOB, footage, DSR, drilled formation, drillbit's type, diameter, tooth, and bearing wear, mud circulation, and GHD | ANN | - | - |
| Amar and Ibrahim [75] | Formation depth and pore pressure gradient, drillbit's tooth wear, WOB, equivalent circulation density (ECD), DSR, and Reynolds number function | ANN | - | - |
| Bataee and Mohseni [76] | Bit diameter, depth, WOB, DSR, and mud weight | ANN | Shadegan oil field | - |
| Elkatatny [77] | Q, DSR, WOB, torque, SPP, UCS, and the drilling fluid density and plastic viscosity | ANN | - | AAPE of 4.0% |
| Al-AbdulJabbar et al. [78] | Q, DSR, WOB, torque, SPP, and UCS | ANN | | $R$ is greater than 0.94 |
| Ahmed et al. [79] | Ten parameters combine the drilling parameters and mud properties | SVR | - | AAPE of only 2.83% |
| Al-AbdulJabbar et al. [32] | Combination of the petrophysical data with the drilling parameters | ANN | Horizontal drilling through carbonate formations | $R$ of 0.96 |
| Al-AbdulJabbar et al. [33] | Q, DSR, torque, WOB, and SSP | ANN | Horizontal drilling of natural gas-bearing sandstone formation | $R$ of 0.954, AAPE of 8.85%, and RMSE of 0.44 ft/h |
| Alali et al. [80] | Q, DSR, and WOB | ANN | - | - |

formation from which the training data were obtained, and remarks on the accuracy of the optimized models.

## 15.5  ESTIMATION OF PETROPHYSICAL PROPERTIES THROUGH MACHINE LEARNING APPLICATION

Reservoir petrophysical properties such as porosity and permeability are very important for reserve calculation (i.e., porosity) and fluid movability and recovery prediction (i.e., permeability). Therefore, accurate estimation of these parameters is critical to make the development decision for newly discovered reservoirs and maintenance plans for currently producing reservoirs.

### 15.5.1  PREDICTION OF THE RESERVOIR POROSITY

Reservoir porosity represents the percentage of the void spaces within the rock. Prediction of poros-
ity could be through laboratory measurement or by using well-log data. Porosity prediction through laboratory measurement is costly and time-consuming, while evaluation of the porosity based on well-log data is associated with many uncertainties and requires several corrections. Recently, several machine learning models were optimized to evaluate the reservoir porosity.

Anifowose and Abdulraheem [81] used two-hybrid intelligent models to estimate the reservoir porosity as a function of the grain volume, grain density, and interval top. The two-hybrid models are combining the use of the ANFIS, SVR, and FN. The use of the hybrid systems did not considerably improve the porosity estimation compared to the use of a single model, and this is because of the use of inputs that are not highly related to the porosity such as the interval top.

Anifowose et al. [82] compared the performance of the ANN and SVR in estimating the porosity of carbonate formation from the same inputs considered by Anifowose and Abdulraheem [81]. Both ANN and SVR showed low accurate results in predicting the reservoir porosity with RMSE in the range of 2%–14% and R between 0.50 and 0.95 for all training, testing, and validation data. This performance is also attributed to the use of the interval top as input.

In 2018, Elkatatny et al. [83] compared the performance of the ANN, SVR, and ANFIS models in predicting the carbonate reservoir porosity. The authors considered the use of well-log data of the RHOB, neutron porosity, and sonic travel time (DT) to train the machine learning models. An empirical correlation for porosity estimation was derived from the optimized ANN model and its associated weights and biases. The porosity was predicted by the ANN-based empirical equation, ANFIS, and SVR models with $R^2$'s of 0.97, 0.97, and 0.93, respectively, for the testing data.

Al-AbdulJabbar et al. [84] optimized the ANN model for estimating the porosity of carbonate reservoir as a function of the real-time measurable drilling data of ROP, Q, DSR, SPP, torque, and WOB. The optimized ANN model enabled real-time prediction of the reservoir porosity with high accuracy, and the porosity of the testing data was predicted with an R of 0.907.

In another study, Kanfar et al. [85] proposed an estimation of the porosity of sandstone and shale formation by combining the inception-based convolutional neural network (IBCNN) combined with a temporal convolutional network (TCN). For training the intelligent models, the authors used mechanical-specific energy (MSE) in addition to the drilling data used earlier by Al-AbdulJabbar et al. [84]. The models predicted the porosity with moderate accuracy (R = 0.6).

Recently, Gamal et al. [86] evaluated the possibility of optimizing Random forest (RF) and decision tree (DT) models for real-time prediction of the reservoir porosity while drilling through com-
plex reservoir lithology of carbonate, sandstone, and shale formations. The models were optimized using the drilling parameters such as WOB, torque, SPP, DSR, ROP, and Q. The optimized RF and DT models predicted the porosity for the validation data through three formations (i.e., carbonate, sandstone, and shale) accurately with Rs of 0.92 and 0.88, and AAPEs of 6.50% and 8.58%, respectively. Figure 15.4 shows the continuous porosity profile predicted by Gamal et al. [86] along with the whole drilled formation in real time during the drilling process.

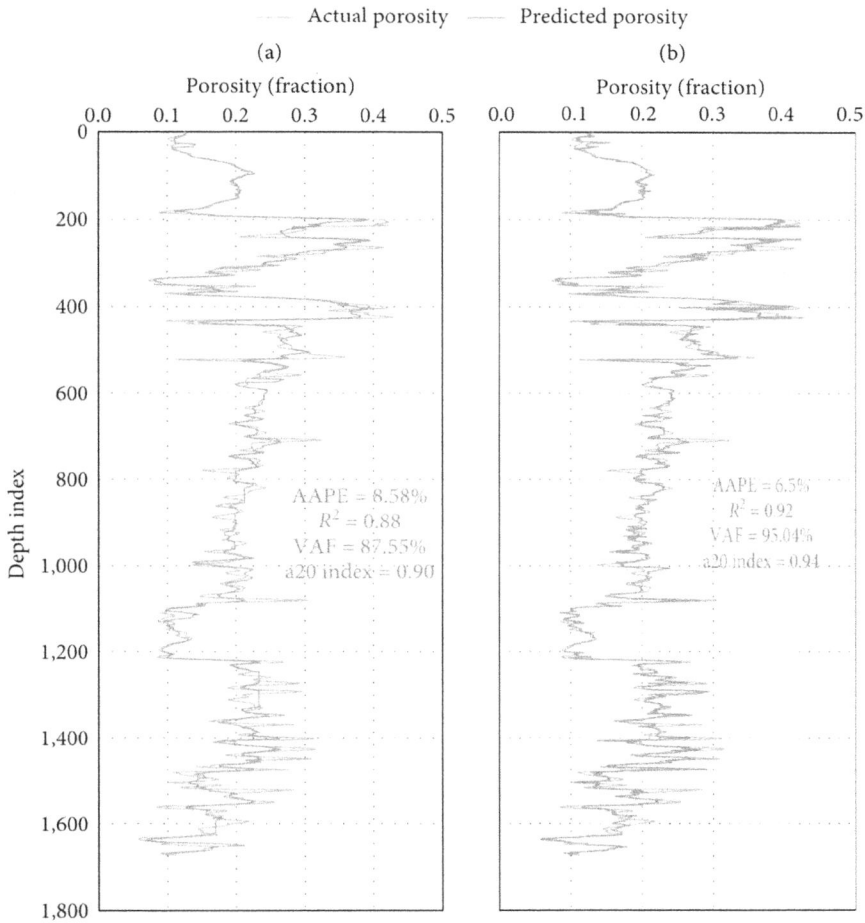

**FIGURE 15.4**  Porosity change as a function of depth. (a) ANN model and (b) RF model (From Ref. [86]. With permission.)

Table 15.2 summarizes some of the optimized machine learning models for reservoir porosity prediction along with the inputs considered for training these models, the field or formation from which the training data were obtained, and remarks on the accuracy of the optimized models.

### 15.5.2  PREDICTION OF RESERVOIR PERMEABILITY

Reservoir permeability is a rock petrophysical property that measures the ability of the formation rocks to transmit fluids. Accurate prediction of reservoir permeability is necessary since this property determines the mobility of the hydrocarbons through reservoir rocks, and hence, the ability to produce hydrocarbons after drilling and completing the drilled wells. Estimation of permeability in the laboratory from retrieved core samples is accurate but costly and time-consuming. Relative permeability is a property that determines how the rock will enable the flow of one fluid in case of the presence of more than a fluid inside the pore spaces.

In 2007, Basbug et al. [87] optimized the ANN model for evaluating the permeability of carbonate formations based on predicted porosity, irreducible water saturation, and specific surface area. The authors reported that the optimized ANN predicted carbonate formation permeability with satisfactory accuracy.

**TABLE 15.2**

**Summary of Some of the Optimized Machine Learning Models for Reservoir Porosity Prediction**

| Authors | Inputs | Machine Learning Models | Field or Formation | Remarks |
|---|---|---|---|---|
| Anifowose and Abdulraheem [81] | Grain volume, grain density, and interval top | Two-hybrid intelligent models combining ANFIS, SVR, and FN | - | Hybrid systems did not improve the porosity estimation because of using interval top as input |
| Anifowose et al. [82] | Grain volume, grain density, and interval top | ANN and SVR | - | ANN and SVR were low accurate because of using interval top as input |
| Elkatatny et al. [83] | RHOB, neutron porosity, and $DT_P$ | ANN, SVR, and ANFIS | Carbonate reservoir | $R^2$s of 0.97, 0.97, and 0.93 for ANN-based empirical equation, ANFIS, and SVR, respectively |
| Al-AbdulJabbar et al. [84] | ROP, Q, DSR, SPP, torque, and WOB | ANN | Carbonate reservoir | R of 0.907 |
| Kanfar et al. [85] | ROP, Q, DSR, SPP, torque, WOB, and MSE | Combining IBCNN and TCN | Sandstone and shale formation | R = 0.6 |
| Gamal et al. [86] | WOB, torque, SPP, DSR, ROP, and Q | RF and DT | Carbonate, sandstone, and shale formations | Rs of 0.92 and 0.88, and AAPEs of 6.50% and 8.58% for RF and DT, respectively |

Later, Al-Fattah and Al-Naim [88] used genetic algorithms to optimize ANN model design parameters for estimation of oil and water relative permeability for a giant carbonate Saudi reservoir. The author developed two ANN models: one for estimation of the oil relative permeability and another one for prediction of the water relative permeability. The ANN models were trained using formation porosity, wettability, residual oil saturation, initial water saturation, and well location. For the testing data, the ANN models showed high accuracy and predicted oil and water relative permeability with Rs of 0.989 and 0.956, respectively.

Anifowose and Abdulraheem [81] suggested the use of two-hybrid intelligent models to estimate reservoir permeability based on gamma-ray log, density log, porosity log, neutron porosity log, water saturation, deep resistivity log, microspherically focused log, and caliper log. The two-hybrid models are combining the use of ANFIS, SVR, and FN. The use of hybrid systems did not considerably improve the permeability estimation compared to the use of a single model, and this is because of the use of inputs which are not highly related to the permeability such as the caliper log.

In another study, Anifowose et al. [82] compared the performance of SVR with that of the ANN model in estimating carbonate reservoir permeability from the same inputs considered by Anifowose and Abdulraheem [81]. Both ANN and SVR showed low accurate results in predicting the reservoir permeability with RMSE of 0.6–1.0 mD and R between 0.60 and 0.83 for all training, testing, and validation data. This is also attributed to the use of the caliper log as input.

To improve permeability prediction, Anifowose et al. [89] suggested combining the use of the seismic data in addition to the well-log data as inputs to learn the ANN, SVR, and ANFIS model. Six well-log data of neutron porosity, porosity log, caliper log, gamma rays, water saturation, and density log, in addition to five seismic attributes of instantaneous frequency, half energy, rms amplitude, arc length, and instantaneous phase, were considered for training the machine learning models. The input data were collected from a giant carbonate reservoir. The results showed that combining well-log and seismic data improved predictability of the reservoir permeability.

**TABLE 15.3**

**Comparison of Some of the Optimized Machine Learning Models for Permeability Prediction**

| Authors | Inputs | Machine Learning Models | Field or Formation | Remarks |
|---|---|---|---|---|
| Basbug et al. [87] | Porosity, irreducible water saturation, and specific surface area | ANN | Carbonate formations | ANN predicted the permeability with satisfactory accuracy |
| Al-Fattah and Al-Naim [88] | Formation porosity, wettability, residual oil saturation, initial water saturation, and well location | ANN | Giant carbonate Saudi reservoir | $R$s of 0.989 and 0.956 for the predicted oil and water relative permeability, respectively |
| Anifowose and Abdulraheem [81] | Gamma-ray log, density log, porosity log, neutron porosity log, water saturation, deep resistivity log, microspherically focused log, and caliper log | Two-hybrid intelligent models combining ANFIS, SVR, and FN | - | Hybrid systems did not improve the permeability estimation because of using interval top as input |
| Anifowose et al. [82] | | ANN and SVR | - | RMSE of 0.6–1.0 mD and R between 0.60 and 0.83 |
| Anifowose et al. [89] | Neutron porosity, porosity log, caliper log, gamma-ray, water saturation, density log, instantaneous frequency, half energy, rms amplitude, arc length, and instantaneous phase | ANN, SVR, and ANFIS | Giant carbonate reservoir | Combining the well-log and seismic data improved predictability of the reservoir permeability |
| Nooruddin et al. [90] | porosity and mercury injection capillary pressure data of displacement pressure, Winland parameter, Swanson parameter, Pittman parameter, Dastidar parameter, and Purcell integration | ANN | - | $R$ of 0.98 |

Later, Nooruddin et al. [90] optimized the ANN model to estimate the reservoir rock permeability based on the measured porosity and mercury injection capillary pressure data of displacement pressure, Winland parameter, Swanson parameter, Pittman parameter, Dastidar parameter, and Purcell integration. The optimized ANN model showed high accuracy in evaluating the permeability with an $R$ of 0.98.

A comparison of some of the optimized machine learning models for permeability prediction is presented in Table 15.3 along with the input parameters for training these models, the field or formation from which the training data were obtained, and remarks on the accuracy of the optimized models.

## 15.6  USE OF MACHINE LEARNING TECHNIQUES IN FORECASTING AND OPTIMIZING THE OIL AND GAS PRODUCTION

The key to the final economic decision of whether to develop the hydrocarbon fields and drill new wells is dependent on the expected hydrocarbon production profile or production rate with time through the life of the reservoir. Therefore, accurate generation and prediction of this profile are critical.

### 15.6.1 History Matching and Production Profile Forecasting

In 2012, Esmaili et al. [91] optimized the neural networks for hydrocarbon forecasting from shale plays. The authors optimized the neural networks to forecast hydrocarbon production based on many parameters which were divided into six groups of data: well location data, static-Marcellus data, completion information, hydraulic fracturing data, production, and operational constraints, and additional parameters. Testing the optimized neural networks showed high accuracy in forecasting oil production, as shown in Figure 15.5; the production profile forecasted by the optimized neural networks (gray line) was able to match the actual production from the reservoir for the last 4 months of production (dark gray dots). The neural network model also forecasted the future production rate for one more year.

Mohamed et al. [92] proposed ANN models for history matching and forecasting future oil, gas, and water production rates for highly faulted, multilayered, and clastic oil reservoirs. The models were trained using static and dynamic data such as well location, porosity, permeability, pay thickness, initial water saturation, gamma-ray, deep resistivity, shale volume, clay volume, water and gas injection rates, reservoir pressure, completion footage, wellhead pressure, and temperature. The trained ANN model was successful in matching the historical oil, gas, and water production rates and forecasting the future rates with high accuracy.

Recently, Khan and Louis [93] proposed the use of ANN model for hydrocarbon production forecasting in an unconventional reservoir. The authors trained the ANN model to forecast oil production based on well performance data such as bottom-hole pressure, tubing head pressure, production rate, and constraints. The optimized model was highly accurate in forecasting the production rate with $R^2$ of 0.996.

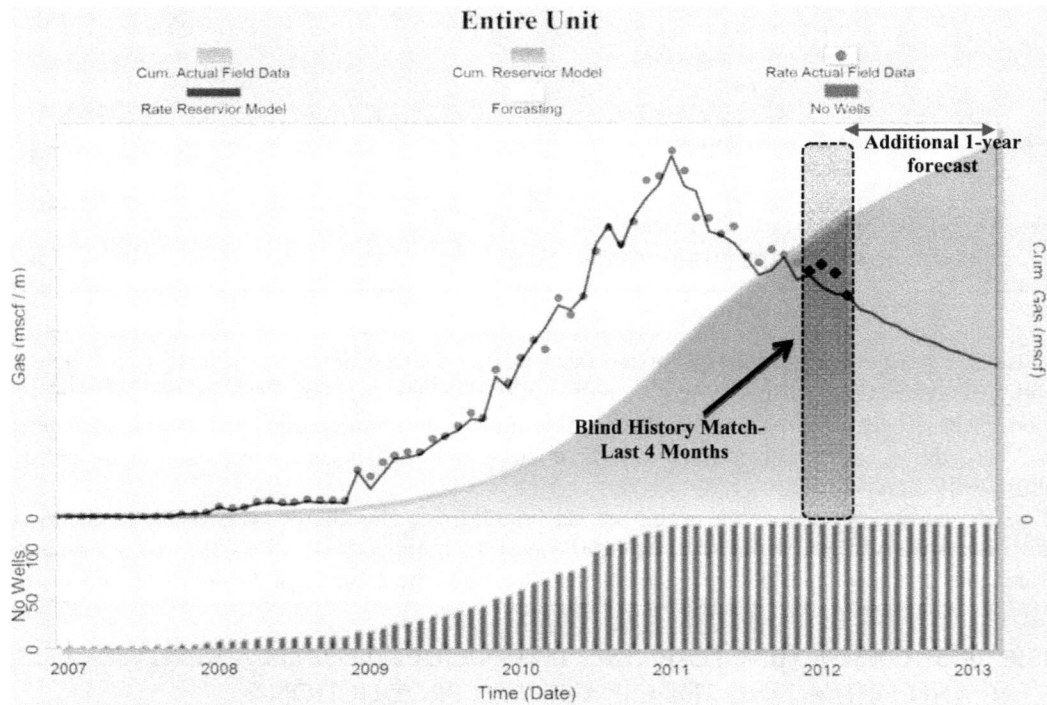

**FIGURE 15.5** Blind history matching of last 4 months of production and forecasting of additional 1 year (From Ref. [91]. With permission.)

### 15.6.2 Hydrocarbon Production Optimization

Park et al. [94] proposed the use of an advanced polynomial neural networks (PNN) model for optimizing the integrated production system, which includes a gas reservoir, well (tubing) flow, and surface production facilities. The proposed PNN exhibited significant improvement in the prediction of reservoir performance with a maximum error of 1.0%.

In 2009, Kandziora [95] suggested a machine learning-based model for optimizing oil and gas production rate based on monitoring the performance of electrical submersible pumps (ESP) and detecting the failure of the pumps in advance to enable operators to preempt costly ESP failures. The developed model was able to determine the probability of an ESP failure 12 days before its actual failure.

Akanji et al. [96] deployed an enhanced neuro-fuzzy technique for optimizing the oil production in the Oredo oil field in Nigeria. The model was developed to optimize oil production as a function of production history, properties of the reservoir rock and fluid, well geometry, completion profile, architecture, and surface data. The optimized enhanced neuro-fuzzy model was able to optimize oil production accurately with a nondimensional error index of 1.14.

## 15.7 CONCLUSION

This chapter summarizes the application of different machine learning models for estimating different essential parameters required while evaluating drillability of hydrocarbon wells, assessment of the rock geomechanical and petrophysical properties, optimizing hydrocarbon production from developed reservoirs, and forecasting future hydrocarbon production. Based on this study, the following observations can be made:

- The use of machine learning models enabled optimizing the drillability of the hydrocarbon wells, prediction of reservoir rock properties, and optimization of hydrocarbon production accurately.
- The use of the machine learning models helped in obtaining accurate continuous profiles of properties such as the ROP, reservoir porosity, reservoir permeability, Young's modulus, Poisson's ratio, and UCS. These profiles are obtained at a low cost.
- Training machine learning models using real-time surface-measurable drilling mechanical parameters enabled the real-time estimation of petroleum-related properties.
- Advance detection of the failure of different equipment and tools used in the hydrocarbon system was enabled by the use of different machine learning models. This is important to enable preempt costly equipment failures.
- The use of the machine learning techniques has led to big transition in techniques and procedures for reducing the high uncertainty associated with the high-value investments in the oil and gas upstream, and it also minimized the time needed for uncertainty handling.

## REFERENCES

1. US EIA. Technically Recoverable Shale Oil and Shale Gas Resources: An Assessment of 137 Shale Formations in 41 Countries Outside the United States. Independent Statistics & Analysis. U.S. Department of Energy. Washington, DC 20585. June 2013.
2. Onwukwe, S. I., Izuwa, N. C., Ileaboya, E. E., Ihekoronye, K. K.,2019. Optimizing production in brown fields using re-entry horizontal wells. *Advances in Petroleum Exploration and Development*. 18(1), 27-35. Doi: 10.3968/11299.
3. US EIA, Incorporating international petroleum reserves and resource estimates into projections of production. Independent Statistics & Analysis. U.S. Department of Energy. Washington, DC 20585. June, 2011.
4. Koroteev, D., Tekic, Z., 2021. Artificial intelligence in oil and gas upstream: Trends, challenges, and scenarios for the future. *Energy and AI* 3, 100041. Doi: 10.1016/j.egyai.2020.100041.

5. Hassanvand, M., Moradi, S., Fattahi, M., Zargar, G., Kamari, M., 2018. Estimation of rock uniaxial compressive strength for an Iranian carbonate oil reservoir: Modeling vs. artificial neural network application. *Petroleum Research* 3(4), 336–345. Doi: 10.1016/j.ptlrs.2018.08.004.

6. Zhang, T., 2021. An estimation method of the fuel mass injected in large injections in common-rail diesel engines based on system identification using artificial neural network. *Fuel* 122404. Doi: 10.1016/j.fuel.2021.122404.

7. Mahmoud, A.A., Elkatatny, S., Al-Abduljabbar, A., 2021. Application of machine learning models for real-time prediction of the formation lithology and tops from the drilling parameters. *Journal of Petroleum Science and Engineering* 108574. Doi: 10.1016/j.petrol.2021.108574.

8. Mahmoud, A.A., Elkatatny, S., and Al-Shehri, D., 2020. Application of machine learning in evaluation of the static young's modulus for sandstone formations. *Sustainability* 12(5). Doi: 10.3390/su12051880.

9. Siddig, O., Mahmoud, A.A., Elkatatny, S.M., Soupios, P., 2021. Utilization of artificial neural network in predicting the total organic carbon in devonian shale using the conventional well logs and the spectral gamma-ray. *Computational Intelligence and Neuroscience* 2021, 2486046. Doi: 10.1155/2021/2486046.

10. Cortes, C., Vapnik, V., 1995. Support-vector networks. *Machine Learning* 20, 273–297.

11. Balabin, R.M., Safieva, R.Z., Lomakina, E.I., 2011. Near-infrared (NIR) spectroscopy for motor oil classification: From discriminant analysis to support vector machines. *Microchemical Journal* 98(1), 121–128. Doi: 10.1016/j.microc.2010.12.007.

12. Vikara, D., Khanna, V., 2021. Machine learning classification approach for formation delineation at the basin-scale. *Petroleum Research*, 7(2), 165-176. Doi: 10.1016/j.ptlrs.2021.09.004.

13. Dargahi-Zarandi, A., Hemmati-Sarapardeh, A., Shateri, M., Menad, N.A., Ahmadi, M., 2020. Modeling minimum miscibility pressure of pure/impure $CO_2$-crude oil systems using adaptive boosting support vector regression: Application to gas injection processes. *Journal of Petroleum Science and Engineering* 184, 106499. Doi: 10.1016/j.petrol.2019.106499.

14. Hu, G., Xu, Z., Wang, G., Zeng, B., Liu, Y., Lei, Y., 2021. Forecasting energy consumption of long-distance oil products pipeline based on improved fruit fly optimization algorithm and support vector regression. *Energy* 224, 120153. Doi: 10.1016/j.energy.2021.120153.

15. Serfidan, A.C., Uzman, F., Türkay, M., 2020. Optimal estimation of physical properties of the products of an atmospheric distillation column using support vector regression. *Computers & Chemical Engineering* 134, 106711. Doi: 10.1016/j.compchemeng.2019.106711.

16. Mahmoud, A.A., Elkatatny, S., Abouelresh, M., Abdulraheem, A., Ali, A., 2020. Estimation of the total organic carbon using functional neural networks and support vector machine. *Proceedings of the 12th International Petroleum Technology Conference and Exhibition*, Dhahran, Saudi Arabia, 13–15 January; IPTC-19659-MS. Doi: 10.2523/IPTC-19659-MS.

17. Mahmoud, A.A., Elkatatny, S., Alsabaa, A., Al Shehri, D., 2020. Functional neural networks-based model for prediction of the static young's modulus for sandstone formations. *Proceedings of the 54th US Rock Mechanics/Geomechanics Symposium*, 28 June–1 July.

18. Bello, O., Asafa, T.A Functional networks softsensor for flowing bottomhole pressures and temperatures in multiphase production wells. In *Proceeding of the SPE Intelligent Energy Conference & Exhibition*, Utrecht, The Netherlands, 1–3 April 2014. Doi: 10.2118/167881-MS.

19. Anifowose, F., Abdulraheem, A. 2011. Fuzzy logic-driven and SVM-driven hybrid computational intelligence models applied to oil and gas reservoir characterization. *Journal of Natural Gas Science and Engineering* 3(3), 505–517. Doi: 10.1016/j.jngse.2011.05.002.

20. Brieman, L., 2001. Random forests. *Machine Learning* 45, 5e32.

21. Ji, X., Yang, B., Tang, Q., 2020. Seabed sediment classification using multibeam backscatter data based on the selecting optimal random forest model. *Applied Acoustics* 167, 107387. Doi: 10.1016/j.apacoust.2020.107387.

22. Marins, M.A., Barros, B.D., Santos, I.H., Barrionuevo, D.C., Vargas, R.E.V., Prego, T.M., de Lima, A.A., de Campos, M.L.R., da Silva, E.A.B., Netto, S.L., 2021. Fault detection and classification in oil wells and production/service lines using random forest. *Journal of Petroleum Science and Engineering* 197, 107879. Doi: 10.1016/j.petrol.2020.107879.

23. Osman, H., Ali, A., Mahmoud, A.A., Elkatatny, S., 2021. Estimation of the rate of penetration while horizontally drilling carbonate formation using random forest. *ASME. Journal of Energy Resources Technology*. Doi: 10.1115/1.4050778.

24. Tran, T.V., Ngo, H.H., Hoang, S.K., Tran, H.N.T., Lambiase, J.J., 2020. Depositional facies prediction using artificial intelligence to improve reservoir characterization in a mature field of Nam con son basin, offshore Vietnam. *Paper Presented at the Offshore Technology Conference Asia*, Kuala Lumpur, Malaysia, 2–6 November. Doi: 10.4043/30086-MS.

25. Carpenter, C., 2019. Artificial intelligence improves seismic-image reconstruction. *Journal of Petroleum Technology* 71, 65–66. Doi: 10.2118/1019-0065-JPT.

26. Mahmoud, A.A., Elkatatny, S., Abdulraheem, A., Mahmoud, M., Ibrahim, O., Ali, A., 2017. New technique to determine the total organic carbon based on well logs using artificial neural network (White Box). *Paper SPE-188016-MS Presented at the 2017 SPE Kingdom of Saudi Arabia Annual Technical Symposium and Exhibition*, Dammam, Saudi Arabia, 24–27 April. Doi: 10.2118/188016-MS.

27. Mahmoud, A.A., Elkatatny, S., Ali, A., Abouelresh, M., Abdulraheem, A., 2019. Evaluation of the total organic carbon (TOC) using different artificial intelligence techniques. *Sustainability* 11(20), 5643. Doi: 10.3390/su11205643.

28. Mahmoud, A.A., Elkatatny, S., Ali, A., Abouelresh, M., Abdulraheem, A., 2019. New robust model to evaluate the total organic carbon using fuzzy logic. *Paper SPE-198130-MS Presented at the SPE Kuwait Oil & Gas Show and Conference*, Mishref, Kuwait, 13–16 October. Doi: 10.2118/198130-MS.

29. Goda, H.M., Maier, H.R., Behrenbruch, P., 2007. Use of artificial intelligence techniques for predicting irreducible water saturation - Australian hydrocarbon basins. *Paper Presented at the Asia Pacific Oil and Gas Conference and Exhibition*, Jakarta, Indonesia, October 30–November 1. Doi: 10.2118/109886-MS.

30. Mahmoud, A.A., Elkatatny, S., Chen, W., Abdulraheem, A., 2019. Estimation of oil recovery factor for water drive sandy reservoirs through applications of artificial intelligence. *Energies* 12(19), 3671. Doi: 10.3390/en12193671.

31. Mahmoud, A.A., Elkatatny, S., Abdulraheem, A., Mahmoud, M., 2017. Application of artificial intelligence techniques in estimating oil recovery factor for water drive sandy reservoirs. *Paper SPE-187621-MS Presented at the 2017 SPE Kuwait Oil & Gas Show and Conference*, Kuwait City, Kuwait, 15–18 October. Doi: 10.2118/187621-MS.

32. Al-Abduljabbar, A., Elkatatny, S., Mahmoud, A.A., Moussa, T., Al-shehri, D., Abughaban, M., Al-yami, A., 2020. Prediction of the rate of penetration while drilling horizontal carbonate reservoirs using the self-adaptive artificial neural networks technique. *Sustainability* 12, 1376. Doi: 10.3390/su12041376.

33. Al-Abduljabbar, A., Mahmoud, A.A., Elkatatny, S., 2021. Artificial neural network model for real-time prediction of the rate of penetration while horizontally drilling natural gas-bearing sandstone formations. *Arabian Journal of Geosciences* 14, 117. Doi: 10.1007/s12517-021-06457-0.

34. Aliyev, R., Donald Paul, D., 2019. A novel application of artificial neural networks to predict rate of penetration. *Paper Presented at the SPE Western Regional Meeting*, San Jose, CA, 23–26 April. Doi: 10.2118/195268-MS.

35. Mahmoud, A.A.; Elkatatny, S.; Abduljabbar, A., Moussa, T., Gamal, H., Al Shehri, D., 2020. Artificial neural networks model for prediction of the rate of penetration while horizontally drilling carbonate formations. *Proceedings of the 54rd US Rock Mechanics/Geomechanics Symposium*, 28 June–1 July.

36. Ashena, R., Rabiei, M., Rasouli, V., Mohammadi, A.H., 2020. Optimization of drilling parameters using an innovative GA-PS artificial intelligence model. *Paper Presented at the SPE Asia Pacific Oil & Gas Conference and Exhibition*, Virtual, November 17–19. Doi: 10.2118/202325-MS.

37. Ahmed, S.A., Mahmoud, A.A., Elkatatny, S., 2019a. Fracture pressure prediction using radial basis function. *Paper AADE-19-NTCE-061 Presented at the 2019 AADE National Technical Conference and Exhibition*, Denver, CO, 9–10 April.

38. Ahmed, S.A., Mahmoud, A.A., Elkatatny, S., Mahmoud, M., Abdulraheem, A., 2019b. Prediction of pore and fracture pressures using support vector machine. *Paper IPTC-19523-MS Presented at the 2019 International Petroleum Technology Conference*, Beijing, China, 26–28 March. Doi: 10.2523/IPTC-19523-MS.

39. Rashidi, M., Asadi, A., 2018. An artificial intelligence approach in estimation of formation pore pressure by critical drilling data. *Paper Presented at the 52nd U.S. Rock Mechanics/Geomechanics Symposium*, Seattle, Washington, DC 17–20 June.

40. Abdelgawad, K.Z., Elzenary, M., Elkatatny, S., Mahmoud, M., Abdulraheem, A., Patil, S., 2019. New approach to evaluate the equivalent circulating density (ECD) using artificial intelligence techniques. *Journal of Petroleum Exploration and Production Technology* 9, 1569–1578. Doi: 10.1007/s13202-018-0572-y.

41. Elzenary, M., Elkatatny, S., Abdelgawad, K.Z., Abdulraheem, A., Mahmoud, M., Al-Shehri, D., 2018. New technology to evaluate equivalent circulating density while drilling using artificial intelligence. *SPE Kingdom of Saudi Arabia Annual Technical Symposium and Exhibition*. Doi: 10.2118/192282-MS.

42. Abbas, A.K., Assi, A.H., Abbas, H., Almubarak, H., Al Saba, M., 2019. Drill bit selection optimization based on rate of penetration: Application of artificial neural networks and genetic algorithms. *Paper Presented at the Abu Dhabi International Petroleum Exhibition & Conference*, Abu Dhabi, UAE, 11–14 November. Doi: 10.2118/197241-MS.

43. Alsaihati, A., Elkatatny, S., Mahmoud, A.A, Abdulraheem, A., 2021. Use of machine learning and data analytics to detect downhole abnormalities while drilling horizontal wells, with real case study. *Journal of Energy Resources and Technology* Doi: 10.1115/1.4048070.

44. Elkatatny, S., Al-AbdulJabbar, A., Mahmoud, A.A., 2019. New robust model to estimate formation tops in real-time using artificial neural networks (ANN). *Petrophysics* 60(06), 825–837. Doi: 10.30632/PJV60N6-2019a7.

45. Ali, A., Aïfa, T., Baddari, K., 2014. Prediction of natural fracture porosity from well log data by means of fuzzy ranking and an artificial neural network in Hassi Messaoud oil field, Algeria. *Journal of Petroleum Science and Engineering* 115, 78–89. Doi: 10.1016/j.petrol.2014.01.011.

46. Khalifah, H., Al Glover, P.W.J., Lorinczi, P., 2020. Permeability prediction and diagenesis in tight carbonates using machine learning techniques. *Marine and Petroleum Geology* 112, 104096. Doi: 10.1016/j.marpetgeo.2019.104096.

47. Moussa, T., Elkatatny, S., Mahmoud, M., Abdulraheem, A. 2018. Development of new permeability formulation from well log data using artificial intelligence approaches. *ASME. Journal of Energy Resources Technology* 140(7), 072903-072903-8. Doi: 10.1115/1.4039270.

48. Mahmoud, A.A., Elkatatny, S., Ali, A., Moussa, T., 2019. Estimation of static young's modulus for sandstone formation using artificial neural networks. *Energies* 12(11), 2125. Doi: 10.3390/en12112125.

49. Ahmed, A., Elkatatny, S., Abdulraheem, A., 2021. Real-time static Poisson's ratio prediction of vertical complex lithology from drilling parameters using artificial intelligence models. *Arabian Journal of Geosciences* 14, 436. Doi: 10.1007/s12517-021-06833-w.

50. Elkatatny, S., 2018. Application of artificial intelligence techniques to estimate the static poisson's ratio based on wireline log data. *Journal of Energy Resources Technology* 140(7), 072905. Doi: 10.1115/1.4039613.

51. Gowida, A., Elkatatny, S., Moussa, T., 2020. Comparative analysis between different artificial based models for predicting static Poisson's ratio of sandstone formations. *Paper Presented at the International Petroleum Technology Conference*, Dhahran, Kingdom of Saudi Arabia, 13–15 January. Doi: 10.2523/IPTC-20208-MS.

52. Tariq, Z., Elkatatny, S.M., Mahmoud, M.A., Abdulraheem, A., Abdelwahab, A.Z., Woldeamanuel, M., Mohamed, I.M., 2017. Development of new correlation of unconfined compressive strength for carbonate reservoir using artificial intelligence techniques. *Paper Presented at the 51st U.S. Rock Mechanics/ Geomechanics Symposium*, San Francisco, California, USA, 25–28 June.

53. Al-Azani, K., Elkatatny, S., Abdulraheem, A., Mahmoud, M., Al-Shehri, D., 2018. Real time prediction of the rheological properties of oil-based drilling fluids using artificial neural networks. *Paper Presented at the SPE Kingdom of Saudi Arabia Annual Technical Symposium and Exhibition*, Dammam, Saudi Arabia, 23–26 April. Doi: 10.2118/192199-MS.

54. Abdelgawad, K., Elkatatny, S., Moussa, T., Mahmoud, M., Patil, S., 2019. Real time determination of rheological properties of spud drilling fluids using a hybrid artificial intelligence technique. *Journal of Energy Resources Technology*, 141(3). Doi: 10.1115/1.4042233.

55. Alsabaa, A., Gamal, H., Elkatatny, S,, Abdulraheem, A., 2020. Real-time prediction of rheological properties of invert emulsion mud using adaptive neuro-fuzzy inference system. *Sensors* 20(6), 1669. Doi: 10.3390/s20061669.

56. Gowida, A., Elkatatny, S., Ramadan, E., Abdulraheem, A., 2019. Data-driven framework to predict the rheological properties of $CaCl_2$ brine-based drill-in fluid using artificial neural network. *Energies* 12(10),1880. Doi: 10.3390/en12101880.

57. Elkatatny, S., 2017. Real-time prediction of rheological parameters of KCl water-based drilling fluid using artificial neural networks. *Arabian Journal for Science and Engineering* 42, 1655–1665. Doi: 10.1007/s13369-016-2409-7.

58. Elkatatny, S.M., 2016. Determination the rheological properties of invert emulsion based mud on real time using artificial neural network. *Paper Presented at the SPE Kingdom of Saudi Arabia Annual Technical Symposium and Exhibition*, Dammam, Saudi Arabia, 25–28 April. Doi: 10.2118/182801-MS.

59. Elkatatny, S., Mahmoud, S., 2018. Development of new correlations for the oil formation volume factor in oil reservoirs using artificial intelligent white box technique, *Petroleum* 4(2), 178–186. Doi: 10.1016/j.petlm.2017.09.009.

60. Mahdiani, M.R., Norouzi, M., 2018. A new heuristic model for estimating the oil formation volume factor. *Petroleum* 4, 300–308. Doi: 10.1016/j.petlm.2018.03.006.
61. Oloso, M.A., Hassan, M.G., Bader-El-Den, M.B., Buick, J.M., 2017. Hybrid functional networks for oil reservoir PVT characterisation. *Expert Systems with Applications* 87, 363–369. Doi: 10.1016/j.eswa.2017.06.014.
62. Ahmadi, M.A., Pournik, M., Shadizadeh, S.R., 2015. Toward connectionist model for predicting bubble point pressure of crude oils: Application of artificial intelligence. *Petroleum* 1, 307–317. Doi: 10.1016/j.petlm.2015.08.003.
63. Wood, D.A., Choubineh, A., 2018. Transparent open-box learning network and artificial neural network predictions of bubble-point pressure compared. *Petroleum*. Doi: 10.1016/j.petlm.2018.12.001.
64. Moussa, T., Elkatatny, S., Abdulraheem, A., Mahmoud, M., Rami Alloush, R., 2017. A hybrid artificial intelligence method to predict gas solubility and bubble point pressure. *Paper Presented at the SPE Kingdom of Saudi Arabia Annual Technical Symposium and Exhibition*, Dammam, Saudi Arabia, 24–27 April. Doi: 10.2118/188102-MS.
65. Mohmad, N.I., Mandal, D., Amat, H., Sabzabadi, A., Masoudi, R., 2020. History matching of production performance for highly faulted, multi layered, clastic oil reservoirs using artificial intelligence and data analytics: A novel approach. *Paper Presented at the SPE Asia Pacific Oil & Gas Conference and Exhibition*, Virtual, 17–19 November. Doi: 10.2118/202460-MS.
66. Khan, H., Louis, C., 2021. An artificial intelligence neural networks driven approach to forecast production in unconventional reservoirs – comparative analysis with decline curve. *Paper Presented at the International Petroleum Technology Conference*, Virtual, March 23–April 1. Doi: 10.2523/IPTC-21350-MS.
67. Panja, P., Velasco, R., Pathak, M., Deo, M., 2018. Application of artificial intelligence to forecast hydrocarbon production from shales. *Petroleum* 4(1), 75–89. Doi: 10.1016/j.petlm.2017.11.003.
68. Zendehboudi, S., Ahmadi, M.A., James, L., Chatzis, I., 2012. Prediction of condensate-to-gas ratio for retrograde gas condensate reservoirs using artificial neural network with particle swarm optimization. *Energy & Fuels* 26(6), 3432–3447. Doi: 10.1021/ef300443j.
69. Chen, C., Yang, M., Han, X., Zhang, J., 2019. Water flooding performance prediction in layered reservoir using big data and artificial intelligence algorithms. *Paper Presented at the Abu Dhabi International Petroleum Exhibition & Conference*, Abu Dhabi, UAE, 11–14 November. Doi: 10.2118/197585-MS.
70. Sierra, D.M., Rojas, A.A., Araque, V.S., 2020. Low salinity water injection optimization in the namorado field using compositional simulation and artificial intelligence. *Paper Presented at the SPE Latin American and Caribbean Petroleum Engineering Conference*, Virtual, 27–31 July. Doi: 10.2118/198995-MS.
71. Dang, C., Nghiem, L., Fedutenko, E., Gorucu, E., Yang, C., Mirzabozorg, A., 2018. Application of artificial intelligence for mechanistic modeling and probabilistic forecasting of hybrid low salinity chemical flooding. *Paper Presented at the SPE Annual Technical Conference and Exhibition*, Dallas, TX, 24–26 September. Doi: 10.2118/191474-MS.
72. Hassan, A., Elkatatny, S., Mahmoud, M., Abdulraheem, A., 2018. A new approach to characterize $CO_2$ flooding utilizing artificial intelligence techniques. *Paper Presented at the SPE Kingdom of Saudi Arabia Annual Technical Symposium and Exhibition*, Dammam, Saudi Arabia, April 23–26. Doi: 10.2118/192252-MS.
73. Mahmoud, A.A., Elkatatny, S., Mahmoud, M., Abouelresh, M., Abdulraheem, A., Ali, A., 2017b. Determination of the total organic carbon (TOC) based on conventional well logs using artificial neural network. *International Journal of Coal Geology* 179, 72–80. Doi: 10.1016/j.coal.2017.05.012.
74. Bilgesu, H.I., Tetrick, L.T., Altmis, U., Ameri, M.S., 1997. A new approach for the prediction of rate of penetration (ROP) values. *Proceedings of the SPE Eastern Regional Meeting*, Lexington, KY, 22–24 October; SPE-39231-MS. Doi: 10.2118/39231-MS.
75. Amar, K., Ibrahim, A., 2012. Rate of penetration prediction and optimization using advances in artificial neural networks, a comparative study. *Proceedings of the 4th International Joint Conference on Computational Intelligence*, Barcelona, Spain, 5–7 October, pp 647–652. Doi: 10.5220/0004172506470652.
76. Bataee, M., Mohseni, S., 2011. Application of artificial intelligent systems in ROP optimization: A case study in shadegan oil field. *Paper Presented at the SPE Middle East Unconventional Gas Conference and Exhibition*, Muscat, Oman, 31 January–2 February 2011. Doi: 10.2118/140029-MS.
77. Elkatatny, S., 2018. New approach to optimize the rate of penetration using artificial neural network. *Arabian Journal for Science and Engineering* 43(11), 6297–6304. Doi: 10.1007/s13369-017-3022-0.
78. Al-AbdulJabbar, A., Elkatatny, S., Mahmoud, M., Abdulraheem, A., 2018. Predicting rate of penetration using artificial intelligence techniques. *Proceeding of the SPE Kingdom of Saudi Arabia Annual Technical Symposium and Exhibition*, Dammam, Saudi Arabia, April. Doi: 10.2118/192343-MS.

79. Ahmed, S.A., Elkatatny, S., Abdulraheem, A., Mahmoud, M., Ali, A.Z., Mohamed, I.M., 2018. Prediction of rate of penetration of deep and tight formation using support vector machine. *Proceeding of the SPE Kingdom of Saudi Arabia Annual Technical Symposium and Exhibition*, Dammam, Saudi Arabia, April. Doi: 10.2118/192316-MS.

80. Alali, A.M., Abughaban, M.F., Aman, B.M., Ravela, S., 2021. Hybrid data driven drilling and rate of penetration optimization. *Journal of Petroleum Science and Engineering* 200, 108075. Doi: 10.1016/j.petrol.2020.108075.

81. Anifowose, F., Abdulazeez Abdulraheem, A., 2010. Prediction of porosity and permeability of oil and gas reservoirs using hybrid computational intelligence models. *Paper Presented at the North Africa Technical Conference and Exhibition*, Cairo, Egypt, 14–17 February 2010. Doi: 10.2118/126649-MS.

82. Anifowose, F.A., Ewenla, A.O., Eludiora, S.I., 2011. Prediction of oil and gas reservoir properties using support vector machines. *Paper Presented at the International Petroleum Technology Conference*, Bangkok, Thailand, 7–9 February 2012. Doi: 10.2523/IPTC-14514-MS.

83. Elkatatny, S., Tariq, Z., Mahmoud, M., Abdulraheem, A., 2018. New insights into porosity determination using artificial intelligence techniques for carbonate reservoirs. *Petroleum* 4(4), 408–418. Doi: 10.1016/j.petlm.2018.04.002.

84. Al-AbdulJabbar, A., Al-Azani, K., Elkatatny, S., 2020. Estimation of reservoir porosity from drilling parameters using artificial neural networks. *Petrophysics* 61, 318–330. Doi: 10.30632/PJV61N3-2020a5.

85. Kanfar, R., Shaikh, O., Yousefzadeh, M., Mukerji, T., 2020. Realtime well log prediction from drilling data using deep learning. *Petroleum Engineering*.

86. Gamal, H., Salaheldin Elkatatny, S., Alsaihati, A., Abdulraheem, A., 2021. Intelligent prediction for rock porosity while drilling complex lithology in real time. *Computational Intelligence and Neuroscience* 2021, 9960478. Doi: 10.1155/2021/9960478.

87. Basbug, B., Karpyn, Z.T., 2007. Estimation of permeability from porosity, specific surface area, and irreducible water saturation using an artificial neural network. *Paper Presented at the Latin American & Caribbean Petroleum Engineering Conference*, Buenos Aires, Argentina, 15–18 April. Doi: 10.2118/107909-MS.

88. Al-Fattah, S.M., Al-Naim, H.A., 2009. Artificial-intelligence technology predicts relative permeability of giant carbonate reservoirs. *SPE Reservoir Evaluation & Engineering* 12 (2009): 96–103. Doi: 10.2118/109018-PA.

89. Anifowose, F.A., Abdulraheem, A., Al-Shuhail, A., Schmitt, D.P., 2013. Improved permeability prediction from seismic and log data using artificial intelligence techniques. *Paper Presented at the SPE Middle East Oil and Gas Show and Conference*, Manama, Bahrain, March 2013. Doi: 10.2118/164465-MS.

90. Nooruddin, H.A., Anifowose, F., Abdulraheem, A., 2013. Applying artificial intelligence techniques to develop permeability predictive models using mercury injection capillary-pressure data. *Paper Presented at the SPE Saudi Arabia Section Technical Symposium and Exhibition*, Al-Khobar, Saudi Arabia, 19–22 May. Doi: 10.2118/168109-MS.

91. Esmaili, S., Kalantari-Dahaghi, A., Mohaghegh, S.D., 2012. Forecasting, sensitivity and economic analysis of hydrocarbon production from shale plays using artificial intelligence & data mining. *Paper Presented at the SPE Canadian Unconventional Resources Conference*, Calgary, Alberta, Canada, 30 October–1 November. Doi: 10.2118/162700-MS.

92. Mohmad, N.I., Mandal, D., Amat, H., Sabzabadi, A., Masoudi, R., 2020. History matching of production performance for highly faulted, multi layered, clastic oil reservoirs using artificial intelligence and data analytics: A novel approach. *Paper Presented at the SPE Asia Pacific Oil & Gas Conference and Exhibition*, Virtual, 17–19 November 2020. Doi: 10.2118/202460-MS.

93. Khan, H., Louis, C., 2021. An artificial intelligence neural networks driven approach to forecast production in unconventional reservoirs – comparative analysis with decline curve. *Paper Presented at the International Petroleum Technology Conference*, Virtual, 23 March–1 April. Doi: 10.2523/IPTC-21350-MS.

94. Park, H.J., Lim, J.S., Kang, J.M., Roh, J., Min, B.H., 2006. A hybrid artificial intelligence method for the optimization of integrated gas production system. *Paper Presented at the SPE Asia Pacific Oil & Gas Conference and Exhibition*, Adelaide, Australia, 11–13 September. Doi: 10.2118/100997-MS.

95. Kandziora, C., 2019. Applying artificial intelligence to optimize oil and gas production. *Paper Presented at the Offshore Technology Conference*, Houston, TX, 6–9 May 2019. Doi: 10.4043/29384-MS.

96. Akanji, L.T., Dala, J., Bello, K., Olalekan, O., Jadhawar, P., 2019. Application of artificial intelligence in well screening and production optimisation in oredo oilfields, Niger Delta, Nigeria. *Paper Presented at the SPE Nigeria Annual International Conference and Exhibition*, Lagos, Nigeria, 5–7 August 2019. Doi: 10.2118/198877-MS.

# Section III

Energy Transition:
Policies and Prospects

# 16 Energy and Sustainable Development

*Tri Ratna Bajracharya*
Tribhuvan University

*Shree Raj Shakya*
Tribhuvan University
Institute for Advanced Sustainability Studies (IASS)

*Anzoo Sharma*
Tribhuvan University
Center for Rural Technology (CRT/N)

## CONTENTS

DOI: 10.1201/9781003315353-19

## 16.1 EVOLUTION OF SUSTAINABLE DEVELOPMENT

The concept of sustainable development was launched in 1987 by World Commission on Environment and Development, in an announcement published in "Our Common Future" (Brundtland Report). The report by World Commission on Environment and Development (1987) states that "development involves a progressive transformation of economy and society along with the satisfaction of essential human needs and aspirations for an improved living standard" and defines sustainable development (SD) as "the development that meets the needs of the present without compromising the ability of future generations to meet their own needs." The definition implies that the earth and its natural resources should not be depleted by a small group of affluent societies; people in developing countries and future generations, everywhere in the world should be able to meet their basic needs (Schrijver, 2008). SD is a concept along with a development paradigm that demands improvement of wellbeing without harming the earth's ecologies; such as climate change and demolition of species. Four primary dimensions of SD can be derived from the Brundtland Report; they are: safeguarding long-term ecological sustainability, satisfying basic human needs, and promoting intragenerational and intergenerational equity (Holden et al., 2014). The report also points out that the present state of technology and the social organization of environmental resources, together with the limited ability of the biosphere to absorb the effects of human activities impose limitations on a new era of economic growth. Many problems of resource depletion and environmental stress arise due to disparities in economic and political power. Thus, the strategies for SD must aim to promote harmony among human beings and between humanity and nature (World Commission on Environment and Development, 1987). According to Kolk (2016), integrating economic, environmental, and social concerns in every executive process is very important from which SD is achievable.

The rise of the idea of SD relates to the industrial revolution. The Malthusian "Theory of Population" (1798) hypothesizes that the human population tends to grow exponentially while subsistence (food) will grow in only an arithmetic progression. Thus, population grew faster than food production and tended to outstrip it in a short time (Coomer, 1981; Dixon & Fallon, 1989; Rostow, 1980). Pigou (1929) noted that "the divergence between marginal private costs and benefits and marginal social cost and benefits create externalities; hence a tax on those activities that produce negative externalities at a rate equal to that external cost will accurately reflect the comprehensive cost and benefits of the activity." The concept of SD is said to be derived from economics as a discipline (Emas, 2015; Mensah, 2019). Now it became a subject of interest in every cluster.

Basically, in the 19th century, industrial countries started to discover that the population growth, the increase in consumption, the disparity in economic and political power, and the danger that crucial resources such as forest products and petroleum resources and water could be depleted, which in due course might threaten the wellbeing of mankind boosted the awareness of the need to use resources sustainably (Jevons, 1865; Marsh, 1864; Mill, 1870; Wallace, 1898). The 20th century began with unlimited optimism but was soon shattered by economic turmoil, destructive world wars, and ecological crises, which unveiled a downside to the scientific and technological advances. Post-World War II, an unprecedented economic boom paved the way for renewed optimism about the prospects of rising living standards worldwide. From the 1960s, the idea of an imminent ecological crisis was published in spine-chilling scientific articles such as Rachel Carson's *The Silent Spring* (1962), Paul Ehrlich's *The Population Bomb* (1968), Edward Goldsmith's *A Blueprint for Survival* (1972) and Fritz Schumacher's *Small Is Beautiful* (1973) and publicized through media. The UN conference in Stockholm (1972) was the first world conference on the environment with the need for a common outlook and for common principles to inspire and guide the peoples of the world in the preservation and enhancement of the human environment. The worldwide recession following the two major oil crises in history: 1973 and 1979, demonstrated the potential consequences of resource shortage and led to an awareness of the limits to economic growth (Du Pisani, 2006). These realizations necessitated a paradigm shift to a new notion of development. In 1983 the United Nations Commission on Environment and Development was created and in 1987, the

Commission issued the Brundtland Report. The report highlighted that humanity can make development sustainable and emphasized three fundamental components of SD; they are environmental protection, economic growth, and social equity. Previously development and conservation had been regarded as conflicting ideas because conservation was understood as the protection of resources and development as the exploitation of resources (Redclift, 1988). The 1992 Earth Summit held in Rio de Janeiro brought the world's governments to rethink economic development and find ways to stop polluting the planet and depleting its natural resources and unanimously adopted the Agenda 21 (1992): a comprehensive blueprint of actions toward SD, to be taken globally, nationally and locally by organizations of UN, Governments and major groups in every area in which human impacts on the environment. In September 2000, political leaders from around the world came together for formulating concrete 2015 targets for millennium development goals (MDGs) related to the priority challenges of SD (Shah, 2008). The Earth Summit (2002) brought tens of thousands of participants for building partnerships for SD. The participants at the Earth Summit lacked to organize the financial resources for the execution of the agenda into action and to realize the targets. Together with scientific and technological capabilities, education at all levels is essential to support SD. After two decades, the 1992 Rio Earth Summit, the 2012 United Nations Conference on Sustainable Development launched a process that outgrew MDGs to develop a set of SDGs for people, for the planet, and documented clear and practical steps for implementation of SD as followings (Table 16.1).

The three pillars of SD are economic growth, environmental stewardship, and social inclusion that carries across all segments of development, from agriculture to urbanization, water accessibility, energy development and use, transportation, and infrastructure (Muralikrishna & Manickam, 2017). As of today, various actors are operating with diverse stakeholders in various areas of

## TABLE 16.1
### SDG Goals and the Corresponding Number of Targets

| Goal No. | Goal | No. of Targets |
|---|---|---|
| 1 | End poverty in all its forms everywhere | 7 |
| 2 | End hunger, achieve food security and improved nutrition and promote sustainable agriculture | 8 |
| 3 | Ensure healthy lives and promote wellbeing for all at all ages | 13 |
| 4 | Ensure inclusive and equitable quality education and promote lifelong learning opportunities for all | 10 |
| 5 | Achieve gender equality and empower all women and girls | 9 |
| 6 | Ensure availability and sustainable management of water and sanitation for all | 8 |
| 7 | Ensure access to affordable, reliable, sustainable and modern energy for all | 5 |
| 8 | Promote sustained, inclusive and sustainable economic growth, full and productive employment and decent work for all | 12 |
| 9 | Build resilient infrastructure, promote inclusive and sustainable industrialization and foster innovation | 8 |
| 10 | Reduce inequality with and among countries | 10 |
| 11 | Make cities and human settlements inclusive, safe, resilient and sustainable | 10 |
| 12 | Ensure sustainable consumption and production patterns | 11 |
| 13 | Take urgent action to combat climate change and its impacts | 5 |
| 14 | Conserve and sustainably use the oceans, seas and marine resources for sustainable development | 10 |
| 15 | Protect, restore and promote sustainable use of terrestrial ecosystems, sustainably manage forests, combat desertification and halt and reverse land degradation and hat biodiversity loss | 12 |
| 16 | Promote peaceful and inclusive societies for sustainable development, provide access to justice for all and build effective, accountable and inclusive institutions at all levels | 12 |
| 17 | Strengthen the means of implementation and revitalize the global partnership for sustainable development. | 19 |

sustainability with the shared objective of raising awareness on this topic and creating conditions for it to grow and develop. Their economic and industrial activities had a significant impact on the environment and the social balance caused by the unprecedented global scale exploitation of raw materials and uneven distribution of wealth between the rich and poor societies (Boyden, 1993; Goudie, 1981).

## 16.2  PRINCIPLES AND TYPES OF SUSTAINABLE DEVELOPMENT

SD paradigm has attracted a wide range of both governmental and non-governmental actors in contemporary development discourse. SD emphasizes a positive revolution trajectory anchored essentially on social inclusion, environmental protection, and economic growth.

### 16.2.1  SALIENT PRINCIPLES OF SUSTAINABLE DEVELOPMENT

The salient principles, which inhibit the concept of SD are unanimously recognized by all nations since they were stated in Brundtland Commission Report (1987), Rio Declaration (1992), and Agenda 21. The principles are as follows:

i. **Environmental protection:** for development to be sustainable, an effective environmental protection mechanism is to be formulated by each nation within its territories.
ii. **Intergenerational equity:** the right of each generation of human beings to benefit from the cultural and natural resources as well as the obligation to preserve such heritage for generations to come.
iii. **Use and conservation of natural resources:** the present generation must utilize natural resources such that the survival of future generations is safeguarded.
iv. **The precautionary principle:** Any human behavior or developmental activity, which bears an adverse effect on the environment, must be prevented at all costs. According to this principle, "where there is the threat of serious or irreversible environmental damage, lack of full scientific certainty should not be used as a reason for postponing measures to prevent environmental degradation."
v. **The polluter pays principle:** National authorities should aim to encourage the internationalization of environmental costs and the use of economic instruments taking into account the method that the polluter should in principle bear the cost of pollution with due regard to the public interest and without misrepresenting international trade and investment.
vi. **Principle of liability to help and cooperate:** The state should cooperate to strengthen indigenous capacity building for SD.
vii. **Poverty eradication:** It is the worst contributing factor for polluting and destroying the environment as it reduces people's capacity to use resources sustainably. Elimination is poverty is a major goal for SD, particularly in developing countries.
viii. Principle of "public trust."

### 16.2.2  TYPES OF SUSTAINABLE DEVELOPMENT

#### 16.2.2.1  Economic Sustainability

Economies consist of markets where transactions occur based on activities related to production, distribution, and consumption. Traditionally economists, supposing that the abundant supply of natural resources allocated resources inefficiently. Further, they presumed that economic growth accompanied by technological and scientific advancement would also fill up natural resources destroyed during the process. However, it has been realized that the natural resources are not unlimited; besides not all of them can be refilled up. This has prompted a rethink of the traditional economic postulations and has questioned the viability of uncontrolled growth and consumption (Basiago, 1996). Economic sustainability refers to practices that support long-term economic growth without negatively impacting the social, environmental, and cultural aspects of the community. It makes use of the necessary prerequisite for economic growth, capital maintenance and extends the (produced) capital concept to include non-produced natural capital (Bartelmus, 2013).

### 16.2.2.2 Social Sustainability

According to Daly (1992), social sustainability is the understanding of concepts of six terms; accessibility, cultural identity, equity, empowerment, institutional stability, and participation (Daly, 1992). Sustainability at the social level raises the communities, culture, and development of people to help achieve a meaningful life, education, attention to gender equality, peace, proper healthcare, and stability across the globe (Saith, 2006). However, social sustainability does not mean guaranteeing that everyone's requirements are fulfilled. Rather, it is objective at providing favorable conditions for everyone to have the capacity to realize their needs if they so desire (Kolk, 2016). Anything that hinders this ability is considered a barrier and needs to be addressed for progressing toward SD (Brodhag & Taliere, 2006; Pierobon, 2019)

### 16.2.2.3 Environmental Sustainability

Environmental sustainability ensures ecosystem integrity and carrying capacity of nature such as it remains productively stable and resilient to support human life and development (Brodhag & Taliere, 2006). This necessitates the natural wealth to be sustainably utilized as a source of economic inputs and as a sink for waste; which implies natural equilibrium is maintained by not harvesting natural resources faster than they can be regenerated and not emitting wastage faster than they can be assimilated by the environment (Diesendorf, 2000; Goodland & Daly, 1996).

## 16.3 ENERGY TYPES AND SOURCES

The most common definition of energy is the ability to do work or to produce heat. Human civilization has evolved because humans have learned to change energy from one form to another. Energy can manifest in various forms including heat, light, motion, electrical, chemical, and gravitational. The sun, directly or indirectly, is the source of all the energy available on earth.

### 16.3.1 ENERGY TYPES

Energy is broadly classified as:

### 16.3.1.1 Primary Energy

This energy is the energy epitomized in natural resources before undertaking any man-made conversions or transformations. Examples of primary energy resources are coal, crude oil, sunlight, wind, flowing rivers, vegetation, and uranium. Most of the primary energy sources, which we get from the earth, such as coal and crude oil, cannot be used directly by the consumer in raw form. Principally, these primary energy sources must be converted to secondary energy sources such as electricity or gasoline as stated in Figure 16.1.

The supply-side management (SSM) deals with the efficient conversion of energy systems that denote actions taken to ensure the generation, transmission, and distribution of energy. This includes transmission and distribution of electricity including grid systems, substations, and on-site generation. In the case of primary fuel, it includes the transport of fuels; liquid, gaseous and solid fuels (Figure 16.2).

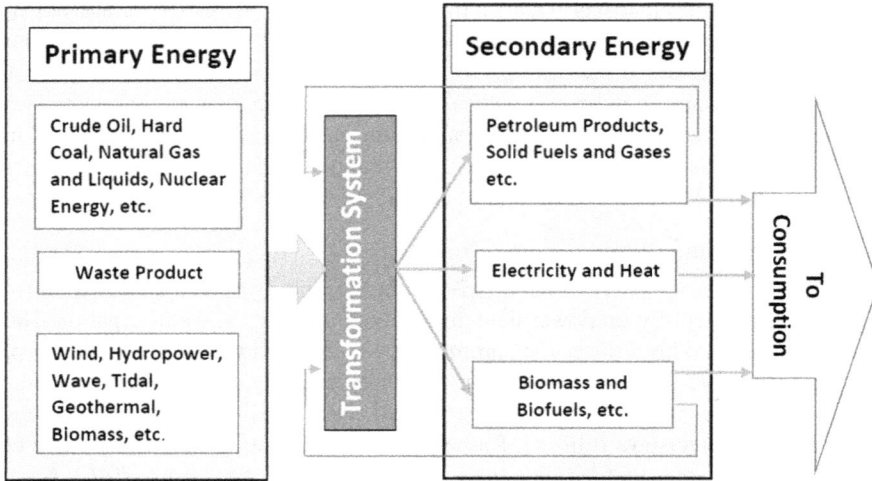

**FIGURE 16.1**   Transformation of energy.

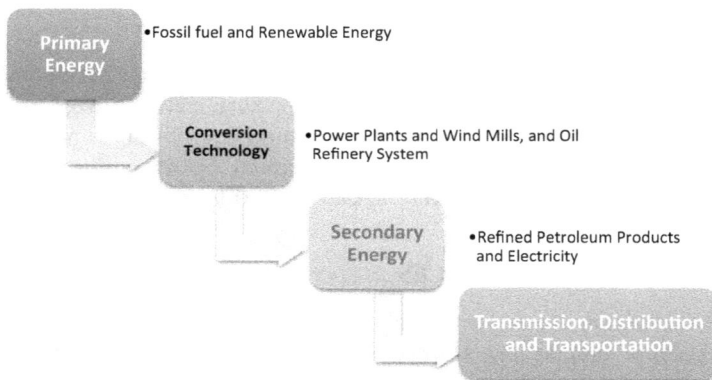

**FIGURE 16.2**   Energy system: supply-side.

### 16.3.1.2   Secondary Energy

As stated above, the energy in the primary (natural state) cannot be used in primary form. It is necessary to convert it into a secondary form. Such energy sources artificially produced or converted through a certain process are called secondary sources. While converting from primary to secondary, there will be some amount of losses and it is based on the efficiency of the conversion system. The processes of conversion from primary sources into secondary sources. The organization of their distribution to the final users is much more complex and requires many people and much knowledge to be managed in the best possible way. Figure 16.3 describes the energy system of the demand side. The term Demand-Side Management refers to a group of actions designed to manage and optimize the end-use energy consumption and to cut costs, from grid charges to general system charges, including taxes.

These sources are further divided into commercial and noncommercial; any energy source that has a fixed price tag in the market is known as a commercial source of energy. They are like petrol, diesel, kerosene, and cooking gas. Usually, noncommercial energy sources do not have a fixed price in the market, and the cost is local and varies according to the place, for example, firewood. The overall energy sources are specifically classified, as shown in Figure 16.4.

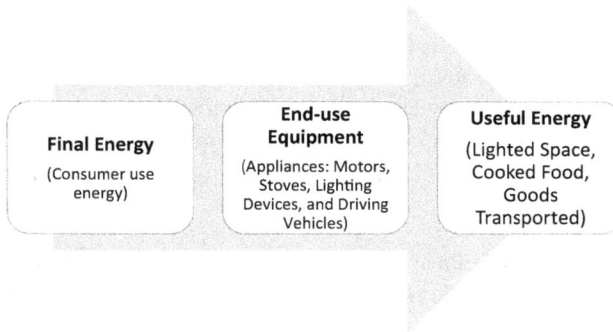

**FIGURE 16.3** Energy system: demand side.

**FIGURE 16.4** Classification of energy resources.

## 16.3.2 ENERGY SOURCES

The sources of energy are diverse. They can be found in various physical states, and with varying degrees of ease or difficulty in capturing their potential energies. Broadly, the energy can be divided into two basic categories.

### 16.3.2.1  Nonrenewable Energy

If the supply of energy from a source is limited, then the energy is said to be nonrenewable energy. Nonrenewable resources, such as fossil fuels and nuclear material, are received from the earth. These nonrenewable sources of energy can be depleted that will ultimately drive up overall energy costs. Most energy sources supporting modern life are non-renewables such as petroleum, hydrocarbon, coal, and nuclear energy.

### 16.3.2.2  Renewable Energy

If the source is a continuously available flow of energy or the source is naturally replenished over time, the energy is known as renewable energy. There are five major renewable energy sources: solar, geothermal, wind, biomass from plants, and hydropower from flowing water. The terms "renewable energy," "green energy," "clean energy," and "alternative energy" are commonly used to describe energy derived from renewable resources. These sources are renewable, sustainable, abundant, and environmentally friendly. Countries have responded to the threat of exhaustion of nonrenewable energy by stepping up campaigns to embrace renewable forms of energy.

## 16.4  ROLE OF ENERGY IN SUSTAINABLE DEVELOPMENT

SD is taken as a guiding strategy for the world's social, economic, and environmental transformation and in which energy is kept at the center of any SD discussion. Energy is a basic requirement to achieve many basic human needs and services. It is a driver of the macroeconomic growth of the country and also a major source of environmental degradation if it is not properly used. Since energy plays a crucial role in economic and social advancement, apart from a production input, it is comprehensively considered a strategic commodity.

### 16.4.1  Energy and Economy

The industrial revolution was fueled by the unprecedented supply of coal and since then, the steady growth in energy consumption has been linked to increasing levels of production and economic opportunities. However, as per the standard model and the neoclassical theory of economic growth, only two important factors affects the production (GDP); they are capital (K) and labor (L), and that energy and other natural resources inputs have very little contribution to the economy due to their negligible role in the national accounts (Ayres & Voudouris, 2014). Treatment of energy as an explicit factor of production along with capital and labor was a retort to economic alarms raised by the Arab oil embargo that led to the "energy crises" of 1973–1974 and by the Iranian Revolution in 1979–1980. Both the events created oil price spikes followed by deep recessions (Ayres et al., 2013). According to Lindenberger and Kümmel (2011), economic productivity (Y) depends on production factors, viz. capital (K), labor (L), energy (E), and time (t).

$$GDP(Y) = fx(K, L, E)$$

$$k(t) = \frac{k(t)}{K_o}, \quad l(t) = \frac{L(t)}{L_o}, \quad E(t) = \frac{E(t)}{E_o}, \quad k, l, e \text{ are dimensionless quantities}$$

$$y(k, l, e, t) = \frac{Y\left(kK_o, lL_o, eE_o\right)}{Y_o}$$

$$\frac{dy}{y} = \alpha \frac{dk}{k} + \beta \frac{dl}{l} + \gamma \frac{de}{e} + \delta \frac{dt}{t - t_o} \quad \alpha, \beta, \gamma \text{ are output elasticities.}$$

α,β,γ represents the weights with which the growth rates of capital, labor, and energy to the growth of outputs, productive power of the production factors.

The output elasticity is much higher for energy and much lesser for labor than the cost-shares of these factors. Hence, energy and its conversion into end-use attributed as "technological progress," determines the economic state of nations (Lindenberger & Kümmel, 2011). Though globalization and market liberalization have attributed to the growth of emerging economies, energy is the oxygen of the economy as it is an input for nearly all goods and services. Without adequate and affordable energy, heat, light, and power, no one can build or run factories and cities that provide goods, jobs, homes, or life comforting amenities.

Energy consumption and per capita GDP or development, are highly correlated over time and space. An empirical study carried out by Esen and Bayark (2017) found that in the long-run contribution of energy consumption to the economic growth of low-income economies like Benin, Ethiopia, and Nepal is statistically insignificant but the energy consumption has a positive and statistically significant effect on the economic growth of lower to medium income economies indicating that the energy consumption can promote the development of economy (Lee, 2005; Mahalik & Mallick, 2014; Zhang-wei & Xun-gang, 2012), while the effect decreases as the level of development (income) increases, indicating that countries use their energy resources for production more effectively and more efficiently as their level of development increases. One of the examples, we can take from Nepal's case. There was a massive power shortage till 2020. Now, dramatic improvement has been made in Nepal producing surplus power, which is directly linked to the improvement of GDP. Nepal is going to be an upgraded developing nation in 2022. The result echoes the findings of a study done by Coers and Sanders (2013) on 30 OECD countries that offers a strong unidirectional causality running from GDP to energy usage and suggests policies aimed at reducing energy consumption or adopting energy efficiency are not likely to degrade economic growth.

Two major energy efficiency indicators are energy intensity, and economic energy efficiency.

Energy intensity is defined as the quantity of energy consumed to produce a unit GDP. This is a measure of the energy inefficiency of an economy. The higher the energy intensities, the higher the price or cost of converting energy into GDP. A nation that is highly economically productive with fuel-efficient technologies has lower energy intensity. Energy intensity in various region of the world is given in (Table 16.2). The inverse way of studying energy intensity is economic energy

**TABLE 16.2**
**Energy Indicators (2017)**

| Region/Country | Energy-Intensity Toe/$ 2010 PPP) |
|---|---|
| World | 119 |
| Developing countries | 123 |
| Europe | 109 |
| Central Asia | 201 |
| Eastern Asia | 136 |
| Western Asia | 108 |
| Southeast Asia | 95 |
| Southern Asia | 111 |
| Latin America and the Caribbean | 92 |
| Northern America | 129 |
| Northern Africa | 94 |
| Sub-Saharan Africa | 173 |

*Source:* IEA, https://www.iea.org/reports/sdg7-data-and-projections/energy-intensity#abstract.

efficiency or the economic rate of return on its consumption of energy, that is, the quantity of GDP produced per unit of energy consumed. India has the highest economic energy efficiency as compared to Nepal, China, and the world average. Target 7.3 of SDG "by 2030, double the global rate of improvement in energy efficiency" requires an average annual energy-intensity improvement of 2.7% (UN DESA, 2020)

### 16.4.2 ENERGY AND SOCIETY

Quality and quantity of energy available at any location determine how and how much crops are cultivated, how food is cooked, the health implications of cooking and heating practices, gender-based work division, what kinds of income-generating activities are adopted, and so on. Consequently, several social indicators posit a strong relationship with per capita energy use. The human development index measures human wellbeing in a society in terms of three dimensions; they are long and healthy life, being knowledgeable, and having a decent standard of living. Energy influences all these dimensions as it is undeniably essential to meet the basic human need as food and water, and acquire acceptable living conditions as to health care, education, shelter, and employment. Many rural populations of developing countries depend for survival on noncommercial energy sources, principally biomass, which is collected at a negligible monetary cost but huge environmental, labor, and health cost. Among different forms of energy, the most important form for human wellbeing and society is electricity. Table 16.3 shows the correlation between per capita energy consumption and HDI. The top five countries with the highest per capita electricity consumption, that is, Iceland, Bahrain, Canada, USA, and Japan have very high HDI (0.8–1.0) and the countries with a higher growth rate of per capita electricity consumption like China, Angola, and Laos have performed better in increasing HDI.

**TABLE 16.3**

**Per Capita Electricity Consumption vs HDI, 2018**

| Country | Per Capita Electricity Consumption (MWh) | | | Human Development Index | | |
|---|---|---|---|---|---|---|
| | 1990 | 2018 | Growth Rate (%) | 1990 | 2018 | Growth Rate (%) |
| Iceland | 16.14 | 54.6 | 4.4 | 0.807 | 0.946 | 0.57 |
| Bahrain | 15.62 | 18.6 | 0.6 | 0.749 | 0.852 | 0.46 |
| Canada | 16.17 | 15.4 | -0.2 | 0.850 | 0.928 | 0.31 |
| USA | 11.69 | 13.1 | 0.4 | 0.865 | 0.925 | 0.24 |
| Japan | 6.71 | 8.0 | 0.6 | 0.818 | 0.917 | 0.41 |
| Russian Federation | 6.67 | 6.9 | 0.1 | 0.735 | 0.823 | 0.40 |
| **China** | **0.51** | **4.9** | **8.4** | **0.499** | **0.755** | **1.49** |
| Mexico | 1.14 | 2.3 | 2.5 | 0.659 | 0.776 | 0.59 |
| India | 0.27 | 0.97 | 4.7 | 0.429 | 0.642 | 1.45 |
| **Angola** | **0.05** | **0.34** | **7.1** | **0.400**[a] | **0.582** | **2.11** |
| Equatorial Guinea | 0.07 | 1.05 | 10.2 | 0.525[a] | 0.582 | 0.57 |
| Nepal | 0.04 | 0.23 | 6.4 | 0.387 | 0.596 | 1.55 |
| Ethiopia | 0.02 | 0.08 | 5.1 | 0.292[a] | 0.478 | 2.78 |
| **Laos** | **0.12[a]** | **0.71** | **10.4** | **0.405** | **0.609** | **2.29** |
| Niger | 0.03[a] | 0.07 | 4.8 | 0.22 | 0.391 | 2.08 |

*Source:* From IEA, 2021, *IEA Energy Atlas*, International Energy Agency. With permission; UNDP, 2021, *Human Development Index*. United Nations Development Program. With permission.

[a] 2000.

*Note:* Countries in bold have higher growth rate of per capita electricity consumption and have performed better in increasing HDI.

### 16.4.3 ENERGY AND ENVIRONMENT

The production, refinement, transportation, and consumption of energy have significant consequences on local soil, water, and air. Of the total energy supply, 14.282 Mtoe, fossil fuels constitute 81.3% (IEA, 2021). The combustion of fossil fuels is accompanied by carbon emissions. Two third of greenhouse gas emissions come from energy extraction and usage. On one hand, energy is seen as an important factor supporting all socio-economic activities; on the other hand, it is the principal cause of higher average global temperature, acidic rain, environmental quality degradation, and severe weather events, which are harmful to the ecosystem. Climate change is the greatest challenge to sustainable development as it can potentially undermine equality, food security, healthy ecosystems, poverty alleviation, and other dimensions of SD. Limiting global warming to 1.5°C above the pre-industrial level is necessary to avoid catastrophic consequences and irreversible changes. However, this will result in a trade-off with efforts to achieve SDGs. The challenge is to formulate and implement policies that alleviate poverty, avoid ecosystem degradation while also cutting emissions, reducing climate change impacts, reducing poverty, and, facilitating adaptation.

So many developmental goals hinge on access to clean, affordable energy—from creating new economic opportunities to securing gender equality to addressing climate change (shown in Figure 16.5). Knowing where the world stands on SDG7 is especially important this year in the context of COVID-19 and the role energy access plays in fighting the pandemic. Health centers need reliable electricity to treat patients. Powered cold chains are needed to keep vaccines at the required temperatures as they are distributed to communities, both urban and rural. And access to clean cooking solutions reduces exposure to indoor air pollution that increases vulnerability to COVID-19 and other illnesses.

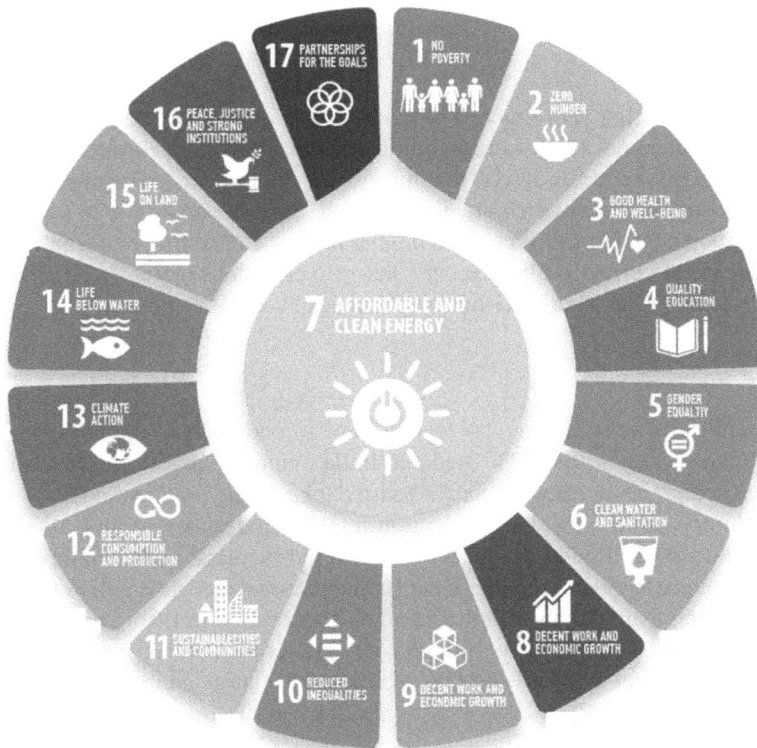

**FIGURE 16.5** Impact of clean and affordable energy on other SDG goals. (From Adhikari, M., Adhikari, N., Karanjit, S., Gautam, S., Maskey, M., Mallik, J. K., … Aryal, M. B (2020). *Interventions in Renewable Energy for Sustainable Development.* Retrieved from https://www.aepc.gov.np/rerl/public/uploads/documents/3YDLLh8AGKhgAjnab9cW5y4psuftriw8Hf9yes27.pdf. With permission.)

## 16.5  SUSTAINABLE DEVELOPMENT IN DEVELOPING AND DEVELOPED COUNTRIES

Sustainability is the process or actions through which humankind avoids the exhaustion of natural resources, meanwhile does not allow the quality of modern societies to decrease. John Elkington, often designated as the "godfather of sustainability" in his book *Green Swans: Regenerative Capitalism* (2020) addresses precisely the need to re-design business and economy since creating economic growth simply by increasing the consumption of materials goods is no longer a viable option. Several initiatives are been taken all around the world as preparedness for the next generation. Some of those inspiring the reluctance ones are as follows.

### 16.5.1  SWEDEN

Sweden tops among 149 countries that have adopted SDG targets that need to be met by 2030. The political environment of Sweden is characterized as peaceful and democratic for a long time and has incubated a tradition of collaboration between different performers in society – economic, political, and social. The country has been able to develop a welfare model with ambition and the ability to guarantee all inhabitants access to education, housing, employment, health care, and school. As regards the environment, Sweden has adopted a generational goal, which implies the conditions for solving environmental issues are to be met within one generation. Consequently, the GHG emission was reduced by 25% between 1990 and 2015 (Sustainable Development Knowledge Platform, 2017). Sweden aims to become the world's first fossil fuel-free welfare state, reaching net-zero emission by the year 2045 (Springtime–Intellecta, 2020).

### 16.5.2  CANADA

In Canada, SD is progressing very much. The country has attained an overall high score in social and economic development. Canada is responding to some challenges through concrete actions to advance gender equality, reduce poverty, and narrow the socio-economic gaps that exist between different groups, foster inclusion, celebrate diversity and improve equality of opportunity for all. Its geographical location contributes to relatively higher energy consumption, despite the government having made policies to curb GHG by placing a net-zero emissions target by 2050, carbon pricing, and commitment to ban sales of all fuel-burning vehicles by 2040.

### 16.5.3  INDIA

India is a geographically very big country where home to one-sixth of the human population and holds the key to the success of the 2030 Agenda. India's commitment to the SDGs can be sensed on the national development agenda as "Sabka Saath Sabka Vikas" (Collective Efforts for Inclusive Growth). The country achieved a sharp reduction in child and maternal mortality rate, provided clean cooking fuel to 80 million poor households, fostered financial inclusion through "Jan Dhan-Adhar-Mobile" (finance-national unique identity number-mobile phone) trinity for developing capabilities for entrepreneurship and employment, and succeeded in lifting more than 271 million people out of multidimensional poverty through economic growth and empowerment. The government of India has set a target to install 450 GW of renewable energy and restore 26 million hectares of degraded land by 2030 (HLPF, 2020).

### 16.5.4  EUROPE'S MOMENT

With the onset of Coronavirus, the characteristics of modern societies such as healthcare and welfare system, societies and economies, and the technique of living and working together have been shaken to its core. As to protect lives and livelihoods, repair and revitalize the Single Market and build a lasting prosperous recovery, the European Commission has launched a €1.85 trillion Green Recovery Program expected to be achieved through green transition and digital transformation. The recovery strategies resonate with European Green Deal:

a. Rolling out renewable energy and energy projects, especially wind, solar and kick-starting a clean hydrogen economy in Europe;
b. Cleaner transport/production and logistics or services, including the installation of 1 million charging points for electric vehicles and a boost for rail travel and clean mobility in European cities and regions;
c. Strengthening the Just Transition Fund to support re-skilling, helping businesses create new economic opportunities;
d. A massive restoration of infrastructure and buildings and a more circular economy, bringing local jobs;
e. Investing in more and better connectivity, especially in the rapid deployment of 5G networks;
f. A stronger industrial and technological presence in strategic sectors, including artificial intelligence, cybersecurity, supercomputing, and cloud;
g. Building a real data economy as a motor for innovation and job creation;
h. Increased cyber resilience.

Table 16.4 shows the top, medium, and least performers toward SDG and major SDG challenges for the respective countries. European countries are among the best performer but are still struggling to meet major two SDG goals: responsible consumption and production (12) and climate action (13). Interestingly the countries with the least SDG performance index have already achieved goal 13: climate actions.

**TABLE 16.4**
**SDG Performance Index of the UN Member States, 2020**

| Rank | Country | Score | SDG Achieved | SDG Major Challenges Remain |
|---|---|---|---|---|
| 1 | Sweden | 84.72 | 1, 3, 5, 7 | 12, 13 |
| 2 | Denmark | 84.56 | 1, 10 | 12, 13, 14 |
| 3 | Finland | 83.77 | 1, 4, 6, 7, 16 | 12, 13 |
| 4 | France | 81.13 | 1 | 12, 13 |
| 5 | Germany | 80.77 | | 12, 13, 14 |
| 81 | Panama | 69.19 | 7 | 2, 9, 10, 16 |
| 82 | Bahrain | 68.83 | | 2, 13 |
| 83 | Egypt | 68.79 | | 2, 3, 5, 8, 9, 10, 16 |
| 84 | Jamaica | 68.66 | | 2, 10, 14 |
| 85 | Nicaragua | 68.66 | 13 | 2, 3, 6, 8, 9, 10, 16 |
| 162 | Liberia | 47.12 | 12, 13 | 1, 2, 3, 4, 5, 6, 7, 8, 9, 11, 16, 17 |
| 163 | Somalia | 45.21 | 13 | 1, 2, 3, 5, 6, 7, 8, 9, 15, 16 |
| 164 | Chad | 43.75 | 13 | 1, 2, 3, 4, 5, 6, 7, 8, 9, 10, 11, 16 |
| 165 | South Sudan | 43.66 | 13 | 1, 2, 3, 4, 6, 7, 8, 9, 10, 11, 16 |
| 166 | The central African Republic | 38.54 | 13, 15 | 1, 2, 3, 4, 5, 6, 7, 8, 9, 10, 11, 12, 16, 17 |

*Source:* Sustainable Development Report (2020).

## 16.6  CONCLUSIONS

Energy is seen as an important factor supporting all socio-economic activities; but nonrenewable energy is the principal cause of higher average global temperature, acidic rain, environmental quality degradation, and severe weather events, which are harmful to the ecosystem. Energy and development are correlated with each other, which is the main driver of SD. To support SD, the world's population needs access to modern energy services that are both reliable, affordable and renewable. Thus, effective policy measures are required to kick-start the effective implementation of SD. In most of the countries, especially in the least developed and even in developing countries policies are available; but not in practice. There is a need for the effective implementation of these policies along with the supportive regulatory role of the government.

## REFERENCES

Adhikari, M., Adhikari, N., Karanjit, S., Gautam, S., Maskey, M., Mallik, J. K., … Aryal, M. B. (2020). *Interventions in Renewable Energy for Sustainable Development.* Retrieved from https://www.aepc. gov.np/rerl/public/uploads/documents/3YDLLh8AGKhgAjnab9cW5y4psuftriw8Hf9yes27.pdf.

AGENDA 21. (1992). *1992 Earth Submit.* Retrieved from https://sustainabledevelopment.un.org/content/ documents/Agenda21.pdf.

Ayres, R. U., van den Bergh, J. C. J. M., Lindenberger, D., & Warr, B. (2013). The underestimated contribution of energy to economic growth. *Structural Change and Economic Dynamics, 27,* 79–88. Doi: 10.1016/J. STRUECO.2013.07.004.

Ayres, R., & Voudouris, V. (2014). The economic growth enigma: Capital, labour and useful energy? *Energy Policy, 64,* 16–28. Doi: 10.1016/J.ENPOL.2013.06.001.

Bartelmus, P. (2013). Green accounting and energy. In *Reference Module in Earth Systems and Environmental Sciences.* Elsevier. Doi: 10.1016/b978-0-12-409548-9.01331-2.

Basiago, A. D. (1996). The search for the sustainable city in 20th century urban planning. In *Environmentalist* (Vol. 16(2), 135–155). Kluwer Academic Publishers. Doi: 10.1007/BF01325104.

Boyden, S. (1993). The human component of ecosystems. In M. J. McDonnell & S. T. A. Pickett (Eds.), *Humans as Components of Ecosystems: The Ecology of Subtle Human Effects and Populated Areas* (pp. 72–77). Springer New York. Doi: 10.1007/978-1-4612-0905-8_7.

Brodhag, C., & Taliere, S. (2006). Sustainable development strategies: Tools for policy coherence. *Natural Resources Forum, 30*(2), 136–145. Doi: 10.1111/j.1477–8947.2006.00166.x.

Coers, R. J., & Sanders, M. (2013). The energy-GDP nexus; addressing an old question with new methods. *Energy Economics, 36,* 708–715. Doi: 10.1016/j.eneco.2012.11.015.

Coomer, J. C. (1981). Quest for a sustainable society. In *Quest for a Sustainable Society.* Elsevier. Doi: 10.1016/ c2013-0-03517-7.

Daly, H. E. (1992). U.N. conferences on environment and development: Retrospect on Stockholm and prospects for Rio. *Ecological Economics, 5*(1), 9–14. Doi: 10.1016/0921–8009(92)90018-N.

Diesendorf, M. (2000). Chapter 2: *Sustainability and sustainable development.* In D. Dunphy, J. Benveniste, A. Griffiths, & P. Sutton (Eds.), Sustainability: The Corporate Challenge of the 21st Century (19-37). Allen & Unwin, Sydney.

Dixon, J. A., & Fallon, L. A. (1989). The concept of sustainability: Origins, extensions, and usefulness for policy. *Society and Natural Resources, 2*(1), 73–84. Doi: 10.1080/08941928909380675.

Du Pisani, J. A. (2006). Sustainable development-historical roots of the concept. *Environmental Sciences, 3*(2), 83–96. Doi: 10.1080/15693430600688831.

Elkington, J. (2020). *Green Swans The Coming Boom in Regenerative Capitalism.* Fast Company Press, New York.

Emas, R. (2015). *The Concept of Sustainable Development: Definition and Defining Principles.* Doi: 10.13140/ RG.2.2.34980.22404.

Esen, Ö., & Bayrak, M. (2017). Does more energy consumption support economic growth in net energy-importing countries. *Journal of Economics, Finance and Administrative Science, 22*(42), 75–98.

Goodland, R., & Daly, H. (1996). Environmental sustainability: Universal and non-negotiable. *Ecological Applications, 6*(4), 1002–1017. https://doi.org/10.2307/2269583.

Goudie, A. S. (1981). The human impact on the natural environment. In *Bulletin of Science, Technology & Society* (Issue 1). MIT Press. Doi: 10.1177/027046769101100141.

HLPF. (2020). *India:Voluntary National Review 2020*. High-Level Political Forum on Sustainable Development, Government of India, New Delhi. Retrieved from https://sustainabledevelopment.un.org/content/documents/26279VNR_2020_India_Report.pdf.

Holden, E., Linnerud, K., & Banister, D. (2014). Sustainable development: Our Common Future revisited. *Global Environmental Change*, *26*(1), 130–139. Doi: 10.1016/j.gloenvcha.2014.04.006.

IEA. (2021). *IEA Energy Atlas*. International Energy Agency. Retrieved from http://energyatlas.iea.org/#!/tellmap/-1118783123/1.

Jevons, W. S. (1865). *The Coal Question; An Inquiry Concerning the Progress of the Nation, and the Probable Exhaustion of our Coal-Mines* (1st ed.). Macmillan and Co, New York.

Kolk, A. (2016). The social responsibility of international business: From ethics and the environment to CSR and sustainable development. *Journal of World Business*, *51*(1), 23–34. Doi: 10.1016/j.jwb.2015.08.010.

Lee, C. C. (2005). Energy consumption and GDP in developing countries: A cointegrated panel analysis. *Energy Economics*, *27*(3), 415–427. Doi: 10.1016/j.eneco.2005.03.003.

Lindenberger, D., & Kümmel, R. (2011). Energy and the state of nations. *Energy*, *36*(10), 6010–6018. Doi: 10.1016/J.ENERGY.2011.08.014.

Mahalik, M. K., & Mallick, H. (2014). Energy consumption, economic growth and financial development: Exploring the empirical linkages for India. *The Journal of Developing Areas*, *48*(4), 139–159.

Malthus, T. R. (1798). *An Essay on the Principle of Population*. J. Johnson, London.

Marsh, G. P. (1864). *Man and Nature or Physical Geography as Modified by Human Actions*. Sampson Low, Son and Marston, London.

Mensah, J. (2019). Sustainable development: Meaning, history, principles, pillars, and implications for human action: Literature review. *Cogent Social Sciences*, *5*(1). Doi: 10.1080/23311886.2019.1653531.

Mill, J. S. (1870). Chapter VI: Of the stationary state. In W. J. Ashley (Ed.), *Principles of Political Economy: With Some of Their Applications to Social Philosophy* (7th ed.). Longmans, Green and Co, London.

Muralikrishna, I. V., & Manickam, V. (2017). Sustainable development. In *Environmental Management* (pp. 5–21). Elsevier. Doi: 10.1016/B978-0-12-811989-1.00002-6.

Pierobon, C. (2019). Promoting sustainable development through civil society: A case study of the EU's NSA/LA thematic programme in Kyrgyzstan. *Development Policy Review*, *37*(S2), O179–O192. Doi: 10.1111/dpr.12411.

Pigou, A. C. (1929). *The Economis of Welfare*. Macmillan and Co, Limited. New York.

Redclift, M. (1988). Sustainable development; Exploring the contradictions. In *Methuen & Co. Ltd* (Vol. 23, Issue 2). Doi: 10.1093/cdj/23.2.130.

Rostow, W. W. (1980). *The World Economy: History & Prospect*. University of Texas Press, Austin.

Sachs, J., Schmidt-Traub, G., Kroll, C., Lafortune, G., Fuller, G., & Woelm, F. (2020). *Sustainable Development Report*. Cambridge University Press, Cambridge.

Saith, A. (2006). From universal values to Millennium Development Goals: Lost in translation. *Development and Change*, *37*(6), 1167–1199. Doi: 10.1111/j.1467-7660.2006.00518.x.

Schrijver, N. J. (2008). The evolution of sustainable development in international law: Inception, meaning and status. In *The Evolution of Sustainable Development in International Law: Inception, Meaning and Status*. Martinus Nijhoff Publishers. Doi: 10.1163/9789047444466.

Shah, M. M. (2008). Sustainable development. In *Encyclopedia of Ecology, Five-Volume Set* (pp. 3443–3446). Elsevier Inc. Doi: 10.1016/B978-008045405-4.00633-9.

Springtime–Intellecta. (2020). *Sweden and the Leaving No One Behind Principle Nationally and Globally*. Retrieved from https://sustainabledevelopment.un.org/content/documents/26672Sweden_and_LNOB_2020.pdf.

Stephenson, M. H. (2021). Affordable and clean energy. In *Geosciences and the Sustainable Development Goals*. *Sustainable Development Goals Series* (pp. 159–182). Springer. Doi: 10.1007/978-3-030-38815-7_7.

Sustainable Development Knowledge Platform. (2017). *Sweden: Voluntary National Review 2017*. Government of Sweden. Retrieved from https://sdgtoolkit.org/wp-content/uploads/2017/06/VNR-Sweden-Full-report.pdf.

UN DESA. (2020). *Goal 7 | Department of Economic and Social Affairs*. United Nations. Retrieved from https://sdgs.un.org/goals/goal7.

UNDP. (2021). *Human Development Reports*. United Nations Development Programme. Retrieved from https://hdr.undp.org/data-center/human-development-index#/indicies/HDI.

United Nations Conference on the Human Environment. (1972). *Report on the United Nations Conference on the Human Environment*. Retrieved from https://digitallibrary.un.org/record/523249?ln=en.

Wallace, A. R. (1898). *The Wonderful Century* (1st ed.). Cambridge University Press. Doi: 10.1017/CBO9781139095013.

World Commission on Environment and Development. (1987). *Our Common Future*. Retrieved from https://sswm.info/sites/default/files/reference_attachments/UN%20WCED%201987%20Brundtland%20Report.pdf.

World Summit on Sustainable Development. (2002). *Sustainable Development Goals Knowledge Platform*. United Nations. Retrieved from: https://documents-dds-ny.un.org/doc/UNDOC/GEN/N02/636/93/PDF/N0263693.pdf?OpenElement.

Zhang-wei, L., & Xun-gang, Z. (2012). Study on relationship of energy consumption and economic growth in China. *Physics Procedia*, *24*, 313–319. Doi: 10.1016/j.phpro.2012.02.047.

# 17 Economics of Energy and Green Growth
## *Decoupling Debate*

*Meltem Ucal*
Kadir Has University

## CONTENTS

## 17.1 INTRODUCTION

Since economic growth enhances people's standard of living, it has been made the ultimate goal of the economy for a long time to solve unemployment, poverty and equity issues. To achieve economic growth, energy is considered a key factor because heating, lighting, transport, and the transformation of inputs into outputs are facilitated by energy. Therefore, it is understood that energy is important for economic development. However, ever-increasing energy use has negative environmental consequences such as increase in greenhouse gas emissions and the issue of climate change (Ackah and Kizys, 2015). At this point, green growth is considered a key element to achieving sustainable development. The reason is that in contrast to conventional economic growth, green growth is expected to protect the environment and allow economic growth simultaneously (Capasso et al., 2019). To achieve these objectives, international collaboration among developing and developed countries is needed (Sertyesilisik and Sertyesilisik, 2017).

Since the economic growth of the last century is inefficient because of the close and inverse relationship between the progress of humankind and the global environmental quality, anthropogenic pressures on the environment have increased, which in turn contributes to the problem of global warming. To face environmental degradation and to improve resource efficiency, 12th goal of the Sustainable Development Goals (SDG 12) aims to achieve sustainable production and consumption. Furthermore, target 8.4 of the SDGs aims to improve global resource efficiency in consumption and production and makes an effort to decouple economic growth from environmental degradation. It is obvious that a shift in policy-making is required toward acknowledging more the need for an integrated approach to production and consumption when environmental impacts are addressed. In this way, decoupling the economic growth from resource use and associated environmental impacts may be rendered easier (Sanyé-Mengual et al., 2019).

In the next section, the concept of green growth will be examined within the context of energy economics. In the subsection of it, some of the headline green growth indicators namely, production-based carbon dioxide productivity, GDP per unit of energy-related carbon dioxide emissions; demand-based carbon dioxide productivity, GDP per unit of energy-related carbon dioxide emissions, and environmentally adjusted multifactor productivity growth will be presented considering various OECD countries. In the third section, the concept of decoupling will be introduced. Next, statistical indicators related to decoupling will be shown regarding various countries in the fourth section. The fifth section concludes the study.

## 17.2 CONCEPTUAL ANALYSIS OF ENERGY ECONOMICS AND GREEN GROWTH

Energy demand has increased rapidly on a global scale, especially in emerging market economies. This results from population and economic growth in these economies. By 2030, it is expected that the global middle-class will double (Hennicke et al., 2014). If the situation continues in this way, demand for energy will rise by 1.3% each year by 2040 (IEA, 2019). Rising energy demand will create new challenges for the world when greater prosperity accompanies the population and economic growth. Since the number of consumers and their need for energy increase every day, the issue of energy security comes into prominence.

The availability of abundant stocks of fossil fuels with high energy density, which were the crucial factors for the Industrial Revolution, led to the decline of exhaustible resource stocks and the overuse of sinks for the uptake of waste products. This situation puts considerable pressure on natural ecosystems (Jakob and Edenhofer, 2014). The current economic growth model was a success in the 20th century; however, today it is required to have the necessary inexpensive raw materials to be able to sustain the model, which seems impossible. In addition, the earth has reached its limit because of increasing emissions and waste (Jänicke, 2012). Greenhouse gas emissions increased as a result of the rise in fossil fuel consumption levels and this situation exacerbated global warming (Fankhauser and Jotzo, 2018). The rise of the global middle-class led to unsustainable consumption and production patterns (Hennicke et al., 2014). Nevertheless, there is a solution to every problem. New thinking and new systems offer transformative solutions for producing, delivering, and consuming energy in the context of sustainable energy future. At this point, the idea of green growth becomes important (OECD, 2012).

Defining green growth can be possible in three ways. For instance, if one follows the OECD's definition for green growth, it is defined as "fostering economic growth and development while ensuring that natural assets continue to provide the resources and environmental services on which our well-being relies." According to the United Nations Environment Program (UNEP), green growth is defined as "green economy that simultaneously grows income and improves human well-being while significantly reducing environmental risks and ecological scarcities." World Bank defines green growth as "economic growth that is efficient in its use of natural resources, clean in that it minimizes pollution and environmental impacts, and resilient in that it accounts for natural hazards and the role of environmental management and natural capital in preventing physical disasters" (Hickel and Kallis, 2019: 2). The concept of green growth is also directly related to the concept of sustainable development. SDGs adopted by United Nations (UN) 2030 Agenda for Sustainable Development express the ambition to end poverty and create a sustainable economic growth path as well as to protect the planet from degradation (Van Vuuren et al., 2017).

The concepts of green economy and green growth come into prominence in environmental and economic discourse especially after the 2008 global financial crisis erupted and sustainable development failed to reconcile conflicting global economic, development, and ecological imperatives (Ferguson, 2015). Unfortunately, society's desire for prosperity through economic growth without consideration of its costs also aggravates the situation (Gupta, 2015). Although economic growth is considered as necessary to alleviate poverty and advance living standards by most economists and policymakers throughout the world, the growth comes at the cost of fast-growing natural resource use and carbon emissions, rising and increasingly volatile natural resource prices, and climate change (Scrieciu et al., 2013; Schandl et al., 2016). Therefore, the concept of green growth has become a dominant policy response to climate change and ecological breakdown. According to the green growth theory, continued economic expansion can be aligned with planet's ecology (Hickel and Kallis, 2019). It helps reduce environmental load while continuing the growth and facilitates related technological and structural change (Hoffmann, 2011).

Transforming one economy from currently prevailing socially and ecologically unsustainable economic situation (Ferguson, 2015) to a sustainable one is possible through developing green growth strategies. Adopting green growth strategies is helpful in reducing pollution, waste and greenhouse gas emissions, natural resource depletions, as well as strengthening energy efficiency, preserving biodiversity, etc. For instance, Lin et al. (2015) argue that the share of renewable energy can be expanded and green energy technology can be promoted in South Africa to increase the efficiency of energy consumption and to reduce carbon dioxide emissions. Economy and environment will be harmonized thanks to green growth and climate change will be prevented through it with the help of conservation of resources and energy (Dinda, 2014).

Environmental sustainability may have an impact on economic performance through four main channels identified by green growth literature. These are investment/capital accumulation, productivity/innovation, market efficiency, and economic resilience. Although green growth is a much wider concept than low-carbon energy, these channels help structure the clean energy growth discussion (Fankhauser and Jotso, 2018).

Providing concrete recommendations and measurement tools such as indicators is also very important to support countries' efforts to achieve economic growth and development. Therefore, Organization for Economic Cooperation and Development (OECD) studies these subjects and suggests a flexible policy framework for various circumstances and stages of development. In the next section, the values of green growth indicators will be addressed taking various countries into account.

### 17.2.1  GREEN GROWTH INDICATORS

There are different statistical indicators for measuring green growth; some of them are: (i) production-based carbon dioxide productivity, GDP per unit of energy-related carbon dioxide emissions; (ii) demand-based carbon dioxide productivity, GDP per unit of energy-related carbon dioxide emissions; and (iii) environmentally adjusted multifactor productivity growth. These are also considered as the headline indicators of green growth. Their values can be seen in Table 17.1.

**TABLE 17.1**
**Production-Based $CO_2$ Productivity, GDP per Unit of Energy-Related $CO_2$ Emissions by Selected OECD Countries**

| Country | Year | Production-Based $CO_2$ Productivity, GDP per Unit of Energy-Related $CO_2$ Emissions (Number, 2010) |
|---|---|---|
| Austria | 2010 | 5.12 |
| | 2011 | 5.40 |
| | 2012 | 5.70 |
| | 2013 | 5.63 |
| | 2014 | 6.00 |
| | 2015 | 5.95 |
| | 2016 | 6.09 |
| | 2017 | 5.97 |
| | 2018 | 6.38 |

*(Continued)*

**TABLE 17.1** (*Continued*)
**Production-Based CO$_2$ Productivity, GDP per Unit of Energy-Related CO$_2$ Emissions by Selected OECD Countries**

| Country | Year | Production-Based CO$_2$ Productivity, GDP per Unit of Energy-Related CO$_2$ Emissions (Number, 2010) |
|---|---|---|
| Belgium | 2010 | 4.20 |
| | 2011 | 4.79 |
| | 2012 | 4.82 |
| | 2013 | 4.77 |
| | 2014 | 5.17 |
| | 2015 | 4.95 |
| | 2016 | 5.07 |
| | 2017 | 5.25 |
| | 2018 | 5.35 |
| Czech Republic | 2010 | 2.58 |
| | 2011 | 2.70 |
| | 2012 | 2.77 |
| | 2013 | 2.87 |
| | 2014 | 3.04 |
| | 2015 | 3.17 |
| | 2016 | 3.19 |
| | 2017 | 3.31 |
| | 2018 | 3.44 |
| Denmark | 2010 | 5.05 |
| | 2011 | 5.75 |
| | 2012 | 6.57 |
| | 2013 | 6.33 |
| | 2014 | 7.22 |
| | 2015 | 7.96 |
| | 2016 | 7.78 |
| | 2017 | 8.52 |
| | 2018 | 8.54 |
| Finland | 2010 | 3.35 |
| | 2011 | 3.91 |
| | 2012 | 4.30 |
| | 2013 | 4.19 |
| | 2014 | 4.53 |
| | 2015 | 4.91 |
| | 2016 | 4.74 |
| | 2017 | 5.18 |
| | 2018 | 5.13 |
| France | 2010 | 6.86 |
| | 2011 | 7.40 |
| | 2012 | 7.36 |
| | 2013 | 7.40 |
| | 2014 | 8.29 |
| | 2015 | 8.20 |
| | 2016 | 8.23 |
| | 2017 | 8.30 |
| | 2018 | 8.83 |

(*Continued*)

**TABLE 17.1 (*Continued*)**
**Production-Based CO$_2$ Productivity, GDP per Unit of Energy-Related CO$_2$ Emissions by Selected OECD Countries**

| Country | Year | Production-Based CO$_2$ Productivity, GDP per Unit of Energy-Related CO$_2$ Emissions (Number, 2010) |
|---|---|---|
| Germany | 2010 | 4.22 |
| | 2011 | 4.54 |
| | 2012 | 4.48 |
| | 2013 | 4.39 |
| | 2014 | 4.74 |
| | 2015 | 4.78 |
| | 2016 | 4.85 |
| | 2017 | 5.07 |
| | 2018 | 5.41 |
| Greece | 2010 | 3.75 |
| | 2011 | 3.46 |
| | 2012 | 3.43 |
| | 2013 | 3.71 |
| | 2014 | 3.91 |
| | 2015 | 3.96 |
| | 2016 | 4.05 |
| | 2017 | 4.10 |
| | 2018 | 4.29 |
| Hungary | 2010 | 4.57 |
| | 2011 | 4.72 |
| | 2012 | 5.04 |
| | 2013 | 5.46 |
| | 2014 | 5.72 |
| | 2015 | 5.56 |
| | 2016 | 5.54 |
| | 2017 | 5.52 |
| | 2018 | 5.78 |
| Italy | 2010 | 5.29 |
| | 2011 | 5.43 |
| | 2012 | 5.53 |
| | 2013 | 5.90 |
| | 2014 | 6.25 |
| | 2015 | 6.11 |
| | 2016 | 6.25 |
| | 2017 | 6.44 |
| | 2018 | 6.64 |
| Poland | 2010 | 2.60 |
| | 2011 | 2.77 |
| | 2012 | 2.88 |
| | 2013 | 2.96 |
| | 2014 | 3.21 |
| | 2015 | 3.29 |
| | 2016 | 3.27 |
| | 2017 | 3.28 |
| | 2018 | 3.41 |

(*Continued*)

**TABLE 17.1** (*Continued*)
**Production-Based CO$_2$ Productivity, GDP per Unit of Energy-Related CO$_2$ Emissions by Selected OECD Countries**

| Country | Year | Production-Based CO$_2$ Productivity, GDP per Unit of Energy-Related CO$_2$ Emissions (Number, 2010) |
|---|---|---|
| Spain | 2010 | 5.67 |
| | 2011 | 5.56 |
| | 2012 | 5.49 |
| | 2013 | 5.97 |
| | 2014 | 6.14 |
| | 2015 | 5.97 |
| | 2016 | 6.41 |
| | 2017 | 6.19 |
| | 2018 | 6.58 |
| Sweden | 2010 | 8.59 |
| | 2011 | 9.62 |
| | 2012 | 10.32 |
| | 2013 | 10.90 |
| | 2014 | 11.28 |
| | 2015 | 11.84 |
| | 2016 | 11.84 |
| | 2017 | 12.24 |
| | 2018 | 13.02 |
| Switzerland | 2010 | 9.60 |
| | 2011 | 10.76 |
| | 2012 | 10.49 |
| | 2013 | 10.43 |
| | 2014 | 11.75 |
| | 2015 | 12.08 |
| | 2016 | 12.12 |
| | 2017 | 12.58 |
| | 2018 | 13.44 |
| Turkey | 2010 | 4.71 |
| | 2011 | 4.88 |
| | 2012 | 4.92 |
| | 2013 | 5.58 |
| | 2014 | 5.45 |
| | 2015 | 5.57 |
| | 2016 | 5.41 |
| | 2017 | 5.20 |
| | 2018 | 5.46 |
| United Kingdom | 2010 | 4.74 |
| | 2011 | 5.23 |
| | 2012 | 5.05 |
| | 2013 | 5.32 |
| | 2014 | 5.99 |
| | 2015 | 6.36 |
| | 2016 | 6.85 |
| | 2017 | 7.24 |
| | 2018 | 7.45 |

*Source:*  OECD, OECD Green Growth Indicators: Database Documentation, OECD Publishing, 2019. With permission.

This indicator can be calculated as the real gross domestic product (GDP), which is generated per unit of carbon dioxide emitted (US$/kg). GDP is expressed at constant 2010 US dollars ($) using purchasing power parity (PPP) (OECD, 2019).

The other headline indicator of green growth is demand-based carbon dioxide production. Its values can be seen in Table 17.2.

This indicator can be calculated as GDP that is generated per unit of carbon dioxide emitted from final demand (US$/kg). The carbon dioxide from energy use emitted during the various stages of production of goods and services consumed in domestic final demand, irrespective of where the stages of production occurred is reflected in demand-based emissions (OECD, 2019) (Table 17.3).

**TABLE 17.2**

**Demand-Based CO$_2$ Productivity, GDP per Unit of Energy-Related CO$_2$ Emissions by Selected OECD Countries**

| Country | Year | Demand-Based CO$_2$ Productivity, GDP per Unit of Energy-Related CO$_2$ Emissions |
|---|---|---|
| Austria | 2010 | 3.91 |
| | 2011 | 4.00 |
| | 2012 | 4.22 |
| | 2013 | 4.21 |
| | 2014 | 4.42 |
| | 2015 | 4.44 |
| Belgium | 2010 | 3.42 |
| | 2011 | 3.63 |
| | 2012 | 3.78 |
| | 2013 | 3.79 |
| | 2014 | 4.02 |
| | 2015 | 3.90 |
| Czech Republic | 2010 | 2.65 |
| | 2011 | 2.75 |
| | 2012 | 2.96 |
| | 2013 | 3.04 |
| | 2014 | 3.20 |
| | 2015 | 3.43 |
| Denmark | 2010 | 3.19 |
| | 2011 | 3.35 |
| | 2012 | 3.75 |
| | 2013 | 3.65 |
| | 2014 | 3.97 |
| | 2015 | 4.28 |
| Finland | 2010 | 3.05 |
| | 2011 | 3.37 |
| | 2012 | 3.61 |
| | 2013 | 3.57 |
| | 2014 | 3.70 |
| | 2015 | 3.98 |
| France | 2010 | 4.58 |
| | 2011 | 4.82 |
| | 2012 | 5.04 |
| | 2013 | 5.06 |
| | 2014 | 5.42 |
| | 2015 | 5.52 |

(*Continued*)

**TABLE 17.2** (*Continued*)
**Demand-Based $CO_2$ Productivity, GDP per Unit of Energy-Related $CO_2$ Emissions by Selected OECD Countries**

| Country | Year | Demand-Based $CO_2$ Productivity, GDP per Unit of Energy-Related $CO_2$ Emissions |
|---|---|---|
| Germany | 2010 | 3.46 |
| | 2011 | 3.65 |
| | 2012 | 3.79 |
| | 2013 | 3.73 |
| | 2014 | 3.96 |
| | 2015 | 4.09 |
| Greece | 2010 | 2.89 |
| | 2011 | 2.74 |
| | 2012 | 2.96 |
| | 2013 | 3.26 |
| | 2014 | 3.28 |
| | 2015 | 3.51 |
| Hungary | 2010 | 3.88 |
| | 2011 | 4.12 |
| | 2012 | 4.49 |
| | 2013 | 4.77 |
| | 2014 | 4.96 |
| | 2015 | 4.92 |
| Italy | 2010 | 3.94 |
| | 2011 | 4.03 |
| | 2012 | 4.36 |
| | 2013 | 4.57 |
| | 2014 | 4.77 |
| | 2015 | 4.76 |
| Poland | 2010 | 2.59 |
| | 2011 | 2.75 |
| | 2012 | 2.91 |
| | 2013 | 3.05 |
| | 2014 | 3.23 |
| | 2015 | 3.39 |
| Spain | 2010 | 4.48 |
| | 2011 | 4.51 |
| | 2012 | 4.75 |
| | 2013 | 5.07 |
| | 2014 | 5.09 |
| | 2015 | 5.02 |
| Sweden | 2010 | 4.68 |
| | 2011 | 4.93 |
| | 2012 | 5.26 |
| | 2013 | 5.37 |
| | 2014 | 5.66 |
| | 2015 | 6.26 |
| Switzerland | 2010 | 4.32 |
| | 2011 | 4.19 |
| | 2012 | 4.36 |
| | 2013 | 4.42 |
| | 2014 | 4.90 |
| | 2015 | 4.79 |

(*Continued*)

**TABLE 17.2 (*Continued*)**
**Demand-Based CO$_2$ Productivity, GDP per Unit of Energy-Related CO$_2$ Emissions by Selected OECD Countries**

| Country | Year | Demand-Based CO$_2$ Productivity, GDP per Unit of Energy-Related CO$_2$ Emissions |
|---|---|---|
| Turkey | 2010 | 3.76 |
| | 2011 | 3.82 |
| | 2012 | 4.11 |
| | 2013 | 4.32 |
| | 2014 | 4.53 |
| | 2015 | 4.74 |
| United Kingdom | 2010 | 3.51 |
| | 2011 | 3.81 |
| | 2012 | 3.74 |
| | 2013 | 3.88 |
| | 2014 | 4.15 |
| | 2015 | 4.35 |

*Source:*   OECD, *OECD Green Growth Indicators: Database Documentation*, OECD Publishing, 2019. With permission.

**TABLE 17.3**
**Environmentally Adjusted Multifactor Productivity Growth by Selected OECD Countries**

| Country | Year | Environmentally Adjusted Multifactor Productivity Growth |
|---|---|---|
| Austria | 2000 | 2.26 |
| | 2001 | 0.78 |
| | 2002 | 1.21 |
| | 2003 | 0.08 |
| | 2004 | 2.54 |
| | 2005 | 1.83 |
| | 2006 | 2.53 |
| | 2007 | 3.12 |
| | 2008 | 0.89 |
| | 2009 | −1.38 |
| | 2010 | 0.98 |
| | 2011 | 2.13 |
| | 2012 | 0.38 |
| | 2013 | 0.17 |
| Belgium | 2000 | 1.71 |
| | 2001 | 0.05 |
| | 2002 | 1.81 |
| | 2003 | 0.86 |
| | 2004 | 2.99 |
| | 2005 | 1.51 |
| | 2006 | 1.70 |
| | 2007 | 2.32 |
| | 2008 | −0.13 |

(*Continued*)

**TABLE 17.3 (*Continued*)**
**Environmentally Adjusted Multifactor Productivity Growth by Selected OECD Countries**

| Country | Year | Environmentally Adjusted Multifactor Productivity Growth |
|---|---|---|
| | 2009 | −1.13 |
| | 2010 | 1.10 |
| | 2011 | 1.73 |
| | 2012 | 0.12 |
| | 2013 | 0.50 |
| Czech Republic | 2000 | 3.14 |
| | 2001 | 3.44 |
| | 2002 | 1.05 |
| | 2003 | 2.38 |
| | 2004 | 3.95 |
| | 2005 | 4.25 |
| | 2006 | 4.83 |
| | 2007 | 4.19 |
| | 2008 | 0.52 |
| | 2009 | −2.81 |
| | 2010 | 0.36 |
| | 2011 | 1.92 |
| | 2012 | −0.66 |
| | 2013 | 0.03 |
| Denmark | 2000 | 2.61 |
| | 2001 | −0.25 |
| | 2002 | 0.17 |
| | 2003 | 0.31 |
| | 2004 | 2.88 |
| | 2005 | 2.11 |
| | 2006 | 1.95 |
| | 2007 | 0.66 |
| | 2008 | −0.98 |
| | 2009 | −3.41 |
| | 2010 | 2.63 |
| | 2011 | 1.40 |
| | 2012 | 0.05 |
| | 2013 | 0.04 |
| Finland | 2000 | 4.60 |
| | 2001 | 1.49 |
| | 2002 | 1.01 |
| | 2003 | 1.50 |
| | 2004 | 3.49 |
| | 2005 | 2.92 |
| | 2006 | 1.80 |
| | 2007 | 3.63 |
| | 2008 | 0.06 |
| | 2009 | −5.70 |
| | 2010 | 1.98 |
| | 2011 | 2.69 |
| | 2012 | −1.53 |
| | 2013 | −0.06 |

(*Continued*)

**TABLE 17.3 (*Continued*)**
**Environmentally Adjusted Multifactor Productivity Growth by Selected OECD Countries**

| Country | Year | Environmentally Adjusted Multifactor Productivity Growth |
|---------|------|----------------------------------------------------------|
| France | 2000 | 3.31 |
| | 2001 | 1.22 |
| | 2002 | 2.18 |
| | 2003 | 0.82 |
| | 2004 | 1.83 |
| | 2005 | 1.48 |
| | 2006 | 2.82 |
| | 2007 | 1.34 |
| | 2008 | −0.26 |
| | 2009 | −1.13 |
| | 2010 | 1.44 |
| | 2011 | 2.48 |
| | 2012 | 0.41 |
| | 2013 | 0.20 |
| Germany | 2000 | 2.85 |
| | 2001 | 2.21 |
| | 2002 | 1.11 |
| | 2003 | 0.47 |
| | 2004 | 1.61 |
| | 2005 | 1.71 |
| | 2006 | 2.76 |
| | 2007 | 2.91 |
| | 2008 | 0.51 |
| | 2009 | −2.38 |
| | 2010 | 2.21 |
| | 2011 | 3.22 |
| | 2012 | 0.37 |
| | 2013 | 0.16 |
| Greece | 2000 | 2.07 |
| | 2001 | 2.55 |
| | 2002 | 2.37 |
| | 2003 | 3.98 |
| | 2004 | 3.36 |
| | 2005 | −1.30 |
| | 2006 | 4.55 |
| | 2007 | 2.08 |
| | 2008 | −0.40 |
| | 2009 | −4.51 |
| | 2010 | −4.11 |
| | 2011 | −7.01 |
| | 2012 | −4.46 |
| | 2013 | −2.32 |
| Hungary | 2000 | 3.05 |
| | 2001 | 2.79 |
| | 2002 | 2.94 |
| | 2003 | 2.05 |
| | 2004 | 4.30 |

(*Continued*)

**TABLE 17.3 (*Continued*)**
**Environmentally Adjusted Multifactor Productivity Growth by Selected OECD Countries**

| Country | Year | Environmentally Adjusted Multifactor Productivity Growth |
|---|---|---|
| | 2005 | 3.81 |
| | 2006 | 2.49 |
| | 2007 | −0.57 |
| | 2008 | 0.89 |
| | 2009 | −4.59 |
| | 2010 | −0.14 |
| | 2011 | 1.49 |
| | 2012 | 0.19 |
| | 2013 | 0.61 |
| Italy | 2000 | 2.73 |
| | 2001 | 1.25 |
| | 2002 | −0.19 |
| | 2003 | −0.39 |
| | 2004 | 1.69 |
| | 2005 | 0.72 |
| | 2006 | 1.74 |
| | 2007 | 1.00 |
| | 2008 | −0.28 |
| | 2009 | −3.03 |
| | 2010 | 2.12 |
| | 2011 | 1.42 |
| | 2012 | −0.21 |
| | 2013 | −0.93 |
| Poland | 2000 | 3.62 |
| | 2001 | 0.97 |
| | 2002 | 1.74 |
| | 2003 | 2.91 |
| | 2004 | 3.22 |
| | 2005 | 1.50 |
| | 2006 | 2.68 |
| | 2007 | 3.42 |
| | 2008 | −0.22 |
| | 2009 | 0.81 |
| | 2010 | 1.23 |
| | 2011 | 2.85 |
| | 2012 | 0.72 |
| | 2013 | −0.96 |
| Spain | 2000 | 2.23 |
| | 2001 | 1.51 |
| | 2002 | 1.04 |
| | 2003 | 1.25 |
| | 2004 | 1.38 |
| | 2005 | 1.79 |
| | 2006 | 2.25 |
| | 2007 | 1.83 |
| | 2008 | 1.17 |
| | 2009 | −1.25 |

(*Continued*)

**TABLE 17.3 (*Continued*)**
**Environmentally Adjusted Multifactor Productivity Growth by Selected OECD Countries**

| Country | Year | Environmentally Adjusted Multifactor Productivity Growth |
|---------|------|---------------------------------------------------------|
|  | 2010 | 1.06 |
|  | 2011 | 0.03 |
|  | 2012 | −0.41 |
|  | 2013 | −0.32 |
| Sweden | 2000 | 3.29 |
|  | 2001 | 0.17 |
|  | 2002 | 2.36 |
|  | 2003 | 2.67 |
|  | 2004 | 3.34 |
|  | 2005 | 2.39 |
|  | 2006 | 3.15 |
|  | 2007 | 1.04 |
|  | 2008 | −1.59 |
|  | 2009 | −3.06 |
|  | 2010 | 2.91 |
|  | 2011 | 1.61 |
|  | 2012 | −0.07 |
|  | 2013 | 0.82 |
| Switzerland | 2000 | 3.27 |
|  | 2001 | 1.23 |
|  | 2002 | 0.55 |
|  | 2003 | −0.20 |
|  | 2004 | 1.68 |
|  | 2005 | 2.42 |
|  | 2006 | 2.92 |
|  | 2007 | 3.07 |
|  | 2008 | 0.34 |
|  | 2009 | −1.90 |
|  | 2010 | 1.81 |
|  | 2011 | 1.53 |
|  | 2012 | 0.49 |
|  | 2013 | 1.19 |
| Turkey | 2000 | 2.90 |
|  | 2001 | −4.80 |
|  | 2002 | 4.01 |
|  | 2003 | 2.25 |
|  | 2004 | 4.88 |
|  | 2005 | 3.34 |
|  | 2006 | 1.48 |
|  | 2007 | −1.25 |
|  | 2008 | −0.91 |
|  | 2009 | −6.65 |
|  | 2010 | 4.34 |
|  | 2011 | 4.33 |
|  | 2012 | −1.51 |
|  | 2013 | ------ |

(*Continued*)

**TABLE 17.3 (*Continued*)**
**Environmentally Adjusted Multifactor Productivity Growth by Selected OECD Countries**

| Country | Year | Environmentally Adjusted Multifactor Productivity Growth |
|---|---|---|
| United Kingdom | 2000 | 3.32 |
| | 2001 | 2.16 |
| | 2002 | 2.86 |
| | 2003 | 3.77 |
| | 2004 | 2.67 |
| | 2005 | 1.92 |
| | 2006 | 2.44 |
| | 2007 | 2.11 |
| | 2008 | 0.29 |
| | 2009 | −2.13 |
| | 2010 | 1.68 |
| | 2011 | 2.21 |
| | 2012 | −0.66 |
| | 2013 | 0.30 |

*Source:* OECD, *OECD Green Growth Indicators: Database Documentation*, OECD Publishing, 2019. With permission.

The change in productivity at the country level encompassing technological change, institutional and organizational improvements is measured through environmentally adjusted multifactor productivity growth. A growth accounting approach is used when it is calculated. According to this approach, when GDP increases or pollution decreases, environmentally adjusted multifactor productivity growth (EAMFP) increases for a given growth of input use.

## 17.3  INTRODUCING THE CONCEPT OF DECOUPLING

According to OECD, decoupling has been defined as "breaking the link between environmental bads and economic goods" (OECD, 2003). This concept is also one of the five interlinked objectives identified by OECD so as to enhance cost-effective and practical environmental policies in the context of sustainable development (Chen et al., 2017). In another article, they also argue that the realization of energy savings and emission-reduction targets is only possible through the degree of the decoupling between carbon dioxide emissions and economic growth (Chen et al., 2018). Decoupling economic growth and human well-being from increasing resource use and negative environmental impacts plays a crucial role in securing long-term sustainability for humankind (Schandl et al., 2016).

Environmental problems emerge because of increasing global energy use. As a result of this, greenhouse gas and air pollution increase and eventually climate change occurs (Csereklyei and Stern, 2015; Shuai et al., 2019). Global warming has been an important issue since the late 1980s. Energy analysts and policymakers have been concerned about the adverse effects of energy-related carbon dioxide emissions because of increasing energy use (Wang et al., 2013). These are important warning signs, which show the world that continuous use of fossil fuels is not sustainable. Therefore, the concept of decoupling is very important in the sense that it helps break the link between economic growth and carbon emission levels (Andreoni and Galmarini, 2012). Decoupling GDP growth from resource use and carbon emissions through technological change and substitution helps continued economic expansion aligned with the planet's ecology (Hickel and Kallis, 2019). It is seen as an imperative in promoting low-carbon economy (Shuai et al., 2019).

There are also three decoupling factors, which affect the relationship between energy and GDP, namely the technological changes, shifts in the composition of the energy inputs, and finally shifts in the composition of outputs. This means that the improvement in energy conservation, which leads to the rate of GDP growth to be higher than the rate of growth of inputs may be possible through new inventions and innovations. In addition, the composition of energy inputs can be changed by a general shift from low-quality fuels to higher-quality fuels. In this way, reducing the energy intensity could be possible. Finally, energy-income relationships may change regarding the shift in the composition of outputs because energy intensities are different in industries (Stern, 2004a).

Reducing energy intensity is very important for well-developed countries because, at one point, it does not negatively affect economic growth. Once the link between energy and output can be broken, initiating energy conservation policies will become possible without creating negative economic side-effects (Climent and Pardo, 2007).

Achieving economic development without causing environmental degradation is an important aspect of the environmental management literature. On one hand, some studies in the literature support the existence of the Environmental Kuznets Curve (EKC), which suggests that as economies grow, environmental degradation increases at first, but then, it reduces thanks to more affluence in the economy. On the other hand, some studies have doubts about countries' growth without causing environmental problems (Conrad and Cassar, 2014).

There is no agreement on precise emission-reduction targets for each country, however, it is known that reducing emissions is possible through different metrics. Targeting carbon intensity is important because it guarantees emissions decoupling. There are three quantitative indicators of decoupling. The first one is decoupling factor and it is based on the rate of growth of emissions intensity. The second one is an emissions-to-economic activity elasticity. Finally, the third one includes the emissions' intensity decreasing rate as well as GDP growth (Grand, 2016).

Gross (2017) examines the relationship between energy and growth in West Germany in 1973–1986 period. During this period, a new energy paradigm has risen in the country because of the oil shocks. The experts found a way for reducing energy consumption while expanding gross domestic production, namely decoupling. This concept helps clarify the divergence of Germany from other large, industrialized states in the 1980s and explains its increased focus on energy conservation rather than on increasing its energy supply.

Parrique et al. (2019) mention decoupling debate in their report. According to the findings of the report, a rigorous analytical framework is necessary to discuss decoupling. Decoupling can be global or local, relative or absolute, territorial or footprint-based. Decoupling of natural resources consumption and environmental impacts from economic growth can be possible through technological progress and structural change. When both variables develop in the same direction but not at the same speed, then relative decoupling occurs. When the two variables go in opposite directions, absolute decoupling occurs. Absolute and relative decoupling can also be used interchangeably with the terms "strong" and "weak" decoupling in the literature (Shi et al., 2016).

## 17.4  COMPARISON WITH STATISTICAL INDICATORS

Portugal has achieved a 20% decline in primary energy intensity for two decades. Furthermore, electricity and natural gas were produced by renewable energy sources this year. The country's renewable share of total primary energy use was 23.5%, which was above the European target of 20% by 2020. It is considered as an important achievement in energy decoupling (Guevara and Domingos, 2017).

The United States is the world's second-largest country that contributes to carbon dioxide emissions. In the 2006–2011 period, energy-related carbon emissions were reduced by 7.7%, which was a percentage more than any country or region. During this time, the United States economy also continued to grow (Jiang et al., 2016). According to this information, the authors examine the factors and systems, which influence the energy-related carbon emissions and discuss the effective measures and policies for decoupling economic growth and carbon emissions in the United States. According to their findings, carbon dioxide emissions increased between 1990–2005 period, while they declined between 2006–2014 period. From 1990 to 2014, per capita, carbon dioxide emissions generally declined by 14.58%. Economic activities and population growth were important factors that negatively affect carbon dioxide emissions. Energy intensity curbed the increase of carbon dioxide emissions (Jiang et al., 2016).

Jiang et al. (2019) combine decoupling analysis with the EKC modeling to explore the interactions between carbon dioxide emissions and economic growth in Guangdong, which is one of the most representative provinces in China. Within the period of 1995–2014, GDP and carbon emissions increased in Guangdong. However, the increase in carbon emissions (from 154.2 to 521.3 Mt $CO_2$) was slower than the increase in economic output (from 191.6 to 1465.9 billion RMB). These statistics indicate that there exists a state of expansive weak decoupling.

Lu et al. (2015) used a decomposition technique while decoupling industrial growth from carbon emissions in Jiangsu province, which is located in China. Within the period of 2005–2012, industrial outputs, energy consumption, and carbon emissions increased in the province with the average annual growth rates of 9.74%, 3.38%, and 3.70%, respectively. The speed of increase slowed down, especially for carbon emissions in this period. In the last 3 years (from 2009 to 2012), the increase in carbon emissions dropped from 12.35% to 2.69%. This means that energy-saving and emission-reduction came to the fore in the province.

Lin et al. (2015) evaluated the occurrence of decoupling of carbon dioxide emissions from GDP in South Africa. In the period of 2010–2012, a strong decoupling occurred in South Africa. In this period, the increase in GDP was 3.07%, however, carbon dioxide emissions declined by 2.83%. On the other hand, in the period of 1994–2010, a weak decoupling occurred in the country. To be more precise, in the period of 1994–2000, the increase in GDP was 2.90% while the increase in carbon dioxide emissions was 1.08%. In the period of 2000–2005, the increase in GDP was 3.89% while the increase in carbon dioxide emissions was 1.34%. Finally, in the period of 2005–2010, the increase in GDP was 3.61%, while the increase in carbon dioxide emissions was 1.48%. It is important to note that an expansive negative decoupling state occurred in South Africa within the period of 1990–1994, which means the increase in carbon dioxide emissions (1.97%) was higher than the increase in GDP (0.33%) in the country.

Marques et al. (2018) examined the impact of economic growth on carbon dioxide emissions in Australia using the EKC and—decoupling index. The decoupling index was 0.3367. This result indicated that there was a weak decoupling for the period of 1965–2016. According to this result, both economic growth and carbon dioxide emissions increased in the period, but the increase in carbon dioxide emissions was slower than that of economic growth. In the first 10-year interval (1965–1975), they found a negative decoupling (−0.0978), which means that the increase in carbon dioxide emissions was faster than that of economic growth. This is not the case for the following decades. The values of the decoupling index were 0.0686 in the period of 1975–1985, 0.0355 in the period of 1985–1995, 0.1728 in the period of 1995–2005, and finally 0.1869 in the period of 2005–2016.

Meng et al. (2018) did a decoupling, decomposition, and forecasting analysis of China's fossil energy consumption from industrial output in the 2001–2014 period. According to their results, China's non-fossil electricity generation increased up to 12.74% within the period of 2001–2010. However, this situation changed in 2011. In this year, China's non-fossil electricity generation was prevented by the drought. Therefore, the growth rate of non-fossil electricity generation was only 0.47%. The drought also occurred in 2013 and this caused an increase in fossil energy consumption in the thermal power sector. In the 2013–2014 period, the decoupling result of fossil energy consumption from value-added of the industry was −0.148.

Pao and Chen (2019) proposed a strategy for decoupling environmental pressure from economic growth in G20 countries. According to the descriptive statistical analysis, they found an absolute decoupling effect because environmental pressure dropped and economy continued to grow in these countries. G20 countries reduced their carbon dioxide emissions rates (both for the 5 (−0.962), 10 (−0.587) and 15 (−0.048) years), fossil fuel consumption (both for the 5 (−0.686) and 10 (−0.335) years), and nuclear energy consumption (both for the 5 (−0.732), 10 (−1.076) and 15 (−0.351) years). At the same time, renewable energy (for 5 years (13.637) and for 10 years (14.480)) and economic growth levels (for 5 years (0.940), for 10 years (0.834) for 15 years (1.209)) increased in G20 countries.

Roinioti and Koroneos (2017) aimed to find the driving forces of carbon dioxide emissions related to energy consumption in Greece using the complete decomposition technique. According to the results, carbon dioxide emissions increased by 2.09 Mt before the crisis period (2003–2008). Carbon dioxide emissions of the energy sector declined by 8.8 Mt in the crisis period (2008–2013). In the entire analyzed period (2003–2013), carbon dioxide emissions declined by 6.7 Mt. In addition, a strong decoupling occurred in 2005–2006, 2007–2008, and finally 2008–2009 in Greece. The values of this effect were 1.153, 1.960, and 5.081, respectively. A weak decoupling occurred in 2003–2004, 2004–2005, 2006–2007, and 2009–2010 in the country. The values of this effect were 0.892, 0.064, 0.404, and 0.210, respectively.

Román-Collado et al. (2018) developed a decoupling elasticity analysis and a two-level decomposition analysis of energy consumption in Columbia from 2000 to 2015. According to their findings, strong decoupling occurred in 2001–2002 with a decline in energy consumption (−0.02) and an increase in GDP (0.03); in 2004–2005 with a decline in energy consumption (−0.02) and an increase in GDP (0.05); in 2006–2007 with a decline in energy consumption (−0.09) and an increase in GDP (0.06); and in 2009–2010 with a decline in energy consumption (−0.07) and an increase in GDP (0.04). In addition, a weak decoupling occurred in 2002–2003 with energy consumption change (0.00) and GDP change (0.04); in 2005–2006 with energy consumption change (0.01) and GDP change (0.06); in 2011–2012 with energy consumption change (0.03) and GDP change (0.04); and finally in 2012–2013 with energy consumption change (0.02) and GDP change (0.05). Expansive negative decoupling occurred in 2000–2001 with energy consumption change (0.02) and GDP change (0.01); in 2007–2008 with energy consumption change (0.22) and GDP change (0.03); in 2008–2009 with energy consumption change (0.04) and GDP change (0.02); in 2013–2014 with energy consumption change (0.09) and GDP change (0.04); and finally in 2014–2015 with energy consumption change (0.05) and gross domestic product change (0.03).

Ru et al. (2012) studied the relationship between economic growth and carbon emissions based on decoupling theory both in developed and developing countries. Their comparative findings can be found in Table 17.4.

Besides the information found in Table 17.4, expansive negative decoupling occurred in Canada in 1960–1965 (0.96); in Japan in 1966–1971 (1.33); in the United States (US) in 1966–1971 (1.58); in China in 1966–1971 (1.76) and in 1972–1978 (1.44); in Brazil in 1990–1995 (1.86) and in 1996–2001 (1.33); and in India in 1960–1965 (1.83), in 1972–1978 (1.72), in 1979–1983 (1.79) and in 1984–1989 (1.39).

## 17.5   DISCUSSION AND CONCLUSION

Although the prices of natural resources have increased, the global consumption of them has also increased continuously. The problem is that these resources are limited on the earth and also damage natural environment. Therefore, conventional economic growth is not sustainable. It is important to understand the urgency to shift toward low-carbon economic model and lifestyle from the current one and to achieve sustainable development that can be supported by the green economy. According to the understanding of green economy, the economy depends on the environment and the coexistence of people and nature. Green growth contributes to resource productivity and provides a way for increasing efficiency through environmental and resource-saving processes and products (Sertyesilisik and Sertyesilisik, 2017).

National, regional, and international institutions such as the OECD, European Union (EU), and the UN use decoupling as a policy objective. For example, the Green Economy Initiative of the UNEP attaches great importance to decoupling human well-being from resource consumption. Since the consumption of natural resources has increased rapidly and this situation has a negative impact on the natural environment, it is of the utmost importance to decouple material and energy flows from social and economic progress urgently. At this point, significant changes in government policies, corporate behavior, and consumption patterns are required for achieving decoupling (UNEP, 2011).

**TABLE 17.4**

**A Comparison between Tapio Decoupling Indicators of Nine Countries**

| Countries | Period | Strong Decoupling | Period | Weak Decoupling |
|---|---|---|---|---|
| Canada | 1979–1983 | −0.23 | 1972–1978 | 0.21 |
| | | | 1984–1989 | 0.42 |
| | | | 1990–1995 | 0.25 |
| | | | 1996–2001 | 0.58 |
| | | | 2002–2007 | 0.48 |
| France | 1979–1983 | −1.81 | 1972–1978 | 0.00 |
| | 1984–1989 | −0.24 | | |
| | 1990–1995 | −0.25 | | |
| | 1996–2001 | −0.38 | | |
| | 2002–2007 | −0.23 | | |
| Japan | 1979–1983 | −0.36 | 1984–1989 | 0.38 |
| | 1996–2001 | −1.53 | 2002–2007 | 0.13 |
| Sweden | 1979–1983 | −3.63 | 1966–1971 | 0.80 |
| | 1984–1989 | −0.14 | | |
| | 1996–2001 | −1.00 | | |
| | 2002–2007 | −0.90 | | |
| United Kingdom | 1972–1978 | −0.70 | 1960–1965 | 0.41 |
| | 1979–1983 | −2.14 | 1966–1971 | 0.52 |
| | 1990–1995 | −0.12 | 1984–1989 | 0.49 |
| | 1996–2001 | −0.30 | 2002–2007 | 0.11 |
| United States | 1979–1983 | −1.49 | 1960–1965 | 0.72 |
| | | | 1972–1978 | 0.12 |
| | | | 1984–1989 | 0.55 |
| | | | 1990–1995 | 0.57 |
| | | | 1996–2001 | 0.31 |
| | | | 2002–2007 | 0.18 |
| China (Mainland) | 1960–1965 | −26.31 | 1979–1983 | 0.33 |
| | | | 1984–1989 | 0.59 |
| | | | 1990–1995 | 0.50 |
| | | | 1996–2001 | 0.02 |
| India | – | – | 1966–1971 | 0.76 |
| | | | 1996–2001 | 0.70 |
| | | | 2002–2007 | 0.63 |
| Brazil | 1979–1983 | −0.75 | 2002–2007 | 0.49 |

*Source:* Ru, X., Chen, S. and Dong, H., J. Sustain. Dev., 5(8), 43, 2012. With permission.
Strong decoupling: $m < 0$. Weak decoupling: $0 \leq m < 0.8$. $m$ indicates decoupling elasticity.

Because the energy system has a negative impact on climate change in the form of increasing carbon dioxide emissions resulted from fossil fuel combustion, decoupling energy use from economic growth is essential. European Environment Agency (EEA) points out that almost all EU countries either decrease energy intensity levels steadily or reduce the ratio of final energy consumption and GDP (Moreau and Vuille, 2018). Further studies will help expand the knowledge of the green economy, the impact of increasing energy use, and decoupling methods.

## REFERENCES

Ackah, I. and Kizys, R. (2015). Green growth in oil producing African countries: A panel data analysis of renewable energy demand. *Renewable and Sustainable Energy Reviews*, 50, pp. 1157–1166. https://doi.org/10.1016/j.rser.2015.05.030.

Andreoni, V. and Galmarini, S. (2012). Decoupling economic growth from carbon dioxide emissions: A decomposition analysis of Italian energy consumption. *Energy*, 44(1), pp. 682–691. https://doi.org/10.1016/j.energy.2012.05.024.

Capasso, M., Hansen, T., Heiberg, J., Klitkou, A. and Steen, M. (2019). Green growth–A synthesis of scientific findings. *Technological Forecasting and Social Change*, 146, pp. 390–402. https://doi.org/10.1016/j.techfore.2019.06.013.

Chen, B., Yang, Q., Li, J.S. and Chen, G.Q. (2017). Decoupling analysis on energy consumption, embodied GHG emissions and economic growth—The case study of Macao. *Renewable and Sustainable Energy Reviews*, 67, pp. 662–672. https://doi.org/10.1016/j.rser.2016.09.027.

Chen, J., Wang, P., Cui, L., Huang, S. and Song, M. (2018). Decomposition and decoupling analysis of $CO_2$ emissions in OECD. *Applied Energy*, 231, pp. 937–950. https://doi.org/10.1016/j.apenergy.2018.09.179.

Climent, F. and Pardo, A. (2007). Decoupling factors on the energy–output linkage: The Spanish case. *Energy Policy*, 35(1), pp. 522–528. https://doi.org/10.1016/j.enpol.2005.12.022.

Conrad, E. and Cassar, L.F. (2014). Decoupling economic growth and environmental degradation: reviewing progress to date in the small island state of Malta. *Sustainability*, 6(10), pp. 6729–6750. https://doi.org/10.3390/su6106729.

Csereklyei, Z. and Stern, D.I. (2015). Global energy use: Decoupling or convergence?. *Energy Economics*, 51, pp. 633–641. https://doi.org/10.1016/j.eneco.2015.08.029.

Dinda, S. (2014). A theoretical basis for green growth. *International Journal of Green Economy*, 8(2), 177–189. https://doi.org/10.1504/IJGE.2014.065851.

Fankhauser, S. and Jotzo, F. (2018). Economic growth and development with low-carbon energy. *Wiley Interdisciplinary Reviews: Climate Change*, 9(1), p. e495. https://doi.org/10.1002/wcc.495.

Ferguson, P. (2015). The green economy agenda: Business as usual or transformational discourse?. *Environmental Politics*, 24(1), pp. 17–37. https://doi.org/10.1080/09644016.2014.919748.

Grand, M.C. (2016). Carbon emission targets and decoupling indicators. *Ecological Indicators*, 67, pp. 649–656. https://doi.org/10.1016/j.ecolind.2016.03.042.

Gross, S.G. (2017). Reimagining energy and growth: Decoupling and the rise of a new energy paradigm in West Germany, 1973–1986. *Central European History*, 50(4), pp. 514–546. https://doi.org/10.1017/S0008938917001017.

Guevara, Z. and Domingos, T. (2017). Three-level decoupling of energy use in Portugal 1995–2010. *Energy Policy*, 108, pp. 134–142. https://doi.org/10.1016/j.enpol.2017.05.050.

Gupta, S. (2015). Decoupling: A step toward sustainable development with reference to OECD countries. *International Journal of Sustainable Development & World Ecology*, 22(6), 510–519. https://doi.org/10.1080/13504509.2015.1088485.

Hennicke, P., Khosla, A., Thakur, M.S. and Wilts, H. (2014). *Decoupling Economic Growth from Resource Consumption*. Berlin: Deutsche Gesellschaft für Internationale Zusammenarbeit (GIZ) GmbH.

Hickel, J. and Kallis, G. (2019). Is green growth possible?. *New Political Economy*, pp. 1–18. https://doi.org/10.1080/13563467.2019.1598964.

Hoffmann, U. (2011). *Some Reflections on Climate Change, Green Growth Illusions and Development Space*. Geneva: United Nations Conference on Trade and Development.

IEA. (2019). *World Energy Outlook 2019*, Paris: OECD Publishing, https://doi.org/10.1787/caf32f3b-en.

Jakob, M. and Edenhofer, O. (2014). Green growth, degrowth, and the commons. *Oxford Review of Economic Policy*, 30(3), pp. 447–468. https://doi.org/10.1093/oxrep/gru026.

Jänicke, M. (2012). "Green growth": From a growing eco-industry to economic sustainability. *Energy Policy*, 48, pp. 13–21. https://doi.org/10.1016/j.enpol.2012.04.045.

Jiang, J.J., Ye, B., Zhou, N. and Zhang, X.L. (2019). Decoupling analysis and environmental Kuznets curve modelling of provincial-level $CO_2$ emissions and economic growth in China: A case study. *Journal of Cleaner Production*, 212, pp. 1242–1255. https://doi.org/10.1016/j.jclepro.2018.12.116.

Jiang, X.T., Dong, J.F., Wang, X.M. and Li, R.R. (2016). The multilevel index decomposition of energy-related carbon emission and its decoupling with economic growth in USA. *Sustainability*, 8(9), p. 857. https://doi.org/10.3390/su8090857.

Lin, S.J., Beidari, M. and Lewis, C. (2015). Energy consumption trends and decoupling effects between carbon dioxide and gross domestic product in South Africa. *Aerosol and Air Quality Research*, 15, pp. 2676–2687. https://doi.org/10.4209/aaqr.2015.04.0258.

Lu, Q., Yang, H., Huang, X., Chuai, X. and Wu, C. (2015). Multi-sectoral decomposition in decoupling industrial growth from carbon emissions in the developed Jiangsu Province, China. *Energy*, 82, pp. 414–425. https://doi.org/10.1016/j.energy.2015.01.052.

Marques, A.C., Fuinhas, J.A. and Leal, P.A. (2018). The impact of economic growth on $CO_2$ emissions in Australia: The environmental Kuznets curve and the decoupling index. *Environmental Science and Pollution Research*, 25(27), pp. 27283–27296. https://doi.org/10.1007/s11356-018-2768-6.

Meng, M., Fu, Y. and Wang, X. (2018). Decoupling, decomposition and forecasting analysis of China's fossil energy consumption from industrial output. *Journal of Cleaner Production*, 177, pp. 752–759. https://doi.org/10.1016/j.jclepro.2017.12.278.

Moreau, V. and Vuille, F. (2018). Decoupling energy use and economic growth: Counter evidence from structural effects and embodied energy in trade. *Applied Energy*, 215, pp. 54–62. https://doi.org/10.1016/j.apenergy.2018.01.044.

OECD. (2003). *OECD Environmental Indicators: Development, Measurement and Use*. https://www.oecd.org/env/indicators-modelling-outlooks/24993546.pdf.

OECD. (2012). *OECD Green Growth Studies Energy*. OECD Publishing, Berlin.

OECD. (2019). *OECD Green Growth Indicators: Database Documentation*. OECD Publishing.

Pao, H.T. and Chen, C.C. (2019). Decoupling strategies: $CO_2$ emissions, energy resources, and economic growth in the Group of Twenty. *Journal of Cleaner Production*, 206, pp. 907–919. https://doi.org/10.1016/j.jclepro.2018.09.190.

Parrique, T., Barth, J., Briens, F., Kerschner, C., Kraus-Polk, A., Kuokkanen, A. and Spangenberg, J.H. (2019). *Decoupling Debunked: Evidence and Arguments against Green Growth as a Sole Strategy for Sustainability*. European Environmental Bureau, Austria.

Roinioti, A. and Koroneos, C. (2017). The decomposition of $CO_2$ emissions from energy use in Greece before and during the economic crisis and their decoupling from economic growth. *Renewable and Sustainable Energy Reviews*, 76, pp. 448–459. https://doi.org/10.1016/j.rser.2017.03.026.

Román-Collado, R., Cansino, J.M. and Botia, C. (2018). How far is Colombia from decoupling? Two-level decomposition analysis of energy consumption changes. *Energy*, 148, pp. 687–700. https://doi.org/10.1016/j.energy.2018.01.141.

Ru, X., Chen, S. and Dong, H. (2012). An empirical study on relationship between economic growth and carbon emissions based on decoupling theory. *Journal of Sustainable Development*, 5(8), p. 43. https://doi.org/10.5539/jsd.v5n8p43.

Sanyé-Mengual, E., Secchi, M., Corrado, S., Beylot, A. and Sala, S. (2019). Assessing the decoupling of economic growth from environmental impacts in the European Union: A consumption-based approach. *Journal of Cleaner Production*, 236, p. 117535. https://doi.org/10.1016/j.jclepro.2019.07.010.

Schandl, H., Hatfield-Dodds, S., Wiedmann, T., Geschke, A., Cai, Y., West, J., Newth, D., Baynes, T., Lenzen, M. and Owen, A. (2016). Decoupling global environmental pressure and economic growth: Scenarios for energy use, materials use and carbon emissions. *Journal of Cleaner Production*, 132, pp. 45–56. https://doi.org/10.1016/j.jclepro.2015.06.100.

Scrieciu, S., Rezai, A. and Mechler, R. (2013). On the economic foundations of green growth discourses: The case of climate change mitigation and macroeconomic dynamics in economic modeling. *Wiley Interdisciplinary Reviews: Energy and Environment*, 2(3), pp. 251–268. https://doi.org/10.1002/wene.57.

Sertyesilisik, B. and Sertyesilisik, E. (2017). Ways of fostering green economy and green growth. In: *Sustainable Economic Development* (pp. 49–65). Cham: Springer.

Shi, L., Vause, J., Li, Q., Tang, L. and Zhao, J. (2016). Decoupling analysis of energy consumption and economic development in China. *Energy Sources, Part B: Economics, Planning, and Policy*, 11(9), pp. 788–792. https://doi.org/10.1080/15567249.2011.585372.

Shuai, C., Chen, X., Wu, Y., Zhang, Y. and Tan, Y. (2019). A three-step strategy for decoupling economic growth from carbon emission: Empirical evidences from 133 countries. *Science of the Total Environment*, 646, pp. 524–543. https://doi.org/10.1016/j.scitotenv.2018.07.045.

Stern, D.I. (2004a). Energy and economic growth. In: Cleveland, C.J. (Ed.). *Encyclopedia of Energy*. San Diego, CA: Academic Press, pp. 35–51.

United Nations Environment Programme. (2011). UNEP/Earthprint, Switzerland.

Van Vuuren, D.P., Stehfest, E., Gernaat, D.E., Doelman, J.C., Van den Berg, M., Harmsen, M., de Boer, H.S., Bouwman, L.F., Daioglou, V., Edelenbosch, O.Y. and Girod, B. (2017). Energy, land-use and greenhouse gas emissions trajectories under a green growth paradigm. *Global Environmental Change*, 42, pp. 237–250. https://doi.org/10.1016/j.gloenvcha.2016.05.008.

Wang, W., Liu, R., Zhang, M. and Li, H. (2013). Decomposing the decoupling of energy-related $CO_2$ emissions and economic growth in Jiangsu Province. *Energy for Sustainable Development*, 17(1), pp. 62–71. https://doi.org/10.1016/j.esd.2012.11.007.

Ward, J.D., Sutton, P.C., Werner, A.D., Costanza, R., Mohr, S.H. and Simmons, C.T. (2016). Is decoupling GDP growth from environmental impact possible?. *PLoS One*, 11(10). https://doi.org/10.1371/journal.pone.0164733.

# 18 Investigation into the Effects of Energy Transition in Terms of Economic Growth

*Guller Sahin*
Kütahya Health Sciences University

## CONTENTS

## 18.1 INTRODUCTION

Transition thought assumes a flow from a movement place to a destination. The energy transition is a process with multiple meanings that follow different ways. These ways explain two opposing views; a central strategy that depends on rigid and non-renewable sources and a decentralized strategy that can flexibly meet an end user's demands and is based on renewable sources. It requires several theoretical, methodological, and investigations linking interdisciplinary perspectives. Different approaches in energy transition studies tend to determine a flow from a past situation to a predicted or desired future and also comment on the current situation within a historical theme. Even if a starting point that is based on technological, normative, and economic innovation was defined in previous energy transition experiences, it is difficult to determine the historical length of a way based on a time scale selected by researchers. Another difficulty is to define the nature of available energy transitions; namely, to estimate a common destination. The European scenario is an obvious example of how hard it is to draw a method for the boundaries and direction of such a process, and also how it can be combined with culturally permanent meanings and applications (Sarrica et al., 2016).

Energy transitions are attributed to a persistent thought in which energy systems have continuously changed. However, transitions from primitive forms of power, such as animals, water, wind, and firewood to coal, and from coal to liquid and gaseous hydrocarbons, took time, with gradual processes that have occurred over decades or centuries. Modern energy transitions are unique, as well as being characterized by the "velocity" property that is short compared to the past, allowing us to evaluate the global challenges of today. Today's energy transitions have socio-economic, ecological and geopolitical dimensions, given their strong ties to poverty, inequality, climate change, national security, economic growth, prosperity, innovations, sustainable settlements and global energy trade. The thing that contributes to the chaos is the "dynamic balance." This is because energy demand necessitates balancing with a better supply, while it continues to change and grow between countries and sectors (Adewuyi et al., 2020; Singh et al., 2019).

Sustainable energy transitions are explained by the concepts of "strong, effective, and efficient developing energy sectors," without compromising current and future socio-environmental security. A number of countries have attempted to reach the sustainable energy transition goal in compliance with energy needs and procurement requirements. This is because the economic and social growth of a country is mostly based on efficiency in the energy sector (Adewuyi et al., 2020). The energy transition of developed countries does not only ride on their financial power, but also on a policy structure that provides broader possibilities, appropriate legislative action, and legally established competitive and regulated market environments. Sustainable energy transitions are basically motived in a political manner across the globe (Burke and Stephens, 2017).

For Vidadili et al. (2017), targets for sustainable energy generation are reduced or zero carbon emissions; harmless ecological impact; increased energy transition safety; reduced energy generation costs; and the use of improved green energy technologies. Integration of these five targets with each other is required for a sustainable future. All the related targets both increase economic safety and promote environmental quality. Therefore, it is mandatory for any country to put cheap, reliable, and clean energy at the center of its socio-economic growth, environmental sustainability, and technological revolution.

Grubler (2012) answers the question, Can sustainable energy transitions serve as an engine for a future whose sustainability is currently seen as questionable? According to his emphasis on social, economic, and environmental criteria, available energy systems are unsustainable; there is an urgent need for the energy transition. Miller et al. (2013) accept that the future of energy systems is one of the main policy challenges facing developed countries. According to Fri and Savitz (2014), neither government agencies nor private markets encourage energy transition. Transition to newer and cleaner energy systems, such as renewable electric vehicles or sources of electricity, do not merely necessitate remarkable changes in technology but also in tariffs and pricing policies, political regulations, and consumer behavior (Nielsen et al., 2015; Sovacool, 2009; Sovacool and Hirsh, 2009). This is because the dynamics of energy transitions are crucial.

According to the 2019 Report of the International Energy Agency, scientific evidence requires a reduction of global greenhouse gas emissions faster than ever; emissions arising from energy ranked first in 2018. Giddens (2009), within this context, mentioned the climate paradox, the point of no return will already have been reached when countries gain a clear understanding of how quickly and for how long they need to switch to energy forms with low-carbon emissions. Therefore, multi-dimensional and rapid energy transitions, including decarbonization, digitalization, decentralization, and decreasing usage, are mandatory.

Energy transition in this study is scrutinized from the economic growth perspective. This chapter consists of seven titles. After the introduction part, the energy transition in the second chapter is evaluated within the framework of historical developments and trends, opportunities, and challenges. The third chapter explains the effects of the energy transition in terms of economic growth. A case study follows the literature review. Uncertainty problems related to energy and action priorities are discussed within the scope of the sample set. This chapter is completed with a conclusion and policy implications.

## 18.2 FRAMEWORK OF ENERGY TRANSITION

Although there is no common or standard definition of energy transition in the literature, there are a number of common themes within those definitions (Bromley, 2016; Sovacool, 2016). Fouquet and Pearson (2012) define energy transition as a transition from one economic system to another that is dependent on one or more energy sources and technologies. Another definition belongs to Hirsh and Jones (2014), who state that energy transitions are the changes in fuels and fuel technology. According to Miller et al. (2015), energy transitions are the technological changes in a fuel source to produce energy and made it possible to use this fuel. O'Connor (2010) describes remarkable changes that affect sources, carriers, and converters, especially in energy consumption patterns in society. The energy transition is also called "conventional energy," "conventional power," and "conventional fuels" (Le and Sarkodie, 2020). A broad perspective emphasizes the differences in an energy system that covers changes in technology and a specific fuel source in general.

Effective modern energy transitions include passing to a more sustainable, affordable, inclusive, and secure energy system at the right time by offering solutions for difficulties regarding global energy (Singh et al., 2019). It is expected that an energy system serves three main difficulties that can be used as leverage for global comparisons. These challenges that show system performance are "safety and access," "environmental sustainability," and "economic development and growth." The success of country-specific actions (socio-economic growth priorities, dependence on energy, depth in capital markets, demography, natural resource equipment, business environment, and others) to improve energy system performance is based on a large provider cluster. "Being ready to transition" is the degree to which the energy system of a country has the economic, political, and social structures that will allow the transition to a safer, inclusive, reliable, and sustainable energy system that fosters economic development. It is important to understand the energy system performance shown in Figure 18.1

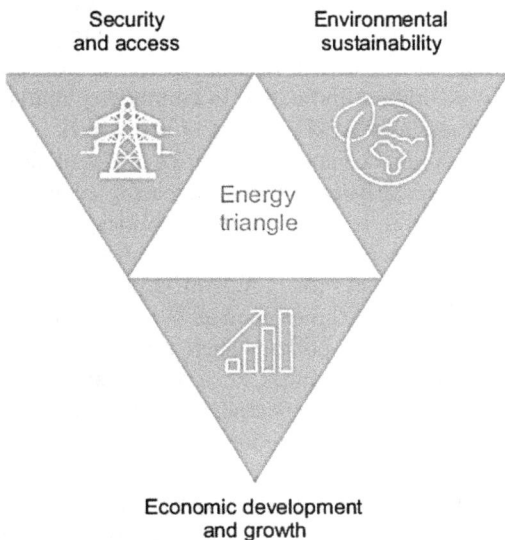

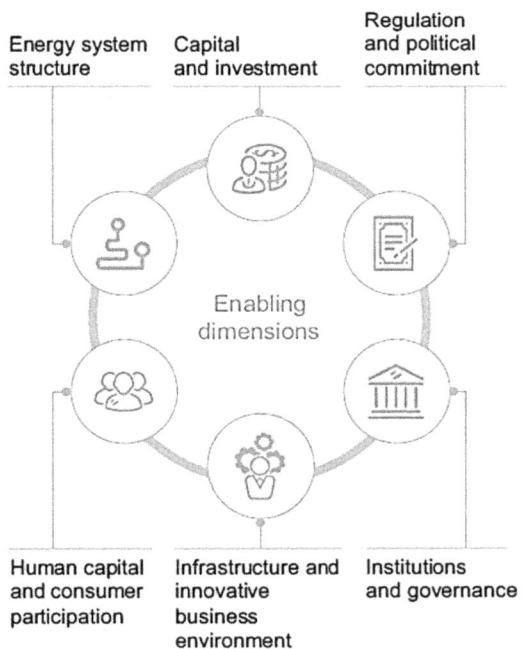

**FIGURE 18.1** Energy transition framework. (From Singh, H.V., Bocca, R., Gomez, P., et al., Energy Strat. Rev., 26, 100382, 2019. With permission.)

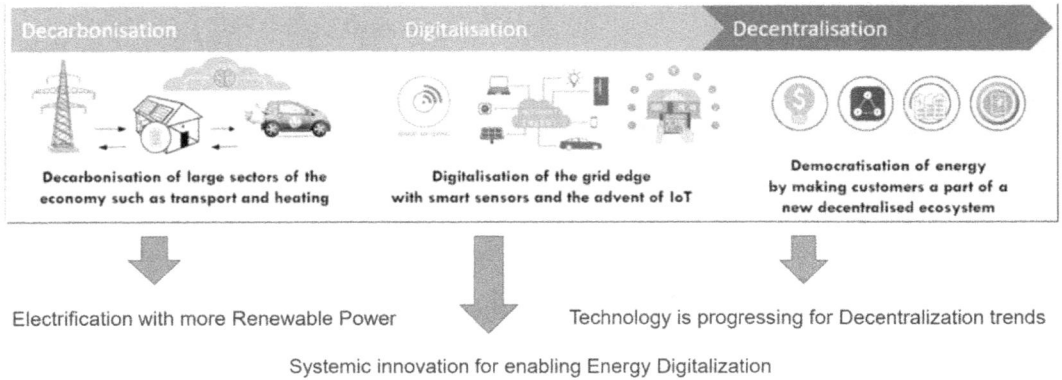

**FIGURE 18.2** Main drivers of energy transitions. (From Arifin, Z., Speech presented at the Seminar Nasional: Energi, telekomunikasi dan optimasi 2019 in Bandung, Indonesia (December 14, 2019), 2019. With permission.)

and the structures required for energy transition preparation, as well as the improvement potential degrees within these structures, to ensure an effective energy transition (World Economic Forum, 2018).

Global energy transitions are managed by downward trends in storage, decentralization, alternative transport, renewable energy, and traditionally centralized energy system components; upward trends in innovation problems (World Energy Council, 2019). Accordingly, the main driving forces of energy transitions are explained as decarbonization, digitization, and decentralization (as shown in Figure 18.2). Related forces represent basic interactions that are sustainable energy solutions. In this regard, there is a need to create technology trends that lead to more connectivity and decentralization between dimensions.

The World Energy Council determines the basic strategic priorities of countries as follows: updating market design rules for the purpose of decentralization; using electrification as a tool to purify the energy sector from carbon; being affected by the velocity of energy transitions by global strategic competition and nationalism; and the increasing role of lithium and cobalt in commodity market risks that are under the control of oil and gas price fluctuations (World Energy Council, 2019).

The BP Energy Outlook Report (2019) considers different scenarios to identify different sides of energy transitions. These scenarios have common characteristics, such as increasing economic growth and passing to reduced carbon fuel mixtures. However, scenarios vary by politics, technology, or behavioral assumptions. Common characteristics of energy transitions are highlighted in all of the scenarios. "Evolving transition (ET)" in Figure 18.3 assumes that government policies, technologies, and social preferences have continued to develop at the same speed as in the recent past. A number of the scenarios, for example, focus on specific rules and policies, such as prohibiting the use of disposable plastics. Others emphasize possible behavioral changes, such as increased trade disputes or large oil producers regulating their economies faster than expected. Moreover, "less carbon" and "more energy" require challenges, including a contribution to reducing carbon emissions, and are considered in distinct sectors of energy facing energy systems.

Figure 18.4, in which energy transitions are analyzed from three different scopes, shows the sectors, regions, and fuels where energy is used. According to the ET scenario, global energy demand will increase by almost one-third by 2040. This ratio represents a lower growth rate compared to previous years. An increase in energy consumption has a broad base in all the main sectors of the economy. Three-quarters of the increase in energy demand consists of industry and buildings. The total increase in energy demand is according to regional results from developing economies under the leadership of China and India. Different regional trends cause remarkable changes in global energy trade flows as well. Renewable energy is the fastest-growing energy source, constituting

Primary energy consumption by fuel
Billion toe

CO₂ emissions
Gt of CO₂

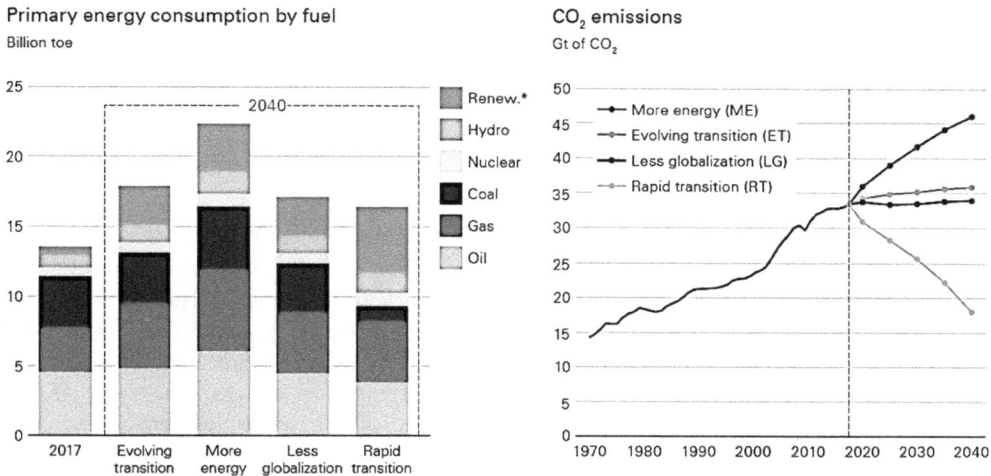

*Note.* Renewables include wind, solar, geothermal, biomass, and biofuels.

**FIGURE 18.3** Different aspects of energy transition. *Note:* Renewables include wind, solar, geothermal, biomass, and biofuels. (From BP Energy Outlook, 2019. With permission.)

Primary energy demand
Billion toe

End-use sector    Region    Fuel

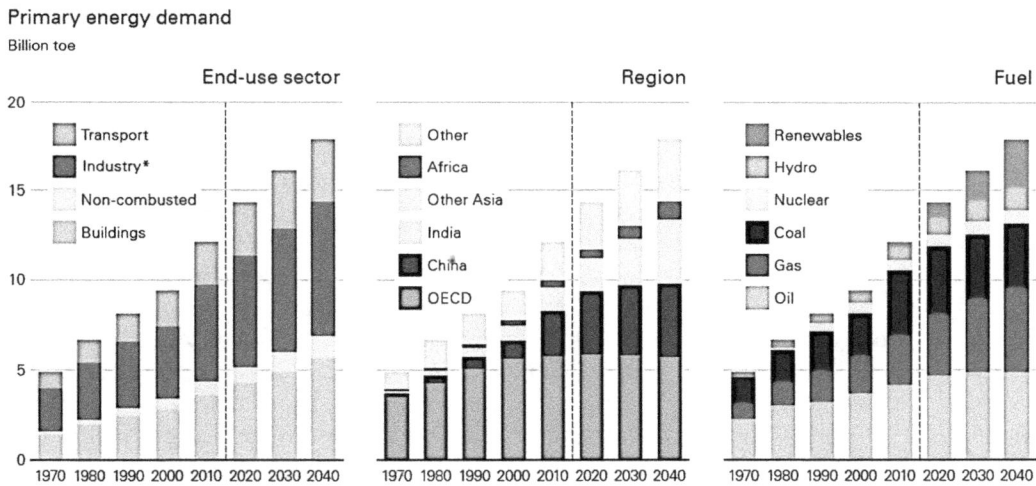

*Note.* Industry excludes non-combusted use of fuels.

**FIGURE 18.4** The energy transition: sectors, regions and fuels. *Note:* Industry excludes non-combusted use of fuels. (From BP Energy Outlook, 2019. With permission.)

almost half of the increase in energy. Demand for natural gas is growing faster than the demand for petroleum and coal. The increased availability of energy sources plays an increasing role in shaping global energy markets (BP Energy Outlook, 2019).

## 18.2.1 Historical Developments and Trends

Energy system transitions are long-term change processes that occur over decades or centuries (Fouquet and Pearson, 2012; Global Energy Assessment, 2012). Conventional energy transitions have been historically directed by innovations. Coal replaced wood, electric lights replaced oil lamps, and internal combustion engines replaced steam engines. Such transitions occurred in separate ways and at their own pace, determined by the socio-economic forces of the time. Modern energy transitions have different dynamics and are multi-dimensional: they are determined by a policy agenda aimed at ending the devastating consequences of fossil fuel burning; they must occur in all energy sectors and be global; and these transitions should be as fast as possible to avoid the devastating effects on society, the environment, and the world (Bromley, 2016). For Tahvonen and Salo (2001), the transition between energy forms is a smooth transition from renewable resources to non-renewable ones first and then to renewable resources again. It is impossible to change the world's energy system based on fossil fuels in the short term. Changing non-renewable energy resources with comprehensive and reliable alternatives is a decades-old task that requires expensive commitment; in this direction, energy systems can be called "a slow-maturing resource" (Sovacool, 2016).

Technological and related institutional/organizational transitions in the end-use of energy constitute the main driving forces of historical energy transitions. For example, steam power revolutionized manufacturing and transportation, which also increased the demand for coal. Similarly, internal combustion engines, automobiles, and petrochemistry directed the growth of the oil industry. The electrification of lighting, industrial drivers, and transportation caused electric services to arise (Grubler, 2012).

The transition process from conventional energy resources to fossil fuels involved several services and sectors in former times, from 1500 to 1920. The basic economic driving forces for transitions were better or cheaper energy services generation opportunities. The presence of a market that wants to pay more for these properties enables energy resources and technology to be gradually refined until they compete for available energy resources. The related process means that the innovation chain took more than a century, with its dissemination stage taking almost 50 years. Similarly, low-carbon energy sources and technologies bring low-carbon impacts that can develop in a proper market while competing for fossil fuels (Fouquet, 2010).

Fouquet and Pearson (2012) highlight that the process from technological innovation to dominating the proper market takes at least 40 years for a single system, with a transition that covers the whole economy taking centuries. The prevalence of conventional transitions has arisen as a relatively rare event. Grubler (2012) emphasizes the importance of services and end-use of energy, the long process of the energy transition, and models that characterize the successful scale of technologies and industries driving historical energy transitions.

According to Myhrvold and Caldeira (2012), a rapid recovery choice is next to impossible since energy transitions are intrinsically slow. Allen (2012), Grubb et al. (2015), and Rubio and Folchi (2012) also express energy transitions as gradual and slow processes, taking as long as 75–130 years. Fast energy transitions are accepted as anomalies limited to countries with extremely small populations or unique contextual conditions that are rarely seen elsewhere.

Hard coal became important when the steam engine was discovered in 1760, water power resources came into prominence when the dynamo was invented in 1873, and oil gained great importance when internal combustion engines were invented in the 1900s, along with internal combustion diesel engines in the 1910s (Energy Portal, 2020). Figure 18.5 shows the direction of the evolution of energy during energy transition processes. Accordingly, the relative global significance of each of the dominant energy fuels either reached a peak or was overtaken by another type of fuel (Le and Sarkodie, 2020). It can be seen that there have been quite long processes in global-scale energy transitions.

Coal is the oldest energy resource after wood, and it was first used to heat houses in England in the 9th century. Coal exceeded the 25% threshold value more than 500 years after the first commercial coal mines were developed in England in 1871. Crude oil, too, exceeded a similar threshold within 90 years after the first commercial well was drilled in 1859. The total of natural gas, hydroelectricity, nuclear energy, and other resources has now exceeded the 25% threshold (Energy Portal, 2020; Sovacool, 2016). Energy service prices fell faster and more dramatically compared to energy

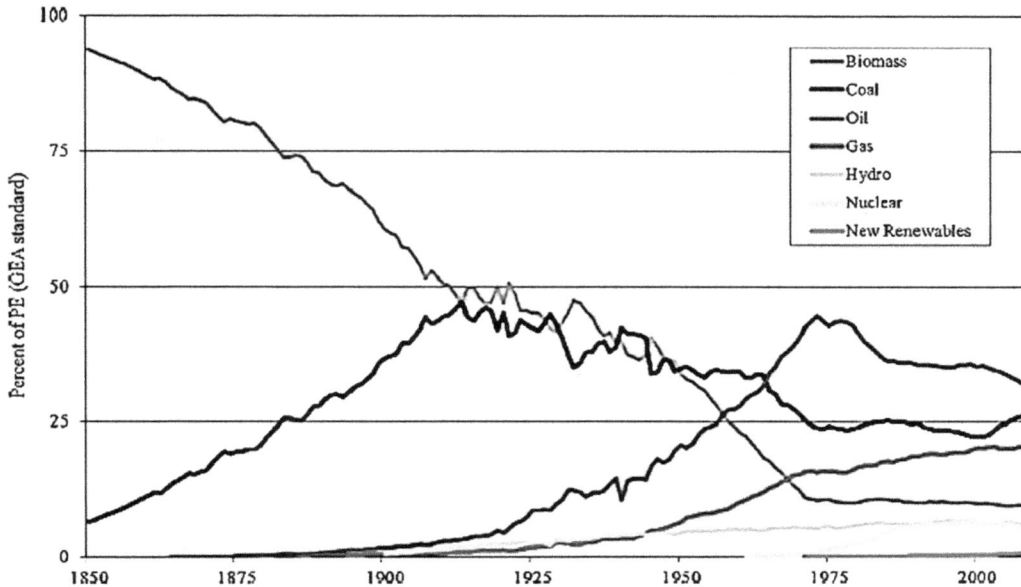

Note. Other renewable resources are wind, solar and geothermal energy.

**FIGURE 18.5** Structural change in world energy supply. *Note:* Other renewable resources are wind, solar and geothermal energy. (From Foell, W.K., Energy Res. Soc. Sci., 54, 96–112, 2019. With permission.)

input prices in Britain's energy history. Moreover, there have been much faster transitions in ultimate energy applications compared to energy supply systems (Grubler, 2012).

The current position of natural gas began with the lighting of street lamps using natural gas in Baltimore, USA in 1816 (Energy Portal, 2020). Increasing natural gas from 1% to 20% of the market has taken 70 years. Following its discovery in the 1860s, crude oil in the USA captured 10% of the national market in the 1910s. It reached 25% after 30 years. It took an extra 103 years to generate only 5% of the total energy used, and 26 years to reach that 25%. It took 38 years for nuclear-powered electricity to reach a 20% share in 1995. It has taken 50–70 years for this resource to reach such great potency (Sovacool, 2016).

England and Germany experienced huge increases in energy density. Energy density gradually increased with industrialization in France at the end of the 19th century. The industrialization process of Italy, in the meantime, was carried out without increasing energy density. For quite a long time, conventional energy carriers in Italy constituted a large part of the energy consumption for quite a long period. England and Germany were the first countries that industrialized by exporting a significant amount of their production using inefficient technology. Moreover, they had large, cheap coal reserves. After this, France and Italy industrialized using more effective technologies without exporting or accessing large, cheap coal reserves (Fouquet, 2016; Gales et al., 2007).

Germany is known as one of the first countries to adopt a strong energy policy aiming to obtain a high share of the modern energy resources within the energy matrix. It followed a long route in which priorities and technology have significantly changed. The country began developing energy transition strategies in the 1970s and has continued to strengthen them to this day. The first law allowing for a decentralized renewable energy grid supply was created in 1990. The Renewable Energy Law came into force in 2000. Germany has attracted huge attention since 2000 when it introduced policy plans toward "Energiewende," which is a pioneering energy transition by the government. A policy framework was presented covering the energy and economic sectors. Energy efficiency measures and on-site backup policies were also implemented. It was decided to phase out nuclear power in 2011. There was a two-stage process in legislative activities that are character-ized by a commitment to developing more open demand response mechanisms with the Electricity Market 2.0 and that started with an energy concept in 2010; the related process focuses on energy efficiency measures, which are currently under development. Germany's success was largely due to a mix of financial programs, voluntary and mandatory policies, and new regulations implemented by numerous economic agents, from consumers to national and regional institutions (Chen et al., 2019; Leiren and Reimer, 2018; Valdes et al., 2019).

Arguments showing that energy transitions go through long processes due to their nature have found more support from energy analysts who analyze the diffusion or innovation of major carriers or specific technologies. Lund (2006) shows in his study, which examines the market penetration rates of energy production and new energy technologies, the exponential penetration rates of new energy technologies can range from 4% to 40% in a year. Changeover durations, corresponding to 1%–50% of the assumed market potential, are between 10 and 70 years. The law ratio is associated with a greater energy impact; changeover durations of less than 25 years appear to be associated with end-use technologies. Sovacool (2016), in addition to all of this, emphasizes that reimburse-ment of R&D activities related to energy technologies takes years. Within this scope, he assumes that changes and alterations will occur over a long-term period when the temporal dynamics of conventional and modern energy transitions are conceptualized. He also attributed commonality in energy transitions to the fact that the larger the scale of widespread uses and transformations, the longer their substitution would take.

Characteristics change in primary energy over time occurs between 80 and 130 years on a global scale. A surprising observation is that change averages have significantly decelerated from the mid-1970s. Scale or market size, technological interrelation, and infrastructure needs are among the factors that explain slow energy transition rates. Comparative advantages can include multiple dimensions of efficiency, costs, and performance; the presence of suitable markets for testing and developing new technologies are factors that can accelerate transitions (Grubler, 2012).

Sovacool (2016) has a main flow perspective for energy transitions as prolonged events usually last for centuries. Moreover, he also explains that a dominant timing view will not always be supported by pieces of evidence. Related evidence is as follows: observing rapid shifts in terms of end-of-energy utilization and key movers; confirming historical records by examples of rapid national-scale tran-sitions in energy supply; and the possibility that modern transition drivers are distinct from tradi-tional transition drivers. Examples, such as the adoption of stoves, air conditioners, and flexible fuel vehicles, show that transition has rapidly occurred between only a few years and 10 years, or within a single generation in varying scales and industries. Air conditioning in the USA is an example of short energy transitions. The transition to nuclear power in France also followed a rapid course. Fifty-six reactors were built between 1974 and 1989 after the oil shock. Nuclear energy increased from 4% in 1970 to 10% of the national electric supply in 1978; it was almost 40% in 1982. According to Grubler (2010), the reasons for this success can be aligned as follows: the presence of a perfect institutional environment that allows for centralized decision-making; standardized reactor design; regulatory stability; in-house engineering resources; and a strong nationalized utility company that allows them to act as the main representative of reactor construction.

### 18.2.2 OPPORTUNITIES AND CHALLENGES

Energy transitions have always provided the basis for several variables. First of all, energy flows associated with energy generation and consumption, whose coordination was improved by energy markets, have developed. Changes have been created to find, use, and transform energy. Energy transitions present changes related to policies that organize the socio-political pole of energy sys-tems, such as modernizing a country, increasing its independence, or reducing poverty.

Figure 18.6 shows the global perspective regarding energy transitions arising from different regional and national priorities. The figure brings a remarkable perspective that allows energy

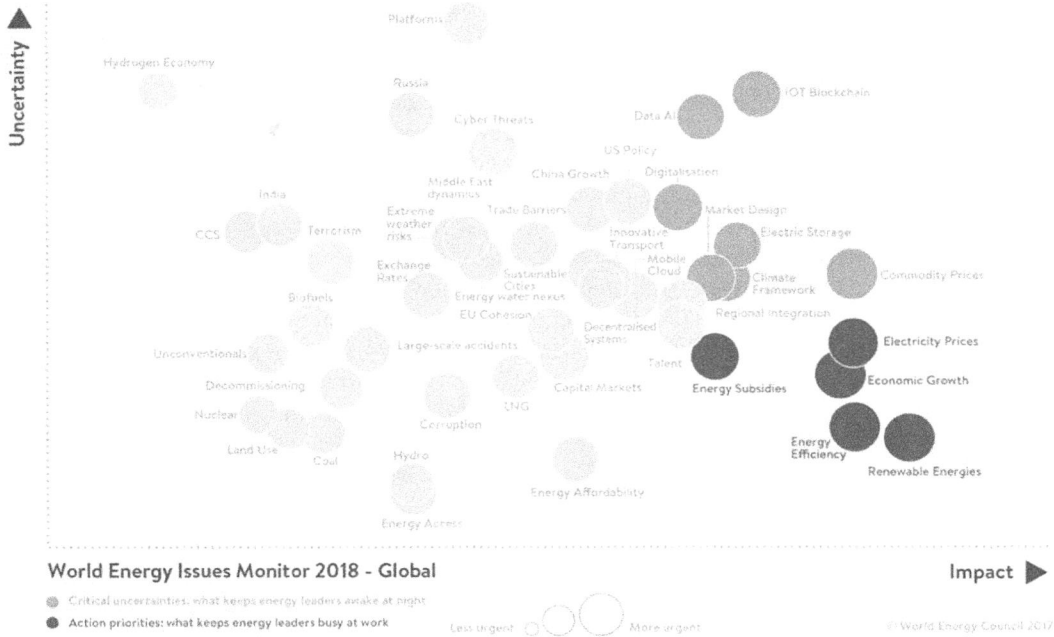

**FIGURE 18.6** The global perspective on energy transition. (From World Energy Council, World Energy Issues Monitor 2018, 2018. With permission.)

leaders to distinguish important change signals. Global problems emphasize that innovations are the main source of worry. Issues such as digitalization, market design, electricity storage, renewable energy, and decentralized systems have attracted interest as effects in the energy industry increase. Moreover, it can be seen that attention regarding central technology has decreased; there is more certainty in electricity prices and in the affordability of energy. An increasing effect of digitalization combines with alternative technologies, such as renewable energy, blockchain, and artificial intelligence (World Energy Council, 2018).

The global energy system faces many interrelated difficulties. These difficulties, basically, are the increasing global population, the need to provide access to modern energy services for more than a billion people without access to electricity, and the available supply resources to reduce carbon emissions. Decentralized renewable energy systems are anticipated to provide electrical access for more than two-thirds of the rural population, helping to move to globally reliable and low-carbon forms of electricity (Katre and Tozzi, 2019).

Although the use of renewable energy has rapidly increased, the rates of increase in ultimate energy consumption are below the desired level. Because the energy demand in developed countries increases slowly, it takes time to change current infrastructure and energy consumption habits. Energy demand in developed countries is increasing, and fossil fuels play a significant role in meeting this demand. In addition, in the current situation, it is impossible for energy arising from renewable energy resources to compete with fossil fuels in terms of pricing. In this context, it is extremely difficult for renewable energy to increase its share of total energy consumption (Karagöl and Kavaz, 2017).

Changes in the 21st century in energy and economic systems will be characterized by a much greater dependence on renewable energy than on fossil fuels. It is foreseen that increasing electrification and digitalization in the energy sector, as well as decentralization in energy supply, will accompany this rise in renewable energy. This multi-dimensional, low-carbon energy transition will fundamentally change energy geopolitics, including those between energy producers and customers, and also power relations between countries. Hydrocarbon exporting countries experience potentially negative economic and political effects because of decreasing energy exports, while energy-importing countries can benefit from increased energy supplies and clean energy technology exports in certain cases. As such opportunities develop and difficulties increase, management of international relations through diplomacy will become an increasingly important tool of foreign policy while countries make efforts at strategic positioning in the future energy environment. Different types of multilateral diplomacy will be required to simultaneously harmonize the energy transition interests of multiple stakeholders (Griffiths, 2019). Increasing energy efficiency can offer a solution to the climate crisis, encourage employment and create job opportunities (Ayres et al., 2013; Kaya, 2020). However, the main energy resource in the modern world is substantially based on oil and natural gas; this situation complicates the transition to renewable and low-carbon energy systems (Boulogiorgou and Ktenidis, 2020).

Available estimations of methods aimed at reducing emissions show a renewed focus on heat, productivity, and transportation. There is also a need for demand-side elasticity to harbor new generation capacity in wind energy. However, it is a slow process even if there is stable growth in both electric vehicle purchases and residential solar photovoltaics. In the meantime, it is accepted that the targets and policies currently in place are insufficient to meet the Paris climate commitments and that countries have struggled with a range of socio-economic and socio-technical challenges that ensure the necessary political momentum for a coordinated national energy strategy (Berka et al., 2020).

Oil-exporting countries have significantly relied on their oil revenues to meet their budget needs. Within this context, they will have a telling future if there is any remarkable decrease in global oil demand arising from the low-carbon energy transition. Related countries will have to establish bilateral relations that provide for the safety of energy demand along with economic opportunities stemming from a low-carbon energy transition to cope with this difficulty (Griffiths, 2019).

Technological progress, financial developments, and new market opportunities create a cost-reducing effect regarding power generation using renewable energy resources. Using solar and wind energy in coastal regions can compete with fossil fuels in terms of cost without calculating externalities. For example, wind energy is accepted as one of the most cost-effective options in terms of grid-based energy in many countries, such as Canada, China, Australia, and the United States. Furthermore, power generation from geothermal, biomass, and water energy sources may be more cost-effective than fossil-sourced production (Karagöl and Kavaz, 2017). There have been innovations and growth opportunities stemming from rapid technological developments and cost reductions in the areas such as solar power, wind power, electric vehicles, smart buildings, bidirectional power flows, and microgrids due to developed energy technologies (Arifin, 2019).

Energy policies that successfully evaluate energy problems in a particular country or region should be developed by considering the wide inequality in energy per capita anywhere in the world. Some of the main difficulties that encircle the applications of policies in energy transition, as in developing countries, are non-supportive political dynamics, superiority of bad law, inadequate networks, and market infrastructure (Adewuyi et al., 2020).

## 18.3  EFFECTS OF ENERGY TRANSITION IN TERMS OF ECONOMIC GROWTH

The transition of the energy consumption structure from non-renewable energy to clean and renewable energy should also be accepted as a factor that contributes to the sustainable improvement of the economy, in addition to being a factor that reduces $CO_2$ emissions. Predictions about the relationship between various power use and economic growth toward sustainable development help policymakers and interested parties in designing and implementing effective environmental and energy policies (Le and Sarkodie, 2020). Social behavior has changed under the pressure of both the consumption of natural resources and the use of energy to produce and use materials and technology over many years. It is highlighted that the relationships between energy consumption, the economy, and welfare differentiate poor countries from rich countries and that indicators, such as infant mortality, life expectancy, and economic development, are used to represent these changes (Santos and Balestieri, 2018).

Spreading new energy technologies and resources has an important impact on the economy. Just as with the industrial revolution, this has played a crucial role in rescuing the economy from the boundaries imposed by soil. There is proof of increased economic output due to the increasing productivity of energy technologies. Energy transitions, at the same time, have transformed economic activities, and also made decentralized production processes possible. Moreover, demographic and social transformations have been encouraged and political structures have changed (Fouquet, 2016).

Historical evidence supports the view that energy consumption per capita has increased in tandem with income levels. It is traditionally assumed that energy density reached a peak through the industrialization process and that it follows an inverted-U shape. Moreover, bringing incorporated biomass and traditional energy sources into the analysis of a wider range of historical experiences shows that this stylized fact is not always valid or dependent on the economic properties of this situation. There is strong historical evidence of energy density convergence because industrialized economies have become less energy-intensive, and industrializing economies have become more energy-intensive (Fouquet, 2016).

Growth in global energy demand is spreading over a broad base in all the main sectors of the economy. Different trends in how the energy in the sector is used and consumed have significant effects on the energy transition. The industry sector today consumes almost half of global energy and raw-material fuels; the remainder of the fuel is used in buildings (29%) and transportation (21%). As the gains in energy efficiency increase, the growth of energy use in all sectors slows down. On the other hand, as improvements in equipment efficiency develop, the slowdown in demand will be more apparent in the transportation sector. The growth of energy demand that is used in the industry also slows down. However, the non-combustible use of fuels in the industry is the fastest-growing source of increasing demand. The importance of energy that is used in buildings grows because increasing prosperity in developing economies leads to significant increases in power demand for cooling, lighting, and electrical equipment (BP Energy Outlook, 2019).

## 18.4 LITERATURE REVIEW

Conceptual approaches in the literature evaluate energy transitions in terms of a broad perspective. Related approaches include studies that focus on socio-technical transitions, road dependency from existing systems, governance and environmental reform, changing end-user values and behavior, and political economy constraints of justice and equity (Singh et al., 2019). Energy transition dynamics are analyzed, and transition to a more sustainable energy system process is evaluated in the literature that develops from different perspectives and disciplines (Valdes et al., 2019). The scope of research is national in general and gathered according to regions and sectors (Cherp et al., 2017, Foell, 2019).

Gales et al. (2007) examine the energy transition processes in Holland, Italy, and Spain over more than 200 years. They present a revision of the popular idea that there is an inverse-U curve in energy density in the long run. Lindmark and Andersson (2010), who focus on technological factors behind the energy use of households in Sweden, state that although coal has been substituted for wood in a number of urban areas, firewood consumption in rural areas did not decrease in the 19th century. Households with higher living standards consume more fuel, and rural households consume more fuel compared to urban households.

For Ayres et al. (2013), rerouting global economic policies continues to be a source of worry because of the peak in oil; climate change is an economic problem that needs a serious response. According to Kander and Stern (2014), although the technical change rate for modern energy is higher in Sweden, innovation in the use of conventional energy carriers contributed more to growth from 1850 to 1890; and modern energy has contributed to economic growth much more than conventional energy after the 1890s.

For Bromley (2016), there is a need for a rapid energy transition framework; this framework needs to focus on the main triggers of climate change. Fouquet (2016) emphasizes that consumers' reactions to energy markets have changed as a result of economic development. Regarding England's experience, the income elasticity of demand for energy services tends to follow an inverse-U shaped curve; therefore, it is important to formulate integrated energy service policies to reduce carbon-intensive infrastructure risks for developing countries. Haarstad (2016) shows how governance processes in Europe come together to direct urban low-carbon energy transitions. Sarrica et al. (2016) offer a cultural approach as an attempt to overcome the dilemmas between technical and humanitarian, and individual and social accounts of energy transitions in Europe.

Burke and Stephens (2017) focus on the democratization of renewable energy development, the energy democracy agenda, and theorizing on the relationship between renewable energy and political power. Cherp et al. (2017) contribute to an understanding of the differences in the use of low-carbon electricity sources by comparing the evolution of wind, solar, and nuclear energy in Germany and Japan. Differences are associated with the faster growth of energy insecurity and electricity demand and the easier diffusion of wind energy technology in Japan. Differences in Germany are associated with a weakening of the nuclear energy regime caused by stagnation and competition based on coal and renewable resources. Kuzemko et al. (2017) emphasize the management of sustainable energy transitions through innovation on the demand side in Germany. They reveal that critical policy debates occur in real time and are associated with changes in energy markets in which the political decision-making process is affected. Moreover, there is also a governance imperfection in regard to making demand-side reaction and local energy markets possible.

For Leiren and Reimer (2018), the success of feed-in tariffs for the use of renewable energy sources changes coalitions of interest in Germany, based on the historical institutionalization perspective. According to Mori's (2018) findings, renewable energy creates feedback effects in China, benefits from the sources and powers of the government to form alliances agreeable to government policy tendencies while blocking regime actors, and institutional reforms for the energy transition. For Mori, resources and power are obtained from oligopolistic or monopolistic electricity supply systems, as well as state price control.

Bohlmann et al. (2019) confirm that the impact of transition to an energy supply mix with a smaller share of coal production in South Africa is responsive to other economic and political conditions, notably, the global coal market and the reaction to coal exporting by the country. Chen et al. (2019) compare energy transitions in China and Germany and observe that infrastructure, policy instruments, and market reform have played a key role in the energy transition processes of both countries. Foell (2019) conducts two centuries of analysis of household energy transitions in Europe and the USA. According to his study findings, the energy transition process in homes, in line with global trends, changes from wood to coal, from coal to natural gas, and from natural gas to renewable energy. Griffiths (2019) submits an analysis regarding how bilateral energy diplomacy supports the interests of the Gulf Cooperation Council countries during a low-carbon energy transition within the scope of the United Arab Emirates case study. The results highlight the importance of developing dual-energy diplomacy with countries that can ensure the safety of domestic energy supply and also support the markets and economic diversity for hydrocarbon sources. Katre and Tozzi (2019) analyze the role to establish and manage decentralized renewable energy systems in India, evaluating the government as a system, and developing a new conceptual framework. Nolden (2019) concludes that energy governance in England has gradually been shaped by decentralization and digitalization that can ease or avoid value creation via decarbonization and democratization. Overland et al. (2019) present the geopolitical gains and losses index of 156 countries after transitioning to renewable energy. They attach the indicators of fossil fuel production and reserves, renewable energy resources, governance, and conflict to the energy transition index. Perez et al. (2019) argue that priorities and different energy security perspectives among the member states of the European

Union constitute two country groups. The first country group is the one that focuses on renewable energy and uses it as an industrial opportunity and a way to reduce import dependency. The other group includes countries that perceive renewable energy resources as too volatile and expensive to replace fossil fuels; they also prioritize reliable supplies. Valdes et al. (2019) perform a comparative analysis of developments in energy policies in Germany and Chile. In addition, they evaluated difficulties to solve the potential of the productive industry to respond to full demand in Chile and also made suggestions for the promotion of demand response. They also emphasize on lessons that are taken to organize the broad demand response potential in the German electricity markets.

Berka et al. (2020) contribute to an understanding of energy transition methods peculiar to the country by focusing on grassroots energy innovation practices in New Zealand following an established energy transition path. Van Boxstael et al. (2020) analyze energy transitions along spatial boundaries in terms of systemic agents in the cases of Sweden and Spain. The findings show how a systemic transnational agent eases energy transition within two scopes. Chhetri et al. (2020) reveal that the public's attitude toward energy and climate are situational rather than global. For them, policy reaction against energy transition is shaped by political, social, and institutional conditions, despite worries regarding climate change. Dietzenbacher et al. (2020) research the effects of global energy transition on renewable energy use using a structural segregation analysis. The energy transition is measured by substituting non-renewable energy for renewable energy. Johnstone et al. (2020) review breakdown waves in clean energy transition in the cases of Germany and England. It can be seen that, despite the comparable renewable energy shares in electric systems of the two countries, the power cut scale in Germany significantly exceeds the scales of technology, market, and business models; ownership; and actors, as well as regulation in England. Le and Sarkodie (2020) present evidence from emerging markets and developing economies for the dynamic connection between renewable and non-renewable energy, economic growth and environmental quality. The long-term effects reveal that the consumption of non-renewable and renewable energy resources remarkably contributes to the economic growth of selected countries. Li and Taeihagh (2020) research the policy mix of China to ease the energy transition and how this policy mix has developed. Chronological analysis accepts two dimensions of temporal changes in policy mix; policy intensity and changes in intensity, and changes in political tool combinations. Mahjabeen et al. (2020) highlight that renewable and non-renewable energy consumption in D-8 countries has a positive impact on environmental quality and economic growth. Matschoss et al. (2020) conduct network analyses of energy transition area experiments in Finland. Data visualization of networks helps to review how arena participants have built energy transitions in the energy sector. Tvinnereim et al. (2020) research to what extent the public in Norway is familiar with the energy transition concept. According to the results, issues related to energy transition are not popular in society, and therefore, related issues are far from the daily concerns of the people (especially, regarding employment). Wang and Wang (2020) analyze whether energy transitions separate economic growth from an increase in emissions with evidence from 186 countries. The results show our upper middle income and lower-middle-income countries will take more responsibility to actively speed up the energy transition process and reduce carbon emissions. Therefore, economic growth will powerfully separate from carbon emissions.

## 18.5  CASE STUDY

This study scrutinizes the effects of energy growth in terms of economic growth within the G-7 countries' sample via panel statistics for the 1970–2018 period.

### 18.5.1  MODEL AND DATA

The model has been built on the Cobb–Douglas function that was extended by considering the studies of Dietzenbacher et al. (2020), Kander and Stern (2014), Le and Sarkodie (2020), Mahjabeen et al. (2020), and Tahvonen and Salo (2001). Per capita output is defined as follows:

$$y_t = A_t k_t^a, \ \alpha > 0 \tag{18.1}$$

where $y$ is the gross domestic product, $A$ is the technology, and $k$ is the capital. Technology may change over time, and energy, as endogenously, is determined by the government spending and trade openness index. Moreover, $CO_2$ emissions are integrated into the model to evaluate the effect of the technological variable on economic growth. In this context, the linear form of equality in Eq. 18.1 is as follows:

$$\text{GDP}_{it} = \partial_0 + \partial_1 C_{it} + \partial_2 \text{CE}_{it} + \partial_3 \text{ME}_{it} + \partial_4 \text{GE}_{it} + \partial_5 \text{TO}_{it} + \partial_6 \text{CO}_{2it} + \mu_{it} \tag{18.2}$$

The GDP, C, CE, ME, GE, TO, and $CO_2$ variables in model respectively explain the following: GDP (per capita, constant US\$ 2010); capital formation (gross, constant US\$ 2010); conventional energy consumption (non-renewable energy,[1] million tonnes oil equivalent); modern energy consumption (renewable energy,[2] million tonnes oil equivalent); government spending (final consumption, fixed US\$ 2010); the difference between the export and import of goods and services (constant US\$ 2010); and carbon-dioxide emissions (million tonnes of $CO_2$). All of the variables are in a model with logarithmic $(\ln)$ structure. $\partial_0$ is the constant parameter, $\partial_1 \ldots \ldots \partial_6$ is slope parameters, and $\mu$ is the error term. The sub-symbol $i$ represents Germany, the USA, the UK, France, Italy, Japan, and Canada, while $t$ is the 1970–2018 time period. Data on the GDP, C, GE, and TO variables were obtained from the World Development Indicators of the World Bank, while data on the CE, ME, and $CO_2$ variables were obtained from British Petroleum.

Capital and technology are components of economic growth, while energy consumption (conventional and modern energy), government spending, and the trade openness index are technology variable. Moreover, $CO_2$ emissions are included in the model for the decarbonization goal. Since the anthropogenic effect of conventional energy resources is strategically important for economic growth, energy is discussed as conventional and modern energy use with reference to the idea of encouraging modern energy use. The sample set in this study is the G-7 countries that make up about 45% of the global economy. Relations between economic growth, energy transitions, and $CO_2$ emissions should be considered in the sample selection. According to 2018 data, conventional and modern energy consumption of the G-7 countries can be seen in Figure 18.7. The amount of $CO_2$ emissions can be seen in Figure 18.8.

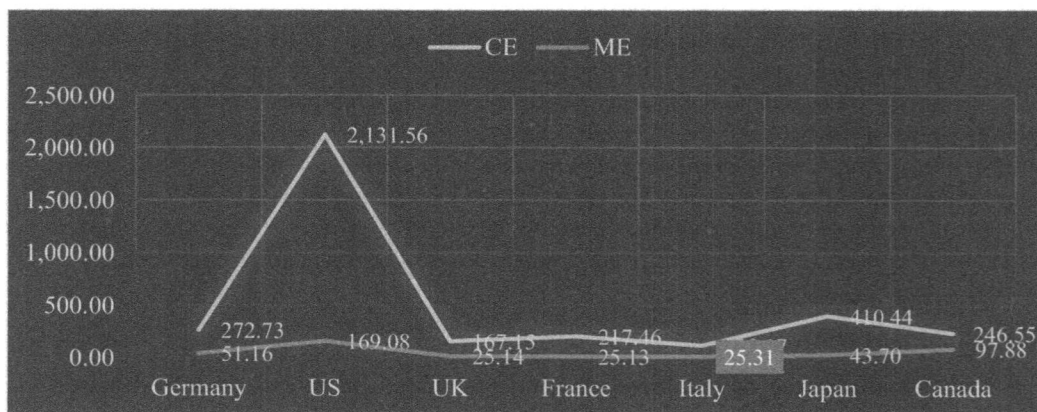

**FIGURE 18.7**  Conventional and modern energy consumption in 2018 (million tonnes oil equivalent). (This was created by the researcher using data from the BP Energy Outlook Report, 2019.)

---

[1] Non-renewable sources including oil, gas, coal, and nuclear energy.
[2] Renewable sources including wind, geothermal, solar, biomass and waste, and hydro.

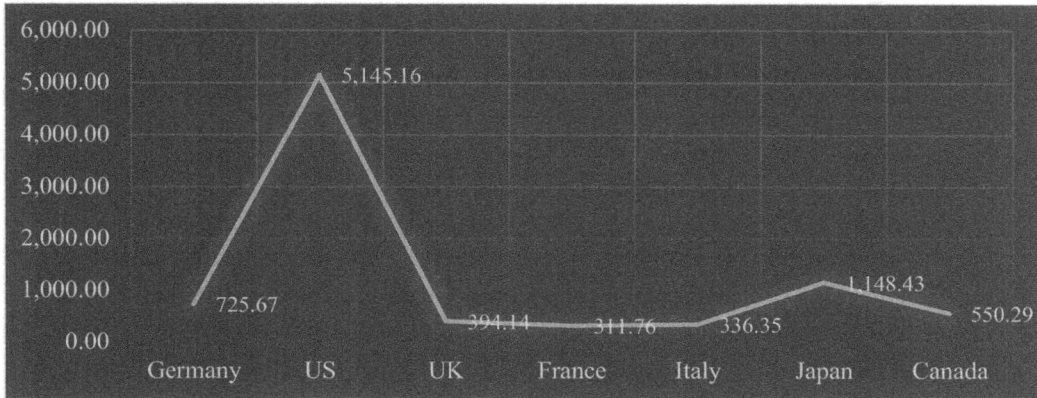

**FIGURE 18.8** $CO_2$ Emissions in 2018 (million tonnes of $CO_2$). (This was created by the researcher using Data from the BP Energy Outlook Report, 2019.)

## 18.5.2 EMPIRICAL STRATEGY

Estimations of the model within Eq. 18.2 were performed by cointegration and causality analyses.[3] A two-stage empirical methodology toward the statistical reliability of tests was followed. We applied the correlation test between units in the first stage, while unit root tests were conducted in the second stage.

### 18.5.2.1 Cross-Sectional Dependence and Panel Unit Root Tests

The type of panel unit root test is determined by whether there is cross-sectional dependency, namely, a correlation between units. A CD Lagrange multiplier (LM) test that was developed by Breusch and Pagan (1980) was used if the time dimension was greater than the number of cross-sections $(T > n)$ in the 343 observed $(N = T * n)$ matrix. Breusch–Pagan $CD_{LM}$ test findings within Table 18.1 support the evidence of cross-sectional dependence between units by denying the $H_0$ hypothesis. In these circumstances, the second-generation panel unit root tests that are used in the case of existing correlation between units should be preferred.

Levin et al. (2002) with Harris and Tzavalis (1999) panel unit root tests were applied in series with a difference from the cross-sectional mean to reduce the effect of correlation between the units. According to the stationarity findings in Table 18.1, the GDP, C, CE, ME, GE, and $CO_2$ variables were inserted in the cointegration model at their first differences; the TO variable was inserted in the same model at its level value.

### 18.5.2.2 Panel Cointegration Test and Long-Term Estimates

Panel cointegration analyses were conducted to check the possible long-term balance relationship between the variables after cross-sectional dependence and unit root tests were applied. Westerlund (2007) panel cointegration analysis findings within Table 18.2 confirm the presence of a long-term cointegration relationship between the variables according to (Gt-Ga) and panel (Pt-Pa) average statistics by denying the $H_0$ hypothesis.

As can be seen in Table 18.3, heterogeneous parameter estimates (Mean Group estimator) in the long-term relationship in the panel cointegration model give evidence that the capital, conventional energy consumption, government spending, and $CO_2$ emissions increase GDP in the long

---

[3] Methodology regarding panel cointegration and causality analyses is not explained because of its frequent use in academic literature.

**TABLE 18.1**

**Cross-Sectional Dependence and Unit Root Tests**

| | CD Test | Unit Root Tests | |
|---|---|---|---|
| Variable | Statistic | Levin-Lin-Chu | Harris-Tzavalis |
| $\ln \text{GDP}$ | 969.5545 (0.0000)[a] | 0.9494 (0.8288) | 1.3443 (0.9106) |
| $\Delta \ln \text{GDP}$ | | −11.1562 (0.0000)[a] | −26.7361 (0.0000)[a] |
| $\ln C$ | 809.8334 (0.0000)[a] | −0.9298 (0.1762) | 0.1657 (0.5658) |
| $\Delta \ln C$ | | −15.4566 (0.0000)[a] | −35.7634 (0.0000)[a] |
| $\ln \text{CE}$ | 443.3433 (0.0000)[a] | −1.0053 (0.1574) | 1.2866 (0.9009) |
| $\Delta \ln \text{CE}$ | | −16.8163 (0.0000)[a] | −40.1118 (0.0000)[a] |
| $\ln \text{ME}$ | 620.9608 (0.0000)[a] | 4.6849 (1.0000) | 2.5889 (0.9952) |
| $\Delta \ln \text{ME}$ | | −20.4604 (0.0000)[a] | −48.8982 (0.0000)[a] |
| $\ln \text{GE}$ | 947.6166 (0.0000)[a] | −2.5713 (0.0051)[b] | 0.6364 (0.7377)[a] |
| $\Delta \ln \text{GE}$ | | −7.7422 (0.0000)[a] | −20.9769 (0.0000)[a] |
| $\ln \text{TO}$ | 82.72699 (0.0000)[a] | −4.9764 (0.0000)[a] | −12.5134 (0.0000)[a] |
| $\Delta \ln \text{TO}$ | | −21.6134 (0.0000)[a] | −48.8534 (0.0000)[a] |
| $\ln \text{CO}_2$ | 484.4824 (0.0000)[a] | −0.8668 (0.1930) | 1.1933 (0.8836) |
| $\Delta \ln \text{CO}_2$ | | −15.0981 (0.0000)[a] | −40.0195 (0.0000)[a] |

*Note:* $\Delta$ notation is the difference processor; p-values are in parentheses; a and b are 1% and 10% statistical significance levels, respectively.

**TABLE 18.2**

**Panel Cointegration Test**

| Test | Value of Test | z-Value |
|---|---|---|
| Gt | −3.875 | −3.757 (0.000) |
| Ga | −21.800 | −2.519 (0.006) |
| Pt | −10.125 | −3.771 (0.000) |
| Pa | −21.204 | −3.383 (0.000) |

*Note:* p-values are in parentheses; the model has a constant term, but no trend, and according to the Akaike information criterion, the lag length is within the range of 1–2.

term under the statistical significance of the numbers. Modern energy consumption and the trade openness index are statistically insignificant.

The formula below shows the reorganization of the model in Eq. 18.2, based on long-term slope parameters:

$$\text{GDP}_{it} = 0.23 C_{it} + 0.11 \text{CE}_{it} + 0.25 \text{GE}_{it} + 0.03 \text{CO}_{2it} \tag{18.3}$$

For the findings, government spending (+0.25) is the variable that affects GDP the most; the variable that least affects GDP is $\text{CO}_2$ emissions (+0.03).

**TABLE 18.3**
**Panel Long-Run Elasticity Estimates**

| Variable | Coefficient | z-Statistic | Prob. |
|---|---|---|---|
| $\Delta \ln C$ | $0.2333225^a$ | 14.23 | (0.000) |
| $\Delta \ln CE$ | $0.1058761^a$ | 4.34 | (0.000) |
| $\Delta \ln ME$ | 0.0096192 | 0.81 | (0.416) |
| $\Delta \ln GE$ | $0.246763^a$ | 6.44 | (0.000) |
| $\ln TO$ | $-0.0001413$ | $-0.37$ | (0.714) |
| $\Delta \ln CO_2$ | $0.0307369^b$ | 1.72 | (0.085) |

*Note:* p-values are in parentheses; a and b are 1% and 5% statistical significance levels, respectively.

**TABLE 18.4**
**Granger Causality Test**

| Variables | ln GDP | ln C | ln CE | ln ME | ln GE | ln TO | ln CO$_2$ |
|---|---|---|---|---|---|---|---|
| ln GDP | - | $6.5215^a$ | $4.6610^a$ | $3.8899^a$ | $12.3105^a$ | 0.4231 | 0.0469 |
|  |  | (0.0000) | (0.0000) | (0.0001) | (0.0000) | (0.6722) | (0.9626) |
| ln C | $5.8337^a$ | Direction of causality: | | | | | |
|  | (0.0000) | $\ln GDP \Leftrightarrow \ln C$ | | | | | |
| ln CE | 0.2564 (0.7976) | Direction of causality: | | | | | |
|  |  | $\ln GDP \Rightarrow \ln CE$ | | | | | |
| ln ME | $6.8337^a$ (0.0000) | Direction of causality: | | | | | |
|  |  | $\ln GDP \Leftrightarrow \ln ME$ | | | | | |
| ln GE | $18.1651^a$ (0.0000) | Direction of causality: | | | | | |
|  |  | $\ln GDP \Leftrightarrow \ln GE$ | | | | | |
| ln TO | 1.1328 (0.2573) | Direction of causality: | | | | | |
|  |  | $\ln GDP \not\Leftrightarrow \ln TO$ | | | | | |
| ln CO$_2$ | $2.7971^b$ (0.0052) | Direction of causality: | | | | | |
|  |  | $\ln GDP \Leftarrow \ln CO_2$ | | | | | |

*Note:* p-values are in parentheses; a and b are 1% and 5% statistical significance levels, respectively.

### 18.5.2.3 Panel Causality Test

As can be seen in Table 18.4, according to the heterogeneous panel causality test findings that were developed by Dumitrescu and Hurlin (2012), there is a feedback relationship between GDP with capital, modern energy consumption, and government spending; a one-way causality relationship from GDP to conventional energy consumption, and also from $CO_2$ emissions to GDP.

## 18.6  DISCUSSION

This study scrutinizes the effects of energy transition in terms of economic growth within a theoretical framework. We also conducted a case study in the G-7 countries' sample set. Since energy transitions include radical changes in the regulatory systems of countries, there were actual policy implications for the case studies. Countries in the sample set contribute to almost 45% of net global wealth (38 trillion US$/2019) (World Bank, 2020). National perspectives regarding energy problems of countries are based on implications in the 2020 World Energy Issues Monitor Report, prepared by the World Energy Council. National perspectives, at the same time, signal regional uncertainty problems and action priorities, including countries (World Energy Council, 2020).

Germany's energy landscape in 2020 reflects uncertainty and action areas. It should be noted that perceptions related to the urgency and impact of problems have varied. Integration of discontinuous renewable energies causes priorities and worries in both the public and private sectors. England, nowadays, focuses on climate framework, market design, and hydrogen economy by redefining lower uncertainty and effects that are attributed to issues such as EU harmonization and IoT/blockchain following Brexit. Its action priorities have taken shape around renewable energy, capital markets, and China. France displays a focused energy environment profile in which the main certainties concern climate framework and Middle Eastern dynamics. Electricity storage, extreme weather risks, energy pricing, and renewable energy are among other uncertainties. Action priorities are energy efficiency, relevant cost, and access. Uncertainties and the main sources of worry for Italy regarding the evolution of sustainable cities are EU harmonization, USA policy, and China. Its action priorities are digitalization and technological problems, renewable energy, energy efficiency, and market design.

The United States displays an energy landscape in which uncertainties are first reduced around energy geopolitics and regional issues; macroeconomic risks are then emphasized. Its action priorities focus on economic growth and are shaped around energy technologies. Sources of uncertainty in Canada include the US policy and energy geopolitical issues regarding China. Moreover, uncertainties concerning the climate framework have increased. Action priorities are regional integration, economic growth, and modern energy resources.

The critical uncertainty issues in Japan are nuclear and excess air risks and large-scale accidents. The hydrogen economy is seen as a new critical uncertainty issue. China shows its priorities as being the climate framework and artificial intelligence.

## 18.7  CONCLUSION AND POLICY IMPLICATIONS

The effects of the energy transition in terms of economic growth are analyzed in this study by way of panel statistics for the 1970–2018 period. According to the cointegration results, capital, conventional energy consumption, government spending, and $CO_2$ emissions all increase GDP in the long term. Creating an increase in GDP through government spending and conventional energy consumption that is used instead of technology empirically confirms that technology supports economic growth. The causality results support the feedback hypothesis between GDP with capital, modern energy consumption, and government spending. One-way causality relationships from GDP to conventional energy consumption can be seen; $CO_2$ emissions to GDP. Our study results agree with the studies of Bromley (2016), Dietzenbacher et al. (2020), Kander and Stern (2014), Le and Sarkodie (2020), and Mahjabeen et al. (2020).

The results suggest that, when we look from a political perspective, effective policies that promote energy and economic structural arrangements should be implemented to reduce atmospheric $CO_2$ emissions. The most important strategy for reducing climate risk and traversing the current unsustainable spiral of the economy has obliged us to make huge investments in energy efficiency and modern energy technologies in the medium term. Related investments ought to be long-term and attractive to investors. As mentioned by Ayres et al. (2013), reducing resource costs by taxing energy use is a good policy choice. It is assumed that increasing taxes regarding natural resource use will make environmental goals more achievable. Moreover, preventing waste by governments by way of taxing and eliminating current organizations that avoid productivity, especially in the electricity sector, is another policy recommendation. Policies that increase economic growth and social welfare, decrease negative environmental results and pollution costs, and those that prevent greenhouse gas accumulation need to be implemented. Boulogiorgou and Ktenidis (2020) highlight the increasing energy efficiency at all stages of the energy chain, from production to final consumption. Transition to energy systems with low-carbon will guarantee improved energy access that directly contributes to social issues, such as accelerating decentralized clean energy services and reducing poverty. In addition, an "intense decarbonization" goal must focus on national and international efforts. This thesis is an implicit part of all other theses. We should accept that cheap fossil fuels (particularly liquid fuels) are about to deplete. Further delay will only sharpen the costly results of climate change. Dependence on oil in particular, and carbon-based fossil fuels in general, should immediately be abandoned. There is a need for financial innovation to drive private investment into decarbonization and energy independence. Specific financial techniques, such as securing of assets and the use of credit default derivatives may be used to market the securities, based on renewable resources and energy efficiency.

## BIBLIOGRAPHY

Adewuyi, O.B., Kiptoo, M.K. Afolayan, A. F. et al. (2020). Challenges and prospects of Nigeria's sustainable energy transition with lessons from other countries' experiences. *Energy Reports* 6: 993–1009. https://doi.org/10.1016/j.egyr.2020.04.022.

Allen, R.C. (2012). Backward into the future: The shift to coal and implications for the next energy transition. *Energy Policy* 50: 17–23. https://doi.org/10.1016/j.enpol.2012.03.020.

Arifin, Z. (2019). Energy transition and digitalization in power system. *Speech presented at the Seminar Nasional: Energi, telekomunikasi dan optimasi 2019 in Bandung, Indonesia (14 December 2019).*

Ayres, R.U., Campbell, C.J., Casten, T.R. et al. (2013). Sustainability transition and economic growth enigma: Money or energy?. *Environmental Innovation and Societal Transitions* 9: 8–12. http://dx.doi.org/10.1016/j.eist.2013.09.002.

Berka, A.L., MacArthur, J.L., Gonnelli, C. (2020). Explaining inclusivity in energy transitions: Local and community energy in Aotearoa New Zealand. *Environmental Innovation and Societal Transitions* 34: 165–182. https://doi.org/10.1016/j.eist.2020.01.006.

Bohlmann, H.R., Horridge, J.M., Inglesi-Lotz, R. et al. (2019). Regional employment and economic growth effects of South Africa's transition to low-carbon energy supply mix. *Energy Policy* 128: 830–837. https://doi.org/10.1016/j.enpol.2019.01.065.

Boulogiorgou, D., Ktenidis P. (2020). TILOS local scale technology innovation enabling low carbon energy transition. *Renewable Energy* 146: 397–403. https://doi.org/10.1016/j.renene.2019.06.130.

BP Energy Outlook (2019). https://www.bp.com/content/dam/bp/business-sites/en/global/corporate/pdfs/energy-economics/energy-outlook/bp-energy-outlook-2019.pdf (accessed 14 May 2020).

Breusch, T.S., Pagan, A.R. (1980). The LaGrange multiplier test and its applications to model specification in econometrics. *The Review of Economic Studies* 47(1): 239–253. https://doi.org/10.2307/2297111.

Bromley, P.S. (2016). Extraordinary interventions: Toward a framework for rapid transition and deep emission reductions in the energy space. *Energy Research & Social Science* 22: 165–171. https://doi.org/10.1016/j.erss.2016.08.018.

Burke, M.J., Stephens, J.C. (2017). Political power and renewable energy futures: A critical review. *Energy Research & Social Science* 35: 78–93. https://doi.org/10.1016/j.erss.2017.10.018.

Chen, C., Xue, B., Cai, G. et al. (2019). Comparing the energy transitions in Germany and China: Synergies and recommendations. *Energy Reports* 5: 1249–1260. https://doi.org/10.1016/j.egyr.2019.08.087.

Cherp, A., Vinichenko, V., Jewell, J. et al. (2017). Comparing electricity transitions: A historical analysis of nuclear, wind and solar power in Germany and Japan. *Energy Policy* 101: 612–628.

Chhetri, N., Ghimire, R., Wagner, M. et al. (2020). Global citizen deliberation: Case of world-wide views on climate and energy. *Energy Policy* 147: 111892. https://doi.org/10.1016/j.enpol.2020.111892.

Dietzenbacher, E., Kulionis, V., Capurro, F. (2020). Measuring the effects of energy transition: A structural decomposition analysis of the change in renewable energy use between 2000 and 2014. *Applied Energy* 258: 114040. https://doi.org/10.1016/j.apenergy.2019.114040.

Dumitrescu, E.-I., Hurlin, C. (2012). Testing for granger non-causality in heterogeneous panels. *Economic Modelling* 29(4): 1450–1460. https://doi.org/10.1016/j.econmod.2012.02.014.

Energy Portal (2020). https://www.enerjiportali.com/fosil-yakitlar-nelerdir/ (accessed 5 December 2020).

Foell, W.K. (2019). A two-century analysis of household energy transitions in Europe and the United States: From the Swiss Alps to Wisconsin. *Energy Research & Social Science* 54: 96–112. https://doi.org/10.1016/j.erss.2019.03.009.

Fouquet, R. (2010). The slow search for solutions: Lessons from historical energy transitions by sector and service. *Energy Policy* 38(11): 6586–6596. https://doi.org/10.1016/j.enpol.2010.06.029.

Fouquet, R. (2016). Lessons from energy history for climate policy: Technological change, demand and economic development. *Energy Research & Social Science* 22: 79–93. https://doi.org/10.1016/j.erss.2016.09.001.

Fouquet, R., Pearson, P.J.G. (2012). Past and prospective energy transitions: Insights from history. *Energy Policy* 50: 1–7. https://doi.org/10.1016/j.enpol.2012.08.014.

Fri, R. W., Savitz, M. L. (2014). Rethinking energy innovation and social science. *Energy Research & Social Science* 1: 183–187. https://doi.org/10.1016/j.erss.2014.03.010.

Gales, B., Kander, A., Malanima P. et al. (2007). North versus south: Energy transition and energy intensity in Europe over 200 years. *European Review of Economic History* 11(2): 219–253. https://doi.org/10.1017/S1361491607001967.

Giddens, A. (2009). *The Politics of Climate Change*. New York: Polity Press.

Global Energy Assessment (2012). *Toward a Sustainable Future*. Cambridge, UK: Cambridge University Press and New York, NY: International Institute for Applied Systems Analysis.

Griffiths, S. (2019). Energy diplomacy in a time of energy transition. *Energy Strategy Reviews* 26: 100386. https://doi.org/10.1016/j.esr.2019.100386.

Grubb, M., Hourcade, J.-C., Neuhoff, K. (2015). The three domains structure of energy-climate transitions. *Technological Forecasting and Social Change* 98: 290–302. https://doi.org/10.1016/j.techfore.2015.05.009.

Grubler, A. (2010). The costs of the French nuclear scale-up: A case of negative learning by doing. *Energy Policy* 38(9): 5174–5188. https://doi.org/10.1016/j.enpol.2010.05.003.

Grubler, A. (2012). Energy transitions research: Insights and cautionary tales. *Energy Policy* 50: 8–16. https://doi.org/10.1016/j.enpol.2012.02.070.

Haarstad, H. (2016). Where are urban energy transitions governed? Conceptualizing the complex governance arrangements for low-carbon mobility in Europe. *Cities* 54: 4–10. https://doi.org/10.1016/j.cities.2015.10.013.

Harris, R.D.F., Tzavalis, E. (1999). Inference for unit roots in dynamic panels where the time dimension is fixed. *Journal of Econometrics* 91(2): 201–226. https://doi.org/10.1016/S0304-4076(98)00076-1.

Hirsh, R.F., Jones, C.F. (2014). History's contributions to energy research and policy. *Energy Research & Social Science* 1: 106–111. https://doi.org/10.1016/j.erss.2014.02.010.

International Energy Agency (2019). *$CO_2$ Emissions from Fuel Combustion*. https://webstore.iea.org/co2-emissions-from-fuel-combustion-2019 (accessed 20 June 2020).

Johnstone, P., Rogge, K.S., Kivimaa, P. et al. (2020). Waves of disruption in clean energy transitions: Sociotechnical dimensions of system disruption in Germany and the United Kingdom. *Energy Research & Social Science* 59: 101287. https://doi.org/10.1016/j.erss.2019.101287.

Kander, A., Stern, D.I. (2014). Economic growth and the transition from traditional to modern energy in Sweden. *Energy Economics* 46: 56–65. https://doi.org/10.1016/j.eneco.2014.08.025.

Karagöl, E.T., Kavaz, İ. (2017). Dünyada ve Türkiye'de yenilenebilir enerji. *Analiz* 197. https://setav.org/assets/uploads/2017/04/YenilenebilirEnerji.pdf (accessed 22 November 2020).

Katre, A., Tozzi, A. (2019). Using hugs, carrots and sticks: How agents exercise power in the transition to community-owned energy systems in remote India. *Energy Research & Social Science* 54: 129–139. https://doi.org/10.1016/j.erss.2019.04.008.

Kaya, H. (2020). Yenilenebilir enerji istihdamında küresel durumun değerlendirilmesi. *Sosyal Bilimler Araştırmaları Dergisi* I(II): 10–21. https://dergipark.org.tr/tr/download/article-file/1115151 (accessed 13 May 2020).

Kuzemko, C., Mitchell, C., Lockwood, M. et al. (2017). Policies, politics and demand side innovations: The untold story of Germany's energy transition. *Energy Research & Social Science* 28: 58–67. https://doi.org/10.1016/j.erss.2017.03.013.

Le, H.P., Sarkodie, S.A. (2020). Dynamic linkage between renewable and conventional energy use, environmental quality and economic growth: Evidence from Emerging Market and Developing Economies. *Energy Reports* 6: 965–973. https://doi.org/10.1016/j.egyr.2020.04.020.

Leiren, M.D., Reimer, I. (2018). Historical institutionalist perspective on the shift from feed-in tariffs towards auctioning in German renewable energy policy. *Energy Research & Social Science* 43: 33–40. https://doi.org/10.1016/j.erss.2018.05.022.

Levin, A., Lin, C.-F., Chu, C.-S.J. (2002). Unit root tests in panel data: Asymptotic and finite-sample properties. *Journal of Econometrics* 108(1): 1–24. https://doi.org/10.1016/S0304-4076(01)00098-7.

Li, L., Taeihagh, A. (2020). An in-depth analysis of the evolution of the policy mix for the sustainable energy transition in China from 1981 to 2020. *Applied Energy* 263: 114611. https://doi.org/10.1016/j.apenergy.2020.114611.

Lindmark, M., Andersson, L.F. (2010). Household firewood consumption in Sweden during the nineteenth century. *Journal of Northern Studies* 2: 55–78. http://urn.kb.se/resolve?urn=urn:nbn:se:umu:diva-39640 (accessed 27 November 2020).

Lund, P. (2006). Market penetration rates of new energy technologies. *Energy Policy* 34(17): 3317–3326. https://doi.org/10.1016/j.enpol.2005.07.002.

Mahjabeen, S.Z.A., Chughtai, S. et al. (2020). Renewable energy, institutional stability, environment and economic growth nexus of D-8 countries. *Energy Strategy Reviews* 29: 100484. https://doi.org/10.1016/j.esr.2020.100484.

Matschoss, K., Repo, P., Lukkarinen, J. (2020). Network analysis of energy transition arena experiments. *Environmental Innovation and Societal Transitions* 35: 103–115. https://doi.org/10.1016/j.eist.2020.03.003.

Miller, C.A., Iles, A., Jones, C.F. (2013). The social dimensions of energy transitions. *Science as Culture* 22(2): 135–148. https://doi.org/10.1080/09505431.2013.786989.

Miller, C.A., Richter, J., O'Leary, J. (2015). Socio-energy systems design: A policy framework for energy transitions. *Energy Research & Social Science* 6: 29–40. https://doi.org/10.1016/j.erss.2014.11.004.

Mori, A. (2018). Socio-technical and political economy perspectives in the Chinese energy transition. *Energy Research & Social Science* 35: 28–36. https://doi.org/10.1016/j.erss.2017.10.043.

Myhrvold, N.P., Caldeira, K. (2012). Greenhouse gases, climate change and the transition from coal to low-carbon electricity. *Environmental Research Letters* 7(1): 014019. https://doi.org/10.1088/1748-9326/7/1/014019.

Nielsen, J.R., Hovmøller, H., Blyth, P.-L. et al. (2015). Of "white crows" and "cash savers:" A qualitative study of travel behavior and perceptions of ridesharing in Denmark. *Transportation Research Part A: Policy and Practice* 78: 113–123. https://doi.org/10.1016/j.tra.2015.04.033.

Nolden, C. (2019). Transaction cost analysis of digital innovation governance in the UK energy market. *The Journal of Energy Markets* 12(2): 1–21. https://doi.org/10.21314/JEM.2019.194.

O'Connor, P. A. (2010). Energy transitions (The Pardee Papers No.12.). *Speech presented at the Boston University the Frederick S. Pardee Center for the Study of the Longer-range future in Boston, USA (November 2010)*.

Overland, I., Bazilian, M., Uulu, T.I. et al. (2019). The GeGaLo Index: Geopolitical gains and losses after energy transition. *Energy Strategy Reviews* 26: 100406. https://doi.org/10.1016/j.esr.2019.100406.

Pérez, M.d.l.E.M., Scholten, D., Stegen, K.S. (2019). The multi-speed energy transition in Europe: Opportunities and challenges for EU energy security. *Energy Strategy Reviews* 26: 100415. https://doi.org/10.1016/j.esr.2019.100415.

Rubio, M.d.M., Folchi, M. (2012). Will small energy consumers be faster in transition? Evidence from the early shift from coal to oil in Latin America. *Energy Policy* 50: 50–61. https://doi.org/10.1016/j.enpol.2012.03.054.

Santos, H.T.M.d., Balestieri, J.A.P. (2018). Spatial analysis of sustainable development goals: A correlation between socioeconomic variables and electricity use. *Renewable & Sustainable Energy Reviews* 97: 367–376. https://doi.org/10.1016/j.rser.2018.08.037.

Sarrica, M., Brondi, S., Cottone, P. et al. (2016). One, no one, one hundred thousand energy transitions in Europe: The quest for a cultural approach. *Energy Research & Social Science* 13: 1–14. https://doi.org/10.1016/j.erss.2015.12.019.

Singh, H.V., Bocca, R., Gomez, P. et al. (2019). The energy transitions index: An analytic framework for understanding the evolving global energy system. *Energy Strategy Reviews* 26: 100382. https://doi.org/10.1016/j.esr.2019.100382.

Sovacool, B.K. (2009). Rejecting renewables: the socio-technical impediments to renewable electricity in the United States. *Energy Policy* 37(11): 4500–4513. https://doi.org/10.1016/j.enpol.2009.05.073.

Sovacool, B.K. (2016). How long will it take? Conceptualizing the temporal dynamics of energy transitions. *Energy Research & Social Science* 13: 202–215. https://doi.org/10.1016/j.erss.2015.12.020.

Sovacool, B.K., Hirsh, R.F. (2009). Beyond batteries: An examination of the benefits and barriers to plug-in hybrid electric vehicles (PHEVs) and a vehicle-to-grid (V2G) transition. *Energy Policy* 37(3): 1095–1103. https://doi.org/10.1016/j.enpol.2008.10.005.

Tahvonen, O., Salo, S. (2001). Economic growth and transitions between renewable and nonrenewable energy resources. *European Economic Review* 45(8): 1379–1398. https://doi.org/10.1016/S0014-2921(00)00062-3.

Tvinnereim, E., Lægreid, O.M., Fløttum, K. (2020). Who cares about Norway's energy transition? A survey experiment about citizen associations and petroleum. *Energy Research & Social Science* 62: 101357. https://doi.org/10.1016/j.erss.2019.101357.

Valdes, J., González, A.B.P., Camargo, L.R. et al. (2019). Industry, flexibility, and demand response: Applying German energy transition lessons in Chile. *Energy Research & Social Science* 54: 12–25. https://doi.org/10.1016/j.erss.2019.03.003.

Van Boxstael, A., Meijer, L.L.J., Huijben, J.C.C.M. et al. (2020). Intermediating the energy transition across spatial boundaries: Cases of Sweden and Spain. *Environmental Innovation and Societal Transitions* 36: 466–484. https://doi.org/10.1016/j.eist.2020.02.007.

Vidadili, N., Suleymanov, E., Bulut, C. et al. (2017). Transition to renewable energy and sustainable energy development in Azerbaijan. *Renewable & Sustainable Energy Reviews* 80: 1153–1161. https://doi.org/10.1016/j.rser.2017.05.168.

Wang, Q., Wang, S. (2020). Is energy transition promoting the decoupling economic growth from emission growth? Evidence from the 186 countries. *Journal of Cleaner Production* 260: 120768. https://doi.org/10.1016/j.jclepro.2020.120768.

Westerlund, J. (2007). Testing for error correction in panel data. *Oxford Bulletin of Economics and Statistics* 69(6): 709–748. https://doi.org/10.1111/j.1468-0084.2007.00477.x.

World Bank (2020). *World Development Indicators*. https://databank.worldbank.org/reports.aspx?source=World%20Development%20Indicators# (accessed 22 October 2020).

World Economic Forum (2018). *Fostering Effective Energy Transition 2018*. http://www3.weforum.org/docs/WEF_Fostering_Effective_Energy_Transition_report_2018.pdf (accessed 15 May 2020).

World Energy Council (2018). *World Energy Issues Monitor 2018*. https://www.worldenergy.org/assets/downloads/Issues-Monitor-2018-HQ-Final.pdf (accessed 27 November 2020).

World Energy Council (2019). *World Energy Issues Monitor 2019*. https://www.worldenergy.org/assets/downloads/1.-World-Energy-Issues-Monitor-2019-Interactive-Full-Report.pdf (accessed 27 November 2020).

World Energy Council (2020). *World Energy Issues Monitor 2020*. https://www.worldenergy.org/assets/downloads/World_Energy_Issues_Monitor_2020_-_Full_Report.pdf (accessed 22 October 2020).

# 19 Energy Security
## *Role of Renewable and Low-Carbon Technologies*

*Liliana N. Proskuryakova*
National Research University Higher School of Economics

## CONTENTS

## 19.1 INTRODUCTION. TRENDS AND FACTORS THAT FACILITATE OR HINDER THE APPLICATION OF RES AND LOW-CARBON TECHNOLOGIES

Global energy consumption has significantly changed over the past 40 years: the OECD countries have managed to slow down energy demand and partially substituted coal with natural gas and renewables (RES), while the economic boom in Asia and Latin America resulted in rapidly accelerating demand for energy resources and an increased share of coal in final energy consumption. Forecasted energy demand over the next 20 years varies significantly. However, many experts agree that the demand for nuclear energy will remain unchanged or even slightly drop, demand for oil and coal will drop more significantly, while demand for natural gas and renewables will grow. The changed energy balances will assure the new energy security architecture that is based on a wider modern concept of energy security.

Renewable and low-carbon technologies have seen a rapid development over the past decades due to price decline, environmental and climate concerns. Affordable renewable energy technologies have made many countries self-sufficient, allowed some economies to evolve from energy poor to energy rich. Even in countries that do not have high insolation or space for wind farms, technological and economic benefits of renewable energy have contributed to higher energy security. Natural gas, the cleanest of all fossil fuels, is considered to be a good complimentary fuel in countries that are switching to green growth. The fluctuating nature of renewables is addressed by more efficient and faster energy distribution through smart grids and cheaper high-capacity energy storage (Proskuryakova and Ermolenko, 2019).

The recognition of the green growth concept and related policy at the international and national levels, notwithstanding local and regional authorities, plays a key role in their implementation. The energy policy of local and regional institutions promotes investments in a sustainable decentralized energy sector, the efficient utilization of resources, the creation of green jobs, and the development of regional clusters and partnerships.

DOI: 10.1201/9781003315353-22

For example, Italy is planning to accomplish the 20–20–20 objectives[1] through implementation of regional renewable energy development programs (EEA, 2018). The Maltese government promoted the construction of RES-based generation facilities and the installation of energy-efficient equipment through direct investments, by attracting additional investments on the municipal level. In Romania, energy-efficient solutions, and local decentralized heating systems are being promoted via relevant policies pursued by local administrations. The government of Scotland intends to hand over 2.4 billion pounds' worth of renewable energy facilities with a capacity of least 500 MW to local authorities or communities by 2020.

The introduction of stricter environmental and climate-related standards and requirements represents a challenge for fossil fuel exporting countries, such as Saudi Arabia and Russia, due to the need to bring their national legislation in line with that of trading counterparts—importers of energy resources (Day and Day 2017). With the projected introduction of carbon tax in the EU and elsewhere, of all carbon intensive sectors extractive industries will suffer the most.

Investments in renewables across the globe are increasingly profitable and assure the dominance of renewable energy in total new installed capacities since 2015. Before the pandemic, in 2019 global renewable generation capacity reached 2.537 GW with the largest share belonging to hydropower −1.190 GW. In 2020, the first year of pandemics investments in renewables moved up 2%. Capital costs continued the downward trend and facilitated the installation of record volumes of both solar (132 GW) and wind (73 GW) (BNEF, 2021). In the past years, 90% of the new renewable capacities have been represented by widely available solar and wind facilities. The predominant R&D and the commercial focus on solar and wind limit chances for other, new renewable energy sources to advance.

## 19.2   AN OVERVIEW OF THE DOMINANT RENEWABLE AND LOW-CARBON TECHNOLOGIES

Worldwide until 2020 the dominant renewable and low-carbon technologies were hydro, solar PV, and wind power, along with digital transformation of the industry. On the one hand, these technologies have had positive social, climate, and environmental implications, and, thus, positive impact on social and environmental security. On the other hand, for other countries, especially fossil fuel poor economies, these technologies have also contributed to energy security by diversifying energy mix and increasing the share of local energy sources.

The focus on renewable and low-carbon energy sources has opened up discussion on their classification, application, and measurement. Researchers outline a number of renewable, low- and zero-carbon technologies that have diverse implications for energy security. Followed by the summary of their security implications in Tables 19.1–19.3, there is a more detailed description of the main technologies.

> Power and heat generation technologies: Small hydro power plants are most often constructed at "run-of-river." They are among the most cost-effective and environmentally friendly energy technologies suitable both for developed and developing countries, in particular for rural electrification. Unlike large hydro systems they do not require the construction of a dam or water storage. Small hydro power has better energy security characteristics than solar or wind power, as it is a much more concentrated energy resource, its availability is more predictable, it is continuously available and may be used in hard-to-reach territories without access to the central grid (such as mountainous regions). It has all the benefits of renewables, including zero fossil fuel consumption, very low environmental and climate impact. However, this technology requires a well-suited site (river) that is at

---

[1]The EU program to reduce greenhouse gases' emissions by 20% compared with 1990; increase RES share in total energy consumption to 20%; and reduce primary energy consumption by 20% by 2020 through increased energy efficiency.

**TABLE 19.1**

**Renewable and Low-Carbon Technologies for Electricity Generation with Implications for Energy Security**

| | Electricity Only | Implications for Energy Security | |
|---|---|---|---|
| | | Risk/LCOE[a] | Power Supply[b]/Technology Maturity[c] |
| Low carbon | Gas peaking | Medium risk<br>High LCOE | Uninterrupted supply<br>Mature |
| | Large-scale energy storage | Medium risk<br>Medium LCOE | Uninterrupted supply<br>Mature |
| Renewable and zero carbon | Nuclear energy | High risk<br>Medium to high LCOE | Uninterrupted supply<br>Mature |
| | Large-scale solar PV farms | Medium risk<br>Low to medium LCOE | Variable supply<br>Mature |
| | Small-scale solar PV | Low risk<br>High LCOE | Variable supply<br>Mature |
| | Offshore and onshore wind farms | Medium risk<br>Low LCOE | Variable supply<br>Mature |
| | Small-scale wind | Low risk<br>Low LCOE | Variable supply<br>Mature technology |
| | Large-scale hydro | High risk | Variable supply<br>Mature |
| | Small hydro | Low risk | Variable supply<br>Mature |
| | Tidal range power | High risk | Variable supply<br>Mature |
| | Tidal stream | Medium risk | Variable supply<br>Mature |
| | Wave power | Medium risk | Variable supply<br>Mature technology |
| | Geothermal | Low risk<br>Low LCOE | Uninterrupted supply<br>Mature |
| | Hydrogen fuel cells | Medium risk<br>Low to medium LCOE | Uninterrupted supply<br>Ascent |

*Source:* Author's analysis based on RSPB, The RSPB's 2050 Energy Vision Meeting the UK's Climate Targets in Harmony with Nature, RSPB, 2016 and Neill et al., (2018). With permission; Lazard (2020), IEA (2021).

[a] High, medium, or low.
[b] Uninterrupted or variable.
[c] R&D, ascent, mature, or decline.

the same time close to power consumers, the maximum useful power output is limited, and river flows may have seasonal variations. Additionally, there may be a conflict of interest with fisheries and agricultural water use (Paish, 2002).

Solar energy technologies are dominated by two types: solar photovoltaics (solar PV) and concentrated solar power (CSP). Solar power is used to generate electricity and thermal energy (heating or cooling), to directly meet lighting needs, and to produce other fuels, such as hydrogen. Many technologies are modular in nature, which makes their application possible in both centralized and decentralized energy systems. The factors that limit energy security of solar energy include

**TABLE 19.2**

**Renewable and Low-Carbon Technologies for Combined Heat and Power Generation with Implications for Energy Security**

| | Combined Heat and Power, CHP | Risk/LCOE[a] | Heat and Power Supply[b]/ Technology Maturity[c] |
|---|---|---|---|
| | | **Implications for Energy Security** | |
| Low carbon | Gas-fired CHP in centralized heat supply systems; | Medium risk<br>Low LCOE | Uninterrupted supply<br>Mature technology |
| | Gas-fired micro CHP in decentralized (including individual) heat supply systems | Medium risk<br>Low LCOE | Uninterrupted supply<br>Mature technology |
| Renewable and zero carbon | Waste or biomass CHP in centralized heat supply systems | Medium risk<br>Low LCOE | Uninterrupted supply<br>Mature technology |
| | Biomass and micro CHP in decentralized (including individual) energy systems | Medium risk<br>High LCOE | Uninterrupted supply<br>Mature technology |
| | Energy from hot dry rocks | Medium risk<br>Low LCOE | Uninterrupted supply<br>R&D phase (demonstration projects) |
| | Nuclear energy CHP | Medium risk<br>High LCOE | Uninterrupted supply<br>R&D phase (demonstration projects) |

*Source:* Author's analysis based on Hansen, K., Energy Strat. Rev., 24, 68–82, 2019; Lazard, Levelized Cost of Energy and of Storage, Lazard.com, 2020; IEA, Net Zero by 20150; IAEA, 2019. A Roadmap for the Global Energy Sector, OECD/IEA, Paris, 2021.

[a] High, medium, or low.
[b] Uninterrupted or variable.
[c] R&D, ascent, mature, or decline.

**TABLE 19.3**

**Renewable and Low-Carbon Technologies for Heat Generation with Implications for Energy Security**

| | Heat Only | Risk/LCOE[a] | Power Supply[b]/ Technology Maturity[c] |
|---|---|---|---|
| | | **Implications for Energy Security** | |
| Low carbon | | | |
| | Air source and ground source heat pumps | Medium risk<br>Medium LCOE | Uninterrupted supply<br>Mature technology |
| Renewable and zero carbon | Solar hot water | Medium risk<br>High LCOE | Variable supply<br>Mature technology |
| | Biomass boiler | Medium risk<br>High LCOE | Uninterrupted supply<br>Mature technology |

*Source:* Author's analysis based on Zongxian, Z, et al., Systems Eng. Proc. 4, 99–106, 2012; Hansen, K., *Energy Strat. Rev.*, 24, 68–82, 2019; Fang, Y., Zhao, S., Solar Energy, 208, 937–948, 2020; Ecclesiastical, Safe Use of Biomass Boilers. Risk Management, Ecclesiastical, 2021; IEA, *Net Zero by 20150. A Roadmap for the Global Energy Sector*, OECD/IEA, Paris, 2021.

[a] High, medium, or low.
[b] Uninterrupted or variable.
[c] R&D, ascent, mature, or decline.

its variability and limited predictability, the dependence of generation on solar irradiance that varies significantly depending on geographic location (Edenhofer et al., 2011; Lange, 2013). The factors that contribute to energy security are wide availability of this resource, advanced technologies with increasing capacity factor and ability to capture sunlight even in cloudy days, a wide array of applications (large-scale solar power plants, rooftop panels, solar windows and many more).

Wind power generation depends on strong and consistent winds that vary depending on geographic location and are hard to harvest within cities. Turbines have to be placed in open areas or well above any nearby objects (trees, buildings, etc.) in order to have access to high quality wind. Similar to solar power plants, large-scale wind farms and turbines drive the constant reduction of electricity costs. Onshore wind power generation has already become cost-competitive with newly constructed conventional power plants. Offshore wind power plants require higher capital investments and grid connectivity, while at the same time are more efficient due to strong winds. Increasing efficiency and lowering cost are the factors that contribute to better energy security of wind power generation. Wind turbines require equipment weight reduction, better aerodynamic efficiency, new lighter and stronger materials (IRENA, 2016).

Biomass that is used for energy generation (organic matters extracted from plants, animals) is widely available and includes residues and waste that would otherwise be stored or disposed of (in forestry, agriculture, housing and utilities, and industrial processes). It also includes traditional fuels for cooking and heating (such as logs/wood fuel) widely used in developing countries. The technologies allow for production of biofuels and bioenergy (electricity and heat, CHP) through biomass combustion, gasification, ultrasonic and microwave intensification, catalytic cracking, hydrodeoxygenation, and anaerobic digestion (Goh, 2019). Biomass may provide for power supply in the periods of non-availability of other renewables, thus contributing to security of energy supply. Moreover biomass, unlike other renewables, may be used for heat in CHP generation.

Energy generation from various types of waste is growing all over the world, including generation by manufacturing companies (in-house generation). Increased amount of solid domestic waste (SDW), agricultural and animal farming waste, water treatment sediments, and other types of waste make up the rationale for energy-from-waste solutions. The global volume of SDW amounted to about 2.01 billion tons a year in 2016 and this figure is projected to increase to 3.40 billion tonnes in 2050. Specific waste generation figures vary for countries and cities depending on the level of their urbanization and wealth (Kaza et al., 2018). In the medium to long term, waste may become a profitable quasi-renewable energy source comparable with fossil fuels.

As solar, wind, and other forms of variable renewable generation increase, grid operators face the need to balance supply and demand, which is essential for assuring stable and secure energy supply. The required flexibility in low-carbon energy systems may be provided by energy storage, demand response solutions, and digital technologies. The former include batteries, flywheels, compressed air, and other options. Demand response programs offer consumers a possibility to reduce or shift their electricity usage during peak hours in exchange for special rates or other financial incentives (US Office of Electricity, 2021).

Another major trend that will affect future energy security is the development of clean and efficient nuclear energy technologies (Karakosta et al., 2013). New multipurpose nuclear energy systems may be able to generate heat and power. Closed nuclear fuel cycle technologies will be applied in the next (fourth) generation reactors that ensure up to 50 times more efficient use of nuclear fuel and significantly reduced nuclear waste resulting in higher reliability and safety (Russian Nuclear Community, 2012). In addition to zero greenhouse gas emissions, the new reactors will be much more environment friendly. Such reactors are currently being developed in Russia, the European Union, India, China, and Japan (IAEA, 2012; European Commission et al., 2014). The deployment of fourth-generation reactors is expected after 2030. The main uncertainty factor here is the project costs affecting their competitiveness with other types of power generation.

- **Green infrastructure solutions:** Digital transformation is underway in most energy systems and segments. Digital technologies contribute to a more efficient use of resources and extend the use of renewables (Kamble et al., 2018; Menzel and Teubner, 2020). Electric power industry is pioneering these changes through smart grids; virtual power plants; smart substations; active and passive smart homes; smart electricity, heating and cooling systems; energy-efficient lighting; and other. The digitalization of energy infrastructure based on cloud computing and Big Data systems is expected to reduce energy costs and create backup end-user capacities (Moyer and Hughes, 2012).

Smart grids of different size may assure centralized, distributed, and individual energy supply of industrial and household consumers. Many countries are constructing national smart grids, where all energy market players (generating and distributing companies, and consumers) actively participate in energy transmission and distribution. Smart grids encompass computers, controls, automation, and other new technologies and equipment. They are instrumental for energy security as they seek to assure sustainable, reliable, safe and high quality electricity to all consumers by providing a digital response to quickly changing electric demand (Bayindir et al., 2016; US Department of Energy, 2019). The most known, and arguably, the most widespread elements of smart grids are new meters and similar appliances capable of sending and receiving information in real time mode to manage electricity demand. Smart grid is also instrumental in integrating various energy storage options and assuring rapid redistribution of variable renewable power from the location where it is available to the location with electricity demand at any given time (Hyams, 2012).

The deployment of renewables and digital transformation contribute to a wider adoption of decentralized energy systems that are not connected to the national transmission grid and operate independently. Smart mini- and microgrids may run on renewables (solar, wind, and/or biomass) or various hybrids, whereby a limited number of consumers and prosumers and interconnected. Such small-scale grids include power plants with capacity ranging from few kW to 10 MW, virtual power plants, and energy storage systems. The customers that benefit from mini- and microgrids are households, small businesses, and service centers. Modern small-scale grids are comparable to larger-scale grids in terms of reliability and security of supply. They may be set up by companies, utilities, communities, non-governmental organizations, or a combination of different actors. The trend toward creating autonomous distributed energy supply systems is noticeable both in developed and developing countries.

Digital transformation reinforces the more mature trends related to decentralization of energy systems, green growth and cross-sectoral solutions, such as vehicle-to-grid. Digital technologies also increase the traceability of goods such oil or green power, thus confirming the quality and the reliability or supplier. Other technologies widely used for digital transformation of energy systems are digital platforms for energy contracts and marketplaces (including distributed ledger and blockchain), Internet of Energy, and those that provide for integrated full-cycle energy services (Energy-as-a-Service) (Dellermann et al., 2017; Ketter et al., 2018; Adeyemi et al., 2020). In exploration and extraction remote sensing and new geoinformation systems based on 3D modeling are applied for a faster and more comprehensive assessment of new deposits. In the oil and gas sector, digital technologies are used to sustain the volume of fossil fuel extraction and for increasing marginality and lowering processing costs. In the coal industry, digital solutions are applied for optimizing the supply chain and increasing safety, for instance, with the use of geolocation and gas-metering appliances, personnel monitoring devices, and air quality meters in coalmines (Dprom.online, 2020). Overall, digital technology solutions in extractive industries may be grouped into digital upstream projects (such as cognitive systems to support expert decisions), digital deposites (with a focus on unconventional reserves), and digital downstream (including integrated value chain management).

Digitalization contributes to energy security by ensuring higher efficiency of the power industry and reliability of power supply, including improvements in efficiency (power losses, outages, labor productivity, and costs), faster power transmission and redistribution of energy across regions, and cross-sectoral implications (i.e., for transport sector in vehicle-to-grid technology). However, digital technologies also incur security risks by increasing the probability of cyberattacks on power plants and assuring personal information safety.

- **Hydrogen:** Many countries have announced their intention to move on toward a hydrogen economy, made possible by the advances in hydrogen energy generation, the development of new fuel cell types, and the significant progress achieved in related fields. The invention of hydrogen fuel cells led to discovering a clean, safe, and economically viable alternative to fossil fuels, making it possible to store electricity on a large-scale basis. However, before hydrogen will be widely adopted several barriers have to be addressed: its low cost-effectiveness, explosion hazards, and lack of distribution infrastructure. The worldwide hydrogen generation industry is expected to reach US$201 billion by 2025 (Globalnewswire, 2021), up from US$143 billion in 2019 (Research and Markets 2021).

Before the invention of hydrogen fuel cells, the application of hydrogen for energy-related purposes was limited by low profit margins, high explosion hazards, and a lack of economically viable storage and distribution systems. Hydrogen fuel cells provided solutions for all these problems. They provide large volumes of electricity and make it available to users not connected to the grid. Due to very high energy density, hydrogen is safe in mixtures with liquids, which could make its transportation possible via already existing liquid fossil fuel distribution pipelines. However, there are several barriers to large-scale application of hydrogen that include insufficient safety and durability of fuel cells and hydrogen storage systems; the lack of the decentralized infrastructure that would make hydrogen cars attractive to consumers; and the high costs of electrolyzers and hydrogen production. The possibility and time required to overcome these barriers remain uncertain.

There are several types of hydrogen that are defined based on production type, which is a crucial consideration from the perspective of energy and environment security. Similar to other energy sources, the most "dirty" hydrogen production technology is usually the cheapest. White, colorless hydrogen may be found in nature, mostly in gaseous form ($H_2$). Brown/black hydrogen is produced of coal through gasification processes at very high temperatures (over 700°C). $CO_2$ and carbon monoxide are emitted in the atmosphere as byproduct. Gasification may also be used to produce hydrogen from biomass.

Gray hydrogen is produced from natural gas through steam reforming with some $CO_2$ as byproduct. This technology accounts for most of hydrogen generation worldwide. Should carbon capture and storage be added in the process whereby $CO_2$ is stored underground, hydrogen will be referred to as blue, and the technology will be referred to as low-carbon.

Pink hydrogen is obtained through electrolysis with the use of nuclear energy. Yellow hydrogen is produced through the same technology with the use of a mixture of various energy sources. Green, or clean, hydrogen accounts for 1% of all hydrogen generation worldwide and is produced with the use of renewables (Energy Cities, 2020). The major obstacle to wider green hydrogen use is high production cost: it has to fall by more than 50% to US$2.0–2.5 per kg to make this energy source a viable alternative to fossil fuels (Edwardes-Evans, 2020). This cost decrease is projected by 2030.

- **Energy storage:** Major energy security factors also include the development of energy storage technologies (new types of accumulators and batteries, etc.), and relevant infrastructure (charging and power output systems) for providing centralized, distributed, and individual energy supply to transport systems. The already available stationary and portable energy storage systems can provide additional advantages for the low-carbon energy systems of the future. Given the increased efficiency and service life, reduced production and running costs, and the reduced demand for passive power capacities, energy storage systems could

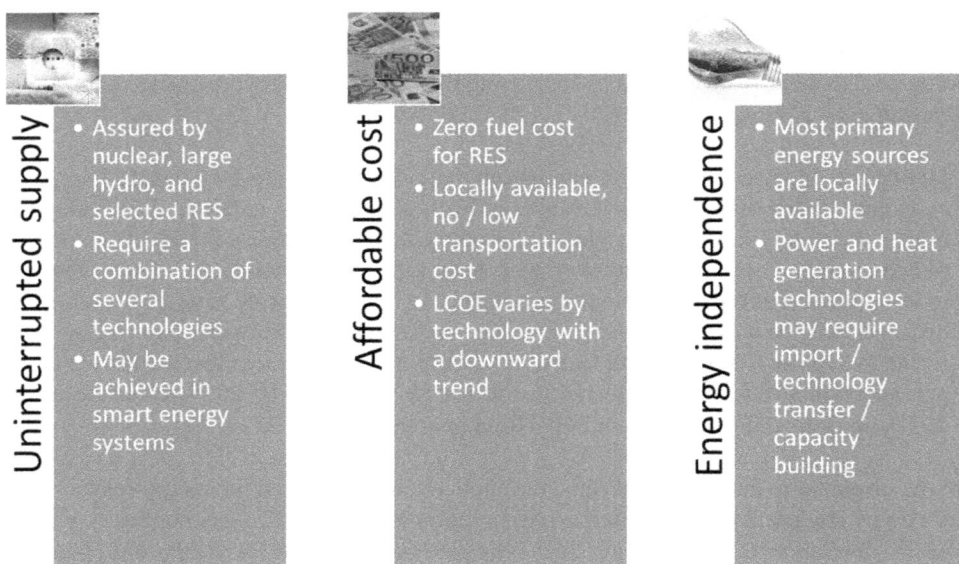

**FIGURE 19.1** Summary of the main RES and low-carbon technologies contribution to energy security (author's analysis).

significantly increase the efficiency of numerous centralized and decentralized generation systems, including solar-, nuclear-, wind-, geothermal-based, etc. The emergence of new electrochemical power sources, with increased safety, higher capacity, and lower production costs could accelerate the electrification of transport systems. There is a particular interest in chemical batteries and electrochemical capacitors (Skoltech, 2016). For example, Tesla, the largest producer of electric cars in the world, is already making lithium-ion batteries. They allow car manufacturers to cut costs by 30%, leading to reduced car prices. Potentially this may lead to a shortage of lithium, and reduced demand for platinum and palladium, which are currently used in internal combustion cars' catalytic converters. Major factors hindering the development of energy storage systems include their high costs, projected material (lithium, cobalt, nickel) deficit, and the legal uncertainty surrounding their use in the energy system.

Summing up, the main contributions of RES and low-carbon technologies to energy security are toward uninterrupted supply, affordable cost, and energy independence (Figure 19.1). While the contributions to greater environmental security and climate action is beyond doubt, low input in energy security of some of technologies is attributable to their insufficient maturity or small market share. New energy technologies that are at the R&D phase (i.e. energy from hot dry rocks, hydrogen, closed nuclear fuel cycle and other) have the potential to make a significant contribution to energy security.

Renewable technologies are charactrized by zero fuel cost, while LCOE depends on the power plant capacity, location, and other factors. The energy supply based on renewable sources varies depending on weather, time, and area, therefore additional technologies are necessary to assure uninterrupted supply (i.e., smart grid, storage systems, combination of various types of power generation). Low-carbon technologies like nuclear or hydrogen have high LCOE and high fuel cost, but assure uninterrupted supply. All of these technologies are based on locally available primary energy sources (with the exception of nuclear) and, thus, contribute to higher energy independence.

## 19.3 NATIONS LEADING IN RENEWABLE AND LOW-CARBON TECHNOLOGIES

Multiple rankings are aimed at identification of world leaders in renewable and low-carbon technologies. For instance, the Green Future Index by MIT Technology Review ranks 76 countries and territories by carbon emissions, energy transition, social green consumer practices, clean innovations—patents and investments, and climate policy. Europe is leading the Index with 15 countries from this region among top 20. Iceland, Denmark, and Norway occupy top positions. Other countries included in top 10 are France, Ireland, Finland, Costa Rica, New Zealand, Belgium, Netherlands, and Germany. The world largest economies USA and China are assessed as lagging behind taking 40th and 45th positions accordingly (MIT Technology Review, 2021).

The national leaders in domestic and international renewable energy investments are the USA, Germany and China (IHS Markit, 2021). If these investments are assessed as a share of GDP, China comes third after Chile and South Africa with 0.9% in 2015 and 1.4%, respectively (O'Meara, 2020). China also holds technological leadership with over 150,000 renewable energy patents in 2016, which made up 29% of the world total. The United States comes second with over 100,000 patents, Japan and the EU follow with around 75,000 patents each (Geopolitics of Renewables, 2021). Technological leadership expressed in terms of patents is a good measure of technological independence that is an important energy security component.

Paraguay, Democratic Republic of Congo, and Ethiopia have the highest share of renewables in their primary energy supply, followed by Iceland, Zambia, and Tanzania. Africa is a clear leader in this indicator followed by Latin America (OECD, 2021). The leader, Paraguay, assured 113% of its primary energy supply from renewables, exporting the excessive amounts of electricity. For comparison, in 2019, only 23% of China's energy consumption is from "clean" sources (including natural gas), almost 58% is assured by coal, and the rest comes from oil and nuclear power. The country's 13th 5-year plan for social and economic development (2016–2020) included a target to raise the share of domestic renewable and non-fossil-fuel energy consumption to 20% by 2030 (O'Meara, 2020).

The countries leading in installed renewable energy capacity in 2020 were China, the United States, and Brazil. China has installed capacity of around 895 GW, the US around 292 GW, and Brazil 150 GW. The three leaders are followed by India (134 GW), Germany (132 GW), Canada (101 GW), Japan (101 GW), Italy (55 GW), and France (55 GW) (Statista, 2021). In terms of new renewable capacity, China accounted for more than half of the global non-hydro renewables additions in 2020.

The leaders change from year to year, and this is illustrated in Figure 19.2. In 2018, Russian Federation was ranked 9th among countries with the highest installed renewable capacity, and in 2019, Spain occupied this place. In 2020, neither Russia nor Spain could make it to the top 10 in this indicator.

The different measures described above clearly demonstrate that world leaders in renewable and low-carbon energy depend on the indicators selected for comparison. Simple and composite indicators need to take account of the countries' energy systems, infrastructure, social and economic context, international relations, weather conditions and climate, size, and geographic location. These considerations are important in assessing the contributions of renewable and low-carbon energy technologies to national energy security.

## 19.4 THE FUTURE REVOLUTIONARY TECHNOLOGIES THAT WILL ALTER SECURITY ARCHITECTURE

There are a number of technologies that may significantly alter existing energy systems, if and when they are invented and appear on the market. Besides technical issues that have to be addressed, the new revolutionary technologies have to be economically feasible. At present, a number of them are at the stage of research and development and associated with either making the use of traditional fuels cleaner or energy transmission faster. Some other technologies aim at exploiting new energy sources.

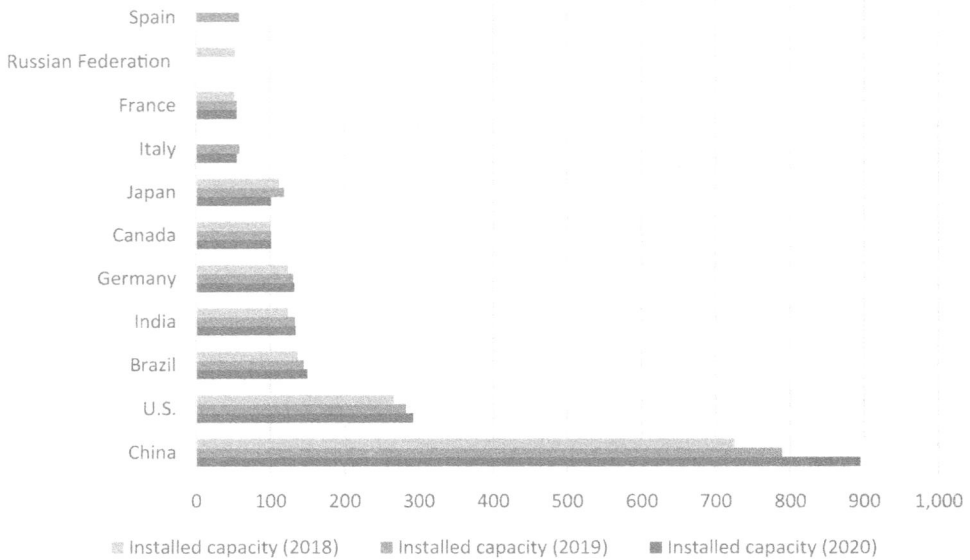

**FIGURE 19.2**  Top 10 countries in installed renewable capacity, 2018–2020, GW. (From IRENA, IRENA Country Rankings, 2021a; IRENA, Renewable Capacity Statistics 2021, 2021b. With permission.)

The active development of infrastructure for the capture, storage, and use of carbon dioxide (CCSU) will make a significant contribution to decreasing the anthropogenic impact on climate, which is mainly caused by the production and use of hydrocarbons (WEC, 2019). Technologies for capturing $CO_2$ and subsoil storage in the course of hydrocarbon extraction are already used by companies (e.g., Statoil), but solutions for capturing $CO_2$ released in the course of fossil fuels combustion have not been implemented yet. The main barriers and, at the same time, uncertainty factors, are the high costs of existing technologies and infrastructure deployment, as well as the lack of legislation regulating their application. Australia, the United Kingdom, the European Union, Canada, Norway, and the United States are applying certain policy tools for promoting the accelerated responses to existing challenges (such as $CO_2$ emission trading schemes and emission taxes).

The chances for achieving energy storage breakthroughs, including power-to-gas (P2G) technologies, remain vague. Currently existing chemical energy storage units have a limited number of charge-discharge cycles. Supercapacitors are much more durable but their capacity does not yet match industrial users' requirements. Superconductive induction energy storage systems are developing at a very high rate. Their mass application could happen when their unit price (per 1 kWh of stored energy) drops to US$100–150. This will allow for a wider application of power generation technologies that cannot operate uninterrupted (i.e., solar, wind, and tidal energy) (D'Aprile et al., 2016), and will substantially increase the energy security of renewable energy systems.

Superconductivity technologies allow for transmitting large volumes of electricity at long distances, and are expected to open new opportunities for the energy industry and the entire economy. The global superconductor-based products market is forecasted to grow from US$4.85 billion in 2020 to $8.78 billion in 2025 (PR Newswire, 2021). The growth will be largely determined by the rate of technological development, a reduction in the cost of superconductors, and, lately, by the post-COVID-19 recovery. Between 2005 and 2014, the cost of 1 kA-m has already decreased from US$10,000 to US$100. When it reaches US$25 per 1 kA-meter, large-scale commercial applications of superconductors will be possible. In commercial energy generation, they would be particularly useful for making cables and power electrical machinery, and for energy storage purposes (induction storage systems). Superconductive cables with super-low energy loss would elevate energy systems' efficiency (and security) to a much higher level, opening up new options for positioning generating facilities and exporting electricity.

There is a chance that nuclear fusion technologies may emerge in the next 30 years. One of the biggest R&D programs in this area is the International Thermonuclear Experimental Reactor (ITER), a joint research project implemented by Russia, India, the EU, China, the United States, the Republic of Korea, and Japan (ITER, 2021). Despite the delays and the high costs, the ITER remains the most promising nuclear fusion project. The results of these experiments are expected in the medium to long term. The reactor's launch and operation will allow one to test the viability of the suggested engineering solutions and to assess the economic feasibility of their commercialization. If the research turns out to be successful, the ITER-based technological solutions may be implemented at nuclear fusion reactors.

Systems that use energy from hot dry rocks (or petrothermal energy) have been explored in the United States, Russia, and some European counties for decades. The R&D have been completed and the systems entered the feasibility studies phase. Several demonstration projects have been completed to test this technology, including Felton Hill in the United States, Soultz in France, the Rosemanowes quarry in England, the Hijiori and Ogachi in Japan, and in the Cooper Basin in Australia. These technologies focus on the extraction and use of heat accumulated in hot dry rocks of the Earth's crust to generate affordable power and heat with stable adjustable parameters. Petrothermal systems could provide a steady energy supply in most locations, while the highest demand may occur in remote and poorly developed regions, as well all those areas that have power deficit. This low-carbon energy resource is attractive because of low energy production costs, close to zero emissions, and an opportunity to recuperate excessive heat (through establishing a closed-cycle system) (Gnatus et al., 2011). Among the risks of these "engineered" energy systems that were identified in the demonstration phase are high water losses observed during pressurized circulation and probability of induced seismic activity.

Researchers had been working on increased efficiency of solar energy use as early as in the 1940s, when it was suggested that an automated space station be launched to redirect solar energy to Earth using microwaves or laser beams (Asimov, 1941). India, China, the United States, and Japan have been developing their own satellite-based robotic solar power stations that would wirelessly transfer huge volumes of clean renewable energy to the Earth. The main barrier hindering the construction of such stations is the high cost of space launches required to place the satellites into orbit. Accordingly, the first such space-based solar power plant is estimated to cost up to US$20 billion. Taking into account the declining space launch costs due to competition by private companies, this amount may be revisited.

Dark matter remains the least researched potential energy source. This topic is currently at the basic research stage. According to one of the hypotheses, the visible mass of the Universe accounts for about 5% of common matter, 70% is "vacuum energy," and 25% is dark matter that is invisible, does not emit either light or other electromagnetic waves, and does not huddle when affected by gravity. Experiments to discover this matter are conducted at the LHC proton accelerator at the European Organization for Nuclear Research (Switzerland). The potential for using dark matter as an energy source for spacecraft on long missions is being discussed (Liu, 2009). If relevant hypotheses are confirmed, a unit of dark matter mass could emit 5 billion times more energy than a mass unit of dynamite does (Khel, 2016).

## 19.5 CONCLUSIONS

Renewable and low-carbon technologies are becoming more widespread due to environmental security and climate change concerns. They have already become instrumental in satisfying the demand for electricity. However, when it comes to heat generation and CHP there still are very limited options to assure them through renewables and low-carbon solutions. Advances in new energy technologies (including those based on new energy sources) and technologies that are in the R&D phase could make a significant input in energy security.

In addition to technologies used for heat and power generation, there are a number of solutions required to connect renewable and low-carbon facilities with each other, with energy storage, and the grid (consumers). Fast and efficient transmission of energy from the locations where it is generated to the places where it is consumed is also necessary to assure high energy security. Therefore, the following complimentary technologies are being developed: CCSU, smart grid, energy storage, superconductivity, and digital solutions.

New energy sources may have a revolutionary effect for the energy industry and energy security. Similar to the way energy poor countries could become energy rich through wide application of renewables, a discovery a new abundant energy source that could provide uninterrupted supply at an affordable price could revolve the energy poverty problem at the global level. In addition to assuring energy security, new renewable and low-carbon energy sources could contribute to environmental security, social and economic stability, and climate action.

The country leaders in renewable and low-carbon energy security vary depending on the indicators and timeframe selected for comparison. Among the leaders are the USA, China, the European Union and African countries. Besides increased security, the aim of some countries is to export excessive energy or energy technologies, thus ensuring additional source of revenue and/or technological leadership.

## ACKNOWLEDGMENT

This chapter is based on the study funded by the Basic Research Program of the National Research University Higher School of Economics.

## REFERENCES

Adeyemi A et al. (2020) Blockchain technology applications in power distribution systems. *The Electricity Journal* 33(8): 106817.
Asimov I (1941) *Reason. A Short Study.* Astounding Science Fiction, USA.
Bayindir R. et al. (2016) Smart grid technologies and applications. *Renewable and Sustainable Energy Reviews* 66: 499–516.
BNEF (2021) Energy transition investment hit $500 billion in 2020 – For first time, *BloombergNEF*, 19 January. https://about.bnef.com/blog/energy-transition-investment-hit-500-billion-in-2020-for-first-time/. Accessed 28 May 2021.
D'Aprile P, Newman J, Pinner D (2016) The new economics of energy storage. *McKinsey Sustainability*, McKinsey. https://www.mckinsey.com/business-functions/sustainability/our-insights/the-new-economics-of-energy-storage. Accessed: 3 Feb 2021.
Day C, Day G (2017) Climate change, fossil fuel prices and depletion: The rationale for a falling export tax. *Economic Modelling* 63: 153–160.
Dellermann D, Fliaster A, Kolloch M (2017) Innovation risk in digital business models: The German energy sector. *Journal of Business Strategy* 38(5): 35–43.
Dprom.online (2020) Digital mines – Technologies moved underground? *Extractive Industries*. https://dprom.online/mtindustry/tsifrovye-shahty-tehnologii-poshli-vglub/. Accessed: 9 June 2021.
Ecclesiastical (2021) *Safe Use of Biomass Boilers. Risk Management*, Ecclesiastical. https://www.ecclesiastical.com/risk-management/biomass/. Accessed: 9 June 2021.
Edenhofer O, et al. (2011) *Renewable Energy Sources and Climate Change Mitigation: Special Report of the Intergovernmental Panel on Climate Change.* Cambridge University Press, Cambridge, England.
Edwardes-Evans H (2020) *Green Hydrogen Costs Need to Fall over 50% to Be Viable: S&P Global Ratings.* S&P Global Platts, SP Global. https://www.spglobal.com/platts/en/market-insights/latest-news/electric-power/112020-green-hydrogen-costs-need-to-fall-over-50-to-be-viable-sampp-global-ratings. Accessed: 17 May 2021.
EEA (2018) *Overall Progress towards the European Union's '20-20-20' Climate and Energy Targets*, European Environment Agency. Luxemburg. Publications Office of the European Union. https://www.eea.europa.eu/themes/climate/trends-and-projections-in-europe/trends-and-projections-in-europe-2017/index. Accessed: 28 May 2021.

Energy Cities (2020) 50 Shades of (grey and blue and green) hydrogen. *Energy Cities*, 13 November. https://energy-cities.eu/50-shades-of-grey-and-blue-and-green-hydrogen/. Accessed: 10 June 2021.

European Commission, Joint Research Centre, and Institute for Energy and Transport (2014) *2013 Technology Map of the European Strategic Energy Technology Plan (Set-Plan) Technology Descriptions.* Luxembourg, Publications Office of the European Union.

Fang Y, Zhao S (2020) Risk-constrained optimal scheduling with combining heat and power for concentrating solar power plants. *Solar Energy* 208: 937–948.

Geopolitics of Renewables (2021) *A New World: The Geopolitics of the Energy Transformation*, http://geopoliticsofrenewables.org/report. http://geopoliticsofrenewables.org/report. Accessed: 9 June 2021.

GlobalNewswire (2021) *The Worldwide Hydrogen Generation Industry Is Expected to Reach $201 Billion by 2025.* GlobeNewswire News Room. https://www.globenewswire.com/news-release/2021/03/01/2184177/28124/en/The-Worldwide-Hydrogen-Generation-Industry-is-Expected-to-Reach-201-Billion-by-2025.html. Accessed: 9 June 2021.

Gnatus NA, Khutorskoy MD, Khmelevskoi VK (2011) Petrothermal energy and geophysics. *Moscow University Geology Bulletin*, 66(3): 151–157.

Goh CS (2019) *Chapter 6 – Energy: Current approach.* In: Tam VWY, Le KN (eds), *Sustainable Construction Technologies.* Butterworth-Heinemann, Oxford, UK, pp. 145–159.

Hansen K (2019) Decision-making based on energy costs: Comparing levelized cost of energy and energy system costs. *Energy Strategy Reviews* 24: 68–82.

Hyams MA (2012) 20 – Wind energy in the built environment. In: Zeman F (ed), *Metropolitan Sustainability.* Woodhead Publishing (Woodhead Publishing Series in Energy), Oxford, UK, pp. 457–499.

IAEA (2012) *Assessment of Nuclear Energy Systems Based on a Closed Nuclear Fuel Cycle with Fast Reactors.* IAEA. https://www.iaea.org/publications/8919/assessment-of-nuclear-energy-systems-based-on-a-closed-nuclear-fuel-cycle-with-fast-reactors. Accessed: 9 June 2021.

IAEA (2019) *Guidance on Nuclear Energy Cogeneration.* IAEA. https://www.iaea.org/publications/13385/guidance-on-nuclear-energy-cogeneration. Accessed: 28 May 2021.

IEA (2021) *Net Zero by 20150. A Roadmap for the Global Energy Sector.* OECD/IEA, Paris.

IHS Markit (2021) *News Release.* IHS Markit Online Newsroom, IHS Markit. https://news.ihsmarkit.com/prviewer/release_only/slug/bizwire-2021-5-5-ihs-markit-rankings-show-united-states-already-the-worlds-most-attractive-market-for-renewables-investment. Accessed: 9 June 2021.

IRENA (2016) *Wind Power: Technology Brief.* https://www.irena.org/publications/2016/Mar/Wind-Power. Accessed: 9 June 2021.

IRENA (2021a) *IRENA Country Rankings.* https://www.irena.org/Statistics/View-Data-by-Topic/Capacity-and-Generation/Country-Rankings. Accessed: 9 June 2021.

IRENA (2021b) *Renewable Capacity Statistics 2021.* https://www.irena.org/publications/2021/March/Renewable-Capacity-Statistics-2021. Accessed: 9 June 2021.

ITER (2021) *ITER – The Way to New Energy*, ITER. http://www.iter.org. Accessed: 9 June 2021.

Kamble SS, Gunasekaran A, Gawankar SA (2018) Sustainable Industry 4.0 framework: A systematic literature review identifying the current trends and future perspectives. *Process Safety and Environmental Protection* 117: 408–425.

Karakosta C et al. (2013) Renewable energy and nuclear power towards sustainable development: Characteristics and prospects. *Renewable and Sustainable Energy Reviews* 22(C): 187–197.

Kaza S et al. (2018) *What a Waste 2.0: A Global Snapshot of Solid Waste Management to 2050.* World Bank Publications.

Ketter W et al. (2018) Information systems for a smart electricity grid: Emerging challenges and opportunities. *ACM Transactions on Management Information Systems* 9(3): 1–22.

Khel I. (2016) Which technologies could be developed on the basis of dark matter? *HiNews.Ru.* 13 January 2016. http://hi-news.ru/. Accessed 26 Nov 2020 (in Russian).

Lange MA (2013) 3.11 – Renewable energy and water resources. In: Pielke RA (ed), *Climate Vulnerability.* Academic Press, Oxford, pp. 149–166.

Lazard (2020) *Levelized Cost of Energy and of Storage*, Lazard.com. http://www.lazard.com/perspective/levelized-cost-of-energy-and-levelized-cost-of-storage-2020/. Accessed 28 May 2021.

Liu J (2009) *Dark Matter as a Possible New Energy Source for Future Rocket Technology.* Cornell University Library. Archive: 0908.1429 [astro-ph. co].

Menzel T, Teubner T. (2020) Green energy platform economics – understanding platformization and sustainabilization in the energy sector. *International Journal of Energy Sector Management* 15(3): 456–475.

MIT Technology Review (2021) *The Green Future Index.* MIT Technology Review. https://www.technologyreview.com/2021/01/25/1016648/green-future-index/. Accessed 9 June 2021.

Moyer JD, Hughes BB (2012) ICTs: D they contribute to increased carbon emissions?, *Technological Forecasting and Social Change* 79(5): 919–931.

Neill SP et al. (2018) Tidal range energy resource and optimization – Past perspectives and future challenges. *Renewable Energy* 127: 763–778.

OECD (2021) *Energy – Renewable energy – OECD Data*, OECD. https://data.oecd.org/energy/renewable-energy.htm. Accessed: 9 June 2021.

O'Meara S (2020) China's plan to cut coal and boost green growth. *Nature* 584(7822): S1–S3.

Paish O (2002) Small hydro power: technology and current status. *Renewable and Sustainable Energy Reviews* 6(6): 537–556.

PR Newswire (2021) *Global Superconductors Market Report 2021: Growing Demand for MRI Machines is Expected to Drive the Growth of the Superconductors Market*. PR Newswire. https://www.prnewswire.com/news-releases/global-superconductors-market-report-2021-growing-demand-for-mri-machines-is-expected-to-drive-the-growth-of-the-superconductors-market-301255020.html. Accessed 9 June 2021.

Proskuryakova L, Ermolenko G (2019) The future of Russia's renewable energy sector: trends, scenarios and policies. *Renewable Energy* 143: 1670–1686.

RSPB (2016) *The RSPB's 2050 Energy Vision Meeting the UK's Climate Targets in Harmony with Nature*. RSPB. https://ww2.rspb.org.uk/Images/energy_vision_summary_report_tcm9-419580.pdf. Accessed 28 May 2021.

Russian Nuclear Community (2012) Innovative 4th-generation nuclear systems. 22 June 2012. http://www.atomic-energy.ru/technology/34307. Accessed 11 Jun 2021 (in Russian).

Skoltech (2016) New technologies in the energy sector. *Presentation, given at the XVII April International Academic Conference, April 20, 2016*. National Research University Higher School of Economics, Moscow.

Statista (2021) *Renewable Energy Capacity Worldwide by Country 2020*. Statista. https://www.statista.com/statistics/267233/renewable-energy-capacity-worldwide-by-country/. Accessed 9 June 2021.

US Department of Energy (2019) *Smart Grid: The Smart Grid*. SmartGrid.gov. https://www.smartgrid.gov/the_smart_grid/smart_grid.html. Accessed: 9 June 2021.

US Office of Electricity (2021) *Demand Response*. US Office of Electricity. https://www.energy.gov/oe/activities/technology-development/grid-modernization-and-smart-grid/demand-response. Accessed 9 June 2021.

WEC (2019) *World Energy Trilemma Index 2019*. World Energy Council. https://www.worldenergy.org/publications/entry/world-energy-trilemma-index-2019. Accessed 9 June 2021.

Zongxian Z et al. (2012) Risk assessment of concentrating solar power based on fuzzy comprehensive evaluation. *Systems Engineering Procedia* 4: 99–106.

# 20 Promotion of Decarbonization by Private Sector's Approaches

*Nobuhiro Sawamura*
Asia Pacific Energy Research Centre

## CONTENTS

## 20.1 INTRODUCTION

In recent years, abnormal climate conditions such as extremely large typhoons, hurricanes, cold and heatwaves, floods, and droughts have occurred worldwide. According to many reports and studies by international organizations and research institutes, these severe natural disasters have been caused by climate change. These disasters have caused severe damages to human lives and the ecosystem.

Therefore, international society and public opinion have more focused on decarbonization due to recent movements toward climate change. To promote decarbonization, many actors have taken actions in both the public and private sectors. Some actors individually tackle, while others work in cooperation with partners. As an example of the former, private enterprises advance technological development and innovation for reducing carbon emissions. As an instance of the latter, each government cooperates with its partner government for human resources development and the development of legal systems.

DOI: 10.1201/9781003315353-23

When considering the recent trend of efforts for decarbonization, the private sector is eager to tackle this problem. Primarily, financial institutions appear to promote decarbonization diligently. For instance, funds and institutional investors tend to withdraw their capital from private enterprises in the fossil fuel industry (divestment). Then, private banks announce to plant engineering corporations and big natural resources companies to stop financing new coal-fired power plants and resource development projects (e.g., coal and oil). These financial institutions seem to be under intense pressure from central banks, financial authorities, other financial institutions, international organizations, NGOs, and private investors who are enthusiastic about decarbonization, promoting renewable energy, and climate change. However, they started their voluntary efforts to raise funds and improve their corporate image.

Likewise, private enterprises also work on decarbonization because they have faced difficulties operating businesses due to these disasters. Stopping using electricity generated by fossil fuels, some private enterprises such as Apple provide electricity to their facilities with 100% renewable energy. Manufacturers work on decarbonization by inventing energy-saving products such as home electric appliances to improve their corporate images, increasing sales, and securing investors' funds.

As a different approach, the public and private sectors cooperate and establish many consortia and initiatives worldwide to promote decarbonization. Under Task Force on Climate-related Financial Disclosures (TCFD), private enterprises contribute to decarbonization by disclosing information about their financial affairs affected by climate change risks. Private companies cooperate and start new initiatives such as RE100 to achieve zero carbon and 100% renewable power. Under other initiatives such as Carbon Disclosure Project (CDP) and the Science Based Targets initiatives (SBTi), active efforts with other organizations have been made for greenhouse gas (GHG) emission reduction and achievement of the goals of the Paris Agreement.

In this chapter, how decarbonization has been promoted from the private sector's viewpoint will be analyzed. In addition, how decarbonization efforts in the private sector affect energy transition will be analyzed. Then, through divestment, consortia, and initiatives, the private sector can promote decarbonization and gain various benefits. Furthermore, as divestment, consortia, and initiatives work as both compulsory and voluntary driving forces, the private sector can also contribute to the energy transition from fossil fuels to renewable and new energy.

## 20.2 DIVESTMENT

### 20.2.1 DIVESTMENT BACKGROUND

Divestment is defined as "the disposal of assets in any of a variety of ways, usually for ethical, financial, or political reasons" (White, 2021). In the 20th century, the tobacco, munitions, adult services, gambling industries, and businesses relating to apartheid in South Africa were subject to divestment (Ansar and Caldecott, 2016, p. 68). In the 21st century, the fossil fuel industry has started to be targeted as divestment.

According to Linnenluecke et al. (2015, p.4),

> the fossil fuel divestment movement started as a grassroots movement at Swarthmore College in the US in 2011, with a student group asking their institution to freeze immediately new investments in the fossil fuel industry and to divest of stocks in the top 200 fossil fuel companies with the largest reserves. The student group followed in the footsteps of earlier college and university divestment campaigns in the US.

Furthermore, 350.org, a climate action NGO, has initiated fossil fuels divestment campaigns since 2012 (Ansar and Caldecott, 2016, p. 68).

In recent years, financial institutions, institutional investors, insurance companies, sovereign wealth funds (SWF), and pension funds have started divestment from the fossil fuel industry around the globe. There are several reasons why these institutions divest from the fossil fuel industry. First, they face pressures not to invest in industries that negatively impact the environment from their clients, environmental NGOs, the mass media, and public opinion. These organizations with high consciousness of the environment may change their trustee institutions if trustee institutions such as financial institutions ignore clients' opinions and philosophy about the environment. Secondly, financial institutions may face difficulties in getting returns from investment and lending. It would be assumed that companies cannot make profits and continue their businesses due to impacts caused by climate change. For instance, grain companies cannot harvest abundant and high-quality crops due to droughts. Also, the beach resort industry cannot provide attractive services to customers owing to abnormal climate disasters such as floods. Thirdly, insurance and reinsurance companies have been up grave difficulties with their businesses. The number of severe natural disasters such as floods, cyclones, and heavy rains that believe to be caused by climate change has been increasing. As a result, insurance and reinsurance companies have paid a large amount of insurance money every year. Therefore, they divest their assets from the fossil fuel industry for their business sustainability.

## 20.2.2  THE CURRENT SITUATION OF DIVESTMENT

Under the recent divestment campaign, financial institutions, institutional investors, insurance companies, pension funds, and central banks have taken active actions.

In 2019, Mitsubishi UFJ Financial Group, Inc. (MUFG), the largest Japanese financial group, released its sustainable finance goals and expressed that it will not finance new coal-fired power generation projects (MUFG, 2019).

For institutional investors, Blackrock, the world's largest asset manager, regards investing in thermal coal producers as one of the Environmental, Social, Governance (ESG) risks. Therefore, it has been actively divesting its financing from the thermal coal production industry (BlackRock, 2021).

Many insurance companies have already announced divestment from coal mines utilities having many coal-power plants. For example, major global insurance companies such as AXA and Aviva have announced their divestment from the coal industry (Ansar and Caldecott, 2016, p. 70). Furthermore, Daiichi Life Insurance in Japan has announced not to invest in a new international coal-fired power plant (Webb and Paola, 2018, p. 6). Then, the largest Japanese insurance company, Nippon Life Insurance Company, showed its "ESG Investment and Financing" policy and announced not to invest in new international coal-fired power plant projects in Japan or overseas (Nippon Life Insurance Company, 2017).

Japan's Government Pension Investment Fund (GPIF), the world's largest pension fund, has promoted ESG investment. In addition, according to Uzsoki (2020), "Besides some allocation to ESG indices, GPIF also puts pressure on its portfolio companies to improve their ESG performance and actively manage climate-related risks" (p. 10).

Also, the central banks have enthusiastically worked on divestment and ESG finance. Mark Carney, a former Governor of the Bank of England, has been eager for ESG finance, and the Bank of England released the UK Money Markets Code in 2017. It includes ESG criteria highlighting the increasing importance of ESG (Bank of England, 2021). Then, the French and Dutch central banks have determined to incorporate ESG factors into their investment decisions. Subsequently, Riksbank, the central bank of Sweden, stated that it would not invest in government bonds that governments issue with high carbon footprints (Uzsoki, 2020, p. 11).

## 20.2.3  CRITICISMS AGAINST DIVESTMENT

Although divestment and ESG finance, including sustainable investing, seem to follow current trends, there are criticisms against them. Furthermore, some people point out the difficulties of sustainable investing.

Firstly, the central bank in Switzerland, the Swiss National Bank (SNB), indicates difficulties of divestment and ESG finance. According to Uzsoki (2020), "SNB argues that sustainable investing decreases the universe of tradable assets and therefore makes asset allocation more difficult" (p. 11).

Secondly, others argue that divestment has no significant influence on the fossil fuel industry and funding. They add that divestment cannot broadly impact the fossil fuel industry and financial institutions integrating assets relating to fossil fuels because the amount of divestment money is too tiny (Bergman, 2018, p. 2).

Thirdly, divestment could heighten the risk of unemployment in the worldwide labor market. Beautiful and famous words such as ecologically friendly do not always provide good results. According to Healy and Barry (2017), for example, "Chomsky points out that when the La Guajira coalmines in Colombia eventually close, ten thousand workers and dependent communities will be turned into a new form of 'ecological refugee'" (p. 455).

### 20.2.4   New Business Opportunities Obtained from Divestment

For the private sector, divestment is one of the most intense external pressures in the market and society. If the private sector is reluctant to promote decarbonization, it may lose business opportunities and be kicked out of the marketplace in the world. Even as its fossil fuel resources such as coal, oil, and gas may lose value and become stranded assets, it may suffer many financial losses. The private sector should promote green innovation, energy transition, and decarbonization through divestment as an effective tool. Therefore, the private sector should not regard divestment as a severe risk but as a precious business chance (Figure 20.1).

As Bergman (2018) points out, divestment may have another aspect as "reinvestment in climate change mitigation measures such as renewable energy, energy efficiency, low carbon infrastructure

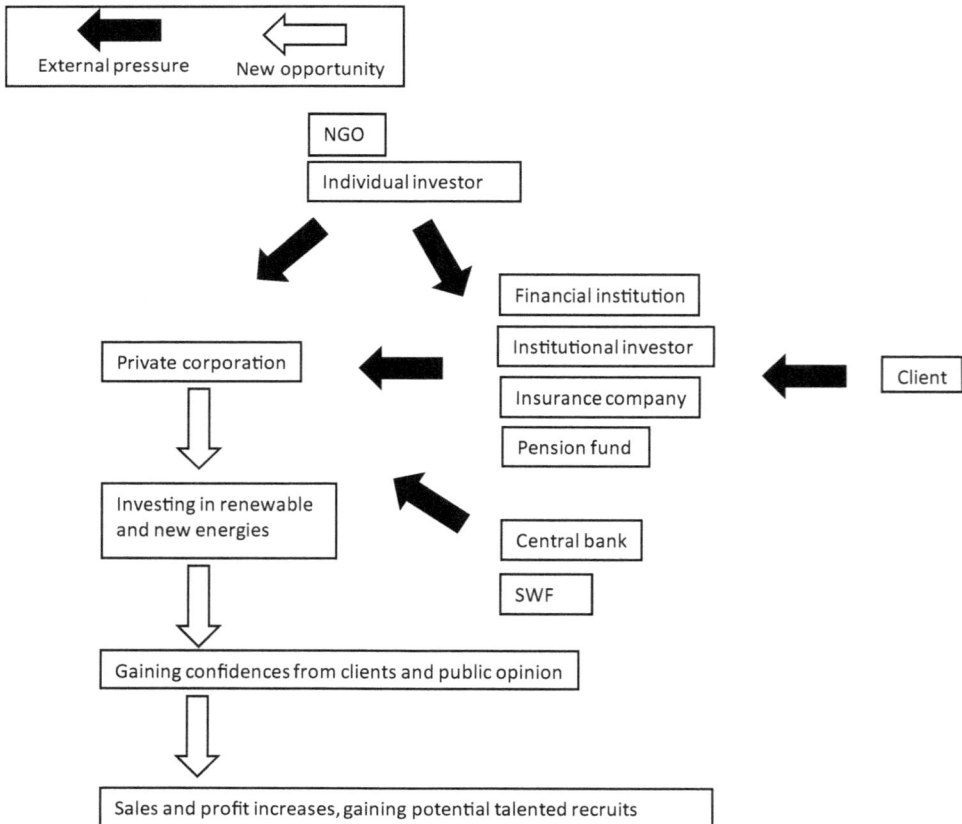

**FIGURE 20.1**   Promotion of decarbonization by divestment.

and demand reduction" (p. 3). Instead of holding on to assets emitting considerable amounts of GHG, the fossil fuel industry and utilities owing thermal power plants can invest their precious resources in renewable energy such as solar power and wind power and new energy such as hydrogen and ammonia. Furthermore, investing in new technology such as perovskite solar cells and carbon capture use and storage can produce new business opportunities.

## 20.3 CONSORTIA AND INITIATIVES

### 20.3.1 THE SCIENCE BASED TARGETS INITIATIVES (SBTi)

#### 20.3.1.1 SBTi Background

There are many consortia and initiatives established for assisting the private sector in meeting the goals of the Paris Agreement. As one of them, the SBTi is a joint initiative with CDP, World Resources Institute, the World Wide Fund for Nature, and the United Nations Global Compact. Then, science-based targets (SBTs) are corporate dedications to reduce GHG emissions regarding how private companies will achieve the goals of the Paris Agreement based on scientific knowledge (Aden, 2019). Companies were encouraged to set carbon reduction targets to restrain global temperature rise to 2°C above pre-industrial levels (Giesekam et al., 2021, p. 2).

#### 20.3.1.2 The Current Situation of SBTi

In response to international movements to climate change and decarbonization, SBTi started to act more ambitiously. From 2019 SBTi has advanced new joining members to restrict to well below 2°C above pre-industrial levels and make efforts to limit to 1.5°C (Giesekam et al., 2021, p. 2). From 2019, it also started the Business Ambition for 1.5°C campaign with other organizations, promoting corporations to adopt 1.5°C-aligned targets (SBT, 2021a). As of August 2021, more than 1,000 corporations have set emission reduction targets under SBTi (SBT, 2021b). Among them, more than 600 corporations worldwide have committed to the 1.5°C targets by way of the Business Ambition for the 1.5°C campaign (SBT, 2021a). Both U.S. and Japanese companies are eager for SBTi. It can be assumed that the number of Japanese companies will continuously increase because its government positively supports SBTi.

#### 20.3.1.3 Benefits of SBTi

For the private sector, there are benefits to adopting SBTs. Firstly, it can gain a good reputation, credibility, and investor confidence because targets are based on scientific evidence and knowledge (Piper and Longhurst, 2021, p. 4). If companies obtain these, they can achieve sales and profit increases. Secondly, they can advance innovation through SBTs (Piper and Longhurst, 2021, p. 4). That is because they can challenge new businesses by using investment capital from investors for innovation.

#### 20.3.1.4 Challenges of SBTi

SBT member corporations are obliged to target Scope 1, Scope 2, and Scope 3 emissions. Scope 1 is direct emissions from its member's factories, offices, and other facilities. Scope 2 is indirect emissions from purchased electricity, heat, and so on. Scope 3 is other indirect emissions from its upstream (emissions from its suppliers) and downstream (emissions from products or services of its customers).

There are challenges regarding Scope 3 emission measurements. Firstly, there are problems with discrepancies with Scope 3 emission measuring. The believability of data over Scope 3 emissions is often suspected as its measurement requires modeling and estimations. Second, the measurement methodology achieving Scope 3 emissions contains issues. If two member companies attain Scope 3 emission targets and the supplier is the same, both companies probably assert their achievement. In this case, it is difficult to decide which one achieved its Scope 3 emission targets (Piper and Longhurst, 2021, p. 8–9). Transparency, data quality, and measurement methodology need to improve to solve these problems.

### 20.3.2  TASK FORCE ON CLIMATE-RELATED FINANCIAL DISCLOSURES (TCFD)

#### 20.3.2.1  TCFD Background

Since the financial sector predicts climate risks will negatively impact their own and borrowers' businesses, it has worked on divestment, especially from the fossil fuel industry and utilities. Furthermore, it also started new campaigns, namely, organizing new consortia and initiatives to mitigate and adapt to climate risks.

In 2015, the Financial Stability Board (FSB), chaired by Mark Carney, a former Governor of the Bank of England, started up TCFD. In 2017, the TCFD chaired by Michael Bloomberg released its financial recommendations for disclosure of climate-related financial risks (Climate Disclosure Standards Board, 2018, p. 4).

The TCFD (2021) shows its goal:

> The Financial Stability Board established the TCFD to develop recommendations for more effective climate-related disclosures that could promote more informed investment, credit, and insurance underwriting decisions and, in turn, enable stakeholders to understand better the concentrations of carbon-related assets in the financial sector and the financial system's exposures to climate-related risks.

Then, the TCFD, the Task Force led by the private sector, promotes private corporations to disclose climate-related risks that may cause potential obstacles to their business, improving business sustainability.

In 2017, the TCFD announced four thematic areas of recommended climate-related financial disclosure: (i) Governance, (ii) Strategy, (iii) Risk management, and (iv) Metrics and Targets (Chiba, Mori, and Shimizu, 2017, p. 3). Based on these areas, it organized and released its proposals (Itadu, 2019, p. 21). Its proposals request companies to introduce scenario analysis and show their business outlook, including their resilience against damages caused by climate change disasters.

The TCFD comprehends that there are generally two categorized risks: (i) transition risks and (ii) physical risks. Transition risks are related to adverse effects and costs that companies face to mitigate climate change impacts. Physical risks are related to burdens, which corporations confront to adapt to natural disasters and environmental changes caused by climate change.

According to Mori and Chiba (2017, p.2),

> There are five main kinds of transition risks: 1) Policy risk: policy actions for promoting climate change efforts; 2) Litigation or legal risk: failure of companies to mitigate climate change impacts, failure to adapt to climate change, and insufficiency of the financial disclosure; 3) Technology risk: technological failure of improvements or innovations to support low-carbon transitions; 4) Market risk: shifts in supply and demand for certain commodities, products, and services as climate risks; and 5) Reputation risk: changes in customer perception to favour lower-carbon goods and services.

As an example of policy risk, governments make laws regulating GHG emissions at a stricter level. Regarding market risk, a rapid transition from gasoline vehicles to electric vehicles (EV) and fuel cell vehicles has occurred. Divestment caused by lower reputation about fossil fuels can be defined as reputation risk.

Regarding physical risks, Mori and Chiba (2017, p.2) point out:

> The physical risks are divided into acute risk and chronic risk: 1) Acute risk: damage caused by extreme weather events, such as cyclones, hurricanes and floods; and 2) Chronic risk: damage caused by slow onset events, such as sea level rise, increasing temperatures, ocean acidification and salinization.

As an instance of an acute risk, the transportation system may be disrupted by natural disasters (e.g., floods and landslides) caused by climate change. As a result, people cannot move on time by transportation and products cannot be delivered punctually.

### 20.3.2.2   The Current Situation of TCFD

After active discussions, the TCFD released the final report in 2017. This report includes climate-related risks and opportunities, scenario analysis, guidelines, and related recommendations (TCFD, 2017).

According to Enokibori (2019, p.244), this report shows two significant points so that investors can use climate-related risks that private companies face or will face for their investment decisions:

> First, corporate managements should recognize climate change as one of significant issues and adopt climate change as main business challenges… Second, this report requests to display the outlook of climate change risks based on scenario analysis as information which private companies disclose.

This report finds out that the first point focuses on the current climate change risks and their counter-measures. On the other hand, the second point notes the future climate change risks and their measures.

As the recent movement, in 2019, TCFD Consortium was founded by leaders from the Japanese industry and academia. The financial and non-financial sectors have had sequential talks under this consortium. Currently, it targets to promote further discourses about the disclosure of climate-related information. Furthermore, for its objective, "virtuous cycle of environment and growth," it cohosted TCFD Summit 2019 and 2020 and will cohost TCFD Summit 2021 with the Ministry of Economy, Trade, and Industry (TCFD Consortium, 2021; TCFD Summit, 2021).

### 20.3.2.3   Benefits of TCFD

The TCFD may be recognized as an obstacle for certain corporations' business activities and regulations on their business plans. However, there are several kinds of benefits brought about by the TCFD to the private sector. They range from finance to investment transition.

In the perspective of finance, Guthrie and Blower (2017, p. 14) point out:

> Disclosures by the financial sector are designed to foster an early assessment of climate-related risks and opportunities, improve the pricing of climate risks and lead to more informed capital allocation decisions.

Therefore, the financial sector can detect climate-related risks earlier, perform risk hedges effectively, and get benefits from the TCFD scheme from the viewpoint of finance.

As another aspect, if the private sector positively discloses climate-related information based on the TCFD recommendations, investors can obtain a broader range of selections of financial assets to decide on their portfolios (Mori and Chiba, 2017, p. 7). If so, investors with high consciousness of the environment may choose bonds and stocks whose companies use more renewable energy and provide greener and more environmentally friendly products and services. Thereby, the stock market gets lively, leading to economic growth.

From the viewpoint of investment transition, more disclosure of financial-related information based on the TCFD recommendations promotes dialogues among companies, banks, and investors. This movement helps corporations raise the recognition of the environment, leading to increased investment in companies and accelerating decarbonizing of their business structures (Mori and Chiba, 2017, p. 7). As a result, it is assumed that if corporations are evaluated and supported by investors in terms of green business models, they invest in renewable and new energies. Then, more investment transits from gray to green.

### 20.3.2.4   Challenges of TCFD

Although the TCFD recommendations are epoch efforts led by the private sector in decarbonization and business, there are challenges to its operation.

Firstly, as its recommendations are voluntary bases and lack a regulative framework and legal enforceability, its effectiveness could be weak (Mori and Chiba, 2017, p. 7). In general, the private sector does not prefer regulations and legislation to restrict its private economic activities.

Secondly, the TCFD recommendations face difficulties in data quality and financial impact. These difficulties contain:

(1) gaps in emissions measurement methodologies and product lifecycle emissions methodologies; (2) lack of robust and cost effective tools to quantify the potential impact of climate related risks and opportunities; (3) variability of climate-related impacts across and within different sectors and markets; and (4) the high degree of uncertainty around the timing and magnitude of climate-related risks.

*Mori and Chiba (2017, p. 8–9)*

It is pretty challenging to measure GHG emissions, climate-related risks, and opportunities in a cross-cutting manner. Measurement methodologies that show accurately how more transparent and qualified data are gained need to be invented.

### 20.3.3   RE100

#### 20.3.3.1   RE100 Background

As another initiative, RE100 plays a significant role in the promotion of decarbonization by the private sector. It is a worldwide initiative to urge private corporations to use 100% renewable energy in their business activities. It shows that "our mission is to accelerate change towards zero carbon grids at scale" (RE100, 2021a). Led by the Climate Group and in partnership with CDP, it has encouraged increased usage of renewable energy and decarbonization.

Becoming a member of RE100 requires private companies to satisfy several criteria, take action, and make reports. Firstly, private firms need to fulfill detailed items in the RE100 joining criteria. Secondly, RE100 requests them to set ambitious target dates to achieve 100% renewable energy (e.g., 100% by 2050). Thirdly, in attaining 100% renewable energy, they must meet technical criteria, the guidance on making credible renewable electricity usage claims, and market boundary criteria. Fourthly, it demands them to do transparent reporting annually (RE100, 2021b).

#### 20.3.3.2   The Current Situation of RE100

As of July 2022, more than 300 RE100 member corporations have committed to achieving 100% renewable energy. There are many industries, such as IT, manufacturing, and service industries, committing to this target (RE100, 2021c).

Google, one of RE100 members, has been enthusiastic about adopting renewable energy and promoting carbon-free energy from an earlier stage. In 2017, it achieved 100% renewable energy for its operations, including its data centers and offices, by purchasing electricity derived from renewable energy and adopting other measures. Now it aims to decarbonize its electricity supply entirely and run its business on 24/7 carbon-free energy by 2030 (Google, 2021).

In 2018, Apple declared that it had accomplished powering its data centers, offices, and facilities in 43 countries with 100% renewable energy (Apple, 2021a). It has been promoting and supporting its manufacturing suppliers to achieve 100% renewable electricity (Apple, 2021b).

#### 20.3.3.3   Benefits of RE100

Although private corporations face numerous severe requirements to join RE100, there are several advantages like the TCFD.

Firstly, the young generation mainly prefers environmentally friendly products and services (Frankfurt School-UNEP Centre/BNEF, 2020, p. 16). If companies diligently adopt more renewable energy and decarbonize their operations for joining RE100, their reputations will improve. As a result, customers, especially the young generation, will be attracted and purchase more products and services than gray ones. In the long run, companies can increase their sales and profits.

Secondly, corporations can improve their image and corporate social responsibility by joining RE100 and announcing their goals at that place (Holman, 2020, p. 11). As mentioned in the section on divestment, many financial institutions, institutional investors, insurance companies, pension funds, and private investors consider the ESG strategies of each corporation as one of the critical factors for their investment. If companies show their excellent ESG performance regarding introducing renewable energy under RE100, they can get more investments.

Thirdly, promoting the usage of renewable energy and joining RE100 can lead to hiring talented mid-career workers and potential recruits (Frankfurt School-UNEP Centre/BNEF, 2020, p. 16). Indeed, one study shows that 42% of workers consider companies' environmental activities when choosing their jobs in the UK (Holman, 2020, p. 11).

### 20.3.3.4 Challenges of RE100

For promoting decarbonization, RE100 is one of the progressive initiatives led by the private sector. However, there are challenges for private corporations to be members of RE100.

Firstly, heavy industries such as steel and cement face challenges to be members of RE100. As these industries have many large plants and emit much more GHG than other industries, they are more challenging to meet several criteria and join RE100. If checking the RE100 member list, it finds out that most of the members are service industries and manufacturing industries with less GHG emissions than heavy industries (RE100, 2021d).

Second, it is relatively harder for companies in emerging markets to join RE100. Generally, infrastructures such as power grids and storage batteries for renewable energy are not built enough to supply electricity derived from renewable energy in emerging countries.

## 20.4 CONCLUSIONS

At first glance, as principal actors, international organizations, and governments have made rules at Conferences of the Parties held each year and have implemented countermeasures for climate change.

However, the private sector has also made efforts for decarbonization and climate change. Firstly, through divestment, the financial sector has worked on decarbonization by itself. Although divestment has both advantages and disadvantages, the private sector (e.g., natural resource industries, utilities, and heavy industries) also recognizes risks regarding climate change and makes efforts for decarbonization. In addition, the private sector can use external pressures called divestment as an opportunity to invest in renewable and new energies. How it deals with divestment forcibly promoting decarbonization predominantly affects its business operations. If it perceives as an opportunity for its growth, it can accelerate decarbonization and make its business grow by investing its resources in new green technology.

Secondly, the private sector can voluntarily promote decarbonization by organizing and managing consortia and initiatives such as SBTi, TCFD, and RE100. As these schemes are not legal and regulative systems binding companies, their effectiveness to mitigate and adapt to climate change depends on how enthusiastically and devotedly each corporation tackles. However, as mentioned previously, joining these consortia and initiatives gives several benefits to the private sector. Therefore, the private sector can use these schemes for its future business sustainability and its business growth.

Although divestment, organizing, and managing consortia and initiatives are contrasting and different approaches, the private sector has its potential as promising as the public sector for decarbonization. It can be hoped that the private sector will increasingly cooperate with the public sector and contribute to achieving the Paris Agreement and other future endeavors for climate change.

## REFERENCES

Aden, Nate. 2019. *Japan Is Leading on Business Climate Engagement. Will Ambitious Policies Follow?* World Resources Institute. https://www.wri.org/insights/japan-leading-business-climate-engagement-will-ambitious-policies-follow.

Ansar, Atif and Ben Caldecott. 2016. Divestment campaigns: Bottom-up geo-economics. *Connectivity Wars: Why Migration, Finance and Trade Are the Geo-Economic Battlegrounds of the Future*, Edited by Mark Leonard, pp. 68–70. https://ecfr.eu/wp-content/uploads/Connectivity_Wars.pdf.

388

Apple. 2021a. Apple now globally powered by 100 percent renewable energy, Press Release, April 9 2018. https://www.apple.com/newsroom/2018/04/apple-now-globally-powered-by-100-percent-renewable-energy/

Apple. 2021b. Environmental progress report. https://www.apple.com/environment/pdf/Apple_Environmental _Progress_Report_2021.pdf.

Bank of England. 2021. Bank of England publishes updated UK money markets code. https://www.bankofengland.co.uk/news/2021/april/boe-publishes-updated-uk-money-markets-code.

Bergman, Noam. 2018. Impacts of the fossil fuel divestment movement: Effects on finance, policy and public discourse. *Sustainability* 10, 2529, 2–3. https://doi.org/10.3390/su10072529.

BlackRock, Inc. 2021. "BlackRock's 2020 Letter to Clients: Sustainability as BlackRock's New Standard for Investing," https://www.blackrock.com/us/individual/blackrock-client-letter.

Chiba, Yohei, Naoki Mori, and Noriko Shimizu. 2017. *Strengthening the Integration of Climate Risks in the Banking Sector.* Policy Brief 38, the Institute for Global Environmental Strategies (IGES), p. 3. https://www.iges.or.jp/en/publication_documents/pub/policy/en/5984/PB_38_E_0516rev_final.pdf.

Climate Disclosure Standards Board (CDSB). 2018. *Ready or Not: Are Companies Prepared for the TCFD Recommendations? Joint CDSB and CDP Report*, p. 4. https://www.cdsb.net/sites/default/files/tcfd_preparedness_report_final.pdf.

Enokibori, Miyako. 2019. Environmental information disclosure accelerated by CDP and the data usages in financial market. *Journal of Life Cycle Assessment, Japan* 15, 3, 244. https://www.jstage.jst.go.jp/article/lca/15/3/15_242/_pdf/-char/ja.

Frankfurt School-UNEP Centre/BNEF. 2020. *Global Trends in Renewable Energy Investment 2020*, p. 16. https://www.fs-unep-centre.org/wp-content/uploads/2020/06/GTR_2020.pdf.

Giesekam, Jannik, Jonathan Norman, Alice Garvey, and Sam Betts-Davies. 2021. Science-based targets: On target? *Sustainability* 13(4), 2. https://www.mdpi.com/2071-1050/13/4/1657/htm.

Google. 2021. *Our Progress Our Efforts Are Designed to Help Us All Get the Most Out of Technology, without Using More Resources.* https://sustainability.google/progress/#.

Guthrie, Lois, and Luke Blower. 2017. *Corporate Climate Disclosure Schemes in G20 Countries after COP 21.* Climate Disclosure Standards Board (CDSB), p. 14. https://www.jstor.org/stable/resrep15540?seq=1#metadata_info_tab_contents.

Healy, Noel, and John Barry. 2017. Politicizing energy justice and energy system transitions: Fossil fuel divestment and a 'just transition.' *Energy Policy*, 455. https://doi.org/10.1016/j.enpol.2017.06.014.

Holman, Dex Jan Loek. 2020. *Green Is Good: Does a Renewable Energy Program Lower the Cost of Debt for Listed EU Companies?.* Erasmus School of Economics, Erasmus University Rotterdam, p. 11. https://www.google.co.jp/url?sa=t&rct=j&q=&esrc=s&source=web&cd=&ved=2ahUKEwih_8mSppnyAhVxHKYKHaBsDE4QFnoECAcQAw&url=https%3A%2F%2Fthesis.eur.nl%2Fpub%2F52696%2FGreen-is-Good-Dex-Holman-414473.pdf&usg=AOvVaw0_u-F8g1fiwQeq-WU9ilqH.

Itadu, Naotaka. 2019. [The Statutory Disclosure Movement based on TCFD proposals: At the Center of the Case of BHP Billiton, a major resource company] TCFD no Teigen ni motoduku hotei kaiji no ugoki -Ote shigen kaisha BHP Billiton no jirei wo chushin ni- (in Japanese), [Nomura Capital Market Quarterly] Nomura shihon shijo quarterly (in Japanese), p. 21. http://www.nicmr.com/nicmr/report/repo/2019/2019win03.pdf.

Linnenluecke, Martina, Cristyn Meath, Saphira Rekker, Baljit Sidhu, and Tom Smith. 2015. "Divestment from fossil fuel companies: Confluence between policy and strategic viewpoints." *Australian Journal of Management*, 4. https://doi.org/10.1177/0312896215569794.

Mitsubishi UFJ Financial Group, Inc. 2019. *MUFG Sets Sustainable Finance Goals and Revises Environmental and Social Policy Framework.* https://www.mufg.jp/dam/pressrelease/2019/pdf/news-20190515-001 _en.pdf.

Mori, Naoki, and Yohei Chiba. 2017. *Impact of Climate Change –Transforming Business Behaviour in Favour of Sustainable Development. IGES Discussion Paper*, The Institute for Global Environmental Strategies (IGES), 2, 7–9. https://www.iges.or.jp/en/publication_documents/pub/discussionpaper/en/6017/ISAP+2017_Discussion+paper_FTF_English.pdf.

Nippon Life Insurance Company. 2017. *ESG Investment and Financing.* https://www.nissay.co.jp/english/esg/.

Piper, Katherine, and James Longhurst. 2021. Exploring corporate engagement with carbon management techniques [Version 1; Peer review: 2 approved with reservations]. *Emerald Open Research*, 4, 8–9. https://doi.org/https://doi.org/10.35241/emeraldopenres.14024.1.

RE100. 2021a. *About Us.* https://www.there100.org/about-us.

RE100. 2021b. *What Are the Requirements to Become a RE100 Member? Technical Guidance.* https://www.there100.org/technical-guidance.

RE100. 2021c. *RE100 Members.* https://www.there100.org/re100-members.

RE100. 2021d. *RE100 Members*. https://www.there100.org/re100-members.

Science Based Targets. 2021a. *Climate Ambition: SBTi Raises the Bar to 1.5°C*. July 15, 2021. https://science-basedtargets.org/news/sbti-raises-the-bar-to-1-5-c.

Science Based Targets. 2021b. *Companies Taking Action*. https://sciencebasedtargets.org/companies-taking-action.

Task Force on Climate-related Financial Disclosures (TCFD). 2017. *Final Report: Recommendations of the Task Force on Climate-Related Financial Disclosures*. https://assets.bbhub.io/company/sites/60/2020/10/FINAL-2017-TCFD-Report-11052018.pdf.

Task Force on Climate-related Financial Disclosures (TCFD). 2021. *Our Goal*. https://www.fsb-tcfd.org/about/#our-goal.

TCFD Consortium. 2021. *About Us*. https://tcfd-consortium.jp/en.

TCFD Summit. 2021. https://tcfd-summit.go.jp/indexEn.html.

Uzsoki, David. 2020. Sustainable investing shaping the future of finance. *International Institute for Sustainable Development*, 10–11. https://mava-foundation.org/wp-content/uploads/2020/02/sustainable-investing-IISD.pdf.

Webb, Matthew, and Paola Parra. 2018. *Japan at an International Crossroads – Seeking a Sunset for Coal*. E3G, 6. https://climateanalytics.org/media/310518_e3g_ca_sunset_for_japanese_coal_1.pdf.

White, Judith A. 2021. *Divestment Economics*. Encyclopedia Britannica. https://www.britannica.com/topic/divestment.

# 21 Electrification and Targets in Developing Countries

*Tri Ratna Bajracharya*
Tribhuvan University

*Shree Raj Shakya*
Tribhuvan University
Institute for Advanced Sustainability Studies (IASS)

*Anzoo Sharma*
Tribhuvan University
Center for Rural Technology (CRT/N)

## CONTENTS

## 21.1 ELECTRIFICATION AND SUSTAINABLE DEVELOPMENT

Economic growth and community prosperity are highly influenced by the availability of reliable and secure energy. Electricity is considered the most convenient and versatile form of energy. It can be generated from primary sources like coal, gas, petroleum, hydropower, wind, and solar, and can be transported at the highest rate. It can be converted to mechanical energy; motors typically used for commercial and industrial purposes have a nominal minimum efficiency of at least 75%. Modern-day services are highly dependent on communication facilities whose life source is electricity. Access to energy, particularly in the form of electricity, is vital for households, enterprises, and community facilities. It assists in raising people's living standards as it offers access to information, enhances teaching-learning activities and quality of education, improves sanitary conditions and facilitates health services, offers newer income-generating opportunities and extended working hours, which in turn reduces poverty, mortality rate, gender inequality and promotes sustainable consumption and production behaviors and reduces the environmental footprint. To achieve these socio-economic and environmental benefits, SDG goal 7 is formulated to "ensure access to affordable, reliable, sustainable, and modern energy for all." The targets and respective indicators of SDG 7 are listed in Table 21.1. The COVID-19 pandemic amplified the need for access to modern energy in the health sector: cold chains to store and distribute vaccines and reliable power to treat patients in health centers.

DOI: 10.1201/9781003315353-24

**TABLE 21.1**
**Targets and Indicators of SDG 7**

| Target | Indicator |
|---|---|
| 7.1 By 2030, ensure universal access to affordable, reliable, and modern energy services | 7.1.1 Proportion of population with access to electricity<br><br>7.1.2 Proportion of population with primary reliance on clean fuels and technology |
| 7.2 By 2030, increase substantially the share of renewable energy in the global energy mix | 7.2.1 Renewable energy share in the total final energy consumption |
| 7.3 By 2030 double the global rate of improvement in energy efficiency | 7.3.1 Energy intensity measured in terms of primary energy and GDP |
| 7.a By 2030, enhance international cooperation to facilitate access to clean energy research and technology, including renewable energy, energy efficiency and advanced and cleaner fossil-fuel technology, and promote investment in energy infrastructure and clean energy technology | 7.a.1 International financial flows to developing countries in support of clean energy research and development and renewable energy production, including in hybrid systems |
| 7.b By 2030, expand infrastructure and upgrade technology for supplying modern and sustainable energy services for all in developing countries, in particular least developed countries, small island developing States, and land-locked developing countries, in accordance with their respective programs of support | 7.b.1 Investments in energy efficiency as a percentage of GDP and the amount of foreign direct investment in financial transfer for infrastructure and technology to sustainable development services |

## 21.2 ENERGY ACCESS IN DEVELOPING COUNTRIES

Although most of the developed countries achieved electrification for large populations from the 1950s to 1970s, at a global level, the number of people without access to electricity is still substantial (Figure 21.1). In 2000, 78.3% of the world population had access to electricity, in 2010, that increased to 83.3%, 89.57% in 2018, and 90% by 2019, with average annual electrification of 0.82% between 2016 and 2018, which implies 1.326 billion, 1.154 billion, 789 million and 759 million people did not have access to electricity in the respective years and estimates that about 620 million people will remain without access in 2030. The population of the developing world are mostly connected to unreliable electricity systems, whereas those with no access to electricity are concentrated in Sub-Saharan Africa and Southern Asia. By the end of 2020, the electrified fraction in Latin America, the Caribbean, Eastern Asia, and Southeastern Asia exceeded 98%, and in Central Asia and Southern Asia, more than 96% of the population had gained access to electricity (IEA et al., 2022).

Figure 21.2 shows the 20 countries with the largest share of electricity access-deficit population. These countries accounted for 76% of 580 million access-deficit people. Bangladesh, Kenya, and Uganda showcase the greatest improvement in the annual electrification growth rate of greater than 3% since 2010 (IEA et al., 2021b). The major disparity in access to electricity is observed between urban and rural areas: the remote localities are untied to the grid for reasons like lack of sufficient power generation capacity, poor transmission, and distribution infrastructure, high cost of supply, or unaffordability. Despite the ease of reach, the poor communities are the unserved ones in urban localities. Access deficits, i.e., the heavy reliance on inefficient and unsustainable use of traditional biomass fuels, are both manifestations and causes of poverty. However, access to modern energy on its own does not guarantee poverty alleviation but is indispensable to meeting other SDG goals.

The rate of electricity connection is improving faster in rural settings through decentralized renewable-based sources. Around the globe in 2019, almost 11 million people are connected to mini-grids, rising from 5 million in 2010, and the population with access to electricity through off-grid solar solutions rose from 85 to 105 million from 2016 to 2019 (GOGLA, 2020). According to Regulatory Indicators for Sustainable Energy (RISE) (ESMAP, 2020), since 2010, the development of policies to support mini-grid and off-grid electricity systems has been overwhelming.

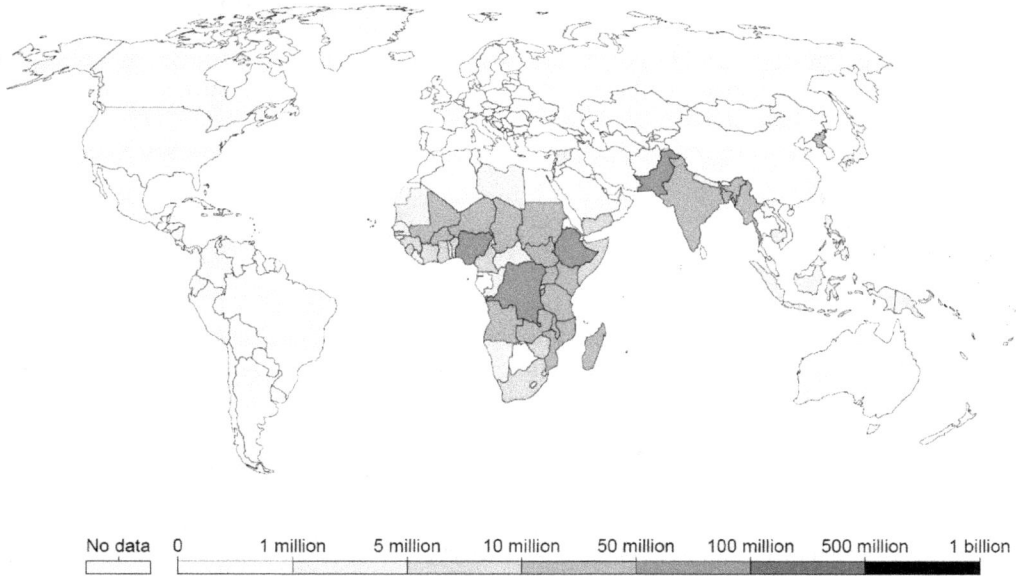

FIGURE 21.1 Share of population without access to electricity, 2019. (From IEA, IRENA, UNSD, World Bank, & WHO, 2021. With permission.).

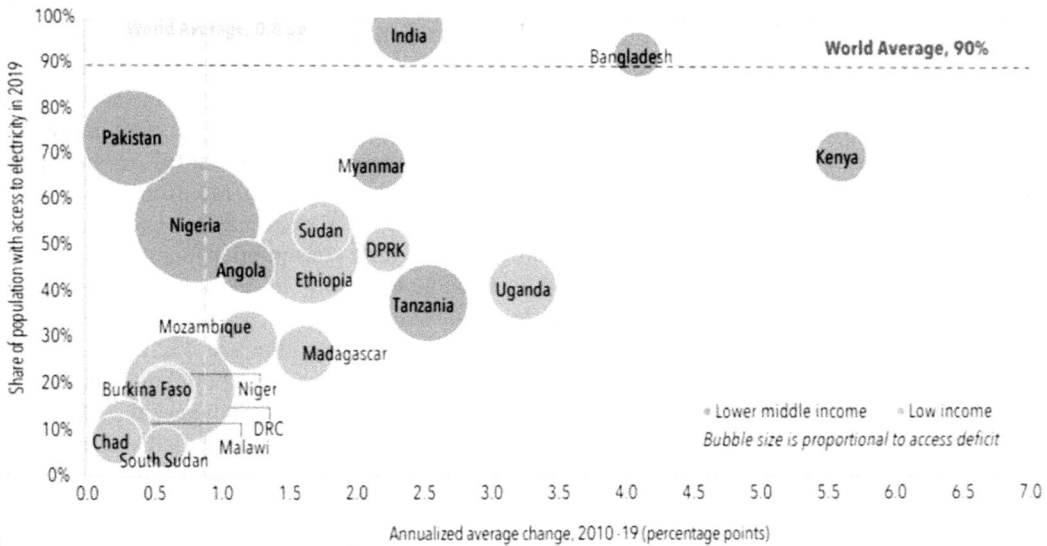

FIGURE 21.2 Electricity access in the top 20 access-deficit[1] countries 2010–2019. (From IEA, IRENA, UNSD, World Bank, & WHO, 2021. With permission.).

Developed countries have reached the threshold for electrification and electrified development. Now, the challenge for them is the transition from dirty fuel to clearer fuel. But the challenge for developing and emerging economies are complex: on the one hand, these countries have to focus on energy infrastructure among access to other basic services like water, health, education, and road, and on the other hand, must fight global climate emergencies through low-carbon approaches.

---

[1] A country's "access deficit" is defined as the number of people without access to electricity in the country.

## 21.3  POLICIES SUPPORTING ELECTRIFICATION IN DEVELOPING COUNTRIES

In recent years, a significant level of progress in expanding electricity access has been observed in many countries. There are some common leading factors for this progress, such as enabling policies and incentives, sustained political commitment and financing, strong institutions, and the use of both off-grid and grid-based electrifications. The journey of electrification of some countries is briefly elaborated below:

### 21.3.1  China

In China, the voyage of electric power began in 1879 with the installation of a 10 hp DC generating unit. By 1947, the nation's power generating capacity was 1,850 MW, and electricity generation was 4.3 billion kWh, which was concentrated majorly in the urban areas. In 1949, 33 small hydropower plants generating merely 3.63 MW of power were operated at the rural level (county and below) (Chinese Society for Electrical Engineering, 2009).

China achieved rural electrification in three phases. The first phase began in 1949–78, the founding year of the People's Republic of China, particularly with the development of small hydropower for improving irrigation and drainage facilities to secure agricultural production. By 1979 the installed capacity reached 6,380 MW at the county and below level, and rural electricity consumption leaped to 59.3 kWh from 0.02 in 1947, and approximately half a billion rural people were given access to electricity within the period (Jiahua et al., 2006). The rapid expansion and large-scale power sector development began in 1979, with the end of the Cultural Revolution and the introduction of economic reform; on average, 5 GW generating units are put into service each year (Chinese Society for Electrical Engineering, 2009). The central government combined electrification with a poverty eradication strategy and formulated a series of policies to support the development of small hydropower that powered the preliminary electrification course and transferred the authority of local electricity system management to local governments. Some major policies include: (i) "the one who invests owns and operates" and "unified construction and operation for unified generation and supply"; (ii) a subsidy and special loans for rural hydropower electrification; (iii) revenue from electricity for development of electricity; (iv) a reduction in value-added tax rate from normal 17% to 6%; and (v) the integration of small local grids to larger grids. Electrification accelerated rural economic expansion, and income growth led to greater electricity consumption, supporting further investment in rural electrification. It facilitated sustainable electricity development through feedback mechanisms rather than counting on government subsidies. By 1998, the total installed generating capacity in the county was 44.15 GW, and total electricity generation was 132.1 billion kWh, contributing 26.7% of total rural electricity consumption, 97% of the Chinese population had access to electricity, and 8.81 million rural households remained to be electrified (Jiahua et al., 2006).

Due to the 1997 Asian financial crisis, the industrial electricity demand decreased and there was a surplus in the supply. Yet, China was battling to connect the final 3% to electricity because of the deteriorating rural electricity system and higher tariffs in rural areas. In 1998 the government declared to invest RMB 290 billion (USD 37 billion) within 5 years to extend and renovate the rural grids. From 1999 to 2004, an additional 776 million population had access to electricity, which made 99.21% rural population electrified. Despite these attempts, everyone could not have access to affordable electricity. In 2012, the central government launched a 3-year action plan, "Electricity for All (2013–2015)," to connect the final 2.73 million people (He and Victor, 2017). With its success, the country was declared "full electrification" in 2015. The geography of China's bulk grid expansion is shown in Figure 21.3. In the final phase of electrification, grid extension and off-grid solar PV were equally divided for providing access depending on terrain and distance from the substation. For some localities, the cost of a highly reliable grid connection was calculated as high as RMB 100,000 (US$16,000) per household, and micro-grids with the cost range of RMB 9,000–20,000 (US$1,400–3,200) per household. The optimum option for such a household was individual solar

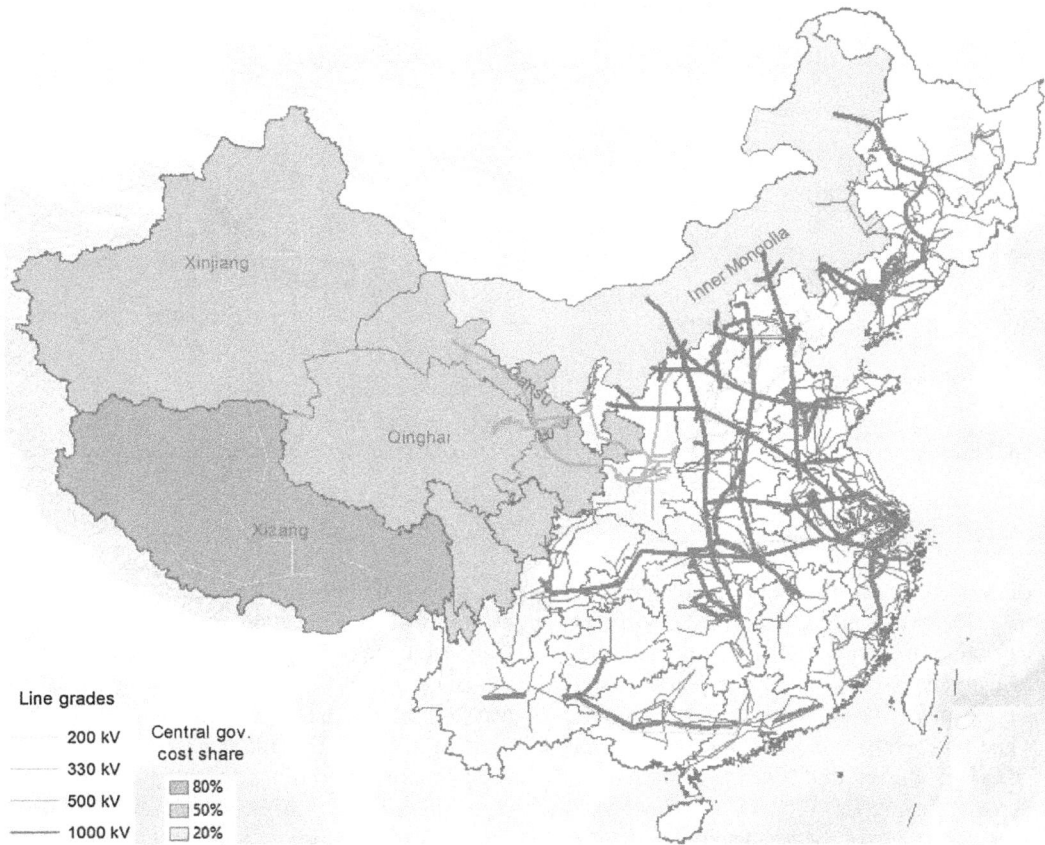

Note: The Geography of China's bulk grid and electrification. Western provinces generally have the weakest grid systems (200 kV and above lines shown in the figure) and have been the focus of final electrification because they have large scattered rural populations and high levels of poverty—for these reasons, they have also attracted the largest share of central government funding (shading).

**FIGURE 21.3** The geography of China's bulk grid and electrification. (From He, G., Victor, D. G., Resour. Conserv. Recycl., 122, 335–338, 2017. With permission.)

PV (0.3–0.4 kW), costing in the range of RMB 7,500–11,200 (US$1,200–1,800) per household (He and Victor, 2017).

The sources of power generation and the sectoral electricity consumption in China is shown in Figure 21.4; industrial sector consumes 61.6% of the electricity generated while 82.5% of the electricity generation is powered by fossil fuel. All three levels of government played well-coordinated roles in the path of 100% electrification: the central government offered policy leadership and investment, provincial governments conducted the local assessment, trials of various approaches and technologies, coordinated project execution, and the local government facilitated the management and operation. Between 2013 and 2015, the central government allotted RMB 24.8 billion (approximately USD 4 billion) to extend the grids and off-grid small PV systems. However, the financing ranged from 20% to 80% of the cost. For example, the project development is relatively expensive in the poor and remote regions like Tibet, and as such, the central government supported 80% of the costs. In case of the settlements that were too expensive to provide service, the government relocated people to a more hospitable terrain within the country.

The authoritarian central government of China can manage large investment funds comparatively easily and mobilize state-owned power firms and local officials than the democratic governments. For instance, not all governments can relocate people from inaccessible locations.

Electricity Generation by Source

Electricity final consumption by sector

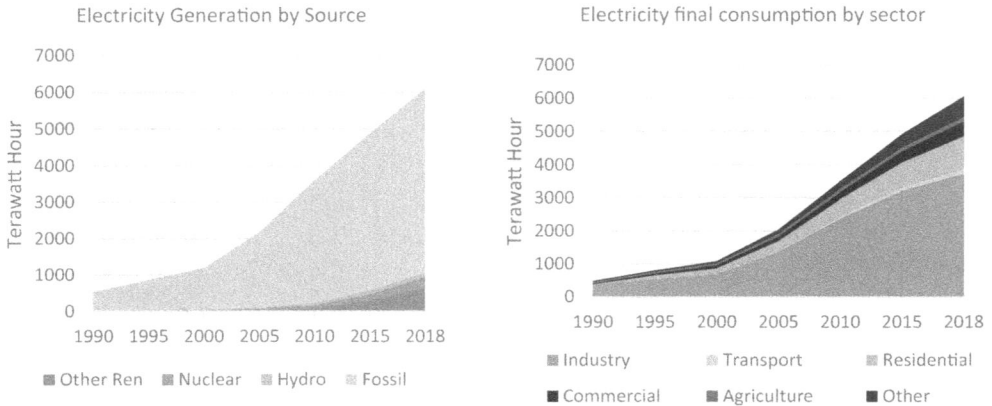

**FIGURE 21.4** Electricity development in People's Republic of China 1990–2018. (Data from IEA, 2021a).

### 21.3.2 INDIA

India was introduced to electricity by the British during the colonial period. The first hydel power station, 13 kW, was commissioned in 1896 to supply power to Darjeeling tea plantations. When the country became independent in 1947, the national generating capacity was 1,362 MW, and the electricity was available only in a few urban centers like Bombay, Calcutta, Mysore, Madras, and Srinagar. Then, the consumers of electricity were public utilities, industries, and few private residences.

The blueprint for electrifying India was formulated by Dr. BR Ambedkar and his team in the 1940s. He believed that "cheap and abundant electricity can only be provided through a centralized system to ensure the success of industrialization and bring about socio-economic development." The endorsement of the Electricity Supply Act, 1948 vesting the creation of State Electricity Boards (SEBs) in each province, an autonomous entity responsible for generation, transmission, and distribution of electricity, to improve the reliability of the system, allow economies of scale and rise geographical coverage had commenced the electrification evolution in the real sense (Palit and Bandyopadhyay, 2017). However, over half a century period (2047–2001), only 55.8% of the total population (87.6% urban and 43.48% rural) had electricity access; the reason being the goal for electricity system development was to support the industries in towns/peri-urban areas and agricultural production by pumping water. With the beginning of the new millennium, it was realized that robust electricity distribution networks and rural electrification were the prime movers of rural development. The then government, led by Honorable Prime Minister Atal Bihari Vajpayee, launched the Rural Electricity Supply Technology Mission in 2002 that opened an enabling environment, and subsequently enacted the Electricity Act in 2003. The 2003 Act set out targets for rural electrification and obligated the federal and provincial government to provide power for all by 2012, and also highlighted the importance of distributed power generation. In 2005, Rural Electricity Infrastructure and Household Electrification Scheme, popularly known as Rajiv Gandhi Grameen Vidyutikaran Yojana (RGGVY), were launched that was dedicated to all rural households living below the poverty line and aimed for at least 1 kWh of electricity per household per day. It was funded with a 90% contribution by the central government and 10% by the Rural Electrification Corporation (REC) (National Portal of India, n.d.-b). Though the RGGVY gave huge momentum to the electrification process (with the coverage of 95% of the targeted un-electrified villages and 76.74% of the below-poverty line households as well as the intensification of the electricity infrastructure in 78.56% of partially electrified villages) the scheme was heavily criticized for the quality of power supply, sustainability of infrastructure, sub-transmission and hence the desired results could not be met (Palit and Bandyopadhyay, 2017). In 2015, Deen Dayal Upadhyaya Gram Jyoti Yojana of

Electricity Generation by Source

Electricity final consumption by sector

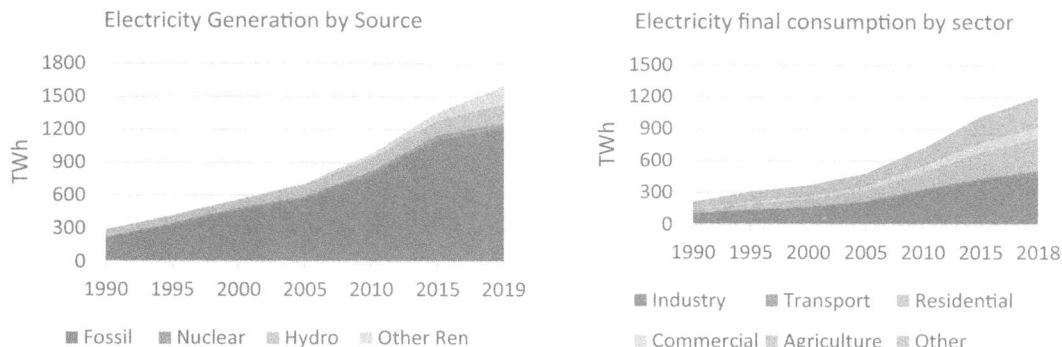

**FIGURE 21.5** Electricity development in India 1990–2018. (Data from IEA, 2021b).

INR43,033 crore (USD 5.8 billion), the newer avatar of RGGVY, was launched with the ambition of providing 24*7 power to rural households and adequate power to agricultural consumers through separate feeders (National Portal of India, n.d.-a). Yet approximately 30 million households were deprived of electricity in 2017. A new, more inclusive, and conceptualized scheme, "Saubhagya," was launched with a total outlay of INR16,320 crores (USD 2.2 billion). The scheme provides a solar home system (SHS) of 200–300 Wp capacity for the dwellers of extremely remote areas.

India has managed to provide almost half a billion people with electricity within a decade despite its geographical size and diverse terrain. According to World Bank Statistics, in 2019, 97.8% of the total population (96.6% in rural and 100% in urban) have access to electricity. However, the quality of the electrification is often under question. Next is the challenge of sustaining electrification by ensuring reliable, round-the-clock electricity at an affordable cost that can serve the anticipated benefits. It is recorded that the Indian distribution companies (DISCOMs) are already under great financial stress (refer to Table 21.2) that is limiting their ability to make payments to the energy generators and invest in the modernization of the grid, which is bound to push back the deployment and dispatchment of renewable energy for the energy transition. In addition, the combined average aggregated technical and commercial (AT&C) losses of all the states were 18.19% in FY19, down from 26% in FY15, while the target for March 2019 was set as 15% overall.

India has set a target of 175 GW renewable energy by 2022 and to reach the non-fossil fuel-based electric power generation capacity of 40% by 2030 (Ministry of Environment Forest and Climate Change, 2015). As of April 2021, the total national installed capacity was 382.73 GW, of which the share of the thermal plant was 234.728 GW, hydro 46.209 GW, nuclear 6.78 GW, and other renewable energy sources were 95.013 GW (Ministry of Power, 2021). The sources of power generation and the sectoral electricity consumption in India is shown in Figure 21.5; industrial sector consumes 40.1% of the electricity generated while 76% of the electricity generation is powered by fossil fuel.

### 21.3.3 MEXICO

Mexican electricity system can be traced back to 1879, when the first thermoelectric plant was operated in the textile industry in the city of Leon (Breceda, 2000). By 1899, 177 small power plants with a total installed capacity of 20 MW had already been installed to operate engines, especially in mining industries, which later expanded to the process of production in manufacturing, flour, and textile mills (Breceda, 2000; Carreón-Rodríguez et al., 2009). The surplus was sold to some elite residential and commercial clients. The electricity market structure was vertical and privately owned, mostly foreigners from Canada, France, the United States, and Germany. Until the mid-1920s, there was no concept of "universal access" to electricity or rural electrification. On May 11,

**TABLE 21.2**

**State-Wise DISCOMs Performance on Key Parameters in FY2018/19**

| Indian | RE Target as per NEP 2018 MW | Energy Mix -RE % | EODB Ranking Nos. | AT&C Loss (2018–2019) % | ACS-ARR Gap (2018–2019) Rs/kWh | Annual Financial Losses as per ARR Rs Crore | Payments Overdue to Generators Rs Crore | Annual Capex Rs Crore | Govt. Subsidy Rs Crore |
|---|---|---|---|---|---|---|---|---|---|
| Uttar Pradesh | 13,221 | 4.8% | 12 | 21.2 | 1.77 | −9,506 | 13,327 | 8,048 | 8,900 |
| Madhya Pradesh | 12,058 | 3.1% | 7 | 27.7 | −0.88 | −3,000 | 594 | 3,945 | 15,000 |
| Maharashtra | 22,045 | 14.8% | 13 | 16.9 | −0.055 | NA | 572 | 5,241 | NA |
| Jharkhand | 2,005 | 1.1% | 4 | 45.9 | 0.54 | −1,470 | 4,141 | 2,494 | 7,800 |
| Punjab | 4,376 | 4.0% | 3 | 19.1 | 0.36 | −906 | 23 | 2,490 | 8,855 |
| Haryana | 5,131 | 10.0% | 20 | 18.3 | 0.04 | −202 | 46 | 2,565 | 7,140 |
| Bihar | 2,762 | 1.2% | 18 | 34.3 | 0.57 | −1,578 | 627 | 5,453 | 2,952 |
| Rajasthan | 14,505 | 38.0% | 9 | 24.3 | 0.87 | NA | 1,955 | 4,632 | 1,709 |
| Chhattisgarh | 2,087 | 3.6% | 6 | 23.3 | 0.04 | −541 | 17 | 997 | 300 |
| Andhra Pradesh | 18,612 | 31.0% | 1 | 10.9 | 0.74 | 1 | 2,517 | 3,771 | 6,030 |
| Karnataka | 14,817 | 52.0% | 8 | 13.9 | 0.05 | 78 | 5,626 | 6,821 | 2,195 |
| West Bengal | 5,735 | 3.0% | 10 | 38.1 | 0.21 | −41 | 24 | 1,076 | 912 |
| Tamil Nadu | 21,793 | 43.0% | 15 | 13.5 | 1.15 | −7,761 | 12,197 | 2,025 | 8,430 |
| Gujarat | 17,133 | 11.8% | 5 | 12.6 | −0.05 | NA | 193 | 4,062 | 1,100 |

*Source:* Garg, V., Shah, K., The Curious Case of India's Discoms: How Renewable Energy Could Reduce Their Financial Distress, 2020. With permission.

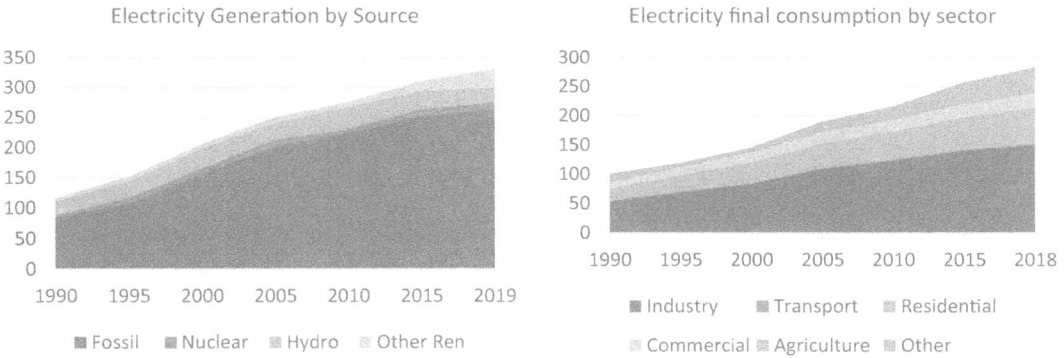

**FIGURE 21.6** Electricity development in Mexico 1990–2019. (Data from IEA, 2021c.)

1926, the "National Electric Code" was published that declared electricity a public good and authorized Congress to legislate affairs associated with the electricity industry. Despite the new framework, the private companies were not interested in investing in the rural areas; hence the government tried to solve the issue by establishing a public company: "Comison Federal de Electricidad" (CFE), in 1937. CFE later led to the consolidation of hundreds of regional monopolies that were creating technical difficulties in the transmission grid and eventually nationalization of the electricity industry in 1960 and over the decade connected millions, including those from adverse social backgrounds to the grid, the access to electricity rose from 25.0% to 58.9% of the total population from 1960 to 1970 (Gutierrez-Poucel, 2007). By 1972, the Mexican electricity grid had expanded almost throughout the country. The benefits of electrification in rural communities were apparent: pumping and irrigation raised the agricultural yield, which raised the incomes, expanded the domestic market, and strengthened the nation-building process. It also improved community life, allowing better education, health services, and public services like lighting and entertainment. Electricity tariffs were set for political benefits, hence were heavily subsidized, backed by oil revenues, with little consideration to the cost of production, whose windfall and pitfall were sensed in the oil price surge of the 1970s and oil price crash of the early 1980s, respectively (Carreón-Rodríguez et al., 2009). Consequently, the reforms in the power sector and the overall economy were implemented by opening Mexican markets to international free trade and forming economic institutions, including the Energy Regulatory Commission (CRE), to regulate the energy market. In 1996, the national Congress and federal government granted greater financial and executive autonomy to local government as a part of decentralization policy, and to date, decisions for infrastructure investment, including rural electrification program (REP) are mostly made at the regional/local level. To attract private utilities in rural electrification projects, the government with CFE decided to reduce initial impedance by providing subsidies in capital cost and tariffs in rural areas with lower population density and inadequate income opportunities.

Mexico was declared 100% electrified in 2017. The sources of power generation and the sectoral electricity consumption in Mexico is shown in Figure 21.6; industrial sector consumes 52.8 % of the electricity generated while 79.5% of the electricity generation is powered by fossil fuel, and as of 2019, the share of renewable in total final energy consumption is limited to 10.3% (IEA et al., 2021a).

Electrification trend in some of the developing countries (2001–2019) are shown in Figure 21.7.

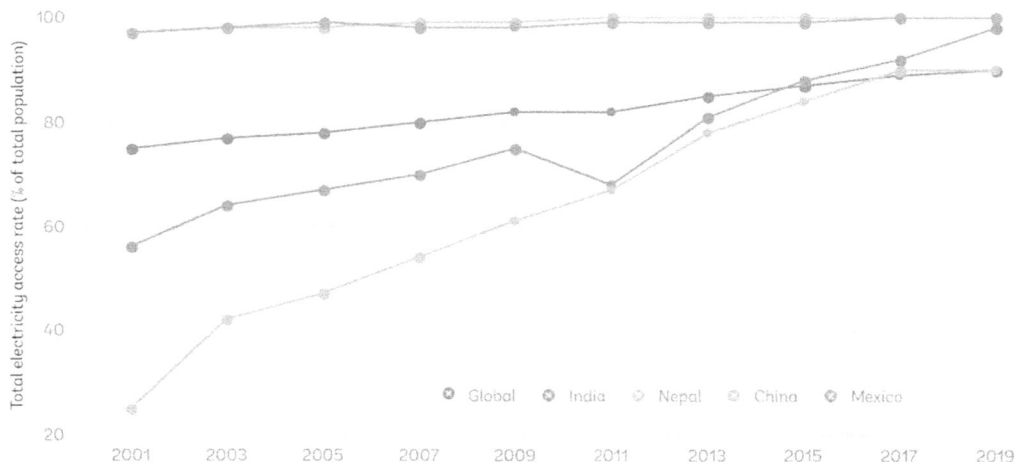

**FIGURE 21.7** Electrification trend in developing countries (2001–2019). (From IEA, IRENA, UNSD, & World Bank, Mexico. Tracking SDG 7, 2021a. With permission.)

## 21.4 FINANCING MECHANISM FOR SUSTAINABLE ELECTRIFICATION

From the examples above, it can be clearly understood that access to the final section of the population, those who are poor, vulnerable, and remote, is the most challenging because reaching the balance between affordability and financial viability is complicated. Hence financing is the biggest hurdle to universal energy access. Extraordinary measures are being designed and implemented to mobilize and scale up investment and therefore, to ramp up universal access to electricity by 2030. In developing countries, the public sector is the critical source of finance in renewable energy investment. Tracking SDG7 Report mentioned that the flow of international public finance has increased threefold over the period of 2010–18 when viewed as a 5-year moving average; however, the 46 least developed countries received only 20% of this amount (IEA et al., 2021a). New connections for power access are needed to provide energy for all by 2030 by connection type and fuel source is shown in Figure 21.8. IEA and IRENA scenarios project that the annual investment requirement in the renewable power sector needs to be increased significantly from the present USD 300 billion to USD 550–850 billion throughout 2019–2030. This level of investment can be anticipated from international public financial flows. Meanwhile, private capital must also be energized through various approaches to minimize investors' risk. The three major financing instruments are:

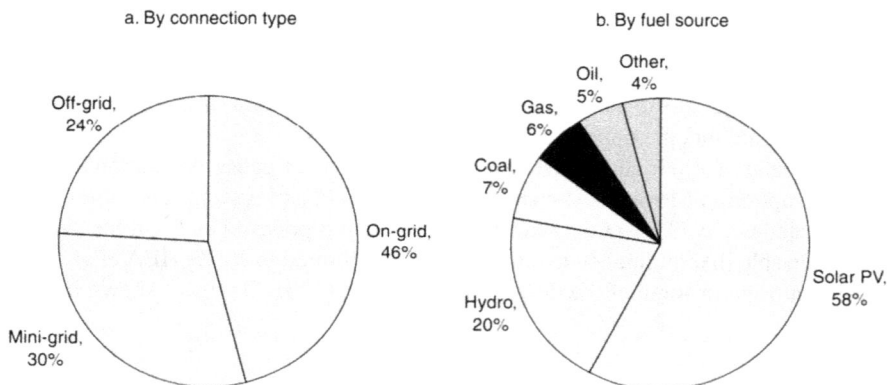

**FIGURE 21.8** New connections for power access are needed to provide energy for all by 2030 by connection type and fuel source. (From IEA, Energy Access Outlook, International Energy Agency, 2017. With permission.)

**TABLE 21.3**
**Micro-Grid Management and Ownership Models**

| Model | Strength | Challenges | Example |
|---|---|---|---|
| Cooperatives | Investment by communities in cash and kind | • Attracting financing<br>• Capacity for O&M<br>• Can't raise tariffs | Sri Lanka, Mali, Nepal, Philippines |
| Private investors | Attracts private investment | • High tariffs, need to earn ROE<br>• Community buy-in and conflicts | Cambodia, Mali, Senegal, Tanzania |
| Utility/REA | Experienced in utility operation | • Low priority<br>• Private investment difficult | Indonesia, Tanzania, Colombia, Ghana |
| Hybrid | Benefit from strengths of multiple models | • Added complexity and increased tariff | Nepal, Pakistan |

*Source:* USAID, Electrification in Developing Countries Utilities, Scaling Renewable Generation, and Access: Challenges and Lessons Learned, 2017. With permission.

i. **Grants:** Grants are nonrepayable funds, products, or services that may consist of matching grants, crowdfunding, and results-based financing (RBF). Grants are especially important in the early stages of the market.
ii. **Debt:** It is used to raise working and/or investment capital mostly by the growth phase companies.
iii. **Equity:** It is the nonrepayable capital investment raised by the project promoters.

Various catalytic tools attract funding by providing incentives or reducing risk to the investors.

Electricity in the areas where the settlement is dense is usually served through the grid and is managed by utilities. The areas that are scarcely populated and are at a longer distance from the grid are mostly electrified by off-grid solar PV, usually solar home systems and portable lanterns. The most challenging areas are where the concentration of settlement is medium, and the grid extension is not economically viable. Such areas are connected through mini-grid based on micro to mini scale wind, hydro, solar, biomass gasifier, diesel or gasoline generators, or hybrid systems and managed privately by the community. This model is set up under government subsidy to make electricity affordable to the local population. In many countries, the electricity is found to be utilized inefficiently because the consumers have little obligation to pay for the service such that if the government fail to implement a cross-subsidy mechanism such model crumble. Hence in recent days, policies are formulated to involve local participation in project installation, operation, and maintenance and emphasize productive uses like local entrepreneurship and industries (Almeshqab and Ustun, 2019; Bhattacharyya, 2013). Over time, these mini-grid are consolidated into an integrated grid system, if financially feasible. Some micro-grid management and ownership models with strengths and challenges are shown in Table 21.3.

## 21.5 WAY FORWARD: BEYOND CONNECTIVITY

However, the binary metric of access to electricity does not illustrate the quality and quantity of connection nor it explains if the "access" is sufficient to satisfy household and enterprise needs and support socio-economic development. Since access to energy is widely accepted as an important ingredient toward achieving SDGs: eliminating poverty and gender inequalities, improving health and education services and food security; its evaluation must comprise both quantitative and qualitative aspects, including the supplied amount, time of supply, voltage, reliability, emissions, affordability and fuel collection time among others (Bhatia and Angelou, 2015a). The Energy Sector Management Assistance Project (ESMAP) Technical Report 008/15 (Bhatia and Angelou, 2015b) has developed a multi-tier

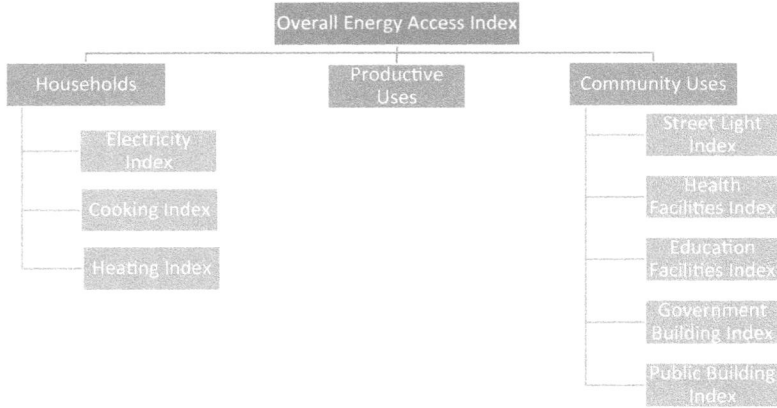

**FIGURE 21.9** Multi-tier framework energy access indices. (From Bhatia, M., Angelou, N., Beyond Connections: Energy Access Redefined. ESMAP Technical Report 008/15, 2015b. With permission.)

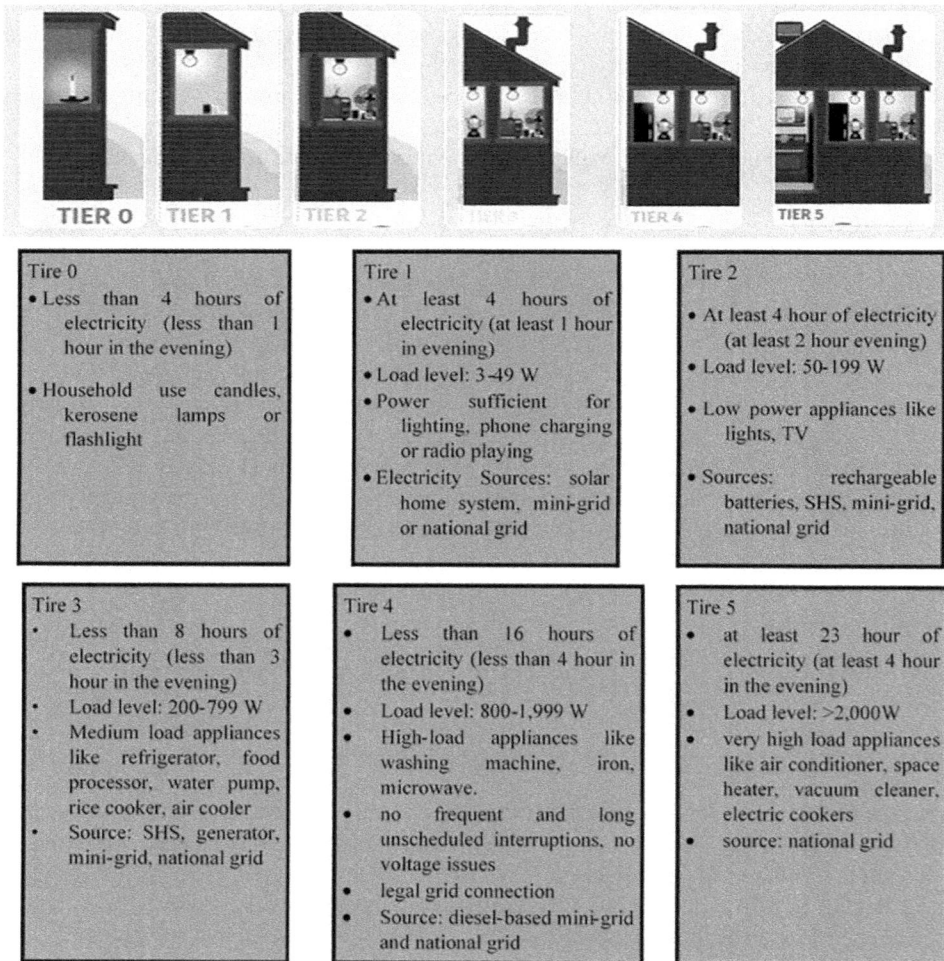

| Tire 0 | Tire 1 | Tire 2 |
|---|---|---|
| • Less than 4 hours of electricity (less than 1 hour in the evening)<br><br>• Household use candles, kerosene lamps or flashlight | • At least 4 hours of electricity (at least 1 hour in evening)<br>• Load level: 3–49 W<br>• Power sufficient for lighting, phone charging or radio playing<br>• Electricity Sources: solar home system, mini-grid or national grid | • At least 4 hour of electricity (at least 2 hour evening)<br>• Load level: 50-199 W<br><br>• Low power appliances like lights, TV<br><br>• Sources: rechargeable batteries, SHS, mini-grid, national grid |
| **Tire 3** | **Tire 4** | **Tire 5** |
| • Less than 8 hours of electricity (less than 3 hour in the evening)<br>• Load level: 200-799 W<br>• Medium load appliances like refrigerator, food processor, water pump, rice cooker, air cooler<br>• Source: SHS, generator, mini-grid, national grid | • Less than 16 hours of electricity (less than 4 hour in the evening)<br>• Load level: 800-1,999 W<br>• High-load appliances like washing machine, iron, microwave.<br>• no frequent and long unscheduled interruptions, no voltage issues<br>• legal grid connection<br>• Source: diesel-based mini-grid and national grid | • at least 23 hour of electricity (at least 4 hour in the evening)<br>• Load level: >2,000W<br>• very high load appliances like air conditioner, space heater, vacuum cleaner, electric cookers<br>• source: national grid |

**FIGURE 21.10** Minimum requirements of electricity in the household by tier. (Adopted from: Pinto, Yoo, Portale, & Rysankova, 2019).

framework (MTF) underlying "Beyond Connections" that is inclusive of the intermediary stages between no access and full energy access, covering the range of improvements that various technologies provide along with three areas of energy access — (i) households, (ii) productive activities and (iii) community institutions (refer to Figure 21.9). The MTF as shown in Figure 21.10 measures not only if users receive energy services but also the service's capacity, duration, reliability, quality, affordability, legality, and safety.

Measuring progress on SDG7 should include looking beyond minimal energy access for basic household services to sufficient job creation and income generation levels. Developing countries need sufficient electricity to power businesses and grow their economies, but this electricity needs to come from low-carbon sources to achieve net-zero emissions by 2050 and deliver on development priorities.

## 21.6   CONCLUSIONS

Access to clean, modern, and efficient energy sources is vital for sustainable economic development and for uplifting the living standard of the people. To achieve socio-economic and environmental benefits related to the sustainable energy transition, United Nations has formulated sustainable development goals (SDG) emphasizing access to energy in the SDG goal 7 stated to "ensure access to affordable, reliable, sustainable and modern energy for all." To realize this goal, electrification plays a crucial role as it is the most versatile form of carbon-free energy and is thus recognized at the top level of the energy ladder. As most of the developing countries still lack 100% electrification and where the connectivity is achieved, it is usually with the poor level of per capita electricity consumption as compared to the developed world, they are striving toward achieving electricity access goals by formulating and implementing various country-specific targets and policies. The energy access index has been increasingly used to monitor the progress toward achieving this target. With access to electricity, the connectivity will be inevitably reliable, affordable, legal, and safe to realize the sustainable development goals.

## REFERENCES

Almeshqab, F., & Ustun, T. S. (2019). Lessons learned from rural electrification initiatives in developing countries: Insights for technical, social, financial and public policy aspects. *Renewable and Sustainable Energy Reviews, 102*, 35–53. https://doi.org/10.1016/j.rser.2018.11.035.

Bhatia, M., & Angelou, N. (2015a). *Beyond Connections: Energy Access Redefined. ESMAP Technical Report 008/15.* Washington DC. Retrieved from https://openknowledge.worldbank.org/bitstream/handle/10986/24368/Beyond0connect0d000technical0report.pdf?sequence=1&isAllowed=y

Bhatia, M., & Angelou, N. (2015b). *Beyond Connections.* https://doi.org/10.1596/24368

Bhattacharyya, S. C. (2013). Financing energy access and off-grid electrification: A review of status, options, and challenges. *Renewable and Sustainable Energy Reviews.* https://doi.org/10.1016/j.rser.2012.12.008

Breceda, M. (2000). *Debate on Reform of the Electricity Sector in Mexico Report on its Background, Current Status and Outlook Prepared for the North American Commission for Environmental Cooperation.* Retrieved from http://www3.cec.org/islandora/en/item/1611-debate-reform-electricity-sector-in-mexico-en.pdf

Carreón-Rodríguez, V. G., Jiménez, A., & Rosellón, J. (2009). The Mexican electricity Sector: economic, legal, and political issues. In *The Political Economy of Power Sector Reform* (pp. 175–214). Cambridge University Press. https://doi.org/10.1017/cbo9780511493287.006

Chinese Society for Electrical Engineering. (2009). *China's Power Industry Milestones 1879–2008.* Retrieved from https://www.ieee-pes.org/images/files/pdf/Chinas_Power_Industry_Milestones_IEEE_PES-20081010.pdf

ESMAP. (2020). *Regulatory Indicators for Sustainable Energy (RISE) 2020: Sustaining the Momentum.* Washington DC. Retrieved from https://www.esmap.org/rise_2020_report

Garg, V., & Shah, K. (2020). *The Curious Case of India's Discoms: How Renewable Energy Could Reduce Their Financial Distress.* Retrieved from https://ieefa.org/wp-content/uploads/2020/08/The-Curious-Case-of-Indias-Discoms_August-2020.pdf

GOGLA. (2020). *Global Off-Grid Solar Market Report: H1 2016 to H2 2020.* Retrieved from https://www.gogla.org/global-off-grid-solar-market-report

Gutierrez-Poucel, L. E. (2007). From central planning to decentralized electricity distribution in Mexico. In D. F. Barnes (ed.), *The Challenge of Rural Electrification: Strategies for Developing Countries* (pp. 132–162). Routledge. Retrieved from https://books.google.com.np/books?id=iOBi17Pr3fIC&pg=PA13 5&lpg=PA135&dq=rural+electrification+momentum+mexico&source=bl&ots=Qvh8KWOjhG&sig=A CfU3U2KKleVo5EBNrOp5Rl6cy0JsDVrMw&hl=en&sa=X&ved=2ahUKEwjO26_28bLxAhWS4nM BHY6tCKcQ6AF6BAgGEAM#v=onepage&q=rural electrification momentum mexico&f=false

He, G., & Victor, D. G. (2017). Experiences and lessons from China's success in providing electricity for all. *Resources, Conservation and Recycling, 122,* 335–338. https://doi.org/10.1016/j.resconrec.2017.03.011

IEA. (2017). *Energy Access Outlook.* International Energy Agency. Retrieved from https://www.iea.org/ data-and-statistics/charts/investment-required-to-provide-electricity-to-all-by-2030-energy-for-all-case

IEA. (2021a). *Data & Statistics - China (People's Republic of China and Hong Kong China) 1990-2018.* Retrieved June 9, 2022, from https://www.iea.org/data-and-statistics/data-browser/?country=CHINAR EG&fuel=Electricity and heat&indicator=ElecGenByFuel

IEA. (2021b). *Data & Statistics - India.* Retrieved June 9, 2022, from https://www.iea.org/data-and-statistics/ data-browser?country=INDIA&fuel=Electricity and heat&indicator=ElecGenByFuel

IEA. (2021c). *Data & Statistics - Mexico.* Retrieved June 9, 2022, from https://www.iea.org/data-and-statistics/data-browser?country=MEXICO&fuel=Electricity and heat&indicator=ElecGenByFuel

IEA, IRENA, UNSD, World Bank, & WHO. (2021). *TRACKING SDG7: The Energy Progress Report.* Washington DC. Retrieved from https://trackingsdg7.esmap.org/data/files/download-documents/2021_ tracking_sdg7_executive_summary.pdf

IEA, IRENA, UNSD, & World Bank. (2021a). *Mexico. Tracking SDG 7.* Retrieved June 26, 2021, from https:// trackingsdg7.esmap.org/country/mexico

IEA, IRENA, UNSD, World Bank, & WHO. (2020). *Tracking SDG7: The Energy Progress Report.* Washington DC. Retrieved from https://www.irena.org/-/media/Files/IRENA/Agency/Publication/2020/May/ SDG7Tracking_Energy_Progress_2020.pdf

IEA, IRENA, UNSD, World Bank, & WHO. (2021b). *Tracking SDG7: The Energy Progress Report.* Washington DC. Retrieved from https://trackingsdg7.esmap.org/data/files/download-documents/2021_ tracking_sdg7_executive_summary.pdf

IEA, IRENA, UNSD, World Bank, & WHO. (2022). *Tracking SDG 7: The Energy Progress Report. World Bank, Washington DC.* © World Bank. License: Creative Commons Attribution—Non Commercial 3.0 IGO (CC BY-NC 3.0 IGO).

Jiahua, P., Wuyuan, P., Meng, L., Xiangyang, W., Lishuang, W., Zerriffi, H., et al. (2006). *Rural Electrification in China 1950–2004: Historical Processes and Key Driving Forces.* Stanford. Retrieved from http:// pesd.stanford.edu

Ministry of Environment Forest and Climate Change. (2015). *India's Intended Nationally Determined Contribution.* Government of India. Retrieved from https://moef.gov.in/wp-content/uploads/2018/08/ Press_Statement__INDC_English.pdf

Ministry of Power. (2021). *Power Sector at a Glance All India.* Retrieved June 25, 2021, from https://power-min.gov.in/en/content/power-sector-glance-all-india

National Portal of India. (n.d.-a). *Deen Dayal Upadhyaya Gram Jyoti Yojana.* National Portal of India. Retrieved June 25, 2021, from https://www.india.gov.in/spotlight/deen-dayal-upadhyaya-gram-jyoti-yojana

National Portal of India. (n.d.-b). *Rajiv Gandhi Grameen Vidyutikaran Yojana.* National Portal of India. Retrieved June 25, 2021, from https://www.india.gov.in/rajiv-gandhi-grameen-vidyutikaran-yojana

Palit, D., & Bandyopadhyay, K. R. (2017). Rural electricity access in India in retrospect: A critical rumination. *Energy Policy, 109,* 109–120. https://doi.org/10.1016/j.enpol.2017.06.025

Pinto, A., Yoo, H. K., Portale, E., & Rysankova, D. (2019). *Nepal Beyond Connections Energy Access Diagnostic Report Based on the Multi-Tier Framework.* Washington DC. Retrieved from https://open-knowledge.worldbank.org/bitstream/handle/10986/35266/Energy-Access-Diagnostic-Report-Based-on-the-Multi-Tier-Framework.pdf?sequence=1

USAID. (2017). *Electrification in Developing Countries Utilities, Scaling Renewable Generation, and Access: Challenges and Lessons Learned.* Retrieved from https://www.eia.gov/conference/2017/pdf/ presentations/dorian_mead.pdf

World Bank. (2021). *Data.* Retrieved April 22, 2021, from https://data.worldbank.org/indicator/SP.POP.TOTL.

# 22 Regional Energy Policies, Trade and Energy Security in South Asia
## *A Quest for Diplomacy*

*Mabroor Hassan*
International Islamic University
Green Environ Sol (Private) Limited

*Manzoor Khan Afridi, and Muhammad Irfan Khan*
International Islamic University

## CONTENTS

## HIGHLIGHTS

- Energy is a basic need and commodity for development in South Asia.
- Regional energy trade can reduce dependence on fossil fuel to foster environmental security.
- Analysis of energy policies is an essence to procure energy across the region.
- Multi-Criteria Decision Analysis provides an analysis of energy trade feasibility.

## 22.1 ENERGY TRADE AND SECURITY IN SOUTH ASIA

Economic growth embarks on energy use, whereby both have potential impacts on energy security, environment, and social wellbeing [1]. Meanwhile, energy security, development, and environmental concerns are rapidly changing the type and intensity of priorities, decision-making, orientation, and current positioning of different countries in the geopolitical context and future of the region [2,3].

DOI: 10.1201/9781003315353-25

**TABLE 22.1**
**Energy Reserves in South Asia**

| Countries | Coal Million tons | Oil Million barrels | Natural Gas TCF | Hydro MW | Biomass Million tons |
|---|---|---|---|---|---|
| Afghanistan | 440 | NA | 15 | 25,000 | 18–27 |
| Bangladesh | 884 | 12 | 8 | 775 | 0.18 |
| Bhutan | 2 | NA | NA | 30,000 | 26.60 |
| India | 90,085 | 5,700 | 39 | 150,000 | 139 |
| Maldives | NA | NA | NA | NA | 0.06 |
| Nepal | NA | NA | NA | 83,000 | 27.04 |
| Pakistan | 17,550 | 324 | 33 | 100,000 | NA |
| Sri Lanka | NA | 150 | NA | 2,000 | 12 |
| **Total** | **108,981** | **5,906** | **95** | **388,775** | **223** |

*Source:* [4,5].

South Asia has an estimated 108,981 million tons reserves of coal, followed by 5,906 million barrels of oil, 95 trillion cubic feet (tcf) of natural gas, the tremendous potential of 388,775 MW of hydropower, 223 million tons of biomass, the massive potential of solar, and wind energy (Table 22.1) that offer opportunities for economic growth [4,5].

Despite these ample energy resources, increasing energy demand and burgeoning population in South Asia have induced rapid changes in the environmental profile of the region, energy security, economic growth patterns, human wellbeing index, and climate change vulnerability. It is anticipated that minor energy and environmental stresses have a high potential for consequential conflicts and civil strife. On the other hand, energy insecurity in South Asia reflects impacts on livelihood, poor economic growth, human insecurity, and frustration among people. However, appropriate planning, decision-making, and policies with energy and environmental security nexus might metamorphose conflict into cooperation and binding mechanism in South Asia [6]. However, energy endowments vary among South Asian countries and lack cooperation among neighboring countries to secure it from each other [7], but energy trade can be a functional area for regional cooperation [8]. Furthermore, regional energy trade can cut the use of fossil fuels for energy production, mitigate climate change, and foster environmental security. Therefore, member states of the South Asian Association for Regional Cooperation (SAARC) signed a SAARC Framework Agreement on Energy Cooperation (Electricity) in November 2014 with the goal to create the SAARC Market for Electricity (SAME) and sustainable procurement of energy at an affordable cost. The core objectives of this agreement are to enable cross-border trade of electricity among member states on a voluntary basis through mutual agreements and negotiate to buy and selling entities in economic terms. Nevertheless, SAARC countries are still facing problems to develop SAME which necessitates significant diplomatic efforts and policy reforms. The lack of policy reforms and collective efforts is hindering the regional energy trade, economic development, and energy security. It seems that there is a knowledge gap on technical, political, and sustainability assessment of energy policies in the context of regional energy trade and environmental sustainability. Hence, energy policies were analyzed with two objectives of assessing the adequacy of energy policies of member states (i) for trade and development and (ii) energy and environmental security. It is hoped that this academic pursuit will offer a new paradigm for assessing the need for environmental diplomacy for sustainable development in South Asia.

## 22.2  BACKGROUND

Su et al. [9] stressed that energy has become a scarce global strategic resource and power generation is distributed geographically. Meanwhile, many countries are generating costly and environmentally unfriendly energy from fossil fuels due to scarcity and poor capacity to utilize renewable energy resources [10]. Such countries also have an option of regional energy trade to cut the use of fossil fuels, cost, and environmental damages. But lack of cooperation, poor planning, inadequate decision-making, and poorly decided energy generation and procurement are invoking environmental damages, poor financial returns, and stakeholder reactions [11]. However, detailed technical, economic, environmental, social, and political considerations in decision-making are needed prior to regional energy trade to manage financial returns, enhance trade feasibility, stakeholder acceptability, and environmental sustainability.

The technical aspects such as reliability, efficiency, cost-effectiveness, mutual understanding between investors and policymakers, speed of implementation, and technology penetration in the energy mix at national, regional, and international levels for safe procurement of energy can make regional energy trade more feasible and can contribute in energy security [12]. Similarly, Büyüközkan and Güleryüz [13] referred to consider the identification of potential and availability of resources, reliability, maturity, safety, security, and import or export of energy as technical aspects to foster regional energy trade and security.

Economic aspect is fundamental to energy and environmental sustainability. The competitive investment includes the cost of labor, equipment, engineering services, infrastructure construction work, technological installations, supply, and trade [14]. According to Mahmood et al. [15], the regional energy trade might be an economically viable option for improving regional cooperation and prosperity. Behera and Dash [16] are of the view that the trade of energy will pledge economic development, raise income, reduce taxes, alleviate poverty, reduce dependence on fossil fuels, and control carbon emissions.

Fonseca and Rosen [17] believed that energy security is directly linked with environmental security. Both are among important issues on the national security agenda in different regions of the world. This is perhaps, due to the relationship of recent energy trends in exploration, consumption of oil and natural gas, water security, air pollution, global warming, and climate change with environmental security. Seter [18] linked water security and climate change with conflicts, political concerns, and environmental security as well as intensifying the governance capacity, regional trade, migration, and national security. Similarly, Wang et al. [19] argued that power generated from fossil fuels (oil, gas, and coal) emits carbon dioxide, NOx, and other gases, which eventuated into temperature rise resulting in climate change, threatened social structure, and environment. Bangladesh, Nepal, and Sri Lanka are facing the largest share to influence on precipitation due to $CO_2$ from the power sector, while India and Pakistan have an average influence on precipitation [20]. Additionally, water and energy are supposed to be independent and interlinked valuable resources [21]. Water is a ubiquitous source for hydropower generation and cooling during thermal power generation. Similarly, water utilization including collection, treatment, and distribution to end-user requires energy. The mutual vulnerability of energy and water is amplified by the growing demand and climate change [22]. According to Chatterjee et al. [23], there is an urgent need for individual response and regional cooperation to complement energy trade and environmental security. Meanwhile, Sovacool [24] argued that the interconnected regional energy grids could potentially reduce greenhouse emissions, cut reliance on fossil-fueled sources of energy and be helpful in reducing the impacts of climate change.

Mühlemeier et al. [25] refer to ponder social considerations for faultless functioning of technical aspects, and sustainable energy procurement from the environmental, economic, and political points of view through public support and involvement. Social criteria cover access to energy as a basic human right, social acceptability, job creation, social benefits, and cooperation [13]. It is being used as a tool for stakeholder consensus on energy trade through addressing the social benefits including access to affordable energy and employment opportunities. According to Dvořák et al. [26], 1,114,210 people were directly or indirectly employed in the energy sector of the European Union in 2010, but 2 years later, the number increased to 1,218,230 (10%) in 2012 due to the approval of the European Energy Strategy (Energy 2020) and regional cooperation as well as interconnection.

The long-term development, procurement of energy, and sustainability have a challenge of long-term commitment and cooperation but identification of political aspects can aspire collaboration over a long-time perspective [27]. Khan et al. [28] identified the constraints for energy trade including geopolitical, social, economic, variable prices, reliability constraints, security, variable resource availability, environmental concerns, interconnection, lack of cooperation, poor political insight, low institutional capacity, and regulatory enforcement, as well as the right of ownership in terms of fiscal, social, political, and environmental benefits are restricting the feasibility of regional energy. However, some countries in the world have successfully overcome these constraints and procured energy in regions such as Nordic Pool and European Union. The successful energy transitions in the Nordic region are the outcome of strong political will and policies [24].

From the above discussion, it is synthesized that the feasibility of regional energy trade and benefits to some extent extents were highlighted in previous studies, but the capacity of energy policies in the region, policy reforms criteria for successful energy procurement and neutral tools to reduce constraints were seldom discussed. Therefore, critical analysis of energy policies of countries in South Asia in terms of technical, economic, environmental, social, and political aspects to assess criteria for policy reforms for future regional energy trade, and linked region energy trade with development, energy, and environmental security is presented. Meanwhile, negotiation and environmental diplomacy as a neutral tool through intergovernmental organizations have been identified to resolve disputes regarding energy trade and environmental security, formulate, and implement the agreements by providing technical assistance and information.

## 22.3  MULTI-CRITERIA DECISION ANALYSIS (MCDA)

The sustainable procurement of energy at a regional level is gaining importance due to low cost and low carbon energy goals. Strong policies with a clear vision, stakeholder engagement, and cooperation are key governance structures for connecting the actors [29]. The regional planners require multi-criteria analysis to determine the pertinent energy policies for secure supply and environmental sustainability [30]. MCDA offers a suitable methodology with a requisite structure that stays on track toward objectives and requirements [31]. MCDA provides a basis to policymakers for long-term sustainable decisions by analyzing the different perspectives of multiple sustainability criteria for regional energy transitions [32]. MCDA applied to energy policies in the region (Table 22.2) to assess the energy trade feasibility, environmental diplomacy, development, and environmental security nexus focused on successful cooperation for procurement of energy to ensure regional energy security and environmental sustainability.

The evaluation criteria were selected from numerous research studies which focused on regional energy transitions and joint publication by the United Nations (UN) and the International Atomic Energy Agency [33]. The technical, economic, environmental, social, and political criteria (Figure 22.1) were used to assess the requirements for regional energy trade, security, and environmental sustainability [32].

The simple multi-attribute rate technique (SMART) is well-known for assigning criteria scores, weighting, and normalization for energy policies [34]. The SMART was used because it is easy to use and presents a clear picture of the scenario. The participants from the entire region being a representative of different stakeholders, from different professional backgrounds (technical, policy experts, energy experts, academia, NGOs, and environmentalists) were asked to rank the selected criteria. The data from 40 experts were collected. The experts ranked significance ranking (how much each criterion is important for regional energy trade, development, energy security, and environment) and existing ranking (how much each criterion is considered in existing policies) (out of 100) for each criterion, dispensed minimum ten points to least considered criteria and increased rating with improving significance [35]. The Pearson correlation was applied to results for assessment of the statistical relationship between significance ranking and consideration of selected criteria in energy policies in regions.

**TABLE 22.2**

**An Overview of Regional Energy Policies and Institutional Framework**

| Countries | Energy Policies | Ministries/Departments |
|---|---|---|
| Afghanistan | Energy Sector Strategy 2008, Afghanistan Rural Renewable Energy Policy 2013 | Ministry of Energy and Water (MEW) Ministry of Rural Rehabilitation and Development (MRRD) Inter-Ministerial Commission for Energy (ICE) |
| Bangladesh | National Energy Policy 2004, Renewable Energy Policy of Bangladesh 2008 | Ministry of Power, Energy and Mineral Resources (MoPEMR) Sustainable Energy Development Agency (SEDA) |
| Bhutan | Bhutan Sustainable Hydropower Development Policy 2008 Alternative Renewable Policy 2013 | Ministry of Economic Affairs (MEA) Department of Energy |
| India | National Electricity Policy 2005 Strategic Plan for New and Renewable Energy Sector for the Period 2011–17 Policy for Repowering of the Wind Projects 2015 | Ministry of Power (MOP) The Central Electricity Regulatory Commission (CREC) Ministry of Petroleum and Natural Gas (MoPNG) Ministry of Coal Ministry of New and Renewable Energy (MoNRE) Department of Atomic Energy (DAE) |
| Maldives | Maldives National Energy Policy and Strategy 2010 | Ministry of Housing and Environment (MHE) Maldives Energy Authority (MEA) |
| Nepal | The Hydropower Development Policy 2001 The Hydropower Development Policy 2049 Rural Energy Policy 2006 | Ministry of Water Resources (MoWR) Ministry of Environment, Science, and Technology (MoEST) Department of Electricity Development Alternative Energy Promotion Centre (AEPC) |
| Pakistan | Nation Power Policy 2013 Power Generation Policy 2015 Alternative and Renewable Energy Policy (Short-term) 2006 Alternative and Renewable Energy Policy (Mid-term) 2011 Alternative and Renewable Energy Policy, 2019 (Draft) | Ministry of Energy (MoE) Private Power and Infrastructure Board (PPIB) Alternative Energy Development Board (AEDB) National Electric Power Regulatory Authority (NEPRA) Pakistan Atomic Energy Commission (PAEC) Water and Power Development Authority (WAPDA) Ministry of Energy Oil and Gas Regulatory Authority (OGRA) |
| Sri Lanka | National Energy Policy & Strategies of Sri Lanka 2008 | Ministry of Power and Energy (MPE) Ministry of Petroleum and Petroleum Resources Development (MPPRD) Ceylon Electricity Board |

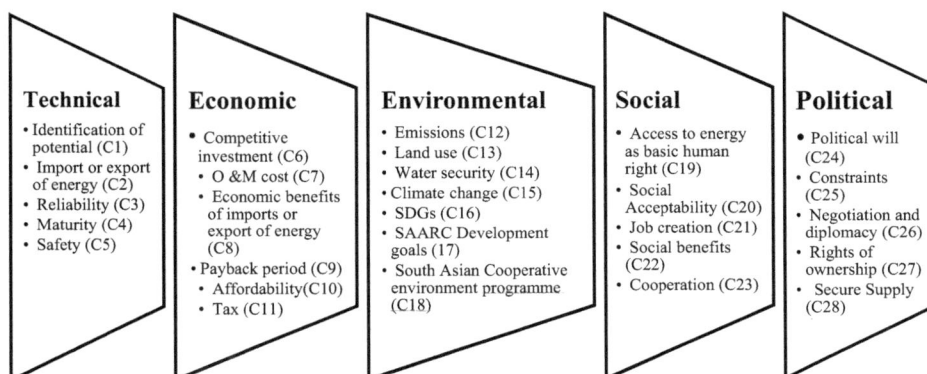

**Technical**
- Identification of potential (C1)
- Import or export of energy (C2)
- Reliability (C3)
- Maturity (C4)
- Safety (C5)

**Economic**
- Competitive investment (C6)
- O &M cost (C7)
- Economic benefits of imports or export of energy (C8)
- Payback period (C9)
- Affordability (C10)
- Tax (C11)

**Environmental**
- Emissions (C12)
- Land use (C13)
- Water security (C14)
- Climate change (C15)
- SDGs (C16)
- SAARC Development goals (17)
- South Asian Cooperative environment programme (C18)

**Social**
- Access to energy as basic human right (C19)
- Social Acceptability (C20)
- Job creation (C21)
- Social benefits (C22)
- Cooperation (C23)

**Political**
- Political will (C24)
- Constraints (C25)
- Negotiation and diplomacy (C26)
- Rights of ownership (C27)
- Secure Supply (C28)

**FIGURE 22.1** Evaluation criteria for analysis of energy policies in the region.

## 22.4  RATIONAL OF REGIONAL ENERGY POLICIES FOR ENERGY TRADE, SECURITY, AND ENVIRONMENTAL SUSTAINABILITY IN SOUTH ASIA

The economic growth and burgeoning population in South Asia are inducing rapid changes in the energy and environment profile of the region, energy security, environmental security, and human vulnerability. These minor energy and environmental stresses have a high ability of consequential conflicts and civil strife. On the other hand, appropriate planning, decision-making, and policies of energy and energy trade might metamorphose conflict into cooperation and binding mechanism in South Asia.

### 22.4.1  TECHNICAL ASPECTS

The South Asian countries are facing different environmental concerns and priorities which have significant transboundary impacts and dimensions. The experts assigned the exceptional significance ranking to the identification of potential (C1) (83). South Asian countries comprehend the importance of the significance of installed capacity identification and availability of energy resources (Figures 22.2 and 22.3). They contemplate the potential and availability of both renewable and nonrenewable energy resources in their energy policies except in the Maldives.

The experts believe that the energy trade is not only feasible in South Asia but has the potential to provide affordable energy, reduce the dependence on fossil fuels, and protect the environment. The energy trade (C2) was ascribed significance ranking (79) in terms of feasibility, affordability, reliance on fossil fuel, and mitigation of climate change. Although all South Asian nations are intended to import or export energy, some countries demonstrated it as part of their foreign policy rather than their energy policies. Afghanistan, Bangladesh, and Sri Lanka contemplated the import of energy being deficit countries. Bhutan and Nepal had pondered the energy export as a source of

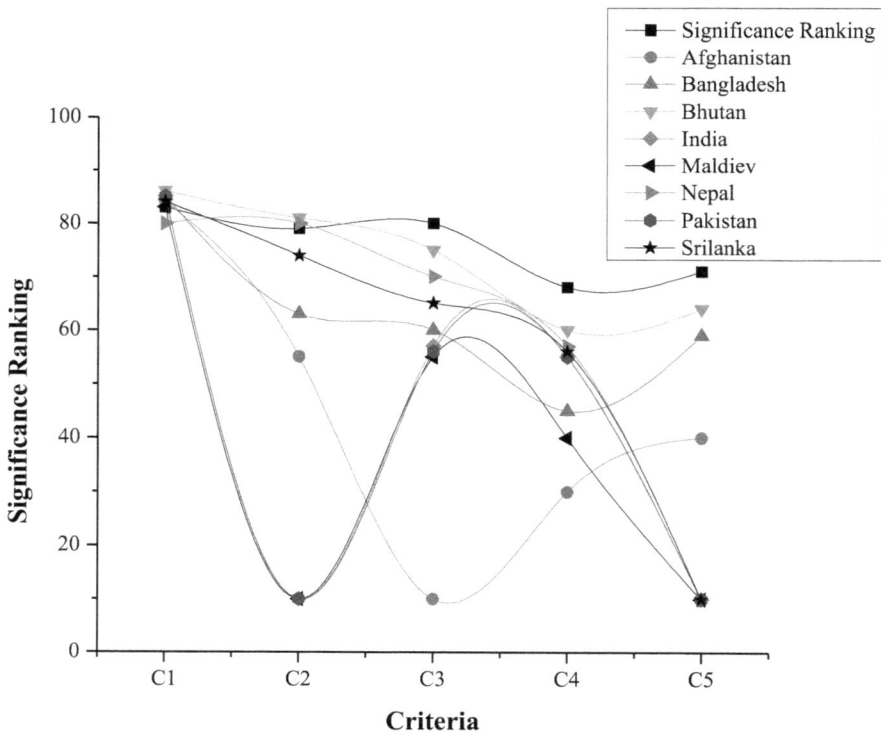

**FIGURE 22.2**  Ranking of technical criteria in energy policies in the region.

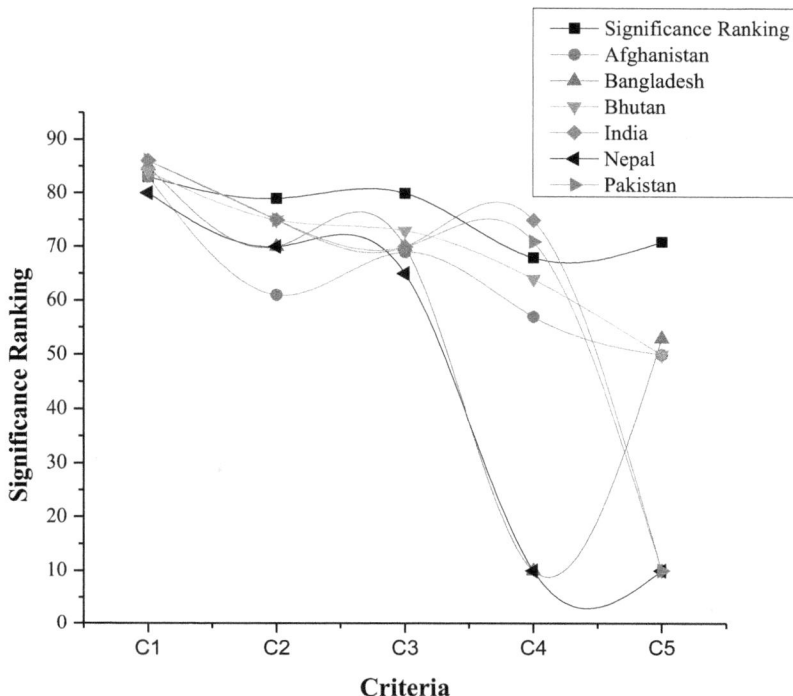

**FIGURE 22.3**   Ranking of technical criteria in ARE policies in the region.

revenue. However, the energy policies of India, Maldives, and Pakistan did not consider the import or export of energy but they apprehend it as a pillar of their foreign policy. The renewable energy policies in the region mainly focused on energy trade in terms of decentralized energy and technology transfer from developed countries to reduce dependence on fossil fuels (Figures 22.2 and 22.3). An integrated energy market can enhance cross-border electricity trade (import or export) from countries abundant with energy resources to countries with less renewable energy resources as well as encourage foreign investment, technology transfer, reduce the reliance on fossil fuels and protect the environment.

It had been noted that all alternative and renewable policies consider reliability (C3) at a national level. However, the reliability of energy systems at a regional level is not mediated in the energy policies and did not comprehend regional energy security. Meanwhile, non-reliable energy systems had adverse impacts on environmental security (Figures 23.2 and 23.3). Securing reliable and environmentally sound supply is among the geopolitical challenges of the current era and seeking strategic bilateral and regional cooperation.

The maturity of technology and energy projects is an effective criterion in decision-making and investment perspective. The experts placed the significance ranking (68) to maturity (C4). The existing ranking of maturity varies in energy policies in the region. Alike other criteria, the concept of maturity is also limited to the national level, but the regional maturity of energy projects is uncommon (Figures 23.2 and 23.3).

A systematic view of the trilemma of safety, security, and sustainable development of energy systems is getting importance due to growing attention to energy and environmental issues. The results revealed the significance ranking (71) to safety (C5), while Afghanistan, Bangladesh, and Bhutan had contemplated the internal safety concerns while energy policies of other regional states are silent on safety concerns, posing divesting impact on sustainable development. Safety is particularly demanding due to the involvement of multiple stakeholders, preferences, intensive investment of

**TABLE 22.3**

**The Pearson Correlation between Significance Ranking and Existing Ranking of Selected Criteria in Regional Energy Policies**

| Criteria | AFG | BGD | BTN | IND | MDV | NPL | PAK | SKL |
|---|---|---|---|---|---|---|---|---|
| Technical | 0.44 | 0.83[c] | 0.96[b] | 0.38 | 0.53 | 0.69 | 0.37 | 0.7 |
| Economic | 0.03 | 0.59 | 0.94[a] | 0.5 | 0.42 | 0.57 | 0.51 | 0.8[b] |
| Environmental | 0.00 | −0.03 | −0.09 | 0.51 | −0.03 | −0.03 | 0.21 | −0.03 |
| Social | 0.46 | 0.6 | 0.16 | 0.62 | 0.5 | 0.9[b] | −0.55 | 0.13 |
| Political | 0.38 | 0.91[b] | 0.15 | 0.91[b] | 0.91[b] | 0.59 | 0.91[c] | 0.63 |

[a] $p < 0.01$.
[b] $p < 0.05$.
[c] $p < 0.1$.

**TABLE 22.4**

**Correlation between Significance Ranking and Existing Ranking of Selected Criteria in the Regional Alternate and Renewable Energy Policies**

| Criteria | AFG | BGD | BTN | IND | NPL | PAK |
|---|---|---|---|---|---|---|
| Technical | 0.83[c] | 0.92[b] | 0.83[c] | 0.5 | 0.98[a] | 0.55 |
| Economic | 0.12 | 0.2 | 0.5 | 0.43 | 0.56 | 0.51 |
| Environmental | −0.03 | −0.03 | −0.09 | 0.48 | −0.03 | 0.12 |
| Social | 0.67 | 0.33 | −0.11 | 0.45 | 0.71 | 0.31 |
| Political | 0.48 | 0.91[b] | 0.43 | 0.76 | 0.61 | 0.75 |

[a] $p < 0.01$.
[b] $p < 0.05$.
[c] $p < 0.1$.

capital, and safety concerns in fragile geopolitical situations of South Asia. The Pearson correlation between significance and existing ranking of technical criteria in energy policies in the region was presented in Table 22.3. Furthermore, Table 22.4 demonstrates the Pearson correlation between significance and the existing ranking of technical criteria in alternative and renewable energy policies.

The experts consider technical aspects ineluctable in energy policies from a regional perspective for energy trade, sustainable development, and environmental security. However, it is corroborated by energy policies analysis that South Asian countries accounted for technical aspects in their energy policies at national but they omit different technical aspects required for regional trade. However, alternative and renewable energy policies had pondered the technical aspects for attracting foreign investment at a local level but still away from the export of alternative and renewable energy. This less assiduity would not only reduce the regional energy trade feasibility but harnessing of energy, without technical aspects that threatened environmental security and sustainability.

## 22.4.2 ECONOMIC ASPECTS

Economic aspects investment cost, operation and maintenance cost, trade and supply cost, economic benefits of trade, payback period, affordability, and tax in regional context had been covered to rationalize regional energy trade and security. Competitive investment refers to the total expenditures required to make energy projects operational. The SAARC Framework Agreement on Energy

Cooperation mandated the planning of cross-border interconnection (Article 7), build, operate, and maintain (Article 8), transmission service agreement (TSA) (Article 9), system operation and settlement (Article 11) being a part of the competitive investment. The experts dispensed significance ranking (80) to competitive investment (C6). The competitive investment (trade cost, connection cost, infrastructure cost, operation, and maintenance cost) for regional energy procurement has been incorporated in only Bhutan Sustainable Hydropower Development Policy 2008 with an existing ranking (77). It is a shallow type of energy policy in which investors or developers will be responsible for all costs. Unfortunately, other South Asian countries did not define competitive investment. However, most of the alternative and renewable policies in the region have contemplated investment costs in case of foreign investment in respective countries. The experts assigned the significance ranking (71) to operation and maintenance costs (C7). Similar to competitive investment, Bhutan Sustainable Hydropower Development Policy 2008, Alternative Renewable Policy 2013 of Bhutan, the alternative and renewable policies of Bangladesh, India, and Pakistan speculates the supply cost (Figures 22.4 and 22.5). The Royal Government of Bhutan has the authority to accede to the supply cost with the investor or the importer. Furthermore, other South Asian states have different trends to acquiesce supply costs in different projects. There are mainly two trends, every partner country bear supply cost within their territory like India, Iran, and Pakistan gas pipeline projects, and the second trend is sponsorship against more supply like India and Bhutan, India and Nepal, Central Asia and South Asia (CASA-1000) electricity transmission between Afghanistan, Pakistan, India, and Central Asia. Therefore, it is an urgent need to define supply costs in the energy policies of the region.

The threat of energy security may result in dependence on imported fossil fuels to meet the energy demand of the growing population which is contributing to climate change. Transmission access (Article 12) of the SAARC Framework Agreement on Energy Cooperation covers cross-border trade and non-discriminatory access to respective transmission grids. Meanwhile, facilitating buying and selling entities (Article 13) edict the member states in cross-border energy trade and create a rational market. The experts believe in addressing the economic benefits of trade in

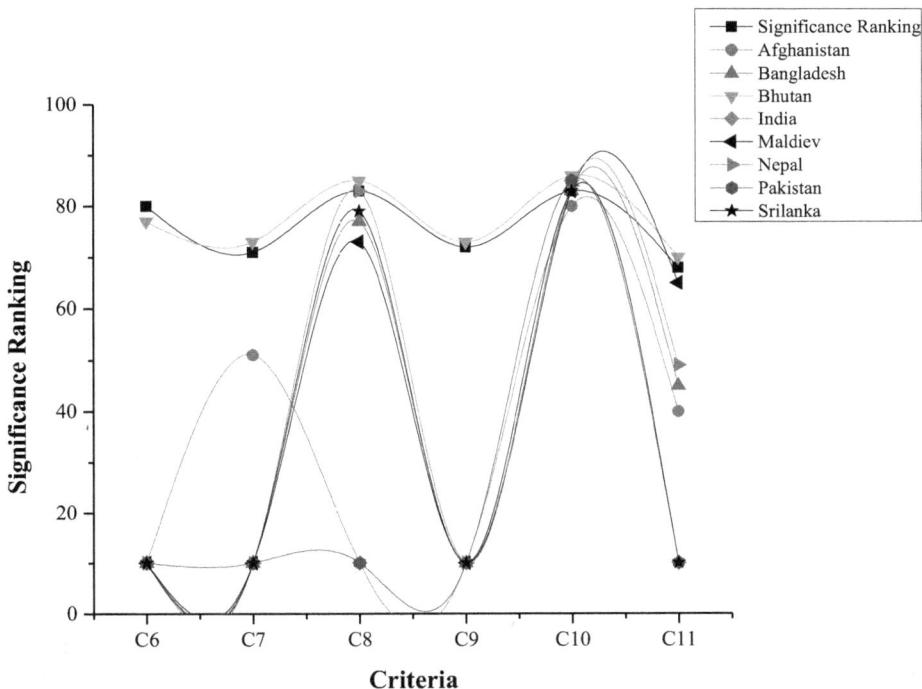

**FIGURE 22.4**   Ranking of economic criteria in energy policies in the region.

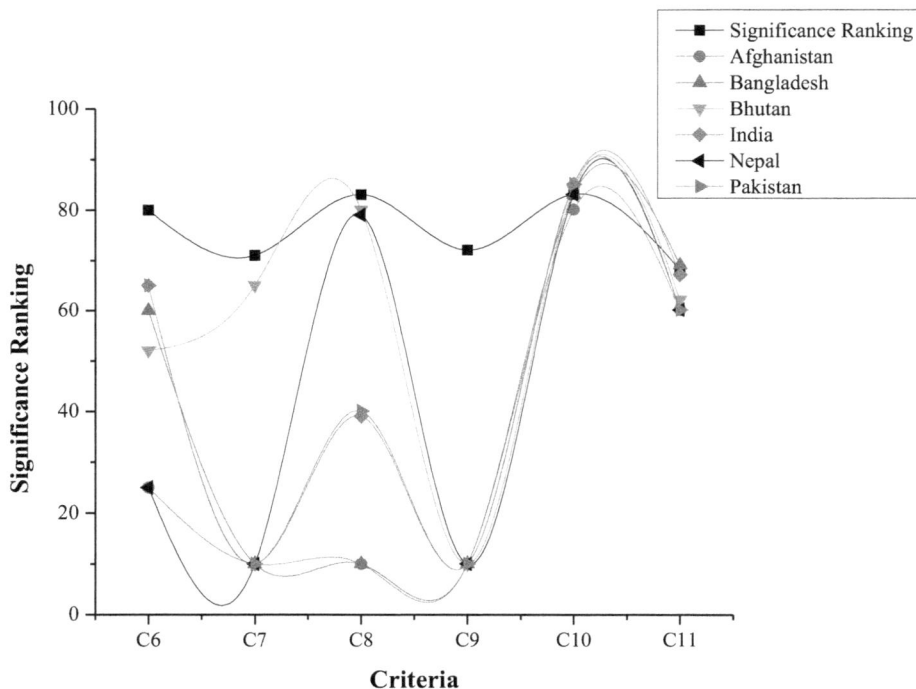

**FIGURE 22.5**    Ranking of economic criteria in ARE policies in the region.

energy (C8) and ascribed significance ranking (83). National Energy Policy 2004 of Bangladesh, Maldives National Energy Policy and Strategy, and National Energy Policy & Strategies of Sri Lanka 2008 have visualized the economic benefits of energy import due to limited indigenous resources (Figures 22.4 and 22.5). On the other hand, Bhutan Sustainable Hydropower Development Policy 2008 and The Hydropower Development Policy of Nepal are reflecting the economic benefits of energy export. Both countries commensurate energy export as a valuable source of revenue due to a limited industry. The energy policies of Afghanistan, India, and Pakistan did not visualize the economic benefits of energy trade. Thus, different countries are gradually connected by this network; these inter-regional flows are helpful in reducing carbon, greenhouse gas emissions, and combating climate change. The growing energy demand, outstripping domestic supply, and supply gap in India, Pakistan, Bangladesh, Afghanistan, and Sri Lanka are rationalizing the energy trade. Meanwhile, Tajikistan, Kyrgyzstan, Nepal, Bhutan, Myanmar, Turkmenistan, and Iran can export energy due to excessive energy, and endowment of resources [36].

The project payback period adduces the time span (number of years) required to return the total investment. The experts believe in considering the payback period from an investment perspective as well as trade and affordability. They accredited the significance ranking (72) to the payback period (C9). It is a notable economic criterion in energy policies. The Royal Government of Bhutan has contemplated the payback period in their policy with an existing ranking of 73. Bhutan Sustainable Hydropower Development Policy 2008 solely vouches for the payback period for rationalizing regional energy trade. However, all other energy policies are silent on the payback period, limiting the trade opportunity and feasibility.

The energy trade may have to deal with two types of taxes including the tax on energy trade and a carbon tax. The tax on energy trade could be a potential source of revenue generation. SAARC Framework Agreement on Energy Cooperation decree duties and taxes (Article 4), the essence of trade duty, fees, and levies exemption on cross-border energy trade. The experts allotted significance ranking (68) to tax (C11). The South Asian countries riveted on tax on trade as a source of revenue but contempt the carbon tax. However, the existing ranking of tax is comparatively better in alternative and renewable energy policies due to tax rebates on the import of ARE technologies (Figures 22.4 and 22.5). There is a need to mediate a clear tax ratio and direction to facilitate regional energy trade. Furthermore, a carbon tax can promote renewable energy, decrease emissions, mitigate climate change, and environmental security. The economic criteria narrated the different values of Pearson's correlation between significance ranking and existing ranking (Tables 22.3 and 22.4).

The economic parameters ought to adeptly address for successful procurement of energy, environmental sustainability, and security. The cost of sustainable energy should not be higher than nonrenewable and fossil fuel energy. The affordable long term is imperative to continue development, industrial operation, and sustainable development. However, there were gaps in economic criteria, poorly or moderately addressed except in Bhutan, rendering contemptible potential to viable energy trade, and environmental security.

## 22.4.3 ENVIRONMENTAL ASPECTS

Environmental criteria are the backbone to evaluate the relationship between energy security, economic growth, and environmental security. This study appended the climate change, water security, Sustainable Development Goals (SDGs), and South Asian Cooperative Environment Programme (SACEP) to adeptly comprehend the relationship of regional energy trade with environmental security and development. The experts confer significance ranking (68) to greenhouse gas emissions (C12) from energy processes and threats to environmental security. They expected that the energy trade can reduce greenhouse gas emissions by revoking unsustainable energy in South Asia. The energy policies in the region fiat the greenhouse gas emissions, partially at the national level except for Afghanistan and Pakistan (Figures 22.6 and 22.7), while, on the other hand, all renewable energy policies in South Asia consider greenhouse gas emissions.

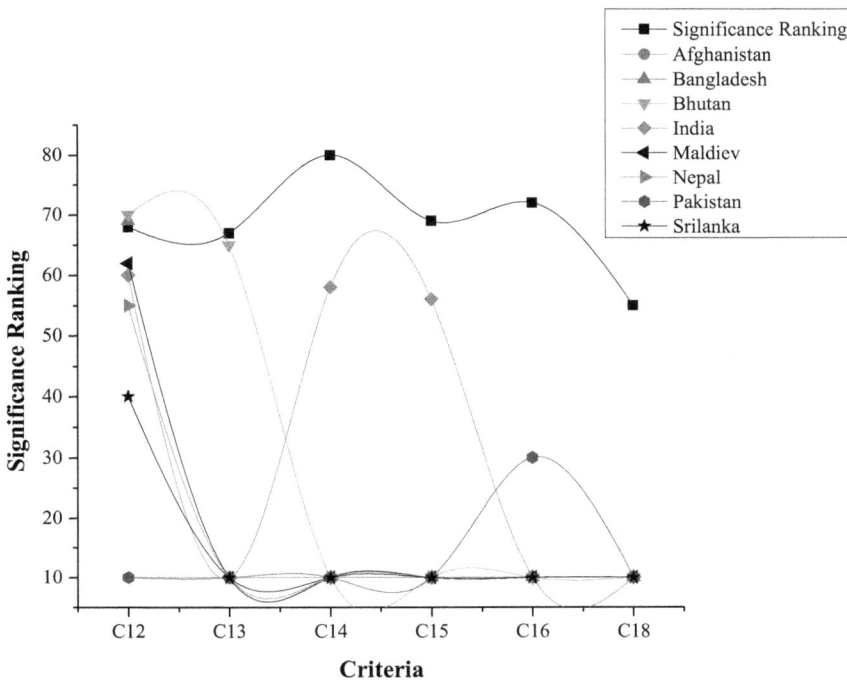

**FIGURE 22.6**  Ranking of environmental criteria in energy policies in the region.

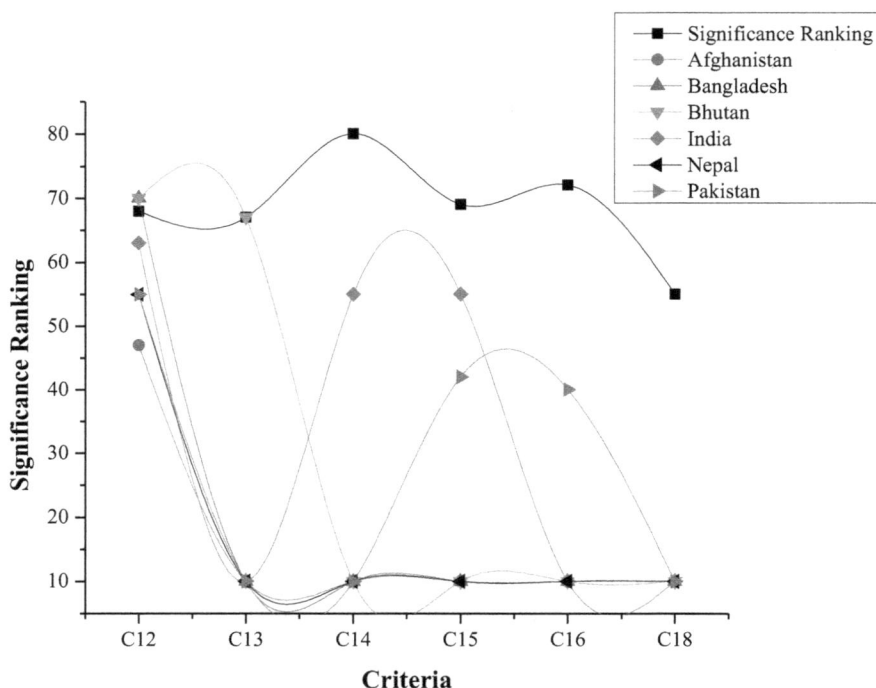

**FIGURE 22.7**  Ranking of environmental criteria in ARE energy policies in the region.

The increasing energy demand and expansion of energy systems require land. The land is required for every stage of energy trade including generation, transmission, terminals, and grids. The experts envisage the significance of land use (C13) for energy trade and rank significance ranking (67). The results anticipated the land use criteria only in Bhutan Sustainable Hydropower Development Policy 2008 and Alternative Renewable Policy 2013 Bhutan (Figures 22.4 and 22.5). While land use criteria were overlooked in all energy policies in the region. The abstraction of these criteria poses an imminent threat to energy trade feasibility and environmental security. It deems likely to misuse land, damage arable land, and stresses the natural ecosystem.

Energy-water-climate change nexus has attained peculiar attention after COP21. Therefore, the energy, water, and climate change integration are substantial for sustainable energy procurement. The experts consign significance ranking to water security (C14) (80) and climate change (C15) (69). The South Asian states did not envision water security and climate change in energy policies except Alternative and Renewable Energy Policy of Pakistan which partially consider climate change. However, the Government of India has formulated an integrated strategy for energy, water security, and climate change (Figures 22.4 and 22.5). The South Asian countries are intended to generate hydropower for energy trade after the natural gas. Water security and climate change are increasing the vulnerability of the energy sector and reducing the feasibility of energy trade. The inapt attention to climate change and water security is a threat to energy security and environmental security and is likely to cause regional conflicts.

The SDGs present a unique approach to global commitment, strengthen governance, integrate sectoral policies, and foster global sustainable development [37]. Goal 7 is energy security, and it stresses ensuring affordable, reliable, sustainable, and modern energy for all. The burgeoning energy demand and industrial growth are important requirements for this goal. The South Asian regions are energy-deficit. The power generation, distribution, and secure supply across borders could be a gateway to achieving SDG 7. The results reveal significance ranking (72) to the SDGs

(C16) for energy trade, sustainable development, and environmental security. It is important to motion that all regional energy policies did not attribute Sustainable Development Goals, perhaps due to the formulation of energy policies before the induction of SDGs (Figures 22.4 and 22.5). The experts highly recommended affordability (C10) as an economic criterion in energy policy. They tiered the significance ranking (83) to affordability. The existing ranking of affordability is meeting or close to the expected ranking in all energy policies. However, all South Asian nations had adopted the SDGs. Pakistan had particularly adopted SDGs as Pakistan Development Goals in policy visions. The scantiness of SDGs in energy policies can derail the global pathway to energy security and sustainable development.

The South Asian countries have reiterated the obligation of strong and intensify regional cooperation through successive summits at the SAARC platform since 1987. The electric power consumption and generation of SAARC countries show unidirectional causality runs to environmental pollution and damage. Therefore, regional environmental security is at risk, and looking for an urgency to mediate the concerns. SAARC Development Goals and the provisions of SACEP are attempts to protect, manage, and preserve diverse and fragile regional ecosystems. They also address the threat induced by climate change and natural disasters. Although SAARC Development Goals (C17) did not consider energy security, they are now invalid and replaced by SDGs. The experts ascribed the significance ranking (55) to South Asia Cooperative Environment Programme (C18). The results indicated that provisions of the South Asia Cooperative Environment Programme are not part of all energy policies in the region (Figures 22.4 and 22.5). The slight attention to provisions of the South Asia Cooperative Environment Programme is likely to foster the damaged environment and ecosystem. The results deposed the diverse Pearson's correlation between significance ranking and existing ranking of environmental criteria in regional energy policies and renewable energy policies (Tables 22.3 and 22.4).

The diminutive incorporation of environmental criteria in regional energy policies presents apprehension to environmental security and sustainable development. It would not only stress natural resources, leading to conflicts and reducing the capacity to settle disputes, but also threaten national security. Furthermore, it is affecting the basic requirement for sustainable development, environment-friendly technologies, compliance with emission limits, waste management, and long-term public satisfaction. Meanwhile, it is reducing the capacity to adhere to global agreements like the Paris Agreement and Clean Development Mechanism as well as reducing the feasibility of energy trade is a technical barrier to trade.

### 22.4.4 SOCIAL ASPECTS

Social criteria rationalize the successful regional procurement of energy through the involvement of social actors in visions, foresight, institutionalization, and development of energy material and flow. Access to affordable energy is a basic human right being a fundamental driver of human needs, development, and sustainability. The experts rated commendable significance ranking (81) to access to energy as a basic human right (C19). The experts believe that regional energy trade can stimulate to meet access to energy as a basic human right. Overall, all South Asian countries are committed to providing accessible, affordable, and sustainable energy for all.

Social acceptance and benefits are undeniable aspects of regional energy trade due to several geopolitical disputes in South Asia. The experts visualize the worth of social acceptance (C20) with a significance ranking of 74. The energy policies of Bangladesh, Bhutan, and India were intended to least focus on social acceptance. Social acceptance usually relies on the public and reiterates historical narrative, practices, and socio-political context. The embodiment of social benefits can foster the cross-border procurement of energy due to comprehensive stakeholder involvement. The experts reveal significance ranking (72) to social benefits (C22). The comprehension of social benefits was observed to be less significant in Bangladesh, India, and Pakistan energy policies (Figures 22.8 and 22.9). Moreover, energy policies in the region grasp the social benefits in the national context only,

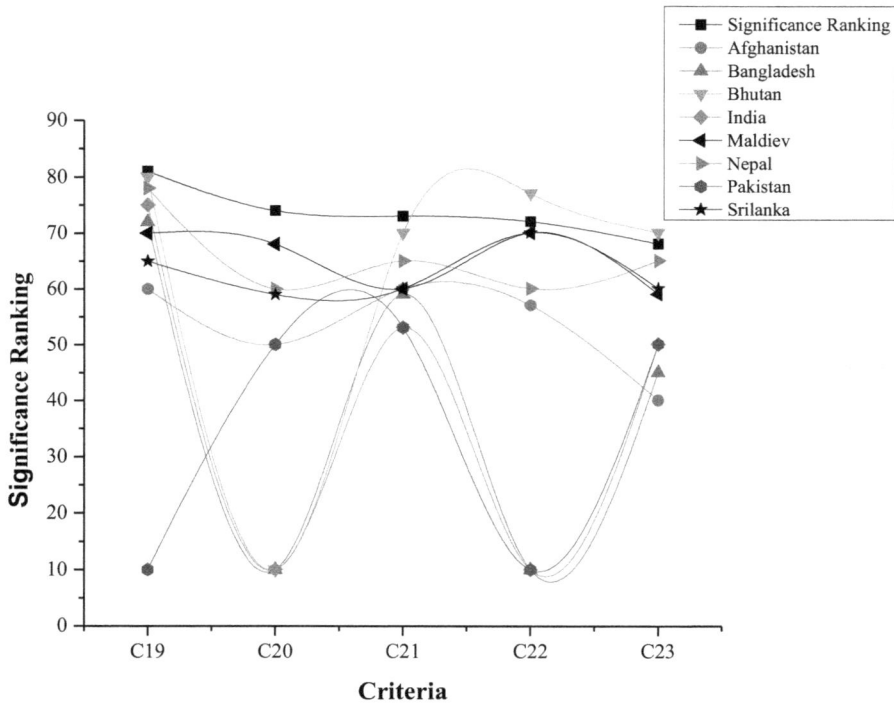

**FIGURE 22.8** Ranking of social criteria in energy policies in the region.

but the notion of social benefit at the regional level can explicit the energy trade, settle disputes, and environmental protection. Social acceptance highlights the responses and representation of people toward the energy projects for the deployment of energy, ranging from construction of infrastructure, and generation to distribution. Unfortunately, South Asia had numerous disputes due to historical narratives and socio-political context which are threatening energy trade feasibility and environmental security. Batel and Devine-Wright [38] recommend reproducing the way of thinking and eradicating social inequity to create larger and interconnected energy systems for social and environmental sustainability at local and regional levels.

Job creation is among the main energy policy goals. A significant number of new employment opportunities from the deployment of projects in the energy sector represent the success of energy policies. The SAARC energy grid and SAME had a significant potential to create new employment opportunities, particularly in the renewable energy sector. The experts allotted the significance ranking (73) to job creation (C21). It is important to mention that all energy policies in the region had contemplated the employment opportunities at a national level (Figures 22.8 and 22.9). There is a need to assess the employment potential in the SAARC energy grid and SAME. It would be helpful to enhance social acceptance and cooperation.

The energy, economic growth, and environmental concerns nexus are complicating the paradigm of development in both developing and developed countries. Articles 5 and 14 of the SAARC Framework Agreement on Energy Cooperation emphasized cooperation through data updating, sharing of information, and knowledge as well as joint research. The experts believe that regional cooperation is imperative for regional energy trade, development, and environmental security. They dispensed the significance ranking (72) to cooperation (C23). The regional energy integrated the cooperation expect National Energy Policy 2004 of Bangladesh, National Energy Policy of India 2005, and National Power Policy 2013 of Pakistan. These countries consolidated the cooperation as part of their foreign policy. While all alternative and renewable energy policies in the region soak

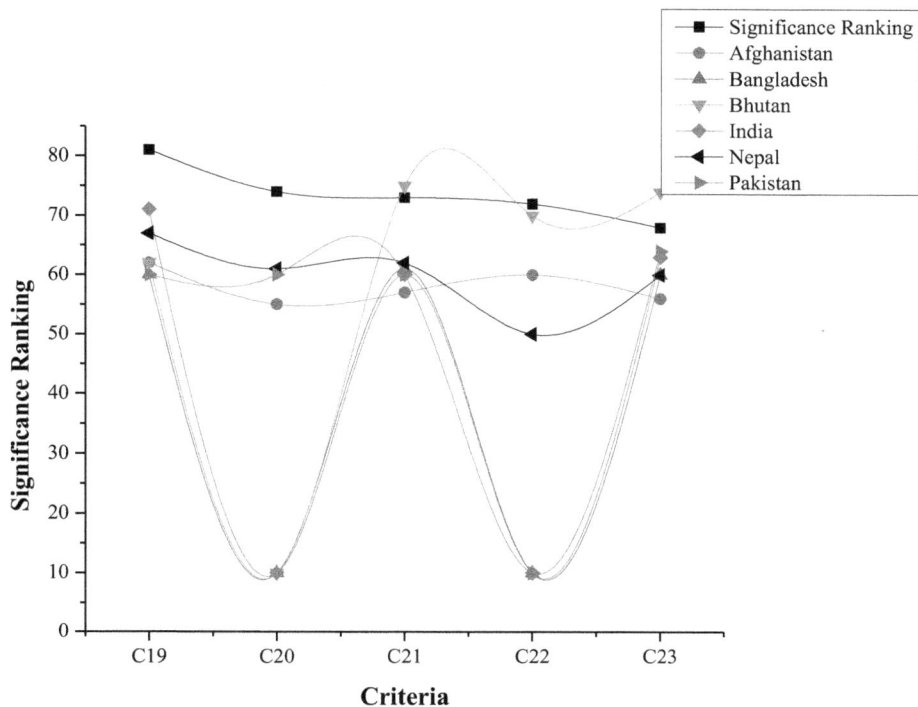

**FIGURE 22.9** Ranking of social criteria in ARE policies in the region.

up the local, regional, and global cooperation for the successful deployment of energy (Figures 22.8 and 22.9). The energy policies of South Asian states have a comparatively better capacity for social aspects than other criteria. The results represented different values of Pearson's correlation between significance ranking and existing ranking of social criteria (Tables 22.3 and 22.4). The cooperation between entities (local and regional) may play a significant role in the deployment of energy projects. Meanwhile, it will improve the institutional capacities of governments, technological assistance, exchange finance, and economic development. These factors have the ability to a substantial reduction in emissions which eventually protects the environment.

### 22.4.5 POLITICAL ASPECTS

The political criteria for regional energy trade include political will, identification of constraints, rights of ownership, secure supply, negotiation, and diplomacy. The strong political will and commitment at national and regional levels determine the pathways of the energy transition. The experts underpin political will, support, and commitment as a substantial criterion for energy trade, development, and environmental security. They imparted the highest significance ranking (92) to political will (C24). It is notable to adduce that all South Asian countries are politically committed to energy trade, development, and environmental security. Despite diverse challenges, the South Asian states are demonstrating their commitment to either energy policies or part of foreign policy and eventually signed different energy projects.

The identification of constraints increases the energy trade feasibility and implication of multilateral environmental agreements. Article 15 of the SAARC Framework Agreement on Energy Cooperation mandated to structure and function of institutional mechanisms for regulatory issues related to energy trade. The experts attributed the significance ranking (78) to identification constraints (C25). However,

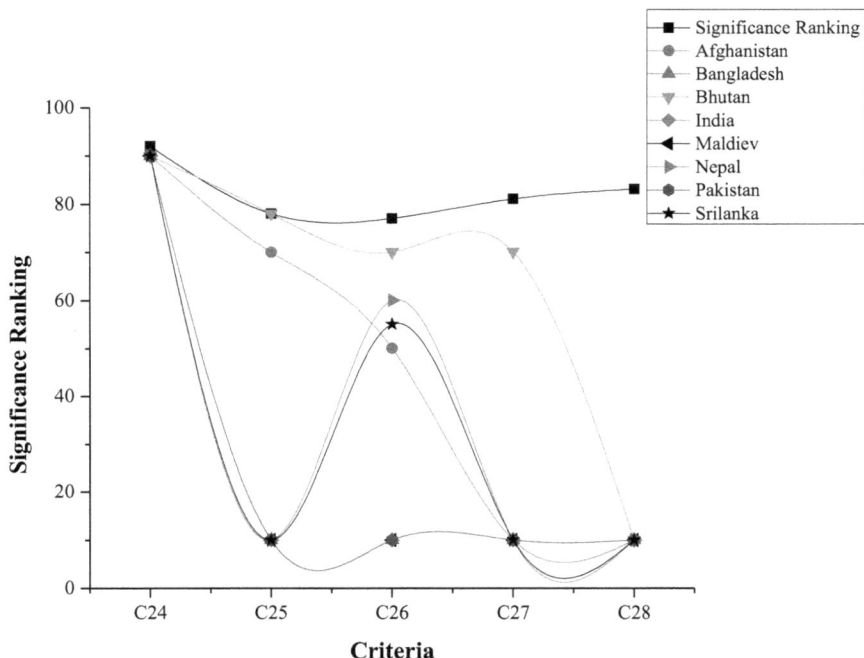

**FIGURE 22.10**  Ranking of political criteria in energy policies in the region.

Afghanistan Energy Sector Strategy 2008, Bhutan Sustainable Hydropower Development Policy 2008, Afghanistan Rural Renewable Energy Policy, and Bhutan Alternative Renewable Policy 2013 had unfolded the constraints in energy generation, and procurement to or from neighboring countries (Figures 22.10 and 22.11). While all other energy policies in the region were not considering the identification of constraints, the energy trade, and environmental security. The sustainable procurement of energy and environmental security is not possible without the identification of constraints in synergy to lower social acceptance, poor economic situation, low technical capacity, and environmental challenges.

Article 16 of the SAARC Framework Agreement on Energy Cooperation behests member states to negotiate on the disputes arising from the interpretation and implementation of this agreement. If the member states are unable to settle the dispute, they can refer it to SAARC Arbitration Council. The experts accredited the significance ranking (77) to negotiation and diplomacy (C26). While all regional states have presented the negotiation and diplomacy for regional energy trade. South Asia is known for various complex disputes (Figures 22.10 and 22.11). The assessment of social, political, historical consideration, geopolitical narratives, and the role of the intergovernmental, international, and transitional regulatory regime is required for being agreed upon complex reforms to harness regional energy trade, climate change mitigation, development, and environmental security. Therefore, negotiation and diplomacy are imperative for balancing the energy, environment, and development nexus.

The right of ownership is among the most controversial issues in the case of regional energy transitions [39]. The experts assigned significance ranking (81) to right of ownership (C27). The Royal Government of Bhutan only anticipated the right of the owner in their energy policies (Bhutan Sustainable Hydropower Development Policy 2008 and Bhutan Alternative Renewable Policy 2013). While other energy policies in the region had no criteria for the right of ownership, which limited the trade feasibility (Figures 22.10 and 22.11), the decisions of states can immediately impinge and potentially affect the entire region in the globalized world. It can also affect the threat to energy use, water use, physical security, lifestyle, and environmental security. The right of ownership is considered to be critical for the successful procurement of energy.

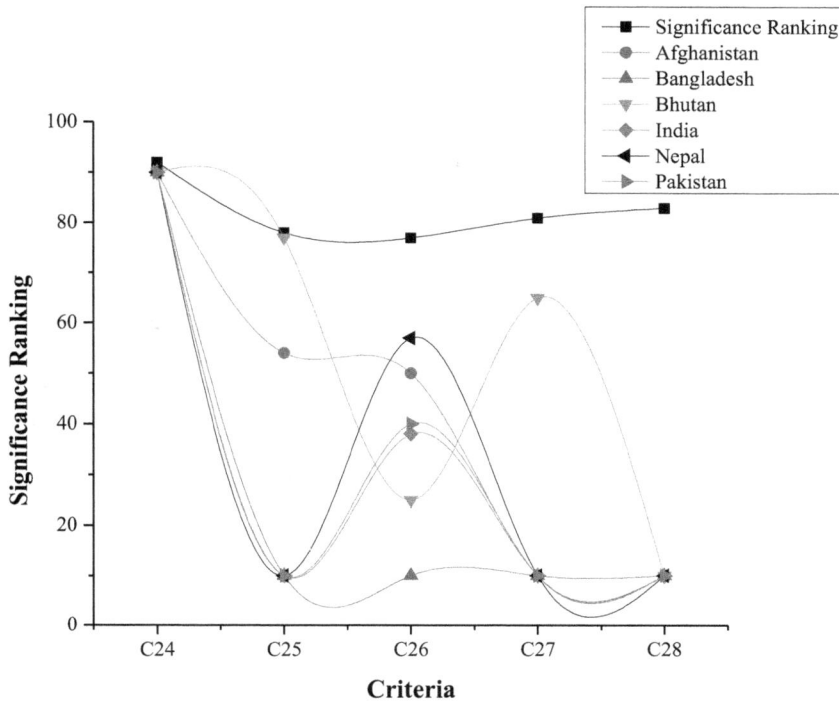

**FIGURE 22.11**   Ranking of political criteria in ARE policies in the region.

The regional energy security and trade depend on a secure energy corridor from resource-rich states to resource-deficit states [40]. However, Article 10 of the SAARC Framework Agreement on Energy Cooperation enacts the protection of grid systems and supply. This article directed the member states to develop and coordinate joint protection systems, and secure cross-border inter-connection, and grid. The experts confer a high significance ranking (83) to secure supply (C28) but unfortunately, all energy policies in the region did not define the measures for safe and secure supply (Figures 22.10 and 22.11). The secure supply is a consequential concern for regional trade in South Asia due to geopolitical tensions between India and Pakistan as well as Afghanistan and Pakistan from both India and Pakistan's perspectives. The scantiness of this criterion not only reduces trade feasibility but also abates social acceptance, economic viability, and environmental security. The results exhibited a positive Pearson's correlation between significance ranking and existing ranking of political criteria in most of the regional energy policies (Tables 22.3 and 22.4).

Political criteria shape the preferences of countries, policy choices, production choices, and relations with various interest groups as well as interlink the primary energy supply, climate change, economic, and environmental benefits. Furthermore, political aspects are interconnected with geography, natural resource endowment, economic wealth, industry, and global energy prices. The South Asian region has unique and complex political circumstances. Meanwhile, staid political hiatus in regional energy policies have endured threats to energy trade, development, regional energy, and environmental security.

## 22.5   ENVIRONMENTAL DIPLOMACY AND DISPUTE SETTLEMENT

Environmental diplomacy can play an effective role in transforming energy security, regional energy trade, development, and environmental security nexus. It can enable states to cope with environmental challenges, sovereignty concerns, and disputes arising from regional energy trade as well as reap the benefits of cooperation, fostering economic development, and energy security

[6]. The results highlight many gaps in technical, economic, environmental, social, and political aspects of regional energy policies. These are reducing the energy trade feasibility, hindering the implementation of the SAARC Framework Agreement on Energy Cooperation, arising disputes which likely to complicate the complicated geopolitical circumstances, and reducing the opportunity to secure energy. There is a lack of regional cooperation due to geopolitical tension which has squashed the capacity of political forums to settle the disputes and barriers in regional energy trade. Therefore, circumstances require a negotiating tool, non-aggressive in nature, emphasizing sustainable energy as a basic human right as well as a commodity for development beyond political interests. Environmental diplomacy through the UN Environment Programme (UNEP), International Union for Conservation of Nature (IUCN), South Asian Network for Development and Environmental Economics, International Centre for Integrated Mountain Development, the World Bank, and the Asian Development Bank in collaboration with political forums SAARC, the Economic Cooperation Organization (ECO), Association of South East Nations (ASEAN), and Shanghai Cooperation Organization could produce effective results, settle disputes, and long-term cooperation. Then, regional would lead toward the utilization of SAARC energy rings through SAARC energy grid, development, energy, and environmental security like Nordic Power Pool (NORDPOOL) and Greater Mekong Sub-region (GMS).

## 22.6   CONCLUSIONS

The South Asian countries are facing challenges of energy and environmental security, which are hampering development. The energy policies in the region have exalted gaps in technical, economic, environmental, social, and political criteria. The results reveal that Bhutan Sustainable Hydropower Development Policy, 2008 tames most of the selected criteria, and looks more technical, suitable, and compatible for regional energy trade. However, consideration of selected criteria varies in other energy policies in the region. They have modest adequacy for regional trade and compatibility with environmental security and sustainability. The energy policies are omitting cornerstone criteria for successful regional energy transitions, development, and environmental protection including safety, competitive investment, operation, and maintenance cost, interconnection cost, taxes, integration with water security and climate change, SDGs, land use, SACEP, social acceptance, political will, constraints, diplomacy, right of ownership, and supply security. Overall, energy policies in South Asia have a demure capacity for regional procurement of energy and sustainability in the social-geopolitical context of South Asia. In sum, energy policies in the region, regulatory mechanisms, cooperation, and political interest postulate to rethink and review energy security, and environmental sustainability.

### 22.6.1   POLICY CONSIDERATIONS

The promotion of regional energy trade will create significant opportunities for economic growth and regional cooperation. This will also reduce dependence on fossil fuels, reduce emissions, and combat climate change. The following consideration in policies can inherent optimal development of natural resources, energy, and environmental security in the region.

- Formulate and implement the regional energy policy like European Energy Strategy.
- Since complex geopolitical circumstances may demote the success of regional energy policy. A review of the SAARC Framework Agreement on Energy Cooperation and the development of regional energy guidelines will be convenient for member states. The member states have an opportunity to develop guidelines by the following mechanisms of South Africa Energy Pool, East Africa Energy Pool, NORDPOOL, and GMS.
- Review the regulatory mechanisms and harmonize policies' regional requirements for energy trade and SDGs.

- Develop a mutual administrative mechanism for duties, taxes, land use, and secure supply.
- Develop a competitive market for regional energy trade.
- Change the traditional political mindset and rethink historical mistrust.
- Integrate energy policies with water security and climate change.
- Identify requirements for dispute settlement, initiate negotiation, and diplomacy.
- Environmental diplomacy through global institutions (UNEP, IUCN, World Bank) might be useful in the geopolitical context of South Asia.

## ACKNOWLEDGMENTS

This chapter is based on research carried out by the first author for his Ph.D. dissertation, entitled "A Study on the Relationship of Development with Environmental Security in a National and Regional Context: The Case of Environmental Diplomacy for Pakistan," submitted to International Islamic University, Islamabad.

## REFERENCES

[1] Radovanović, M., Filipović, S., & Pavlović, D. (2017). Energy security measurement – A sustainable approach. *Renewable and Sustainable Energy Reviews, 68,* 1020–1032.

[2] Winzer, C. (2012). Conceptualizing energy security. *Energy Policy, 46,* 36–48.

[3] Hassan, M., Afridi, M. K., & Khan, M. I. (2018). An overview of alternative and renewable energy governance, barriers, and opportunities in Pakistan. *Energy & Environment, 29*(2), 184–203.

[4] SAARC Energy Centre. (2010). *Study for a Development of Potential Regional Hydropower Plant in South Asia.* The South Asian Association for Regional Cooperation Energy Centre, Islamabad, Pakistan.

[5] SAARC Secretariat. (2010). *SAARC Regional Energy Trade Study (SRETS).* The South Asian Association for Regional Cooperation Secretariat, Katmandu, Nepal.

[6] Ali, S.H., Zia, A. (2017). Transboundary Data Sharing and Resilience Scenarios: Harnessing the Role of Regional Organizations for Environmental Security. In: Adeel, Z., Wirsing, R. (eds) Imagining Industan. Water Security in a New World. Springer, Cham.

[7] Hassan, M., Khan Afridi, M., & Irfan Khan, M. (2019). Energy policies and environmental security: A multi-criteria analysis of energy policies of Pakistan. *International Journal of Green Energy, 16*(7), 510–519.

[8] Tripathi, D. (2012). Energy security: The functional area of regional cooperation for South Asia. *Eurasia Border Review, 3*(2), 91–102.

[9] Su, M., Zhang, M., Lu, W., Chang, X., Chen, B., Liu, G., Hao, Y & Zhang, Y. (2017). ENA-based evaluation of energy supply security: Comparison between the Chinese crude oil and natural gas supply systems. *Renewable and Sustainable Energy Reviews, 72,* 888–899.

[10] Ahmed, T., Mekhilef, S., Shah, R., Mithulananthan, N., Seyedmahmoudian, M., & Horan, B. (2017). ASEAN power grid: A secure transmission infrastructure for clean and sustainable energy for South-East Asia. *Renewable and Sustainable Energy Reviews, 67,* 1420–1435.

[11] Büyüközkan, G., & Karabulut, Y. (2017). Energy project performance evaluation with sustainability perspective. *Energy, 119,* 549–560.

[12] Amer, M., & Daim, T. U. (2011). Selection of renewable energy technologies for a developing country: A case of Pakistan. *Energy for Sustainable Development, 15*(4), 420–435.

[13] Büyüközkan, G., & Güleryüz, S. (2017). Evaluation of renewable energy resources in Turkey using an integrated MCDM approach with linguistic interval fuzzy preference relations. *Energy, 123,* 149–163.

[14] Şengül, Ü., Eren, M., Shiraz, S. E., Gezder, V., & Şengül, A. B. (2015). Fuzzy TOPSIS method for ranking renewable energy supply systems in Turkey. *Renewable Energy, 75,* 617–625.

[15] Mahmood, A., Javaid, N., Zafar, A., Riaz, R. A., Ahmed, S., & Razzaq, S. (2014). Pakistan's overall energy potential assessment, comparison of LNG, TAPI and IPI gas projects. *Renewable and Sustainable Energy Reviews, 31,* 182–193.

[16] Behera, S. R., & Dash, D. P. (2017). The effect of urbanization, energy consumption, and foreign direct investment on the carbon dioxide emission in the SSEA (South and Southeast Asian) region. *Renewable and Sustainable Energy Reviews, 70,* 96–106.

[17] Fonseca, B., & Rosen, J. D. (2017). Energy and Environmental Security. In: *The New US Security Agenda* (pp. 135–159). Springer International Publishing.

[18] Vesco, P., Dasgupta, S., De Cian, E., & Carraro, C. (2020). Natural resources and conflict: A meta-analysis of the empirical literature. *Ecological Economics, 172*, 106633.

[19] Wang, T. R., Mousseau, V., Pedroni, N., & Zio, E. (2017). An empirical classification-based framework for the safety criticality assessment of energy production systems, in presence of inconsistent data. *Reliability Engineering & System Safety, 157*, 139–151.

[20] Akhmat, G., Zaman, K., Shukui, T., Irfan, D., & Khan, M. M. (2014). Does energy consumption contribute to environmental pollutants? Evidence from SAARC countries. *Environmental Science and Pollution Research, 21*(9), 5940–5951.

[21] Jin, S. W., Li, Y. P., Yu, L., Suo, C., & Zhang, K. (2020). Multidivisional planning model for energy, water and environment considering synergies, trade-offs and uncertainty. *Journal of Cleaner Production, 259*, 121070.

[22] Teotónio, C., Rodríguez, M., Roebeling, P., & Fortes, P. (2020). Water competition through the 'water-energy' nexus: Assessing the economic impacts of climate change in a Mediterranean context. *Energy Economics, 85*, 104539.

[23] Chatterjee, R., Mehra, M., & Banerjee, S. (2000). *Environmental Security in South Asia.* Indian Energy Sector, the Energy Research Institute, New Delhi, India.

[24] Sovacool, B. K. (2017). Contestation, contingency, and justice in the Nordic low-carbon energy transition. *Energy Policy, 102*, 569–582.

[25] Gregg, J.S., Nyborg, S., Hansen, M., Schwanitz, V.J., Wierling, A., Zeiss, J.P., Delvaux, S., Saenz, V., Polo-Alvarez, L., Candelise, C. and Gilcrease, W. (2020). Collective action and social innovation in the energy sector: A mobilization model perspective. *Energies, 13*(3), 651.

[26] Dvořák, P., Martinát, S., Van der Horst, D., Frantál, B., & Turečková, K. (2017). Renewable energy investment and job creation: A cross-sectoral assessment for the Czech Republic with reference to EU benchmarks. *Renewable and Sustainable Energy Reviews, 69*, 360–368.

[27] Mikkola, M., Jussila, A., & Ryynänen, T. (2015, November). Collaboration in regional energy-efficiency development. In *International conference on Smart and Sustainable Planning for Cities and Regions* (pp. 55–66). Springer, Cham.

[28] Khan, Z., Linares, P., & García-González, J. (2017). Integrating water and energy models for policy-driven applications. A review of contemporary work and recommendations for future developments. *Renewable and Sustainable Energy Reviews, 67*, 1123–1138.

[29] Binder, C. R., Mühlemeier, S., & Wyss, R. (2017). An indicator-based approach for analyzing the resilience of transitions for energy regions. Part I: Theoretical and conceptual considerations. *Energies, 10*(1), 36.

[30] Parkinson, S. C., Makowski, M., Krey, V., Sedraoui, K., Almasoud, A. H., & Djilali, N. (2017). A multi-criteria model analysis framework for assessing integrated water-energy system transformation pathways. *Applied Energy, 210*, 477–486

[31] Seager, T., Gisladottir, V., Mancillas, J., Roege, P., & Linkov, I. (2017). Inspiration to operation: Securing net benefits vs. zero outcome. *Journal of Cleaner Production, 148*, 422–426.

[32] Volkart, K., Weidmann, N., Bauer, C., & Hirschberg, S. (2017). Multi-criteria decision analysis of energy system transformation pathways: A case study for Switzerland. *Energy Policy, 106*, 155–168.

[33] IAEA. (2005). *Energy Indicators for Sustainable Development: Guidelines and Methodologies.* International Atomic Energy Agency, Vienna, Austria.

[34] Edwards, W. (1977). How to use multiattribute utility measurement for social decision making. *IEEE Transactions on Systems, Man, and Cybernetics 7*(5), 326–340.

[35] Wang, J. J., Jing, Y. Y., Zhang, C. F., & Zhao, J. H. (2009). Review on multi-criteria decision analysis aid in sustainable energy decision-making. *Renewable and Sustainable Energy Reviews 13*(9), 2263–2278.

[36] The World Bank. (2008). *Potential and Prospects for Regional Energy Trade in the South Asia Region.* The International Bank for Reconstruction and Development/the World Bank Group, Washington, DC.

[37] Biermann, F., Kanie, N., & Kim, R. E. (2017). Global governance by goal-setting: The novel approach of the UN Sustainable Development Goals. *Current Opinion in Environmental Sustainability, 26*, 26–31.

[38] Batel, S., & Devine-Wright, P. (2017). Energy colonialism and the role of the global in local responses to new energy infrastructures in the UK: A critical and exploratory empirical analysis. *Antipode, 49*(1), 3–22.

[39] Ipek, P. (2017). Oil and intra-state conflict in Iraq and Syria: Sub-state actors and challenges for Turkey's energy security. *Middle Eastern Studies, 53*(3), 406–419.

[40] Hassan, M., Afridi, M. K., & Khan, M. I. (2017). Environmental diplomacy in South Asia: Considering the environmental security, conflict and development nexus. *Geoforum, 82*, 127–130.

# 23 Electricity Sector Reforms, Private Sector Participation and Electricity Sector Performance in Sub-Saharan Africa

*Chigozie Nweke-Eze*
Integrated Africa Power (IAP)
University of Bonn

*Chinedu Miracle Nevo*
The Open University Business School

*Ewere Evelyn Anyokwu*
University of Bonn

## CONTENTS

DOI: 10.1201/9781003315353-26

## 23.1  INTRODUCTION

Before the early 1990s, the public sector financed and managed a significant part of all Sub-Saharan African (SSA) countries' electricity sector, and their power systems were vertically integrated. However, from the early 1990s, electricity sector reforms and policies began to sweep over most of the countries in the SSA. The reform included the corporatization and commercialization of state-owned utilities; adoption of new electricity sector legislation which approves the liberalization of the utilities and provides for the establishment of independent regulatory bodies; unbundling of utilities; introduction of competition through private sector participation (PSP); and provision of incentives for investment (Newbery, 2002; Pollitt, 2012; Jamasb et al., 2015; Maulidia et al., 2019).

Several studies argue that these reforms would be essential in bringing the electricity sector of these countries out of its under-financed and inefficient state (Bhattacharya and Kojima, 2012; Pollitt, 2012; Maulidia et al., 2019; Dertinger and Hirth, 2020). Specifically, it is envisaged to improve the technical and financial performance of the industry; improve access to electricity; and make up for the government's inability to mobilize sufficient capital investment for electricity sector development and expansion in the implemented countries (Zhang et al., 2008; Sen and Jamasb, 2012; Kessides, 2012; Erdogdu, 2013a).

By the late 2000s, virtually all SSA countries had implemented at least one form of the electricity sector reforms, most commonly private sector participation (PSP). The PSP adopted by the individual countries was in the form of management or lease contracts, concession contracts, independent power projects, and divestitures (World Bank, 2021). As of 2021, 586 electricity sector projects with PSP reached financial closure with a total investment of more than 93.6 billion United States Dollars (USD) in the studied 33 SSA countries between 1990 and 2021 (World Bank, 2021).

Despite this widespread implementation of PSP in the region, there has been little or no improvement in the performance of the power sector. The privatization reform process has been complex, inefficient, and slow in the SSA countries where reforms have begun (Prasad, 2008; 2011; Vagliasindi et al., 2010; Nepal and Jamasb, 2012; Eberhard et al., 2011). State-owned utilities still retain most ownership and management of unbundled electricity utilities, limiting most private sector participation to mainly the generation sector—where the private investors are Independent Power Producers (IPPs) (Gratwick and Eberhard, 2008; Eberhard et al., 2011). Many private management contracts have been awarded, yet there is no visible effect on the performance of the electricity sectors (Gboney, 2009; Eberhard et al., 2011). Only a few out of the many issued private leases and concessions have survived (Nellis, 2008; Eberhard et al., 2011). These trends question the effectiveness of the currently implemented liberalized model of electricity sector reform in improving the performance of the electricity sector in SSA.

However, there has being little efforts at empirically assessing the contributions or otherwise of the current progress of reforms in the SSA electricity sector. Earlier works that have attempted to explain the effects of the sector reforms have at best being based on case studies (Oseni, 2011; Fritsch and Poudineh, 2015; Peng and Poudineh, 2017) with limited external validation. Other works have utilized basic descriptive statistics based on scanty data (Imam et al., 2018), while a bulk of the literature on electricity sector reforms in SSA were based on conceptual analysis and opinion reports (Joskow, 1998, 2008; Plane, 1999; Briceno-Garmendia et al., 2004). Although these studies already provides a peak view into the state of electricity sector reforms and the potential effects, they have so far made limited contribution to the electricity sector reforms in terms of evidence-based recommendations that could be universally considered.

This chapter thus bridges the existing gap in the literature by exploring and analyzing the impacts of electricity sector reform in Sub-Saharan Africa. It specifically examines the impacts of private sector participation (as an important and most universally implemented tenet of the electricity sector reforms in SSA) on electricity sector performance in SSA. This chapter adds to the SSA electricity reform literature on three grounds. First, it is the first chapter known to the authors that explores the impacts of electricity sector reforms in the SSA using robust empirical models.

This gives the advantage that the study results can be revalidated given changing circumstances in the region. Second, unlike earlier studies that take the case study approach, the current research covers a significant number of countries in the SSA region. By implication, the study results provide a valid estimate of the effects of PSP in the African continent, given the different contexts in which countries exist. Lastly, the study contributes to the growing body of studies assessing the interactions between sectoral reforms and sectoral outcomes, which is scarce on the African continent.

The following section reviews empirical studies on the impact of electricity sector reform on SSA. Section 23.3 presents and explains the model and methodology of analyses for measuring the effects of PSP on electricity sector performance in SSA. Sections 23.4 and 23.5 present the results and discuss the findings of the empirical analyses. Section 23.6 concludes by discussing the policy implications and recommendations.

## 23.2 LITERATURE REVIEW

This section reviews empirical studies on effects of electricity sector reform and the impacts of private sector participation, particularly on the electricity sector performance and development in SSA. They are discussed under the sub-sections of productivity and economic efficiency; quality of service and system efficiency; electricity pricing; and electricity access.

### 23.2.1 PRODUCTIVITY AND ECONOMIC EFFICIENCY

The extensive privatization efforts of the electricity sector of SSA countries were driven by the need to improve productivity and economic efficiency in electricity production and delivery (Jamasb et al., 2015). Impacts of the reforms on productivity and efficiency have, however, been sparsely studied in SSA. Most studies have been on other regions or have been restricted on a global scale. There is, however, evidence that countries with privatized electricity sectors have better-performed utilities than those without privatization (Megginson and Netter, 2001). These evidence manifests in the increased labor productivity, and technical scale efficiency as found in the research carried out by Politt (2008), Fischer et al. (2003), Cullman and von Hirschhausen (2008), and Celen (2013).

However, the extent of performance improvement differs across countries and economies (Ramos-Real et al., 2009; Galan and Pollitt, 2014). In a non-parametric study of 27 least developing countries, Yuno and Hawdon (1997) found that competition is an important prerequisite for privatization to lead to higher efficiency and productivity. They write that simple exchange of ownership and management of assets does not lead to improved productivity and efficiency in itself (Yuno and Hawdon, 1997). Competition in privatized utilities in SSA has been proven to be often hampered because of undue government interference in the contract awarding process and undue influences on regulatory agencies (Politt, 1995). Yuno and Hawdon (1997) also noted significant differences in productivity and efficiency of small and large-scale providers, with higher productivity found in large-scale investments. This implies that the mostly smaller-scale investments in SSA electricity utilities also mean relatively lower average productivity and efficiency. ESMAP (2011) identifies the existence of independent regulators as another vital prerequisite.

Plane (1999), in his parametric analysis of the impact of the privatization of the electricity sector in Cote d'Ivoire, reported a significant increase in utility productivity and performance after implementing privatization. However, he noted that the increase in performance, especially technical efficiency and productivity, has been irregular over the post-privatization years. Estache et al. (2008), using a non-parametric analysis on a sample of 12 operators from 12 Southern African countries to measure the impact of reforms on total factor productivity (TFP), record a similar performance and efficiency level across the examined countries but notes an unclear relationship between adoption of the liberalized electricity sector reform and increased efficiency.

### 23.2.2 Quality of Service and System Efficiency

One of the main aims of the widespread reforms of the electricity sector of SSA countries were to increase the efficiency of existing capacities, reduce energy losses, increase the quality of service of utilities, add new capabilities to existing ones; and by so doing, improve access to quality electricity supply (Joskow, 1998; Briceno-Garmendia et al., 2004). A panel data is analysis of impacts on technical and distribution (T&D) loses in SSA countries carried out by ESMAP (2009) proves otherwise. Average T&D losses in SSA stood at about 28% in 2009, with huge variations across countries—with Angola having an average system loss of 14.5% and Swaziland 68% (ESMAP, 2009).

Cubbin and Stern (2004, 2006) and Zhang et al. (2005, 2008), in their panel analyses of a set of developing countries (including SSA countries), show that the existence of independent regulatory agency and competition is a condition that must be satisfied for a meaningful increase in per capita electricity generation, capacity and quality of service of utilities. Nagayama (2010) supported these findings in his multi-country panel data analyses of developing and developed countries. Nagayama (2010) found that IPPs, unbundling, wholesale market, and regulatory bodies, together, lead to a reduction in transmission and distribution losses.

Nepal and Jamasb (2012) confirmed that privatization on its own did not lead to improvement in electricity sector performance. Yunos and Hawdon (1997) also observe that privatization without accompanying competition does not affect the efficiency of electricity supply. Many of these conditions are yet to be fulfilled in most of the countries analyzed in this study. There are often no or irregular efficiency gains resulting from privatized management in most developing economies (Plane, 1999; Estache et al., 2008). In some developing countries cases, public utilities were found to be more efficient in the management of electric utilities than private utilities (Jamasb et al., 2015; Erdogdu, 2013b; AICD, 2008).

Progress in reforms has led to increased system losses in many developing countries due to the simultaneous decrease in national government research and utility development investments (Hall, 1999; Erdogdu, 2013b). This may also contribute to the increase in system losses in some of the implementing SSA countries.

### 23.2.3 Electricity Pricing

The liberalized model of electricity sector reform is expected to introduce competition in the industry by attracting many new players into the market (Fan et al., 2007). The competition introduced through the reforms is expected to increase efficiency and decrease the cost of electricity production, thereby bringing electricity prices to an optimal lower level (Yang and Sharma, 2012). However, the impacts of the reforms on electricity tariffs in most developing countries are not significant. In other cases, the outcomes are mixed. Pollitt (1995) and Nagayama (2007), in their analyses of these impacts, write that country-level political interference in investment decision making, awarding of contracts and IPPs, and undue influences of independent regulatory agencies, limits and prevent the significant positive impact of the electricity sector reforms on the electricity tariff. ESMAP's (2011) study using a panel data of 20 countries with varied system sizes shows that vertical unbundling has reduced electricity tariff by an average of just 10%.

Electricity subsidies often make the capturing of the actual impacts of reforms on the tariffs unclear (Jamasb et al., 2015). Nagayama (2009), in his panel data analyses of 75 developed, developing, and transitioning economies, shows a decline in national electricity subsidies after the commencement of reforms in many developing countries, including in some SSA countries. Situations like this make it difficult to measure the impact of the reforms on electricity prices if the subsidies are not taken into consideration. There have always been relatively high tariffs prior to the reforms in SSA countries (Jamasb et al., 2015). Electricity prices have remained constant or have slightly increased regardless of the reforms in most SSA countries, but this has also been accompanied by the reduction in electricity tariff subsidies in some cases (Eberhard et al., 2011). However, in SSA countries such as Angola, Malawi, South Africa, Zambia, Nigeria, and Zimbabwe, high tariffs exist with high electricity tariff subsidies (Eberhard et al., 2011).

## 23.2.4 ELECTRICITY ACCESS

Increasing national population access to electricity was also one of the major reasons for the widespread liberalization of the electricity sector in SSA (Jamasb et al., 2015). Participation of the private sector was expected to increase the number of investors in electricity generation and capacity thereby increasing access to electricity (Zhang et al., 2008). Several findings, however, prove that this is not always the case in SSA (Zhang et al., 2008; Eberhard et al., 2011). There is also scarce information on the impact on access to the poor and coverage for people who live in rural areas. Prasad (2008), in his multi-country analysis of SSA countries, finds that access is only improved when reforms are specifically targeted toward improving access to the poor. Multi-country analyses by ESMAP (2011) show that the existence of independent regulatory agencies helps to strengthen public electricity access by 50%. This finding was supported by Erdogdu's (2014) finding that reforms can lead to system supply self-sufficiency only in the presence of necessary prerequisites—independent regulatory agencies and competition.

The study by Gassner et al. (2008) investigates whether private sector participation improves electricity and water distribution performance, using a dataset from about 1,200 utilities drawn from 71 developing and transition economies. The study findings reveal that the private sector delivers higher labor productivity and operational efficiency expectations, convincingly outperforming a set of comparable companies that remained state owned and operated. Specifically, a comparison of the annual values for performance indicators from pre- and post-PSP periods highlights that the following could be attributed to PSP: (i) a 29% increase for electricity distribution companies, (ii) a 32% increase in electricity sold per worker, (iii) a 45% increase in bill collection rates in electricity, and (iv) an 11% reduction in distribution losses for electricity (following divestiture). All these changes, which occurred within 5 years, were significantly above those recorded for state-owned companies. The study, however, was silent on the role of regulations on private sector performance with respect to electricity generation.

A study by Aslani (2014) argued that government increased support for renewable energy utilization will further raise the private sector contribution in electricity generation in Sub-Saharan Africa, also supported by Rafique et al. (2019). The author suggested a framework that consists of five main tasks for renewable energy upscaling to achieve this goal. The first is the correct identification of consumer behavior in the RE industry. The second is to understand the different levels of investment in the RE supply. The third strategy is also to expand the subjects related to investment and business in academic engineering programs and research institutes of developing countries. The fourth strategy is introducing renewable energy future centers (REFC) that will serve as an environment for professionals and investors debates. According to the study, the last strategy is a re-definition of concepts related to sustainability, which is essential, especially for investors in private sectors.

With an approximated 6,000 MW of electricity and only 40% of it generated annually, Oseni (2011) believes that the electricity sector's performance is grossly short of expectation and requires a quick and adequate policy intervention for a stable and efficient electricity market to be achieved in the country. Oseni (2011) further quipped that power inadequacies in the country limit private sector involvement, thereby undermining the country's potential to industrialize. In a way, as their study portrays, the private sector's impact on electricity sector performance is still limited because of public sector policy anomalies. These public sector issues also increase private sector business overhead costs (African Economic Research Consortium, 2005). Maulidia et al. (2019) recommended that the government initiate a clear policy framework that facilitates private sector investment while simultaneously addressing the monopolized power market system that oversees a complex, changing anomalies of electricity pricing regulations that make investment uncertain and risky.

For many developing countries of the world, the opening of the electricity industries for private sector participation is driven by three main reasons: public sector budget constraints; the need to increase the capacity or enhance "systems" reliability, or both; and the positive outcomes of the early investigations with private participation in Chile and the United Kingdom with respect to driving up electricity sector performance (Izaguirre, 1998). Given these perceived positive effects of private participation on electricity sector development, about 62 developing countries initiated private participation in the electricity sector in different proportions—ranging from the privatization of most sector operations in Bolivia, Hungary and Argentina, to management contracts for the state-owned utility in Mali, all between 1990 and 1997.

A significant part of the extant literature reviewed suggests a need for the state's role toward electricity sector development to be redefined to give the private sector more opportunity to participate. This can help in increasing the stability and performance of the electricity industry in developing countries and ultimately heighten their contributions to overall energy access, especially in Sub-Saharan Africa. Jamasb (2006) further suggest that countries should reform the electricity sector at intervals and that this would be more effectively done with the adoption simpler reform models and gradual implementation. However, these recommendations would gain more weight with additional empirical-based evidence, one of the contributions of this current chapter.

## 23.3  METHODOLOGY

### 23.3.1  Data

Our dataset is composed of 39 African countries[1] between 1990 and 2017. This limit was defined by the availability of data and private sector participation at any time within the sample period. Thus, only countries with at least one private sector involvement in the reference period were included in the study.[2] Moreover, most of the electricity sector reforms in Africa started after 1990. Before then, almost all the countries held a monopolistic electricity sector where only the government invests, generates, distributes, and manages the "sector's activities." Given this timeframe, our observation size was 1,092. Due to some missing variables, some of our models utilize fewer observations.[3] The study utilized the most comprehensive dataset on private sector involvement in the electricity sector across the globe. We constructed our electricity sector reform variables from data provided by the World Bank private sector investment portfolios. This dataset provides details of the private sector activities across various sectors including the transportation and agricultural sectors. One of the advantages of the dataset is that it provides complete details of the nature, size, and intensity of investments. It also supplies relevant details of the status of the investments and the extent of current private sector activity and future involvement in the sector.

We complemented the participation data with other datasets from the World Bank Development Indicators. We obtained institutional data from the World Bank Governance Indicators and complemented these with data from the fourth Polity project (see Table A2 in the Appendix for further information). Lastly, we utilized the data from the International Energy Agency (IEA) in defining the performance of the energy sector.

---

[1] The sample of countries included in the study are: Algeria, Angola, Botswana, Burkina Faso, Cameroon, Cape Verde, Chad, Comoros, Congo Republic, Côte d'Ivoire, Egypt, Ethiopia, Gabon, Gambia, Ghana, Guinea, Guinea-Bissau, Kenya, Lesotho, Liberia, Madagascar, Malawi, Mali, Mauritius, Morocco, Mozambique, Namibia, Nigeria, Rwanda, Sao Tome and Principe, Senegal, Sierra Leone, South Africa, Tanzania, Togo, Tunisia, Uganda, Zambia and Zimbabwe.

[2] Due to lack of sufficient data, it was not possible to determine the inclusion or exclusion of countries based on the relative share of the private sector in the overall electricity sector activity. Although, this would have been a more appropriate way to capture the overall relevance of the reform process.

[3] See Tables 23.3–23.5 for further information.

## 23.3.2   MEASUREMENT OF VARIABLES

To measure the performance of the electricity sector, we utilized two main variables: (i) the net generated electricity per country for a year, and (ii) the percentage of electricity lost in transmission and distribution to total electricity produced. These variables constitute the bulk of electricity sector performance studies (Davis and Wolfram, 2012; Jamasb et al., 2005; Polemis, 2016; Steggals et al., 2011). A few studies deviate from this norm by utilizing electricity pricing as performance indicators (see Bacchiocchi et al., 2015).

Our model follows the model specifications of patterns of Bacchiocchi et al. (2015) and Polemis (2016). However, our measures differ in several ways. First, we focus on private sector participation as our definition of the electricity sector reform, while previous studies have used measures that were reflective of the entire regulation process. This involved the use of reform indices based on existing energy laws, regulatory body and independence of the electricity regulatory authority. One of the problems associated with the use of index numbers is that the results are largely driven by the unknown weights of the components of the index. Second, an extensive database for electricity sector laws and other components of the reform index is unavailable for developing countries and African countries in particular. Hence studies of this nature have been limited to more developed economies and the European continent (Polemis, 2016).

To circumvent this problem, we adopt a superior source of actual private sector participation data computed and published by the World Bank with better African coverage. Our proxy provides an adequate measure of electricity sector reform. According to Bacon and Besant-Jones (2001), full-scale power sector reform composes of six main elements, including (i) enhancing commercialization of electricity by operational enterprises in the generation, transmission, systems control, and services supply both by governments owned and privately operated enterprises, (ii) encouraging efficiency, innovation, and responsiveness improving competitive market structure within the chain of electricity production to use systems, (iii) full sector restructuring to engage non-actors to become actors in the sector and breaking down bundles activities into multiple unbundled processes, (iv) encouraging and ensuring privatization which is the bedrock of competition and commercialization (an aspect that has been mostly hindered in state-owned monopolistic electricity markets in developing countries), (v) economic regulation of the electricity market by independent sector decoupled from the production and supply chain activities, and (vi) limiting state actions to policy development. Going from the above, our measure captures three main components of the restructuring process (namely: i, iii, and iv).

While Bacchiocchi et al. (2015) measured their regulatory reforms using dummy variables, we adopt a complete set of information to characterize our proxy. However, the components of reforms are not the same.[4] One problem with the dummy variable approach is that it fails to account for the intensity of the level of reform. Therefore, the sizes of reform scores, for example, do not matter. For our case, we use the data on investment portfolio and volume of installed electricity as a measure of our reform. The major problem encountered with this measure is that there are sufficient gaps in that data, limiting our methodological approach (Jamasb et al., 2005). We overcome this problem by generating a measure of cumulative investment and installed capacity by the private sector using data provided by the World Bank (2019). This indicator builds up by adding the current year's value of investment or installed capacity to the existing values of the investments.[5] This proxy seems reasonable since assets usually accumulate over time especially hard investments such as electricity (Richard and Mullen, 2009; Leeper et al., 2010).

---

[4] While previous studies have tried to measure reform using direct reform scores of levels of monopoly in the electricity sector for example, our study uses the participation of private sectors as the proxy for reform. The advantage of our proxy is that it goes beyond the policy setting and records that actual action of reformation (Jamasb et al., 2005).

[5] For studies that have used similar estimates see (Allcott and Keniston, 2018; Richard and Mullen, 2009; Leeper et al., 2010).

Some may argue that our measure automatically introduces serial correlation to our model, we claim otherwise since we already correct for that using an appropriate econometric model. To affirm our position, we test the robustness of our model using alternative measures of the variable, including a dummy variable approach that takes the value of one for any year there is an investment or private sector-led electricity installation. One concern with this dummy is that we cannot explain the perpetuation of the effects of the participants across the years. We also test for the robustness of our measure using the direct value of investment and installation in a given year. Lastly, we ascertain the difference in levels of investments by accounting for the number of investments or installations led by the private sector as well as the cumulative numbers of these variables. Overall, our results do not differ significantly based on the choice of variables.

The performance of the electricity sector may be determined by other factors besides the reformation activities in the industry. The omission of these factors may bias our model. Following existing literature, especially the broad analysis presented by Jamasb et al. (2005) and Polemis (2016), we control for differences in economic characteristics of countries and level of economic development using the values of Gross Domestic Product (GDP) per capita. The economic strength of a country largely determines the extensive capabilities for that country to carry out sufficient reforms as well as the performance of its economic sectors. Second, we control for the population size and the proportion of the population with access to electricity. According to Jamasb et al. (2005), these factors determine the average electricity demand and thus the extent to which reforms may be required as well as the carrying capacity of the electricity infrastructure. In some of our specifications, we include a country-specific time trend to capture the development in the country's technology. Lastly, we control for the effects of institutional factors[6] that may explain the reliability of the electricity industry.

### 23.3.3 EMPIRICAL SPECIFICATION

Our study attempts to tease out the impact of electricity sector reform (with specific emphasis on private sector participation) on the performance and efficiency of the electricity sector for some African countries. Following previous related studies such as Bacchiocchi et al. (2015), Davis and Wolfram (2012), and Polemis (2016), we estimate a reduced form equation in the following order:

$$Y_t^c = \beta P_t^c + \gamma X_t^c + \delta_t + \mu^c + \varepsilon_t^c ss \tag{23.1}$$

where $Y_t^c$ represents the performance of the electricity sector in country $c$ in year $t$. $P_t^c$ is the measure of private sector reform (in our case, private sector performance) for the same country in the current year? The vector of other factors that influence the performance of the electricity sector is denoted by $X_t^c$. $\delta_t$ and $\mu^c$ are sets of time and country-specific fixed effects. These variables capture the impact of global shocks that affect all the sample countries simultaneously, for example, the global economic downturn in 2008–2009. The country-fixed effects capture country-specific time-invariant shocks such as geographical factors and institutional characteristics. The $\alpha$, $\beta$, and $\gamma$ are the parameters of interest which shows the magnitude and direction of the relationship between the variables of interest and the outcome variable while $\varepsilon$ is the error term which explains the part of the outcome variable that cannot be explained by the factors included in the model. Table 23.1 reports the complete set of variables included in the model while Tables 23.2 and 23.3 show the breakdown of the summary statistics of each variable.

Economic variables and investment variables, in particular, most often exhibit a dynamic characteristic. For example, investment in 1 year remains active for some years to come. In extreme cases,

---

[6] We argue in our main specifications that institutions in Africa have remained mostly enduring in the time period under analysis. However, there may have been some changes across countries, for this reason, we include these factors as robustness to our main specification.

## TABLE 23.1
### Summary Statistics

| Variable | Number of Observation | Mean | Standard Deviation | Minimum | Maximum |
|---|---|---|---|---|---|
| Electricity distribution losses | 1,092 | 1.463 | 3.735 | 0 | 24.47 |
| Electricity installed capacity | 1,090 | 2.830 | 7.761 | 0 | 55.14 |
| Net electricity generation | 1,092 | 12.153 | 37.3530 | 0 | 245.58 |
| Cumulative private sector installed capacity | 1,091 | 528.806 | 1080.79 | 0 | 6325 |
| Cumulative private sector investment | 1,091 | 443.8427 | 1573.83 | 0 | 15429.33 |
| Private sector investment dummy (0/1) | 1,092 | 0.165 | 0.371 | 0 | 1 |
| Private sector installed capacity | 1,092 | 33.79 | 158.89 | 0 | 3,000 |
| Private sector investment | 1,092 | 27.43 | 139.11 | 0 | 2621.58 |
| Private sector investment (number) | 1,091 | 0.304 | 1.44 | 0 | 30 |
| GDP per capita | 1,069 | 1959.71 | 2211.19 | 164.33 | 11949.28 |
| Population density | 1,092 | 86.08528 | 113.0556 | 1.740462 | 622.9621 |
| Access to electricity (% population) | 1,001 | 29.86578 | 15.86432 | 4.90249 | 90.78317 |
| Voice and accountability score | 741 | −0.463764 | 0.6564918 | −1.674941 | 1.007172 |
| Political stability score | 741 | −0.4238405 | 0.7992985 | 2.436677 | 1.219244 |
| Governance effectiveness score | 741 | −0.610505 | 0.5842014 | 1.884888 | 1.056994 |
| Regulatory quality score | 741 | −0.5518675 | 0.5636701 | −2.236245 | 1.12727 |
| Rule of law score | 741 | −0.5621717 | 0.6186422 | −2.008507 | 1.07713 |
| Control of corruption score | 741 | −0.5348506 | 0.5923351 | −1.701552 | 1.216737 |
| Polity2 | 1,062 | 1.29 | 5.39 | −9 | 10 |
| Democracy | 1,062 | 11.61 | 12.19 | 0 | 81 |

## TABLE 23.2
### Effects of Private Sector Participation on Electricity Sector Performance

| Dependent Variable | Net Electricity Generation (MW) | | | Electricity Distribution Losses (MW) | | |
|---|---|---|---|---|---|---|
| | 1 | 2 | 3 | 4 | 5 | 6 |
| | Panel A: Cumulative Private Sector Investments (USD) | | | | | |
| Lagged dependent variable | 1.028*** | 1.028*** | 0.976*** | 1.031*** | 1.034*** | 0.981*** |
| | (0.00712) | (0.00712) | (0.0332) | (0.00830) | (0.00857) | (0.0192) |
| Cumulative investment | 0.000296** | 0.000267** | 0.000218* | −6.69e−03** | −7.13e−03** | −6.33e−03* |
| | (0.000126) | (0.000124) | (0.000124) | (3.16e−03) | (3.58e−03) | (3.64e−03) |
| GDP per capita (log) | | | 3.056 | | | 0.706*** |
| | | | (3.260) | | | (0.185) |
| Population density | | | −0.0351 | | | −0.00226 |
| | | | (0.0270) | | | (0.00207) |
| Access to electricity (% population) | | | 0.0411 | | | −0.00376 |
| | | | (0.0454) | | | (0.00451) |
| Constant | −0.0465 | −0.300 | −19.54 | 0.0546** | 0.00190 | −4.546*** |
| | (0.0471) | (0.306) | (21.12) | (0.0230) | (0.0926) | (1.279) |
| Observations | 1,052 | 1,052 | 1,033 | 1,052 | 1,052 | 1,033 |

*(Continued)*

**TABLE 23.2** (*Continued*)
**Effects of Private Sector Participation on Electricity Sector Performance**

| Dependent Variable | Net Electricity Generation (MW) | | | Electricity Distribution Losses (MW) | | |
|---|---|---|---|---|---|---|
| | 1 | 2 | 3 | 4 | 5 | 6 |
| Number of countries | 39 | 39 | 39 | 39 | 39 | 39 |
| Country-fixed effects | Yes | Yes | Yes | Yes | Yes | Yes |
| Year fixed effects | No | Yes | Yes | No | Yes | Yes |
| AR (1) (*p-values*) | 0.0300 | 0.0231 | 0.0192 | 0.0598 | 0.0541 | 0.0466 |
| AR (2) (*p-values*) | 0.3297 | 0.3091 | 0.3018 | 0.2397 | 0.2045 | 0.2118 |

*Note:* Two-step systems generalized method of the moment. A robust finite sample corrected the standard error. Statistical significance of the p-values is reported with stars. * Significant at 10%, ** significant at 5% and ***significant at 1% level respectively (*p < 0.10, **p < 0.05, ***p < 0.01).

the effects of an investment decision are not observed in the first years of the investment but become essential after a few years (Ogunniyi et al., 2020). For this purpose, we employ a dynamic estimation approach to exploit the dynamic relationships between our variables of interest. We also exploit the panel dimensions of our data by adopting a dynamic panel model framework. Not only does our model satisfy the need for the dynamic structure of our data, but it also helps to cater for potential endogeneity problems that may arise from specifying a dynamic model of the form:

$$Y_t^c = \alpha Y_{t-1}^c + \beta P_t^c + \gamma X_t^c + \delta_t + \mu^c + \varepsilon_t^c \tag{23.2}$$

where $Y_{t-1}^c$, on the other hand, is the previous year's value of electricity sector performance and all other variables maintain the same description as in Eq. (23.1). Within our model, we treat electricity sector reform (proxied by private sector participation) as endogenous because of the potential issues of reverse causality that may occur between the current reforms and the past and present performance of the electricity sector. According to Jamasb et al. (2005), countries respond to the performance or quality of the given sector, in the presence of institutional and adequate policy setup by regulating the electricity or energy sector.

A major estimation issue occurs with specifying lagged dependent variable $Y_{t-1}^c$ while controlling for time-fixed effects $\delta_t$. Potential correlation between the unobserved time-invariant heterogeneity and the time-varying control variable could introduce some element of serial correlation and ultimately biases the estimation results. Estimating Eq. (23.2) with the ordinary least square model could yield an upward-biased estimate of the variables of interest since $Y_{t-1}^c$ is correlated with the error term $\varepsilon_t^c$ (this is because, as $Y_t^c$ is correlated with unobserved variables, so also is $Y_{t-1}^c$ correlated these variables which are captured in the error term). Earlier studies suggest handling these problems using a within estimator with a fixed effect (Ogunniyi et al., 2020). However, while the within estimation with fixed effects may correct for the unobserved heterogeneity problem with the base outcome variable, it does not correct the correlation between the errors and the lagged dependent variable. For our study, we follow the recommended approach of Woodridge (2010) and Arellano and Bond (1991) in using the first difference instrumental variable approach in the following manner:

$$\Delta Y_t^c = \alpha \Delta Y_{t-1}^c + \beta \Delta P_t^c + \gamma \Delta X_t^c + \Delta \delta_t + \Delta \mu^c + \Delta \varepsilon_t^c \tag{23.3}$$

where $\Delta Y_t^c$ is the year on change in the value of the outcome variable (electricity sector performance) given as the difference between current year value and the previous year's value $(Y_t^c - Y_{t-1}^c)$. The first right-hand-side (RHS) variable is the year on the lagged difference of the lagged dependent

**TABLE 23.3**

**Effects of Private Sector Participation on Electricity Sector Performance**

| Dependent Variable | Net Electricity Generation (MW) | | | Electricity Distribution Losses (MW) | | |
|---|---|---|---|---|---|---|
| | 1 | 2 | 3 | 4 | 5 | 6 |
| Panel B: Cumulative Private Sector Installed Capacity (MW) | | | | | | |
| Lagged dependent variable | 1.022*** | 1.021*** | 0.973*** | 1.038*** | 1.045*** | 0.992*** |
| | (0.00418) | (0.00333) | (0.0311) | (0.00891) | (0.00929) | (0.0199) |
| Cumulative installed capacity | 0.000913*** | 0.00128*** | 0.000919* | −0.000171*** | −0.000265*** | −0.000221*** |
| | (0.000259) | (0.000385) | (0.000479) | (5.56e−05) | (7.26e−05) | (7.58e−05) |
| GDP per capita (log) | | | 3.701 | | | 0.591*** |
| | | | (3.148) | | | (0.192) |
| Population density | | | −0.0297 | | | −0.00261 |
| | | | (0.0242) | | | (0.00207) |
| Access to electricity (% population) | | | 0.0254 | | | −0.00167 |
| | | | (0.0389) | | | (0.00454) |
| Constant | −0.351** | −0.0124 | −24.20 | 0.108*** | 0.300** | −3.792*** |
| | (0.142) | (0.0834) | (20.38) | (0.0319) | (0.132) | (1.328) |
| Observations | 1,052 | 1,052 | 1,033 | 1,052 | 1,052 | 1,033 |
| Number of countries | 39 | 39 | 39 | 39 | 39 | 39 |
| Country-fixed effects | Yes | Yes | Yes | Yes | Yes | Yes |
| Year fixed effects | No | Yes | Yes | No | Yes | Yes |
| AR (1) (*p-values*) | 0.0326 | 0.0282 | 0.0207 | 0.0605 | 0.0542 | 0.0473 |
| AR (2) (*p-values*) | 0.3388 | 0.3230 | 0.3104 | 0.2360 | 0.1999 | 0.2065 |

*Note:* Two-step systems generalized method of the moment. A robust finite sample corrected the standard error. Statistical significance of the p-values is reported with stars.

*Significant at 10%, **significant at 5% and ***significant at 1% level, respectively (*p < 0.10, **p < 0.05, ***p < 0.01).

variable ($Y_{t-1}^c - Y_{t-2}^c$). The other RHS variables are defined as the outcome variable. The instruments consist of the linear and non-linear lagged values of the RHS and time variables, respectively. We test for the validity of our instruments using the Arelleno and Bond (1991) AR (1) and AR (2).

## 23.4   RESULTS

### 23.4.1   Overview of the Data

Table 23.1 shows the summary statistics for the panel data (composing of the mean and the correlation matrix). The table shows that the average electricity distribution losses for all countries in all years were 1.5% of the total electricity output. The standard deviation of the data was more than three times above the mean. This is reflective of the differences and high variability of electricity sector efficiency over time and across countries. These differences could have also been attributed to changes in technological progress over time (which we capture within our model using a time trend). The value of installed electricity generation capacity followed a similar trend to the loss values. We find that net electricity generation was also spatially and temporally erratic. On average, only about 10% of the sample countries recorded at least one investment portfolio from the private sector in each. These values do not substantially differ from the overall investments in the electricity sector for the countries under review.

Generally, investments in electricity are quite low in Africa, where the average access to electricity is only about 30% of the total population. Although low, the private sector installed electricity capacity was on average 34 MW but reached up to 3,000 MW in large projects. Most private sector-led investments are relatively small and could have been constrained by access to required capital for massive projects. Bacon and Besant-Jones (2001) show that electricity projects are usually capital intensive. Despite the relatively small sizes of these projects, they could significantly influence the electricity sector, especially in Africa where general access and production of energy remains low.

The correlation matrix presented in Table A1 in the Appendix reports the correlation between the variables included in the RHS of the model. The results presented using Spearman's pairwise correlation model and corrected with the Bonferonni sample size correction show that almost all the variables included are weakly correlated. The maximum correlation value was about 45%. These weak correlation coefficients imply that our model does not suffer from serious multi-collinearity problems. A variance inflation factor (VIF) analysis proves further that our results are robust to cross-correlation problems.[7]

### 23.4.2   Effects of Private Sector Participation on Electricity Sector Performance

We present a set of results from our empirical analysis in this section. We first run regression model (3), exploiting our data's panel and dynamic structure. Results are reported for the two proxies of our electricity sector performance indicators in columns 1–3 and 4–6, respectively. We also report the different measures of our private sector participation variable in Tables 23.2 and 23.3 for the pairs of outcome variables. In columns 1 and 4 of Tables 23.2 and 23.3, we do not control for time-fixed effects. We include year effects in columns 2 and 5, respectively, and in columns 3 and 6, we exploit the entire model using a complete set of controls, including country and year effects.

Our results show that the model suffers from first-order residual serial correlation in all the specifications, but these correlations are corrected in the second order. These are shown by the statistically significant AR (1) p-value and the not significant AR (2) p-values. Overall, our model is stable with small means square errors and log-likelihood values.[8]

In all the different models presented in Tables 23.2 and 23.3, we find that electricity sector reform improves both the performance and efficiency of the electricity sector in SSA. We find that a unit

---

[7] Results of the VIF for each model are available on request.
[8] Estimates available in full analysis can be reported at request.

change (growth) in investment (that corresponds to an additional million-dollar investment) results in the increase in net electricity generation by about 300,000 kWh. In a similar vein, the same investment unit saves about 70,000 kWh of electricity in the distribution process. The results are comparable across all the models provided in the table. Including additional controls in columns 2, 3, 5 and 6, reduced the magnitude of our parameter and the standard errors. For example, the results in columns 3 and 6 were about 27% and 5% lower than their counterparts in columns 1 and 4, respectively. More so, while investment was statistically significant at a 5% significance level in columns 1, 2, 4, and 5, they were only weakly significant at 10% in columns 4 and 6. This shows that ignoring other controls would have resulted in an upwardly biased estimate.

Similar to the investment variable, we find that the physical value of the private sector activity (installed capacity from the private sector) improves the net value of electricity generation. While we take the results in Table 23.3 as indicative,[9] we find that increasing the capacity installation from the private sector strongly decreased the overall loss in electricity distribution. This implies that in general, private sector-led activities promote overall efficiency in the electricity sector.

### 23.4.3 Heterogeneous Effects

To examine the heterogeneity of impacts of private sector participation in the electricity sector by the point of investment and the technology type, we compare the performance of the electricity sector for investments made in the renewable electricity and conventional and mixed investment types. We also compare the outcomes for greenfield investments, brownfield, divestiture, and management activities carried out by the private sector.

Table 23.4, columns 1–4, shows the regression estimates for the effects of private investments on the performance of the electricity sector at the different stages of the electricity value chain. We restrict our analysis at this point to the estimation of the impacts on net electricity generation. In columns 4–7, we report the results for the technology types invested. We find a positive and statistically significant impact of private sector participation on the net electricity generation for both greenfield and brownfield investments and divested investment and investment in management activities. The results revealed that investment in management activities yields the highest positive net generation impact. Greenfield investments, on the other hand, showed a smaller positive impact. The results reflect the capacity of each activity type to increase overall efficiency in the electricity sector. While the management-based investments may yield quick results in improving the outcomes of the electricity sector, investment in new projects may take more time to have significantly large impacts. Concerning the technology types, we find a negative but insignificant effect of investment in conventional fossil-based technologies on the electricity sector. Investment portfolios in renewable electricity technologies were reported to increase the net generation of electricity, although marginally. One possible reason for these results is that renewable energy investment portfolios are often smaller than their conventional electricity investment counterparts.

### 23.4.4 Robustness Checks

We ran several regressions to subject our earlier estimates to a robustness test. First, with the understanding that the electricity sector reforms and the performance of the electricity sector do not happen in isolation, for example Jamasb et al. (2005) noted that both reform and performance might be driven by the institutional performance and capability of the responsible bodies. Although we already for country-fixed effects that capture the trend in institutional development, at this point, we control for potential time-variant institutional performance of the country. We controlled for indices for control of corruption, regulatory quality, voice and accountability, the rule of law, governance

---

[9] We expect a strong correlation between installed capacity and generation which may not be related to the private sector participation but the base value of measurement (even though our correlation coefficients were quite low).

**TABLE 23.4**

**Heterogeneous Effects of Investment and Technology Types on the Electricity Sector**

Dependent Variable: Net Electricity Generation

| | Investment Type | | | | Technology Type | | |
|---|---|---|---|---|---|---|---|
| | Brownfield | Divestiture | Greenfield | Management | Conventional | Renewable | Mixed |
| Lagged dependent variable | 1.019*** | 1.022*** | 0.871*** | −0.439 | 1.017*** | 0.692*** | −0.120 |
| | (0.00609) | (0.00975) | (0.110) | (0.386) | (0.00888) | (0.122) | (0.325) |
| Cumulative investment | 0.000415*** | 0.00147 | 0.000368*** | 0.00607*** | −0.000116 | 0.000920* | −0.000471 |
| | (0.000109) | (0.00101) | (0.000113) | (0.00177) | (0.000328) | (0.000552) | (0.000730) |
| GDP per capita (log) | 0.0987 | 0.959*** | 2.765 | −0.0357 | −0.0151 | 2.261 | 4.138 |
| | (1.038) | (0.264) | (2.026) | (0.525) | (0.961) | (2.791) | (3.484) |
| Population density | 0.00241 | −0.0140 | −0.0133 | 0.00255 | −0.00611 | −0.000602 | −0.109** |
| | (0.00346) | (0.00881) | (0.00882) | (0.00421) | (0.00680) | (0.0173) | (0.0542) |
| Access to electricity (% population) | −0.0235 | 0.0594 | −0.0626 | −0.0207 | 0.0636** | −0.0363 | 0.0401 |
| | (0.0277) | (0.0540) | (0.0400) | (0.0156) | (0.0278) | (0.0545) | (0.0283) |
| Constant | 2.588 | −7.960*** | 403.5 | 9.893* | −1.334 | 6.454 | 9.750 |
| | (6.515) | (1.540) | (652.3) | (5.871) | (6.538) | (16.83) | (31.91) |
| Observations | 125 | 55 | 379 | 140 | 358 | 191 | 137 |
| Number of countries | 14 | 5 | 30 | 14 | 32 | 28 | 15 |
| Country-fixed effects | Yes | Yes | Yes | Yes | Yes | Yes | Yes |
| Year fixed effects | Yes | Yes | Yes | Yes | Yes | Yes | Yes |
| AR (1) p-value | 0.0132 | 0.0598 | 0.0285 | 0.0458 | 0.0313 | 0.0877 | 0.1299 |
| AR (2) p-value | 0.2130 | 0.1090 | 0.2598 | 0.1879 | 0.4133 | 0.1356 | 0.5089 |

*Note:* Two-step systems generalized method of the moment. A robust finite sample corrected the standard error. Statistical significance of the p-values is reported with stars. * Significant at 10%, ** significant at 5% and ***significant at 1% level respectively (*p < 0.10, **p < 0.05, ***p < 0.01).

effectiveness, and political stability (with data from the WGI). We also used the democracy score and government durability (both from the Polity IV project) as a measure of the institution. We analyze our regression models with these additional control variables for our two outcome variables (net electricity generation and electricity distribution losses) and the two measures of private sector participation (investment and installed capacity).

We find that our results are robust to the addition of the institutional variables for the net electricity generation outcome. We find a positive statistically significant effect on net generation for both the cumulative investment and cumulative installed capacity regression. Our results confirm the previous results even though our parameter estimates are smaller after the additional controls (Tables 23.5 and 23.6). As before, we find that including institutional controls to our regression models for the electricity distribution losses analysis, our parameter estimates are still negative. However, they are no longer significant for both the WDI and Polity IV data sets.

Second, we used dummies to represent our measure of private sector participation. The use of dummies allows us to reproduce our estimates without introducing bias by potential measurement errors. As an additional test in this direction, we also use the number of investments in a country in any year as our measure of PSP. In addition to helping eliminate potential endogeneity bias for measurement errors, this measure also helps differentiate the intensity of private sector influence at any point in a given country. We also measure the cumulative number and functional PSP to further capture the impact of additional investment over the years for a given country in our sample. These results are presented in Table 23.7. Columns 1–3 report the regression results for net electricity generation, while columns 4–6 show the electricity distribution losses estimates. For all the measures, we find that our previous results are robust, although of a smaller magnitude. Private sector investments in the electricity sector were also found to reduce the electricity waste in transmission (see columns 4–6 of Table 23.7).

Finally, we subject our model to a different model specification. We ignore the potential endogeneity within our model and the dynamic nature of our dataset. We provide estimates of the private sector participation using a static fixed-effects model. Overall, we find that our initial results continue to perform well in estimates for the net generation model. While we find both positive and significant impacts of private sector-led investments and installed capacity in the electricity sector on the net electricity generation, we also find a positive significant impact of the two measures of PSP for the distribution losses. This contrasts the initial results we have using the differenced systems generalized method of the moment model (see results in Table A3 in the Appendix).

## 23.5  DISCUSSION OF FINDINGS

This sub-section discusses the findings of the empirical analyses in relation to the findings of other empirical studies on the subject matter. The results show that increase in total investment cost of electricity projects with "PSP" leads to increase in installed capacity and in electricity generation per capita. This finding is in line with the theoretical expectation that the relationship between private investment and electricity generation and installed capacity is expected to be positive, in the case of an effective liberalized model of electricity sector reform. To buttress this finding, the results also indicate that the positive relationships among the variables are significant, although the nominal effects was small. The small nominal effect observed in the results is attributed to the fact that most of the analyzed countries do not have fully independent regulatory institutions and unhampered competition as found and indicated by Zhang et al. (2005) and Cubbin and Stern (2004) to be a very important condition for higher electricity capacity and generation. Also on this note, Zhang et al. (2008) further reiterate that privatization, in isolation, has no significant impact on electricity generation potentials. Further highlighting the importance of independent regulation and governance, Cubbin and Stern (2006) found that even when privatization and competition are controlled for, effective and independent regulatory law and governance positively correlates with an increase in electricity capacity and generation. The gradual improvement of positive effects found in our result is backed by the study of Erdogdu (2014) which indicated that electricity supply level increased as electricity sector reforms progressed over time. On the other hand, electricity

**TABLE 23.5**
**Inclusion of Institutional Controls**

| Dependent Variable | Net Electricity Generation | | | |
|---|---|---|---|---|
| PSP Measure | Cumulative Investment | | Cumulative Installed Capacity | |
| | (1) | (2) | (3) | (4) |
| Lagged dependent variable | 0.998*** | 1.060*** | 0.988*** | 1.051*** |
| | (0.0484) | (0.0495) | (0.0905) | (0.0498) |
| Cumulative investment/installed capacity | 0.000171* | 0.000250* | 0.000752 | 0.00105** |
| | (9.85e−05) | (0.000140) | (0.000532) | (0.000497) |
| GDP per capita (log) | 7.753*** | 3.452 | 7.752 | 3.982 |
| | (0.980) | (3.390) | (5.744) | (3.332) |
| Population density | −0.0341*** | −0.0406 | −0.0312 | −0.0361 |
| | (0.00800) | (0.0308) | (0.0252) | (0.0289) |
| Access to electricity (% population | 0.0269* | 0.0255 | 0.0197 | 0.0112 |
| | (0.0143) | (0.0414) | (0.0372) | (0.0364) |
| VA | 0.376 | | 0.644 | |
| | (0.548) | | (1.146) | |
| PV | 0.0452 | | 0.0947 | |
| | (0.343) | | (0.282) | |
| GE | 0.430 | | 0.322 | |
| | (0.813) | | (0.701) | |
| RQ | −0.549 | | −0.202 | |
| | (0.757) | | (1.123) | |
| RL | 0.160 | | 0.0246 | |
| | (0.869) | | (1.010) | |
| CC | 0.0824 | | −0.0447 | |
| | (0.795) | | (0.961) | |
| Polity2 | | −0.00614 | | 0.0286 |
| | | (0.0632) | | (0.0556) |
| Durable | | 0.0106 | | 0.0195 |
| | | (0.0305) | | (0.0325) |
| Constant | −53.08*** | −22.24 | −52.67 | −25.80 |
| | (6.955) | (21.95) | (39.01) | (21.56) |
| Institution database | WGI | PolityIV | WGI | PolityIV |
| Observations | 734 | 977 | 734 | 977 |
| Number of countries | 39 | 38 | 39 | 38 |
| Country-fixed effects | Yes | Yes | Yes | Yes |
| Year fixed effects | Yes | Yes | Yes | Yes |
| AR (1) *p-value* | 0.0355 | 0.0205 | 0.0401 | 0.0220 |
| AR (2) *p-value* | 0.3218 | 0.3306 | 0.3323 | 0.3432 |

*Note:* Two-step systems generalized method of the moment. A robust finite sample corrected the standard error. Statistical significance of the p-values is reported with stars. * Significant at 10%, ** significant at 5% and ***significant at 1% level respectively (*p < 0.10, **p < 0.05, ***p < 0.01).

**TABLE 23.6**
**Inclusion of Institutional Controls**

| Dependent Variable | Electricity Distribution Losses | | | |
|---|---|---|---|---|
| PSP Measure | Cumulative Investment | | Cumulative Installed Capacity | |
| | (5) | (6) | (7) | (8) |
| Lagged dependent variable | 0.739*** | 0.765*** | 0.759*** | 0.776*** |
| | (0.0510) | (0.0652) | (0.0623) | (0.0727) |
| Cumulative investment/installed capacity | −4.89e−05 | −5.65e−05 | −0.000169 | −0.000169 |
| | (5.08e−05) | (5.56e−05) | (0.000105) | (0.000149) |
| GDP per capita (log) | 0.240 | 0.564* | 0.272 | 0.501 |
| | (0.464) | (0.341) | (0.468) | (0.381) |
| Population density | −0.00481 | −0.00421 | −0.00482 | −0.00437 |
| | (0.00690) | (0.00617) | (0.00704) | (0.00585) |
| Access to electricity (% population) | 0.0257 | 0.00674 | 0.0258 | 0.00797 |
| | (0.0183) | (0.0123) | (0.0182) | (0.0128) |
| VA | 0.756 | | 0.684 | |
| | (0.528) | | (0.495) | |
| PV | −0.432* | | −0.438* | |
| | (0.225) | | (0.227) | |
| GE | −0.404 | | −0.412 | |
| | (0.714) | | (0.729) | |
| RQ | 0.371 | | 0.316 | |
| | (0.419) | | (0.444) | |
| RL | 0.549 | | 0.610 | |
| | (0.359) | | (0.386) | |
| CC | −0.227 | | −0.176 | |
| | (0.357) | | (0.335) | |
| Polity2 | | 0.00668 | | 0.00552 |
| | | (0.0104) | | (0.0108) |
| Durable | | −0.0104 | | −0.0105 |
| | | (0.00653) | | (0.00663) |
| Constant | 1.861 | −3.594* | 1.616 | −3.185 |
| | (3.207) | (2.050) | (3.226) | (2.187) |
| Institution database | WGI | PolityIV | WGI | PolityIV |
| Observations | 734 | 977 | 734 | 977 |
| Number of countries | 39 | 38 | 39 | 38 |
| Country-fixed effects | Yes | Yes | Yes | Yes |
| Year fixed effects | Yes | Yes | Yes | Yes |
| AR (1) p-value | 0.0646 | 0.0499 | 0.0663 | 0.0520 |
| AR (2) p-value | 0.1248 | 0.1349 | 0.1318 | 0.1345 |

*Note:* Two-step systems generalized method of the moment. A robust finite sample corrected the standard error. Statistical significance of the p-values is reported with stars. * Significant at 10%, ** significant at 5% and *** significant at 1% level respectively ($*p < 0.10$, $**p < 0.05$, $***p < 0.01$).

**TABLE 23.7**
**Alternative Definition of PSP**

| Dependent Variable | Net Electricity Generation | | | Electricity Distribution Losses | | |
|---|---|---|---|---|---|---|
| | (1) | (2) | (3) | (4) | (5) | (6) |
| Investment dummy | 2.202* | | | −0.249* | | |
| | (1.167) | | | (0.147) | | |
| Number of investments | | 1.661*** | | | −0.197*** | |
| | | (0.465) | | | (0.0720) | |
| Cumulative number of investments | | | 0.524*** | | | −0.0607*** |
| | | | (0.116) | | | (0.0179) |
| GDP per capita (log) | 6.521 | 6.635 | 6.192 | 0.802 | 0.813 | 0.763 |
| | (4.262) | (4.115) | (4.099) | (0.583) | (0.568) | (0.566) |
| Population density | −0.0731** | −0.0696** | −0.0735** | −0.0146** | −0.0142*** | −0.0147*** |
| | (0.0340) | (0.0311) | (0.0314) | (0.00545) | (0.00517) | (0.00525) |
| Access to electricity (% population) | −0.230 | −0.222 | −0.250 | −0.0239 | −0.0230 | −0.0262 |
| | (0.278) | (0.260) | (0.276) | (0.0392) | (0.0371) | (0.0389) |
| Constant | −27.12 | −28.33 | −24.15 | −3.330 | −3.459 | −2.979 |
| | (25.95) | (25.39) | (24.99) | (3.718) | (3.676) | (3.637) |
| Observations | 1,070 | 1,070 | 1,069 | 1,070 | 1,070 | 1,069 |
| R-squared | 0.186 | 0.235 | 0.233 | 0.195 | 0.226 | 0.224 |
| Number of countries | 39 | 39 | 39 | 39 | 39 | 39 |
| Country-fixed effects | Yes | Yes | Yes | Yes | Yes | Yes |
| Year fixed effects | Yes | Yes | Yes | Yes | Yes | Yes |

*Note:* Static FE model. Standard errors clustered at country level in parentheses. Statistical significance of the p-values is reported with stars. * Significant at 10%, ** significant at 5% and ***significant at 1% level respectively ($*p < 0.10$, $**p < 0.05$, $***p < 0.01$).

sector reforms in most developed and transitioning economies, with strong institutions and regulatory framework, have led to strong positive improvements in the level of electricity sector productivity and efficiency in the concerned countries (Goto and Sueyoshi, 2009; Cullman and von Hirschhausen, 2008; Jain et al., 2010).

Furthermore, with respect to the efficiency of the electricity sector, the results show that increase in total investment cost of electricity projects with "PSP" does initially lead to increase in distribution losses. Theoretically, the relationship between private investment and distribution loss is expected to be negative. Our results, however, further show that improvements in private sector participation marginally results in improved efficiency performance of the electricity sector in the SSA, only after controlling for the relevant economic and institutional factors. These findings are also similar to the results by ESMAP (2009) who found that electricity sector reforms in analyzed SSA countries did not positively influence efficiency in electricity supply – reporting increased power outages and increased delay in electrical connections. Hall (1999) and Smith (2004) also found similar results in their global multi-country analyses. The findings were also similar to those of Nagayama (2009) who found that electricity sector liberalization does not always lead to improved electricity access or reduced electricity prices. Our results show that the inconsistencies and differences in impact of liberalization on electricity sector efficiency levels across the studies could be accounted for by the differences in the different stages and maturity of the reform process—a factor highly determined by the degree of institutional quality. Studies by Erdogdu (2011), Kundu and Mishra (2011), Haselip and Potter (2010), and Nagayama (2009) also observed differences in types of reforms, for example, variations between electricity reform stages; reforms in the industrial and agricultural electricity industries; existence of electricity subsidies to consumers; and at different macroeconomic situations across countries that were analyzed.

Overall, the findings of our study lead to the acknowledgment of the findings by Sen and Jamasb (2012) who posits that the benefits from investment from the private sector are often blocked by political economy factors, which often manifest at the initial stages of the reform in most of the analyzed countries. Many studies believe that efficient and independent regulatory agencies can effectively channel private investments to improve the performance of the electricity sector, by ensuring that improvement in generation and capacity leads to improved electricity consumption (Jamasb et al., 2015; Zhang et al., 2008). This stand is also backed by the findings of ESMAP (2011) that with the introduction of a regulatory agency that supports healthy competition in the sector, electricity access and consumption is increasing by an average of 50%. Prasad (2008) concluded his findings by stating that electricity sector reforms impact electricity access and consumption and efficiency only when tailored to meet the unique local conditions of the poor.

## 23.6  CONCLUSIONS, POLICY IMPLICATIONS AND RECOMMENDATIONS

Our study provides interesting evidence that successful power sector reforms, especially with the inclusion of the private sector, in prior state-owned monopolistic industries like the electricity industry, could generate overall gains in terms of improving power supply and reducing inefficiencies in the sector. However, these prospects are significantly hinged on effective power sector reforms, underpinned by the existence of independent regulatory agencies, institutional quality, and competition. In fact, our study show that these factors are major determinants in the success of the electricity sector reforms in many developing countries, particularly in SSA countries (also see Akampurira et al., 2009; Jamasb et al., 2015; Nepal, 2013; Eberhard et al., 2011). Several of the reviewed literature in our study found that they have a significant positive impact on electricity access, electricity installation and generation as well as on the efficiency of supply (for example, Jamasb et al., 2015; Nepal, 2013; Eberhard et al., 2011; Vagliasindi et al., 2010). They have also been shown to be important in keeping the balance between increased investment, welfare and equality of access (Jamasb et al., 2015; ESMAP, 2005; Davies et al., 2006).

Furthermore, it is also important to chart specific and tailored reform routes for individual countries according to existing political, social and institutional circumstances in the countries instead of relying on international paradigm or ideologies of electricity sector reform will go a long way to improving reform outcome in SSA countries (Nepal and Jamasb, 2012; Nepal, 2013; Eberhard et al., 2011; Acemoglu and Robinson, 2010). Regarding the slow and hindered speed of reform processes in the analyzed countries and the resulting minimal impacts of the reforms in SSA, we follow Eberhard and Gratwick (2016), Jamasb et al. (2015), and Bhattacharya (2007) to recommend explanation of the reform processes, objectives, and importance as well as transparency in the implementation of the reform. These will go a long way to counter the problems of theft, reluctant and non-payment of electricity tariffs in many SSA countries. The price increase burden as a result of reform implementation due to accompanying removal of electricity subsidies should be well planned and mapped out in phases in order to reduce adverse impacts on electricity consumers (Eberhard and Gratwick, 2016; Jamasb et al., 2015).

The review of literature on the subject matter reveals two practical ways for improving the effectiveness of reforms that suit the conditions of countries in SSA, namely: implementing hybrid electricity markets, and ensuring the effectiveness of regulatory institutions through complementary regulatory measures.

### 23.6.1  HYBRID ELECTRICITY MARKETS

Hybrid electricity market implies that the state has control over electricity utilities, but simultaneously creates room for potential PSP for complementarity, in a Public-Private Partnership framework (Eberhard, 2007; Eberhard and Gratwick, 2016). The hybrid electricity market structure is particularly beneficial in the African context because it can curtail corruption on the part of government while at the same time curtailing capitalist accumulation on the part of private investors. In this way, optimal outcomes can be achieved through embedded checks and balances. This market structure, however, comes with challenges such as uncertainty on whose responsibility it is to ensure adequate and reliable electricity supply or to make plans for procuring new electric utility infrastructure (a typical case for Tanzania and Rwanda) (AICD, 2008). To resolve these challenges, Eberhard and Gratwick (2016), Eberhard (2007), and Kessides (2012) suggest that plans for procurement of new electric utility infrastructure should be left in hands of state-owned utilities—who usually have relatively more resources and professional staff. Furthermore, procurement plans should also consider least possible cost options through competitive bidding processes (Eberhard and Gratwick, 2016; Kessides, 2012). There should also be a strong monitoring oversight for detecting potential and viable opportunities for new capacity investment; setting clear and transparent criteria for building new IPPs among existing state-owned utilities as well as for takeoff of operations of completed projects; ensure adequate and reliable electricity supply; all in a manner that suits national interest as opposed to utilities interests (Jamasb et al., 2015; Eberhard and Gratwick, 2016; Kessides, 2012; Eberhard, 2007). These monitory functions can also be integrated into the functions of already existing regulatory agencies in the concerned countries (Eberhard and Gratwick, 2016; Eberhard, 2007; Kessides, 2012).

## 23.6.2 COMPLEMENTARY REGULATORY FRAMEWORKS

Notwithstanding the existence of independent regulatory agencies in the electricity sector of some SSA countries, the extent of their independency is still questionable (Zhang et al., 2008; Jamasb et al., 2015). The agencies have been criticized for being weak and for lacking in capacity and experiences for making such important decision that has social and political consequences—thereby exacerbating existing regulatory problems (Eberhard, 2007; Eberhard and Gratwick, 2016). The regulatory agencies were initially state-owned and set up to: (i) ensure financial viability and fair tariff, (ii) attract new investments, and (iii) encourage cost and performance efficiency (Newbery, 2002; Pollitt, 2012). Following the reforms, these regulatory institutions were later made independent to ensure the actualization of these duties without undue political influences (Newbery, 2002; Pollitt, 2012). However, unhampered independency has not yet been achieved. To improve the independency of their power sector regulatory oversights, some developing countries have resorted to: contracting regulatory functions to foreign agencies usually located in their former colonial countries (Cubbin and Stern, 2004; Eberhard and Gratwick, 2016); outsourcing regulatory functions to technical consultants and expert panels funded by the state ministries (Brown et al., 2006; Eberhard and Gratwick, 2016); and establishing panels which serve as arbitrators in issues or conflicts between regulators and utilities (Eberhard and Gratwick, 2016). These strategies have also been recorded to be effective in the SSA context. Efficiency of regulatory institution in SSA have been shown to improve significantly when there are co-existence of different forms of regulatory agencies—such as the co-existence of established independent regulatory agency in a country with foreign regulatory agencies inform of contracts; or the co-existence of established independent regulatory agency in a country with regulatory technical consultants and expert panels (Zhang et al., 2008; Jamasb et al., 2015; Nepal, 2013; Nepal and Jamasb, 2012).

# APPENDIX

## TABLE A1
## Correlation Matrix for Variables Included in the Study

| | C_capacity | PSP_dum | C_Invest | PSP_number | Investments | Capacity | Elect_access | GDP_PC | Pop_density |
|---|---|---|---|---|---|---|---|---|---|
| C_capacity | 1 | | | | | | | | |
| PSP_dum | 0.4292* | 1 | | | | | | | |
| C_invest | 0.8606* | 0.4315* | 1 | | | | | | |
| PSP_number | 0.4328* | 0.9964* | 0.4369* | 1 | | | | | |
| Investments | 0.3997* | 0.8429* | 0.4420* | 0.8435* | 1 | | | | |
| Capacity | 0.4575* | 0.9553* | 0.4409* | 0.9547* | 0.8296* | 1 | | | |
| Elect_access | 0.4063* | 0.1009 | 0.4529* | 0.0978 | 0.1566* | 0.1186* | 1 | | |
| GDP_PC | 0.2161* | 0.1013* | 0.3074* | 0.1003 | 0.1264* | 0.1008 | 0.1821* | 1 | |
| Pop_density | −0.1038* | 0.0194 | −0.1398* | 0.0227 | −0.0115 | 0.0262 | −0.4960* | −0.3867* | 1 |

*Note:* Estimation uses Spearman's pairwise correlation model with Bonferonni correction. p-values: * significant at 5%.

## TABLE A2
## Definition and Sources of Variables Included in the Study

| Variable | Description of Variable | Source of Data |
|---|---|---|
| Electricity distribution losses | The volume of electricity lost (% of total generation) | International Energy Agency, 2020 |
| Electricity installed capacity | Total electricity installed capacity (billion kWh) | |
| Net electricity generation | Total electricity generated less loss volume (billion kWh) | |
| Cumulative private sector installed capacity | Year on cumulative installed electricity capacity by the private sector (MW) | Author's calculation |
| Cumulative private sector investment | Year on cumulative private sector investment in electricity (million dollars) | Author's calculation |
| Private sector investment dummy (0/1) | Private sector investment (yes or no) where 1 = yes; 0, otherwise | Author's definition |
| Private sector installed capacity | Installed electricity capacity by the private sector (MW) | |
| Private sector investment | Private sector investment in electricity (million dollars) | |
| Private sector investment (number) | Number of private sector investment made per year | Author's calculation |
| GDP per capita | Gross domestic product per person (proxy for the size of the economy) | World Bank Development Indicators |
| Population density | Number of person per square kilometer (a proxy for country size) | |
| Access to electricity (% population) | Percentage of population with access to electricity (proxy for future electricity demand) | |
| Voice and accountability score | Voice and accountability score | World Governance Indicators |
| Political stability score | Political stability score | |
| Governance effectiveness score | Governance effectiveness score | |
| Regulatory quality score | Regulatory quality score | |
| Rule of law score | Rule of law score | |
| Control of corruption score | Control of corruption score | |
| Polity2 | The measure of the quality of governance | Polity IV project |
| Democracy | Whether a country experiences a democratic rule | |

## TABLE A3
## Results from Static Fixed-Effects Model[10]

| | Net Electricity Generation (MW) | | | Electricity Distribution Losses (MW) | | |
|---|---|---|---|---|---|---|
| | (1) | (2) | (3) | (4) | (5) | (6) |
| Panel A: Cumulative Private Sector Investments (USD) | | | | | | |
| Cumulative investment | 0.00313*** | 0.00238*** | 0.00244*** | 0.000474*** | 0.000356*** | 0.000353*** |
| | (0.00107) | (0.000830) | (0.000814) | (0.000144) | (0.000102) | (0.000122) |
| GDP per capita (log) | | | 6.158 | | | 0.738 |
| | | | (3.861) | | | (0.527) |
| Population density | | | −0.0532* | | | −0.0118** |
| | | | (0.0264) | | | (0.00471) |
| Access to electricity (% population) | | | −0.294 | | | −0.0333 |
| | | | (0.284) | | | (0.0402) |
| Constant | 10.55*** | 7.060*** | −23.99 | 1.234*** | 0.642** | −2.790 |
| | (0.476) | (1.928) | (23.40) | (0.0639) | (0.270) | (3.417) |
| Observations | 1,091 | 1,091 | 1,070 | 1,091 | 1,091 | 1,070 |
| R-squared | 0.144 | 0.208 | 0.252 | 0.149 | 0.219 | 0.259 |
| Number of countries | 39 | 39 | 39 | 39 | 39 | 39 |
| Country-fixed effects | Yes | Yes | Yes | Yes | Yes | Yes |
| Year fixed effects | No | Yes | Yes | No | Yes | Yes |
| Panel B: Cumulative Installed Capacity (MW) | | | | | | |
| Cumulative capacity | 0.00549*** | 0.00450*** | 0.00447*** | 0.000794*** | 0.000620** | 0.000601** |
| | (0.00194) | (0.00166) | (0.00163) | (0.000282) | (0.000253) | (0.000257) |
| GDP per capita (log) | | | 5.685 | | | 0.681 |
| | | | (3.723) | | | (0.514) |
| Population density | | | −0.0543* | | | −0.0122** |
| | | | (0.0275) | | | (0.00559) |
| Access to electricity (% population) | | | −0.276 | | | −0.0302 |
| | | | (0.270) | | | (0.0376) |
| Constant | 9.038*** | 7.014*** | −21.11 | 1.024*** | 0.636** | −2.462 |
| | (1.024) | (1.891) | (22.62) | (0.149) | (0.265) | (3.331) |
| Observations | 1,091 | 1,091 | 1,070 | 1,091 | 1,091 | 1,070 |
| R-squared | 0.222 | 0.241 | 0.279 | 0.210 | 0.236 | 0.272 |
| Number of countries | 39 | 39 | 39 | 39 | 39 | 39 |
| Country-fixed effects | Yes | Yes | Yes | Yes | Yes | Yes |
| Year fixed effects | No | No | No | No | No | No |

*Note:* Static FE model. Standard errors clustered at country level in parentheses. Statistical significance of the p-values is reported with stars. * Significant at 10%, ** significant at 5% and ***significant at 1% level respectively (*$p < 0.10$, **$p < 0.05$, ***$p < 0.01$).

# REFERENCES

Acemoglu, D. and J. L. Robinson. 2010. The role of institutions in growth and development. *Review of Economics and Institutions* 1(2): 1–33.

AERC – African Economic Research Consortium. 2005. *Analysis of the Cost of Infrastructure Failures in a Developing Economy: The Case of Electricity Sector in Nigeria*, Research Paper *148*.

AICD – Africa Infrastructure Country Diagnostic. 2008. *Power Sector Database*. Washington, DC: World Bank.

Akampurira, E., D. Root, and W. Shakantu. 2009. Stakeholder perceptions in the factors constraining the development and implementation of public private partnerships in the Ugandan electricity sector. *Journal of Energy in Southern Africa* 20(2): 2–9.

Arellano, M., and S. Bond. 1991. Some tests of specification for panel data: Monte Carlo evidence and an application to employment equations. *Review of Economic Studies* 58: 277–297.

Aslani, A. 2014. Private sector investment in renewable energy utilization: Strategic analysis of stakeholder perspectives in developing countries. *International Journal of Sustainable Energy* 33(1): 112–124.

Bacchiocchi, E., M. Florio, and G. Taveggia. 2015. Asymmetric effects of electricity regulatory reforms in the EU15 and the new member states: Empirical evidence from residential prices 1990–2011. *Utilities Policy* 35: 72–90.

Bacon, R.W. and J. Besant-Jones. 2001. Global electric power reform: Privatization and liberalization of the electric power industry in developing countries. *Annual Reviews Energy & the Environment* 26: 33–59.

Bhattacharya, A. and S. Kojima. 2012. Power sector investment risk and renewable energy: A Japanese case study using portfolio risk optimization method. *Energy Policy* 40: 69–80.

Bhattacharya, S. 2007. Power sector reform in south Asia: Why slow and limited so far? *Energy Policy* 35(1): 317–332.

Briceno-Garmendia, C., A. Estache, and S. Nemat. 2004. Infrastructure services in developing countries: Access, quality, costs and policy reform. Policy Research Working Paper Series *3468*, The World Bank.

Brown, A., J. Stern, B. Tenenbaum, and D. Gencer. 2006. *Handbook for Evaluating Infrastructure Regulatory Systems*. Washington D.C.: The World Bank.

Cubbin, J. and J. Stern. 2004. Regulatory effectiveness: The impact of good regulatory governance on electricity industry capacity and efficiency in developing countries. Policy Research Working Paper *3536*, The World Bank, Washington D.C.

Cubbin, J. and J. Stern. 2006. The impact of regulatory governance and privatization on electricity industry generation capacity in developing economies. *World Bank Economic Review* 20(1): 115–141.

Cullmann, A. and C.V. Hirschhausen. 2008. From transition to competition. *Economics of Transition* 16(2): 335–357.

Davis, L. W., and C. Wolfram, C. 2012. Deregulation, consolidation, and efficiency: Evidence from US nuclear power. *American Economic Journal: Applied Economics* 4(4): 194–225.

Dertinger, A. & Hirth, L. (2020) Reforming the electric power industry in developing economies, evidence on efficiency and electricity access outcomes. *Energy Policy* 139: 111348.

Eberhard, A and Gratwick, K.N. (2016). *New Models to Scale up Investment in Power Generation in Sub-Saharan Africa*. World Bank, Washington, D.C.

Eberhard, R., Shkaratan, V. (2011). *Africa's Power Infrastructure: Investment, Integration, Efficiency*. Washington D.C.: World Bank.

Erdogdu, E. 2013a. *Essays on Electricity Market Reforms: A Cross-Country Applied Approach*. Ph.D. Thesis, University of Cambridge, UK.

Erdogdu, E. 2013b. Implications of liberalization policies on government support to R&D: Lessons for electricity markets. *Renewable and Sustainable Energy Reviews* 17(1): 110–118.

Erdogdu, E. 2014. Investment, security of supply and sustainability in the aftermath of three decades of power sector reform. *Renewable and Sustainable Energy Reviews* 31(3): 1–8.

ESMAP. 2011. *Revisiting Policy Options on the Market Structure in the Power Sector. The Energy Sector and Mining Assistance Program (ESMAP)*. Washington D.C.: The World Bank.

Estache, A., B. Tovar, and L. Trujillo. 2008. How efficient is African electricity companies? Evidence from the southern African countries. *Energy Policy* 36(6): 1969–1979.

Fan, Y., H. Liao, and Y.-M. Wei. 2007. Can market-oriented economic reforms contribute to energy efficiency improvement? Evidence from China. *Energy Policy* 35(4): 2287–2295.

Fischer, R., R. Gutierrez, and P. Serra. 2003. The effects of privatization on firms and on social welfare: The Chilean case. *Inter-American Development Bank Research Network Working Paper R-456, May*.

Fritsch, J. and R. Poudineh. 2015. *Gas-to-power Market and Investment Incentive for Enhancing Generation Capacity: An Analysis of 'Ghana's Electricity Sector.* Oxford: Oxford Institute for Energy Studies.

Galan, J.E. and M. Pollitt. 2014. *Inefficiency Persistence and Heterogeneity in Colombian Electricity Distribution Utilities. EPRG Working Paper 1403*, Faculty of Economics, University of Cambridge.

Gassner, K., A. Popov, and N. Pushak. 2008. *Does Private Sector Participation Improve Performance in Electricity and Water Distribution?* The World Bank: PPIAF.

Goto, M. and T. Sueyoshi. 2009. Productivity growth and deregulation of Japanese electricity distribution. *Energy Policy* 37(8): 3130–3138.

Hall, D. 1999. *Electricity Restructuring, Privatisation and Liberalisation: Some International Experiences.* Public Services International Research Unit: University of Greenwich, October.

Haselip, J. and C. Potter. 2010. Post-neoliberal electricity market 're-reforms' in Argentina: Diverging from market prescriptions? *Energy Policy* 38(2): 1168–1176.

Imam, M.I., T. Jamasb, and M. Llorca. 2018. *Sector Reforms and Institutional Corruption: Evidence from Electricity Industry in Sub-Saharan Africa, EPRG Working Paper 1801.*

Izaguirre, A.K. 1998. *Private Participation in the Electricity Sector-Recent Trends. Viewpoint Note No. 154.*

Jamasb, T., R. Mota, D. Newbery, M. Pollitt, and E. Directors. 2005. *Electricity Sector Reform in Developing Countries: A survey of Empirical Evidence on Determinants and Performance. World Bank Policy Research Working Paper, 3549 (March).*

Jamasb, T. 2006. Between the state and market: Electricity sector reform in developing countries. *Utilities Policy* 14: 14–30.

Jain, S., T. Thakur, and A. Shandilya. 2010. A parametric and non-parametric approach for performance appraisal of Indian power generating companies. *International Journal of Advanced Engineering Science and Technologies* 1(2): 64–78.

Joskow, P. L. 1998. Electricity sectors in transition. *Energy Journal* 19(2): 25–28.

Joskow. P. L. 2008. Lessons learned from electricity market liberalization. *Energy Journal*, Special issue. *The Future of Electricity: Papers in Honor of David Newbery.*

Kessides, I. N. 2012. The impacts of electricity sector reforms in developing countries. *The Electricity Journal* 25(6): 79–88.

Kundu, G.K., and B. B. Mishra. 2011. Impact of reform and privatisation on consumers: A case study of power sector reform in Orissa, India. *Energy Policy* 39(6): 3537–3549.

Leeper, E., T. B. Walker, S. S. Yang. 2010. Government investment and fiscal stimulus. *Journal of Monetary Economics* 57(8): 1000–1012.

Megginson, W.L. and J.M. Netter. 2001. From state to market: A survey of empirical studies on privatization. *Journal of Economic Literature* 39: 321–389.

Maulidia, M., P. Dargusch, P. Ashworth, and F., Ardiansyah. 2019. Rethinking renewable energy targets and electricity sector reform in Indonesia: A private sector perspective. *Renewable and Sustainable Energy Reviews* 101: 231–247.

Nagayama, H. 2007. Effects of regulatory reforms in the electricity supply industry on electricity prices in developing countries. *Energy Policy* 35(6): 3440–3462.

Nagayama, H. 2009. Electric power sector reform liberalisation and electric power prices in developing countries: An empirical analysis using international panel data. *Energy Economics* 31(3): 463–472.

Nagayama, H. 2010. Impacts on investments, and transmission/distribution loss through power sector reforms. *Energy Policy* 38(7): 3453–3467.

Nepal, R. and T. Jamasb. 2012. Reforming the power sector in transition: Do institutions matter? *Energy Economics* 34(5): 1675–1682.

Newbery, D. 2002. *Issues and Options for Restructuring Electricity Supply Industries. Cambridge Working Papers in Economics No. 0210.*

Oseni, M.O. 2011. An analysis of the power sector performance in Nigeria. *Renewable and Sustainable Energy Reviews* 15: 4765–4774.

Ogunniyi, A. I., G. Mavrotas, K. O. Olagunju, O. Fadare, and R. Adedoyin. 2020. Governance quality, remittances and their implications for food and nutrition security in Sub-Saharan Africa. *World Development* 127: 104752.

Peng, D. and R. Poudineh. 2017. An appraisal of investment vehicles in the 'Tanzania's electricity sector. *Utilities Policy* 48: 51–68.

Plane, P. 1999. Privatisation, technical efficiency and welfare consequences: The case of the Côte d'Ivoire electricity company (CIE). *World Development* 27(2): 343–360.

Polemis, M. L. 2016. New evidence on the impact of structural reforms on electricity sector performance. *Energy Policy* 92: 420–431.

Pollitt, M. 2008. Electricity reform in Argentina: Lessons for developing countries. *Energy Economics*, 30(4): 1536–1567.

Pollitt, M.G. 1995. *Technical Efficiency in Electrical Power Plants. Cambridge Working Papers in Economics 9422*, Faculty of Economics, University of Cambridge.

Pollitt, M. G. 2012. The role of policy in energy transitions: Lessons from the energy liberalisation era. *Energy Policy* 50: 128–137.

Prasad, G. 2008. Energy sector reform, energy transitions and the poor in Africa. *Energy Policy* 36(8): 2806–2811.

Ramos-Real, F.J., B. Tovar, E.F. de Almeida, and H. Q. Pinto. 2009. The evolution and main determinants of productivity in the Brazilian electricity distribution 1998–2005: An empirical analysis. *Energy Economics* 31(2): 298–305.

Richards, P and M.R. Mullen. 2009. Some evidence of the cumulative effects of corporate social responsibility on financial performance. *Journal of Global Business Issues* 3(1): 1–14.

Sen, A. and T. Jamasb. 2012. Diversity in unity: An empirical analysis of electricity deregulation in Indian states. *The Energy Journal* 33(1): 83–130.

Smith, T.B. 2004. Electricity theft: A comparative analysis. *Energy Policy* 32(18): 2067–2076.

Steggals, W., R. Gross, and P. Heptonstall. 2011. Winds of change: How high wind penetrations will affect investment incentives in the GB electricity sector. *Energy Policy* 39(3): 1389–1396.

Woodridge, J.M. 2010. *Econometric Analysis of Cross Section and Panel Data*, 2nd ed. Cambridge, MA: MIT Press.

World Bank 2021. *Private Participation in Infrastructure (PPI) Database*. Washington, DC.

World Bank. http://ppi.worldbank.org/data. Accessed July 2022.

Yang, M. and D. Sharma. 2012. *The Impacts of Electricity Reforms on Electricity Prices. 3rd IAEE Asian Conference, Tokyo*, February.

Yunos, J. M., and D. Hawdon. 1997. The efficiency of the national electricity board in Malaysia: An inter-country comparison using DEA. *Energy Economics* 19(2): 255–269.

Zhang, Y.F., D. Parker, and C. Kirkpatrick. 2005. Competition, regulation, and privatisation of electricity generation in developing countries: Does the sequencing of the reforms matter? *The Quarterly Review of Economics and Finance* 45(2–3): 358–379.

Zhang, Y-F, D. Parker, and C. Kirkpatrick. 2008. Electricity sector reform in developing countries: An econometric assessment of the effects of privatization, competition, and regulation. *Journal of Regulatory Economics* 33: 159–178.

# 24 Electrification and Energy Transition in Sub-Saharan Africa

## Status, Issues and Effects, from a Source-to-Household Perspective

*Vikrant Katekar*
S. B. Jain Institute of Technology, Management and Research

*Sandip Deshmukh*
Birla Institute of Technology & Science, Pilani

## CONTENTS

## 24.1 INTRODUCTION

Home to over 1 billion people, half of whom will be under the age of 25 by 2050, sub-Saharan Africa is a vibrant continent offering human and natural resources that have the potential to deliver sustainable growth and eliminate poverty in the region, enabling Africans to live healthier and more productive lives across the continent [1]. Table 24.1 shows the population, area, and gross domestic product (GDP) of sub-Saharan countries. This region's total population is more than 87,41,12,479 living on a land area of 2,46,56,134 km². The highest population has been recorded for Nigeria as 17,45,07,539. The lowest population is recorded for Seychelles, 87,476. The highest area is occupied by the Democratic Republic of the Congo (6,86,92,542 km²), and the lowest area is occupied by Seychelles (455 km²) [2].

The total market value of all final products and services generated within a nation in a given time is GDP [3]. The per cent change in real GDP, which accounts for inflation, is called economic growth (GDP growth) [4]. GDP per capita is calculated by dividing a country's economic production by its people. It's a strong indicator of a country's living standards [5]. It also explains how often the people of a nation get profit from the economy [6]. GDP per capita (PPP based) is GDP separated by the total population in foreign dollars using purchasing power parity rates [7]. The buying power of an overseas dollar is equal to that of a US dollar in terms of GDP [8]. Except for Somalia GDP of all other countries are tabulated in Table 24.1. As of 2017, the average GDP per capita (PPP) was found as $5566. Concerning the world GDP per capita (PPP) of $17,100 as of 2017, this continent's GDP is 67.45% less than the world GDP. The lowest and highest GDP per capita (PPP) was recorded for the Central African Republic ($700) and Equatorial Guinea ($37400), respectively. However, this continent is developing an entirely new direction of growth, harnessing its wealth and citizens' potential, with the world's largest free trade area and a 1.2 billion-person market. Figure 24.1 shows the comparison of GDP per capita for different sub-Saharan African countries [9].

**TABLE 24.1**
**Biographical Details of Sub-Saharan African Countries [2]**

| Country | Population | Area (km²) | GDP per Capita (PPP) |
|---|---|---|---|
| Angola | 18,498,000 | 1,246,700 | 6,800 |
| Burundi | 8,988,091 | 27,830 | 700 |
| Democratic Republic of the Congo | 68,692,542 | 2,345,410 | 800 |
| Cameroon | 18,879,301 | 475,440 | 3,700 |
| Central African Republic | 4,511,488 | 622,984 | 700 |
| Chad | 10,329,208 | 1,284,000 | 2,300 |
| Republic of the Congo | 3,700,000 | 342,000 | 800 |
| Equatorial Guinea | 1,110,000 | 28,051 | 37,400 |
| Gabon | 1,514,993 | 267,667 | 18,100 |
| Kenya | 39,002,772 | 582,650 | 3,500 |
| Nigeria | 174,507,539 | 923,768 | 5,900 |
| Rwanda | 10,473,282 | 26,338 | 2,100 |
| Sao Tome and Principe | 212,679 | 1,001 | 3,200 |
| Tanzania | 44,928,923 | 945,087 | 3,200 |
| Uganda | 32,369,558 | 236,040 | 2,400 |
| Sudan | 31,894,000 | 1,886,068 | 4,300 |
| South Sudan | 8,260,490 | 619,745 | 1,600 |
| Djibouti | 516,055 | 23,000 | 3,600 |
| Eritrea | 5,647,168 | 121,320 | 1,600 |
| Ethiopia | 85,237,338 | 1,127,127 | 2,200 |
| Somalia | 9,832,017 | 637,657 | NA |
| Botswana | 1,990,876 | 600,370 | 17,000 |
| Comoros | 752,438 | 2,170 | 1,600 |
| Lesotho | 2,130,819 | 30,355 | 3,300 |
| Madagascar | 19,625,000 | 587,041 | 1,600 |
| Malawi | 14,268,711 | 118,480 | 1,200 |
| Mauritius | 1,284,264 | 2,040 | 22,300 |
| Mozambique | 21,669,278 | 801,590 | 1,300 |
| Namibia | 2,108,665 | 825,418 | 11,200 |
| Seychelles | 87,476 | 455 | 29,300 |
| South Africa | 59,899,991 | 1,219,912 | 13,600 |
| Eswatini | 1,123,913 | 17,363 | 11,089 |
| Zambia | 11,862,740 | 752,614 | 4,000 |
| Zimbabwe | 11,392,629 | 390,580 | 2,300 |

*(Continued)*

**TABLE 24.1 (*Continued*)**
**Biographical Details of Sub-Saharan African Countries [2]**

| Country | Population | Area (km²) | GDP per Capita (PPP) |
|---|---|---|---|
| Benin | 8,791,832 | 112,620 | 2,300 |
| Mali | 12,666,987 | 1,240,000 | 2,200 |
| Burkina Faso | 15,730,977 | 274,200 | 1,900 |
| Cape Verde | 499,000 | 322,462 | 7,000 |
| Ivory Coast | 20,617,068 | 322,463 | 3,900 |
| Gambia | 1,782,893 | 11,295 | 2,600 |
| Ghana | 24,200,000 | 238,535 | 4,700 |
| Guinea | 10,057,975 | 245,857 | 2,200 |
| Guinea-Bissau | 1,647,000 | 36,125 | 1,900 |
| Liberia | 4,128,572 | 111,369 | 1,300 |
| Mauritania | 3,359,185 | 1,030,700 | 4,500 |
| Niger | 17,129,076 | 1,267,000 | 1,200 |
| Senegal | 12,855,153 | 196,712 | 3,500 |
| Sierra Leone | 6,190,280 | 71,740 | 1,600 |
| Togo | 7,154,237 | 56,785 | 1,700 |

NA, not available.

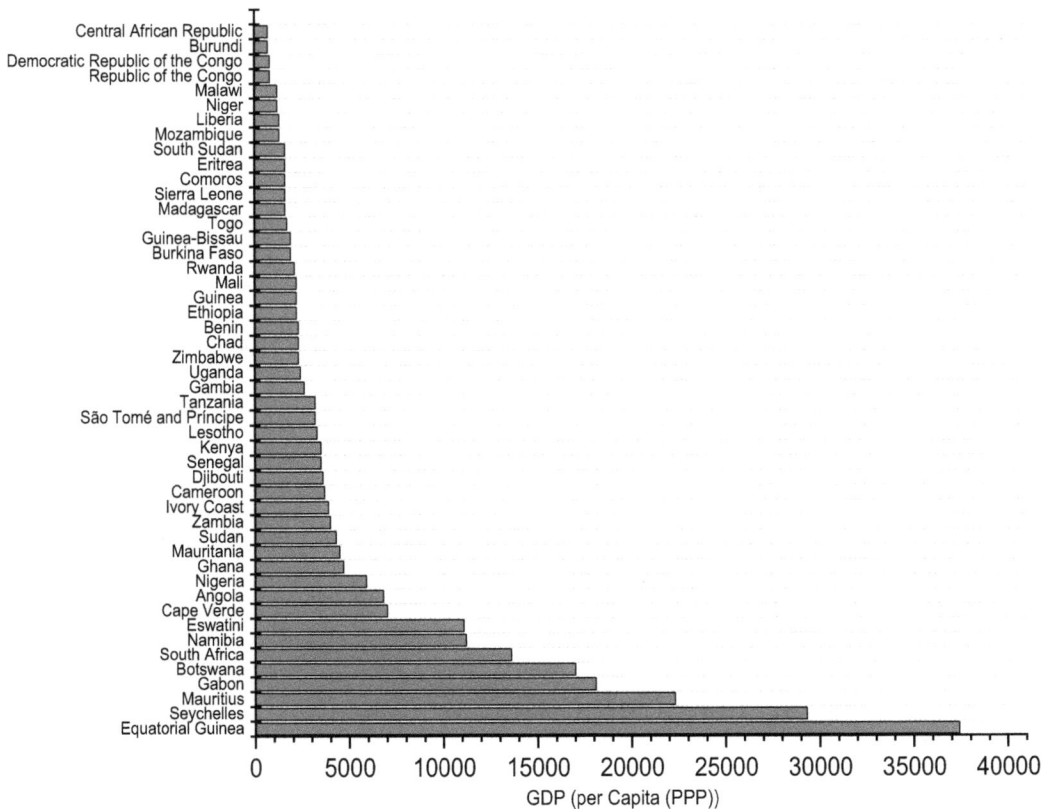

**FIGURE 24.1** GDP per capita (PPP) of sub-Saharan countries. (From Ref. [9]. With permission.)

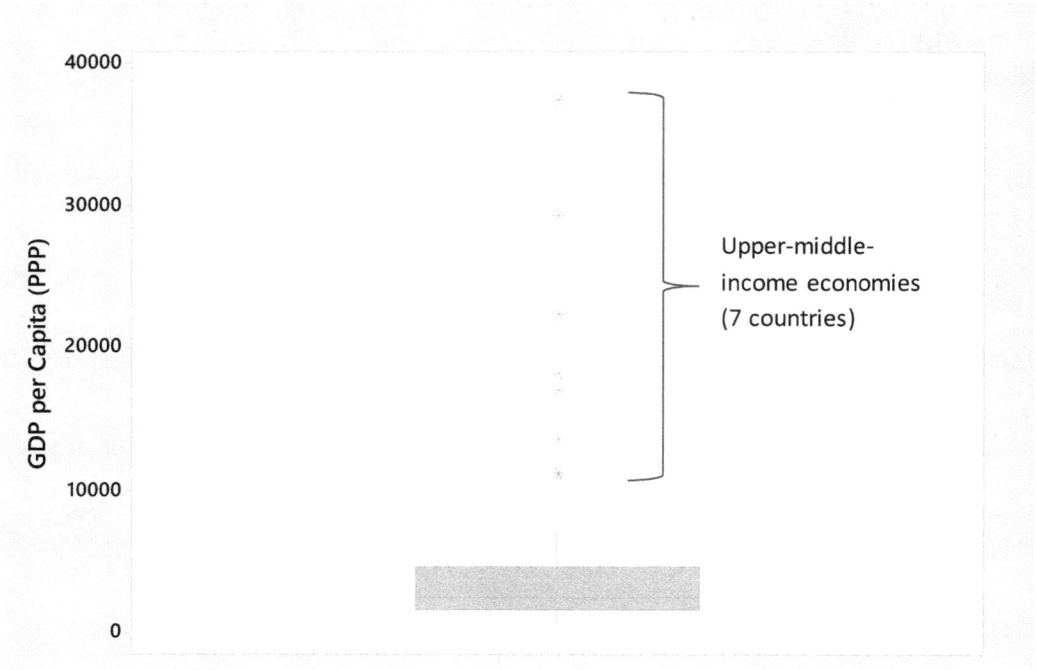

**FIGURE 24.2** Box plot of GDP per capita (PPP) showing outliers, indicating upper-middle-income countries.

There are a total of 49 countries in the sub-Saharan region. The region consists of low-income, lower-middle, upper-middle and high-income nations, 18 vulnerable or affected by violence. There are also 13 small states in this region, marked by a small population, minimal human resources, and a restricted land area. Most of the countries in this region fall under low-income countries [10]. Only seven countries have satisfactory financial progress; they are included in upper-middle-income counties, as shown in Figure 24.2 (outliers). Figure 24.3 shows the map of the sub-Saharan Africa region. Following is the broad classification of sub-Saharan African countries based on their income level [11]:

a. Low-income countries ($1,035 or less) (total 27 countries) [11]

| **Benin** | **The Gambia** | **Niger** |
|---|---|---|
| Burkina Faso | Guinea | Rwanda |
| Burundi | Guinea-Bissau | Sierra Leone |
| Central African Republic | Kenya | Somalia |
| Chad | Liberia | South Sudan |
| Comoros | Madagascar | Tanzania |
| Congo, Dem. Rep. | Malawi | Togo |
| Eritrea | Mali | Uganda |
| Ethiopia | Mozambique | Zimbabwe |

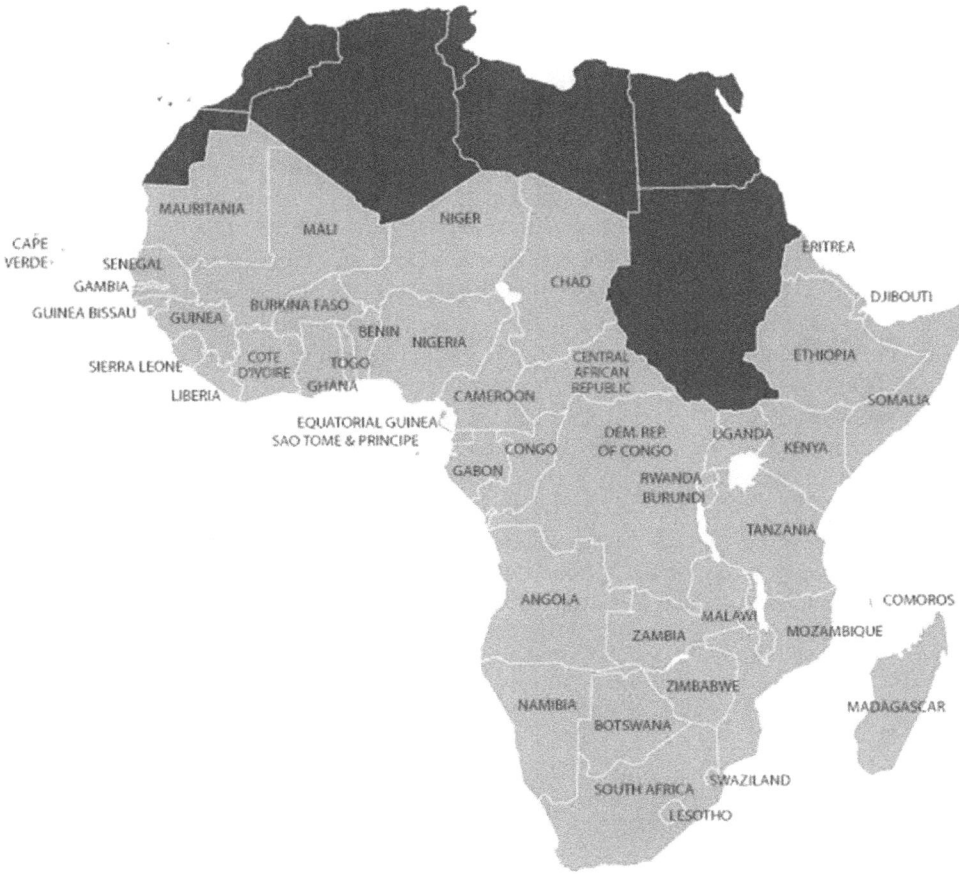

**FIGURE 24.3** Map of sub-Saharan Africa. (From https://images.app.goo.gl/Dik3oMBoF35wANey7. With permission.)

b. Lower-middle-income economies ($1,036–$4,085) (total 13 countries) [11]

| Cameroon | Lesotho | Sudan |
|---|---|---|
| Cape Verde | Mauritania | Swaziland |
| Congo, Rep | Nigeria | Zambia |
| Cote d'Ivoire | Sao Tome and Principe | |
| Ghana | Senegal | |

c. Upper-middle-income economies ($4,086–$12,615) (seven countries) [11]

| Angola | Mauritius | South Africa |
|---|---|---|
| Botswana | Namibia | |
| Gabon | Seychelles | |

As per the United Nations Educational, Scientific, and Cultural Organization (UNESCO), the literacy rate is the proportion of a population who can read and write in a specific age group [12]. The adult literacy rate applies to those aged 15 and over, the youth literacy rate to those aged 15–24, and the elderly literacy rate to those aged 65 and up. The literacy rate is usually defined by how well one can understand a brief, clear sentence about daily life [13]. As per the United Nations International Children's Emergency Fund, the global literacy rate for all people aged 15 and above is 86.3%. The global literacy rate for all males is 90.0%, and the rate for all females is 82.7%. The average literacy rate of the sub-Saharan region is 67.69. The highest literacy rate is recorded for Seychelles (91.85%). This country has male and female literacies are 91.4% and 92.3%, respectively. This may be because this country has the smallest area (only 455 km$^2$) and the lowest population (87,476). It is also interesting to note that the female literacy rate is more significant than the male literacy rate for this country. There are 21 countries (42.85%) whose literacy rate is unknown [14].

Another point to note is that, for Burkina Faso, the female literacy rate is not available, and this country has a meagre male literacy rate (25.30%). Figure 24.4 shows the comparison of male and female literacy rates for sub-Saharan countries. Exact numeric details about literacy are also presented in Table 24.3 [14].

By extension, life expectancy is focused on predicting the average age at which members of a demographic community may pass [15]. The life expectancy for the world in 2006 was 68.17 years, while life expectancy for the sub-Saharan region was stated as 52.47 years in 2006, which was 23.03% less than the world. Table 24.2 shows a list of countries where life expectancy is less

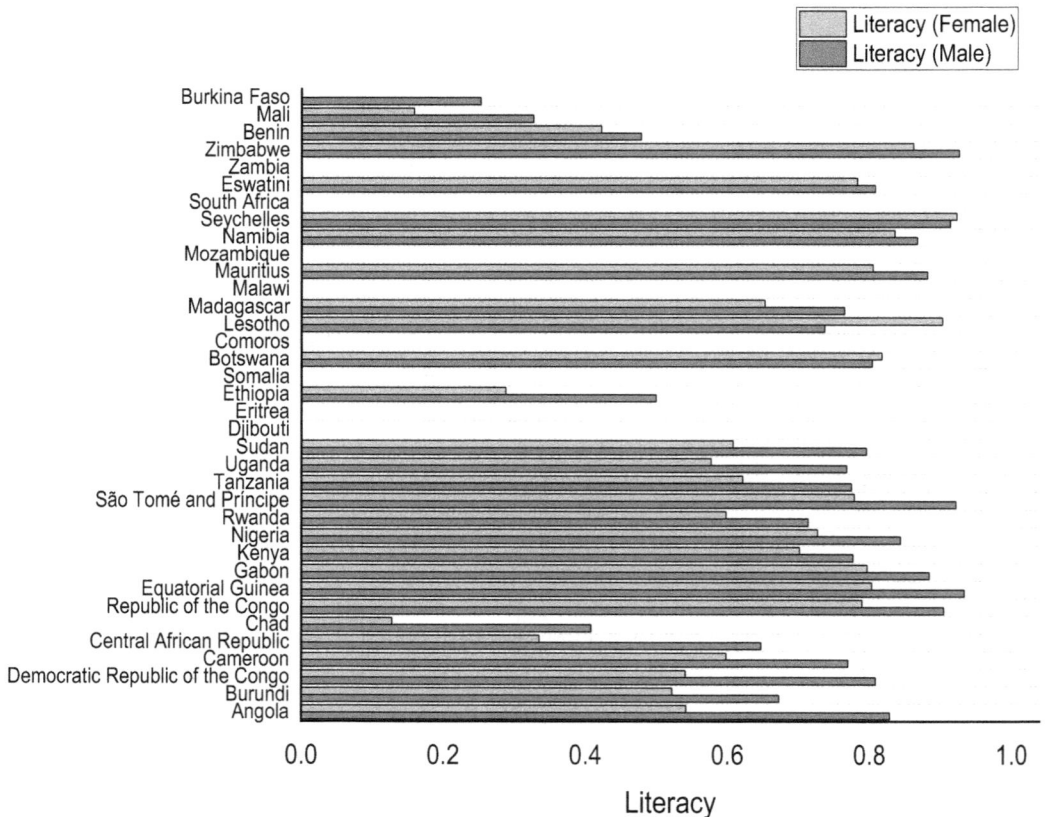

**FIGURE 24.4**  Literacy rate of different countries. (From Ref. [14]. With permission.)

**TABLE 24.2**
**Countries with Less than 50 years of Life Expectancy [16]**

| Country | Life Expectancy (years) |
| --- | --- |
| Eswatini | 40.8 |
| Zambia | 41.7 |
| Angola | 42.4 |
| Mozambique | 42.5 |
| Zimbabwe | 42.7 |
| Lesotho | 42.9 |
| Central African Republic | 44.4 |
| Democratic Republic of the Congo | 46.1 |
| Rwanda | 46.8 |
| Malawi | 47.6 |
| Somalia | 47.7 |
| Burundi | 49 |
| Botswana | 49.8 |

than 50 years. There are 13 countries (26.53% of the total) where life expectancy is less than 50. The detailed information about countries whose life expectancy is less than 50 years is given in Table 24.2 [16]. The lowest life expectancy is recorded for Eswatini (40.8 years).

Similarly, the Human Development Index (HDI) is a statistical index of life expectancy, education, and per capita income used to define human development [17]. As per the UNDP report, the average HDI for the sub-Saharan region was 0.468 as of 2008. The highest and lowest HDIs were recorded for Seychelles (0.773) and the Democratic Republic of the Congo (0.286). Figure 24.5

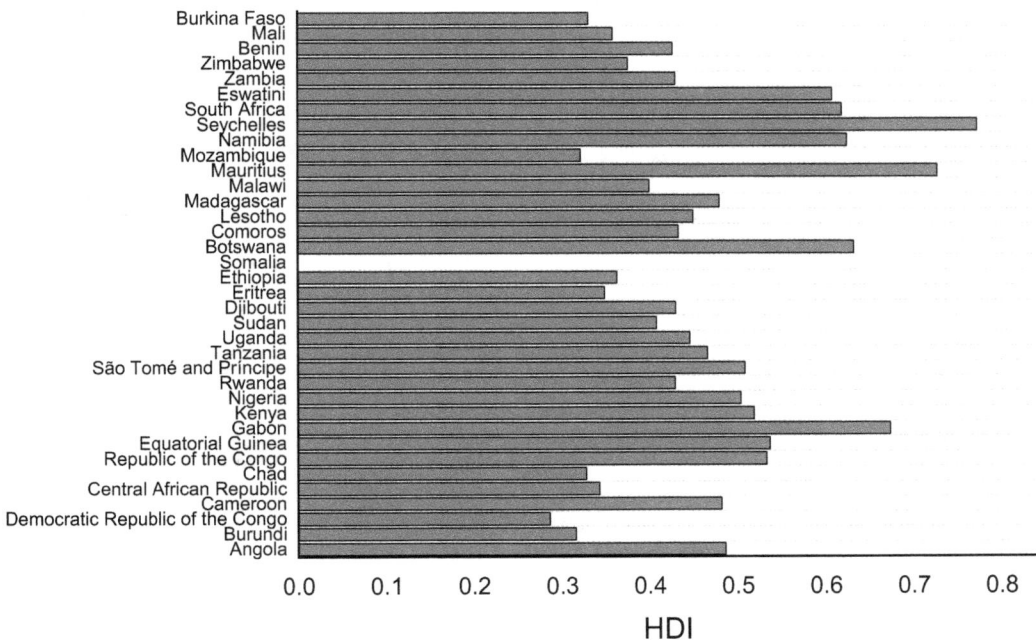

**FIGURE 24.5** Human Development Index for different countries.

shows HDIs for different sub-Saharan countries. HDI value for each sub-Saharan country is given in Table 24.3 [18].

TABLE 24.3
Literacy Rate, Life Expectancy and Human Development Index for Different
Sub-Saharan African Countries [11,14,18]

| Country | Literacy (Male) | Literacy (Female) | Life Exp. (Years) | HDI |
|---|---|---|---|---|
| Angola | 82.90% | 54.20% | 42.4 | 0.486 |
| Burundi | 67.30% | 52.20% | 49 | 0.316 |
| Democratic Republic of the Congo | 80.90% | 54.10% | 46.1 | 0.286 |
| Cameroon | 77% | 59.80% | 50.3 | 0.482 |
| Central African Republic | 64.80% | 33.50% | 44.4 | 0.343 |
| Chad | 40.80% | 12.80% | 50.6 | 0.328 |
| Republic of the Congo | 90.50% | 79.00% | 54.8 | 0.533 |
| Equatorial Guinea | 93.40% | 80.30% | 51.1 | 0.537 |
| Gabon | 88.50% | 79.70% | 56.7 | 0.674 |
| Kenya | 77.70% | 70.2% | 57.8 | 0.519 |
| Nigeria | 84.40% | 72.70% | 57 | 0.504 |
| Rwanda | 71.40% | 59.80% | 46.8 | 0.429 |
| Sao Tome and Principe | 92.20% | 77.90% | 65.2 | 0.509 |
| Tanzania | 77.50% | 62.20% | 51.9 | 0.466 |
| Uganda | 76.80% | 57.7% | 50.7 | 0.446 |
| Sudan | 79.60% | 60.80% | 62.57 | 0.408 |
| South Sudan | NA | NA | NA | NA |
| Djibouti | NA | NA | 54.5 | 0.43 |
| Eritrea | NA | NA | 57.3 | 0.349 |
| Ethiopia | 50% | 28.80% | 52.5 | 0.363 |
| Somalia | NA | NA | 47.7 | NA |
| Botswana | 80.40% | 81.80% | 49.8 | 0.633 |
| Comoros | NA | NA | 63.2 | 0.433 |
| Lesotho | 73.70% | 90.30% | 42.9 | 0.45 |
| Madagascar | 76.50% | 65.30% | 59 | 0.48 |
| Malawi | NA | NA | 47.6 | 0.4 |
| Mauritius | 88.20% | 80.50% | 73.2 | 0.728 |
| Mozambique | NA | NA | 42.5 | 0.322 |
| Namibia | 86.80% | 83.60% | 52.5 | 0.625 |
| Seychelles | 91.40% | 92.30% | 72.2 | 0.773 |
| South Africa | NA | NA | 50.7 | 0.619 |
| Eswatini | 80.90% | 78.30% | 40.8 | 0.608 |
| Zambia | NA | NA | 41.7 | 0.43 |
| Zimbabwe | 92.70% | 86.20% | 42.7 | 0.376 |
| Benin | 47.90% | 42.30% | 56.2 | 0.427 |
| Mali | 32.70% | 15.90% | 53.8 | 0.359 |
| Burkina Faso | 25.30% | NA | 51 | 0.331 |

NA, Not available.

Countries whose literacy rate, life expectancy, and HDI is unknown even if today are as follows (total 21 countries, 42.85% of total sub-Saharan countries) [18].

| **Cape Verde** | **Guinea-Bissau** | **Senegal** |
|---|---|---|
| Comoros | Ivory Coast | Sierra Leone |
| Djibouti | Liberia | Somalia |
| Eritrea | Malawi | South Africa |
| Gambia | Mauritania | South Sudan |
| Ghana | Mozambique | Togo |
| Guinea | Niger | Zambia |

## 24.2　ENERGY SECURITY IN SUB-SAHARAN COUNTRIES

### 24.2.1　ELECTRIFICATION IN SUB-SAHARAN AFRICAN COUNTRIES

Electrification provides electrical energy from the source of its generation to the point of application [19]. Electrification, this term refers to providing necessary infrastructure and supplying electricity to the community [20]. Electrification is the indicator of the socio-economic growth of societies [21]. Table 24.4 illustrates year-wise progress in percentage access to electricity in people of sub-Saharan African countries. In 2018, 47.67% population had access to electricity, of which 31.55% rural and 78.09% urban people have access to electricity. On average, 39.14% of people have access to electricity, which encompasses 22.17% rural and 73.32% urban people [22].

Figure 24.6 shows that the percentage of access to electricity to people is increasing year by year. There is a sharp increase in access in 2015–16. As compared with urban electrification access, the rate of rural electricity access is more progressive. During 2017–18, the percentage increases in overall and rural electricity access are 6.71% and 17.28%, respectively. However, urban electricity access was decreased during this period [23].

Table 24.5 illustrates the year-wise percentage utilisation of energy sources of sub-Saharan countries for electricity generation. 54.86% of the electricity is generated by coal. Its contribution is 61% out of all the fuel sources used. Henceforth, carbon emission is also very significant (55.19%) [24]. Followed by coal, the contribution of renewable energy is highest (24.41%); this is a remarkable achievement of this region. Hydropower plays a vital role in electricity generation (19.79%), i.e., 22%

## TABLE 24.4
**Year-Wise Progress in Percentage Access to Electricity in sub-Saharan African Countries [22]**

| Year | Access to Electricity (% of the Population) | Access to Electricity, Rural (% of Rural Population) | Access to Electricity, Urban (% of Urban Population) |
|---|---|---|---|
| 2000 | 26.00 | 11.35 | 64.17 |
| 2011 | 35.82 | 18.05 | 72.30 |
| 2012 | 37.53 | 21.12 | 71.85 |
| 2013 | 38.38 | 21.45 | 72.35 |
| 2014 | 38.28 | 19.69 | 72.53 |
| 2015 | 39.27 | 20.51 | 73.82 |
| 2016 | 44.67 | 28.87 | 76.39 |
| 2017 | 44.67 | 26.90 | 78.35 |
| 2018 | 47.67 | 31.55 | 78.09 |
| Avg. | 39.14 | 22.17 | 73.32 |

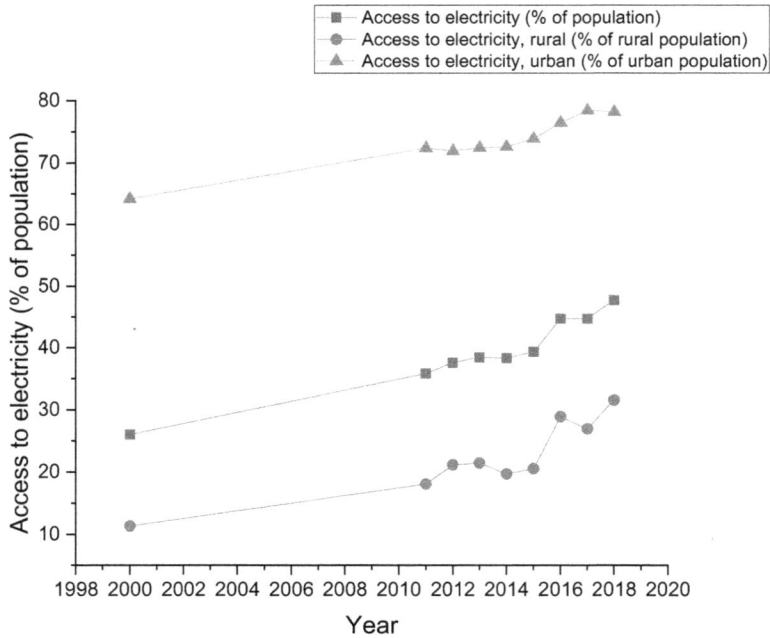

**FIGURE 24.6** Percentage access to electricity for people of sub-Saharan Africa. (From Ref. [23]. With permission.)

**TABLE 24.5**

**Year-Wise Percentage Utilisation of Energy Sources of Sub-Saharan Countries [25]**

| Year | CO$_2$ Emissions from Electricity and Heat Production, Total (% of Total Fuel Combustion) | Electricity Production from Coal Sources (% of Total) | Electricity Production from Hydroelectric Sources (% of Total) | Electricity Production from Natural Gas Sources (% of Total) | Electricity Production from Nuclear Sources (% of Total) | Electricity Production from Oil Sources (% of Total) | Electricity Production from Renewable Sources, Excluding Hydroelectric (% of Total) | Renewable Electricity Output (% of Total Electricity Output) |
|---|---|---|---|---|---|---|---|---|
| 2000 | 56.60 | 62.00 | 17.15 | 3.85 | 4.06 | 2.37 | 0.42 | 20.60 |
| 2011 | 55.79 | 56.03 | 19.93 | 7.35 | 3.04 | 2.85 | 0.67 | 23.80 |
| 2012 | 54.93 | 55.27 | 19.84 | 7.69 | 2.68 | 3.38 | 0.73 | 24.09 |
| 2013 | 53.88 | 53.64 | 20.27 | 7.92 | 3.10 | 3.66 | 0.86 | 24.76 |
| 2014 | 54.77 | 51.61 | 21.10 | 8.52 | 2.96 | 3.81 | 1.70 | 26.58 |
| 2015 | NA | 50.61 | 20.45 | 9.51 | NA | 3.99 | 2.39 | 26.62 |
| Avg. | 55.19 | 54.86 | 19.79 | 7.47 | 3.17 | 3.34 | 1.13 | 24.41 |

NA, not available.

of the total fuel's contribution. But utilisation of other renewable forms of energies (excluding hydro) is highly negligible (approximately 1%) [25].

As illustrated in Figure 24.7, from 2013 onwards, carbon emissions increased, indicating that electricity generation raised in the region and progressed. Electricity generation was decreasing

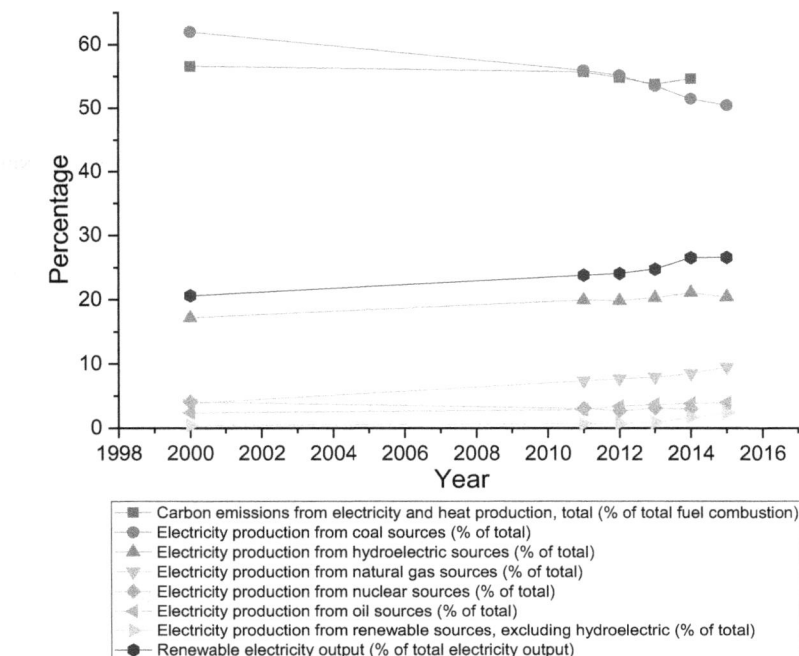

**FIGURE 24.7**  Fuel consumption for electricity generation.

using coal, and the share of renewable energy was growing since 2011. In place of coal, natural gas was used mainly for electricity generation. There was no variation found in electricity generation using oils.

As mentioned earlier, in 2015, most of the electricity generation was by hydropower (Table 24.6) at 38.58%. After that, oil, natural gas, and coal were used. The utilisation of nuclear and other renewable energy for electricity generation was negligible. As shown in Figure 24.8, only one country uses nuclear power for electricity generation (i.e., South Africa), six countries use coal, ten countries use natural gas, 16 countries use renewable, 21 countries use hydro, and 21 countries use oil electricity generation [26].

**TABLE 24.6**
**Electricity Production, Sources, and Access (2015) [26]**

| | Coal | Natural Gas | Oil | Hydro Power | Renewable Sources | Nuclear Power |
|---|---|---|---|---|---|---|
| | \multicolumn{6}{c}{Sources of Electricity Production} | | | | | |
| **Country** | % of Total | % of Total | % of Total | % of Total | % of Total | % of Total |
| Angola | 0 | 0 | 46.8 | 53.2 | 0 | 0 |
| Benin | 0 | 0 | 94.4 | 4.1 | 1.5 | 0 |
| Botswana | 96.4 | 0 | 3.6 | 0 | 0 | 0 |
| Cameroon | 0 | 6 | 17.9 | 75 | 1.1 | 0 |
| Congo, Rep. | 0 | 46.7 | 0 | 53.3 | 0 | 0 |
| Cote d'Ivoire/Ivory Coast | 0 | 78.1 | 5.2 | 15.5 | 1.2 | 0 |
| Dominican Republic | 12.9 | 22 | 53.5 | 6.4 | 5.3 | 0 |
| Eritrea | 0 | 0 | 99.5 | 0 | 0.5 | 0 |

*(Continued)*

**TABLE 24.6 (*Continued*)**
**Electricity Production, Sources, and Access (2015) [26]**

| | Sources of Electricity Production | | | | | |
|---|---|---|---|---|---|---|
| | Coal | Natural Gas | Oil | Hydro Power | Renewable Sources | Nuclear Power |
| **Country** | **% of Total** | **% of Total** | **% of Total** | **% of Total** | **% of Total** | **% of Total** |
| Ethiopia | 0 | 0 | 0 | 92.7 | 7.3 | 0 |
| Gabon | 0 | 46 | 10.2 | 43.2 | 0.6 | 0 |
| Ghana | 0 | 38.8 | 0 | 50.9 | 0 | 0 |
| Kenya | 0 | 0 | 12.5 | 39.2 | 48.3 | 0 |
| Mauritius | 39.4 | 0 | 37.8 | 4.1 | 18.7 | 0 |
| Mozambique | 0 | 12.8 | 0.8 | 86.4 | 0 | 0 |
| Namibia | 0.5 | 0 | 1.8 | 97.8 | 0 | 0 |
| Niger | 41.6 | 0 | 57.6 | 0 | 0.8 | 0 |
| Nigeria | 0 | 81.8 | 0 | 18.2 | 0 | 0 |
| Senegal | 0 | 4.2 | 83.6 | 8.6 | 1.8 | 0 |
| South Africa | 92.7 | 0 | 0.1 | 0.3 | 1.9 | 5.5 |
| South Sudan | 0 | 0 | 99.4 | 0 | 0.6 | 0 |
| Sudan | 0 | 0 | 22.6 | 64.5 | 0 | 0 |
| Tanzania | 0 | 43.9 | 21.9 | 33.5 | 0.7 | 0 |
| Togo | 0 | 0 | 24.7 | 69.1 | 6.2 | 0 |
| Zambia | 0 | 0 | 3 | 97 | 0 | 0 |
| Zimbabwe | 46.8 | 0 | 0.5 | 51.4 | 1.3 | 0 |
| Total | 13.21 | 15.21 | 27.90 | 38.58 | 3.91 | 0.22 |

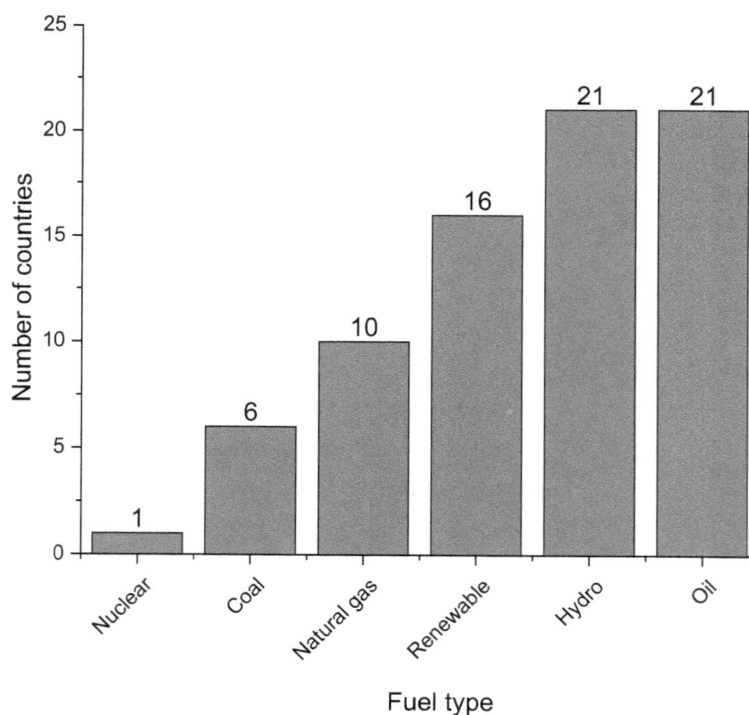

**FIGURE 24.8** Fuel sources used for electricity generation. (From Ref. [26]. With permission.)

As shown in Figure 24.9, six countries use 91%–100%, five countries use 61%–90%, four countries use 31%–60%, and nine countries use less than 30% fossil fuel source for electricity generation. In 2015, South Africa, Benin, Niger, South Sudan, Eritrea, and Botswana were the topmost countries generating electricity by utilising mostly fossil fuels (Figure 24.10). Tables 24.7 and 24.8 illustrate electricity production from oil, gas and coal sources.

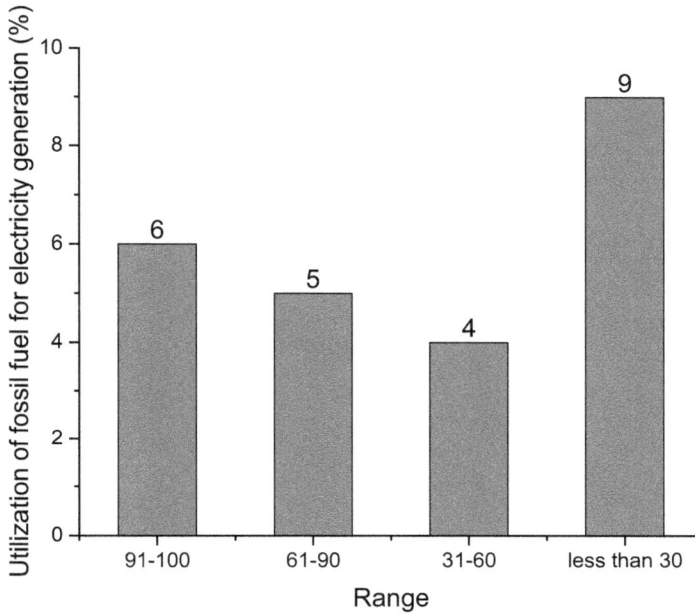

**FIGURE 24.9**  Total number of countries using fossil fuel for electricity generation. (From Ref. [26]. With permission.)

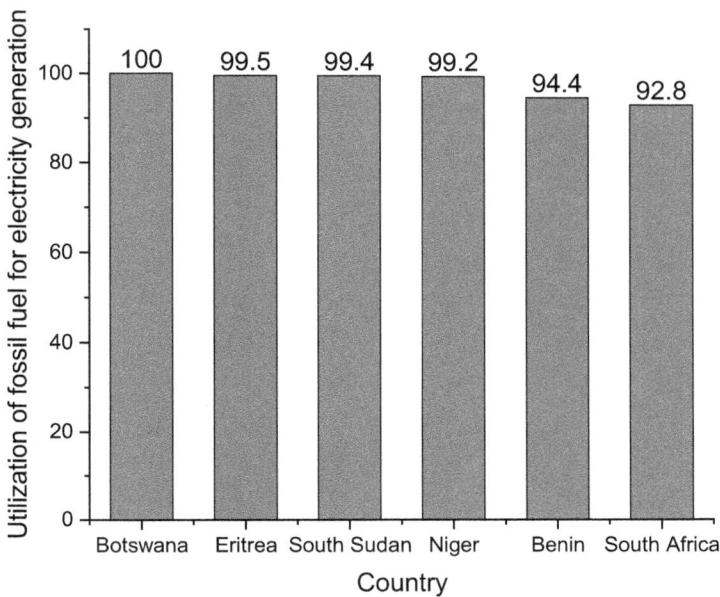

**FIGURE 24.10**  Top countries utilising fossil fuel for electricity generation. (From Ref. [26]. With permission.)

**TABLE 24.7**

**Electricity Production from Oil, Gas and Coal Sources [27]**

| Electricity Production from Oil, Gas and Coal Sources (% of Total) (2015) | % |
|---|---|
| Angola | 46.8 |
| Benin | 94.4 |
| Botswana | 100 |
| Cameroon | 23.9 |
| Congo, Dem. Rep. | 0.2 |
| Congo, Rep. | 46.7 |
| Cote d'Ivoire | 83.3 |
| Eritrea | 99.5 |
| Gabon | 56.3 |
| Ghana | 38.8 |
| Kenya | 12.5 |
| Mauritius | 77.3 |
| Mozambique | 13.6 |
| Namibia | 2.2 |
| Niger | 99.2 |
| Nigeria | 81.8 |
| Senegal | 87.8 |
| South Africa | 92.8 |
| South Sudan | 99.4 |
| Sudan | 22.6 |
| Tanzania | 65.8 |
| Togo | 24.7 |
| Zambia | 3 |

**TABLE 24.8**

**Electricity Production from Oil Sources [27]**

| Electricity Production from Oil Sources (% of Total) (2015) | % |
|---|---|
| Angola | 46.8 |
| Benin | 94.4 |
| Botswana | 3.6 |
| Cameroon | 17.9 |
| Congo, Dem. Rep. | 0.1 |
| Congo, Rep. | 0.7 |
| Cote d'Ivoire | 5.2 |
| Eritrea | 99.5 |
| Gabon | 10.2 |
| Ghana | 0.5 |
| Kenya | 12.5 |
| Mauritius | 37.8 |
| Mozambique | 0.8 |
| Namibia | 1.8 |
| Niger | 57.6 |
| Nigeria | 20.4 |
| Senegal | 83.6 |

*(Continued)*

**TABLE 24.8 (*Continued*)**
**Electricity Production from Oil Sources [27]**

| Electricity Production from Oil Sources (% of Total) (2015) | % |
|---|---|
| South Africa | 0.1 |
| South Sudan | 99.4 |
| Sudan | 22.6 |
| Tanzania | 21.9 |
| Togo | 24.7 |
| Zambia | 3 |
| Zimbabwe | 0.5 |

Table 24.7 electricity generation using fossil fuels [27]. Botswana, Cote d'Ivoire, Niger, and South Sudan are almost generating energy using fossil fuels. However, Congo, Dem. Rep, Namibia, and Zambia are independent of fossil fuel. They use even less than 5% of fossil fuel for electricity generation.

In fossil fuel utilisation, the use of oil is predominant [28]. Table 24.8 enlists the countries with their percentage contribution of oil in electricity generation. Benin, Gabon, and South Sudan mostly use oil for energy electricity production. Botswana, Congo, Dem. Rep, Congo, Rep., Cote d'Ivoire, Ghana, Mozambique, Namibia, South Africa, South Sudan, Zambia, and Zimbabwe are independent on oil for electricity generation. They use less than 5% oil for electricity generation, as shown in Table 24.8.

Table 24.9 illustrates countries with their electricity generation using renewable energy, excluding hydropower. Cote d'Ivoire, Zimbabwe, Mauritius, Ethiopia, Kenya, and South Africa are the

**TABLE 24.9**
**Electricity Production from Renewable Sources, Excluding Hydroelectric Power [29]**

| Electricity Production from Renewable Sources, Excluding Hydroelectric (kWh) – Sub-Saharan Africa (2015) | kWh |
|---|---|
| Benin | 5,000.00 |
| Botswana | 1,000.00 |
| Cameroon | 76,000.00 |
| Congo, Dem. Rep. | 10,000.00 |
| Cote d'Ivoire | 105,000.00 |
| Eritrea | 2,000.00 |
| Ethiopia | 759,000.00 |
| Gabon | 12,000.00 |
| Ghana | 3,000.00 |
| Kenya | 4,659,000.00 |
| Mauritius | 559,000.00 |
| Niger | 4,000.00 |
| Senegal | 70,000.00 |
| South Africa | 4,763,000.00 |
| South Sudan | 2,000.00 |
| Tanzania | 42,000.00 |
| Togo | 5,000.00 |
| Zimbabwe | 129,000.00 |

top countries generating electricity using renewable sources. South Africa is the leading country to generate electric power using renewable energy sources.

Getting electricity at home quickly after applying for it is vital to show the efficiency of electricity infrastructure [30]. As per the World Bank report 2019, 82.59 days are required to get the electricity at the customer's home at the global level [31]. In 2019, 109.57 days were needed in sub-Saharan countries, as shown in Table 24.10. On average, 124.18 days, i.e., 4.13 months, are required, which shows that electrification infrastructure development is slower in sub-Saharan countries. But Figure 24.11 shows a fierce cut shot of this period from 139.49 to 109.57 days (21.44% improvement) for supplying electricity to the customer's home.

As per Figure 24.12, only five countries have 91%–100%, 11 countries have 61%–90%, 20 countries have 31%–60%, and 14 countries have less than 30% electricity access to people. As shown in Table 24.11, almost 50% of sub-Saharan countries do not have electricity in their homes. Only South Africa, Gabon, Cabo Verde, Mauritius, Seychelles countries could provide more than 90% of the people with electricity access, as shown in Figure 24.13 [32].

**TABLE 24.10**
**Time Required to Get the Electricity in Sub-Saharan Countries [32]**

| Year | The Time Needed to Get Electricity (days) |
|------|-------------------------------------------|
| 2011 | 139.49 |
| 2012 | 141.65 |
| 2013 | 136.29 |
| 2014 | 132.14 |
| 2015 | 117.97 |
| 2016 | 115.57 |
| 2017 | 113.45 |
| 2018 | 111.52 |
| 2019 | 109.57 |

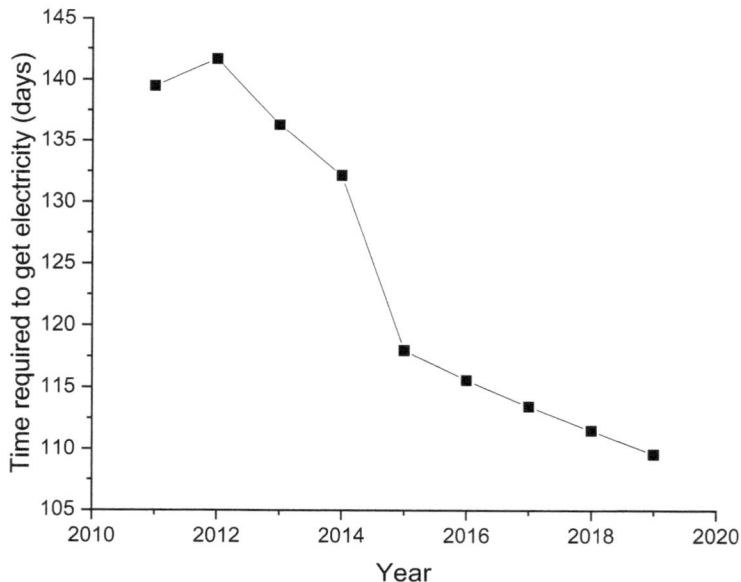

**FIGURE 24.11**   Time required to get electricity. (From Ref. [32]. With permission.)

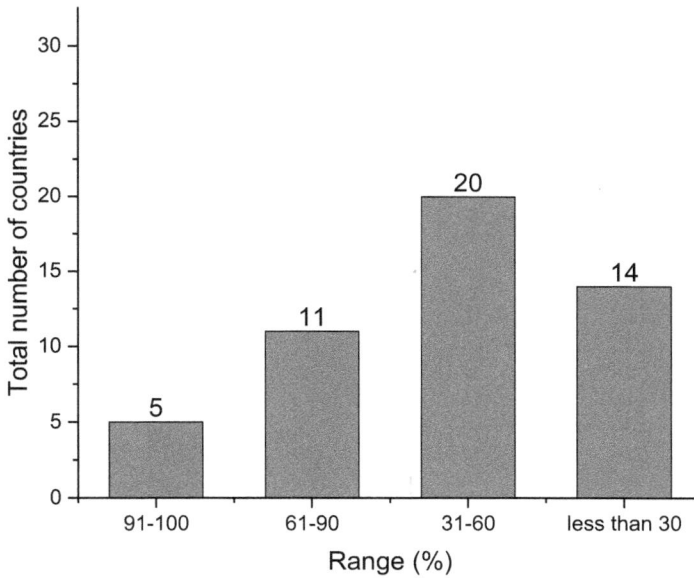

**FIGURE 24.12** Electricity access to people (combined rural and urban). (From Ref. [32]. With permission.)

**TABLE 24.11**
**Access to Electricity (% of the Population) [23]**

| Access to Electricity (% of the Population) (2018) | % |
|---|---|
| Angola | 43.3 |
| Benin | 41.5 |
| Botswana | 64.9 |
| Burkina Faso | 14.4 |
| Burundi | 11 |
| Cabo Verde | 93.6 |
| Cameroon | 62.7 |
| Central African Republic | 32.4 |
| Chad | 11.8 |
| Comoros | 81.9 |
| Congo, Dem. Rep. | 19 |
| Congo, Rep. | 68.5 |
| Cote d'Ivoire | 67 |
| Equatorial Guinea | 67 |
| Eritrea | 49.6 |
| Eswatini | 76.5 |
| Ethiopia | 45 |
| Gabon | 93 |
| Gambia | 60.3 |
| Ghana | 82.4 |
| Guinea | 44 |
| Guinea-Bissau | 28.7 |
| Kenya | 75 |
| Lesotho | 47 |
| Liberia | 25.9 |

*(Continued)*

**TABLE 24.11** (*Continued*)
**Access to Electricity (% of the Population) [23]**

| Access to Electricity (% of the Population) (2018) | % |
|---|---|
| Madagascar | 25.9 |
| Malawi | 18 |
| Mali | 50.9 |
| Mauritania | 44.5 |
| Mauritius | 97.5 |
| Mozambique | 31.1 |
| Namibia | 53.9 |
| Niger | 17.6 |
| Nigeria | 56.5 |
| Rwanda | 34.7 |
| Sao Tome and Principe | 71 |
| Senegal | 67 |
| Seychelles | 100 |
| Sierra Leone | 26.1 |
| Somalia | 35.3 |
| South Africa | 91.2 |
| South Sudan | 28.2 |
| Sudan | 59.8 |
| Tanzania | 35.6 |
| Togo | 51.3 |
| Uganda | 42.6 |
| Zambia | 39.8 |
| Average | 50.74 |

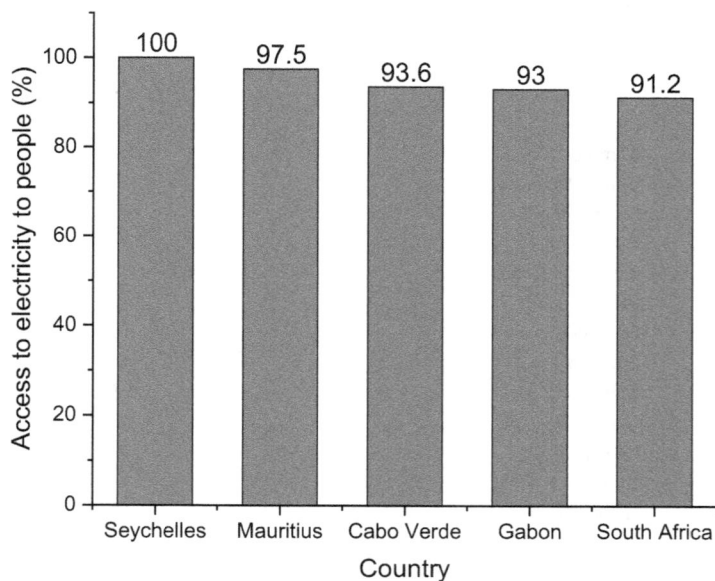

**FIGURE 24.13**  Top five countries providing access to electricity to people.

Table 24.11 presents access to electricity to the people of sub-Saharan countries. On an average basis, 50.74% population have access to electricity. The government of Seychelles successfully providing 100% of people with electricity access. However, only 11% of people in Burundi are getting electricity at their homes.

In rural areas, electricity access is inferior [33]. Only one third (33.33%) of people have electricity in their homes (refer to Table 24.12). Comoros, South Africa, Cabo Verde, Mauritius, and Seychelles can provide electrical connections to more than 75% of rural people. Mauritius and Seychelles deliver 100% electricity access to their rural people, as shown in Figure 24.14. To address the issue of the fulfilment of energy demand for rural communities, governments of sub-Saharan African countries must promote the use of devices working on renewable energy such as windmills [34], smokeless biomass stoves [35], solar water heaters [36], solar cookers [37–39], solar distillation systems [40–47], and solar dryer or air heater [48–51].

**TABLE 24.12**
**Access to Electricity, Rural [23]**

| Access to Electricity, Rural (% of Rural Population) (2018) | % |
|---|---|
| Benin | 18.3 |
| Botswana | 27.9 |
| Burkina Faso | 4.7 |
| Burundi | 3.4 |
| Cabo Verde | 96.9 |
| Cameroon | 23 |
| Central African Republic | 16.3 |
| Chad | 2.7 |
| Comoros | 77 |
| Congo, Dem. Rep. | 1.8 |
| Congo, Rep. | 20.2 |
| Cote d'Ivoire | 32.9 |
| Equatorial Guinea | 6.6 |
| Eritrea | 34.6 |
| Eswatini | 70.2 |
| Ethiopia | 32.7 |
| Gabon | 62.5 |
| Gambia | 35.5 |
| Ghana | 67.3 |
| Guinea | 19.7 |
| Guinea-Bissau | 10 |
| Kenya | 71.7 |
| Lesotho | 37.7 |
| Liberia | 7.4 |
| Malawi | 10.4 |
| Mali | 25.4 |
| Mauritania | 0.6 |
| Mauritius | 100 |
| Mozambique | 8 |
| Namibia | 35.5 |
| Niger | 11.7 |
| Nigeria | 31 |
| Rwanda | 23.4 |

*(Continued)*

**TABLE 24.12 (*Continued*)**
**Access to Electricity, Rural [23]**

| Access to Electricity, Rural (% of Rural Population) (2018) | % |
|---|---|
| Sao Tome and Principe | 55.7 |
| Senegal | 44.2 |
| Seychelles | 100 |
| Sierra Leone | 6.4 |
| Somalia | 14.6 |
| South Africa | 89.6 |
| South Sudan | 23.7 |
| Sudan | 47.1 |
| Tanzania | 18.8 |
| Togo | 22.4 |
| Uganda | 38 |
| Zambia | 11 |
| Average | 33.3 |

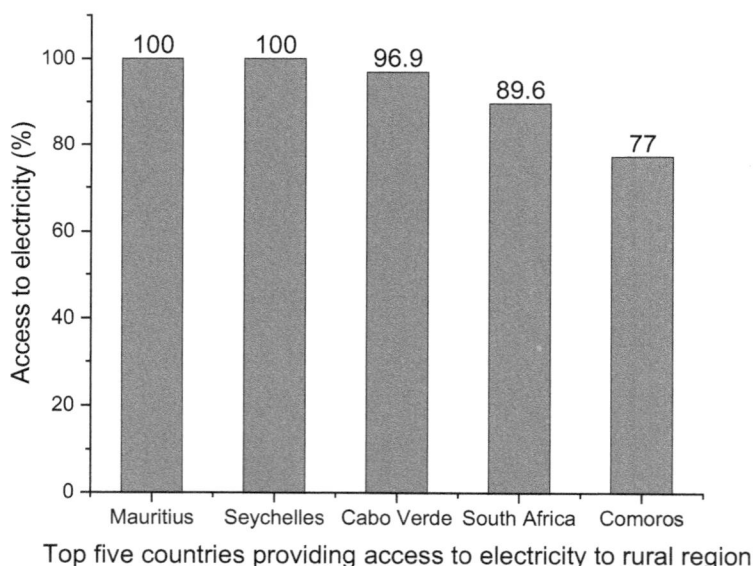

Top five countries providing access to electricity to rural region

**FIGURE 24.14**  Top five countries providing access to electricity to the rural region. (From Ref. [23]. With permission.)

Table 24.12 provides the percentage access of electricity to the rural sub-Saharan region. Only 33.3% of the rural area is electrified till 2018. Electricity access to 100% of rural people is provided by Seychelles and Mauritius, as shown in Figure 24.15.

Electrification in urban areas of sub-Saharan countries is better than in rural regions. As demonstrated in Table 24.13, 75.98% of electrification is done in rural areas. Fourteen countries provide more than 91%, 17 countries offer 71%–90%, 12 countries provide 50%–70% urban electrification. Six governments provide less than 50% of people with electricity accessibility, as shown in Figure 24.16.

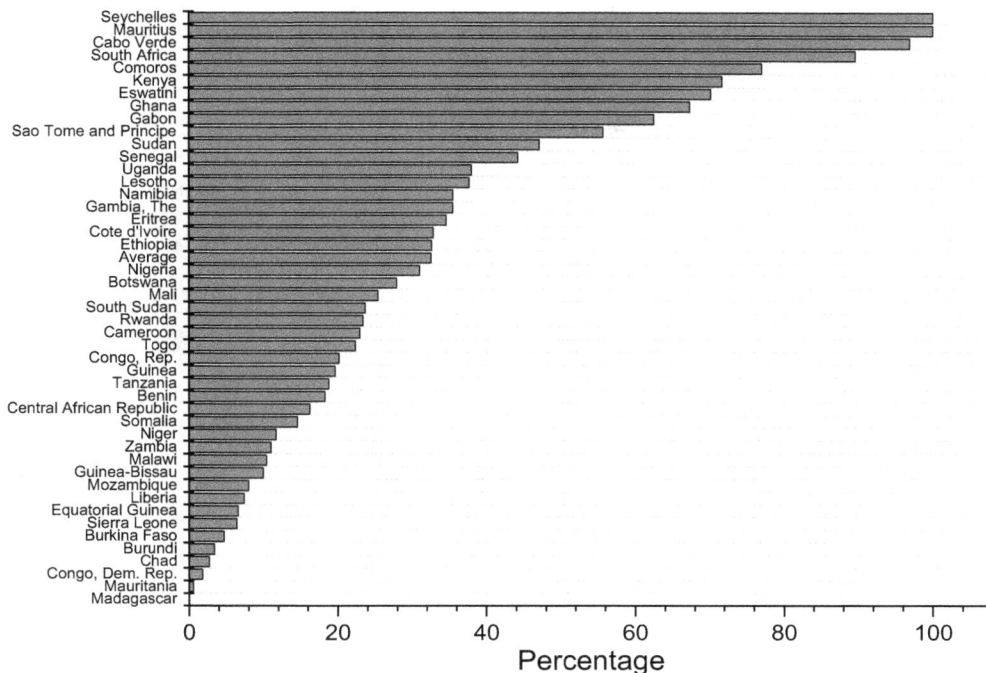

**FIGURE 24.15** Percentage access of electricity to rural people. (From Ref. [23]. With permission.)

---

**TABLE 24.13**

**Access to Electricity, Urban [52]**

| Access to Electricity, Urban (% of Urban Population) (2018) | % |
|---|---|
| Angola | 73.7 |
| Benin | 67.4 |
| Botswana | 81.1 |
| Burkina Faso | 62.3 |
| Burundi | 61.7 |
| Cabo Verde | 91.9 |
| Cameroon | 93.3 |
| Central African Republic | 55.2 |
| Chad | 41.8 |
| Comoros | 94 |
| Congo, Dem. Rep. | 50.7 |
| Congo, Rep. | 92.4 |
| Cote d'Ivoire | 100 |
| Equatorial Guinea | 90.4 |
| Eritrea | 77.1 |
| Eswatini | 96.7 |
| Ethiopia | 92 |
| Gabon | 96.7 |
| Gambia | 76 |
| Ghana | 94.2 |
| Guinea | 87 |

*(Continued)*

**TABLE 24.13** (*Continued*)
**Access to Electricity, Urban [52]**

| Access to Electricity, Urban (% of Urban Population) (2018) | % |
|---|---|
| Guinea-Bissau | 53.1 |
| Kenya | 84 |
| Lesotho | 70.7 |
| Liberia | 43.6 |
| Madagascar | 69.6 |
| Malawi | 55.2 |
| Mali | 85.6 |
| Mauritania | 82.4 |
| Mauritius | 88.5 |
| Mozambique | 72.2 |
| Namibia | 72.2 |
| Niger | 47.6 |
| Nigeria | 81.7 |
| Rwanda | 89.1 |
| Sao Tome and Principe | 76.7 |
| Senegal | 92.4 |
| Seychelles | 99.6 |
| Sierra Leone | 53.2 |
| Somalia | 60.5 |
| South Africa | 92.1 |
| South Sudan | 46.8 |
| Sudan | 83.8 |
| Tanzania | 68.3 |
| Togo | 91.9 |
| Uganda | 57.5 |
| Zambia | 77.2 |
| Average | 75.98 |

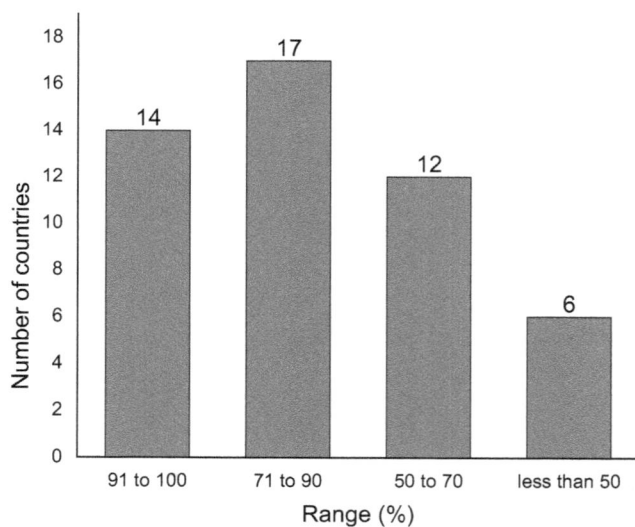

**FIGURE 24.16**  Electricity access to urban people.

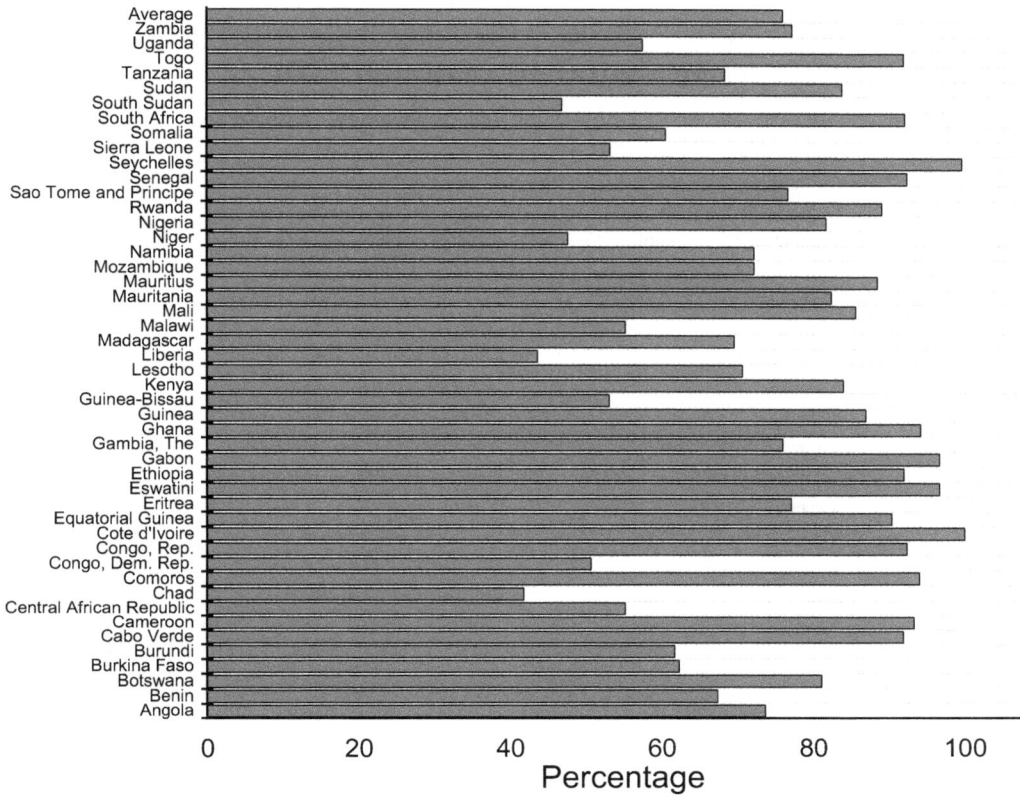

**FIGURE 24.17** Percentage access to electricity by urban people. (From Ref. [52]. With permission.)

Table 24.13 shows country-wise electricity access to urban people. Cote d'Ivoire can provide 100% electricity access to urban people; however, the urban people of Chad have the lowest access to electricity, as shown in Figure 24.17.

Table 24.14 states countries that do not have the infrastructure to generate electricity. They usually purchase electricity from nearby countries. But still, Seychelles can provide 100% electricity access to its population, as shown in Figure 24.18.

## 24.2.2 NEXUS BETWEEN ELECTRIFICATION AND PROGRESS INDICATORS OF SUB-SAHARAN AFRICA

A nexus is a link or sequence of links that connect two or more items. Table 24.15 establishes the nexus between GDP and access to electricity. The bottommost 20 countries are compared to investigate the nexus. Similar colors are given to the names of the country present in both columns. For example, Benin is present in both the columns and is indicated by the same color. Table 24.15 illustrates that the people of Benin have poor access to electricity; hence its GDP is also lower. Similar conclusions are drawn for other countries also. The exact values of GDP and percentage access to electricity are displayed in Table 24.16. From Table 24.15, it has been concluded that 15 countries out of the bottom 20 countries have unquestionably established the nexus between electricity access and GDP.

From Table 24.15, it has also been found that Comoros, Eritrea, Ethiopia, Mali, and Togo (a total of five countries) are having low GDP per capita, but still, they are capable of supplying electric access to their people; however, Angola, Chad, Tanzania, Uganda, Zambia (total five countries) are

**TABLE 24.14**
**Countries Not Generating Electricity (2015) [52]**

| Country | Access to Electricity (%) |
| --- | --- |
| Burkina Faso | 16.3 |
| Burundi | 8.6 |
| Cabo Verde/Cape Verde | 87.1 |
| Central African Republic | 24.1 |
| Chad | 7.7 |
| Comoros | 74.4 |
| Djibouti | 57.5 |
| Equatorial Guinea | 66.3 |
| Eswatini | 65.8 |
| Gambia | 54.4 |
| Guinea | 33.9 |
| Guinea-Bissau | 20.3 |
| Lesotho | 31.8 |
| Liberia | 16.1 |
| Madagascar | 20.2 |
| Malawi | 10.8 |
| Mali | 37.6 |
| Mauritania | 39.5 |
| Rwanda | 22.8 |
| Sao Tome and Principe | 66.2 |
| Seychelles | 100 |
| Sierra Leone | 19.7 |
| Uganda | 18.5 |

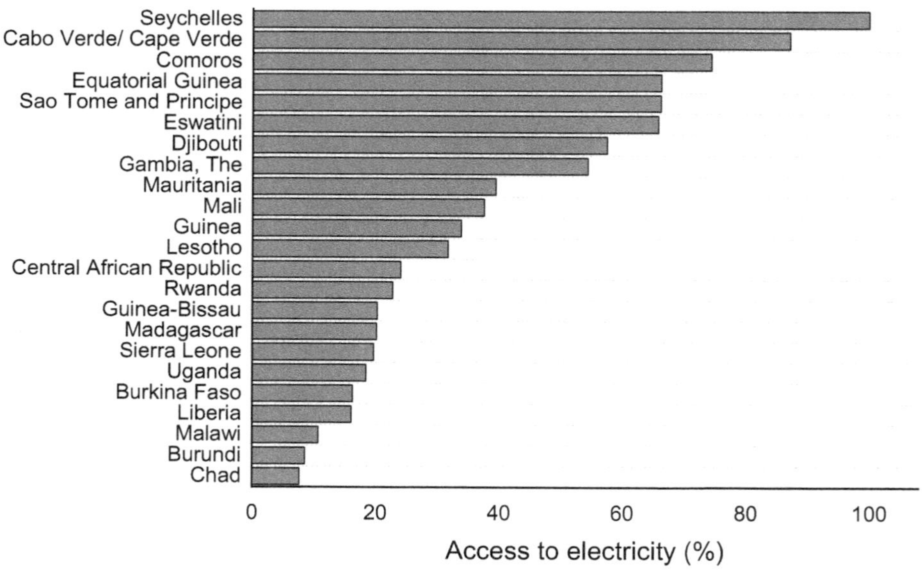

FIGURE 24.18 Electricity access. (From Ref. [52]. With permission.)

**TABLE 24.15**

**Nexus between GDP and Electricity Access to People**

| S. No. | Lower GDP (Bottom 20) | Poor Access to Electricity (Bottom 20) |
|---|---|---|
| 1 | Benin | Angola |
| 2 | Burkina Faso | Benin |
| 3 | Burundi | Burkina Faso |
| 4 | Central African Republic | Burundi |
| 5 | Comoros | Central African Republic |
| 6 | Democratic Republic of the Congo | Chad |
| 7 | Eritrea | Democratic Republic of the Congo |
| 8 | Ethiopia | Guinea |
| 9 | Guinea | Guinea-Bissau |
| 10 | Guinea-Bissau | Liberia |
| 11 | Liberia | Madagascar |
| 12 | Madagascar | Malawi |
| 13 | Malawi | Mozambique |
| 14 | Mali | Niger |
| 15 | Mozambique | Rwanda |
| 16 | Niger | Sierra Leone |
| 17 | Rwanda | South Sudan |
| 18 | Sierra Leone | Tanzania |
| 19 | South Sudan | Uganda |
| 20 | Togo | Zambia |

having reasonable GDP per capita, but still, they are not capable of supplying adequate electricity access to people.

## 24.2.3 ISSUES ASSOCIATED WITH POOR ELECTRICITY ACCESS TO THE PEOPLE

Many issues emerge when people have poor access to electricity. As shown in Table 24.16, poor access to electricity leads to poor GDP of countries. The reasons behind this are that people cannot do their business effectively and hence slower industries growth. In the absence of electricity, education in school and students' study at home are ineffective. The school could not use modem educational aids and internet facilities. Henceforth, educational attainment is low for such countries. Finally, the literacy rate of the country is marginal. Due to the lack of electricity access and its availability and reliability, business and industries are challenging to start; hence, the local employment rate is inferior. Consequently, the migration of people from their native places to urban areas increases the burden on urban people and resources. The effect of poor access to electricity on other development indicators is shown in Figure 24.19.

Agriculture production many times depends on electricity accessibility. Much agricultural equipment needs electricity that is useful to increase the crop yield and to earn more money. To run a water pump at the appropriate time, electricity is essential. In the interrupted or absence of electricity, agricultural production decreases. That increases inflation and the hunger rate. Survival is found to be difficult for many people. Lack of money reserve, people, cannot take care of their family members; hence, there is a challenging issue of undernourished people, high child mortality, low life expectancy, and low HDI (refer to Table 24.16); consequently, there are more issues of people exploitation. Table 24.16 shows that poor access to electricity, GDP, literacy rate, HDI, educational attainment, employment index, and crop production index is less; however, the child death rate and percentage of undernourished people are significantly higher.

**TABLE 24.16**

**Nexus between Access to Electricity and Its Effect on Other Development Indicators [2,9,11,14,22,23,26]**

| S. No. | Name of the Country (Bottom 15) | Access to Electricity (%) | GDP per Capita (PPP) (USD) | Literacy Rate (M/F) (%) | HDI | Child Death (under 5 years) per 1,000 Birth | Educational Attainment | Employment Index | Crop Production Index | Undernourished People (%) |
|---|---|---|---|---|---|---|---|---|---|---|
| 1 | Benin | 41.5 | 2300 | 47.9/42.3 | 0.427 | 90 | 1.8 | 2.5 | 161.4 | 9.58 |
| 2 | Burkina Faso | 14.4 | 1900 | 25.3/NA | 0.331 | 88 | 4.3 | 5 | 145 | 19.24 |
| 3 | Burundi | 11 | 700 | 67.3/52.2 | 0.316 | 57 | 7.6 | 0.8 | 106.2 | 4.25 |
| 4 | Central African Republic | 32.4 | 700 | 64.8/33.5 | 0.343 | 110 | 3.1 | 4.3 | 108.5 | 60 |
| 5 | Democratic Republic of the Congo | 19 | 800 | 80.9/54.1 | 0.286 | 85 | 38.6 | 4.6 | 109.5 | 40.04 |
| 6 | Guinea | 44 | 2200 | NA | NA | 99 | 10.9 | 4.3 | 128.2 | 16.92 |
| 7 | Guinea-Bissau | 28.7 | 1900 | NA | NA | 79 | NA | 3.2 | 140.2 | 26.68 |
| 8 | Liberia | 25.9 | 1300 | NA | NA | 85 | 8.2 | 3.3 | 103.1 | 37.35 |
| 9 | Madagascar | 25.9 | 1600 | 76.5/65.3 | 0.48 | 51 | NA | 1.9 | 120.3 | 43.41 |
| 10 | Malawi | 18 | 1200 | NA | 0.4 | 42 | 8.1 | 6 | 157.2 | 18.19 |
| 11 | Mozambique | 31.1 | 1300 | NA | 0.322 | 74 | 11.5 | 3.4 | 145.2 | 28.14 |
| 12 | Niger | 17.6 | 1200 | NA | NA | 80 | 5 | 0.7 | 196.8 | 16.04 |
| 13 | Rwanda | 34.7 | 2100 | 71.4/59.8 | 0.429 | 34 | 11.7 | 1.4 | 156.6 | 36.58 |
| 14 | Sierra Leone | 26.1 | 1600 | NA | NA | 109 | NA | 4.6 | 171.7 | 24.84 |
| 15 | South Sudan | 28.2 | 1600 | NA | NA | 96 | 16.2 | 12.7 | NA | 19.62 |

NA, not available.

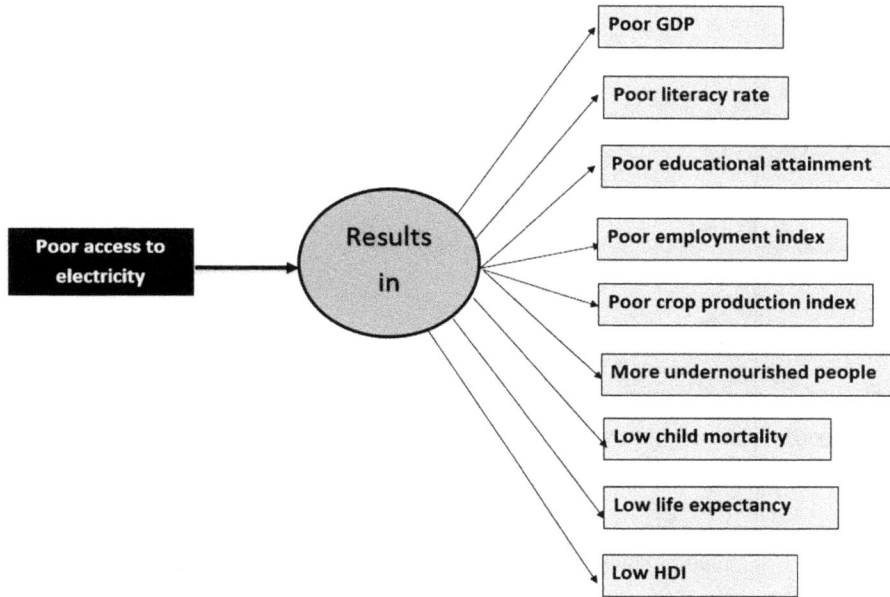

**FIGURE 24.19** Effect of poor access to electricity on other development indicators.

## 24.3 COMPARISON OF SUB-SAHARAN COUNTRIES WITH THE WORLD [53]

This section compares the energy consumption, primary energy mix, energy security, and world energy access in sub-Saharan countries. Table 24.17 presents the global energy mix. It illustrates that as of 2020, global energy fulfilment is mainly done by oil, and its use is increasing year by year. After the oil, coal, then natural gas is primarily used. As shown in Figure 24.20, biofuels, nuclear, hydro, wind, and solar energy utilisation are significantly less. Still, the encouraging point is that their uses day by day are increasing.

Table 24.18 compares the primary energy mix of different parts of the world and sub-Saharan Africa. In 2015, the percentage of access to electricity was the lowest among all continents of the world.

**TABLE 24.17**

**World Energy Consumption Rate in ktoe [53]**

| Year | Coal | Natural Gas | Nuclear | Hydro | Wind, Solar, etc. | Biofuels and Waste | Oil |
|------|------|-------------|---------|-------|-------------------|--------------------|----|
| 1990 | 2,220,587 | 1,662,187 | 525,520 | 184,064 | 36,571 | 904,162 | 3,233,212 |
| 1995 | 2,207,669 | 1,806,624 | 608,098 | 212,766 | 42,391 | 967,469 | 3,373,297 |
| 2000 | 2,317,134 | 2,071,233 | 675,467 | 224,663 | 60,262 | 1,014,659 | 3,669,477 |
| 2005 | 2,990,601 | 2,360,022 | 721,706 | 252,334 | 70,143 | 1,088,960 | 4,010,067 |
| 2010 | 3,649,798 | 2,735,952 | 718,713 | 296,474 | 110,200 | 1,205,287 | 4,127,360 |
| 2015 | 3,842,742 | 2,928,795 | 670,172 | 334,851 | 203,821 | 1,271,235 | 4,328,233 |
| 2018 | 3,838,326 | 3,261,595 | 706,814 | 362,332 | 286,377 | 1,327,127 | 4,496,998 |

*Source:* IEA, World Energy Outlook-2020.

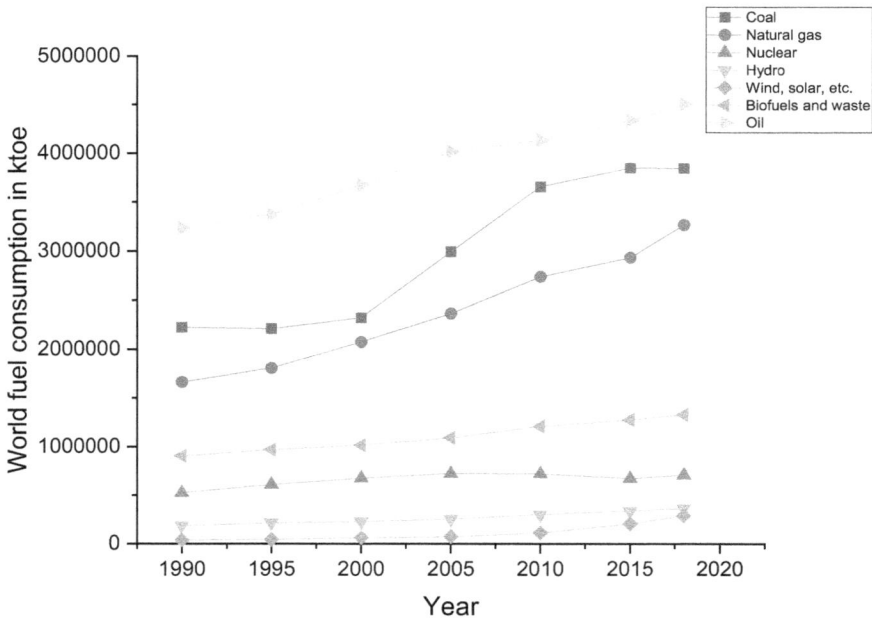

**FIGURE 24.20**   World fuel consumption. (From Ref. [53]. With permission.)

**TABLE 24.18**

**Comparison of the Primary Energy Mix of Sub-Saharan African Countries with the World [54]**

| Region (2015) | Coal % of Total | Natural Gas % of Total | Oil % of Total | Hydro Power % of Total | Renewable Sources % of Total | Nuclear Power % of Total | Access to Electricity % of Population |
|---|---|---|---|---|---|---|---|
| World | 39.2 | 22.8 | 3.3 | 15.9 | 6.8 | 8.1 | 86.6 |
| East Asia & Pacific | 59.4 | 13.5 | 1.7 | 15.2 | 5 | 1.9 | 97.1 |
| Europe & Central Asia | 23.2 | 24.3 | 1.4 | 16.2 | 11.8 | 16.1 | 99.3 |
| Latin America & Caribbean | 6.7 | 27.2 | 9.9 | 44 | 7.6 | 0.7 | 97.3 |
| Middle East & North Africa | 3.2 | 67.1 | 18.6 | 2.3 | 0.5 | 0 | 97 |
| North America | 30.9 | 29 | 0.9 | 12.7 | 7.2 | 18.8 | 100 |
| South Asia | 66.1 | 9.1 | 4.8 | 11.5 | 4.8 | 2.8 | 84.5 |
| Sub-Saharan Africa | 50.6 | 9.5 | 4 | 20.5 | 2.4 | 3 | 39.3 |

As shown in Figure 24.21, in 2015 in sub-Saharan Africa, coal was mainly used for energy generation after hydropower. The use of other energy sources is found to be marginal for this region.

Figure 24.22 describes the rank of sub-Saharan countries in comparison to all continents of the world. In hydropower utilisation, the sub-Saharan region has done an excellent job. This region stood at second number among all continents. In nuclear power and coal utilisation, this region stood at third rank. The use of renewable energy and natural gas is found to be very low among all continents. However, this region has the most down position for access to electricity to the people, which is the leading cause behind why this region is so undeveloped even today.

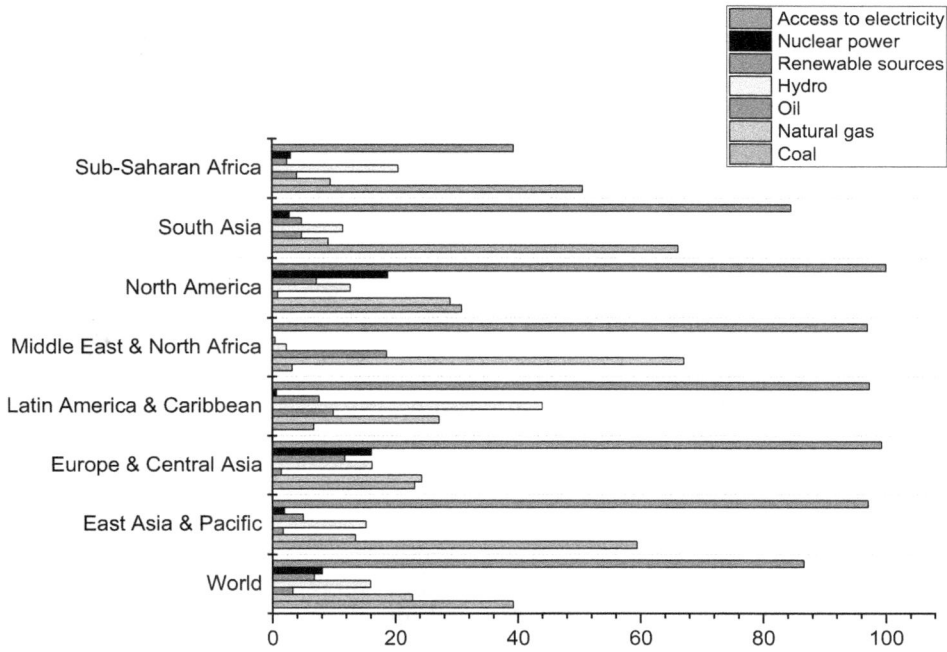

**FIGURE 24.21** Use of different fuel sources for electricity generation in other regions (in percentage). (From Ref. [54]. With permission.)

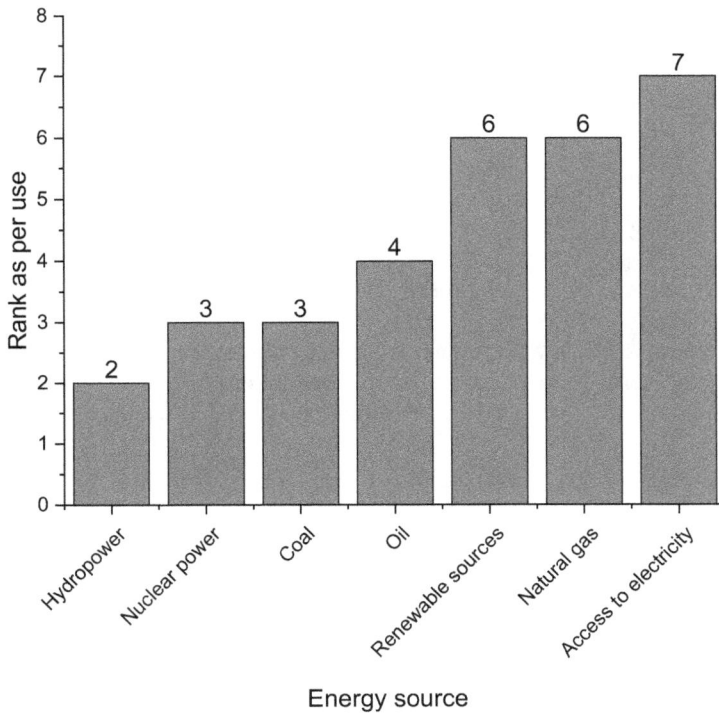

**FIGURE 24.22** Rank of sub-Saharan Africa based on different energy sources utilisation. (From Ref. [55]. With permission.)

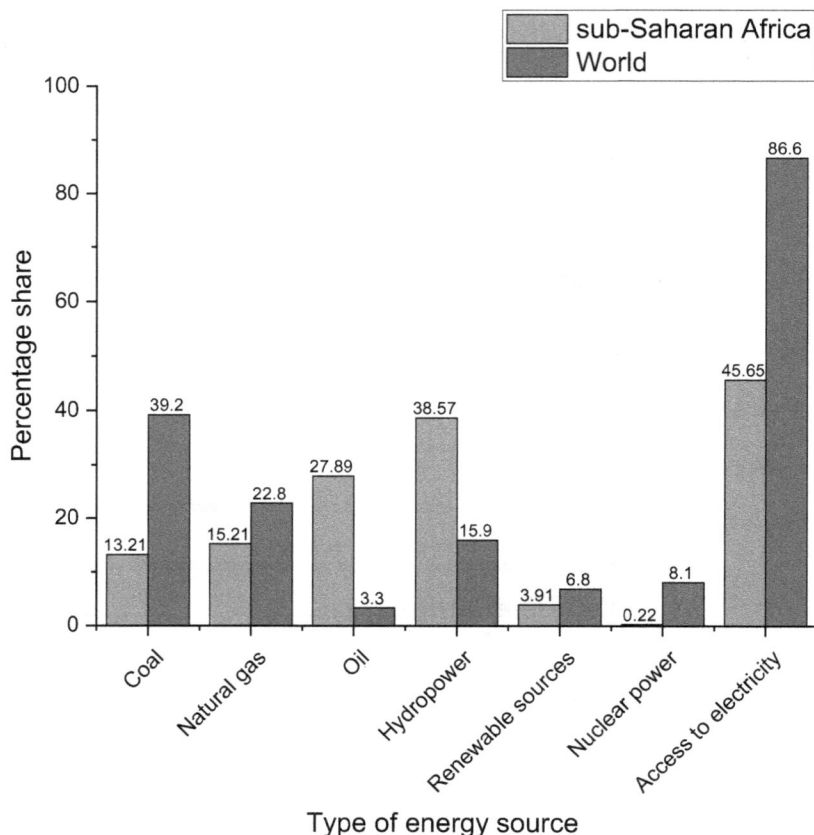

**FIGURE 24.23** Percentage share of fuel utilisation in sub-Saharan countries in comparison with the world. (From Ref. [55]. With permission.)

Figure 24.23 compares the energy mix of sub-Saharan Africa with the world's primary energy mix. It shows that hydropower and oil for electricity generation are significantly higher than in the world. This region needs to reduce oil utilisation for sustainable development growth and to reduce its carbon footprint.

Table 24.19 illustrates how the percentage of access to electricity increased from 2000 to 2019. In the year 2000, only 24% of sub-Saharan people had access to electricity within last 20 years this value is just doubled to 48%. And out of 771 million people living without electricity globally, out

**TABLE 24.19**

**Proportion of the Population with Access to Electricity [54]**

| Region | The Proportion of the Population with Access to Electricity | | | | | | | People without Access (million) |
|---|---|---|---|---|---|---|---|---|
| | National | | | | | Urban | Rural | |
| Year | 2000 | 2005 | 2010 | 2015 | 2019 | 2019 | 2019 | 2019 |
| World | 73% | 77% | 80% | 85% | 90% | 96% | 85% | 771 |
| Sub-Saharan Africa | 24% | 28% | 33% | 40% | 48% | 76% | 29% | 578 |

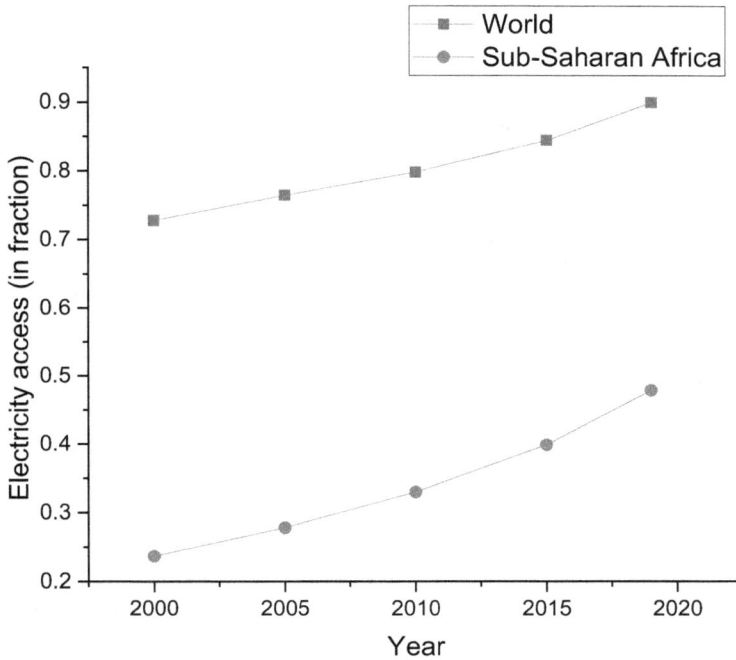

**FIGURE 24.24** Percentage growth in access to electricity in sub-Saharan countries. (From Ref. [54]. With permission.)

of that, 578 million people (72.3%) live in the sub-Saharan region. But as per Figure 24.24, it can be noticed that the progress of electricity access to the people is increasing year by year.

Table 24.20 compares the fuel consumption and percentage access to electricity in sub-Saharan Africa and the world. It shows that the use of hydropower is remarkable for sub-Saharan Africa as compared to the world. To understand the comparison, precisely different indices are defined and estimated. They are also mentioned in Table 24.20. The definition of these indices is given in Table 24.21. The value of these indices indicates a fraction of utilities of different energy sources by sub-Saharan countries.

**TABLE 24.20**
**Comparison of Sub-Saharan Countries with the World with Indices [54]**

| Country (2015) | Coal | Natural Gas | Oil | Hydropower | Renewable Sources | Nuclear Power | Access to Electricity |
|---|---|---|---|---|---|---|---|
| | % of Total | % of Total | % of Total | % of total | % of Total | % of total | % of Population |
| Sub-Saharan Africa | 13.21 | 15.21 | 27.90 | 38.58 | 3.91 | 0.22 | **39.3** |
| World | 39.2 | 22.8 | 3.3 | 15.9 | 6.8 | 8.1 | 86.6 |
| Index | Coal utility index | Gas Utility index | Oil utility index | Hydropower utility index | Renewable energy utility index | Nuclear power utility index | Access to electricity index |
| Value of index | 0.34 | 0.67 | 8.45 | 2.43 | 0.58 | 0.03 | 0.45 |

**TABLE 24.21**
**Characterisation of Indices**

| Index | Definition | Significance |
|---|---|---|
| Coal utility index | It is the ratio of percentage utilisation of coal by sub-Saharan countries to the world's percentage utilisation. | Presently this value is 0.34, indicating that 34% of coal consumption is in the sub-Saharan region. |
| Gas utility index | It is the ratio of percentage utilisation of gas by sub-Saharan countries to the percentage utilisation of gas by the world. | Currently, this value is 0.67, which indicates that 67% of the total gas is used in the sub-Saharan region. |
| Oil utility index | It is the ratio of percentage utilisation of oil by sub-Saharan countries to the percentage utilisation of oil by the world. | The oil utility index is 8.45, which states that oil consumption in this region is 8.45 times more than the world oil consumption, which is significantly high. To control carbon mitigation in this region, oil consumption in this region has to be reduced. |
| Hydropower utility index | It is the ratio of percentage utilisation of hydropower by sub-Saharan countries to the percentage utilisation of hydropower by the world. | The hydropower index is 2.43, which is attractive concerning utility hydro energy for electricity generation, the cleanest energy source. Hydropower utilisation in the sub-Saharan region is 2.43 times more than world hydropower utilisation. |
| Renewable energy utility index | The percentage utilisation of renewable energy by sub-Saharan countries to the world's percentage utilisation of renewable energy. | The renewable energy index encompasses all other renewable energy, excluding hydropower. Its present value is 0.58. It indicates that 58% of the total renewable energy is used in this region than global renewable energy. |
| Nuclear power utility index | The ratio of percentage utilisation of nuclear energy by sub-Saharan countries to the world's percentage utilisation of nuclear energy. | Nuclear energy utilisation is very little less by sub-Saharan region as its index value is 0.03 only. |
| Access to electricity index | It is the ratio of percentage access to electricity by sub-Saharan countries to access electricity access to the world. | This index shows that 45.38% of the global population living in the sub-Saharan region has only electricity access. |

## 24.4   SUMMARY, CONCLUSIONS, AND RECOMMENDATIONS

This article describes sub-Saharan countries' socio-economic development and energy security, the present status of electricity consumption, its future demand, environmental impact, and significant challenges to supplying electricity to all households. Summary and conclusions drawn from this study are as follows:

- Sub-Saharan Africa is home to over 1 billion people. It is a vibrant continent offering human and natural resources that can deliver sustainable growth and eliminate poverty.
- As of 2017, this region's average GDP per capita was $5566, 67.45% less than the world GDP.
- Most of the countries in this region fall under low-income countries. Only seven countries are included in upper-middle-income. Eighteen countries of this region are vulnerable to violence.
- The average literacy rate of the sub-Saharan region is 67.69%.

- The life expectancy for the sub-Saharan region was stated as 52.47 years, which is 23.03% less than the world.
- The average HDI for the sub-Saharan region was 0.468 as of 2008.
- In 2018, 47.67% population had access to electricity, of which 31.55% of rural and 78.09% of urban people have access to electricity.
- On average, 39.14% of people have access to electricity, which encompasses 22.17% of rural and 73.32% of urban people.
- As compared with urban electrification access, the rate of rural electricity access is more progressive.
- During 2017–2018, percentage increases in overall and rural electricity access are 6.71% and 17.28%, respectively. However, urban electricity access was decreased during this period.
- About 54.86% of the electricity is generated by coal. Its contribution is 61% out of all the fuel sources used. Henceforth, carbon emission is also very significant (55.19%).
- After coal, the contribution of renewable energy is highest (24.41%); this is a remarkable achievement of this region.
- Hydropower plays a vital role in electricity generation (19.79%), that is, 22% of the total fuel's contribution.
- In 2019, on an average basis, 124.18 days, that is, 4.13 months, are required to get the electricity at home after applying for it.
- Only five countries have 91%–100%, 11 countries have 61%–90%, 20 countries have 31%–60% access to electricity, and 14 countries have less than 30% electricity access to people.
- In the rural area, 33.33% of people have electricity in their homes.
- Coal, gas, oil, hydropower, renewable energy, and nuclear energy utility indices are 0.34, 0.67, 8.45, 2.43, 0.58, and 0.03, respectively.
- Planning at the micro-level is required for the establishment of electrification infrastructure.
- To handle the issue of energy poverty, more focus must be given to the use of renewable energy for electricity generation.

Even though urban electricity rates in sub-Saharan Africa are much more significant than rural electrification rates, this may not represent reality on the ground. In sub-Saharan Africa, more than 50% of urban people live in slum-like conditions, with many of them being without power. Urbanisation presents a particular obstacle to electrification since increasing energy availability in urban areas is often more difficult than in rural regions.

### 24.4.1  Recommendations for Electricity Planning and Implementation

In terms of electricity planning, most sub-Saharan African nations have given little or no planning attention. A few things can be done:

1. The first is to increase the production and reliability of linked users' current centralised power systems.
   a. The second is to connect the disconnected population, primarily rural, in decentralised systems and mini-grids that may subsequently be linked to wide-area (centralised) grids utilising advances in hardware and digital control systems.
   b. The third step decarbonises the energy mix, focusing on coal, high-carbon local biomass, and diesel (which is now extensively utilised as an energy source in stand-alone generators).
   c. Renewable energy technologies are often seen as a preferred choice. Its abundance on the continent, particularly solar and hydroelectric. Its availability in rural locations is far from the effective energy system. Because renewables are ubiquitous, they are more ecologically benign, provide social advantages, and reduce resource-related political dangers.

## REFERENCES

1. Katekar VP, Asif M, Deshmukh SS (2021) Energy and environmental scenario of south Asia. In: Asif M (ed) *Energy and Environmental Security in Developing Countries, Advanced Sciences and Technologies for Security Applications*, Springer Nature, Switzerland, pp 1–33.
2. World Energy Council (2017) *World Energy Scenarios. 2017 Regional Perspective for Sub-Saharan Africa Regional Perspective for Sub-Saharan Africa Preparing for "The Grand Transition."* World Energy Council, Cornhill, London, United Kingdom. pp 1–12.
3. Katekar VP, Deshmukh SS, Elsheikh AH (2020) Assessment and way forward for Bangladesh on SDG-7: Affordable and clean energy. *Int Energy J* 20:421–438.
4. Deshmukh S, Jinturkar A, Anwar K (2014) Determinants of household fuel choice behavior in rural Maharashtra, India. In: *1st International Congress on Environmental, Biotechnology, and Chemistry Engineering IPCBEE*, Carina Liu (Ed) IACSIT, Singapore. pp 128–133.
5. Katekar VP, Deshmukh SS (2021) Energy-drinking water-health nexus in developing countries. In: Asif M (ed) *Energy and Environmental Security in Developing Countries. Advanced Sciences and Technologies for Security Applications.* Springer, Cham. pp 411–445.
6. Leach M, Deshmukh S (2012) Sustainable energy law and policy. In: Karen E Makuch and Ricardo Pereira (eds), *Environ Energy Law.* Blackwell Publishing Ltd, 1(1):122–138.
7. Speirs J, Gross R, Deshmukh S, et al (2010) *Heat Delivery in a Low Carbon Economy.* BEEE. pp 1–20.
8. Jambhulkar G, Nitnaware V, Pal M, et al (2015) Performance evaluation of cooking stove working on spent cooking oil. *Int J Emerg Sci Eng* 3(4):26–31.
9. The World Bank (2019) *GDP per capita (current US$) – Sub-Saharan Africa.* World Bank. https://data.worldbank.org/indicator/NY.GDP.PCAP.CD?locations=ZG. Accessed 1 Jun 2021.
10. The Royal Scociety (2019) *List of Eligible Countries in Sub-Saharan Africa*, Department of International Development, United Kingdom, p 117.
11. World Bank (2012) Low-income countries in sub-Saharan Africa (World Bank classification). *UK Res Innov* 1, pp 1–463.
12. Katekar VP, Deshmukh SS, Vasan A (2020) Energy, drinking water and health nexus in India and its effects on environment and economy. *J Water Clim Chang* 11: 1–26 https://doi.org/10.2166/wcc.2020.340.
13. Deshmukh S (2011) Energy resource allocation in energy planning. In: Ahmed F Zobaa and Ramesh C Bansal (eds), *Handbook of Renewable Energy Technology.* World Scientific, Singapore. pp 801–846.
14. Smith-Greenaway E (2015) Educational attainment and adult literacy: A descriptive account of 31 sub-Saharan Africa countries. *Demogr Res* 33: 1015–1034. https://doi.org/10.4054/DemRes.2015.33.35.
15. Deshmukh S, Jinturkar A, Anwar K (2014) Determinants of household fuel choice behavior in rural Maharashtra, India. In: Carina Liu (ed), *1st International Congress on Environmental, Biotechnology, and Chemistry Engineering IPCBEE*, IACSIT, Singapore. pp 128–133.
16. UNESCO (2018) *From the UNESCO Science Report, Towards 2030.*, United Nations Educational, Scientific and Cultural Organization, Paris, France.
17. Cheng V, Deshmukh S, Hargreaves A, et al (2011) A study of urban form and the integration of energy supply technologies. In: Bahram Moshfegh (ed), *Proceedings of the World Renewable Energy Congress – Sweden, 6–13 May, 2011*, LiU Electronic Press, Linköping, Sweden. pp 3356–3363.
18. Baigarin K, Al E (2008) *National Human Development Report 2008 – Climate Change and Its Impact on Kazakhstan's Human Development*, Agropromizdat, Astana.
19. Dhurwey AR, Katekar VP, Deshmukh SS (2019) An experimental investigation of thermal performance of double basin, double slope, stepped solar distillation system. *Int J Mech Prod Eng Res Dev* 9:200–206.
20. Leach M, Deshmukh S, Ogunkunle D (2014) Pathways to decarbonising urban systems. In: Tim Dixon, Malcolm Eames, Miriam Hunt, Simon Lannon (eds), *Urban Retrofitting for Sustainability: Mapping the Transition to 2050.* Routledge, Florida, U.S.A. pp 191–208.
21. Bhujade S, Mate A, Katekar V, Sajjanwar S (2017) Biogas plant by using kitchen waste. *Int J Civil, Mech Energy Sci* 64–69. https://doi.org/10.24001/ijcmes.icsesd2017.74.
22. Global Outlook (2019) *Energy Transformation in Sub-Saharan Africa.* International Energy Agency, Paris, France.
23. Blimpo MP, Cosgrove-Davies M (2019) *Electricity Access in Sub-Saharan Africa: Uptake, Reliability, and Complementary Factors for Economic Impact.* International Energy Agency, Paris, France.
24. Bhaisare A, Katekar V, Deshmukh S (2020) Novel energy efficient design of water-cooler for hot and dry climate. *Test Eng Manag* 1:12523–12528.

25. British Petroleum (2020) Energy Outlook 2020 edition explores the forces shaping the global energy transition out to 2050 and the surrounding that. BP energy outlook 2030, *Stat Rev London Br Pet* 81.

26. International Energy Agency (2014) Africa Energy Outlook. A focus on the energy prospects in sub-Saharan Africa. World Energy Outlook Spec Report, *Int Energy Agency Publ* 1–237.

27. Hafner M, Tagliapietra S, Strasser L de (2018) *Energy in Africa Challenges and Opportunities*. Springer Nature, Cham.

28. Katekar VP, Deshmukh SS (2019) Productivity enhancement of solar still using exergy analysis. In: *International conference on Advances in Mechanical and Electrical Engineering (ICAMEE-19), 24–25 August 2019*. G H Raisoni College of Engineering, Nagpur. pp 24–25.

29. Whitepaper (2016) *Renewable Energy in Sub-Saharan Africa*.

30. Jinturkar A, Deshmukh S, Sarode A, et al (2014) Fuzzy-AHP approach to improve effectiveness of supply chain. *Supply Chain Manag Under Fuzziness* 35–59.

31. Khatri V, Katekar V (2012) Helmet cooling with phase change material. *Int J Comput Appl* 2:1–5

32. Sustainable Energy for All (2020) *SEforALL Analysis of SDG7 Progress – 2020*.

33. Saxena A, Deshmukh S, Nirali S, Wani S (2018) Laboratory based experimental investigation of photovoltaic (PV) thermo-control with water and its proposed real-time implementation. *Renew Energy* 115:128–138 https://doi.org/10.1016/j.renene.2017.08.029.

34. Anwar K, Deshmukh S (2020) Parametric study for the prediction of wind energy potential over the southern part of India using neural network and geographic information system approach. *Proc Inst Mech Eng Part A J Power Energy* 234:96–109 https://doi.org/10.1177/0957650919848960.

35. Office A (2013) *Regional Corporation for Energy Access and Energy Security in South and South-West Asia*.

36. Ambade S, Narekar T, Katekar V (2009) Performance evaluation of combined batch type solar water heater cum regenerative solar still. In: *Second International Conference on Emerging Trends in Engineering and Technology (ICETET)*. IEEE, pp 1064–1067.

37. Arunachala UC, Kundapur A (2020) Cost-effective solar cookers: A global review. *Sol Energy* 207: 903–916 https://doi.org/10.1016/j.solener.2020.07.026.

38. Ambade S, Tikhe S, Sharma P, Katekar V (2017) Cram of Novel Designs of Solar Cooker. *Int J Mech Eng Res* 7:109–117.

39. Ukey A, Katekar V (2019) An experimental investigation of thermal performance of an octagonal box type solar cooker. In: *Smart Technologies for Energy, Environment and Sustainable Development. Lecture Notes on Multidisciplinary Industrial Engineering*. Springer, Singapore, pp 769–777.

40. Bhaisare A, Wasnik U, Sakhare A, et al (2021) Performance assessment of improved solar still design with stepped-corrugated absorber plate. In: Reddy A., Marla D., Favorskaya M.N. SSC (ed) *Intelligent Manufacturing and Energy Sustainability. Smart Innovation, Systems and Technologies*. Springer, Singapore, pp 667–674.

41. Katekar VP, Deshmukh SS (2020) A review of the use of phase change materials on performance of solar stills. *J Energy Storage* 30:1–28 https://doi.org/https://doi.org/10.1016/j.est.2020.101398.

42. Elsheikh AH, Katekar VP, Muskens OL, et al (2020) Utilisation of LSTM neural network for water production forecasting of a stepped solar still with a corrugated absorber plate. *Process Saf and Environmental Prot* 1–34. https://doi.org/10.1016/j.psep.2020.09.068.

43. Katekar VP, Deshmukh SS (2020) Thermoeconomic analysis of solar distillation system with stepped-corrugated absorber plate. *Proc Inst Mech Eng Part C J Mech Eng Sci* 1–20 https://doi.org/10.1177/0954406220943227.

44. Katekar VP, Deshmukh SS (2020) A review on research trends in solar still designs for domestic and industrial applications. *J Clean Prod* 257:120544. https://doi.org/10.1016/j.jclepro.2020.120544.

45. Mate A, Katekar V, Bhatkulkar HS (2017) Performance investigation of solar still for batteries of railway engine, Indian Railways, at Ajni Loco Shed, Nagpur. In: *International conference on Advances in Thermal Systems, Materials and Design Engineering (ATSMDE2017)*. VJTI, Mumbai.

46. Katekar VP, Deshmukh SS (2021) Techno-economic review of solar distillation systems: A closer look at the recent developments for commercialisation. *J Clean Prod* 294:126289. https://doi.org/10.1016/j.jclepro.2021.126289.

47. Katekar VP, Deshmukh SS (2019) Performance investigation of solar still by exergy analysis – A review. In: *Advances in Mechanical & Electrical Engineering*. pp 1–9.

48. Madhav Durusoju, Goyal C, Sheik I, et al (2016) Heat transfer enhancement techniques for solar air heater – A review. *Int J Res Appl Sci Eng Technol* 4:451–457.

49. Katekar V, Vithalkar A, Kale B (2009) Enhancement of convective heat transfer coefficient in solar air heater of roughened absorber plate. In: *2009 2nd International Conference on Emerging Trends in Engineering and Technology, ICETET 2009*. pp 1042–1046.

50. Mamulkar C, Katekar V (2012) Performance evaluation of double flow solar air heater. In: *National Conference on Innovative Paradigms in Engineering & Technology (NCIPET-2012)*. pp 5–6.

51. Durusoju M, Goyal C, Sheikh I, et al (2016) An experimental investigation of thermal performance of solar air heater with 'W' wire mesh. *Int J Res Appl Sci Eng Technol* 4:5–14.

52. Zhang YF, Parker D, Kirkpatrick C (2008) Electricity sector reform in developing countries: An econometric assessment of the effects of privatisation, competition and regulation. *J Regul Econ* 33:159–178 https://doi.org/10.1007/s11149-007-9039-7.

53. Terreson D, Richardson S, West J, et al (2020) Global energy outlook 2020 energy transition or energy addition. *Resour Futur* 1–6.

54. International Energy Agency (2020) *World Energy Outlook 2020 – Event*. IEA pp 1–25.

55. World Energy Scenarios (2013) *World Energy Scenarios: Composing Energy Futures to 2050*. pp 1–288. Reproduced from http//www.worldenergy.org/.

# 25 The Role of Nuclear Power under the Energy Transition

## An Unsustainable Approach to Taiwan's Nuclear-Free Homeland By 2025?

Anton Ming-Zhi Gao
National Tsing-Hua University

## CONTENTS

## 25.1 INTRODUCTION

In 2016, the Taiwan government launched the agenda of energy transition and nuclear-free homeland with the target of 20% renewable energy and 0% nuclear power by 2025 (Energia Plus, 2020). Aside from subscribing to the UN's 17 Sustainable Development Goals (SDGs), Taiwan set up an 18th one of becoming nuclear-free by 2025 (Lin Chia-nan, 2018). However, later there were a lot of controversies on the increase of coal-fired power plants and conflictive land use and environmental issues resulting from the development of large-scale photovoltaic and wind power units (Lien, 2021). With Taiwan facing such conflicts, they motivated me to evaluate the compatibility between its unique 18th SDG and sustainable development.

The concept of nuclear-free homeland was introduced in Article 23 of the Basic Environment Act of 2002, which stipulated that "The government shall establish plans to gradually achieve the goal of becoming a nuclear-free country. The government shall also strengthen nuclear safety management and control, protections against radiation, and the management of radioactive materials and monitoring of environmental radiation to safeguard the public from the dangers of radiation exposure" (Ministry of Justice, 2002). However, due to the lack of substantive definition in the provision, every ruling party could claim pursuing such a goal without the massive change of the existing nuclear power policy. For instance, in 2011 after the Fukushima accident, the former ruling party announced the plan to continue using nuclear power plants until the licensing life of 40 years and completion of the construction of the fourth plants (Yueh-Hsun, 2014). This new policy was considered to conform with the nuclear phase-out goal, as even the fourth nuclear power plant would not run forever. In 2016, the then new ruling party claimed to adopt a more ambitious approach to nuclear phase-out by providing that the existing three nuclear power plants (six reactors) could run until the expiry of the 40-year generation licence. The current ruling party also considered such an approach to be one of pursuing a nuclear-free homeland. Therefore, the main purpose of this article is to discover the concept of nuclear-free homeland from legal aspects and from the discussion of legal or policy documents "in the past almost 20 years" and to evaluate if such concept is compatible with the rest of the SDGs.

Our preliminary finding is that there is a serious conflict between Taiwan's SDG 18 and several of the existing 17 SDGs. For instance, under SDG 7, SDG 18 may be helpful in achieving its clean energy goal but may jeopardize affordable energy due to setting an unreasonably high subsidy for the renewable one. Replacing nuclear power with more fossil fuel fired plants may jeopardize SDG 13 (climate change) as well. Not to mention, the lack of law and due process to deliberate the fate of nuclear power and even against the results of the nuclear phase-out referendum in 2018 may violate the meaning of SDG 16. The original delayed progress towards the 2025 nuclear-free homeland goal got a further setback with the outbreak of Covid-19, which held back the development of renewable energy projects.

## 25.2  THE EVOLUTION OF THE CONCEPT OF NUCLEAR-FREE HOMELAND IN TAIWAN

After the change in government in 2000, the opposition and the green party took power and started to follow Germany's approach in promoting the idea of "nuclear phase-out." The Executive Yuan announced the suspension of the construction of Nuclear Power No. 4 at the end of 2000, which suspension led to an issue of political and constitutional law. Even though the issue was dealt with by the Grand Justice of Taiwan, the ambiguous judgement could not resolve this political chaos. There seemed no clear judgement as to whether the suspension decision of the Executive was right or wrong. This issue also led to tension between the Executive Yuan and the Parliament. The functioning of the national machine came to a standstill. In order to resolve this impasse, there were intensive discussions between the ruling party and the opposition party in the Parliament. Finally, a consensus was reached in February 2001 to continue the construction of the Nuclear Power No. 4 plant while pursuing the idea of a "nuclear-free homeland" ("Nuclear-Free Consensus"). It resulted in serious nuclear-free homeland activities in the coming years. However, as there was no concrete context and definition for this new term—nuclear-free homeland—a lot of effort was put into developing this idea.

### 25.2.1  PROPOSAL OF NUCLEAR PHASE-OUT BILL IN 2001

In order to pursue the goal of a nuclear-free homeland, in February 2001, the Executive Yuan adopted a bottom-up approach to develop this objective. It ordered the MOEA to consult with the AEC and propose energy-related bills to implement the idea of a nuclear-free homeland, which would be submitted to the Parliament for its deliberation. Furthermore, the MOEA was also to consider stopping the operation of the existing three nuclear power plants on the condition that it would not affect the electricity supply security of Taiwan (Executive Yuan, 2001).

In view of this call, the MOEAEC started to interpret the idea of a nuclear-free homeland, and proposed the Nuclear Phase-out Bill to the MOEA. The MOEA then approved it immediately in July 2001 and delivered it to the Executive Yuan for its deliberation.

In this bill, the idea of a nuclear-free homeland was only mentioned in the Explanatory Note (the foreword) of the bill. The MOEAEC interpreted the concept of this term as "nuclear phase-out." The bill specified the condition for the phasing out of the existing four nuclear power plants in Art. 5. The follow is the essence of this bill, notwithstanding its very unclear and ambiguous conditions:

1. From the date of phase-out, the annual electricity reserve capacity is able to be maintained at an excess of 15% for the next 7 years.
2. From the date of phase-out, the electricity price does not exceed 15% within 7 years.
3. The compensation to Tai-Power as a result of the phase-out decision would not cause a "significant" financial burden to the government.
4. The phase-out would not cause any infringement of international treaties signed by the government.

This bill is unique in Taiwanese energy law from the following two aspects. First, bearing in mind the usual sequence in which policy comes first, then the law, this was the first time in history that the legal instrument came before the policy, as at that time, the energy policy document promoting nuclear power in Taiwan was still in effect. Second, in terms of nuclear development laws, this was also the first time that Taiwan sought to develop a legal regime that "discouraged" the development of nuclear power. However, in the prevailing political climate, the Executive Yuan seemed to consider this bill, particularly the conditions, too weak, too ambiguous, and too difficult to achieve the objective of nuclear phase-out. The bill was insufficiently discussed; further, the Minister without portfolio at the level of the Executive Yuan began to notice the "German" phase-out approach and decided to adopt a top-down approach, promoting the "Nuclear-free Homeland Bill" idea.

### 25.2.2 Integration into Environmental Basic Act in 2002

The bill related to the nuclear-free homeland idea took some time. Yet, the administrative sector was quite eager in promoting the idea of a nuclear-free homeland. As the concerns related to nuclear energy development were also relevant to the environment-related legislations, the concept of the "nuclear-free homeland" was integrated into the final version of Art. 23 of the Environmental Basic Act and approved by the Parliament on November 19, 2002. Even though the nuclear-free homeland is an empty and open concept in Art. 23, it was helpful in encouraging serious action on the proposal of a Nuclear-free Homeland Bill, the holding of a National Nuclear-free Homeland Conference, and the adoption of a Nuclear-free Homeland Action Plan, etc.

The emptiness of this wording is reflected in the following aspect. In comparison with the Nuclear-free Consensus, Art. 23 does not deal with the issue of whether construction of the Nuclear Power No. 4 plant should continue. Second, unlike the phase-out bill dealing with the fate of Nuclear Power Nos 1–3, this aspect is unclear in the Environmental Basic Act.

A closer look at the development of this concept in the Basic Act reveals its stealthy nature during the legislative process. If the Nuclear-free Consensus were the influence or the inspiration for Art. 23, according to the past legislative practice it should have been mentioned in either the Explanatory Note of the overall bill or in giving the legislative reason for it. Yet, none of that happened in this Act. My guess is that as Art. 23 is the only provision that deals with nuclear issues, the "nuclear-free homeland" wording was surreptitiously introduced into this Act. The mentioning of this concept thus led to the further development of this idea.

### 25.2.3   Proposal of Nuclear-Free Homeland Bill of 2003

The Nuclear-free Homeland Bill of 2003 can be seen as a "shadow" of the German phase-out mechanism. Following an official visit to Germany in February 2003 led by the Minister without portfolio, steps were taken to integrate the German element into an Explanatory Note, and provide legislative explanation of the Nuclear-free Homeland Bill, such as the commercial operation of 32 years as an index to electricity production. This draft bill was then approved by the Executive Yuan on May 7, 2003 and submitted to the Parliament for its deliberation on May 11, 2003. The bill underwent several discussions in the Parliament but was not approved. Even though the bill was formally noted as a "priority bill" of the Executive Yuan, it was ignored in that parliamentary term.

The bill has 20 provisions. The issues of future energy mix adjustment, nuclear plant phase outs, clean energy, public participation in nuclear power policy, safety monitoring of nuclear power plants, energy education, etc., are all mentioned in this bill. Yet, it should be noted that the scheme to phase-out nuclear power plants remains the core of this bill, while the other elements are merely "decorative." In this regard, three provisions regulate the fate of future nuclear power development. Generally, Art. 4 urges the government to adjust the Taiwanese energy mix and gradually reduce nuclear electricity with a view to eliminating gradually the use of nuclear power. Then, more especially, from the implementation date of this Act, there would be no construction licence issued (Art. 5). In this regard, the development of Nuclear Power No. 5 or the expansion of nuclear power plants at the existing sites of Nuclear Power Nos 1–4 is no longer possible. Finally, the German phase-out concept of "scheduling a fixed total electricity amount" and "transferring the remaining electricity among different nuclear power facilities" is stipulated in Art. 6. The nuclear power reactors should stop operation permanently, once they reach the "scheduled fixed total electricity amount." According to Art. 6(2), plant operators are allowed to transfer the unused electricity to other plants after the approval of the Executive Yuan. This mechanism was considered for the less safe and older type reactors of Nuclear Power Nos 1 and 2, and allows the safer reactors to run longer. Finally, this mechanism would prevent any compensation claim from the nuclear plant operators, which was also a very important element of the German-style phase-out.

The Explanatory Note of this bill defines the nature of this bill as Basic Law. Yet, a closer investigation reveals that the development of nuclear power plants would be subject to many limitations, similar to a regular legal regime. For the existing reactors (No. 1–3 plants), it seems unlikely they would operate until the normal decommission date (i.e., in 40 years); not to mention, their life extension for another 20 years. For those reactors under construction (No. 4), their construction would be allowed to continue, but their operation would be limited by this bill as well. The expansion of nuclear generators at the existing sites was clearly prohibited as well. In sum, this bill sets up an inescapable dragnet to avoid potential loopholes. Compared with the previous generation nuclear phase-out bill of 2001, this bill's phase-out conditions are more concrete and workable. Yet, these workable and concrete measures also made the bill difficult to pass in the Parliament, where the opposition party is in the majority (Table 25.1).

### 25.2.4   Proposal of the Nuclear-Free Homeland Bill since 2012

After the Fukushima accident, the leader of the opposition party (i.e., the party in power at the time the nuclear-free homeland was being promoted) again proposed the Vision of A Nuclear-free Homeland of 2025 (Chao, 2011). Despite losing the presidential election in 2012, the anti-nuclear party proposed the Nuclear-free Homeland Bill in 2012 (Legislative Yuan, 2012). In response, the ruling party proposed the Energy Security and Nuclear-free Homeland Bill in March 2013 (Executive Yuan, 2013). Apart from these two main versions, there were also the other three

**TABLE 25.1**

**Transition of the Concept of a Nuclear-Free Homeland in Major Policies and Legislation**

| Contents | Nuclear-Free Homeland Consensus (2001.2.13) | Nuclear Phase-out Bill 2001 | Nuclear-Free Homeland Policy (2002) | Environmental Basic Act of 2002 | Nuclear-Free Homeland Bill (2003) | Nuclear-Free Homeland Action Plan (2003.9) | Energy Policy White Paper 2005 |
|---|---|---|---|---|---|---|---|
| Continuing the construction of nuclear no 4 | O | O | O | O | O | O | X |
| Overall energy supply and demand | O | O | O | O | O | O | X |
| Renewable energy development | X | X | O | O | O | O | O |
| Dealing with the phase-out issue of existing nuclear power Nos 1–3 | X | V | O | X | O | O | X |
| Nuclear safety | X | X | O | O | O | O | X |
| Radioactive protection | X | X | O | O | O | O | X |
| Any nuclear weapon | X | X | O | X | O | X | X |
| Nuclear waste management | X | X | O | X | O | O | X |
| New nuclear power installations (construction of nuclear No. 5 or expansion at existing sites) | X | X | ? | X | O | ? | X |
| Extending the life of the existing nos 1–3 power plants | X | ? | ? | X | O | ? | X |

*Source:* Compiled by the author.
O: The issue is clearly dealt with.
X: The issue is not dealt with.
?: It is unclear whether this issue is dealt with or not.

versions proposed by the non-ruling parties, namely, the Nuclear-free Homeland Bill by Taiwan Solidarity Union (Legislative Yuan, 2013), Nuclear-free Homeland Implementation Bill by 21 DPP MPs (Legislative Yuan, 2012), and No-Nuclear Power Homeland Promotion Bill by the People First Party (Legislative Yuan, 2011). Of the five versions, the government version set criteria before phasing out all nuclear power plants. Furthermore, it provided no fixed timeline for the nuclear phase-out, and reserved the possibilities of extending the life of the Nos 1–3 nuclear power plants and issuing a licence to the No. 4 nuclear power plant (Table 25.2).

**TABLE 25.2**

**Comparison of Different Versions of the Nuclear-Free Homeland Bills in the Post-Fukushima Era**

|  | The Government | DPP | Taiwan Solidarity Union | 21 DPP MPs | People First Party |
|---|---|---|---|---|---|
| Pre-conditions for nuclear phase-out | Yes | No |  |  |  |
| Timeline of a nuclear-free homeland | No | 2025 |  |  | No |
| Prohibition of life extensions or issuing a new licence | No | Yes |  |  | Yes (but excluding the fourth nuclear power plant) |

*Source:* Compiled by the author.

### 25.2.5 After 2016: The Concept of a Nuclear-Free Homeland in the Taiwan Electricity Act Amendment: Toward a Liberalized and Green Energy Market (January 23, 2017)

#### 25.2.5.1 Zero Nuclear Power Generation by 2025

The legislation for a nuclear-free homeland disappears after 2016. After the MOEA's New Energy Policy was implemented in 2016, Taiwan began tackling the soon to be four nuclear power plants. In contrast to the deliberations on whole legislative bills that focused on a nuclear-free homeland during the previous few years (Law Bank, 2003), the DPP took a unique approach. First, the idea of a nuclear-free homeland was integrated into the Amendment (2017) as Article 95. Second, instead of explicitly stating "nuclear-free homeland," Article 95 only provided the operational deadline as follows: "The nuclear energy-based power-generating facilities shall wholly stop operating by 2025." Third, instead of using the word "decommission" (which was in the previous draft about a nuclear-free homeland), it stated, "wholly stop operating." These conceptual changes led to a legal concern about the Amendment's relationship to the legal regime of current atomic energy. The draft Amendment to promote a nuclear-free homeland referred to further revisions of the Nuclear Reactor Facilities Regulation Act (2003); thus, to avoid conflicts, Article 95 would relate to application issues such as whether the three nuclear power plants could meet the "stop operations" requirement while applying for life extension (OGEL, 2019).

Previously, under the "Nuclear-free Homeland Draft Act," a comprehensive approach would handle all aspects of decommissioning the nuclear power plants. For example, in one version of the Energy Security and Nuclear-free Homeland Bill in 2013, an evaluation mechanism would be created to determine whether the phase-out would influence the overall electricity supply. However, a thorough and thoughtful approach is missing from the current Article 95 (Electricity Act) because many unresolved issues remain.

#### 25.2.5.2 Fate of the Three Nuclear Power Plants

Under Article 95 of the Amendment, the entire electricity system gradually moves to a nuclear-free system. However, in mid-2017, the central government approved the operation of two new nuclear reactors: the first reactor of the second nuclear power plant and the second reactor of the third nuclear power plant. Then, on August 15, 2017, due to operational negligence of a gas valve, a chain reaction led to system failure and Taiwan experienced an unprecedented electricity blackout that affected 6.68 million households (Focus Taiwan, 2017). About 1 year later in mid-2018 (just months ago), the government approved a restart of the controversial second reactor at the second nuclear power plant (Focus Taiwan, 2018a). Apparently, the beginning of Taiwan's journey toward a nuclear-free homeland was not very smooth. Instead, it began with a dramatic increase in nuclear power in Taiwan's electricity mix.

### 25.2.5.2.1    Fate of the Fourth Nuclear Power Plant

The fate of the nearly complete fourth nuclear power plant was less positive. Despite the lack of clarity about it in Article 95, the government took deliberate action against it, circumventing the law. Because of the lack of legal measures to deal with sunk assets, the government asked Tai-Power to return the fuel rods to the United States. Tai-Power had already sent back several fuel rods in mid-2018 (Focus Taiwan, 2018a). Some readers might wonder why Tai-Power was so compliant, but this state-owned company could not challenge the government's implicit order.

Dealing with the fourth nuclear power plant has been problematic. The plant is comprised of two advanced boiling water reactors (each 1,350 MWe net). General Electric is the reactor supplier, and the turbines and generators are supplied by Mitsubishi Heavy Industries. The nuclear power plant cost about USD 9.4 billion. As discussed above, the DPP government stopped budgetary funding as a political tactic to halt its construction in 2000 while continuing construction and injecting funding after consensus was reached on a nuclear-free homeland. After the Japanese Fukushima nuclear power plant accident in 2011, the DPP, then the opposition party, again raised the issue. In 2014, the DPP launched a major protest against continuing with the construction, which pressured the KMT government to halt construction of the fourth power plant. Construction of the plant was deferred (WISE, 2014).

As noted, in mid-2018, the government began to return the fuel rods, although many people opposed that action because it could be perceived as a symbol to abandon investment in nuclear power plants.

To avoid a dramatic increase in electricity tariffs caused by the sunk costs of this nuclear power plant, the government tried to have Tai-Power not reflect these costs in their accounts. Perhaps it is understandable that in 2014, the construction works differed although the regular maintenance of the power plants remains. Likely, construction will be completed and the plant will go online. However, the recent step of returning the fuel rods seems to signal the end. It is very difficult to imagine that Taiwan, a country under the rule of law, has no legal remedy for dealing with the USD 9.4 billion in sunk costs. A major debate in the Parliament on this issue is not necessary for such an important decision, which only relies on Tai-Power's decision to return the fuel rods and sink the assets. Further, it seems unnecessary for the government to persuade the public to accept higher electricity prices after abandoning this power plant project.

### 25.2.5.2.2 2018: Referendum to Revoke Article 95

In December 2017, the Referendum Act was largely revised (Law & Regulations Database of the Republic of China, 2018). The major change concerned easing the burden of the threshold for the number of proposers by reducing it from about 1 million to around 280,000 people (Referendum Act, 2018). In March 2018, pro-nuclear groups used this as an opportunity to propose "go green with nuclear power," because municipal elections were set for November 2018 that could include a referendum vote (Freschi, 2018). Thus, in mid-2018, they launched a referendum to remove the nuclear-free homeland clause from Article 95. This referendum was launched by the alumni and a professor of National Tsing-Hua University, Taiwan, on https://www.green-nuclear.vote/.

The proposer documents had to be submitted by mid-September for the referendum to be included on the November 24, 2018 ballot. However, after accepting 315,000 signatures by September 6, the government refused to accept a second attempted delivery of 24,000 signatures on September 13 (Shellenberger, 2018). Shih-Hsiu Huang, cofounder of Nuclear Myth-Busters and the chief organizer of the referendum, staged a hunger strike outside the office of the Central Election Commission (CEC) for several days. The hunger strike ended, but the CEC still refused to accept the signatures. As expected, on October 16, 2018, the referendum failed to pass CEC review because the number of qualified public endorsements for the proposal did not meet the legally required threshold, falling short by 2,326 signatures (Focus Taiwan, 2018a).

In mid-September, the organizer of the referendum brought the CEC's refusal to accept the signatures before the High Court to request a provisional injunction to order the CEC to accept them.

**TABLE 25.3**

**Results of the Nuclear Referendum in 2018**

| | Referendum Results | | | | | | | | |
|---|---|---|---|---|---|---|---|---|---|
| | For | | Against | | Invalid/ | | Registered | | |
| Question | Votes | % | Votes | % | Blank | Total | Voters | Turnout | Outcome |
| Repealing the planned end of nuclear power stations | 5,895,560 | 59.49 | 4,014,215 | 40.51 | 922,960 | 10,832,735 | | 54.83 | Adopted |

*Source:* http://referendum.2018.nat.gov.tw/pc/en/00/m00000000000000000.html.
Central Election Commission "*2018 Referendum*." Nov. 24, 2018. https://web.archive.org/web/20181125031636/http:/referendum.2018.nat.gov.tw/pc/en/00/00000000000000000.html

On October 17, the Administrative court ordered the CEC to accept supplemental signatures for a pro-nuclear power referendum proposal (Focus Taiwan, 2018b). Finally, the CEC approved the referendum on October 24, 2018 (Shellenberger, 2018).

After the vote on November 24, 2018, close to 6 million citizens voted for the revoke of Article 95 of the nuclear-free homeland clause (Table 25.3). The clause was officially revoked on May 22, 2019. Even so, the nuclear-free homeland policy continues as there is no life extension of the existing three power plants. The first (No 1) nuclear power plant was issued a decommission licence in July 2019. The two reactors of the second nuclear power plant are scheduled to cease operations in 2021 and 2023, while the third is scheduled to cease operations in 2025. As the fuel rods of the fourth nuclear power plant have been sent back to the United States, test operations remain unlikely. On December 13, 2019, the CEC approved a proposal to hold a referendum on the revival of the contested fourth nuclear plant (Taiwan News, 2019). December 18, 2021 has been named as the date the vote be held. The voting is invalid due to not reach the voting threshold.

## 25.3  COMPATIBILITY OF NUCLEAR-FREE HOMELAND WITH SDGs

Taiwan is an isolated island with 98% energy being imported and is classified as having an independent electricity system like Japan and South Korea, which is also much different from the nuclear phase-out pioneer countries like Italy and Germany. To support the export of products from high technology industry, the reliance on fossil fuel base and nuclear power generation is inevitable. At the end of 2016, the total installed capacity of electricity in Taiwan was 49.06 GW, which came from coal-based thermal (34.73%), gas-based thermal (32.32%), nuclear (10.49%), renewable energy (9.10%), fuel-based thermal (8.06%), and pumped storage hydro (5.3%) (Energy Market Company, 2017). The installation mix would not reflect the reality of such high reliance on traditional electricity sources. From the electricity production mix of 2016, the role of renewables was reduced to 4.8% and the rest account for 95.2% (Wang, 2017). Thus, how to change such a mix to replace all of nuclear power plus a certain extent of fossil fuel generation by 2025 would be a huge challenge for Taiwan, particularly within such a limited timeline (Figure 25.1).

After four years of efforts in implementing such an energy transition plan, in 2020, 45% of Taiwan's electricity generation came from coal, 35.7% from natural gas, 11.2% from nuclear plants, and 5.4% from renewables, and there was only a 0.6% increase from 2016, while the reliance on traditional electricity remained. However, the expansion of coal- and gas-fired plants and renewable has already led to concerns of air pollution/carbon emission (coal- and gas-fired) and land use conflict. (PV, onshore and offshore wind). This chapter will elaborate on these.

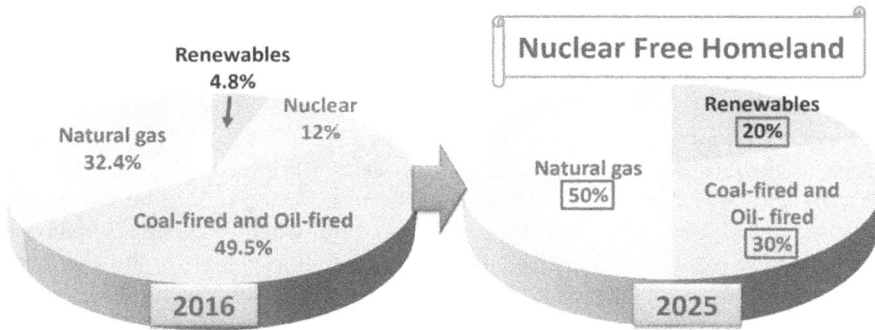

**FIGURE 25.1** Taiwan Energy Transition Plan by 2025. (Data from Current Thrusts to 2025 Renewable Energy Targets in Taiwan, http://www.cieca.org.tw/v_comm/inc/download_file.asp?re_id=2998&fid=34830 (Wang, 2017))

### 25.3.1 DECISION-MAKING PROCESS OF NUCLEAR-FREE HOMELAND

Needless to say, nuclear power has the advantage in achieving SDG 7 and 13; Goal 7: Affordable and clean energy and Goal 13: Climate action from its low carbon, and relatively cheap features. However, this power is also controversial for its nuclear legacy, which makes it less "clean" under the SDG7.

To promote a nuclear-free homeland, "just" measures would be necessary to fulfil the requirement of Goal 16: Peace, justice, and strong institutions or the due process or rule of law concerns of a democratic society. However, it is not the case in Taiwan.

In 2018, before holding the referendum, the government played tricks to send back the fuel rods to the US and also used a lot of tricks to delay the referendum signatures process. After citizens voted for not setting 2025 as the deadline, the government tried to interpret narrowly to include the abolition of Art 95, Electricity Act of 2017, but passively decided not to set up the life extension schedule. For the fourth nuclear power plant, the government continued to send back the fuel rods. For these milestone decisions affecting billions of sunken cost of investment, usually a formal legislative discussion and decision is necessary, but not for the case of nuclear-free homeland policy.

Also, such an approach cannot answer a basic and logical question: If nuclear power were to be so dangerous and uncontrollable, how come the current ruling party decided not to switch off all nuclear power but to squeeze it to the last minute of mid-2025 (the operation licence of the second reactor of the Third Nuclear Power Plant in Taiwan)? Interestingly, most environmental NGOs also buy out such an idea, and it would be interesting to see such SD perception in the world.

### 25.3.2 RENEWABLE DEVELOPMENT AND SDGS

The original government objective was to replace nuclear power with renewables. Thus, there was a large jump of renewable "installation" capacity over the past 4 years, particularly solar PV (from 1,245.06 in 2016 to 5,817.21 MW by the end of 2020). For the rest of the Renewable Energy System (RES), there was only a slight increase (onshore wind: 682.09 → 725.71; offshore wind power: 8.00 → 128.00); such a big increase of PV would contribute to the SDGs of 7 and 13. Perhaps, it is also helpful for the creation of decent work and economic growth (Goal 8) and contributes to Goal 11: Sustainable cities and communities. Certain energy intensive industries or high technology companies could also plan to install solar panels on their roofs, which would meet Goal 12: Responsible consumption and production.

Yet, the large expansion of PV also affects the existing land use and certain SDGs in a negative way. For instance, compared with other type of RES, PV is relatively expensive. Thus, such a large number of PVs would lead to the affordability concerns of SDG 7. Secondly, the changing land use to install PVs also affect life on land or below water (Goal 14: Life below water; Goal 15: Life on land), as they occupy a lot of land or pond space (Taipei Times, 2019). Thirdly, the PV also affects citizens, such as installation in fish farming area upsets the life of original fish farmers (Focus Taiwan, 2021b). This would lead to concerns about farmers having to sacrifice for energy transition (Goal 10: Reduced inequality). Finally, it is still doubtful whether the large expansion of such land use would be the basis for sustainable cities and communities (Goal 11). Several communities oppose such big PV parks in their vicinity in south Taiwan (Ming-sho Ho, 2020).

### 25.3.3 THIRD LNG TERMINAL TURMOIL AND ITS REPLACEMENT

In spite of the official statement of replacing nuclear power with renewables, facing the opposition and slow development in RES lead to the reliance on traditional fossil fuel. To strike a balance among different types of fossil fuel, gas was chosen as the main one and to become the dominant (from the current No. 2) electricity source by 2025. Unlike the ease of importing gas via undersea or land pipelines as European countries or the US do, Taiwan has to build third LNG terminals to import gas to replace nuclear power. Environmental groups have argued that construction would harm the algal reef and endangered species in the area. Rescue Datan's Algal Reefs Alliance initiated a referendum on "Do you agree that CPC Corp., Taiwan's LNG terminal should be relocated from its planned site near the algal reef coast of Datan and its surrounding waters?" It is scheduled to be held in August (Focus Taiwan, 2021a) if this LNG terminal project is vetoed in the referendum, it could be beneficial to SDGs 13–15. However, there is also a worry that the government may change to coal-fired power plants to replace nuclear energy. This may again affect and worsen SDGs 13–15.

### 25.3.4 LACK OF COAL PHASE-OUT SCHEDULE

Around the world, the mainstream of energy transition policy focuses more on coal phase-out, instead of the nuclear one. Not to mention, Taiwan's nuclear-free homeland is not nuclear phase-out in nature but has a normal operation life of 40 years. Of course, there is usually a trade-off between nuclear and coal power in different countries. For instance, in France, the UK and Sweden, the coal phase-out schedule can be done earlier. Yet, for nuclear phase-out countries like Germany, the schedule is quite stretched but at least there is a committed timeline of 2038. There is a legislation in place to achieve the vision: Act to Reduce and End Coal-Powered Energy and Amend Other Laws of 2020 (Goess, 2020). Yet, in Taiwan, there has never been such a commitment, and only a 2025 vision to reduce coal-fired power plants to 30%. Yet albeit the current LNG terminal controversy, it seemed unlikely to meet such targets and may instead see an increase. There is no legislation to push forward such coal phase-out agenda as well. In this way, several SDGs like 3, 6, 7, 13, 14, and 15 would be negatively affected (Figure 25.2).

### 25.3.5 NO PLAN AND LAW TO DEAL WITH NUCLEAR LEGACY

A responsible nuclear-free homeland requires not only a regime for setting up a nuclear power decommission schedule, but also one for both low-level radioactive waste (LRW) and high radioactive waste (HRW). For instance, in 2011, Germany committed to phasing out nuclear power by 2022. A Site Selection Act for controversial HRW was adopted in 2013 (BGR, 2017), and revised in 2017 (BMJV, 2017). However, this responsible attitude is not evident in Taiwan.

Even if Tai-Power has already fulfilled its legal duties to submit a plan for LRW and HRW, a deadlock is evident. In terms of LRW, it is the unresolved passive attitude of holding a referendum

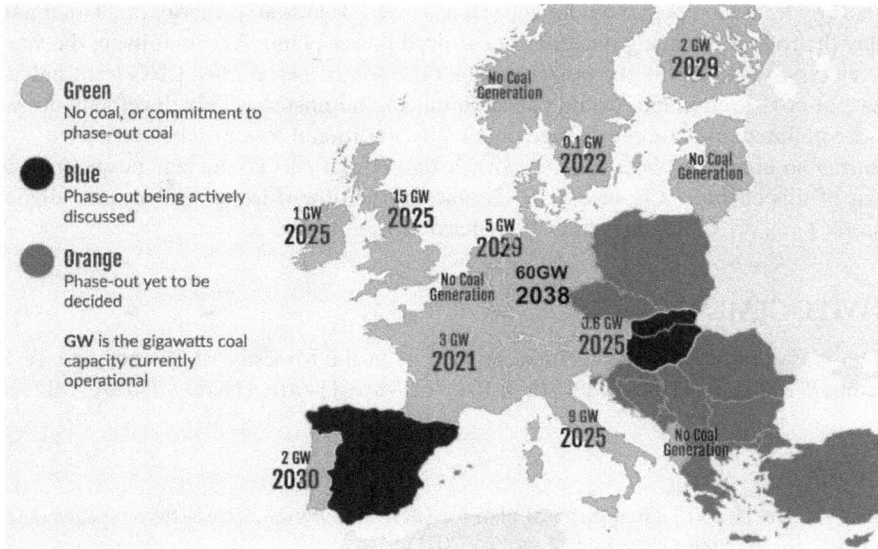

**FIGURE 25.2** Current coal phase-out commitments in Europe. (Data from https://onclimatechange-policydotorg.wordpress.com/2019/06/18/europes-phase-out-of-coal/.)

by the local government. According to Article 29 of the Nuclear Materials and Radioactive Waste Management Act, it is the responsibility of nuclear operators to plan for final disposal (Nuclear Materials and Radioactive Waste Management Act, 2002). Tai-Power proposed the first version of the final disposal plan for low radioactive waste LRW in 2004 (Atomic Energy Council, 2004). The plan has been subject to revision several times: in 2007, 2014, and 2015 (Atomic Energy Council, 2007). However, based on opposition from the local county during the site selection process, Tai-Power was ordered in December 2016 to propose an alternative programme for final disposal of LRW. A centralized LRW storage site is planned (Atomic Energy Council, 2017), which will be used to store it before the final disposal site is available. To continue evaluating technical aspects of the final disposal of LRW, Tai-Power also conducted related research. The research report (Atomic Energy Council, 2017) can be viewed as a soft approach to proceeding with the site selection process.

For HRW, no definite legislative timeline was set for selecting a site, and the current draft bill is badly drafted. The multi-level veto and broader local referendum could be expected to worsen site selection for HRW more than for LRW. The resolution to both deadlocks requires the determination of the Parliament, the president, and cabinet. Otherwise, the only way out seems to be the continued use of existing LRW and HRW sites.

## 25.4 CONCLUSIONS

Taiwan's approach in becoming the world's third nuclear power-free country is problematic from different aspects, particularly under the eye of just transition and the harmonization among different targets of SDGs.

From the just transition aspect, the lack of a thoughtful deliberation process to make this decision and ignorance to the referendum results render this initiative unjust. Not to mention the lack of rule of law in abolishing billions in sunk investment. Finally, the irresponsible attitude of the ruling government in tackling nuclear waste would be unjust as well. It continues to argue that there is no toilet to store nuclear power waste to justify its nuclear power legacy. However, it still wants to switch to nuclear power and continue to produce waste before 2025. Such irresponsible attitude is sure to be an unjust energy transition.

A just energy transition plan would meet SDGs as well. Renewable energy or cleaner natural gas usually play the role of replacing nuclear or coal-fired power plants. Yet in Taiwan, the standstill of renewable electricity evident in the power mix and the controversial third LNG terminal leading to the increase of coal-fired plants would cause an unjust and unsustainable development by moving back to more polluted and climate change unfriendly traditional fossil fuel electricity.

Taiwan has no chance to become the world's first country to go nuclear power-free, but from the analysis of this chapter, it is sure to be eligible for the title of irresponsible, unrealistic, unjust, against people's majority, and unsustainable nuclear-free country.

## ACKNOWLEDGEMENT

This chapter was completed with the funding support of the Ministry of Science and Technology, project number: 109-2410-H-007-038-MY2; 109-2627-M-011-001; 111-NU -E-007 -002 -NU.

## REFERENCES

Atomic Energy Council. 2004. Final disposal plan for LRW. http://www.aec.gov.tw/webpage/control/waste/files/index_07_a-93.pdf (accessed October 24, 2021).
Atomic Energy Council. 2007. For example, the final disposal plan for LRW 2007, 2012, 2014, 2015, etc. https://www.aec.gov.tw/webpage/control/waste/files/index_07_a-96.pdf (accessed October 24, 2021).
Atomic Energy Council. 2017. Technology assessment report on LRW final disposal. http://www.aec.gov.tw/webpage/control/waste/files/index_07_e-1.pdf (accessed October 24, 2021).
Bundesanstalt für Geowissenschaften und Rohstoffe (BGR). 2017. Site selection. https://www.bgr.bund.de/EN/Themen/Endlagerung/Standortauswahl/standortauswahl_node_en.html (accessed October 24, 2021).
Bundesministeriums der Justiz und für Verbraucherschutz. 2017. Bundesamt für Kustiz, Gesetz zur Suche und Auswahl eines Standortes für ein Endlager für hochradioaktive Abfälle. https://www.gesetze-im-internet.de/standag_2017/BJNR107410017.html (accessed October 24, 2021).
Chao, V. Y. 2011. Tsai talks "nuclear-free homeland." *Taipei Times*. https://www.taipeitimes.com/News/taiwan/archives/2011/06/09/2003505340 (accessed October 24, 2021).
Energy Market Company. 2017. Developments in Taiwan's Electricity Market (PDF). August 1, 2017. https://www.emcsg.com/f1671,123955/3_-_Dr_Chuan-Neng_Lin_Bureau_of_Energy_Taiwan.pdf (accessed October 24, 2021).
Energia Plus. 2020. Taiwan's Energy Transition: Big Business for Renewables. https://energeiaplus.com/2020/01/23/taiwans-energy-transition-big-business-for-renewables/ (accessed October 24, 2021).
Executive Yuan. 2001. The decision of the Executive Yuan on February 14, 2001. http://www.ey.gov.tw/ct.asp?xItem=21449&ctNode=232&mp=21 (accessed October 24, 2021).
Executive Yuan. 2013. Energy Security and Nuclear-free Homeland Bill of 2013. https://www.ey.gov.tw/Page/9277F759E41CCD91/9ccacfb2-be1e-421b-a8e8-25192d4e482f (accessed October 24, 2021).
Focus Taiwan. 2017. ROUNDUP: Power outage affects 6.68 million households; supply restored. http://focustaiwan.tw/news/aeco/201708150034.aspx (accessed October 24, 2021).
Focus Taiwan. 2018a. Tai-Power seeks to restart No 2 nuclear power plant reactor. http://focustaiwan.tw/news/aeco/201802050010.aspx (accessed October 24, 2021).
Focus Taiwan. 2018b. CEC ordered to accept more signatures for nuclear power referendum. http://focustaiwan.tw/news/aipl/201810170022.aspx (accessed October 24, 2021).
Focus Taiwan. 2021a. KMT urges government to halt construction on controversial LNG project, 03/11/2021. https://focustaiwan.tw/politics/202103110018 (accessed October 24, 2021).
Focus Taiwan. 2021b. Solar panel fish farm to begin generating electricity in 2021. https://focustaiwan.tw/sci-tech/202011060024 (accessed October 24, 2021).
Freschi, N. 2018. Taiwan's Nuclear Dilemma: Nuclear energy has long been a contentious issue for Taiwan, but there are few good alternatives. https://thediplomat.com/2018/03/taiwans-nuclear-dilemma/ (accessed October 24, 2021).
Goess, S. 2020. The German coal phase-out is law: an overview. https://energycentral.com/c/ec/german-coal-phase-out-law-overview (accessed October 24, 2021).

Ho, M.S. 2020. Why do Taiwan's environmentalists oppose renewable energy facilities? https://taiwaninsight.org/2020/07/13/why-do-taiwans-environmentalists-oppose-renewable-energy-facilities/ (accessed October 24, 2021).

Law Bank. 2003. The early draft act to promote a nuclear-free homeland was proposed in 2003. Draft Act to promote a nuclear-free homeland. https://www.lawbank.com.tw/news/NewsContent.aspx?nid=15938.00 (in Chinese) (accessed October 24, 2021).

Legislative Yuan. 2011. Nuclear-free homeland bill by People First Party. https://misq.ly.gov.tw/MISQ/IQuery/misq5000QueryBillDetail.action?billNo=1020425070200800 (accessed October 24, 2021).

Legislative Yuan. 2012. Nuclear-free homeland bill by Chen. https://misq.ly.gov.tw/MISQ/IQuery/misq-5000QueryBillDetail.action?billNo=1020503070200300 (accessed October 24, 2021).

Legislative Yuan. 2013. Nuclear-free homeland bill by Taiwan Solidarity Union. https://misq.ly.gov.tw/MISQ/IQuery/misq5000QueryBillDetail.action?billNo=1020509070200600 (accessed October 24, 2021).

Lien, W. W. 2021. A landscape approach to the conflicts of greens: Planning for energy and wetland land-use growth in southwestern Taiwan's coastal landscape in a climate-changing era. https://scholarsbank.uoregon.edu/xmlui/handle/1794/26129 (accessed October 24, 2021).

Lin, C. 2018. VP says nuclear-free homeland is nation's 18th SDG. *Taipei Times.* https://www.taipeitimes.com/News/taiwan/archives/2018/08/22/2003698969 (accessed October 24, 2021).

Ministry of Justice. 2002. Basic environment act. https://law.moj.gov.tw/ENG/LawClass/LawAll.aspx?pcode=O0100001 (accessed October 24, 2021).

Nuclear Materials and Radioactive Waste Management Act. 2002. Article 29: The treatment, carriage, storage and(/or) final disposal of radioactive waste shall be done by the producer of radioactive waste itself solely or be entrusted to an entrepreneur who or which has the technical capability to finally dispose of the domestic or foreign. https://law.moj.gov.tw/ENG/LawClass/LawAll.aspx?pcode=J0160015 (accessed October 24, 2021).

Nuclear Reactor Facilities Regulation Act. 2003. https://law.moj.gov.tw/Eng/LawClass/LawAll.aspx?PCode=J0160019 (accessed October 24, 2021).

OGEL. 2019. Taiwan's perhaps irresponsible policy and laws on a nuclear-free homeland by 2025: The missing piece of nuclear assets and legacy. https://www.ogel.org/article.asp?key=3819 (accessed October 24, 2021).

Referendum Act. 2018. Article 12: For the matters prescribed in Paragraph Two of Article 2, the number of proposers shall be not less than 1.5% of the total electors in the most recent election of the President and Vice President. https://law.moj.gov.tw/Eng/LawClass/LawContent.aspx?PCODE=D0020050 (accessed October 24, 2021).

Shellenberger, M. 2018. Momentum builds for nuclear power with referendum approved in Taiwan and "Pride Fest" in Germany. *Forbes.* https://www.forbes.com/sites/michaelshellenberger/2018/10/24/momentum-builds-for-nuclear-power-with-refer endum-approved-in-taiwan-and-pride-fest-in-germany/ (accessed October 24, 2021).

Taipei Times. 2019. Chiayi denies solar panels driving away endangered birds. http://www.taipeitimes.com/News/taiwan/archives/2019/05/14/2003715101 (accessed October 24, 2021).

Taiwan News. 2019. Taiwan approves plans for a referendum to revive contested 4th nuclear plant. https://www.taiwannews.com.tw/en/news/3836618 (accessed October 24, 2021).

Wang, R.C. 2017. Current thrusts to 2025 renewable energy targets in Taiwan. http://www.cieca.org.tw/v_comm/inc/download_file.asp?re_id=2998&fid=34830 (accessed October 24, 2021).

WISE. 2014. Taiwan halts fourth power plant. https://www.wiseinternational.org/nuclear-monitor/786/taiwan-halts-fourth-power-plant (accessed October 24, 2021).

Yueh-Hsun, T. Y. H. 2014. Recent development of new energy policy and legislation in Taiwan, with the focus on promotion of biofuel. *Review of Business and Economics Studies*, 2(2):42–50.

# Index

For Product Safety Concerns and Information please contact our EU
representative  GPSR@taylorandfrancis.com
Taylor & Francis Verlag GmbH, Kaufingerstraße 24, 80331 München, Germany

www.ingramcontent.com/pod-product-compliance
Lightning Source LLC
Chambersburg PA
CBHW060957210326
41598CB00031B/4848

9 781032 324982